Trends in Antiviral Drug Development

Trends in Drug Discovery

Trends in Antiviral Drug Development

Edited By

Dr. Michael J. Sofia
Arbutus Biopharma
701 Veterans Circle
Warminster
PA 18974, USA

Dr. Zhengqiang Wang
University of Minnesota
312 Church St. SE
Minneapolis
MN 55455, USA

WILEY-VCH

Editors

Dr. Michael J. Sofia
Arbutus Biopharma
701 Veterans Circle
Warminster
PA 18974, USA

Dr. Zhengqiang Wang
University of Minnesota
312 Church St. SE
Minneapolis
MN 55455, USA

Library of Congress Card No.: applied for

British Library Cataloguing-in-Publication Data
A catalogue record for this book is available from the British Library.

Bibliographic information published by the Deutsche Nationalbibliothek
The Deutsche Nationalbibliothek lists this publication in the Deutsche Nationalbibliografie; detailed bibliographic data are available on the Internet at <http://dnb.d-nb.de>.

Print ISBN: 9783527353019
ePDF ISBN: 9783527845101
ePub ISBN: 9783527845095
oBook ISBN: 9783527845088

Set in 9.5/12.5 STIX Two Text by Lumina Datamatics

Printed and bound by CPI Group (UK) Ltd, Croydon, CR0 4YY

C9783527353019_270625

The manufacturer's authorized representative according to the EU General Product Safety Regulation is Wiley-VCH GmbH, Boschstr. 12, 69469 Weinheim, Germany, e-mail: Product_Safety@wiley.com.

Contents

Preface

As practitioners in the field of antiviral drug discovery and development, we collectively have observed and contributed to the field for over 47 years, published over 130 papers on antiviral research, introduced 17 antiviral drugs into development, and discovered and developed the first direct-acting antiviral drug to cure a viral disease. We have seen major progress in both treating and curing viral diseases and have observed how advancements in new science and technologies helped expand our knowledge of viruses and provided new insights into how to develop innovative antiviral therapeutics. Yet, we have also seen how cunning viruses are in the way they attack the human host and coopt the human body's own biological processes to survive. We have seen how drug hunters constantly battle to innovate and find ways to combat viruses as they continually change and adapt to the environment around them, and as new viruses emerge by jumping from species to species. Consequently, it was our desire to publish a book that would highlight some of the recent advances in the field of antivirals that occurred over the last 20 years and provide to those who are interested in learning about how antivirals have been developed a window into the thinking and processes used by the scientists who made these discoveries. This book delves into the discovery of drugs to treat HIV, CMV, HSV, VARV, RSV, and SARS-CoV-2, and the breakthrough drugs that cured HCV. It also introduces siRNA therapeutics as a new antiviral modality and the drug candidates in the clinic that are progressing toward achieving an HBV cure. Our hope is that the stories told here in *Trends in Antiviral Drug Development* would expand the readers' understanding of what it takes to develop a drug to treat viral diseases and encourage new ideas and approaches to combat the ever-expanding universe of viral infections.

Michael J. Sofia, PhD, FRSC
Zhengqiang Wang, PhD

Introduction

Successes and Challenges in Antiviral Drug Development

Zhengqiang Wang[1] *and Michael J. Sofia*[2]

[1] *Center for Drug Design, College of Pharmacy, University of Minnesota, Minneapolis, MN 55455, USA*
[2] *Arbutus Biopharma, Inc., 701 Veterans Circle, Warminster, PA 18974, USA*

Introduction

Viral infections have had a profound impact on socioeconomics and life expectancy of human societies throughout human history [1–7]. However, the discovery and development of specific antiviral drugs to treat viral diseases has a relatively short history, largely concomitant with the advent of modern biomedical sciences. Specifically, the FDA approval of idoxuridine (IDU) [8] in 1963, a deoxynucleoside analog first synthesized by Prusoff in 1958 [9], ushered in the era of direct-acting antivirals (DAAs). Since then, many antiviral drugs and drug combinations have been successfully developed and marketed [10], which have largely mitigated many human viral pathogens that historically plagued human societies. Not surprisingly, most of these drugs target viruses that cause chronic infections and/or establish latency. Human immunodeficiency virus (HIV), in particular, has been at the center of antiviral drug development and drug regimen evolution, with dozens of single agents and fixed-dose combinations (FDCs) approved, spanning a wide range of compound classes and molecular targets [11, 12]. The everlasting needs to prevent infection, enhance drug accessibility, and improve adherence have also spurred developments in pre-exposure prophylaxis (PrEP) [13, 14], regimen simplification [15–17], and long-acting injectables (LAIs) [18–20]. Prior to HIV, the earliest antiviral drugs mostly targeted human herpesviruses (HHVs), including herpes simplex virus (HSV), varicella zoster virus (VZV), and human cytomegalovirus (HCMV). Other major chronic viruses targeted by various small-molecule antiviral drugs include hepatitis B virus (HBV) and hepatitis C virus (HCV), both causing viral hepatitis and increasing risk of hepatocarcinoma. Although current antiviral drugs typically do not clear HIV, HBV, or HHVs, the development of DAAs to successfully

Trends in Antiviral Drug Development, First Edition. Edited by Michael J. Sofia and Zhengqiang Wang.

cure HCV represents a major breakthrough [21, 22] in antiviral therapy and provides hope that further development in antiviral drugs could eventually lead to the functional cure of other challenging viral infections. Widespread and seasonal acute viral infections, such as influenza virus, respiratory syncytia virus (RSV), and SARS-CoV-2, have also caused major mortality and morbidity, and have been targeted by successful and ongoing antiviral drug development efforts.

Antiviral Drugs Targeting Human Herpesviruses

HHVs, a family of eight large DNA viruses, are highly prevalent within human populations [23]. While primary infections are generally associated with low risk, these viruses are all capable of establishing lifelong latency. In immunocompromised individuals, the reactivation of the latent infection causes high morbidity and mortality. Among HHVs, α-herpesviruses HSV and VZV, and β-herpesvirus HCMV, have received particular attention in drug discovery.

The vast majority, if not all, of the earliest HHV drugs belonged to four distinct subclasses of nucleos(t)ide analogs (Table 1): (1) C-5 substituted thymidine analogs, including IDU [8, 24, 25], trifluridine (TFT) [26], and brivudine (BVDU) [27], where the C-5 substituent is mechanistically important by disrupting base pairing; (2) marine sponge *tectitethya crypta* spongonucleoside [28] vidarabine (ara-A) [29], which contains D-arabinose rather than D-ribose; (3) acyclic nucleosides [30] aciclovir (ACV), penciclovir (PCV), and ganciclovir (GCV); and (4) acyclic nucleoside phosphonate (ANP) [31] cidofovir. Intracellularly, all nucleoside analogs (the first three subclasses) are converted to monophosphate (MP) by a virally encoded kinase, and then further phosphorylated by cellular kinases to the active form triphosphate (TP). ANP cidofovir bypasses the function of viral kinase as the MP is already chemically installed, but still uses cellular kinases to form the active TP. Early mechanistic studies showed that these TPs can stop viral DNA synthesis via competitive viral polymerase inhibition or by getting incorporated into the DNA and acting as chain terminators. Of these nucleos(t)ide anti-herpes antiviral drugs, ACV [32, 33] is considered a milestone drug due to its excellent selectivity and low toxicity and has been used to effectively treat infections caused by most known HHVs, particularly HSV and VZV. Another acyclic nucleoside drug GCV [34, 35] has been the first-line treatment for HCMV for decades.

Another mechanistically distinct viral polymerase inhibitor is the pyrophosphate (PPi) mimic foscarnet (PFA), which binds to a site similar to but distinct from the PPi binding site of the viral polymerase to block the pyrophosphate exchange [36]. Notably, PFA does not require phosphorylation and thus is kinase-independent.

Although polymerase-targeting nucleos(t)ide analogs and the PPi mimic PFA are largely effective against HHVs, they are limited by dose-related adverse effects and the emergence of drug resistance. In line with expanding the anti-herpes antiviral drug repertoire, efforts in recent years have led to the successful development of three mechanistically novel drug classes: amenamevir (AMNV) and pritelivir (PTV), inhibitors of viral helicase–primase complex, for treating VZV and HSV, respectively; letemovir (LTV), which targets pUL56, a component protein of the

Table 1 Antiviral drugs against HHVs.

Drug class and MOA	Drug name	Note	Year approved	Indication approved
Nucleos(t) ide analogs: phosphorylated intracellularly, incorporated and act as chain terminators	Idoxuridine (IDU)	Thymidine analog. 5-I blocking base paring.	1963	Topical treatment of herpes simplex keratitis
	Trifluridine (TFT)	Thymidine analog. 5-CF_3 blocking base paring	1980	Eye drops for treating keratitis and keratoconjunctivitis caused by HSV
	Brivudine (BVDU)	Thymidine analog. 5-Bromovinyl blocking base paring	2000	Herpes zoster (VZV) in adult patients
	Vidarabine (ara-A)	Marine spongonucleoside analog	1976	Ointment for treating HSV; herpes zoster (VZV) in AIDS patients
	Aciclovir (ACV)	Acyclic nucleoside analog	1981	Treatment of HSV and VZV infections
	Penciclovir (PCV)	Competitive viral pol inhibition and chain termination	1996	Treatment of zoster (VZV) and genital HSV
	Ganciclovir (GCV)	Competitive viral pol inhibition and chain termination	1988	First-line treatment of CMV infections
	Cidofovir (CDV)	Competitive viral pol inhibition and chain termination	1996	ACV-resistant HSV and GCV-resistant CMV infections
Pyrophosphate mimic	Foscarnet (PFA)	Binding to the PPi binding site and blocking PPi release	1991	ACV-resistant HSV and VZV; GCV-resistant CMV infections
Terminase inhibitor	Letermovir (LTV)	Targeting viral terminase component protein pUL56	2017	Prophylaxis of post-transplant HCMV
Viral kinase inhibitor	Maribavir (MBV)	Inhibition of viral kinase (pUL97)	2021	Treating post-transplant HCMV infections resistant to other drugs
Viral helicase–primase inhibitors	Amenamevir (AMNV)	Inhibition of viral helicase–primase complex	2017	Herpes zoster (shingles)
	Pritelivir (PTV)	Inhibition of viral helicase–primase complex	Phase 3	HSV infections in immunocompromised patients

HCMV terminase complex, for HCMV prophylaxis post stem cell transplants; and maribavir (MBV), an HCMV kinase (pUL97) inhibitor for treating post-transplant HCMV resistant to other drugs.

Antiviral Drugs Targeting Human Immunodeficiency Virus

HIV drug discovery [37, 38] has been particularly successful with the approval of dozens of single agents in the past four decades (Table 2). These drugs, comprising a few distinct drug classes and mechanisms of action, coalesce to a large repertoire for effective antiretroviral therapy (ART), termed the highly active antiretroviral therapy (HAART) [39]. HAART regimens typically consist of two nucleoside reverse transcriptase inhibitors (NRTIs), along with an integrase inhibitor (INSTI), a protease inhibitor (PI) or a non-nucleoside reverse transcriptase inhibitor (NNRTI).

Drugs Targeting Reverse Transcriptase

The virally encoded RT [40] is the target of the earliest anti-HIV drug class, nucleos(t)ide reverse transcriptase inhibitors (NRTIs) [41]. Azidothymidine (AZT), initially designed and synthesized by Jerome Horwitz in 1964 [42] as an anticancer drug, was approved by FDA in 1987 as the very first HIV drug [43, 44]. In the following years, numerous other NRTIs were approved, including three other furanonucleoside analogs didanosine (DDI) [45, 46], zalcitabine (DDC) [47], and stavudine (D4T) [48]; the L-nucleoside lamivudine (3TC) [49–51] and emtricitabine (FTC) [52]; the carbocyclic nucleoside abacavir (ABC) [53]; and the acyclic nucleoside phosphonate (ANP) tenofovir (TFV) [54]. These nucleos(t)ide drugs are all converted into triphosphate (TP) intracellularly by cellular kinases, recognized and incorporated by RT into viral DNA, and act as obligate chain terminators by virtue of lacking the 3′-OH. Although generally effective against HIV, NRTIs can cause toxicity by inhibiting human mitochondrial DNA polymerase (Pol γ), leading to depletion of mtDNA and mitochondrial dysfunction [55, 56].

Also targeting RT, 4′-ethynyl-2-fluoro-2′-deoxyadenosine (EFdA) is structurally unique and mechanistically distinct from all other RT inhibitors. As a nucleoside analog, EFdA contains the 3′-OH group, which likely allows more efficient phosphorylation and incorporation. Mechanistically, upon incorporation, EFdA can act as a nonobligate chain terminator as the 4′-ethynyl group distorts the alignment of the primer and the next incoming dNTP, and/or trap RT into a translocation-deficient state where the sliding from N site to P site is inhibited [57, 58]. The latter mechanism renders EFdA the first nucleoside reverse transcriptase translocation inhibitor (NRTTI). Due to the exceptional antiviral potency and the long-lasting PK, EFdA has been developed for PrEP [59].

A conceptually distinct class of RT-targeting drugs is the NNRTIs (Table 2), which allosterically target RT by binding to a largely hydrophobic and well-studied binding pocket ~10 Å away from the active site to induce conformational changes to the active site and distort the positioning of the primer grip to affect its alignment

Table 2 Antiviral drugs against HIV.

Drug class and MOA	Drug name	Note	Year approved
NRTIs: intracellularly phosphorylated, incorporated and act as obligate chain terminators	Zidovudine (AZT)	Furanonucleoside; first HIV drug approved	1987
	Didanosine (DDI)	Furanonucleoside	1991
	Zalcitabine (DDC)	Furanonucleoside	1992
	Stavudine (D4T)	Furanonucleoside	1994
	Lamivudine (3TC)	L-nucleoside	1995
	Abacavir (ABC)	Carbocyclic nucleoside	1998
	Tenofovir (TFV)	ANP. Two prodrug regimens (TDF and TAF) approved	2001
	Emtricitabine (FTC)	L-nucleoside analog	2003
NRTTI: intracellularly phosphorylated, incorporated and inhibits RT translocation	Islatravir (EFdA)	3′-OH present. Long acting	Phase 3
NNRTIs: bind to an allosteric site to impact the conformation of the RT active site	Nevirapine (NVP)	First generation	1996
	Delavirdine (DLV)	First generation; rarely used	1997
	Efavirenz (EFV)	First generation	1998
	Etravirine (ETR)	Second generation	2008
	Rilpivirine (RPV)	Second generation	2011
	Dapivirine (DPV)	Second generation	2022 (Africa)
	Doravirine (DOR)	Second generation	2018
PIs: substrate mimics to inhibit viral protease and block the cleavage of viral precursor proteins	Saquinavir (SQV)	Typically used with PK enhancer RTV	1995
	Ritonavir (RTV)	Used as both an antiviral agent and a PK enhancer (inhibiting CYP3A)	1996
	Indinavir (IDV)	Typically used with PK enhancer RTV	1996
	Nelfinavir (NFV)	Inhibits both HIV-1 and HIV-2	1997
	Lopinavir (LPV)	Used along with PK enhancer RTV	2000
	Atazanavir (ATV)	Inhibits UGT1A1	2003
	Fosamprenavir (FPV)	Prodrug of amprenavir (APV)	2003
	Tipranavir (TPV)	Not a substrate mimic. Used along with PK enhancer RTV	2005
	Darunavir (DRV)	Typically used with PK enhancer RTV. Capable of backbone H-bonding	2006

(Continued)

Table 2 (Continued)

Drug class and MOA	Drug name	Note	Year approved
INSTIs: bind to IN and viral DNA complex (intasome) and selectively inhibit IN strand transfer	Raltegravir (RAL)	First-in-class	2007
	Elvitegravir (EVG)	Enhanced by CYP3A inhibitor Cobicistat	2012
	Dolutegravir (DTG)	Used as a single agent or in FDC along with ABC and 3TC	2013
	Bictegravir (BIC)	Used in FDC along with TAF and FTC	2018
	Cabotegravir (CAB)	Long acting. Approved for PrEP	2021
Entry inhibitors: blocking viral entry at attachment, post-attachment, coreceptor, or fusion	Enfuvirtide (ENF)	Polypeptide (36mer) fusion inhibitor	2003
	Maraviroc (MVC)	Coreceptor CCR5 antagonist	2007
	Ibalizumab	Monoclonal antibody targeting CD4 post-attachment	2018
	Fostemsavir (FTR)	Prodrug of attachment inhibitor temsavir (TMR)	2020
Capsid inhibitor: binds to viral capsid protein (CA) to impact capsid stability and compete against host factors crucial for virus	Lenacapavir (LEN)	Long-acting injectable (every six months)	2022

with the polymerase active site [60, 61]. Since NNRTIs do not interfere with cellular kinases or human mitochondrial DNA polymerase, they are generally better tolerated than NRTIs. Although the FDA approval of the first-generation NNRTIs nevirapine (NVP, 1996), delavirdine (DLV, 1997, discontinued), and efavirenz (EFV, 1998) validated NNRTIs as a clinically important drug class, especially in HAART with the NRTIs, these early NNRTI drugs typically exhibit low generic barrier to resistance [62]. The resistance profile was drastically improved with all second-generation NNRTIs, including FDA-approved etravirine (ETR, 2008) [63], rilpivirine (RPV, 2011) [64], and doravirine (DOR, 2018) [65], as well as dapivirine (DPV) recently approved in Africa for use with monthly vaginal rings (VR) for PrEP [66, 67]. Critical to the success of the second-generation NNRTIs is the rationally designed three-ring pharmacophore to confer high binding flexibility and adaptability [68].

Drugs Targeting Protease

HIV aspartyl protease cuts gag and gag-pol polyproteins to enable the maturation of newly assembled HIV virions. Catalytically, two aspartic acid residues, D25 and D125, are essential for its enzymatic activity [69]. Inhibitors are typically designed to mimick the transition state of viral substrates using nonhydrolyzable dipeptide (Y/F-P) hydroxyethylene isostere substituted for the scissile amide bond [70, 71]. However, peptidomimetics in general are susceptible to liver metabolizing enzyme CYP3A [72]. The only approved nonpeptidomimetic PI [73], tipranavir (TPV), which features a dihydropyrone ring as the central scaffold, strongly induces CYP3A expression. Therefore, all approved PIs are used in combination with a CYP3A inhibitor, such as low-dose ritonavir (RTV), which is both an antiviral agent and a PK enhancer [74], or Cobicistat (COBI), an analog of RTV and a nonantiviral CYP3A inhibitor [75]. Of the many PIs approved, the first-generation PIs saquinavir (SQV), RTV, indinavir (IDV), and nelfinavir (NFV) exhibit low genetic barrier to the development of broad class resistance, whereas the second-generation PIs lopinavir (LPV), atazanavir (ATV), fosamprenavir (FPV), TPV, and darunavir (DRV) demonstrate much improved resistance profiles.

Drugs Targeting Integrase Strand Transfer

Integration of viral DNA into host DNA is a required and signature step in the replication cycle of retroviruses [76, 77]. The stably maintained integrated proviral DNA in resting CD4+ T cells allows persistent infections and poses a major barrier toward eradication via antiviral therapy. Integration is carried out by the virally encoded integrase (IN), which has two distinct functions: the 3′-processing (3′-P) and the strand transfer (ST). The 3′-P nicks the terminal 3′ dinucleotide from viral DNA to generate the free 3′ OH, which serves as the nucleophile for the ST where viral DNA is inserted into cellular DNA [78]. In addition, productive HIV integration also entails IN binding to a cellular cofactor, the lens epithelium-derived growth factor (LEDGF)/p75. Although both enzymatic functions [79] and the allosteric binding to LEDGF/p75 [80] have been targeted for the discovery of IN-targeting antivirals, only selective INST inhibitors (INSTIs) have achieved clinical successes, and validated IN as a viable HIV drug target [81]. The chemical matter of selective INSTIs uniquely and invariably features a chelating triad capable of engaging with two Mg^{2+} ions at the DDE active site [82]. Raltegravir (RAL) [83] was the first-in-class INSTI developed which gained FDA approval in 2007. This milestone success was followed by the successful development and approval of four additional INSTIs, elvitegravir (EVG) [84], dolutegravir (DTG) [85], bictegravir (BIC) [86], and cabotegravir (CAB) [87]. Interestingly, CAB is approved to be used in combination with NNRTI RPV as the first LAI ART option for virologically suppressed adults with HIV-1. In addition, long-acting CAB has the potential for HIV PrEP.

Drugs Targeting Viral Entry

HIV entry consists of three major steps [88]: attachment, coreceptor binding, and fusion, all of which have been successfully targeted by FDA-approved drugs (Table 2). The infection starts with binding of the viral glycoprotein gp120 to the CD4 receptor on host cells. Pharmacologically disrupting this binding interaction led to the development of recently approved fostemsavir (FTR), the prodrug of attachment inhibitor temsavir [89, 90]. Significantly, since FTR works in a step prior to coreceptor binding, it inhibits both CCR5-tropic and CXCR4-tropic HIV.

Upon attachment, the binding to CD4 causes gp120 conformational changes to allow HIV to stably bind to the CCR5 (R5-tropic) or CXCR4 (X4-tropic) coreceptor, which is a required step for viral entry. The approved HIV drug maraviroc (MVC) [91, 92] is a CCR5 antagonist that binds to a CCR5 transmembrane pocket to allosterically alter the conformation of the extracellular domains to block viral gp120 binding. This specific mechanism also renders MVC ineffective against X4-tropic HIV and necessitates a tropism assay prior to treatment. Although resistance via tropism switch is rare due to large genetic barrier, high-level resistance to MVC was still selected, which is due to viral entry using MVC-bound CCR5 [93]. Alternatively, the outgrowth of the preexisting X4-tropic viral population can also confer resistance to MVC [94].

The sequential binding of the receptor and the coreceptor triggers fusogenic conformational changes in viral gp41 to expose a fusion peptide that inserts into the cellular plasma membrane to initiate fusion. The fusion-targeting enfuvirtide (ENF) was the first approved HIV entry inhibitor [95]. A polypeptide (36mer) derived from HIV-1 gp41 C-terminal heptad repeat (CHR), ENF interferes with CHR interaction with the highly conserved N-heptad repeat (NHR) to inhibit the formation of the required six-helix bundle (6HB) and subsequent viral entry. ENF is not orally available due to the intrinsically unfavorable physicochemical properties of peptides [96], especially enzymatic degradation and poor membrane permeability, and is administered via subcutaneous injection.

Ibalizumab [97, 98] is another HIV entry inhibitor approved. It is a humanized monoclonal antibody and a post-attachment inhibitor. Mechanistically, ibalizumab binds to a conformational epitope on domain 2 of the extracellular portion of CD4, and blocks HIV from binding to the CCR5 or CXCR4 coreceptors. Similar to the attachment inhibitor FTR, ibalizumab is active against both R5-tropic and X4-tropic HIV isolates.

Drugs Targeting Viral Capsid

The multifunctional HIV capsid protein (CA) is the target of the most recently approved HIV drug lenacapavir (LEN) [99]. Mechanistically [100, 101], LEN binds to a pocket at a conserved interface between CA protomers to interfere with the binding of host cofactors essential for multiple phases of the viral replication cycle, including NUP153 and CPSF6, which are required for viral nuclear entry. LEN shows antiviral potency at picomolar concentrations against all subtypes of HIV-1 tested, and exhibits high synergy and no cross-resistance with other classes of HIV drugs. The exceptional antiviral profile, along with the low in vivo systemic clearance and slow-release kinetics from the subcutaneous injection site, renders LEN an excellent LAI drug [102]. Remarkedly, LEN is approved to be used via injection every six months.

Antiviral Drugs Targeting Hepatitis B Virus

Hepatitis B virus (HBV) chronically infects an estimated 240–350 million people worldwide and remains a major global health threat [103, 104]. Although an effective vaccine is widely available, each year there are still around 1 million people newly infected. Individuals chronically infected with HBV are at high risk of severe liver diseases typically progressing through fibrosis, cirrhosis, and eventually hepatocellular carcinoma. HBV-associated liver diseases account for approximately 750 000 deaths annually. Fortunately, multiple drugs have been developed and marketed against chronic HBV (Table 3) [105]. Although these drugs rarely lead to functional cure, their lifelong use effectively suppresses HBV replication and reduces the risk of liver cancer. Current HBV drugs consist of two classes (Table 3): the immunomodulatory interferon, including IFN alpha-2b and PegIFN alpha-2a [106]; and direct acting nucleos(t)ide analogs (NAs) targeting reverse transcriptase (RT), the viral polymerase. Similar to HIV, HBV-encoded RT has been successfully targeted by various NAs, including cytidine L-nucleoside analog lamivudine (3TC) [107], adenosine analogs adefovir (AFV) [108] and tenofovir (TFV) [107], guanosine carbocyclic analog entecavir (ETV) [109], and L-thymidine telbivudine (TBV) [110] and its 2′-β-F analog clevudine (CLV) [111]; the latter was approved in South Korea only. Notably, while close analogs AFV and TFV were initially approved as ester prodrugs, the recent ProTide prodrug of TFV, tenofovir alafenamide (TAF) [112, 113],

Table 3 Antiviral drugs against HBV.

Drug class and MOA	Drug name	Note	Year approved
Nucleoside polymerase inhibitors: intracellularly phosphorylated, incorporated and act as chain terminators	Lamivudine (3TC)	L-nucleoside	1998
	Adefovir (AFV) dipivoxil	ANP prodrug	2002
	Entecavir (ETV)	Carbocyclic nucleoside	2005
	Telbivudine (TBV)	Thymidine L-nucleoside	2006
	Tenofovir (TFV) disoproxil fumarate (TDF)	Prodrug of ANP	2008
	Tenofovir (TFV) alafenamide (TAF)	ProTide prodrug of ANP	2016
	Clevudine (CLV)	L-nucleoside; approved in Korea	2006
Interferons (IFNs): immunomodulators	IFN alpha-2b	Three times a week; injection	1991
	PegIFN alpha-2a	Long acting (once weekly, injection)	2005

allows much lower dosing and improved safety profile due to the liver-specific delivery. Interestingly, HBV and HIV share two prominent therapeutic commonalities: (1) many NAs approved for treating HBV are also HIV drugs, including 3TC, tenofovir disoproxil fumarate (TDF) and tenofovir alafenamide TAF and (2) these drugs do not cure the infection, presumably because they do not impact the viral reservoirs. Adefovir was also initially being developed as an HIV drug; however, its toxicity at the doses required for systemic usage limited its use, compared to a much lower dose required for a liver-tropic drug. Current efforts in the field are looking to identify a combination regimen that suppresses viral replication, reduces the production of HBV surface antigen, and reactivates an HBV-specific immune response with the hope of achieving high rates of functional cure with a short duration of therapy.

Antiviral Drugs Targeting Hepatitis Delta Virus

Hepatitis D virus (HDV) is a single-stranded (−)-RNA virus whose genome folds into a rod-like structure with no nucleocapsid. HDV requires the presence of HBV to propagate because it uses the HBV surface antigens (HBsAg) to complete its viral coat. It is the HBsAg in the HDV viral coat that is recognized by the sodium taurocholate cotransporter polypeptide (NTCP) receptor on the surface of an hepatocyte facilitating viral entry. The HD antigen (HDAg) expressed as two isoforms (small and large) is the only protein expressed by HDV. Since HDV produces no other proteins, it uses cellular DNA-dependent RNA polymerases to replicate its genome. Post-translational modification of HDAgs modifies their function and therefore enables modulation of the viral life cycle. Coexistence of HDV and HBV leads to more severe liver disease and a more rapid progression to cirrhosis and liver cancer [114, 115].

Current therapies to treat HDV infection consist of pegylated interferon-α (PEG-IFN) and newly introduced bulevirtide (Myrcludex B, BLV), which blocks viral entry (Table 4). Pegylated interferon has shown limited efficacy [116]. However, BLV, which is a myristolated synthetic lipopeptide mimicking the N-terminal region of the pre-S1 domain of HBsAg, has demonstrated promising efficacy and enhanced efficacy has been observed in combination with PEG-IFN [114].

Table 4 Antiviral drugs against HDV.

Drug class and MOA	Drug name	Note	Year approved
Therapeutic protein Innate immune activation	Pegylated interferon-α (PEG-IFN)	Subcutaneous delivery; limited efficacy	Mid-2000s
Myristolated synthetic polypeptide	Bulevirtide (BLV, Myrcludex B)	Inhibits viral entry by blocking the NTCP receptor; only approved by the EMEA in the European Union and not by the US FDA	2020

Antiviral Drugs Targeting Hepatitis C Virus

The hepatitis C virus (HCV) is a liver tropic virus that chronically infects over 75 million individuals worldwide and is a leading cause of chronic liver disease and liver cancer [117]. This (+)-sense RNA virus replicates in the cytoplasm of hepatocytes, does not integrate into the host genome, nor does it appear to leave a reservoir of genetic material in the infected cell [118].

Drugs targeting the hepatitis C virus (HCV) can be classified into two categories: immunomodulators and direct-acting antivirals (Table 5). The first drug developed for curing HCV was the immunomodulator interferon-α (IFN). The evolution of interferon therapies ultimately led to the development of more effective pegylated interferons, PEG-INTRON and PEGASYS. The use of ribavirin (RBV) in conjunction with IFN was shown to increase the cure rates of IFN therapies to 40%; however, significant adverse side effects and genotype coverage limited the use of IFN regimens [119].

The evolution of HCV curative therapies eventually focused on the development of direct-acting antivirals (DAAs) and the elimination of IFN in clinical practice (Table 5). DAA therapeutic development targeted three viral nonstructural proteins, NS3/4a protease, NS5A replication complex protein, and the NS5B RNA-dependent RNA polymerase [22, 120]. The launch of the NS3/4a PIs telaprevir and boceprevir each in combination with IFN not only led to improved cure rates but also added to the undesirable side effect profile already seen with IFN use. Approval of the liver-targeted NS5B polymerase inhibitor sofosbuvir ushered in the era of IFN-free HCV curative therapies with short 8–12-week treatment duration, pan-genotypic profile, high barrier to resistance, minimal side effect profile, and cure rates surpassing 95% [121, 122]. Combinations that included sofosbuvir with either RBV, the NS3/4a PI, simeprevir or NS5A inhibitors, ledipasvir or daclatasvir soon became the regimens of choice for genotypes 1, 2, and 3 HCV patients [123–126]. Eventually, DAA regimens that included combinations of NS5A inhibitors and NS3/4a PIs with or without a non-nucleoside NS5B inhibitor (paritaprevir/ombitasvir/dasabuvir, grazoprevir/elbasvir) were introduced but had limited uptake because of their poor genotype coverage and resistance profile [127, 128]. However, these were eventually surpassed by the highly potent and pan-genotypic regimes sofosbuvir/valpatasvir and glecaprevir/pibrentasvir which dominate the market providing cure rates of over 95% [22, 129, 130]. To address any patients who exhibited breakthrough resistance or relapse, primarily from the NS5A component of these regimens, the triple combination regimen sofosbuvir/valpatasvir/voxilaprevir was introduced [131].

Antiviral Drugs Targeting Acute RNA Respiratory Viruses

Drugs Targeting Human Influenza Viruses

Seasonal flu epidemics are caused by human influenza A virus (IAV) and influenza B virus (IBV). IAVs are classified into many subtypes based on the two viral glycoproteins

Table 5 Antiviral drugs against HCV.

Drug class and MOA	Drug name	Note	Year approved
Therapeutic protein Innate immune activation	Interferon-α (IFN)	Subcutaneous delivery; efficacy was improved with the development of pegylated IFN	1991
Nucleoside NS5B pol inhibitors: intracellularly phosphorylated, incorporated and act as chain terminators	Ribavirin (RBV)	Additional MOAs include inhibiting IMPDH; inducing Th2 to Th1 switch; lethal mutagenesis; and inhibiting RNA capping. Used in combination with PegIFN	1999
	Sofosbuvir (SOF)	ProTide prodrug; liver-targeted delivery.	2013
Non-nucleoside NS5B pol inhibitor: allosterically inhibit pol	Dasabuvir	Genotype 1; binds to HCV NS5B palm I site	2014
	Beclabuvir (BCV)	Binds to NS5B thumb domain site I	Post marketing
NS3/4 protease inhibitors: bid to the active site inhibit the cleavage of HCV polyproteins	Boceprevir (BOC)	Genotype 1; carbonyl amide warhead; withdrawn in 2015	2011
	Telaprevir (TVR)	Genotype 1; carbonyl amide warhead; discontinued in 2014	2011
	Simeprevir (SMV)	Genotype 1	2013
	Paritaprevir (PTV)	Genotype 1; discontinued in 2018	2014
	Grazoprevir (GZR)	Genotypes 1 and 4	2015
	Asunaprevir (ASV)	Genotype 1	2014
	Voxilaprevir	Pan-genotypic; part of combination with SOF and velpatasvir	2017
NS5A inhibitors: bind to HCV NS5A to impact genome replication and virion assembly	Daclatasvir (DCV)	In combination with ASV; genotype 1	2014
	Ledipasvir (LDV)	In combination with SOF; genotype 1	2014
	Ombitasvir (OBV)	Used in combination therapy for genotype 4	2014
	Elbasvir (EBR)	Used in combination with GZR for genotype 1 or 4	2016
	Velpatasvir (VEL)	Used in combination with SOF for all major genotypes	2020
	Pibrentasvir (PIB)	Used in combination with Glecaprevir for all major genotypes	2017

hemagglutinin (H1-H18) and neuraminidase (N1-N11). Two IAV subtypes, A(H1N1) and A(H3N2), along with the two co-circulating IBV lineages, B/Yamagata and B/Victoria, are the main circulating strains of seasonal flu in human populations [132], causing significant mortality globally [133]. Antiviral drug discovery against flu has a long history [134, 135], with the first drug amantadine (AMT) approved by FDA in 1968. By binding to and inhibiting the IAV M2 proton-selective ion channel to block viral uncoating [136], AMT and the later approved rimantadine (RIM) constituted the first flu drug class. However, mutations such as S31N have rendered seasonal IAV completely resistant to AMT and RIM [137]. Current treatment of flu relies heavily on three approved neuraminidase inhibitors (Table 6) [138], the inhalation drug zanamivir (ZAN) [113], the oral drug oseltamivir (OTV) [139, 140], and the IV drug peramivir

Table 6 Antiviral drugs against acute RNA viral infections.

Virus	Drug name	MOA/note	Year approved
Influenza virus	Amantadine (AMT)	Block influenza A virus M2 channel to inhibit viral uncoating. Not being prescribed any longer as all circulating strains are resistant.	1968
	Rimantadine (RIM)		1993
	Zanamivir (ZAN)	Block viral release via inhibiting viral neuraminidase. Approved for treating influenza A and B. Administrated via inhalation (ZAN), oral (OTV), and IV (PRV).	1999
	Oseltamivir (OTV)		1999
	Peramivir (PRV)		2014
	Favipiravir (FPV)	Intracellularly converted into FPV-RTP, incorporated by viral polymerase (PB1), and causes chain termination and/or lethal mutagenesis. Approved for stockpiling.	2014 (Japan)
	Baloxavir marboxil (BXM)	Approved for treating both influenza A and B. Blocking viral cap-snatching by inhibiting viral PA endonuclease.	2018 (USA and Japan)
SARS-CoV-2	Remdesivir (RDV)	ProTide prodrug of C-nucleoside analog. Phosphorylated, incorporated and acts as a delayed chain terminator. IV infusion.	2020
	Molnupiravir (MPV)	Ester prodrug of cytidine analog. Phosphorylated, incorporated and causes lethal mutagenesis. Oral pill.	EUA 2021
	Mindeudesivir (VV116)	Ester prodrug of 7-D C-nucleoside analog. Phosphorylated, incorporated and acts as a delayed chain terminator. Oral pill.	2023 (China)
	Nirmatrelvir (NMV)	Inhibiting SARS-CoV-2 main protease (Mpro). Used in combination (Paxlovid) with PK enhancer ritonavir.	2023
	Ensitrelvir	Inhibiting SARS-CoV-2 main protease (Mpro)	2022 (Japan)

(PRV) [141]. These drugs block the release of new IAV and IBV viral particles by inhibiting neuraminidase. The most recently successful flu drugs target the viral polymerase complex [142], a heterotrimeric RNA-dependent RNA polymerase complex comprising subunits PA (polymerase acidic protein), PB1 (polymerase basic protein 1), and PB2 (polymerase basic protein 2). PB1 provides the viral polymerase activity and is targeted by nucleobase analog favipiravir (FPV, Table 6), which upon intracellular activation to FPV-TP, induces lethal mutagenesis [143] and/or terminates the viral RNA chain [144]. FPV was approved in Japan for stockpiling but has not been approved by FDA. PB2 and PA work concertedly to carry out the cap-snatching [145] process required for viral mRNA capping, where PB2 binds to the 5′ cap of cellular mRNA while PA cleaves cellular mRNA a few nucleotides downstream of the cap. The endonuclease activity of PA is the target of the most recently approved flu drug baloxavir marboxil (BXM, Table 6) [146].

Drugs Targeting SARS-CoV-2

As the most recent viral pandemic, severe acute respiratory syndrome coronavirus 2 (SARS-CoV-2) has caused mortality and morbidity rivaled by few human viruses. Although multiple highly successful vaccines are widely available, they do not effectively protect against infection, due to the quick evolution of SARS-CoV-2 variants [147]. Antiviral drugs are critical in mitigating SARS-CoV-2-associated diseases and deaths. With the extraordinary resources and efforts [148], two mechanistically distinct classes of small-molecule drugs have been successfully developed and marketed, targeting viral RdRp and main protease (Mpro), respectively (Table 6). The first FDA-approved small-molecule SARS-CoV-2 drug was RdRp inhibitor remdesivir (RDV) [149], an adenosine C-nucleoside and a ProTide prodrug [150] initially developed as an antiviral drug against Ebolavirus [151]. However, RDV is not orally available presumably due to high liver abundance of carboxylesterase 1 (CES1) resulting in high first-pass hepatic extraction [152] and low systemic exposure. Instead, it is administered to hospitalized patients via intravenous (IV) infusion. VV116 [153], a triester prodrug of the C7 deuterated GS-441524, the RDV parent nucleoside, was developed and approved in China as an oral drug. Another nucleoside prodrug, molnupiravir (MPV) [154], an ester prodrug of cytidine analog NHC [155], works via a distinct mechanism of inducing lethal mutagenesis as the incorporated NHC-TP creates base mispairing [156].

The other successful drug class of SARS-CoV-2 targets the viral main protease (Mpro) [157–159], which is a cysteine protease with a Cys145-His41 catalytic dyad to cleave the viral polyproteins pp1a and pp1ab. PF-07321332 (nirmatrelvir), evolved from a SARS-CoV-1 Mpro inhibitor [160], and chemically featuring a cyano warhead and a leucine-mimicking bicyclic noncanonical amino acid, was developed as a Mpro covalent inhibitor and an oral drug [161]. However, since nirmatrelvir (Table 6) is susceptible to oxidative metabolism by CYP3A, it is clinically used in combination with low-dose ritonavir, a CYP3A inhibitor, forming a combination regimen called Paxlovid. In the meantime, S-217622 (ensitrelvir), a noncovalent SARS-CoV-2 Mpro inhibitor, was discovered via virtual screening followed by structure-guided optimization [162, 163]. Ensitrelvir gained approval in Japan as a monotherapy (Table 6).

New Trends and Challenges in Antiviral Drug Development

The success of antiviral drug discovery is epitomized by the success achieved in HIV and HCV drug developed in the past three decades. In addition to dozens of single agents and multiple drug classes, milestone achievements in HIV drugs also included the advent of cocktail therapy to increase the genetic barrier to resistance, and the regimen simplification to once-daily pills for enhanced adherence marked by the approval of Atripla [164]. For HCV, the achievement of curative regimens characterized by single-pill combination regimens that deliver >95% cure rates after only 8–12 weeks of therapy have led to global efforts to achieve HCV elimination. Yet new challenges remain in the field of antiviral drug discovery and development.

An evolving area in HIV antiviral therapy is the LAIs for treatment and PrEP [165]. Currently, there are two FDA-approved LAI regimens: Cabenuva, which consists of INSTI cabotegravir and NNRTI rilpivirine as two separate injectables; and Sunlenca with CA inhibitor lenacapavir as the only active ingredient. Cabenuva is administered monthly or every two months, whereas lenacapavir is given every six months. The extraordinarily long efficacy of these regimens, particularly with lenacapavir, can significantly enhance patient adherence to achieve improved treatment outcome. However, as a monotherapy, lenacapavir can be confronted with the quick development of resistance. Developing a lenacapavir-based dual long-acting regimen represents both a new opportunity and a challenge. EFdA (Table 2), the NRTTI, also exhibits PK properties suitable for long-acting PrEP. Second-generation NNRTI DPV has been approved in Africa as monthly virginal ring for PrEP (Table 2).

HIV functional cure [166] is rarely achieved and will continue to be the subject of intense research and development. Due to the latent HIV infection in resting CD4+ T cells, the persistence is likely lifelong even in patients under effective ART [167–170]. Mechanistically, HIV is integrated in the genome of CD4+ T cells to form the reservoir that cannot be eliminated by the host immune system due to insufficient viral proteins made. Therapeutic strategies to a cure typically aim to induce viral protein production while enhancing HIV-specific immunity [171]. A few immunotherapeutic approaches have been reported, featuring Toll-like receptor 7 (TLR7) agonist [172], broadly neutralizing antibody-derived CAR T cells [173], and anti-HIV-1 antibody therapy [174]. In parallel, HBV functional cure is also rarely achieved with current drugs, which necessitates concerted efforts from the scientific and medical communities to identify new antiviral and immune mechanisms aimed at infection cure. Virological barriers toward HBV functional cure include the persistence of viral covalently closed circular DNA, the integration of viral DNA into the cellular genome and the hugely excessive antigen load, particularly hepatitis B surface antigen (HBsAg) [105]. The major immunological barrier is the exhaustion of host immune response within the liver microenvironment [175]. Multiple targeted therapies are now in human studies toward achieving HBV functional cure [176].

Besides influenza virus and SARS-CoV-2, for which specific antiviral drugs have been developed, other emerging and re-emerging viral infections with pandemic potential continue to draw significant efforts from the antiviral research and development community [177]. Specifically, drug discovery efforts have been directed to target respiratory syncytial virus (RSV) against which numerous small-molecule drugs are in the development pipeline [178, 179], flaviviruses [180–182], filoviruses [183–185], paramyxoviruses [186, 187], alphaviruses [188–190], henipaviruses [191, 192], and bunyaviruses [193]. One particular area of interest in viral pandemic preparedness is to develop broad-spectrum antiviral drugs [194] to provide the initial rapid countermeasures. Ribavirin (RBV) [195, 196] and favipiravir (FPV) [197, 198] are the prototypical broad-spectrum antivirals with approved indications against RSV and HCV (RBV) and flu (FPV), and a wide range of additional antiviral activities reported. Of the emerging DNA viruses, poxviruses, particularly the variola virus (VARV) that causes smallpox and the ongoing outbreak of mpox virus (MPXV), are targets of two approved drugs, tecovirimat [199, 200] and brincidofovir [201, 202], as well as substantial ongoing drug discovery efforts [203–205]. Other double-stranded DNA viruses, such as human adenovirus [206, 207] and human papillomavirus [208, 209], have also received significant research attention.

Overall, the landscape of antiviral drugs has been ever expanding, with many mechanistically distinct drug classes introduced and/or enriched, predominantly targeting major chronic viral infections. The current lack of a cure for these viruses will likely continue to drive major research and development efforts. Looking ahead, the field of antiviral drug discovery is presented with plenty opportunities and massive challenges in developing new drugs, preferably broad-spectrum antivirals, against emerging viruses of pandemic potential.

References

1 Gayle HD, Hill GL. Global impact of human immunodeficiency virus and AIDS. *Clin Microbiol Rev*. 2001;14(2): 327–35.

2 Sheena BS, Hiebert L, Han H, Ippolito H, Abbasi-Kangevari M, Abbasi-Kangevari Z, et al. Global, regional, and national burden of hepatitis B, 1990–2019: a systematic analysis for the Global Burden of Disease Study 2019. *Lancet Gastroenterol Hepatol*. 2022;7(9): 796–829.

3 Yang J, Qi JL, Wang XX, Li XH, Jin R, Liu BY, et al. The burden of hepatitis C virus in the world, China, India, and the United States from 1990 to 2019. *Front Public Health*. 2023;11: 1041201.

4 Maleki F, Welch V, Lopez SMC, Cane A, Langer J, Enstone A, et al. Understanding the global burden of influenza in adults aged 18–64 years: a systematic literature review from 2012 to 2022. *Adv Ther*. 2023;40(10): 4166–88.

5 Thèves C, Crubézy E, Biagini P. History of smallpox and its spread in human populations. *Microbiol Spectr*. 2016;4(4). doi: 10.1128/microbiolspec.PoH-0004-2014.

6 Isasi F, Naylor MD, Skorton D, Grabowski DC, Hernandez S, Rice VM. Patients, families, and communities COVID-19 impact assessment: lessons learned and compelling needs. *NAM Perspect*. 2021; 2021. doi: 10.31478/202111c.

7 Alibek K, Baiken Y, Kakpenova A, Mussabekova A, Zhussupbekova S, Akan M, et al. Implication of human herpesviruses in oncogenesis through immune evasion and supression. *Infect Agents Cancer*. 2014;9(1): 3.

8 Prusoff WH. Idoxuridine or how it all began. In: De Clercq E, editor. Clinical Use of Antiviral Drugs. Boston, MA: Springer US; 1988. p. 15–24.

9 Prusoff WH. Synthesis and biological activities of iododeoxyuridine, an analog of thymidine. *Biochim Biophys Acta*. 1959;32(1): 295–6.

10 De Clercq E, Li GD. Approved antiviral drugs over the past 50 years. *Clin Microbiol Rev*. 2016;29(3): 695–747.

11 Xu SJ, Sun L, Liu XY, Zhan P. Opportunities and challenges in new HIV therapeutic discovery: What is the next step? *Expert Opin Drug Dis*. 2023;18(11): 1195–9.

12 Bowe ZG, Shanmugam S, Hall-Thomsen H, Gubernick SI. The landscape and market for HIV therapies. *Nat Rev Drug Discov*. 2023;23(5): 334–5

13 Chou R, Spencer H, Bougatsos C, Blazina I, Ahmed A, Selph S. Preexposure prophylaxis for the prevention of HIV: Updated evidence report and systematic review for the US preventive services task force. *JAMA*. 2023; 330(8): 746–63.

14 Spinner CD, Boesecke C, Zink A, Jessen H, Stellbrink H-J, Rockstroh JK, et al. HIV pre-exposure prophylaxis (PrEP): a review of current knowledge of oral systemic HIV PrEP in humans. *Infection*. 2016;44(2): 151–8.

15 Truong WR, Schafer JJ, Short WR. Once-daily, single-tablet regimens for the treatment of HIV-1 infection. P & T: a peer-reviewed journal for formulary management. 2015;40(1): 44–55.

16 Oh KS, Seo GH, Choi HK, Han E. Effect of single tablet regimen on prescription trends for treatment-naïve patients with HIV/AIDS in Korea. *Sci Rep*. 2022;12(1): 2031.

17 Harris TN, Buscemi LR. Evaluation of treatment simplification strategies in patients living with HIV with multi-drug resistance. *Ther Adv Infect Dis*. 2023;10: 20499361221149869.

18 Slama L, Porcher R, Linard F, Chakvetadze C, Cros A, Carillon S, et al. Injectable long acting antiretroviral for HIV treatment and prevention: perspectives of potential users. *BMC Infect Dis*. 2023;23(1): 98.

19 Venkatesan P. Long-acting injectable ART for HIV: a (cautious) step forward. *Lancet Microbe*. 2022;3(2): e94.

20 Nachega JB, Scarsi KK, Gandhi M, Scott RK, Mofenson LM, Archary M, et al. Long-acting antiretrovirals and HIV treatment adherence. *Lancet HIV*. 2023;10(5): e332–e42.

21 Sofia MJ. Enter sofosbuvir: the path to curing HCV. *Cell*. 2016;167(1): 25–9.

22 Manns MP, Maasoumy B. Breakthroughs in hepatitis C research: from discovery to cure. *Nat Rev Gastroenterol Hepatol*. 2022;19(8): 533–50.

23 Lan K, Luo MH. Herpesviruses: epidemiology, pathogenesis, and interventions. *Virol Sin*. 2017;32(5): 347–8.

24 Cheong L, Rich MA, Eidinoff ML. Introduction of the 5-halogenated uracil moiety into deoxyribonucleic acid of mammalian cells in culture. *J Biol Chem.* 1960;235(5): 1441–7.

25 Whitley RJ. The past as prelude to the future: history, status, and future of antiviral drugs. *Ann Pharmacother.* 1996;30(9): 967–71.

26 Carmine AA, Brogden RN, Heel RC, Speight TM, Avery GS. Trifluridine: a review of its antiviral activity and therapeutic use in the topical treatment of viral eye infections. *Drugs.* 1982;23(5): 329–53.

27 De Clercq E. Discovery and development of BVDU (brivudin) as a therapeutic for the treatment of herpes zoster. *Biochem Pharmacol.* 2004;68(12): 2301–15.

28 Sagar S, Kaur M, Minneman KP. Antiviral lead compounds from marine sponges. *Mar Drugs.* 2010;8(10): 2619–38.

29 Whitley RJ, Tucker BC, Kinkel AW, Barton NH, Pass RF, Whelchel JD, et al. Pharmacology, tolerance, and antiviral activity of vidarabine monophosphate in humans. *Antimicrob Agents Chemother.* 1980;18(5): 709–15.

30 Freeman S, Gardiner JM. Acyclic nucleosides as antiviral compounds. *Mol Biotechnol.* 1996;5(2): 125–37.

31 Clercq ED, Holý A. Acyclic nucleoside phosphonates: a key class of antiviral drugs. *Nat Rev Drug Discov.* 2005;4(11): 928–40.

32 Schaeffer HJ, Beauchamp L, de Miranda P, Elion GB, Bauer DJ, Collins P. 9-(2-Hydroxyethoxymethyl)guanine activity against viruses of the herpes group. *Nature.* 1978;272(5654): 583–5.

33 Elion GB, Furman PA, Fyfe JA, Miranda Pd, Beauchamp L, Schaeffer HJ. Selectivity of action of an antiherpetic agent, 9-(2-hydroxyethoxymethyl)guanine. 1977;74(12): 5716–20.

34 Faulds D, Heel RC. Ganciclovir. A review of its antiviral activity, pharmacokinetic properties and therapeutic efficacy in cytomegalovirus infections. *Drugs.* 1990;39(4): 597–638.

35 Crumpacker CS. Ganciclovir. *N Engl J Med.* 1996;335(10): 721–9.

36 Crumpacker CS. Mechanism of action of foscarnet against viral polymerases. *Am J Med.* 1992; 92(2a):3s–7s.

37 Landovitz RJ, Scott H, Deeks SG. Prevention, treatment and cure of HIV infection. *Nat Rev Microbiol.* 2023;21(10): 657–70.

38 Broder S. The development of antiretroviral therapy and its impact on the HIV-1/AIDS pandemic. *Antiviral Res.* 2010;85(1): 1–18.

39 Shafer RW, Vuitton DA. Highly active antiretroviral therapy (Haart) for the treatment of infection with human immunodeficiency virus type 1. *Biomed Pharmacother.* 1999;53(2): 73–86.

40 Sarafianos SG, Marchand B, Das K, Himmel DM, Parniak MA, Hughes SH, et al. Structure and function of HIV-1 reverse transcriptase: molecular mechanisms of polymerization and inhibition. *J Mol Biol.* 2009;385(3): 693–713.

41 Uchitsubo K, Masuda J, Akazawa T, Inoue R, Tsukada K, Gatanaga H, et al. Nucleos(t)ide reverse transcriptase inhibitor-sparing regimens in the era of standard 3-drug combination therapies for HIV-1 infection. *Glob Health Med.* 2020;2(6): 384–7.

42 Horwitz JP, Chua J, Noel M. Nucleosides. V. The monomesylates of 1-(2′-Deoxy-β-D-lyxofuranosyl)thymine1,2. *J Org Chem*. 1964;29(7): 2076–8.

43 Brook I. Approval of zidovudine (AZT) for acquired immunodeficiency syndrome: a challenge to the medical and pharmaceutical communities. *JAMA*. 1987;258(11): 1517.

44 Fischl MA, Richman DD, Grieco MH, Gottlieb MS, Volberding PA, Laskin OL, et al. The Efficacy of Azidothymidine (AZT) in the treatment of patients with AIDS and AIDS-related complex. *N Engl J Med*. 1987;317(4): 185–91.

45 Robins MJ, McCarthy JR, Jr., Robins RK. Purine Nucleosides. XII. The Preparation of 2′,3′-Dideoxyadenosine, 2′,5′-Dideoxyadenosine, and 2′,3′,5′-Trideoxyadenosine from 2′-Deoxyadenosine*. *Biochemistry*. 1966;5(1): 224–31.

46 Yarchoan R, Mitsuya H, Thomas RV, Pluda JM, Hartman NR, Perno C-F, et al. In vivo activity against HIV and favorable toxicity profile of 2′,3′-dideoxyinosine. 1989;245(4916): 412–5.

47 Horwitz JP, Chua J, Rooge MAD, Noel M, Klundt IL. Nucleosides. IX. The formation of 2′,3′-unsaturated pyrimidine nucleosides via a novel β-elimination reaction1,2. *J Org Chem*. 1966;31(1): 205–11.

48 Horwitz JR, Chua J, Da Rooge MA, Noel M, Klundt IL. Nucleosides. IX. The formation of 2′,2′-unsaturated pyrimidine nucleosides via a novel beta-elimination reaction. *J Org Chem*. 1966;31(1): 205–11.

49 Soudeyns H, Yao XI, Gao Q, Belleau B, Kraus JL, Nguyen-Ba N, et al. Anti-human immunodeficiency virus type 1 activity and in vitro toxicity of 2′-deoxy-3′-thiacytidine (BCH-189), a novel heterocyclic nucleoside analog. *Antimicrob Agents Chemother*. 1991;35(7): 1386–90.

50 Coates JA, Cammack N, Jenkinson HJ, Mutton IM, Pearson BA, Storer R, et al. The separated enantiomers of 2′-deoxy-3′-thiacytidine (BCH 189) both inhibit human immunodeficiency virus replication in vitro. *Antimicrob Agents Chemother*. 1992;36(1): 202–5.

51 Schinazi RF, Chu CK, Peck A, McMillan A, Mathis R, Cannon D, et al. Activities of the four optical isomers of 2′,3′-dideoxy-3′-thiacytidine (BCH-189) against human immunodeficiency virus type 1 in human lymphocytes. *Antimicrob Agents Chemother*. 1992;36(3): 672–6.

52 Liotta DC, Painter GR. Discovery and development of the anti-human immunodeficiency virus drug, emtricitabine (Emtriva, FTC). *Acc Chem Res*. 2016;49(10): 2091–8.

53 Vince R, Hua M. Synthesis and anti-HIV activity of carbocyclic 2′,3′-didehydro-2′,3′-dideoxy 2,6-disubstituted purine nucleosides. *J Med Chem*. 1990;33(1): 17–21.

54 Kim CU, Luh BY, Misco PF, Bronson JJ, Hitchcock MJM, Ghazzouli I, et al. Acyclic purine phosphonate analogs as antiviral agents. Synthesis and structure-activity relationships. *J Med Chem*. 1990;33(4): 1207–13.

55 Brinkman K, Kakuda TN. Mitochondrial toxicity of nucleoside analogue reverse transcriptase inhibitors: a looming obstacle for long-term antiretroviral therapy? *Curr Opin Infect Dis*. 2000;13(1): 5–11.

56 Koczor CA, Lewis W. Nucleoside reverse transcriptase inhibitor toxicity and mitochondrial DNA. *Expert Opin Drug Metab Toxicol*. 2010;6(12): 1493–504.

57 Michailidis E, Huber AD, Ryan EM, Ong YT, Leslie MD, Matzek KB, et al. 4′-Ethynyl-2-fluoro-2′-deoxyadenosine (EFdA) inhibits HIV-1 reverse transcriptase with multiple mechanisms. *J Biol Chem*. 2014;289(35): 24533–48.

58 Salie ZL, Kirby KA, Michailidis E, Marchand B, Singh K, Rohan LC, et al. Structural basis of HIV inhibition by translocation-defective RT inhibitor 4′-ethynyl-2-fluoro-2′-deoxyadenosine (EFdA). 2016;113(33): 9274–9.

59 Stoddart CA, Galkina SA, Joshi P, Kosikova G, Moreno ME, Rivera JM, et al. Oral administration of the nucleoside EFdA (4′-ethynyl-2-fluoro-2′-deoxyadenosine) provides rapid suppression of hiv viremia in humanized mice and favorable pharmacokinetic properties in mice and the rhesus macaque. *Antimicrob Agents Chemother*. 2015;59(7): 4190–8.

60 Esnouf R, Ren J, Ross C, Jones Y, Stammers D, Stuart D. Mechanism of inhibition of HIV-1 reverse transcriptase by non-nucleoside inhibitors. *Nat Struct Biol*. 1995;2(4): 303–8.

61 Spence RA, Kati WM, Anderson KS, Johnson KA. Mechanism of inhibition of HIV-1 reverse transcriptase by nonnucleoside inhibitors. *Science*. 1995;267(5200): 988–93.

62 Shafer RW. Rationale and uses of a public HIV drug-resistance database. *J Infect Dis*. 2006;194: S51-S8.

63 Andries K, Azijn H, Thielemans T, Ludovici D, Kukla M, Heeres J, et al. TMC125, a novel next-generation nonnucleoside reverse transcriptase inhibitor active against nonnucleoside reverse transcriptase inhibitor-resistant human immunodeficiency virus type 1. *Antimicrob Agents Chemother*. 2004;48(12): 4680–6.

64 Molina J-M, Cahn P, Grinsztejn B, Lazzarin A, Mills A, Saag M, et al. Rilpivirine versus efavirenz with tenofovir and emtricitabine in treatment-naive adults infected with HIV-1 (ECHO): a phase 3 randomised double-blind active-controlled trial. *Lancet*. 2011;378(9787): 238–46.

65 Hwang C, Lai MT, Hazuda D. Rational design of doravirine: from bench to patients. *ACS Infect Dis*. 2020;6(1): 64–73.

66 Baeten JM, Palanee-Phillips T, Mgodi NM, Mayo AJ, Szydlo DW, Ramjee G, et al. Safety, uptake, and use of a dapivirine vaginal ring for HIV-1 prevention in African women (HOPE): an open-label, extension study. *Lancet HIV*. 2021;8(2): e87–e95.

67 Gollub EL, Vaughan R. U.S. Women need the dapivirine ring, too: FDA as structural barrier to HIV prevention for women. *AIDS Educ Prev*. 2022;34(4): 311–24.

68 Das K, Clark AD, Lewi PJ, Heeres J, de Jonge MR, Koymans LMH, et al. Roles of conformational and positional adaptability in structure-based design of TMC125-R165335 (etravirine) and related non-nucleoside reverse transcriptase inhibitors that are highly potent and effective against wild-type and drug-resistant HIV-1 variants. *J Med Chem*. 2004;47(10): 2550–60.

69 Davies DR. The structure and function of the aspartic proteinases. *Annu Rev Biophys Biophys Chem*. 1990;19(1): 189–215.

70 Dreyer GB, Metcalf BW, Tomaszek TA, Carr TJ, Chandler AC, Hyland L, et al. Inhibition of human immunodeficiency virus 1 protease in vitro: rational design of substrate analogue inhibitors. *Proc Natl Acad Sci U S A*. 1989;86(24): 9752–6.

71 Huff JR. HIV protease: a novel chemotherapeutic target for AIDS. *J Med Chem*. 1991;34(8): 2305–14.

72 Wacher VJ, Silverman JA, Zhang Y, Benet LZ. Role of P-glycoprotein and cytochrome P450 3A in limiting oral absorption of peptides and peptidomimetics. *J Pharm Sci*. 1998;87(11): 1322–30.

73 Doyon L, Tremblay S, Bourgon L, Wardrop E, Cordingley MG. Selection and characterization of HIV-1 showing reduced susceptibility to the non-peptidic protease inhibitor tipranavir. *Antiviral Res*. 2005;68(1): 27–35.

74 Kempf DJ, Marsh KC, Kumar G, Rodrigues AD, Denissen JF, McDonald E, et al. Pharmacokinetic enhancement of inhibitors of the human immunodeficiency virus protease by coadministration with ritonavir. *Antimicrob Agents Chemother*. 1997;41(3): 654–60.

75 Lepist EI, Phan TK, Roy A, Tong L, Maclennan K, Murray B, et al. Cobicistat boosts the intestinal absorption of transport substrates, including HIV protease inhibitors and GS-7340, in vitro. *Antimicrob Agents Chemother*. 2012;56(10): 5409–13.

76 Craigie R. The molecular biology of HIV integrase. *Future Virol*. 2012;7(7): 679–86.

77 Engelman A, Mizuuchi K, Craigie R. HIV-1 DNA integration: mechanism of viral DNA cleavage and DNA strand transfer. *Cell*. 1991;67(6): 1211–21.

78 Maertens GN, Engelman AN, Cherepanov P. Structure and function of retroviral integrase. *Nat Rev Microbiol*. 2022;20(1): 20–34.

79 Pommier Y, Johnson AA, Marchand C. Integrase inhibitors to treat HIV/Aids. *Nat Rev Drug Discov*. 2005;4(3): 236–48.

80 Maehigashi T, Ahn S, Kim UI, Lindenberger J, Oo A, Koneru PC, et al. A highly potent and safe pyrrolopyridine-based allosteric HIV-1 integrase inhibitor targeting host LEDGF/p75-integrase interaction site. *PLoS Pathogens*. 2021;17(7): e1009671.

81 Hazuda DJ. HIV integrase as a target for antiretroviral therapy. *Curr Opin HIV AIDS*. 2012;7(5): 383–9.

82 Barreca ML, Ferro S, Rao A, De Luca L, Zappalà M, Monforte A-M, et al. Pharmacophore-based design of HIV-1 integrase strand-transfer inhibitors. *J Med Chem*. 2005;48(22): 7084–8.

83 Summa V, Petrocchi A, Bonelli F, Crescenzi B, Donghi M, Ferrara M, et al. Discovery of raltegravir, a potent, selective orally bioavailable HIV-integrase inhibitor for the treatment of HIV-AIDS infection. *J Med Chem*. 2008;51(18): 5843–55.

84 Shimura K, Kodama E, Sakagami Y, Matsuzaki Y, Watanabe W, Yamataka K, et al. Broad antiretroviral activity and resistance profile of the novel human immunodeficiency virus integrase inhibitor elvitegravir (JTK-303/GS-9137). *J Virol*. 2008;82(2): 764–74.

85 Bailly F, Cotelle P. The preclinical discovery and development of dolutegravir for the treatment of HIV. *Expert Opin Drug Discov*. 2015;10(11): 1243–53.

86 Zeuli J, Rizza S, Bhatia R, Temesgen Z. Bictegravir, a novel integrase inhibitor for the treatment of HIV infection. *Drugs Today* (Barcelona, Spain: 1998). 2019;55(11): 669–82.

87 Bares SH, Scarsi KK. A new paradigm for antiretroviral delivery: long-acting cabotegravir and rilpivirine for the treatment and prevention of HIV. *Curr Opin HIV AIDS*. 2022;17(1): 22–31.

88 Wilen CB, Tilton JC, Doms RW. HIV: cell binding and entry. *Cold Spring Harb Perspect Med*. 2012;2(8): a006866.

89 Nowicka-Sans B, Gong YF, McAuliffe B, Dicker I, Ho HT, Zhou N, et al. In vitro antiviral characteristics of HIV-1 attachment inhibitor BMS-626529, the active component of the prodrug BMS-663068. *Antimicrob Agents Chemother*. 2012;56(7): 3498–507.

90 Nettles RE, Schürmann D, Zhu L, Stonier M, Huang SP, Chang I, et al. Pharmacodynamics, safety, and pharmacokinetics of BMS-663068, an oral HIV-1 attachment inhibitor in HIV-1-infected subjects. *J Infect Dis*. 2012;206(7): 1002–11.

91 Kuritzkes D, Kar S, Kirkpatrick P. Maraviroc. *Nat Rev Drug Discov*. 2008;7(1): 15–6.

92 Dorr P, Westby M, Dobbs S, Griffin P, Irvine B, Macartney M, et al. Maraviroc (UK-427,857), a potent, orally bioavailable, and selective small-molecule inhibitor of chemokine receptor CCR5 with broad-spectrum anti-human immunodeficiency virus type 1 activity. *Antimicrob Agents Chemother*. 2005;49(11): 4721–32.

93 Westby M, Smith-Burchnell C, Mori J, Lewis M, Mosley M, Stockdale M, et al. Reduced maximal inhibition in phenotypic susceptibility assays indicates that viral strains resistant to the CCR5 antagonist maraviroc utilize inhibitor-bound receptor for entry. *J Virol*. 2007;81(5): 2359–71.

94 Westby M, Lewis M, Whitcomb J, Youle M, Pozniak AL, James IT, et al. Emergence of CXCR4-using human immunodeficiency virus type 1 (HIV-1) variants in a minority of HIV-1-infected patients following treatment with the CCR5 antagonist maraviroc is from a pretreatment CXCR4-using virus reservoir. *J Virol*. 2006;80(10): 4909–20.

95 Matthews T, Salgo M, Greenberg M, Chung J, DeMasi R, Bolognesi D. Enfuvirtide: the first therapy to inhibit the entry of HIV-1 into host CD4 lymphocytes. *Nat Rev Drug Discov*. 2004;3(3): 215–25.

96 Wang L, Wang N, Zhang W, Cheng X, Yan Z, Shao G, et al. Therapeutic peptides: current applications and future directions. *Signal Transduct Target Ther*. 2022;7(1): 48.

97 Beccari MV, Mogle BT, Sidman EF, Mastro KA, Asiago-Reddy E, Kufel WD. Ibalizumab, a novel monoclonal antibody for the management of multidrug-resistant HIV-1 infection. *Antimicrob Agents Chemother*. 2019;63(6): e00110–19.

98 Jacobson JM, Kuritzkes DR, Godofsky E, DeJesus E, Larson JA, Weinheimer SP, et al. Safety, pharmacokinetics, and antiretroviral activity of multiple doses of ibalizumab (formerly TNX-355), an anti-CD4 monoclonal antibody, in human immunodeficiency virus type 1-infected adults. *Antimicrob Agents Chemother*. 2009;53(2): 450–7.

99 Link JO, Rhee MS, Tse WC, Zheng J, Somoza JR, Rowe W, et al. Clinical targeting of HIV capsid protein with a long-acting small molecule. *Nature*. 2020;584(7822): 614–8.

100 Bester SM, Wei G, Zhao H, Adu-Ampratwum D, Iqbal N, Courouble VV, et al. Structural and mechanistic bases for a potent HIV-1 capsid inhibitor. *Science*. 2020;370(6514): 360–4.

101 Faysal KMR, Walsh JC, Renner N, Márquez CL, Shah VB, Tuckwell AJ, et al. Pharmacologic hyperstabilisation of the HIV-1 capsid lattice induces capsid failure. *eLife*. 2024;13: e83605.

102 Subramanian R, Tang J, Zheng J, Lu B, Wang K, Yant SR, et al. Lenacapavir: a novel, potent, and selective first-in-class inhibitor of HIV-1 capsid function exhibits optimal pharmacokinetic properties for a long-acting injectable antiretroviral agent. *Mol Pharm*. 2023;20(12): 6213–25.

103 MacLachlan JH, Cowie BC. Hepatitis B virus epidemiology. *Cold Spring Harb Perspect Med*. 2015;5(5): a021410.

104 Hou J, Liu Z, Gu F. Epidemiology and prevention of hepatitis B virus infection. *Int J Med Sci*. 2005;2(1): 50–7.

105 Fanning GC, Zoulim F, Hou J, Bertoletti A. Therapeutic strategies for hepatitis B virus infection: towards a cure. *Nat Rev Drug Discov*. 2019;18(11): 827–44.

106 Woo ASJ, Kwok R, Ahmed T. Alpha-interferon treatment in hepatitis B. *Ann Transl Med*. 2017;5(7): 159.

107 Jenh AM, Thio CL, Pham PA. Tenofovir for the treatment of hepatitis B virus. *Pharmacotherapy*. 2009;29(10): 1212–27.

108 Dando T, Plosker G. Adefovir dipivoxil: a review of its use in chronic hepatitis B. *Drugs*. 2003;63(20): 2215–34.

109 Scott LJ, Keating GM. Entecavir: a review of its use in chronic hepatitis B. *Drugs*. 2009;69(8): 1003–33.

110 Lai CL, Gane E, Liaw YF, Hsu CW, Thongsawat S, Wang Y, et al. Telbivudine versus lamivudine in patients with chronic hepatitis B. *N Engl J Med*. 2007;357(25): 2576–88.

111 Choung BS, Kim IH, Jeon BJ, Lee S, Kim SH, Kim SW, et al. Long-term treatment efficacy and safety of clevudine therapy in naïve patients with chronic hepatitis B. *Gut and Liver*. 2012;6(4): 486–92.

112 Ogawa E, Nomura H, Nakamuta M, Furusyo N, Koyanagi T, Dohmen K, et al. Tenofovir alafenamide after switching from entecavir or nucleos(t)ide combination therapy for patients with chronic hepatitis B. Liver Int: Official Journal of the International Association for the Study of the Liver. 2020;40(7): 1578–89.

113 Li ZB, Li L, Niu XX, Chen SH, Fu YM, Wang CY, et al. Switching from entecavir to tenofovir alafenamide for chronic hepatitis B patients with low-level viraemia. Liver Int: Official Journal of the International Association for the Study of the Liver. 2021;41(6): 1254–64.

114 Caviglia GP, Ciancio A, Rizzetto M. A review of HDV infection. Viruses. 2022;14(8). doi: 10.3390/v14081749.

115 Hughes SA, Wedemeyer H, Harrison PM. Hepatitis delta virus. *Lancet* (London, England). 2011;378(9785): 73–85.

116 Sandmann L, Wedemeyer H. Interferon-based treatment of chronic hepatitis D. Liver Int: Official Journal of the International Association for the Study of the Liver. 2023;43(Suppl 1): 69–79.

117 WHO. Hepatitis C. Geneva: World Health Organization 2000 October. Contract No.: Fact Sheet No. 164.

118 Saeed M BE, Rice CM. HCV molecular virology and animal models. In: Sofia MJ, editor. HCV: The Journey from Discovery to a Cure. Topics in Medicinal Chemistry. 1. Switzerland: Springer Nature; 2019. p. 30–68.

119 Alqahtani SA SM. The role of interferon for the treatment of chronic hepatitis C virus infection. In: Sofia MJ, editor. HCV: The Journey from Discovery to a Cure. Topic in Medicinal Chemistry. 1 Switzerland: Springer Nature; 2019. p. 98–113.

120 Manns MP, Maasoumy B. Breakthroughs in hepatitis C research: from discovery to cure. *Nat Rev Gastroenterol Hepatol*. 2022;19(8): 533–50.

121 Gentile I, Maraolo AE, Buonomo AR, Zappulo E, Borgia G. The discovery of sofosbuvir: a revolution for therapy of chronic hepatitis C. *Expert Opin Drug Discov*. 2015;10(12): 1363–77.

122 Manns MP, Cornberg M. Sofosbuvir: the final nail in the coffin for hepatitis C? *Lancet Infect Dis*. 2013;13(5): 378–9.

123 Gimeno-Ballester V, Simon MA, Trigo C, Mar J, San Miguel R. Sofosbuvir plus simeprevir for the treatment of genotype 1 chronic hepatitis C: a review of evidence. *Expert Rev Gastroenterol Hepatol*. 2016;10(11): 1289–303.

124 Kwo P, Gitlin N, Nahass R, Bernstein D, Etzkorn K, Rojter S, et al. Simeprevir plus sofosbuvir (12 and 8 weeks) in hepatitis C virus genotype 1-infected patients without cirrhosis: OPTIMIST-1, a phase 3, randomized study. *Hepatology* (Baltimore, Md). 2016;64(2): 370–80.

125 Sulkowski MS, Gardiner DF, Rodriguez-Torres M, Reddy KR, Hassanein T, Jacobson I, et al. Daclatasvir plus sofosbuvir for previously treated or untreated chronic HCV infection. *N Engl J Med*. 2014;370(3): 211–21.

126 Lawitz E, Mangia A, Wyles D, Rodriguez-Torres M, Hassanein T, Gordon SC, et al. Sofosbuvir for previously untreated chronic hepatitis C infection. *N Engl J Med*. 2013;368(20): 1878–87.

127 Deeks ED. Ombitasvir/paritaprevir/ritonavir plus dasabuvir: a review in chronic HCV genotype 1 infection. *Drugs*. 2015;75(9): 1027–38.

128 Lawitz E, Gane E, Pearlman B, Tam E, Ghesquiere W, Guyader D, et al. Efficacy and safety of 12 weeks versus 18 weeks of treatment with grazoprevir (MK-5172) and elbasvir (MK-8742) with or without ribavirin for hepatitis C virus genotype 1 infection in previously untreated patients with cirrhosis and patients with previous null response with or without cirrhosis (C-WORTHY): a randomised, open-label phase 2 trial. *Lancet* (London, England). 2015;385(9973): 1075–86.

129 Greig SL. Sofosbuvir/velpatasvir: a review in chronic hepatitis C. *Drugs*. 2016;76(16): 1567–78.

130 Yaras S, Demir M, Barutcu S, Yildirim AE, Gurel S, Ucbilek E, et al. The efficacy and tolerability of glecaprevir/pibrentasvir treatment in a real-world chronic hepatitis C patients cohort. *Hepatol Forum*. 2023;4(3): 92–6.

131 Bourlière M, Gordon SC, Flamm SL, Cooper CL, Ramji A, Tong M, et al. Sofosbuvir, velpatasvir, and voxilaprevir for previously treated HCV infection. *N Engl J Med*. 2017;376(22): 2134–46.

132 Bouvier NM, Palese P. The biology of influenza viruses. *Vaccine*. 2008; 26 (Suppl 4):D49–53.

133 Iuliano AD, Roguski KM, Chang HH, Muscatello DJ, Palekar R, Tempia S, et al. Estimates of global seasonal influenza-associated respiratory mortality: a modelling study. *Lancet*. 2018;391(10127): 1285–300.

134 De Clercq E. Antiviral agents active against influenza A viruses. *Nat Rev Drug Discov*. 2006;5(12): 1015–25.

135 Li Y, Huo S, Yin Z, Tian Z, Huang F, Liu P, et al. The current state of research on influenza antiviral drug development: drugs in clinical trial and licensed drugs. *mBio*. 2023;14(5): e0127323.

136 Jing X, Ma C, Ohigashi Y, Oliveira FA, Jardetzky TS, Pinto LH, et al. Functional studies indicate amantadine binds to the pore of the influenza A virus M2 proton-selective ion channel. *Proc Natl Acad Sci USA*. 2008;105(31): 10967–72.

137 Bright RA, Medina MJ, Xu X, Perez-Oronoz G, Wallis TR, Davis XM, et al. Incidence of adamantane resistance among influenza A (H3N2) viruses isolated worldwide from 1994 to 2005: a cause for concern. *Lancet* (London, England). 2005;366(9492): 1175–81.

138 von Itzstein M. The war against influenza: discovery and development of sialidase inhibitors. *Nat Rev Drug Discov*. 2007;6(12): 967–74.

139 Lew W, Chen X, Kim CU. Discovery and development of GS 4104 (oseltamivir): an orally active influenza neuraminidase inhibitor. *Curr Med Chem*. 2000;7(6): 663–72.

140 Schmitz U, Swaminathan S. Discovery and development of oseltamivir at gilead sciences. *Antiviral Ther*. 2022;27(2): 13596535211067598.

141 Alame MM, Massaad E, Zaraket H. Peramivir: a novel intravenous neuraminidase inhibitor for treatment of acute influenza infections. *Front Microbiol*. 2016;7: 450.

142 Wandzik JM, Kouba T, Cusack S. Structure and function of influenza polymerase. *Cold Spring Harb Perspect Med*. 2021;11(9): a038372.

143 Baranovich T, Wong S-S, Armstrong J, Marjuki H, Webby RJ, Webster RG, et al. T-705 (favipiravir) induces lethal mutagenesis in influenza A H1N1 viruses in vitro. *J Virol*. 2013;87(7): 3741–51.

144 Furuta Y, Komeno T, Nakamura T. Favipiravir (T-705), a broad spectrum inhibitor of viral RNA polymerase. *Proc Jpn Acad Ser B Phys Biol Sci*. 2017;93(7): 449–63.

145 De Vlugt C, Sikora D, Pelchat M. Insight into influenza: a virus cap-snatching. *Viruses*. 2018;10(11): 641.

146 Todd B, Tchesnokov EP, Götte M. The active form of the influenza cap-snatching endonuclease inhibitor baloxavir marboxil is a tight binding inhibitor. *J Biol Chem*. 2021;296: 100486.

147 Markov PV, Ghafari M, Beer M, Lythgoe K, Simmonds P, Stilianakis NI, et al. The evolution of SARS-CoV-2. *Nat Rev Microbiol*. 2023;21(6): 361–79.

148 Li G, Hilgenfeld R, Whitley R, De Clercq E. Therapeutic strategies for COVID-19: progress and lessons learned. *Nat Rev Drug Discov*. 2023;22(6): 449–75.

149 Eastman RT, Roth JS, Brimacombe KR, Simeonov A, Shen M, Patnaik S, et al. Remdesivir: a review of its discovery and development leading to emergency use authorization for treatment of COVID-19. *ACS Cent Sci*. 2020;6(5): 672–83.

150 Mehellou Y, Rattan HS, Balzarini J. The ProTide prodrug technology: from the concept to the clinic. *J Med Chem*. 2018;61(6): 2211–26.

151 Warren TK, Jordan R, Lo MK, Ray AS, Mackman RL, Soloveva V, et al. Therapeutic efficacy of the small molecule GS-5734 against Ebola virus in rhesus monkeys. *Nature*. 2016;531(7594): 381–5.

152 Her L, Zhu HJ. Carboxylesterase 1 and precision pharmacotherapy: Pharmacogenetics and nongenetic regulators. *Drug Metab Dispos. Biol Fate Chem*. 2020;48(3): 230–44.

153 Xie Y, Yin W, Zhang Y, Shang W, Wang Z, Luan X, et al. Design and development of an oral remdesivir derivative VV116 against SARS-CoV-2. *Cell Res*. 2021;31(11): 1212–4.

154 Cox RM, Wolf JD, Plemper RK. Therapeutically administered ribonucleoside analogue MK-4482/EIDD-2801 blocks SARS-CoV-2 transmission in ferrets. *Nat Microbiol*. 2021;6(1): 11–8.

155 Cox RM, Wolf JD, Lieber CM, Sourimant J, Lin MJ, Babusis D, et al. Oral prodrug of remdesivir parent GS-441524 is efficacious against SARS-CoV-2 in ferrets. *Nat Commun*. 2021;12(1): 6415.

156 Malone B, Campbell EA. Molnupiravir: coding for catastrophe. *Nat Struct Mol Biol*. 2021;28(9): 706–8.

157 Zhao Y, Zhu Y, Liu X, Jin Z, Duan Y, Zhang Q, et al. Structural basis for replicase polyprotein cleavage and substrate specificity of main protease from SARS-CoV-2. *Proc Natl Acad Sci U S A*. 2022;119(16): e2117142119.

158 Jin Z, Du X, Xu Y, Deng Y, Liu M, Zhao Y, et al. Structure of Mpro from SARS-CoV-2 and discovery of its inhibitors. *Nature*. 2020;582(7811): 289–93.

159 Anand K, Ziebuhr J, Wadhwani P, Mesters JR, Hilgenfeld R. Coronavirus main proteinase ($3CL^{pro}$) structure: basis for design of anti-SARS drugs. *Science*. 2003;300(5626): 1763–7.

160 Hoffman RL, Kania RS, Brothers MA, Davies JF, Ferre RA, Gajiwala KS, et al. Discovery of ketone-based covalent inhibitors of coronavirus 3CL proteases for the potential therapeutic treatment of COVID-19. *J Med Chem*. 2020;63(21): 12725–47.

161 Owen DR, Allerton CMN, Anderson AS, Aschenbrenner L, Avery M, Berritt S, et al. An oral SARS-CoV-2 M^{pro} inhibitor clinical candidate for the treatment of COVID-19. 2021;374(6575): 1586–93.

162 Unoh Y, Uehara S, Nakahara K, Nobori H, Yamatsu Y, Yamamoto S, et al. Discovery of S-217622, a noncovalent oral SARS-CoV-2 3CL protease inhibitor clinical candidate for treating COVID-19. *J Med Chem*. 2022;65(9): 6499–512.

163 Tyndall JDA. S-217622, a 3CL Protease inhibitor and clinical candidate for SARS-CoV-2. *J Med Chem*. 2022;65(9): 6496–8.

164 Choudhary MC, Mellors JW. The transformation of HIV therapy: one pill once a day. 2022;27(2): 13596535211062396.

165 Nachega JB, Scarsi KK, Gandhi M, Scott RK, Mofenson LM, Archary M, et al. Long-acting antiretrovirals and HIV treatment adherence. *Lancet HIV*. 2023;10(5): e332–e42.

166 Ndung'u T, McCune JM, Deeks SG. Why and where an HIV cure is needed and how it might be achieved. *Nature*. 2019;576(7787): 397–405.

167 Finzi D, Blankson J, Siliciano JD, Margolick JB, Chadwick K, Pierson T, et al. Latent infection of CD4+ T cells provides a mechanism for lifelong persistence of HIV-1, even in patients on effective combination therapy. *Nat Med*. 1999;5(5): 512–7.

168 Crooks AM, Bateson R, Cope AB, Dahl NP, Griggs MK, Kuruc JD, et al. Precise quantitation of the latent HIV-1 reservoir: implications for eradication strategies. *J Infect Dis*. 2015;212(9): 1361–5.

169 Lian X, Seiger KW, Parsons EM, Gao C, Sun W, Gladkov GT, et al. Progressive transformation of the HIV-1 reservoir cell profile over two decades of antiviral therapy. *Cell Host Microbe*. 2023;31(1):83–96.e5.

170 Chomont N, El-Far M, Ancuta P, Trautmann L, Procopio FA, Yassine-Diab B, et al. HIV reservoir size and persistence are driven by T cell survival and homeostatic proliferation. *Nat Med*. 2009;15(8): 893–900.

171 Kim Y, Anderson JL, Lewin SR. Getting the "kill" into "shock and kill": strategies to eliminate latent HIV. *Cell Host Microbe*. 2018;23(1): 14–26.

172 SenGupta D, Brinson C, DeJesus E, Mills A, Shalit P, Guo S, et al. The TLR7 agonist vesatolimod induced a modest delay in viral rebound in HIV controllers after cessation of antiretroviral therapy. *Sci Transl Med*. 2021;13(599): eabg3071.

173 Liu B, Zhang W, Xia B, Jing S, Du Y, Zou F, et al. Broadly neutralizing antibody–derived CAR T cells reduce viral reservoir in individuals infected with HIV-1. *J Clin Investig*. 2021;131(19).

174 Gaebler C, Nogueira L, Stoffel E, Oliveira TY, Breton G, Millard KG, et al. Prolonged viral suppression with anti-HIV-1 antibody therapy. *Nature*. 2022;606(7913): 368–74.

175 Bertoletti A, Le Bert N. Immunotherapy for chronic hepatitis B virus infection. *Gut Liver*. 2018;12(5): 497–507.

176 Naggie S, Lok AS. New therapeutics for hepatitis B: the road to cure. *Annu Rev Med*. 2021;72(1): 93–105.

177 Meganck RM, Baric RS. Developing therapeutic approaches for twenty-first-century emerging infectious viral diseases. *Nature Medicine*. 2021;27(3): 401–10.

178 Cockerill GS, Good JAD, Mathews N. State of the art in respiratory syncytial virus drug discovery and development. *J Med Chem*. 2019;62(7): 3206–27.

179 Zou G, Cao S, Gao Z, Yie J, Wu JZ. Current state and challenges in respiratory syncytial virus drug discovery and development. *Antiviral Res*. 2024;221: 105791.

180 Samrat SK, Xu J, Li Z, Zhou J, Li H. Antiviral agents against flavivirus protease: prospect and Future Direction. *Pathogens* (Basel, Switzerland). 2022;11(3): 293.

181 Bifani AM, Chan KWK, Borrenberghs D, Tan MJA, Phoo WW, Watanabe S, et al. Therapeutics for flaviviral infections. *Antiviral Res*. 2023;210: 105517.

182 Pérez-Pérez M-J, Saiz J-C, Priego E-M, Martín-Acebes MA. Antivirals against (Re) emerging flaviviruses: should we target the virus or the host? *ACS Med Chem Lett*. 2022;13(1): 5–10.

183 Gao Y, Zhao Y, Guyatt G, Fowler R, Kojan R, Ge L, et al. Effects of therapies for Ebola virus disease: a systematic review and network meta-analysis. *Lancet Microbe*. 2022;3(9): e683–e92.

184 Salata C, Calistri A, Alvisi G, Celestino M, Parolin C, Palù G. Ebola virus entry: from molecular characterization to drug discovery. *Viruses*. 2019;11(3): 274.

185 Bixler SL, Duplantier AJ, Bavari S. Discovering drugs for the treatment of ebola virus. *Curr Treat Options Infect Dis*. 2017;9(3): 299–317.

186 Contreras EM, Monreal IA, Ruvalcaba M, Ortega V, Aguilar HC. Antivirals targeting paramyxovirus membrane fusion. *Curr Opin Virol.* 2021;51: 34–47.

187 Cox R, Plemper RK. The paramyxovirus polymerase complex as a target for next-generation anti-paramyxovirus therapeutics. 2015;6: 459.

188 Abdelnabi R, Delang L. Antiviral strategies against arthritogenic alphaviruses. *Microorganisms.* 2020;8(9): 1365.

189 Ogorek TJ, Golden JE. Advances in the development of small molecule antivirals against equine encephalitic viruses. *Viruses.* 2023;15(2): 413.

190 Ching K-C, FP Ng L, Chai CLL. A compendium of small molecule direct-acting and host-targeting inhibitors as therapies against alphaviruses. *J Antimicrob Chemother.* 2017;72(11): 2973–89.

191 Bossart KN, Bingham J, Middleton D. Targeted strategies for henipavirus therapeutics. *Open Virol J.* 2007;1: 14–25.

192 Pigeaud DD, Geisbert TW, Woolsey C. Animal models for henipavirus research. *Viruses.* 2023;15(10): 1980.

193 Toba S, Sato A, Kawai M, Taoda Y, Unoh Y, Kusakabe S, et al. Identification of cap-dependent endonuclease inhibitors with broad-spectrum activity against bunyaviruses. *Proc Natl Acad Sci USA.* 2022;119(36): e2206104119.

194 Karim M, Lo CW, Einav S. Preparing for the next viral threat with broad-spectrum antivirals. *J Clin Invest.* 2023;133(11): e170236.

195 Sidwell RW, Huffman JH, Khare, Lois GP, Allen B, Witkowski, et al. Broad-spectrum antiviral activity of virazole: 1-β-D-ribofuranosyl-1,2,4-triazole-3-carboxamide. *Science.* 1972;177(4050): 705–6.

196 Tam RC, Lau JY, Hong Z. Mechanisms of action of ribavirin in antiviral therapies. *Antiviral Chem Chemother.* 2001;12(5): 261–72.

197 Geraghty RJ, Aliota MT, Bonnac LF. Broad-spectrum antiviral strategies and nucleoside analogues. *Viruses.* 2021;13(4): 667.

198 Furuta Y, Takahashi K, Shiraki K, Sakamoto K, Smee DF, Barnard DL, et al. T-705 (favipiravir) and related compounds: Novel broad-spectrum inhibitors of RNA viral infections. *Antiviral Res.* 2009;82(3): 95–102.

199 Grosenbach DW, Honeychurch K, Rose EA, Chinsangaram J, Frimm A, Maiti B, et al. Oral tecovirimat for the treatment of smallpox. *N Engl J Med.* 2018;379(1): 44–53.

200 Yang G, Pevear DC, Davies MH, Collett MS, Bailey T, Rippen S, et al. An orally bioavailable antipoxvirus compound (ST-246) inhibits extracellular virus formation and protects mice from lethal orthopoxvirus challenge. *J Virol.* 2005;79(20): 13139–49.

201 Olson VA, Smith SK, Foster S, Li Y, Lanier ER, Gates I, et al. In vitro efficacy of brincidofovir against variola virus. *Antimicrob Agents Chemother.* 2014;58(9): 5570–1.

202 Chan-Tack K, Harrington P, Bensman T, Choi S-Y, Donaldson E, O'Rear J, et al. Benefit-risk assessment for brincidofovir for the treatment of smallpox: U.S. Food and drug administration's evaluation. *Antiviral Res.* 2021;195: 105182.

203 Harrison SC, Alberts B, Ehrenfeld E, Enquist L, Fineberg H, McKnight SL, et al. Discovery of antivirals against smallpox. *Proc Natl Acad Sci U S A.* 2004;101(31): 11178–92.

204 Delaune D, Iseni F. Drug development against smallpox: present and future. *Antimicrob Agents Chemother*. 2020;64(4): e01683–19.

205 De Clercq E, Jiang Y, Li G. Therapeutic strategies for human poxvirus infections: monkeypox (mpox), smallpox, molluscipox, and orf. *Travel Med Infect Dis*. 2023;52: 102528.

206 Dodge MJ, MacNeil KM, Tessier TM, Weinberg JB, Mymryk JS. Emerging antiviral therapeutics for human adenovirus infection: Recent developments and novel strategies. *Antiviral Res*. 2021;188: 105034.

207 Martínez-Aguado P, Serna-Gallego A, Marrugal-Lorenzo JA, Gómez-Marín I, Sánchez-Céspedes J. Antiadenovirus drug discovery: potential targets and evaluation methodologies. *Drug Discov Today*. 2015;20(10): 1235–42.

208 Toots M, Ustav M, Jr., Männik A, Mumm K, Tämm K, Tamm T, et al. Identification of several high-risk HPV inhibitors and drug targets with a novel high-throughput screening assay. *PLoS Pathog*. 2017;13(2): e1006168.

209 Messa L, Loregian A. HPV-induced cancers: preclinical therapeutic advancements. *Expert Opin Investig Drugs*. 2022;31(1): 79–93.

Part I

Human Herpes Viruses

1

Letermovir for the Prevention of CMV Infection in Transplant Recipients

Peter Lischka and Holger Zimmermann

AiCuris Anti-infective Cures AG, Department of Virology, Wuppertal, Germany

1.1 Background

1.1.1 Cytomegalovirus

Human cytomegalovirus (CMV; also known as HHV5) is the prototype member of the β-herpesviridae and the largest member of the virus family Herpesviridae. CMV is a DNA virus with an enveloped capsid including a linear, double-stranded DNA genome of ca. 230 kb, encoding approximately 200 viral proteins [1].

CMV is a very common infection in the human population worldwide, with prevalence rates in the United States ranging from 40% to nearly 100%, depending on the region and socioeconomic status [2]. Primary infection results from close interpersonal contact (bodily fluids) [3, 4] and is generally symptomless, often passing unnoticed [5]. Once established, the infection is lifelong as the virus enters latency in different cell types, including leukocytes (lymphocytes, monocytes, dendritic cells) and CD34+ hematopoietic progenitor cells [6]. For healthy, immune-competent individuals, the infection remains asymptomatic. However, primary infection or reactivation from latency poses a significant threat to those with weakened or compromised immune systems, particularly transplant recipients [7]. While other clinical scenarios do occur, e.g. congenital CMV infections or HIV/CMV coinfections, and while CMV infection may exhibit other complex interactions with its host, e.g. in intensive care patients or certain cancer patients, these are not the focus of the development of letermovir (LET) described here and are therefore not discussed further.

Trends in Antiviral Drug Development, First Edition. Edited by Michael J. Sofia and Zhengqiang Wang.

1.1.2 Immunocompromised Patients

The situation is very different in the immunocompromised versus the immune-competent populations, particularly for transplant recipients, including human stem cell transplant (HSCT) and solid organ transplant (SOT) patients [1]. In transplant recipients, the use of immune suppressants, crucial to the prevention of graft organ rejection (or graft vs. host disease [GvHD] in the case of HSCT recipients), also weakens containment of the latent/persistent virus, enabling (i) the reactivation of viral replication, (ii) virus spread, and ultimately (iii) the development of CMV end-organ disease. CMV continues to be the most common opportunistic viral infection in transplant recipients [8]. If not rapidly contained, the infection can be serious or even life-threatening, being associated with increased morbidity and mortality from CMV directly and indirectly through the increased risk of GvHD (HSCT recipients), graft organ failure (SOT patients), or an increased risk of other opportunistic infections [9, 10]. Indeed, it is now understood that any level of active CMV replication may negatively impact the overall all-cause mortality of HSCT patients, even independently of successfully controlling virus replication with preemptive therapy (PET) (see below) [11, 12].

1.1.3 Patient Groups with a High Risk for CMV Complications

Although the risk of a primary CMV infection or a reactivation of latent virus is increased in all patients on immune suppression, that risk is not uniformly distributed. In the case of HSCT, the risk of CMV reactivation and disease is known to be markedly higher in D−/R+ recipients (see Figure 1.1, upper panel). These are highly immunosuppressed, seropositive HSCT recipients with an established/latent CMV infection (R+) given blood-forming stem cells from a CMV-naïve, seronegative donor (D−), meaning that these patients can no longer control the persistent virus and are almost defenseless once the virus fully reactivates from latency [13]. Among HSCT patients, approximately 60% are CMV-seropositive at the time of transplant and thus are particularly vulnerable to CMV reactivation [14]. Due to its cytopathogenic potential, CMV may cause end-organ disease in these patients, most commonly involving the gastrointestinal (GI) tract or the lungs, as a result of uncontrolled viral replication in tissues facilitated by impaired functional virus-specific T-cell responses and a lack of neutralizing antibodies [15, 16].

Conversely, CMV-seronegative SOT recipients (R−) given an organ from a CMV-seropositive donor (D+) become infected via latent virus, reactivating in the transplanted donor organ (see Figure 1.1; lower panel). These patients are initially defenseless against the primary infection due to their own CMV-naïve immune system, which is additionally under medical suppression. As described above for HSCT recipients, CMV infection in SOT recipients can also have both direct and indirect effects, with a significant impact on transplant outcomes. Direct effects in SOT include CMV syndrome and tissue-invasive organ disease, whereas CMV-associated indirect effects include, among others, acute and chronic organ rejection and opportunistic secondary infections [17].

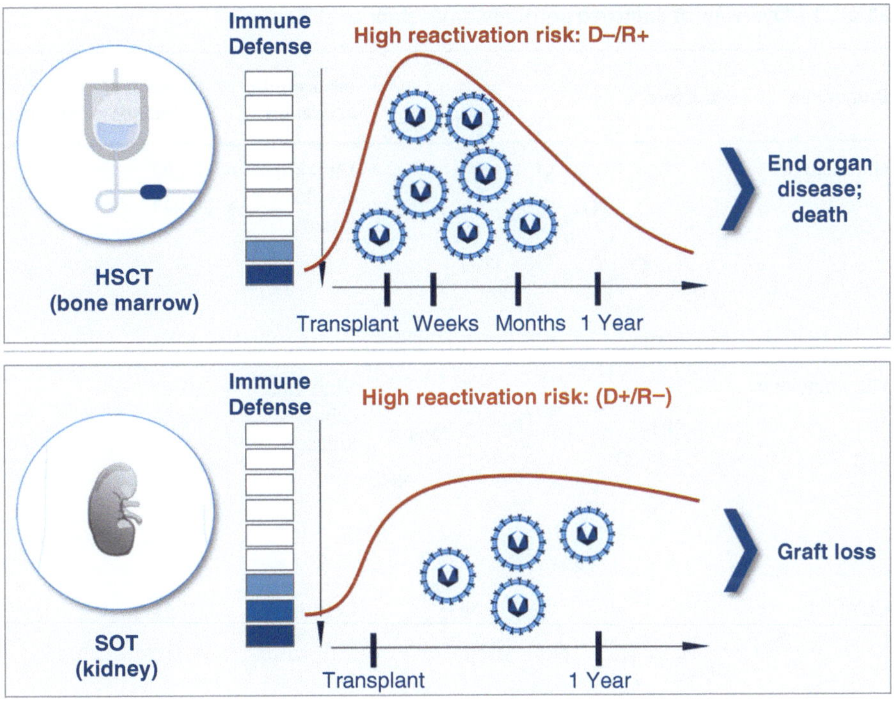

Figure 1.1 High-risk organ donor/organ recipient configurations for CMV reactivation and disease in immunocompromised HSCT and SOT patients.

D, donor; R, recipient; HSCT, hematopoietic stem cell transplantation; SOT, solid organ transplantation

HSCT, D−/R+ describes the situation where a CMV-naïve immune system from a seronegative donor is transplanted into a latently infected, highly immunosuppressed seropositive recipient, leaving the recipient unprotected in the absence of some intervention to prevent/control virus reactivation.

SOT, D+/R− describes the situation where a latently CMV-infected organ from a seropositive donor is transplanted into a seronegative recipient with a pharmacologically suppressed immune system, leaving the recipient unprotected against the virus which has effectively hitch-hiked on the transplanted organ.

Source: Peter Lischka and Holger Zimmermann (2024).

1.1.4 Available CMV Treatments Before Letermovir

1.1.4.1 Antiviral Drugs

When the LET project began, only a few drugs were available for the treatment or prevention of CMV infections, namely ganciclovir (GCV), cidofovir (CDV), and foscarnet (FOS) (see Table 1.1). All drugs were approved in the 1990s (the GCV prodrug valganciclovir [VGCV] was approved in 2001) and all are limited in their use by toxic side effects, pharmacokinetic drawbacks, and resistance development. Moreover, since all of these drugs ultimately share the same primary molecular target, the viral DNA polymerase pUL54, cross-resistance between drugs has been observed in the clinic, in some cases extending to all available polymerase inhibitors, thus eliminating all effective treatment options. In addition to this pharmaceutical armamentarium,

Table 1.1 Overview of approved anti-CMV drugs prior to LET.

Drug name	Structure	Intracellular activation	Location of resistance mutations
Ganciclovir	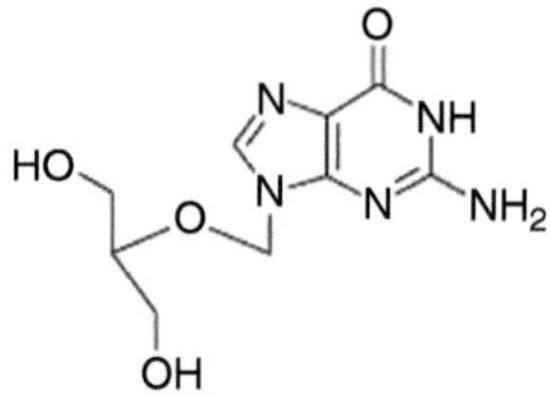	Viral kinase + cellular kinases	UL97 UL54
Valganciclovir	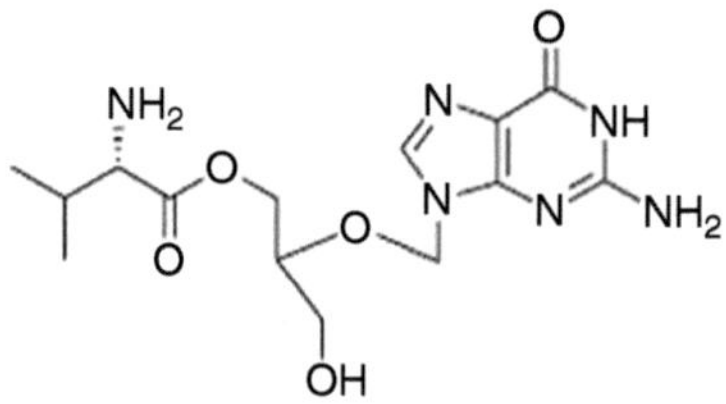	Viral kinase + cellular kinases	UL97 UL54
Cidofovir	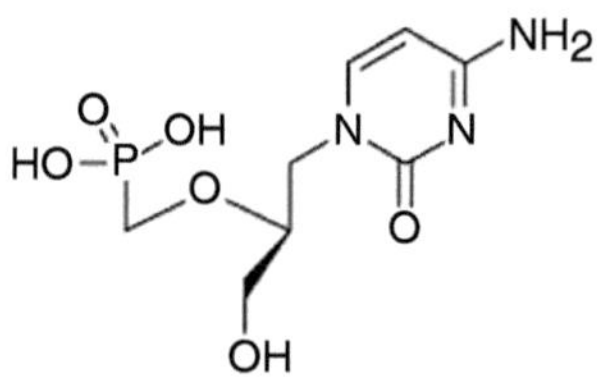	Cellular kinases	UL54
Foscarnet	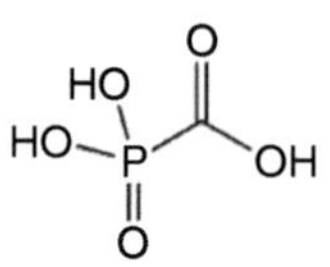	No	UL54

UL54, CMV DNA polymerase; UL97, CMV protein kinase.
All approved anti-CMV drugs prior to LET ultimately target the viral DNA polymerase pUL54 and thus are prone to cross-resistance. Some drugs initially require intracellular activation (phosphorylation) via the viral kinase UL97 and/or cellular kinases (see text for details).

only a handful of new small molecules or monoclonal antibodies were in clinical development at that time. Moreover, there was no effective vaccine on the horizon. This situation has not really changed to date [1, 18].

Ganciclovir (GCV, Cymeven®, Roche) or its oral prodrug valganciclovir (VGCV, Valcyte®, Roche) is an analog of deoxyguanosine [19, 20]. (V)GCV is efficacious against CMV as well as other Herpesviridae and represented the CMV therapy of choice for HSCT and SOT patients at the time [19]. GCV/VGCV requires specific intracellular phosphorylation by the viral protein kinase pUL97 (to the monophosphate), as well as intracellular kinases to form the active triphosphate. Once activated, it inhibits the viral DNA polymerase and is also incorporated into progeny viral DNA, which drastically suppresses the rate of chain extension [21]. For this reason, (V)GCV resistance may be mediated by mutations in either the viral protein kinase pUL97 or the viral polymerase pUL54.

Despite its efficacy, (V)GCV's use is associated with restrictive toxicities, including neutropenia, leucopenia, anemia, and thrombocytopenia, as well as renal and longer-term genetic/reproductive toxicity. The former is particularly relevant in the HSCT recipient population, as the myelotoxic side effects of the drug are not compatible with a prolonged prophylactic treatment regimen, limiting treatment options in HSCT with this drug to so-called PET (see below).

Cidofovir (CDV, Vistide®, Gilead), a phosphonomethoxy analog of cytosine, is a competitive inhibitor of the viral DNA polymerase, acting as a chain terminator. CDV is activated solely by intracellular kinases, making it independent of the viral kinase pUL97. CDV is efficacious and has a long intracellular half-life. However, it also exhibits profound renal toxicity as well as carcino- and teratogenicity.

Foscarnet (FOS, Foscavir®, AstraZeneca), a pyrophosphate analog, does not require intracellular activation for activity. It binds directly to the pyrophosphate binding site of the viral DNA polymerase and inhibits viral DNA replication by blocking the cleavage of the pyrophosphate group from the terminal nucleoside triphosphate and preventing DNA chain extension. While efficacious, the use of FOS is associated with substantial and treatment-limiting toxicities, especially nephrotoxicity and electrolyte disruption, which also restrict or even prohibit its use in combination with other commonly used nephrotoxic drugs in this population such as the immunosuppressive agents tacrolimus or cyclosporine [22, 23]. In addition, FOS requires frequent intravenous (IV) infusions (every 8–12 h). For these reasons, its use was (and is) largely restricted to rescue therapy.

1.1.4.2 Dominant Treatment Strategies

As outlined above, CMV treatment or prevention in the immunocompromised population before LET was entirely restricted to inhibitors of the viral pUL54 DNA polymerase, with substantial and often treatment-limiting toxicities and significant risk of (cross-)resistance. An obvious approach to the problems presented by therapeutic toxicity is simply to wait for clinical signs related to a CMV infection, and only treat once pathology is evident. In the case of CMV, this approach is accompanied by considerable additional risk: (i) a local reactivation of the virus rapidly becomes systemic, resulting in end-organ disease, GvHD in the case of HSCT recipients or organ rejection in the case of SOT recipients, and substantially increases the risk of morbidity and mortality, (ii) CMV is itself immune-modifying, meaning that essentially any degree of active replication may facilitate so-called indirect CMV effects including an increased risk of secondary infections. Therefore, ideally, active CMV replication must be prevented or minimized before it becomes uncontrollable [9, 10, 17, 24].

Considering the toxicity profiles of the available treatments at the time, two preferred strategies were available to combat/prevent CMV infection and disease:

1. Prophylaxis: Antiviral prophylaxis, ideally starting immediately following organ transplantation and continuing during the period of increased risk for CMV reactivation. Although a prophylactic approach is typically easy to coordinate and there is ample evidence demonstrating the efficacy of this strategy for preventing CMV reactivation and disease and positively affecting

CMV-associated indirect effects, prophylaxis could be associated with increased drug costs and, more importantly, the prolonged drug exposure increases the risks of treatment-limiting toxicities. Prophylaxis also carries an elevated risk of late-onset reactivation and disease following cessation of treatment compared with PET (see below). While prophylaxis with (V)GCV for 3–6 months post-transplantation is the dominant strategy for CMV prevention in SOT recipients, in the case of HSCT patients, (V)GCV prophylaxis, even after engraftment, is usually contraindicated due to (i) the myelosuppressive side effects (neutropenia/leucopenia) of the drug and the fact that (ii) reduction of immunosuppression is excluded due to the risk of GvHD [25]. This situation led to the development of the following second option.

2. PET: Antiviral treatment is initiated when viral replication is detected in blood and when viremia/DNAemia reaches a specific threshold. The intention here is to (i) efficiently control virus replication, (ii) prevent symptomatic infection, and (iii) permit a more rapid/effective immune reconstitution by limiting exposure to potentially toxic (myelosuppressive) drugs. However, the risk of direct and indirect CMV effects, such as end-organ disease, rejection, graft loss, and opportunistic infections, is increased. Indeed, due to its immune modulatory activity, the mere presence of CMV replication (including low-level viremia without CMV disease) is associated with increased mortality in the HSCT population [11, 12, 26]. It is also of note that preemptive treatment requires regular time- and resource-intensive monitoring for CMV reactivation in blood.

1.1.4.3 Unmet Medical Need

Clearly, none of the available treatment options at that time could be considered ideal for the prevention or control of CMV in transplant patients, and thus there was a clear and unmet need for an effective CMV therapy that (i) is sufficiently safe to allow prophylaxis and (ii) acts via an alternative mechanism of action (i.e. not mediated by inhibition of pUL54).

1.2 Discovery Phase

The journey of anti-CMV drug development at AiCuris began at the Bayer Anti infective Research Laboratories even before the discovery of LET. Ongoing library screening for anti-CMV substances led to the identification of several new hits, which were selected for lead development, and evaluation of their antiviral and safety properties. It is noteworthy that the applied screening assay set-up (cell culture-based replication assay) appeared to be sensitive to targets late in the virus replication cycle as especially components targeting the viral capsid or the viral terminase complex (see below) were identified, though the reasons for this are unclear. Among these was a member of a new chemical class: the 3,4-dihydroquinazolines.

Structure–activity relationship (SAR)-based studies to improve the candidate's drug characteristics (see Figure 1.2), including efficacy and physical–chemical properties, was based on thousands of derivates and ultimately concluded with BAY 73-6327 (also known as AIC246, LET, and Prevymis®).

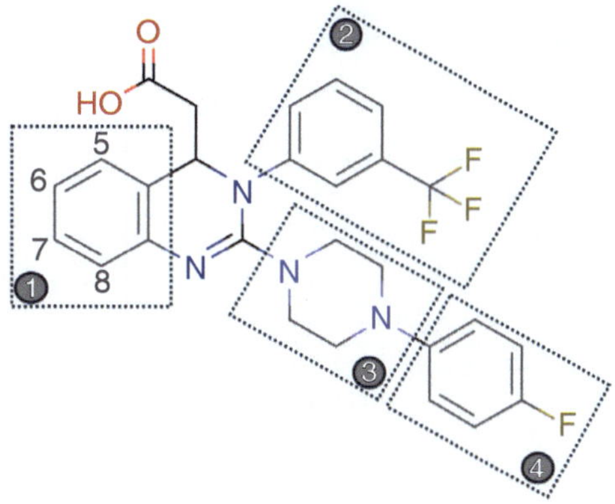

Figure 1.2 SAR for LET.

(1) Lipophilic and polar substituents well tolerated in 6- and 8-position; no improvement by pridyls, bis-substitution favorable.
(2) Lipophilic substitution most active; 3-position better than 4-position better than 2-position; no activity improvement through bis-substitution; benzyl substituents not tolerated.
(3) Loss of activity by replacement with ethylenediamines, piperidines, cyclohexyl, and phenyl.
(4) Substitutions in 3- and 4-positions well tolerated.

Source: Peter Lischka and Holger Zimmermann (2024).

At this early stage, as the initial mechanism of action (MoA) studies and Phase 1 trial planning for LET were commencing, the Anti-infectives Research Division was spun out from Bayer AG and became what is now AiCuris Anti-infective Cures AG, taking a comprehensive early pipeline of anti-infective projects including the anti-CMV drugs program with it.

AiCuris quickly grew from a spin-out of 22 people to an integrated R&D biotech company, including in-house expertise in preclinical, clinical, and chemistry, manufacturing, and controls (CMC) development. Further development of LET up to and including Phase 2b was thus performed and steered by AiCuris.

1.3 Preclinical Characterization

1.3.1 Antiviral Potency/Selectivity/Inhibitory Profile

Initial virology studies revolved around determining the new candidate's potency, first against very well-characterized CMV laboratory strains, then moving on through clinical strains, drug-resistant CMV strains, other herpesviruses, and ultimately to other viral pathogens.

1.3.1.1 *In Vitro* Potency Versus Laboratory CMV Strains

Early *in vitro* potency studies either examined LET's ability to inhibit the viral cytopathic effect on cell cultures [27] or used a recombinant virus expressing the green fluorescent protein and examined LET's effect on the fluorescence of infected cell cultures [28]. CMV laboratory strains exhibit a very narrow cell tropism mainly restricted to fibroblasts; therefore, initial *in vitro* studies used primary fibroblast cells from different tissues (e.g. skin, lung) for efficacy analyses [28, 29]. These studies delivered remarkable results; LET consistently demonstrated both EC_{50} and EC_{90} values in the low nanomolar range and was >400 (based on EC_{50}) to >2000 times (based on EC_{90}) more potent than the contemporary gold standard CMV treatment (i.e. GCV, see Figure 1.3, Part A and Table 1.2).

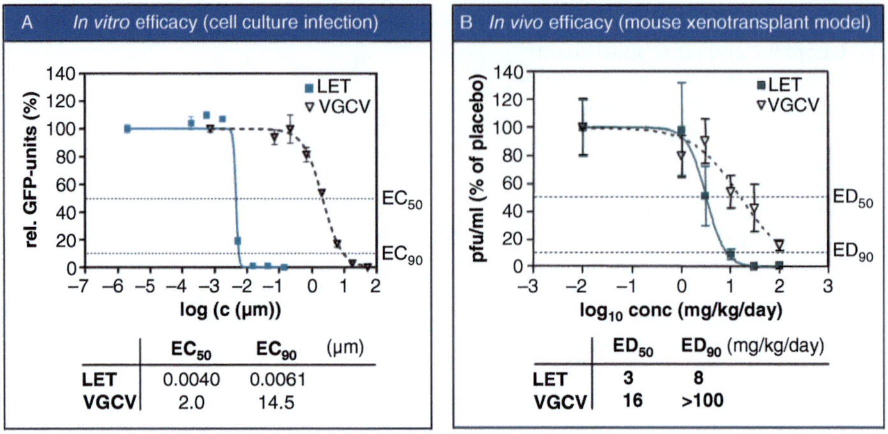

Figure 1.3 *In vitro* and *in vivo* dose-response curves for LET and VGCV.

(A) Comparison of the *in vitro* dose-response curves showing >400- to >2000-fold greater potency of LET compared to VGCV as well as the remarkably steep dose response of LET. Data were obtained from a cell culture infection model using human fibroblast cells and a GFP-expressing CMV laboratory strain. EC_{50} and EC_{90} values for both drugs are indicated.

(B) Comparison of the *in vivo* inhibition curves obtained from a murine xenotransplant model. Data confirm the relevance of the *in vitro* observation and demonstrate a substantially greater efficacy and a steeper dose response for LET vs. VGCV. ED_{50} and ED_{90} values for both drugs are indicated.

Source: Adapted from [28] / American Society for Microbiology.

Perhaps more important than the absolute potency is the selectivity index (SI), i.e. the ratio of cytotoxicity expressed as the 50% cytotoxic concentration (CC_{50}) to EC_{50}, since this underlies the subsequent therapeutic potential *in vivo*. The results showed a consistently high CC_{50} >33 μM (highest concentration tested) in various fibroblast lines, resulting in a selectivity index of at least 12 000 (median 18 000) and indicating good tolerability [28].

Interestingly, at this early stage, it was already recognized that mutant laboratory strains encoding various GCV resistance mutations in UL97 and/or UL54 retained sensitivity to LET (see Table 1.2 and Figure 1.4). This was the first indication that the target site or inhibitory mechanism of AIC246 is different from that of classical polymerase inhibitors like GCV [28, 30].

1.3.1.2 *In Vitro* Potency Versus Clinical CMV Isolates and Resistance-breaking Profile

Potency studies were then extended to clinical CMV isolates including both GCV-susceptible (GCV-S) and GCV-resistant (GCV-R) strains. Ultimately, >70 CMV field isolates were tested for LET sensitivity, out of which 35% demonstrated a GCV-R phenotype due to known GCV resistance mutations that were mapped either to the viral protein kinase UL97 or the viral polymerase UL54. Overall, there was no change in the potency profile between laboratory strains and clinical isolates or between GCV-S and GCV-R strains [28, 30, 31] (Figure 1.4). This demonstrated the broad anti-CMV

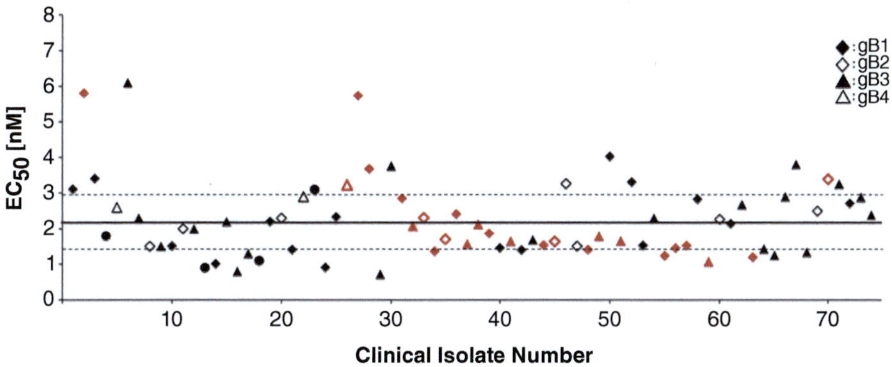

Figure 1.4 LET EC_{50} values from 74 different clinical isolates, including GCV-R strains, by gB subtype.

Black symbols show GCV-S isolates.

Red symbols show GCV-R isolates.

Filled/open diamonds or triangles indicate the individual gB subtype of each virus isolate.

Comparison of individual EC_{50} values from 74 clinical CMV isolates determined by plaque reduction assays. The overall mean EC_{50} value (thick line) ± standard deviation (dashed lines) of the tested viruses is shown. The individual EC_{50} values were shown to be in roughly the same low nanomolar range. The susceptibility of the clinical virus isolates to LET was unaffected by both gB subtype and phenotypic/genotypic GCV resistance.

Source: Adapted from [31] / with permission of Elsevier.

spectrum of this drug and confirmed that LET had an alternative MoA to the currently approved drugs, giving it a potentially resistance-breaking profile.

In addition, the studies with CMV field isolates also revealed that LET susceptibility is not associated with one of the four described CMV glycoprotein B (gB) subtypes (Figure 1.4). This is of importance since due to its involvement in CMV entry and cell-to-cell spread, some studies suggest an impact of different gB genotypes on pathogenesis or disease outcomes in immunocompromised CMV-infected patients [31, 32].

1.3.1.3 *In Vivo* Efficacy (Xenotransplant Model)

As outlined below, LET was found to be inactive against rodent CMVs like mouse- or rat CMV (MCMV; RCMV). Therefore an engineered xenograft mouse model, originally described by Chong et al. [33], was employed for the first *in vivo* evaluation of LET's efficacy. Essentially, gelfoam sponges carrying CMV-infected human fibroblasts were transplanted into mice. After vascularization of the pseudo-organ, mice were treated once daily per os for 9 days, the sponge was explanted, and the CMV titer in the human cells within the sponge was quantified.

Consistent with the prior *in vitro* data, the results showed a significant and dose-dependent reduction of the viral titer in the gelfoam after treatment. Potency was notably greater for LET than GCV, as was the gradient of the dose-dependency curve (Figure 1.3B). While 30 and 100 mg/kg/day LET dose groups achieved a 2-log reduction in viral titer by the end of treatment, 100 mg/day GCV achieved only a 1-log reduction. As seen in the previous *in vitro* studies, LET also appeared to be notably more effective than GCV *in vivo*.

1.3.1.4 *In Vitro* Antiviral Specificity

Having established LET's potency and resistance-breaking potential against CMV, LET was screened against a broad panel of other herpesviruses and other human-pathogenic viruses, as shown in Table 1.2. Interestingly, the data indicated that

Table 1.2 Summary of *in vitro* potency data for LET against CMV laboratory strains and clinical isolates, other herpesviruses, and representative human-pathogenic viruses.

Virus strain	Range of activity EC50 (μM)	
	LET	**GCV**
Potency vs. human CMV		
Laboratory strains	0.0020–0.0051	1.5–3.5
GCV-resistant laboratory strains	0.0016–0.0039	4.5–32
Clinical isolates	0.0001–0.0058	0.6–5
GCV-resistant clinical isolates	0.0014–0.0061	5.9–55
Potency vs. nonhuman CMV (strain)		
MCMV (Smith)	4.5	4.3
RCMV (Maastricht)	>10	0.9
RhCMV (68-1)[a]	>10	0.6
GpCMV (22122)[a]	>10	ND
Potency vs. other herpesviruses (strain)		
Alphaherpesviruses		
VZV (Oka)	>10	0.8
HSV-1 (166v VP22-GFP)	>10	2.2
HSV-2 (01-6332)	>10	2.5
Betaherpesviruses		
HHV-6 (typeA-GS)	>10	ND
Gammaherpesviruses		
EBV (B95-8)	>10	ND
Potency vs. other human-pathogenic viruses		
Adenovirus (HAdV-2)	>10	ND
Hepadnavirus (HBV HepG2.2.15)	>30	ND
Retrovirus (HIV-1 LAI)	>11	ND
Orthomyxovirus (influenza A A/WSN/33)	>10	ND
Flavivirus (HCV replicon)	>32	ND

EBV, Epstein–Barr virus; EC50, concentration at 50% maximum efficacy; GCV, ganciclovir; GpCMV, guinea pig cytomegalovirus; HAdV, human adenovirus; HBV, hepatitis B virus; HCV, hepatitis C virus; HHV-6, human herpes virus 6; HIV, human immunodeficiency virus; HSV, herpes simplex virus; MCMV, murine cytomegalovirus; ND, not determined; RCMV, rat cytomegalovirus; RhCMV, rhesus monkey cytomegalovirus; VZV, varicella-zoster virus.

a. Lischka et al., unpublished data.

Source: Adapted from [28, 30].

LET is remarkably specific for human CMVs since no significant activity was noted against other alpha-, beta-, and gamma-herpesviruses (including CMVs from other species), or unrelated pathogenic viruses (including adeno-, hepadna-, retro-, orthomyxo-, and flaviviruses) [30].

1.3.1.5 Other Characteristics of Letermovir's Inhibitory Profile

In Vitro Inhibition of Both Focal and High Titer Infections Assays examining the cell-to-cell spread of the virus *in vitro* showed that LET could contain an active infection, almost completely preventing the spread to neighboring cells.

In addition, replication assays using varying multiplicities of infection (MOI) showed that LET's potency was more robust against high MOI infections than GCV, exhibiting only a 3-fold increase in EC_{50} across a 300-fold increase in MOI while the EC_{50} of GCV increased by a factor of 5. This interesting drug property was already suspected given the very steep dose-response curve of LET when compared to that of GCV (Figure 1.3) [28].

In Vitro Drug Combination Profile In anticipation of possible clinical comedication scenarios, *in vitro* drug combination assays were performed and showed that LET exhibited additive, i.e. nonantagonistic, effects with all approved anti-CMV drugs, indicating the theoretical potential for combination therapy if needed. Moreover, the same studies showed there were neither agonistic nor antagonistic effects apparent between LET and a large panel of anti-HIV drugs supporting the potential use of LET in CMV-HIV-coinfected individuals undergoing highly active antiretroviral therapy (HAART) [34].

1.3.1.6 Summary

The early *in vitro* and *in vivo* studies of LET's potency, selectivity, and antiviral profile demonstrated a highly potent drug with low toxicity and remarkable specificity for CMV.

While the results of these studies showed that LET may have considerable therapeutic potential and likely had an alternative MoA versus GCV and the other approved anti-CMV drugs, neither these results nor the screening process used to identify LET as a potential anti-CMV drug candidate suggested a likely MoA.

1.4 Mechanism of Action Studies

1.4.1 Target Identification

The replication cycle of CMV is divided into three sequential phases: immediate early (IE), early (E), and late (L). The IE stage begins as soon as viral DNA enters the nucleus through the expression of the IE regulatory genes. During the E phase, viral DNA replication occurs, while viral DNA packing, virus assembly, and egress of progeny virus take place during the L phase of the replication cycle (Figure 1.5, panel C). Accordingly, early MoA studies attempted to identify the point in the replicative cycle affected by LET, in order to narrow the field of likely molecular targets.

In vitro assays were used to examine the effects of drug treatment on viral protein and DNA expression profiles. LET-mediated suppression of viral replication was shown

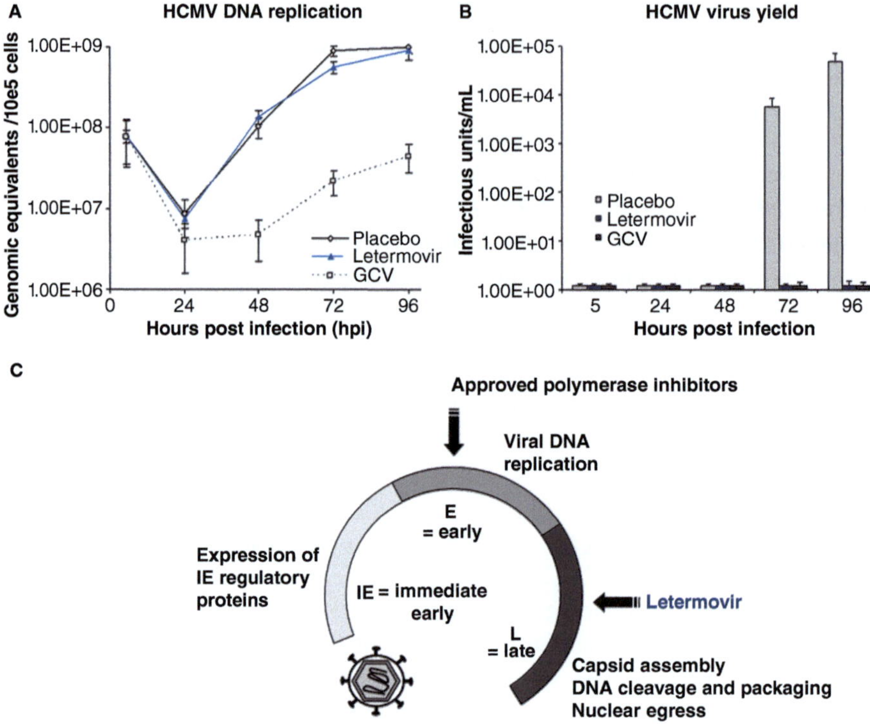

Figure 1.5 *In vitro* inhibition of viral replication without suppression of viral DNA synthesis.

CMV-infected human fibroblasts were treated with LET, GCV, or placebo at 10-fold EC_{50}. (A) Like the placebo treatment, LET had no effect on viral DNA replication; nonetheless, (B) LET treatment effectively inhibited the generation of infectious virus particles. This suggests that LET acts during the L phase of viral replication, i.e. after viral DNA replication and via a mechanism distinct from that of the DNA polymerase inhibitors. (C) CMV replication cycle. *Source:* Adapted from [29] / American Society for Microbiology.

to have no apparent effect on either viral protein expression over the three phases or on viral DNA replication [29]. Thus, LET's MoA appeared to be different from that of GCV or other viral DNA polymerase inhibitors in that LET appeared to act later in the viral replication cycle, targeting a process after DNA replication that is not associated with viral protein expression (Figure 1.5). This assumption was confirmed by time-of-addition studies and is consistent with LET being active against CMV strains that are resistant to the CMV polymerase inhibitors GCV, CDV, or FOS (see above) [28, 29].

The first indication that LET might affect the cleavage and packaging of viral DNA also came from time-of-addition and recovery kinetic studies in which LET showed antiviral kinetics similar to a previously identified viral terminase inhibitor (BAY 38-4766), which was used as a "late-acting" control compound in these experiments [28, 29].

The CMV terminase is an essential multiprotein complex responsible for viral genome maturation and packaging by cleaving the concatenated viral genomic DNA formed in the nucleus from rolling circle viral DNA replication into single genomic units and packaging these genome units into preformed viral capsids (Figure 1.6). The core viral terminase complex is composed of at least three subunits: the proteins pUL56, pUL89, and pUL51, which work together with other packaging proteins and the portal protein pUL104 to complete this process. Recent cryoEM studies suggest that the functional viral terminase complex is a hexameric structure, with each monomer itself comprising the three subunits where pUL51 and pUL56 function as regulator/fixer proteins and pUL89 harbors the ATPase/nuclease function. The terminase complex is highly conserved among different CMV strains and within herpesviruses [35–37].

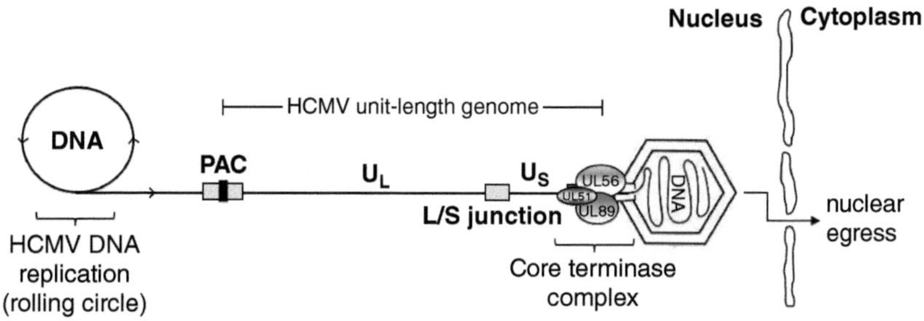

Figure 1.6 Function of the viral terminase complex.

Schematic model of CMV DNA synthesis and packaging. CMV DNA replication, the terminase complex, and the CMV genome structure are illustrated. The CMV genome is comprised of unique long (U_L) and unique short (U_S) regions separated by a region designated the L/S junction. The CMV viral terminase complex cleaves the concatemeric viral DNA at the terminase cleavage site, located within the PAC region at the genome terminus. The resulting individual genomic units are then packaged into preformed viral capsids followed by the nuclear export of mature virions into the cytoplasm (nuclear egress).
Source: Adapted from [29].

The initial idea that LET targets the CMV terminase was confirmed through molecular characterization of LET-resistant strains obtained from *in vitro* drug resistance selection studies. CMV genotyping identified individual mutations in the UL56 terminase gene capable of causing LET resistance as shown by marker transfer experiments (see also Section 1.8) [28, 29, 38]. Since all these data supported the hypothesis that LET exerts its effects via the viral terminase complex, these functions were then directly investigated using a terminase cleavage assay to investigate the processing of viral genome concatemers and electron microscopy to evaluate packaging of viral genomic DNA into nascent pre-capsids. The results showed that LET inhibits the formation of properly processed unit-length CMV genomes (Figure 1.7) and thus prevents packaging of viral genomic DNA into premature viral capsids. As a consequence, empty viral capsids (B-capsids) accumulate in the nucleus of infected cells and are not exported to the cytoplasm for further particle maturation.

Overall, the data confirmed that LET targets the terminase subunit pUL56 and thereby interferes with cleavage and packaging of CMV progeny DNA, i.e. it has a similar MoA to that proposed for the discontinued anti-CMV terminase inhibitors of the benzimidazole-ribonucleoside of sulfonamide class described below.

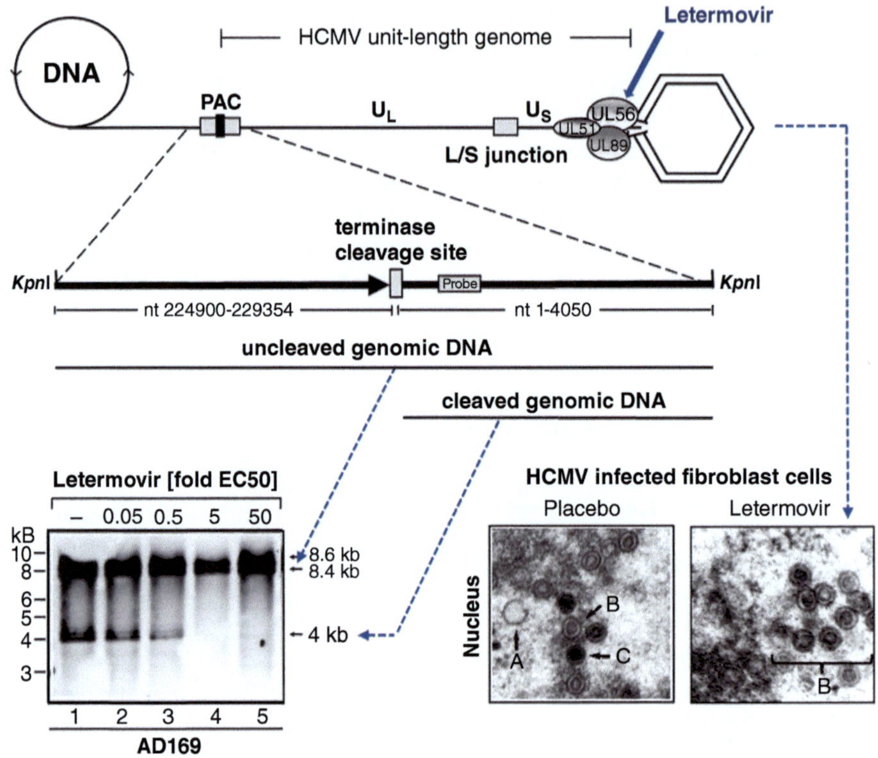

Figure 1.7 Effect of LET on viral DNA processing and packaging.

In concatemeric viral DNA, the PAC region, containing the terminase cleavage site, is flanked by two KpnI restriction sites approximately 8.6 kb apart. Inhibition of the terminase complex by increasing concentrations of LET prevents cleavage of the viral DNA concatemer at the terminase cleavage site, resulting in 8.6 kb fragments after digestion of the isolated viral DNA with KpnI. In the absence of LET, viral DNA is naturally cleaved by the terminase complex at the terminase cleavage site, resulting in ca. 4 kb fragments after digestion with KpnI.

In addition, the failure of genomic viral DNA cleavage due to the presence of LET prevents viral DNA packaging into preformed capsids and the subsequent nuclear egress of mature C-capsids into the cytoplasm. This results in the accumulation of empty B-capsids in the nucleus of infected, LET-treated cells, as shown in the electron micrographs.

Source: Adapted from [29] / American Society for Microbiology.

1.5 Terminase Inhibitors

1.5.1 Previous and Contemporary Drug Candidates Targeting the Terminase Complex

Prior to LET, there were two published chemical classes of terminase inhibitors: the benzimidazole ribonucleosides (e.g. BDCRB/TCRB and GW275175X) and the sulfonamides (e.g. BAY 38-4766) (see Table 1.3).

Table 1.3 Overview of published CMV terminase inhibitors.

Chemical class	Substance	Structure	Range of activity EC_{50} (μM)[a]	Viral genes harboring resistance mutations[b]	*In vitro* drug combination with GCV (antiviral effect)	Activity against nonhuman CMVs
Benzimidazole ribonucleoside	BDCRB	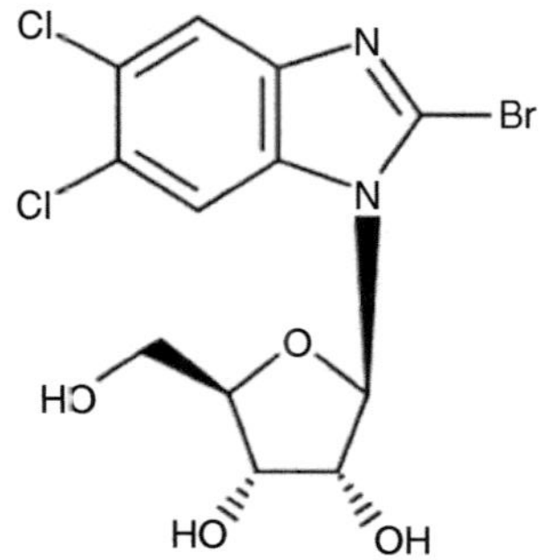	0.3–1.4	UL89 + UL56	Additive-synergistic	Yes
	TCRB	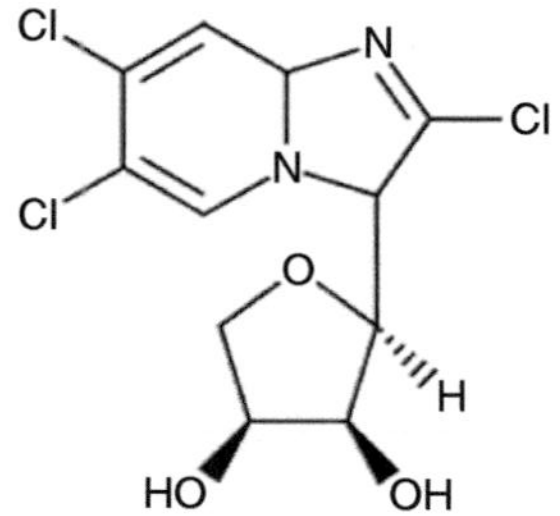	2.9	UL89 + UL56	n.d.	Yes

(*Continued*)

Table 1.3 (Continued)

Chemical class	Substance	Structure	Range of activity EC_{50} (μM)[a]	Viral genes harboring resistance mutations[b]	*In vitro* drug combination with GCV (antiviral effect)	Activity against nonhuman CMVs
	GW275175X	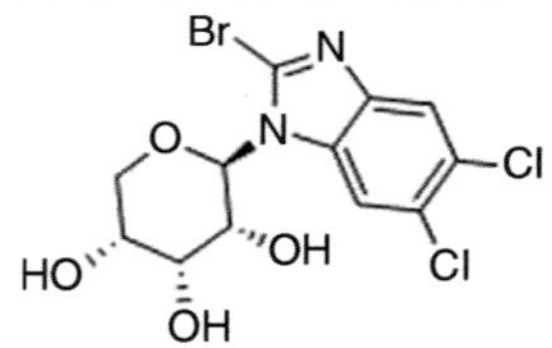	0.3–1.4	UL89 + UL56	n.d.	Yes
Sulfonamide	Tomeglovir (BAY 38-4766)	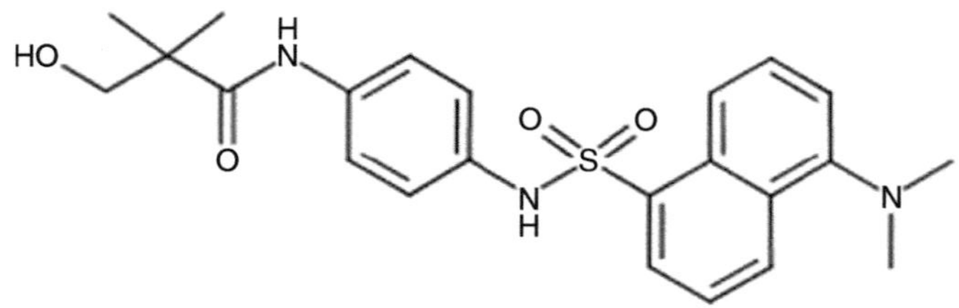	0.3–1.2	UL89 + UL56	Antagonistic	Yes
Quinazoline	LET	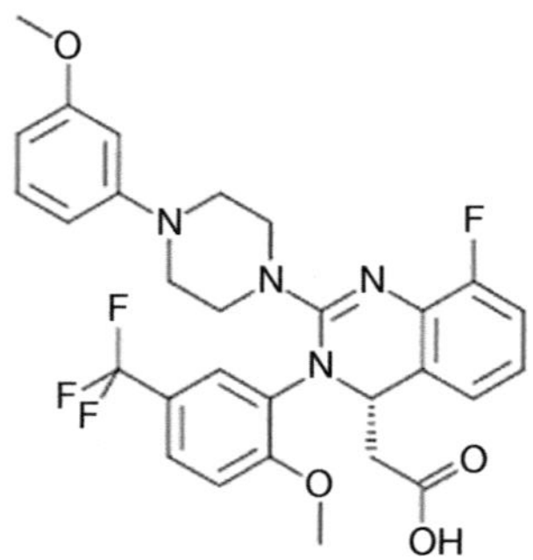	0.002–0.005	UL56	Additive	No

a. Antiviral activity determined against CMV laboratory strains adapted partly from [36].
b. Only genes harboring mutations conferring high-level drug resistance are listed (see text body for references).

The benzimidazole ribonucleosides were originally synthesized by Townsend et al. [39] and later found to exert their anti-CMV activity by interfering with the viral terminase complex [40, 41]. Although clinical development of the first set of these terminase inhibitors (BDCRB and TCRB) was not pursued due to unfavorable pharmacokinetics (PK) properties, medicinal chemistry efforts led to the identification of GW275175X, a D-ribopyranose derivative of BDCRB with improved *in vivo* stability. *In vitro* studies confirmed the MoA of this molecule as a terminase inhibitor, and the drug was selected by GlaxoSmithKline (now GSK plc) as a clinical candidate for the treatment of CMV diseases [42]. The drug progressed to Phase 1, where it demonstrated good safety and tolerability, but further development as an anti-CMV drug was discontinued for strategic decisions [36].

An early terminase inhibitor, identified at Bayer using the same discovery approach as LET, was the sulfonamide BAY 38-4766. It was an effective anti-CMV drug *in vitro*, and it did not inhibit viral DNA synthesis; rather, it interfered with viral genomic processing and packaging [27, 43]. BAY 38-4766 had an EC_{50} in the 0.5–2 μM range and was active against GCV-R strains of CMV [27]. Drug resistance selection studies demonstrated that BAY 38-4766 acted via inhibition of the viral terminase complex, mediated mainly by interaction with the UL89 gene product [43]. Despite its very promising antiviral properties, the development was stopped in Phase 1 because the drug induced CYP enzymes, which led to very low exposure levels in humans.

1.5.2 Letermovir: Same Target, Different Interaction

Although LET shares the overall mode of action with other terminase inhibitor molecules, studies have shown that the molecular interaction between LET and the terminase complex appears to be quite distinct. This is supported by a comparison of the resistance and activity profiles of LET with those of other terminase inhibitors of the structurally distinct sulfonamide or benzimidazole-ribonucleoside classes (see Table 1.3). There is no relevant cross-resistance between LET and other terminase inhibitors [29, 38]. Indeed, at the molecular level, and in contrast to other cleavage-packaging inhibitors [29, 38–41, 43, 44], LET resistance mutations are essentially confined to UL56 and almost entirely to a single and distinct region (see Section 1.8) while the UL89 terminase subunit gene is essentially irrelevant to LET resistance. In addition, the activity spectrum of LET is also highly selective for human CMV and inactive against CMVs of other species, in marked contrast to the broader anti-CMV activity of the other terminase inhibitors [27, 30, 45–47]. Finally, a distinct LET drug combination profile in terms of agonism/antagonism has also been described for LET versus other approved anti-CMV drugs (i.e. polymerase inhibitors) [34].

1.5.3 Advantages of Terminase Inhibitors

To date, LET is the only terminase inhibitor to be granted marketing approval. LET's paradigm-shifting potential is largely a result of its molecular target (pUL56 of the viral terminase complex), or rather, it results from (i) the fact that it does not target the same molecule (viral DNA polymerase) as *all* available anti-CMV drugs at the time and (ii) the fact that LET's target has no direct equivalent in the human host cells. This makes the development of LET resistance independent of resistance to other anti-CMV treatments and thus also enables a genuine rescue therapy option.

In addition, the lack of an equivalent mammalian target also reduces the likelihood of toxicities by minimizing potential interaction with host pathways.

Interestingly, a further potential advantage of LET over polymerase inhibitors has been proposed in the context of prophylaxis: the potential low-level exposure to viral proteins without the release of viable capsids. Depending on the duration of treatment, this low-level exposure may facilitate immune training and reduce the risk of late-onset disease, though the impact of this remains to be confirmed.

1.6 Preclinical Safety Evaluation

To support a long-term prophylactic treatment in patients, a comprehensive evaluation of potential safety pharmacological and toxicological risks was conducted before testing of LET in humans was initiated.

For safety pharmacological investigations, LET was administered to rats and dogs at single doses of up to 45 mg/kg to determine the effect on electrocardiogram (ECG), the cardiovascular system, the respiratory system, the central nervous system, renal function, lipid metabolism, or hematology, blood glucose concentration, and the GI tract. Taking all the data from these safety pharmacological studies together, there were no findings that hindered the clinical development of LET.

Of note, discouraging bioavailability data and emesis noted in the dogs after oral administration of LET disqualified the dog as a model species in this drug development program, and, therefore, only very limited toxicity and toxicokinetic data were generated in dogs. Instead, the monkey was selected as the appropriate nonrodent species for toxicity profiling of LET. The monkey showed high systemic exposure to the compound after oral administration, and it is phylogenetically close to humans.

As a prerequisite for first human studies, subacute (4-week) oral toxicity studies with LET were also performed in a rodent (rat) and a nonrodent (monkey) species. Acute toxicity studies were conducted in rats and mice (oral and IV).

To evaluate the genotoxic potential of LET, *in vitro* and *in vivo* mutagenicity studies were performed.

Subsequently, subchronic to chronic oral toxicity studies in mice (13-week), rats (26-week), and monkeys (39-week), fertility studies in rats and monkeys (sexually mature), carcinogenicity studies in rats and mice, and developmental toxicity studies in rats and rabbits were conducted. In addition, data from a pre- and postnatal toxicity study in rats were evaluated together with appropriate toxicokinetic evaluations.

Moreover, IV dose toxicity studies in rats and monkeys and a local tolerability study in rabbits were carried out to support IV dosing in humans as well.

In summary, LET demonstrated an overall favorable toxicological profile leading to a very positive risk/benefit ratio for the prophylactic treatment in the indicated patients.

1.7 Clinical Development and MAA/NDA Submission

1.7.1 Regulatory Support for Clinical Development

In both the United States and European Union, supportive programs are offered by the regulatory agencies to encourage the development of new drugs where there is a serious unmet medical need (Fast Track) and/or the disease is rare (orphan drug; <200 000 patients in United States, and 5 patients in 10 000 residents in European Union).

These programs facilitate frequent communication with regulatory agencies during drug development for serious or life-threatening diseases. They offer accelerated review and approval times, extended market exclusivity, and cost/tax benefits, which are crucial incentives for drug development in complex niche indications.

The US Food and Drug Administration (FDA) granted Fast Track designation for LET in May 2011. European Medicines Agency (EMA) granted Orphan Drug Designation in April 2011 and the FDA in December 2011.

1.7.2 Phase 1

Phase 1 data confirmed that LET was well suited to IV and oral dosing; oral tablet dosing is particularly advantageous in such a patient population, facilitating simple dosing without the need for an outpatient visit as would be required for IV administrations. However, the anticipated poor condition of the target patient population and the broad range of expected concomitant medications necessitated an extensive Phase 1 program in addition to the standard safety, dose finding, and absorption, distribution, metabolism, and excretion (ADME) characterization studies before LET development could be moved into patients.

1.7.2.1 Drug–Drug Interaction Studies

Early *in vitro* enzyme inhibition studies had suggested that LET may interact with other drugs, including commonly used concomitant medications for immune suppression, such as tacrolimus or cyclosporine [48]. Considering especially the key immune suppressants essential for clinical management of organ or stem cell recipients, appropriate Phase 1 studies were performed.

Drug–drug interaction data show LET to be a substrate and moderate inhibitor of CYP3A and a substrate of 2D6 and is predicted to induce CYP2C9 and 2C19. LET is also a substrate and inhibitor of OATP1B1, 1B3, P-gp, and an inhibitor of OAT3, BCRP, BSEP, and MRP-2.

PK analyses based on the Phase 2b trial (see below) in HSCT patients resulted in LET dose adjustment (50% dose reduction) for patients receiving concomitant cyclosporine [49, 50].

1.7.2.2 Special Populations

In addition, based on the predictable comorbidities in the target population, studies were performed examining the effect of renal and hepatic impairment on LET PK. As may be expected based on LET's predominantly biliary excretion [51, 52], moderate and severe hepatic impairment led to increased exposure [51]. Perhaps less predictably, considering the negligible renal excretion, renal impairment also led to slight but not clinically relevant increases in exposure, but no dose adjustment based on renal function was required [52].

Taken together, this meant that LET could not be recommended for patients with severe hepatic impairment or for those with moderate hepatic impairment with at least moderate renal impairment.

1.7.2.3 IV Formulation

In HSCT recipients, a period of high risk for CMV reactivation is immediately following transplantation and prior to engraftment. LET's safety profile enables administration during this time. However, in the first days post-transplant, patients may be unable to swallow oral formulations. Therefore, an effective CMV prophylaxis in the HSCT population must also be available in a parenteral formulation. Following successful proof-of-concept (PoC), AiCuris developed the hydroxypropyl beta-cyclodextrin formulation of LET to permit parenteral administration [53].

Please note: Phase 2 studies in HSCT subjects proceeded with the oral formulation only, partly because established clinical practices at the time had developed to specifically avoid using anti-CMV drugs pre-engraftment due to their myelosuppressive toxicities. The IV formulation was introduced in Phase 3 when inclusion was allowed directly after transplantation.

1.7.3 Clinical Proof-of-concept

Based on the population of greatest need, the development program was clearly focused on CMV prophylaxis in the HSCT recipient population. Especially HSCT patients (and particularly the R+/D− patients) were effectively unprotected during the window of maximum risk, i.e. before immune reconstitution, and therefore at profound risk from CMV viremia and disease (see Figure 1.1). The well-known toxicities of available drugs prohibited their use in prophylaxis, forcing the use of preemptive strategies. However, although this may lessen the risk of disease, any degree of virus replication, including subclinical, can negatively impact patient outcomes [11, 12]. This population, therefore, showed the greatest probability of achieving a superior risk/benefit for LET versus existing CMV treatments.

1.7.3.1 Phase 2a Clinical Trial (AIC001-2-001)

An initial PoC trial was designed for the preemptive treatment of CMV in HSCT recipients as the intended primary target population, reflecting the need for (i) a like-for-like comparison of LET versus standard of care treatment regimens, (ii) a maximal permitted LET treatment period of just 14 days, and (iii) a maximum dose of 80 mg daily due to limited preclinical and clinical data at the time. Based on these considerations, the AIC001-2-001 trial was initially designed using a preemptive treatment approach in HSCT recipients, i.e. treatment would be initiated upon detection of viremia [54], although the overall goal of the development program remained CMV prophylaxis. In practice, the risk associated with the use of a novel and, therefore, clinically unproven preemptive treatment in a highly vulnerable viremic HSCT patient was perceived as too high, and recruitment failed with only one patient recruited.

The trial population was therefore switched by protocol amendment to the kidney or kidney/pancreas transplant population, again using preemptive treatment with 40 mg LET twice daily (b.i.d.), 80 mg LET once daily (q.d.), or local standard of care determined using a 1 : 1 : 1 randomization scheme. Patients were CMV-positive with no signs of severe systemic infection, end-organ CMV disease, GvHD, liver/renal dysfunction, positive hepatitis B virus (HBV)/HIV infection, diarrhea, severe gastrointestinal (GI) disease, recent treatment with GCV, CDV, or FOS, or treatment with CYP3A4 inducers/inhibitors. This time recruitment progressed well, and 27 subjects were selected and treated.

The primary endpoint for the trial was the decline in CMV DNA versus baseline over a 14-day treatment period. Observations of changes from baseline in viral load at day 15 showed statistically significant reductions in CMV copy numbers in all three treatment groups (LET 40 mg b.i.d.: $P = 0.031$, 80 mg q.d.: $P = 0.018$, standard of care [SoC]: $P = 0.001$); differences between groups were not significant and could reasonably be attributed to baseline differences and the small trial population. Interestingly, the proportion of subjects achieving viral clearance appeared higher for the LET groups (6/12, 50%) compared with the SoC (2/7, 29%). Presumably due to the measure used (viral DNA copy number in blood), the apparent kinetic of viral inhibition was also notably different between LET and SoC (GCV/VGCV), with reported viremia values dropping from day 4 under GCV treatment versus day 11 under LET. This likely reflects the different MoAs of the two drugs (LET does not inhibit DNA replication but does prevent the maturation of infectious virus particles) and provides a caveat in the interpretation of such data.

PoC could, therefore, be considered established. Overall, LET appeared to be at least similarly effective to SoC within the analyzed time frame, and the drug was safe and well tolerated [54].

1.7.3.2 Emergency IND Treatment of a Lung Transplant Patient with Multiresistant CMV Disease

Shortly after the completion of the PoC trial, further successful demonstration of LET's potential was provided via the Emergency Investigational New Drug (eIND) treatment of a lung transplant recipient with severe refractory and multidrug-resistant CMV disease [55]. The patient had been viremic for 5 months and exhibited severe, disseminated, refractory, multidrug-resistant CMV disease of the lungs, GI tract, and retina. By the time of the LET eIND treatment, the patient had already received numerous approved and off-label drugs: GCV, FOS, CDV, leflunomide, CMV hyperimmune globulin, CMX-001 (an experimental, lipid-conjugated derivative of CDV), as well as an artemisinin derivative, all without success.

LET was obtained for emergency treatment since all other treatment options had been exhausted. The subject received 120 mg/day LET for 16 days. Because plasma concentrations were then shown to be toward the lower end of the established safety range based on early Phase 1 data in healthy volunteers, the dose was increased to 240 mg/day.

By the end of the 49-day total treatment, the CMV viral load was below the LLOQ (<600 copies/mL), and clinical evidence indicated ongoing resolution of the lung,

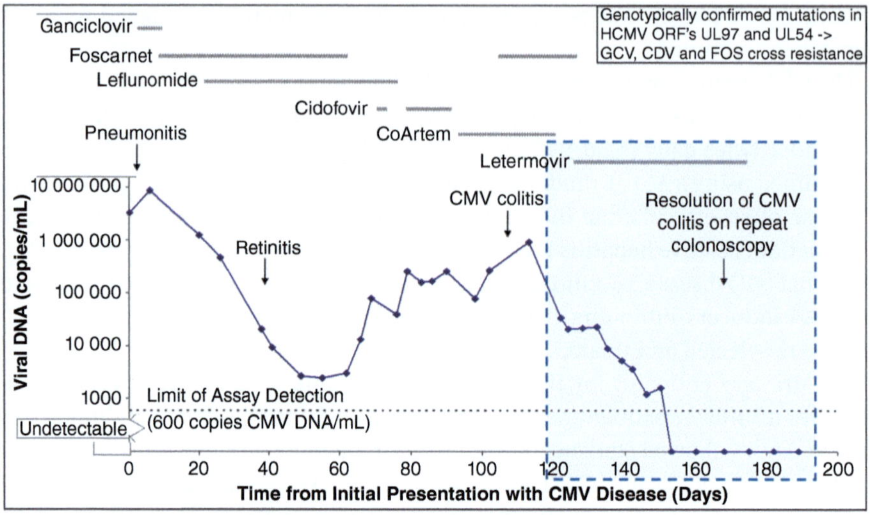

Figure 1.8 Clinical demonstration of LET's efficacy in treatment of an eIND patient with multidrug-resistant CMV.

Viral load in a lung transplant recipient with disseminated, refractory, multidrug-resistant CMV infection and CMV disease in lungs, GI tract, and retina. After exhausting all available approved and off-label anti-CMV treatments, as well as some experimental drugs (top panel), LET treatment was initiated under an emergency IND obtained from the FDA. By the end of a 49-day LET treatment period, the virus was undetectable and the disease was resolving.

Source: Adapted from [55] / with permission of Elsevier.

GI, and retinal disease (see Figure 1.8). Three months after completion of treatment, the subject still had no detectable CMV viremia, and there was no sign of rejection in the lungs.

These results not only supported the efficacy and safety conclusions of the Phase 2a trial, but they also demonstrated the tolerability of the 240 mg/day dose and the enormous potential of LET for life- and organ-saving interventions in such multidrug-resistant cases.

1.7.3.3 Credentials Established

Based on the results from the PoC trial and the additional eIND treatment of a severely ill lung transplant patient with severe, disseminated, drug-resistant CMV disease, the case for LET's potential had been well made. On the clinical side, this appeared to be sufficient to allay remaining concerns relating to LET's use in a highly vulnerable patient population. On the business side however, current offers from out-licensing partners still contained substantial price reductions because they assessed the remaining development risks as high. In contrast, AiCuris shareholders were rather convinced by the existing data and saw further investment into Phase 2b clinical trials as justified taking into account the future potential of the novel drug. Development up to end-of-phase 2 was therefore pursued at AiCuris rather than taking the less advantageous out-licensing deals already available.

1.7.4 Letermovir for CMV Prophylaxis in HSCT Patients

1.7.4.1 Phase 2b: First Prophylaxis Trial in HSCT Patients

Having established PoC, the first clinical trial in the ultimate target HSCT population was a dose-ranging trial (AIC246-01-II-02) designed to compare the efficacy and safety of LET with placebo for CMV prophylaxis [50]. Because the potential for hematotoxicity through LET could not yet be excluded at that time, prophylaxis was started post-engraftment.

A total of 131 R+ HSCT recipients were recruited and randomized 1:1:1:1 to 12 weeks CMV prophylaxis with either 60, 120, or 240 mg/day LET or placebo. Placebo was a justified comparator since there was no alternative prophylaxis available, and patients who developed CMV viremia or disease discontinued and received local SoC preemptive treatment as a rescue therapy.

The primary endpoint in this trial was the rate of all-cause prophylaxis failure. Secondary analyses included specific analyses of virological failure rates and time-to-onset analyses. After 12 weeks of treatment, LET showed a powerful, dose-dependent, and statistically significant reduction versus placebo in all-cause prophylaxis failure (LET: 48% at 60 mg/day, 32% at 120 mg/day, 29% at 240 mg/day compared with placebo: 64%) (Figure 1.9; Panel A). In line with this, the time to onset of prophylaxis failure was also significantly reduced by LET. However, the clearest demonstration of LET's efficacy in this setting was the complete prevention of virological failure at the maximum 240 mg/day LET dose after the exclusion of patients who were already CMV-positive at baseline (Figure 1.9b). Again, safety data showed no findings of concern associated with LET prophylaxis. Especially regarding potential future treatment regimens, comparisons versus placebo showed no indication of either (i) hematotoxicity, i.e. leukopenia or neutropenia, or (ii) renal toxicity. LET's safety profile was, therefore, sufficiently clean to open the way to early initiation of prophylaxis, even before engraftment, meaning LET may even fulfill the clinical ideal of immediate prophylaxis following transplantation, covering the period of maximum risk.

The importance of these results was not lost on the wider anti-infectives industry either. LET's emerging product profile and the lack of a potential alternative CMV prophylactic drug enabled AiCuris to secure the largest German pharmaceutical deal of 2012, by out-licensing LET to Merck & Co., Inc., Rahway, NJ, USA (hereafter: MSD) [56]. AiCuris received a €110 million upfront payment and was eligible for milestone payments of up to €332.5 million based on the successful achievement of development, regulatory, and commercialization milestones. This was a remarkable result for such a young pharmaceutical company and one which earned the company's leadership the German Future Prize of 2018.

Subsequent pivotal trials, marketing approval applications, and indication extensions proceeded through MSD.

1.7.4.2 Phase 3 CMV Prophylaxis Trial in HSCT Patients

The pivotal trial MK-8228-001 compared the safety and efficacy of LET or placebo for CMV prophylaxis in HSCT recipients [49]. Over a period of nearly 2 years, a total of 565 patients from 67 centers in 20 countries were randomized (2:1) and

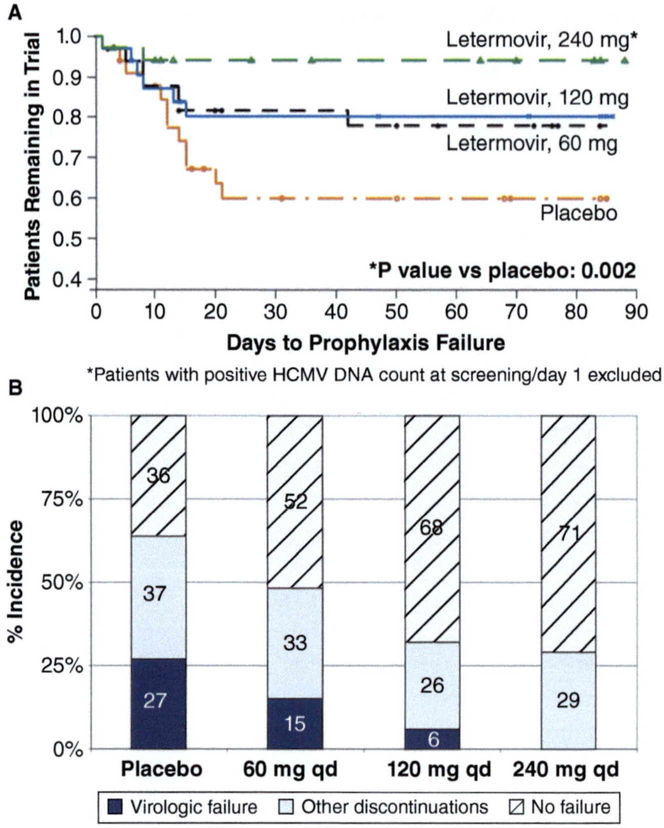

Figure 1.9 Dose-dependent protection of CMV reactivation in HSCT patients (Phase 2b trial).

(A) Kaplan–Meier plot of time-to-failure in the first (Phase 2b) trial of LET prophylaxis in HSCT patients. Subjects who discontinued treatment for reasons other than treatment failure or who were lost to follow-up were censored at the time of discontinuation or at the time of last contact with the center. All subjects who completed treatment were censored at the time of last dose.
LET significantly reduced time-to-failure (all-cause failure including CMV viremia or disease). The effect was dose-dependent and statistically significant ($P = 0.007$) for the highest dose group of 240 mg/day).

(B) Treatment failure rates are shown as virological failure (dark blue; viremia or CMV disease) or any other cause for discontinuation (light blue). LET treatment completely prevented CMV viremia and disease at the highest test dose of 240 mg, once daily ($P = 0.002$). The effect was profound and dose-dependent.

Source: Adapted from [50].

treated with either 480 mg/day LET (240 mg/day if on cyclosporine) or placebo. Treatment was started within 28 days post-transplantation and lasted up to week 14 (ca. 100 days) post-transplantation. Importantly, engraftment was not a prerequisite for treatment initiation in this trial.

The primary endpoint agreed with regulatory agencies and clinicians was clinically significant CMV (csCMV, defined as either CMV disease or CMV viremia warranting preemptive treatment) up to week 24 post-transplantation in patients without viremia at baseline (per central laboratory), thereby including a further 10-week observational period after cessation of prophylaxis. Early discontinuations or missed week 24 data were imputed as meeting the csCMV endpoint (i.e. as prophylaxis failures). Patients were followed up to week 48.

The results again showed LET to be highly effective, this time at reducing the clinically defined CMV endpoint. Of the 495 patients treated in the trial and confirmed as viremia negative at baseline, 37.5% (122/325) of LET-treated patients versus 60.6% (103/170) placebo-treated patients met the primary endpoint; this result was highly statistically significant ($P<0.001$) and did not result from differences in discontinuations/missing data that were comparable between groups. Looking only at the csCMV rates, 17.5% (57/325) of LET-treated patients had csCMV compared with 41.8% (71/170) of placebo subjects (Figure 1.10; Panel A). Importantly, mortality rates were also significantly reduced in the LET group at 24 weeks post-transplantation (10.2% vs. 15.9%; $P=0.03$), and numerically at week 48 ($P=0.12$) (Figure 1.10; Panel B) compared to the placebo group. Subgroup analyses showed that these benefits (differences between treatment groups) were greater for high-risk patients.

Particularly important, nearly 2/3 of patients treated did not have engraftment at the start of treatment. LET treatment had no discernable effect on engraftment times and did not affect the observed rates of GvHD, infections, or relapse (in underlying hematologic disease).

Although minor side effects were noted in this larger Phase 3 population, LET continued to demonstrate a very positive safety profile, with no major side effects.

1.7.4.3 Marketing Approval in HSCT Recipients

The results of the HSCT development program demonstrated effective CMV prophylaxis, with very few side effects and no discernable myelo- or nephrotoxicity, and are suitable for use immediately after transplantation. This was clearly a huge advancement on the only previous option of preemptive intervention with a toxic drug to limit an already active and, therefore, harmful infection.

In November 2017, marketing approval was granted by the US FDA [57], and in January 2018 [58], by the EMA for prophylaxis of CMV infection and disease in adult R+ allogeneic HSCT recipients.

1.7.4.4 Further Clinical Development and Real-world Data

After approval, the number of HSCT patients to benefit from CMV prophylaxis rapidly increased. With the growing number of patients, the amount of real-world data available for meta-analyses also grew, and confirmed the picture already suggested by the Phase 3 controlled clinical trial, that LET primary prophylaxis was a highly effective and safe new drug associated with a significant reduction in (i) CMV reactivation, (ii) csCMV infection, and (iii) CMV disease after allogeneic HSCT. Building on these new data, a recent meta-analysis combining multiple real-world studies, including a large number of outcome events, has now also demonstrated what was already suggested by the Phase 3 study, that primary prophylaxis with LET is also associated with a significant reduction in all-cause mortality and nonrelapse mortality beyond

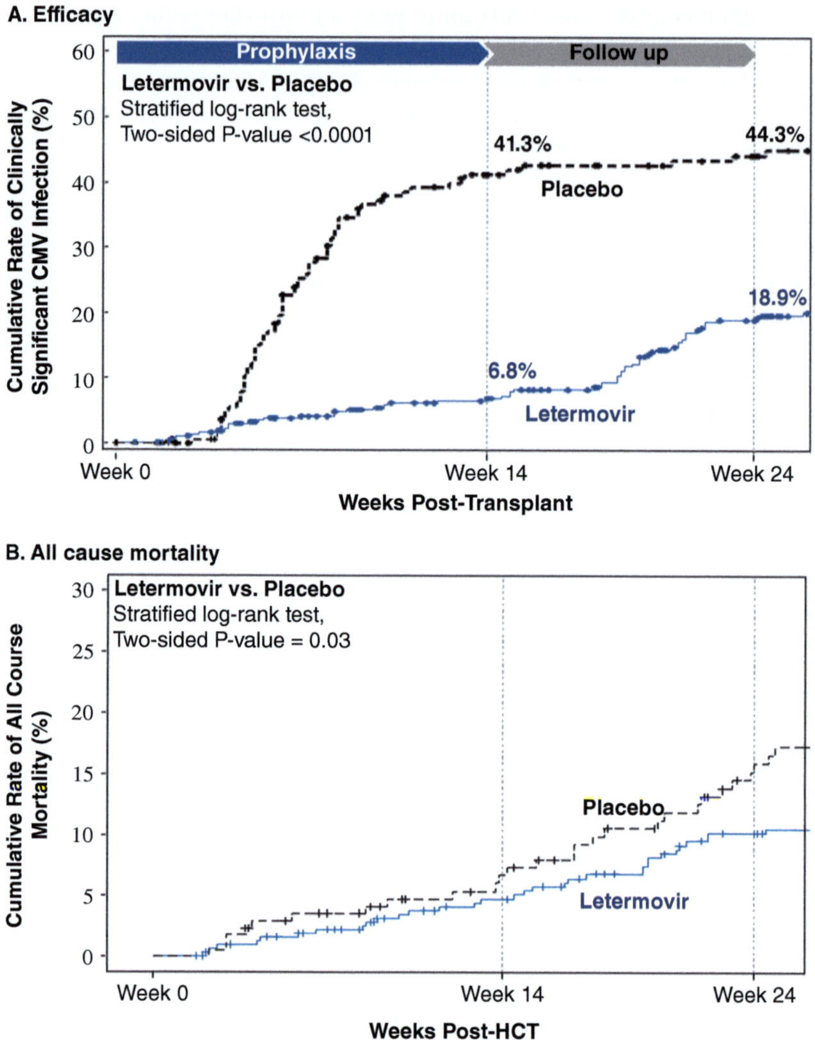

Figure 1.10 Efficacy and all-cause mortality for LET prophylaxis vs. placebo-treated HSCT patients (Phase 3 trial).

(A) The incidence of csCMV was significantly reduced by LET prophylaxis vs. placebo in HSCT recipients, both during prophylaxis and during follow-up after cessation of therapy.

(B) Mortality was significantly reduced in the LET prophylaxis group vs. placebo at week 24.

Source: Adapted from [49].

week 24 post-HSCT [26]. Reasons for the observed mortality benefit of LET prophylaxis compared to the (V)GCV PET control group are (i) the potent antiviral activity of the drug preventing, reducing, or delaying CMV-related complications and (ii) the reduced need to treat patients preemptively with antiviral agents that are associated with serious toxicities.

Though LET is very efficient in suppressing virus reactivation while patients are on therapy, the incidence of csCMV increased again after discontinuation of the drug on day 100 post-HSCT, particularly in patients who remained at high risk of

CMV reactivation (see Figure 1.10). This finding indicates that some patients experience late CMV infections after cessation of the drug [26, 49].

Considering LET's safety profile, along with the aforementioned observations and taking into account results of some observational studies showing that high-risk patients benefited from extended LET prophylaxis [59–61], it was felt likely that extension of the prophylaxis period would be safe and may help reduce the incidence of these so-called late-onset infections [62]. This would also be consistent with the approach taken in other transplant populations (e.g. KT, below), and a new Phase 3 trial was started to compare the safety and efficacy of 100 vs. 200-day prophylaxis periods for safety and efficacy.

1.7.4.4.1 Phase 3 Trial for Extension of the Prophylaxis Period

In this randomized, placebo-controlled trial, 255 HSCT patients who had completed 100 days of LET prophylaxis with no incidence of csCMV and who remained at high risk for CMV reactivation were randomized 2 : 1 to 100 additional days of prophylaxis with either LET or placebo [63]. The primary endpoint was the occurrence of csCMV from randomization (on day 100 after HSCT) up to day 200 (week 28 after HSCT). Secondary comparisons included csCMV up to week 48 as a follow-up.

The results showed that the incidence of csCMV at 28 weeks post-transplantation was significantly reduced by 200 vs. 100 days LET prophylaxis (3% vs. 19% incidence). At the same time, tolerability and safety (including all-cause mortality) were comparable between 100- and 200-day treatment periods.

Interestingly, there was no additional difference in all-cause mortality from randomization to week 28 or week 48 when comparing the 100- and 200-day prophylaxis groups. This suggests that the survival benefit demonstrated for LET prophylaxis (see above) is most likely related to the fact that LET prevents CMV reactivation in the first 100 days post-transplant, when the patients are most vulnerable. Whether there is still a "small" survival benefit beyond 100 days will become apparent when a more extensive database of patients treated for 200 days is available.

In summary, the results of this study suggest that extended-duration LET may be a good option for patients who have multiple risk factors for CMV reactivation. Prolonging LET prophylaxis in HSCT patients further reduces the incidence rates of (i) csCMV infections, (ii) indirect CMV effects like secondary infections (e.g. invasive fungal), (iii) exposure to toxic drugs like GCV, FOS, and CDV, (iv) drug resistance, and (v) is safe and well tolerated.

Later occurrence of CMV infection might mean that reactivation occurs in the setting of a more mature immune system, which could be beneficial, potentially reflected in the early survival benefit of LET compared with the absence of survival benefit seen in this study.

The extension of HSCT treatment to 200 days was granted by the FDA in August 2023 [64] and by the EMA in October 2023 [65].

1.7.4.4.2 Follow-up Trials in Specific Populations

Pediatric HSCT Recipients The pediatric population remained a group in urgent need with no approved treatment. Although publications demonstrated off-label use of LET, no formal approved guidance was available [66, 67].

A further Phase 2b trial was undertaken to investigate the PK, safety, tolerability, and efficacy of 100 days of LET prophylaxis in the pediatric HSCT population [68]. Starting with 12–18-year-olds, the trial progressed stepwise through <12 to 2-year-olds, and then 2-year-olds to newborns. Dosage was determined (and reduced for patients on concomitant cyclosporin A [CsA]) to achieve comparable exposure to the approved adult dose. Safety data and PK models were reviewed and modified as necessary between age cohorts.

The results confirmed the efficacy and safety for all age groups. Incidence rates of csCMV after 100 days of prophylaxis were broadly comparable to previous trials in adult HSCT patients (19.6% at week 14, 25.0% at week 24).

1.7.5 Letermovir for CMV Prophylaxis in KT Patients

As described in the background section, the high-risk population among SOT recipients is different from that of HSCT recipients – but for the same reasons: the elevated risk arises from a CMV infection/reactivation in the setting of a CMV-naïve immune system. In this case, D+/R− [69, 70], representing ca. 20% of all KT recipients [71]. Interestingly, and perhaps somewhat unexpectedly, the proportion of R− recipients and, therefore also, the total number of D+/R− high-risk patients appears to be increasing in the kidney recipient population and is projected to continue to grow for at least the next 20 years [71]. This is of particular importance because a primary CMV infection in these high-risk KT patients is associated with more severe illness and worsening outcomes [72]. Given this, effective measures to prevent or control CMV infections are warranted.

Prevention of CMV disease is not restricted solely to preemptive treatment in this population. According to the current consensus guidelines, VGCV and GCV are the recommended agents of choice for CMV prophylaxis and preemptive treatment in kidney transplantation, whereby high-risk D+/R− patients generally receive universal prophylaxis with VGCV [70, 73]. SOT patients generally have intact hematopoietic systems and are relatively resilient to GCV toxicity compared to HSCT recipients. In this setting, prophylaxis may be extended for up to 200 days, as has been standard for over a decade [74]. This extends protection beyond the period of maximum risk and reduces the likelihood of late-onset CMV thereafter [74, 75]. Despite this, CMV disease still occurs in up to 50% of high-risk D+/R− SOT patients despite prevention and in 17% of R+ patients [17].

Although 200-day prophylaxis with VGCV is quite effective in preventing CMV reactivation, it is nonetheless associated with toxicities, including increased rates of leukopenia/neutropenia (especially when given with toxic concomitant medications) and these can necessitate interruptions in GCV, immunosuppression (IS), or antimicrobial treatments, leading to increased rates of infection, rejection, and organ loss. Indeed, neutropenia is reported by 30–40% of SOT patients in the first year post-transplantation, largely drug-related [17].

Long-term use of GCV also increases the likelihood of resistance to polymerase inhibitors (discussed below). Beyond the risk to the patient, these issues also bring substantial complications to case management, consume more resources, and increase costs.

In the KT population, especially the D+/R− high-risk patients, there was a clear medical need for an effective anti-CMV drug without the issues of myelotoxicity and the need for dose adjustments related to renal function [49, 75]. Based on its safety and efficacy profile in the PoC trial and in the Phase 2b and Phase 3 HSCT trials, LET appeared well placed to offer equivalent efficacy to (V)GCV while avoiding the risks to patients' outcomes from renal- and myelo-toxicity.

1.7.5.1 Phase 3 Noninferiority Trial in KT Recipients

In 2018, following FDA approval of LET for prophylaxis of CMV infection and disease in adult R+ allogeneic HSCT recipients in 2017, MSD began a Phase 3 noninferiority comparison of LET versus VGCV in high-risk (D+/R−) kidney transplant recipients.

Over a 2-year period, the trial recruited a total of 601 patients, randomized 1:1 to 480 mg/day LET (240 mg/day if taking cyclosporine) or 900 mg/day VGCV (adjusted for renal function) beginning not later than day 7 post-transplantation [75]. Consistent with the SoC in this population, the treatment period in this trial was up to 200 days post-transplant, with follow-up at 1 year post-transplantation. The primary comparison was the incidence of CMV disease up to week 52 based on blinded adjudicator assessments, and secondary analyses included CMV disease by 28 weeks post-transplantation.

The results confirmed the noninferiority of 200 days prophylaxis with LET versus VGCV in the prevention of CMV disease with failure rates of 10.4% (LET) versus 11.8% (VGCV) up to 52 weeks post-transplantation. The majority of failures resulted from late-onset disease from a viral rebound in the post-treatment period. Indeed, no LET-treated patients developed CMV disease during the 200-day treatment period, compared to 1.9% of VGCV patients. Incidence rates of viral DNAemia during and post-treatment were consistent with adjudicator-based disease rates. Clearly, LET was at least as effective as VGCV in this population. Nonetheless, the data show some patients had reactivation even after 200 days of prophylaxis, suggesting further extension of prophylaxis may benefit some patients. Ultimately, monitoring of CMV immunity may help in determining the optimal time point for cessation of prophylaxis [72].

As in previous clinical trials, the safety data showed the clear advantage of LET over VGCV: rates of leukopenia or neutropenia were significantly reduced in LET-treated patients versus VGCV (26% vs. 64%, $P < 0.001$) (Figure 1.11) and led to treatment discontinuation in just 1.4% (LET) versus 5.4% (VGCV). Overall drug-discontinuation rates due to adverse events were also notably reduced: 4.1% (LET) versus 13.5% (VGCV).

Overall, the results showed a clear safety advantage for patients receiving LET rather than VGCV. The lack of need for dose adjustment is a further advantage for the clinical management team, particularly in the KT population, resulting in reduced burden and costs.

1.7.5.2 Marketing Approval in KT Recipients

Marketing approval of LET for the prevention of CMV disease in high-risk adult KT recipients was received from the US FDA in June 2023 [64] and EMA in October 2023 [65].

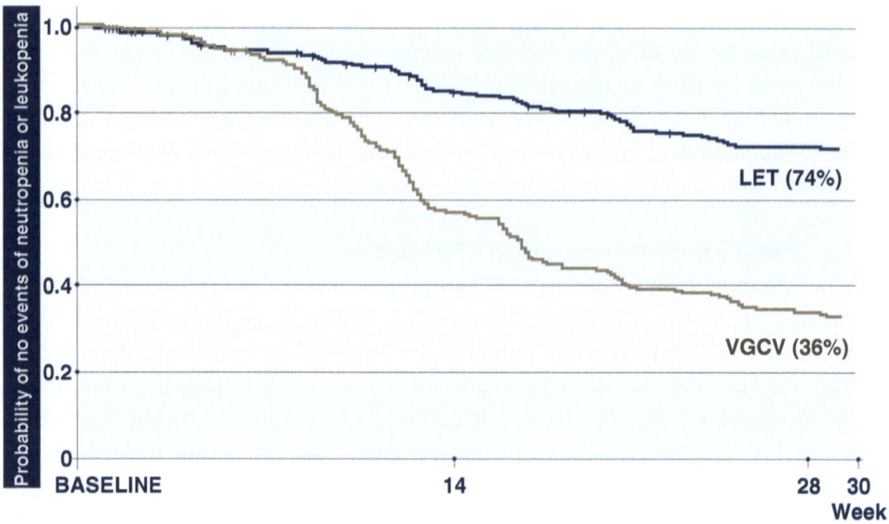

Figure 1.11 CMV prophylaxis with LET greatly reduces the rates of neutropenia and leukopenia compared with VGCV in KT patients (Phase 3 trial).

Event free (neutropenia/leukopenia) probability.

In addition to the comparable efficacy of LET vs. VGCV for prevention of csCMV in KT patients, the rates of neutropenia/leukopenia were significantly reduced in the LET group (26%) vs. GCV (64%), $P < 0.001$.

Source: Adapted from [75] / American Medical Association.

1.7.5.3 Further Clinical Development and Real-world Data

Real-world data examining the risks and benefits of LET versus VGCV in kidney transplantation have already begun to emerge. In addition to at least comparable prophylactic efficacy versus VGCV, LET's safety profile allows maintenance or even increase of mycophenolate dosing without the need for supplemental granulocyte colony-stimulating factor (G-CSF), making the conversion from VGCV particularly useful in cases of VGCV-induced leukopenia [76].

1.8 Drug Resistance

1.8.1 Genetic Characterization of Letermovir Resistance

Knowledge on potential drug resistance is important for any antiviral project. In the context of CMV in transplant recipients, clinical resistance historically meant, e.g. VGCV dose escalation or a transition to an alternative DNA polymerase inhibitor like foscarnet or cidofovir with greater toxicity. Cross-resistance due to the common molecular target, the viral polymerase, was a further complicating issue leading to worsening treatment outcomes. Rapid identification and characterization of amino acid mutations conferring resistance to a novel drug are therefore important to guide and optimize treatment options [77].

As a direct-acting antiviral drug (DAAD), the development of LET resistance is essentially inevitable. Therefore, identification and characterization of potential resistance mutations began early in the development program. Partly, this was essential for understanding LET's MoA, especially considering the target complex comprises three distinct subunits (UL56, UL89, and UL51), and partly because these data are crucial for effective clinical management. Similarly, cataloging natural polymorphisms unrelated to resistance is also important to permit rapid interpretation of sequencing data from patients with CMV viremia [77–79].

First, a number of *in vitro* drug resistance studies were conducted using cell culture selection of LET-resistant virus strains, followed by genotyping of potential target genes involved in cleavage/packing of viral progeny DNA. Amino acid (aa) mutations were identified against a standard genomic reference sequence (CMV AD169) and then transferred to a cloned LET-susceptible wild-type strain for *in vitro* drug susceptibility testing (marker transfer) [38, 79]. This approach, referred to as recombinant phenotyping, has become the standard for attributing a resistant or susceptible phenotype to a given CMV genotype [77, 80]. The work confirmed the principle molecular target as the terminase subunit pUL56 and showed that single-site aa mutations in the UL56 gene could confer LET resistance. Interestingly, all these single-site resistance mutations were seen to cluster in the UL56 region aa 231–369, which was therefore considered the LET-resistance region, a designation that still applies. The different aa resistance mutations in this region were shown to range in their impact from low-level (<20-fold) to high-level (>100-fold) increases in EC_{50}, with only minimal loss of viral growth fitness. The most significant mutations so far are aa exchanges at codon C325, which confer >3000-fold increase in EC_{50}, and thus essentially absolute drug resistance [29, 38, 79, 80].

Following on from this foundational work, a series of subsequent *in vitro* studies using variations of the same approach identified additional unknown resistance mediating mutations or additional aa exchanges at codons already known to confer drug resistance. Interestingly, the results showed that almost all LET-resistance mutations were still located in the UL56 LET resistance region, with the minor exceptions of one low-level resistance mutation at codon 25 in UL56, four low-level resistance mutations in the UL89 gene, and one in the UL51 gene [81–83]. A summary of the known LET resistance mutations is given in Figure 1.12.

Knowledge on LET resistance mutations also enabled screening of wild-type CMV genotype data, either from publicly available databases or from collections of LET-naïve clinical isolates, for the potential presence of preexisting LET-resistant genotypes. In summary, the data showed no evidence of preexisting LET-resistant genotypes circulating in the population [79, 84].

It is of note that in some cases, engineered mutant virus strains lacking the viral polymerase proofreading activity have been used to accelerate the selection of resistant virus mutants and the subsequent identification of novel resistance mutations in cell culture. *In vitro* resistance selection studies using such engineered strains also showed that LET resistance mutations emerged at a lower selection passage compared to foscarnet and GCV, suggesting that LET may have a lower genetic barrier to resistance than polymerase inhibitors [81–83]. However, a caveat to this comparison

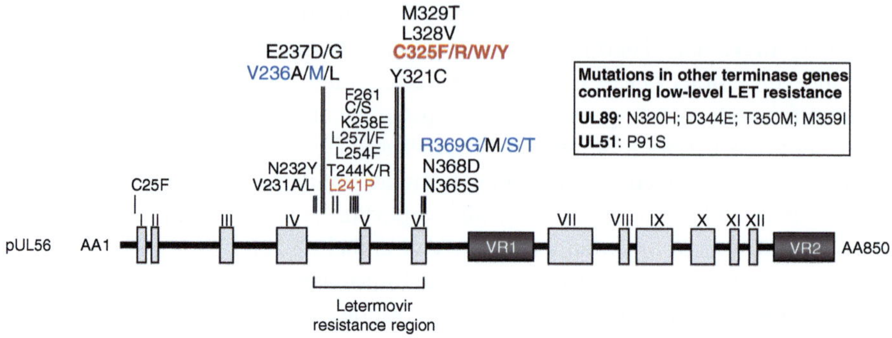

Figure 1.12 Known LET-resistance mutations.

Schematic representation of the UL56 domain organization according to Champier et al. (2008). Conserved regions are indicated as gray boxes (I–XII); variable regions (VR1 and VR2) as black boxes. LET resistance mutations in UL56 are indicated above, resistance mutations in UL89 and UL51 are listed in the box. Resistance mutations are color-coded by the degree of resistance conferred:

Black: EC_{50} ratios <20-fold (most probably not of clinical relevance).

Blue: EC_{50} ratios 20- to 100-fold.

Red: EC_{50} ratios >100-fold.

Note: Mutations at position UL56 C325 confer >3000-fold resistance (i.e. absolute resistance).

Source: References [28, 29, 38, 31, 79, 81–84, 93, 99].

is the inherent bias against drugs that act downstream from DNA replication and thus are less effective than polymerase inhibitors in preventing the formation of DNA mutations generated by a proofreading-deficient viral DNA polymerase. Therefore, the clinical relevance of these data can only be meaningfully interpreted in light of clinical data (see below).

1.9 Letermovir Resistance in Clinical Trials

Resistance genotyping was performed retrospectively for all subjects with csCMV in the Phase 2b, and the Phase 3 HSCT and KT trials. In addition, unknown polymorphisms were characterized by recombinant phenotyping, an approach now considered standard [77].

The first reported clinical case of LET resistance (V236M) was reported in the dose range finding Phase 2b trial in an HSCT patient [78]. It occurred after ~7 weeks of treatment in a subject receiving low-dose LET (60 mg/day), indicating the dose to be at least partially permissive to viral replication and so suboptimal. No cases of resistance were reported in the higher dose groups (120 and 240 mg/day LET).

In the Phase 3 trials, the LET doses used were higher (480 or 240 mg/day with concomitant cyclosporine), and the overall number of resistance events was very low in both HSCT and KT subjects [75, 85].

In total, three subjects in the Phase 3 HSCT trial were shown to have four LET-resistance mutations in UL56. Two of these were already known from preclinical *in vitro* work (V236M and C325W), and two were newly identified mutations, however, at codons already associated with LET resistance (UL56 aa237 and aa369) [85].

In the Phase 3 KT trial, no subjects (of 292 LET-treated) had detectable LET resistance. This is compared to 12 of 297 VGCV-treated subjects with VGCV resistance mutations. However, it is of note that subsequent genetic analyses of VGCV failures, including UL56 sequencing, showed that a single VGCV-treated subject had two low-level LET-resistance mutations that were present with low frequency (<10%). However, re-sequencing of the pUL56 from the same sample failed to confirm the presence of any LET-resistance mutation, which suggests that the detection of such mutations in a LET-naïve patient may have been an artifact of the sequencing process supporting the assumption that no LET-resistant viral strains are circulating in the population [86, 87].

1.10 Real-world Resistance

In the first period after marketing approval, multiple reports of drug resistance were published, which, however, have been largely restricted to the context of secondary prophylaxis or CMV treatment rather than the approved primary prophylaxis indication [88–92].

In the meantime, however, with increasing experience with the clinical use of LET, the resistance rates reported in the clinical trials (see above) appear to be representative of the real world. As concluded by Limaye et al. [75], though resistance to LET is rare in prophylaxis, it appears to be rather more common in cases of therapeutic/rescue off-label use. At an intuitive level, this may reflect the considerable differences between the prophylaxis and treatment settings regarding ongoing virus replication rates (little-to-none vs. potentially high viral loads) and the resulting opportunities for resistance mutations to arise; however, this is not known.

A recent frequency analysis of >1100 clinical samples from suspected LET treatment failures suggested that when LET resistance is detected, single-step mutations conferring high LET resistance (RI ≥3000), especially at the codon 325 position, appear to be more common than low-level resistance mutations (RI <20), confirming the intuitive expectation [93]. This difference can also be explained because in vitro resistance selection experiments often used LET concentrations escalating from a low nanomolar level, whereas therapeutic concentrations of LET at the standard dose are expected to be in the micromolar range, far exceeding the EC_{50} values of many resistant mutants [85]. The occasional appearance of low-grade LET resistance mutations in clinical specimens suggests insufficient drug exposure in those cases [93].

1.11 Outlook for Letermovir

In the high-risk HSCT and kidney recipient populations, LET is now well-established as a safe and effective prophylactic drug to prevent CMV infection/reactivation. Early data reporting the use of LET prophylaxis in other SOT indications likewise suggest good anti-CMV efficacy, safety, and low rates of resistance and thus support the general use of LET as CMV prophylaxis in all transplant patients [76, 94, 95].

LET treatment regimens are specifically optimized for prophylaxis. Nevertheless, an increasing number of literature case reports show that the drug is also being used off-label as a PET or therapy in both HSCT and SOT settings, including cases of resistant/refractory CMV [25, 88, 92, 96–98]. While these are still case reports and small studies, it appears that LET may be efficacious in the treatment setting, though drug resistance rates seem to be higher than in prophylaxis use [75]. The latter observation suggests that dosing regimens may need optimizing for LET to be used as PET or rescue therapy. This cannot be easily addressed without targeted clinical trials but may open new avenues for LET while preventing widespread resistance.

Generally, available data show promise for LET as a first-line agent for universal prophylaxis in high-risk transplant patients. Although this review revolves around the development program that specifically targeted CMV in HSCT/KT transplant recipients, it is also worth remembering the ongoing need for efficacious CMV treatments in the neonatal and HIV populations. Specific investigations should be undertaken to evaluate the potential of this safe and potent drug in these indications.

Acknowledgments

The authors would like to thank Dr. Rob Saunders, biomed context, for his assistance in preparing this manuscript and Dr. Tamara Pfaff and Dr. Dirk Kropeit for their expert critical input.

References

1 Lischka P, Zimmermann H. Antiviral strategies to combat cytomegalovirus infections in transplant recipients. *Curr Opin Pharmacol.* 2008;8(5): 541–8.

2 Cannon MJ, Schmid DS, Hyde TB. Review of cytomegalovirus seroprevalence and demographic characteristics associated with infection. *Rev Med Virol.* 2010;20(4): 202–13.

3 Centers for Disease Control and Prevention. About Cytomegalovirus (CMV) cdc.gov [Available from: https://www.cdc.gov/cmv/overview.html].

4 Dollard SC, Grosse SD, Ross DS. New estimates of the prevalence of neurological and sensory sequelae and mortality associated with congenital cytomegalovirus infection. *Rev Med Virol.* 2007;17(5): 355–63.

5 Bottino P, Pastrone L, Curtoni A, Bondi A, Sidoti F, Zanotto E, et al. Antiviral approach to cytomegalovirus infection: an overview of conventional and novel strategies. *Microorganisms.* 2023;11(10).

6 Forte E, Zhang Z, Thorp EB, Hummel M. Cytomegalovirus latency and reactivation: an intricate interplay with the host immune response. *Front Cell Infect Microbiol.* 2020;10: 130.

7 Cesaro S, Ljungman P, Tridello G, Mikulska M, Wendel L, Styczynski J, et al. New trends in the management of cytomegalovirus infection after allogeneic hematopoietic cell transplantation: a survey of the infectious diseases working pary of EBMT. *Bone Marrow Transplant.* 2023;58(2): 203–8.

8 Siegrist EA, Sassine J. Letermovir and new horizons in prevention of post-transplant CMV. *Expert Rev Clin Pharmacol.* 2023;16(10): 887–9.

9 Razonable R. Direct and indirect effects of cytomegalovirus: can we prevent them? *Enfermedades infecciosas y microbiologia clinica.* 2010;28(1): 1–5.

10 Griffiths P, Baraniak I, Reeves M. The pathogenesis of human cytomegalovirus. *J Pathol.* 2015;235(2): 288–97.

11 Green ML, Leisenring W, Xie H, Mast TC, Cui Y, Sandmaier BM, et al. Cytomegalovirus viral load and mortality after haemopoietic stem cell transplantation in the era of pre-emptive therapy: a retrospective cohort study. *Lancet Haematol.* 2016;3(3): e119–27.

12 Teira P, Battiwalla M, Ramanathan M, Barrett AJ, Ahn KW, Chen M, et al. Early cytomegalovirus reactivation remains associated with increased transplant-related mortality in the current era: a CIBMTR analysis. *Blood.* 2016;127(20): 2427–38.

13 Stern L, Withers B, Avdic S, Gottlieb D, Abendroth A, Blyth E, et al. Human cytomegalovirus latency and reactivation in allogeneic hematopoietic stem cell transplant recipients. *Front Microbiol.* 2019;10: 1186.

14 Staras SA, Dollard SC, Radford KW, Flanders WD, Pass RF, Cannon MJ. Seroprevalence of cytomegalovirus infection in the United States, 1988–1994. *Clin Infect Dis.* 2006;43(9): 1143–51.

15 Boeckh M, Geballe AP. Cytomegalovirus: pathogen, paradigm, and puzzle. *J Clin Invest.* 2011;121(5): 1673–80.

16 Gimenez E, Torres I, Albert E, Pinana JL, Hernandez-Boluda JC, Solano C, et al. Cytomegalovirus (CMV) infection and risk of mortality in allogeneic hematopoietic stem cell transplantation (Allo-HSCT): a systematic review, meta-analysis, and meta-regression analysis. *Am J Transplant.* 2019;19(9): 2479–94.

17 Kotton CN, Kamar N. New Insights on CMV Management in solid organ transplant patients: prevention, treatment, and management of resistant/refractory disease. 2023;12(2): 333–42.

18 Chiuppesi F, Wussow F. The quest for a cytomegalovirus vaccine continues. *Lancet Infect Dis.* 2023;23(12): 1330–2.

19 Summary of *Infect Dis Ther.* product characteristics: Cymevene 2016 [Available from: https://www.ema.europa.eu/documents/referral/cymevene-article-30-referral-annex-iii_en.pdf].

20 Summary of product characteristics: Valcyte 2018 [Available from: http://www.mhra.gov.uk/home/groups/spcpil/documents/spcpil/con1542345229870.pdf].

21 De Clercq E, Li G. Approved antiviral drugs over the past 50 years. *Clin Microbiol Rev.* 2016;29(3): 695 747.

22 Clinigen Healthcare Ltd. Foscavir product information. 2011.

23 Clinigen Healthcare Ltd. Summary of Product Characteristics. 2014.

24 Saullo JL, Miller RA. Cytomegalovirus therapy: role of letermovir in prophylaxis and treatment in transplant recipients. *Annu Rev Med.* 2023;74: 89–105.

25 Limaye AP, Babu TM, Boeckh M. Progress and challenges in the prevention, diagnosis, and management of cytomegalovirus infection in transplantation. *Clin Microbiol Rev*. 2020;34(1).

26 Vyas A, Raval AD, Kamat S, LaPlante K, Tang Y, Chemaly RF. Real-world outcomes associated with letermovir use for cytomegalovirus primary prophylaxis in allogeneic hematopoietic cell transplant recipients: a systematic review and meta-analysis of observational studies. *Open Forum Infect Dis*. 2023; 10(1):ofac687.

27 Reefschlaeger J, Bender W, Hallenberger S, Weber O, Eckenberg P, Goldmann S, et al. Novel non-nucleoside inhibitors of cytomegaloviruses (BAY 38-4766): in vitro and in vivo antiviral activity and mechanism of action. *J Antimicrob Chemother*. 2001;48(6): 757–67.

28 Lischka P, Hewlett G, Wunberg T, Baumeister J, Paulsen D, Goldner T, et al. In vitro and in vivo activities of the novel anticytomegalovirus compound AIC246. *Antimicrob Agents Chemother*. 2010;54(3): 1290–7.

29 Goldner T, Hewlett G, Ettischer N, Ruebsamen-Schaeff H, Zimmermann H, Lischka P. The novel anticytomegalovirus compound AIC246 (Letermovir) inhibits human cytomegalovirus replication through a specific antiviral mechanism that involves the viral terminase. *J Virol*. 2011;85(20): 10884–93.

30 Marschall M, Stamminger T, Urban A, Wildum S, Ruebsamen-Schaeff H, Zimmermann H, et al. In vitro evaluation of the activities of the novel anticytomegalovirus compound AIC246 (letermovir) against herpesviruses and other human pathogenic viruses. *Antimicrob Agents Chemother*. 2012;56(2): 1135–7.

31 Lischka P, Zhang D, Holder D, Zimmermann H. Impact of glycoprotein B genotype and naturally occurring ORF UL56 polymorphisms upon susceptibility of clinical human cytomegalovirus isolates to letermovir. *Antiviral Res*. 2016;132: 204–9.

32 Isaacson MK, Compton T. Human cytomegalovirus glycoprotein B is required for virus entry and cell-to-cell spread but not for virion attachment, assembly, or egress. *J Virol*. 2009;83(8): 3891–903.

33 Chong KT, Mediratta S, Pagano P, Hinshaw R. A novel gelfoarm mouse model for human cytomegalovirus (CMV) replication. 39th Interscience Conference on Antimicrobial Agents and Chemotherapy. 1999;39: 441.

34 Wildum S, Zimmermann H, Lischka P. In vitro drug combination studies of Letermovir (AIC246, MK-8228) with approved anti-human cytomegalovirus (CMV) and anti-HIV compounds in inhibition of CMV and HIV replication. *Antimicrob Agents Chemother*. 2015;59(6): 3140–8.

35 Yang Y, Yang P, Wang N, Chen Z, Su D, Zhou ZH, et al. Architecture of the herpesvirus genome-packaging complex and implications for DNA translocation. *Protein Cell*. 2020;11(5): 339–51.

36 Gentry BG, Bogner E, Drach JC. Targeting the terminase: an important step forward in the treatment and prophylaxis of human cytomegalovirus infections. *Antiviral Res*. 2019;161: 116–24.

37 Neuber S, Wagner K, Goldne T, Lischka P, Steinbrueck L, Messerle M, et al. Mutual interplay between the human cytomegalovirus terminase subunits pUL51, pUL56, and pUL89 promotes terminase complex formation. *J Virol*. 2017.

38 Goldner T, Hempel C, Ruebsamen-Schaeff H, Zimmermann H, Lischka P. Geno- and phenotypic characterization of human cytomegalovirus mutants selected in vitro after letermovir (AIC246) exposure. *Antimicrob Agents Chemother.* 2014;58(1): 610–3.

39 Townsend LB, Devivar RV, Turk SR, Nassiri MR, Drach JC. Design, synthesis, and antiviral activity of certain 2,5,6-trihalo-1-(beta-D-ribofuranosyl)benzimidazoles. *J Med Chem.* 1995;38(20): 4098–105.

40 Underwood MR, Harvey RJ, Stanat SC, Hemphill ML, Miller T, Drach JC, et al. Inhibition of human cytomegalovirus DNA maturation by a benzimidazole ribonucleoside is mediated through the UL89 gene product. *J Virol.* 1998;72(1): 717–25.

41 Krosky PM, Underwood MR, Turk SR, Feng KW, Jain RK, Ptak RG, et al. Resistance of human cytomegalovirus to benzimidazole ribonucleosides maps to two open reading frames: UL89 and UL56. *J Virol.* 1998;72(6): 4721–8.

42 Underwood MR, Ferris RG, Selleseth DW, Davis MG, Drach JC, Townsend LB, et al. Mechanism of action of the ribopyranoside benzimidazole GW275175X against human cytomegalovirus. *Antimicrob Agents Chemother.* 2004;48(5): 1647–51.

43 Buerger I, Reefschlaeger J, Bender W, Eckenberg P, Popp A, Weber O, et al. A novel nonnucleoside inhibitor specifically targets cytomegalovirus DNA maturation via the UL89 and UL56 gene products. *J Virol.* 2001;75(19): 9077–86.

44 Evers DL, Komazin G, Ptak RG, Shin D, Emmer BT, Townsend LB, et al. Inhibition of human cytomegalovirus replication by benzimidazole nucleosides involves three distinct mechanisms. *Antimicrob Agents Chemother.* 2004;48(10): 3918–27.

45 Weber O, Bender W, Eckenberg P, Goldmann S, Haerter M, Hallenberger S, et al. Inhibition of murine cytomegalovirus and human cytomegalovirus by a novel non-nucleosidic compound in vivo. *Antiviral Res.* 2001;49(3): 179–89.

46 North TW, Sequar G, Townsend LB, Drach JC, Barry PA. Rhesus cytomegalovirus is similar to human cytomegalovirus in susceptibility to benzimidazole nucleosides. *Antimicrob Agents Chemother.* 2004;48(7): 2760–5.

47 Nixon DE, McVoy MA. Dramatic effects of 2-bromo-5,6-dichloro-1-beta-D-ribofuranosyl benzimidazole riboside on the genome structure, packaging, and egress of guinea pig cytomegalovirus. *J Virol.* 2004;78(4): 1623–35.

48 McCormick D, Kropeit D, Stoelben S, Budde K, Arns W, Kaul DR, et al. AIC246 (Letermovir) Does not exhibit clinically significant drug-drug interactions when administered with immunosuppressive drugs in transplant recipients. ICAAC; Chicago, USA2011.

49 Marty FM, Ljungman P, Chemaly RF, Maertens J, Dadwal SS, Duarte RF, et al. Letermovir prophylaxis for cytomegalovirus in hematopoietic-cell transplantation. *N Engl J Med.* 2017;377(25): 2433–44.

50 Chemaly RF, Ullmann AJ, Stoelben S, Richard MP, Bornhauser M, Groth C, et al. Letermovir for cytomegalovirus prophylaxis in hematopoietic-cell transplantation. *N Engl J Med.* 2014;370(19): 1781–9.

51 Kropeit D, McCormick D, Erb-Zohar K, Stobernack HP, Zimmermann H, Rübsamen-Schaeff H. Pharmacokinetics and safety of Letermovir, a novel anti-human cytomegalovirus drug, in hepatically impaired patients. ECCMID; April 2015; Copenhagen, Denmark2015.

52 Kropeit D, Scheuenpflug J, Erb-Zohar K, Stobernack HP, Hulskotte EGJ, van Schanke A, et al. Pharmacokinetics and safety of Letermovir, a novel anti-human cytomegalovirus drug, in renally impaired patients. ECCMID; Barcelona, Spain2014.

53 Kropeit D, Scheuenpflug J, Erb-Zohar K, Stobernack HP, Hulskotte EGJ, van Schanke A, et al. Single ascending and multiple dose data from a hydroxypropyl beta cyclodextrin intravenous formulation of letermovir, a novel anti-human cytomegalovirus drug. ICAAC; Denver, CO, USA2013.

54 Stoelben S, Arns W, Renders L, Hummel J, Muhlfeld A, Stangl M, et al. Preemptive treatment of Cytomegalovirus infection in kidney transplant recipients with letermovir: results of a Phase 2a study. Transplant international: official journal of the European Society for Organ Transplantation. 2014;27(1): 77–86.

55 Kaul DR, Stoelben S, Cober E, Ojo T, Sandusky E, Lischka P, et al. First report of successful treatment of multidrug-resistant cytomegalovirus disease with the novel anti-CMV compound AIC246. *Am J Transplant.* 2011;11(5): 1079–84.

56 Merck (NYSE: MRK), AiCuris AG. Aicuris and Merck Enter Exclusive Worldwide License Agreement For Investigational Portfolio Targeting Human Cytomegalovirus [press release]. October 15, 2012.

57 FDA. CDER Summary Review, NDA 209939, 209940 In: Center for Drug Evaluation and Research (CDER), editor. 2017.

58 CHMP EMA. Summary of opinion (initial authorisation). In: CHMP E, editor. 2017.

59 Politikos I, Lau C, Devlin SM, Quach S, Lin A, Perales MA, et al. Extended-duration letermovir prophylaxis for cytomegalovirus infection after cord blood transplantation in adults. *Blood Adv.* 2022;6(24): 6291–300.

60 Hinman B, Cox J, Umoru G, Kamble R, Musick W. Extended duration letermovir in allogeneic hematopoietic stem cell transplant. *Transpl Immunol.* 2023;81: 101936.

61 Chavaz L, Royston L, Masouridi-Levrat S, Mamez AC, Giannotti F, Morin S, et al. CMV infection after letermovir primary prophylaxis discontinuation in allogeneic hematopoietic cell transplant recipients. *Open Forum Infect Dis.* 2023; 10(4):ofad169.

62 Douglas AP, Slavin MA. Extended duration of letermovir prophylaxis: how long is long enough? *Lancet Haematol.* 2023.

63 Russo D, Schmitt M, Pilorge S, Stelljes M, Kawakita T, Teal VL, et al. Efficacy and safety of extended duration letermovir prophylaxis in recipients of haematopoietic stem-cell transplantation at risk of cytomegalovirus infection: a multicentre, randomised, double-blind, placebo-controlled, phase 3 trial. *Lancet Haematol.* 2023.

64 U.S. FDA Approves New Indication for Merck's PREVYMIS® (letermovir) for Prevention of Cytomegalovirus (CMV) Disease in High-Risk Adult Kidney Transplant Recipients [press release]. 2023.

65 EMA. Prevymis - opinion on variation to marketing authorisation ema.europa.eu2023 [Available from: https://www.ema.europa.eu/en/medicines/human/variation/prevymis].

66 Strenger V, Sperl D, Kubesch K, Donnerer J, Schwinger W, Zach K, et al. Letermovir in paediatric HSCT recipients. *J Antimicrob Chemother*. 2019;74(9): 2820–1.

67 Korholz KF, Fuller MA, Hennies M, Holterhus M, Hagedorn S, Ahlmann M, et al. Letermovir for prophylaxis and pre-emptive therapy of cytomegalovirus infection in paediatric allogeneic haematopoietic cell Transplant Patients. *Paediatr Drugs*. 2023;25(2): 225–32.

68 Schulte JH, Gefen A, Groll AH, Fraser CJ, Teal VL, Haber B, et al. Cytomegalovirus (CMV) prophylaxis with letermovir (LET) in pediatric (birth to <<18 years of age) allogeneic hematopoietic cell transplant (HCT) recipients: pharmacokinetics (PK), safety, and efficacy results of a Phase 2b study. IDWeek 2023; Boston, MA, USA2023.

69 Doss KM, Heldman MR, Limaye AP. Updates in cytomegalovirus prevention and treatment in solid organ transplantation. *Infect Dis Clin North Am*. 2023.

70 Kotton CN, Kumar D, Caliendo AM, Huprikar S, Chou S, Danziger-Isakov L, et al. The third international consensus guidelines on the management of cytomegalovirus in solid-organ transplantation. *Transplantation*. 2018;102(6): 900–31.

71 Imlay H, Wagener MM, Vutien P, Perkins J, Singh N, Limaye AP. Increasing proportion of high-risk cytomegalovirus donor-positive/recipient-negative serostatus in solid organ transplant recipients. *Transplantation*. 2023;107(4): 988–93.

72 Stewart AG, Kotton CN. What's new: updates on cytomegalovirus in solid organ transplantation. *Transplantation*. 2023.

73 Razonable RR, Humar A. Cytomegalovirus in solid organ transplant recipients-guidelines of the american society of transplantation infectious diseases community of practice. *Clin Transplant*. 2019;33(9): e13512.

74 Humar A, Lebranchu Y, Vincenti F, Blumberg EA, Punch JD, Limaye AP, et al. The efficacy and safety of 200 days valganciclovir cytomegalovirus prophylaxis in high-risk kidney transplant recipients. *Am J Transplant*. 2010;10(5): 1228–37.

75 Limaye AP, Budde K, Humar A, Vincenti F, Kuypers DRJ, Carroll RP, et al. Letermovir vs valganciclovir for prophylaxis of cytomegalovirus in high-risk kidney transplant recipients: a randomized clinical trial. *JAMA*. 2023;330(1): 33–42.

76 Jorgenson MR, Descourouez JL, Saddler CM, Smith JA, Odorico JS, Rice JP, et al. Real world experience with conversion from valganciclovir to letermovir for cytomegalovirus prophylaxis: letermovir reverses leukopenia and avoids mycophenolate dose reduction. *Clin Transplant*. 2023;37(12): e15142.

77 Chemaly RF, Chou S, Einsele H, Griffiths P, Avery R, Razonable RR, et al. Definitions of resistant and refractory cytomegalovirus infection and disease in transplant recipients for use in clinical trials. *Clin Infect Dis*. 2018.

78 Lischka P, Michel D, Zimmermann H. Characterization of cytomegalovirus breakthrough events in a letermovir (AIC246, MK 8228) phase 2 prophylaxis trial. *J Infect Dis*. 2015.

79 Goldner T, Zimmermann H, Lischka P. Phenotypic characterization of two naturally occurring human Cytomegalovirus sequence polymorphisms located in a distinct region of ORF UL56 known to be involved in in vitro resistance to letermovir. *Antiviral Res.* 2015;116: 48–50.

80 Chou S. Advances in the genotypic diagnosis of cytomegalovirus antiviral drug resistance. *Antiviral Res.* 2020;176: 104711.

81 Chou S. Rapid in vitro evolution of human cytomegalovirus UL56 mutations that confer letermovir resistance. *Antimicrob Agents Chemother.* 2015;59(10): 6588–93.

82 Chou S. A third component of the human cytomegalovirus terminase complex is involved in letermovir resistance. *Antiviral Res.* 2017;148: 1–4.

83 Chou S. Comparison of cytomegalovirus terminase gene mutations selected after exposure to three distinct inhibitor compounds. *Antimicrob Agents Chemother.* 2017;61(11).

84 Lischka P, Michel D, Zimmermann H. Characterization of cytomegalovirus breakthrough events in a phase 2 prophylaxis trial of letermovir (AIC246, MK 8228). *J Infect Dis.* 2016;213(1): 23–30.

85 Douglas CM, Barnard R, Holder D, Leavitt R, Levitan D, Maguire M, et al. Letermovir resistance analysis in a clinical trial of cytomegalovirus prophylaxis for hematopoietic stem cell transplant recipients. *J Infect Dis.* 2020;221(7): 1117–26.

86 Chou S, Watanabe J. Phenotypes of cytomegalovirus genetic variants encountered in a letermovir clinical trial illustrate the importance of genotyping validation. *Antiviral Res.* 2024: 105935.

87 Strizki JM, Diamond TL, Teal VL, Gilbert CL, Wang W, Stauffer N, et al. Cytomegalovirus antiviral resistance among participants in the phase 3 trial of letermovir vs valganciclovir prophylaxis in kidney transplant recipients. *J Infect Dis.* 2024.

88 Cherrier L, Nasar A, Goodlet KJ, Nailor MD, Tokman S, Chou S. Emergence of letermovir resistance in a lung transplant recipient with ganciclovir-resistant cytomegalovirus infection. *Am J Transplant.* 2018;18(12): 3060–4.

89 Jung S, Michel M, Stamminger T, Michel D. Fast breakthrough of resistant cytomegalovirus during secondary letermovir prophylaxis in a hematopoietic stem cell transplant recipient. *BMC Infect Dis.* 2019;19(1): 388.

90 Khawaja F, Batista MV, El Haddad L, Chemaly RF. Resistant or refractory cytomegalovirus infections after hematopoietic cell transplantation: diagnosis and management. *Curr Opin Infect Dis.* 2019;32(6): 565–74.

91 Popping S, Dalm V, Lubke N, Cristanziano VD, Kaiser R, Boucher CAB, et al. Emergence and Persistence of Letermovir-Resistant Cytomegalovirus in a Patient With Primary Immunodeficiency. *Open Forum Infect Dis.* 2019; 6(9):ofz375.

92 Turner N, Strand A, Grewal DS, Cox G, Arif S, Baker AW, et al. Use of Letermovir as Salvage Therapy for Drug-Resistant Cytomegalovirus Retinitis. *Antimicrob Agents Chemother.* 2019;63(3).

93 Chou S, Kleiboeker S. Relative frequency of cytomegalovirus UL56 gene mutations detected in genotypic letermovir resistance testing. *Antiviral Res.* 2022;207: 105422.

94 Saullo JL, Baker AW, Snyder LD, Reynolds JM, Zaffiri L, Eichenberger EM, et al. Cytomegalovirus prevention in thoracic organ transplantation: A single-center evaluation of letermovir prophylaxis. The Journal of heart and lung transplantation: the official publication of the International Society for Heart Transplantation. 2022;41(4): 508–15.

95 Winstead RJ, Kumar D, Brown A, Yakubu I, Song C, Thacker L, et al. Letermovir prophylaxis in solid organ transplant-Assessing CMV breakthrough and tacrolimus drug interaction. *Transpl Infect Dis*. 2021;23(4): e13570.

96 Hofmann E, Sidler D, Dahdal S, Bittel P, Suter-Riniker F, Manuel O, et al. Emergence of letermovir resistance in solid organ transplant recipients with ganciclovir resistant cytomegalovirus infection: A case series and review of the literature. *Transpl Infect Dis*. 2021;23(3): e13515.

97 Paolucci S, Campanini G, Cassaniti I, Tebaldi A, Novazzi F, Fratini A, et al. Emergence of letermovir-resistant CMV UL56 mutant during rescue treatment in a liver transplant recipient with ganciclovir-resistant infection CMV: a case report. *BMC Infect Dis*. 2021;21(1): 994.

98 von Hoerschelmann E, Munch J, Gao L, Lucht C, Naik MG, Schmidt D, et al. Letermovir rescue therapy in kidney transplant recipients with refractory/resistant CMV disease. *J Clin Med*. 2023;13(1).

99 Champier G, Couvreux A, Hantz S, Rametti A, Mazeron MC, Bouaziz S, et al. Putative functional domains of human cytomegalovirus pUL56 involved in dimerization and benzimidazole D-ribonucleoside activity. *Antiviral Therapy*. 2008;13(5): 643–54.

2

Discovery and Development of the Helicase–Primase Inhibitor Pritelivir for the Treatment of Immunocompromised Patients with Resistant HSV Infection

Alexander Birkmann

AiCuris Anti-Infective Cures AG, Wuppertal, Germany

2.1 HSV Virology and Disease

Herpes simplex virus (HSV; comprising HSV type 1 [HSV-1] and type 2 [HSV-2]), belongs to the Herpesviridae family and is characterized by enveloped, double-stranded DNA genomes of approximately 152 kb in size that harbor approximately 90 open reading frames [1]. Historically, HSV-1 has been predominantly associated with "above-the-belt" infections involving the mouth and eyes and persisting within the trigeminal ganglion. In contrast, HSV-2 residing in the sacral ganglion has been linked to "below-the-belt" genital herpes [2]. However, evolving trends indicate a substantial overlap between sites of infection, with increasing cases of genital herpes attributed to HSV-1. This shift may be due to reduced childhood HSV-1 infections resulting from improved hygiene standards, as well as an increase in oral sex practices [3, 4].

Like the closely related Herpesviridae family varicella-zoster virus (VZV), the establishment of latency and persistent infection within peripheral nervous system neurons is intrinsic to HSV [5]. Prior to migration into the peripheral nervous system, initial infection begins when the virus infiltrates the host via skin or mucosal surfaces, which can be associated with or without clinical symptoms, often resulting in local lesions. Once neuronal latency is established, HSV can undergo reactivation driven by still not fully understood molecular triggers, leading to replication in the lower epidermal layers near the original site of infection [6]. The severity of recurrences varies widely, ranging from asymptomatic only with detectable viral particles on the skin, to painful orofacial cold sores or genital lesions. Such outbreaks have a

Trends in Antiviral Drug Development, First Edition. Edited by Michael J. Sofia and Zhengqiang Wang.

significant impact on quality of life, causing distress and psychosocial problems [7, 8]. Additionally, ocular HSV-1 infections can lead to visual impairment, underscoring the clinical importance of effective therapeutic interventions [9].

HSV transmission occurs via infectious viral particles, emanating from active lesions or, especially in the case of HSV-2, via asymptomatic shedding [10, 11]. HSV-1 typically initiates infection during childhood, whereas HSV-2 tends to occur later in adolescence, primarily through sexual transmission [12, 13]. Notably, quantitative polymerase chain reaction (PCR) analysis has shown that asymptomatic viral shedding is common in patients with genital herpes, increasing the risk of transmission even in the absence of visible symptoms. Approximately 70% of HSV-2 transmission occurs during such asymptomatic shedding episodes [14].

Typically, HSV recurrences are self-limiting and promptly cleared by the immune system, primarily via the cellular immune response [15, 16]. However, HSV reactivation can lead to severe and occasionally life-threatening infections. Immunocompromised patients, including transplant recipients and those with suboptimally controlled human immunodeficiency virus (HIV) infections, face a more severe clinical course during HSV infections [17–19]. Unlike otherwise healthy individuals, these patients experience frequent reactivations, prolonged symptoms, extensive lesions, and an elevated risk of systemic virus spread. In those rare cases of a systemic infection, HSV can disseminate to organs, particularly the liver and lungs, resulting in fulminant hepatitis or pneumonitis, respectively [20, 21]. HIV-infected individuals in particular are often co-infected with HSV-2. This alters the course of the HSV-2 infection, causing more frequent and severe reactivations. The presence of HSV-2 infection also increases the risk of HIV acquisition [22].

Neonates born to HSV-positive mothers are another vulnerable patient population. These children are susceptible to congenital infections which can manifest as skin, eye, and mouth infections, herpes simplex encephalitis (HSE), or disseminated disease involving multiple organs. Notably, even immunocompetent individuals between the ages of 6 months and 20 years, as well as those 50 years and older, can develop HSE [23].

The global prevalence of HSV-1 and HSV-2 infection is estimated to be 67% and 11%, respectively, with the highest rates observed in the 45–49 age group, reaching 79% and 17%, respectively [3, 4]. The pervasive burden and severity of these infections, coupled with the stigmatizing nature of labial and especially genital manifestations, underscores the need for ongoing research efforts to discover and develop new therapeutic strategies against HSV, including drugs with a novel mode of action such as helicase primase inhibitors like pritelivir, to combat both resistant and recurrent infections.

2.2 Treatment

HSV poses a significant clinical challenge due to its ability to establish persistent infections and recur intermittently. Therefore, a distinction must be made between primary infections, initial/nonprimary infections (in individuals already infected with one HSV type), or recurrent episodes. While primary and initial infections may present as asymptomatic or with nonspecific symptoms, therapeutic interventions are primarily focused on managing recurrent episodes, which account for the majority of symptomatic cases [24]. It is worth noting, however, that initial infections, particularly in the case of genital herpes, can be prolonged and more painful than subsequent recurrences [12]. Early recognition and administration of antiviral treatment during the initial episode can substantially reduce symptoms and shorten the duration of infection.

Nucleoside analogs, specifically acyclovir and penciclovir, along with their respective prodrugs, valacyclovir and famciclovir, are the cornerstone of treatment for HSV infection (Figure 2.1) [25]. These agents, discovered more than 50 years ago, exhibit

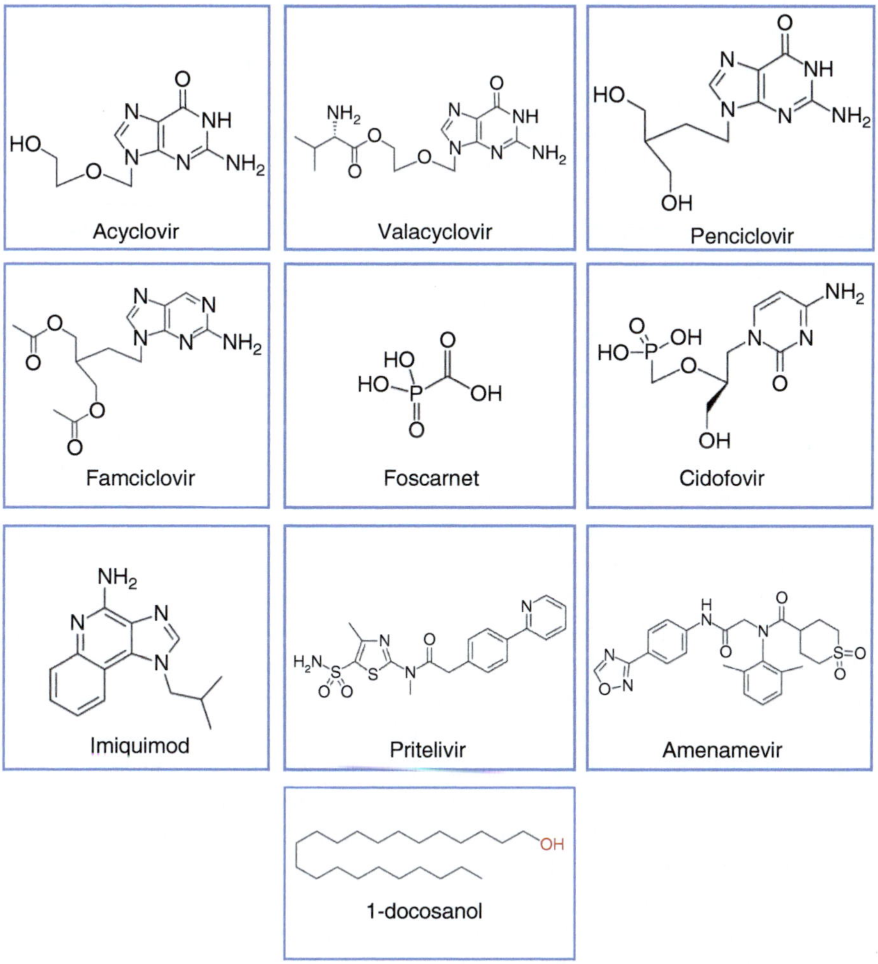

Figure 2.1 Acyclovir and penciclovir are guanosine nucleoside analogs targeting the viral polymerase, while valacyclovir and famciclovir represent prodrugs with better oral bioavailability. Foscarnet is a pyrophosphate analog and cidofovir is an acyclic nucleoside phosphonate, both of which also target the viral polymerase. Imiquimod is an aminoquinoline immune modulator that induces interferon production. Pritelivir and amenamevir are helicase–primase inhibitors. 1-docosanol is considered to be a virus uptake inhibitor.

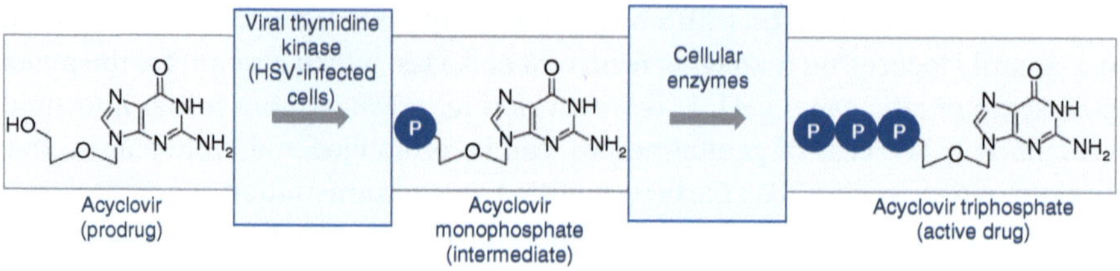

Figure 2.2 The prodrug acyclovir is phosphorylated by the viral thymidine kinase in HSV-infected cells to an intermediate compound, acyclovir monophosphate. The active drug, acyclovir triphosphate, is the product of further cellular enzymatic activity, resulting in additional phosphorylation at the same position.

high specificity for HSV and the closely related VZV causing chickenpox and shingles, respectively, as they rely on activation by the viral enzyme thymidine kinase (TK), a key player in the phosphorylation of thymidine [26]. Viral TK converts guanosine analogs into monophosphates, which are subsequently phosphorylated by cellular kinases into pharmacologically active triphosphates (Figure 2.2). Incorporation of these triphosphates into the growing viral DNA chain, catalyzed by the viral DNA polymerase, culminates in chain termination, thereby disrupting the HSV replication cycle.

Oral acyclovir, valacyclovir, and famciclovir are frequently employed for the short-term management of acute episodes of genital or labial herpes and long-term suppressive therapy, particularly in individuals prone to frequent genital recurrences [2]. Notably, valacyclovir has demonstrated the added benefit of reducing the transmission of genital herpes during suppressive therapy [27]. Topical formulations of acyclovir and penciclovir are primarily used to alleviate symptoms during acute episodes of labial herpes [28]. Acyclovir remains the preferred treatment choice for immunocompromised individuals and is often administered intravenously due to increased efficacy based on higher levels of exposure via this route and the inability of certain patients to swallow [29]. Furthermore, acyclovir or valacyclovir are used as prophylaxis to prevent HSV reactivation in immunosuppressed individuals before or after transplantation, or in HIV patients. Such prophylactic measures have become critical, as HSV recurrence was frequently reported in immunosuppressed patients before the availability of such treatments, with rates reaching 82% in seropositive patients in the months before and after bone marrow transplantation [30].

Although generally considered safe and well tolerated, nucleoside analogs such as acyclovir and valacyclovir have certain shortcomings with respect to efficacy. Long-term suppressive treatment with 500 mg valacyclovir once daily in otherwise healthy individuals with genital herpes resulted in a recurrence-free rate of 40% (or 28%, depending on the statistical method used) [31]. This indicates that more than 50% of patients still experience recurrences despite treatment, consistent with the results of the valacyclovir transmission trial, where transmission among discordant couples is reduced by 50% under valacyclovir compared to placebo but still occurs [27].

In addition to nucleoside analogs, alternative treatment options are available for specific clinical situations. The topical application of 1-docosanol (Figure 2.1), a proposed virus uptake inhibitor, is employed in the management of cold sores and is even available as an over-the-counter treatment [28]. Foscarnet (trisodium phosphonoformate, Figure 2.1) administered intravenously is available or the treatment of infections resistant to nucleoside analogs [32]. Unlike acyclovir or penciclovir, foscarnet selectively interacts with the pyrophosphate binding site of viral DNA polymerases (HSV and cytomegalovirus [CMV]) and does not require activation by TK. It is the only approved treatment for resistant HSV infections in several countries. However, its use is limited due to severe side effects, making it a last resort for cases in which no other treatment options are available. The nucleoside analog cidofovir (intravenous [IV] or topical) is sometimes used off-label as a foscarnet alternative, however, IV administration is also associated with nephrotoxicity and acute renal failure [33]. Another alternative is the immunomodulator imiquimod. Although the efficacy and safety of topical imiquimod on acyclovir-resistant HSV infections have been assessed in several off-label studies, efficacy results remain controversial in the absence of conclusive data from controlled clinical trials [34–36]. However, it is still used by some clinicians to treat these patients [33].

2.3 Resistant Infections

HSV recurrences in immunocompetent individuals are typically self-limited infections that are rapidly controlled by the host immune system, particularly through cellular immunity, i.e. the activity of CD8 cells [16]. Effective management of HSV infections requires the initiation of antiviral therapy at the earliest onset of prodromal signs and symptoms, as delaying treatment until later stages, when viral replication is under the control of the immune system, does not offer additional benefits [37]. Accordingly, the prevalence of resistance to nucleoside analogs, particularly the widely used acyclovir, remains low in immunocompetent patients, ranging from 0.1% to 0.7% [32]. For those with compromised immune responses, HSV replicates more extensively, as viral replication is primarily limited by antiviral treatment in the absence of robust immune control, especially during prolonged therapy [29]. As a result, the prevalence of acyclovir-resistant HSV infection is significantly higher in immunocompromised populations, with variation depending on the type of immunosuppression. More specifically, reports have found acyclovir resistance rates of 3.5–7% in HIV-positive individuals, up to 10% in solid organ transplant recipients, and 14–46.5% in hematopoietic stem cell transplant recipients [38–41].

A common mechanism contributing to resistance to nucleoside analogs is the acquisition of loss-of-function mutations in the viral *TK* gene, often resulting in the introduction of premature stop codons, thereby preventing the expression of the full-length functional enzyme [42]. Additionally, amino acid changes leading to the decreased affinity of TK to nucleoside analogs can occur. Notably, the immune system effectively manages and eliminates such variants in immunocompetent hosts. Conversely, these resistant strains can replicate unhindered in immunocompromised individuals. Another mechanism of resistance to nucleoside analogs involves mutations in the viral polymerase (*pol*) gene that reduce the affinity of the enzyme for the respective triphosphates [43, 44]. Although *pol* mutations are less common than *TK* mutations, they can also result in reduced susceptibility to foscarnet [45].

In cases where resistance to acyclovir is suspected, an initial therapeutic approach involves the administration of high-dose IV acyclovir to overcome resistance

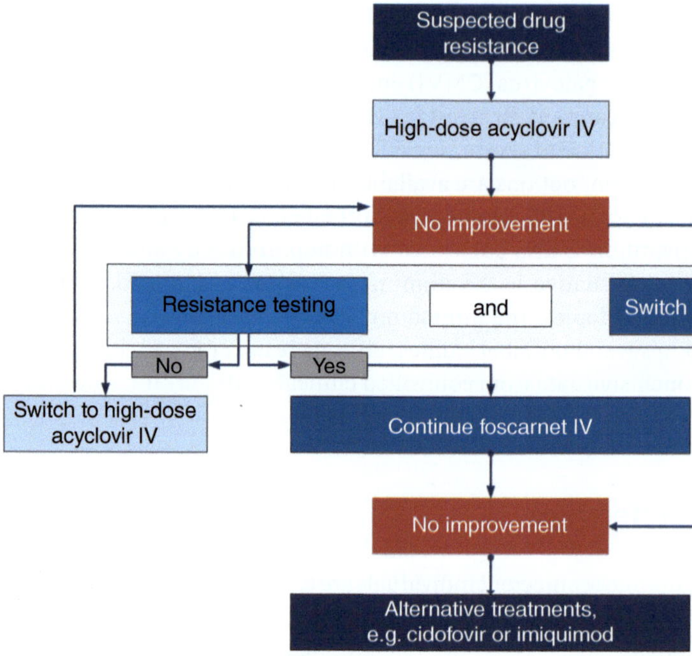

Figure 2.3 The algorithm for treating these patients indicates that options are limited. High-dose IV acyclovir may overcome resistance in some cases, but if there is no improvement, IV foscarnet is the next option. If foscarnet is ineffective, there are currently no proven alternative options.

(Figure 2.3) [46]. This strategy relies on overcoming a potentially reduced affinity of the TK or pol enzyme and the limited extent of initial acyclovir phosphorylation facilitated by cellular kinases in the absence of normal viral TK activity [47]. For patients who do not respond to high-dose acyclovir treatment, foscarnet remains the only US Food and Drug Administration (FDA) approved therapeutic option [48]. Clinical observations have shown that foscarnet can facilitate lesion healing, reduce viral shedding, and relieve pain [49]. However, the efficacy of this treatment is offset by its potential for significant adverse events, primarily affecting serum electrolytes and renal function, occasionally necessitating discontinuation [50]. In addition, several studies have reported the development of foscarnet resistance in acyclovir-resistant strains. Notably, one study reported that 38% of patients treated with foscarnet developed resistance to the drug, highlighting the challenges associated with this therapeutic option [51].

2.4 Pritelivir Discovery and Target Identification

The aforementioned limitations of available treatments, as well as the lack of an effective vaccine, provide a rationale for the development of a new treatment for HSV infections, which ideally should overcome these shortcomings by exploiting a novel mode of action, thereby offering resistance-breaking properties by design, and exhibiting a favorable pharmacological profile in terms of safety, efficacy, and pharmacokinetics (PK).

To identify such a HSV drug, colleagues at Bayer used phenotypic high-throughput screening with HSV-1 infected culture cells and a cell viability readout [52, 53]. The advantage of this approach was that only cells that survived the HSV-1 infection due to the presence of an active inhibitor gave a positive signal, thus also screening for noncytotoxic compounds. Screening of Bayer's small-molecule library, which at the time contained approximately 420 000 compounds, led to BAY 38-9489 as the primary lead, which showed initial activity of 0.5 μM against HSV-1 and 0.7 μM against HSV-2 (Figure 2.4). The subsequent lead optimization process by means of combinational chemistry to improve *in vitro* potency and tolerability, *in vivo* efficacy, and pharmacokinetic properties, which has been recently reviewed [54], resulted in the preclinical and later clinical candidate pritelivir (formerly AIC316, BAY 57-1293).

Since a phenotypic approach was used for screening, the target of the newly discovered compound class was unknown. Through trans-complementation experiments with genome fragments derived from resistant viruses generated *in vitro*, it was shown that fragments carrying mutations in the UL5 helicase could restore the resistant phenotype in wild-type virus-infected cells [53]. Based on this result and additional experiments, the viral helicase–primase complex was identified as the target for this series of lead molecules (Figure 2.5). Since the helicase–primase complex is unique to herpesviruses and some bacteriophages without closely related host cell homologs, high selectivity, and specificity for viral infections could be expected [55]. The validation of the helicase–primase as a target is further supported by the fact that other companies are also developing inhibitors against this complex [56–59] and that one of them, amenamevir (Figure 2.1), had been approved in Japan [60].

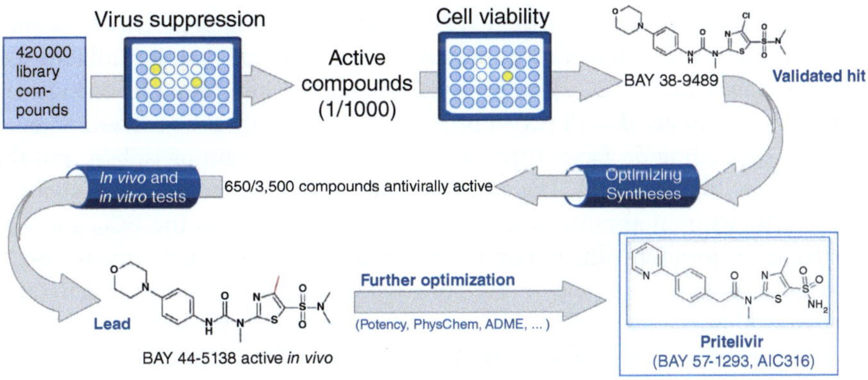

Figure 2.4 Phenotypic high-throughput screening of 420,000 compounds against HSV-1 led to the identification of the initial hit BAY 38-9489. Using a combinational chemistry approach, BAY 38-9489 was optimized to the lead BAY 44-5138, which showed in vivo efficacy, and further optimized to the final clinical candidate pritelivir.

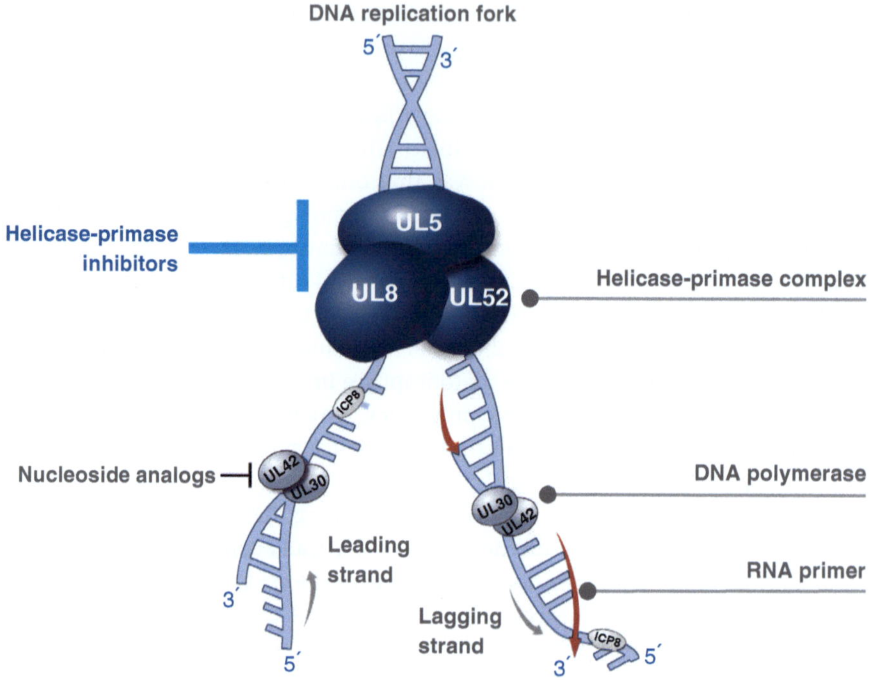

Figure 2.5 Pritelivir is a viral helicase–primase inhibitor that targets a helicase–primase complex mainly present in herpesviruses. Its mechanism of action differs from that of traditionally used nucleoside analogs, which resemble naturally occurring nucleosides, inhibit DNA polymerization, and cause termination of the growing DNA chain.

2.5 Nonclinical Data

2.5.1 *In Vitro* Studies

Pritelivir is selective for HSV-1 and HSV-2 and has only limited activity against other herpesviruses such as VZV or human cytomegalovirus [53]. It exhibits a 50% effective concentration (EC_{50}) range of 0.01–0.07 μM when tested in various cell culture systems infected with HSV-1 or HSV-2 [61]. Notably, the observed potency may vary depending on factors such as cell type, virus strain or isolate, and the specific assay method used. The *in vitro* concentration-response curve shows a steep profile with an approximate twofold difference between the EC_{50} and 90% effective concentration values (Figure 2.6). In these assays, pritelivir was 50–200 times more potent than acyclovir (EC_{50} of 1–4 μM) [53]. In addition, the activity of pritelivir against clinical isolates has been shown to be very similar to that against laboratory strains. In a study of 65 HSV-1 and 19 HSV-2 clinical isolates from Germany, EC_{50} values were consistently at or below 0.05 μM. In another study of 27 HSV-1 and 32 HSV-2 clinical isolates from patients in the United States, the mean EC_{50} was 0.03 μM for both HSV-1 (range 0.011–0.042 μM) and HSV-2 strains (range 0.009–0.065 μM) [53, 61].

To confirm the activity of pritelivir against HSV resistant to nucleoside analogs due to its different mode of action, seven well-characterized acyclovir- and

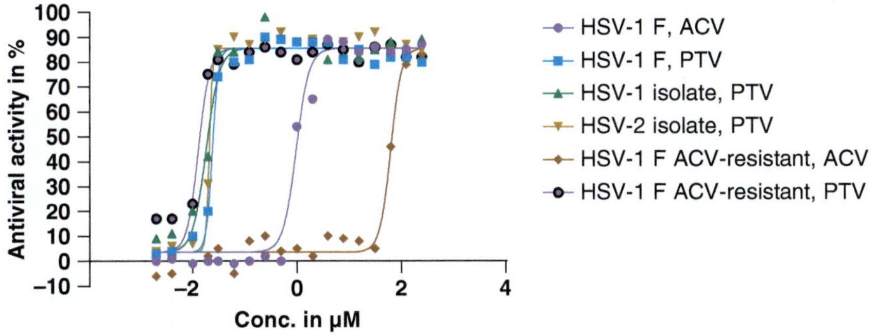

Strain	HSV-1 F	HSV-1 F	HSV-1 isolate	HSV-2 isolate	HSV-1 F ACV-resistant	HSV-1 F ACV-resistant
Treatment	Acyclovir	Pritelivir	Pritelivir	Pritelivir	Acyclovir	Pritelivir
EC_{50} [μM]	0.96	0.02	0.02	0.02	58.99	0.01

Figure 2.6 When plotted as antiviral activity (expressed as percent inhibition) versus concentration in μM, (i.e. concentration-response curve), all of the treatments evaluated were effective in achieving high levels of antiviral activity. However, the concentration required was lower for pritelivir-treated laboratory strains or clinical isolates when compared to acyclovir. Notably, acyclovir treatment of an acyclovir-resistant HSV-1 strain showed activity only at high concentrations, whereas pritelivir was as effective as against wild-type strains.

penciclovir-resistant mutants derived from the HSV-1 laboratory strain KOS were tested for susceptibility to pritelivir [61]. These mutants included four *TK* mutants and three DNA *pol* mutants, all of which were resistant to acyclovir, including cross-resistance to foscarnet and cidofovir for each one of the *pol* mutants. Pritelivir remained active against all seven mutants with EC_{50} values similar to those observed with the wild-type KOS strain. These results underscore the antiviral activity of pritelivir across a broad spectrum of HSV isolates and support its potential as a therapeutic option, including the treatment of resistant infections.

2.5.2 *In Vivo* Studies

A plethora of *in vivo* studies have been conducted in various animal species, including mice, rats, guinea pigs, and rabbits, and also including immunocompromised animals, addressing HSV manifestations such as mucocutaneous infection, genital herpes, HSV keratitis, and HSE [62–67]. The activity of pritelivir in two representative models, a guinea pig model for genital HSV and a mouse model for HSE, is described in more detail. Furthermore, explorative data in immunocompromised mice are summarized.

2.5.2.1 Guinea Pig Model for Genital HSV-2 Infection

The use of HSV-2 infection in guinea pigs is a well-established and predictive model for the study of genital herpes in humans. This model has been used extensively to study the pathogenesis of the disease and the effects of antiviral drugs on its course and outcome [68]. In this paradigm, an investigation was conducted to assess the effects of pritelivir compared to valacyclovir and placebo on symptom severity, viral

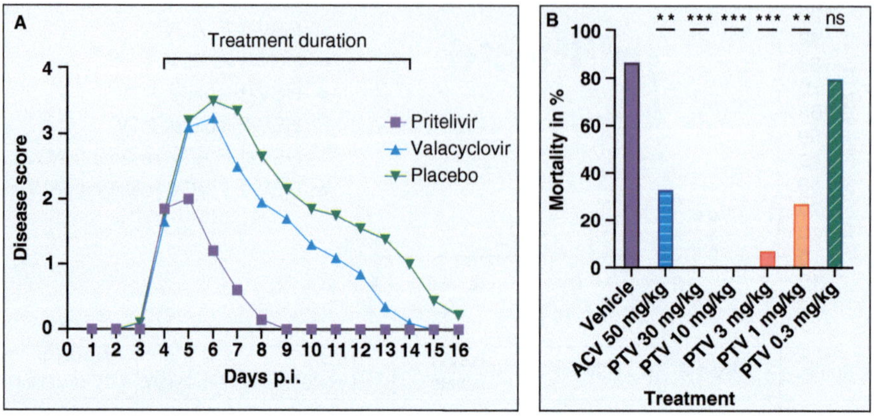

Figure 2.7 The results of the guinea pig model, which aimed to evaluate the effects of delayed postinfection administration on day 4, when lesions were already present, showed that treatment with pritelivir (20 mg/kg twice daily) led to significantly faster healing rates compared to valacyclovir (150 mg/kg twice daily) or placebo initiated on study day 1 (a). In the murine lethal challenge model, treatment with pritelivir resulted in significantly reduced mortality rates when compared with vehicle-treated controls, even when administered as late as 72 h postinfection at twice-daily dosages ranging from 0.3 to 10 mg/kg in mice infected with the HSV-2 MS strain. Acyclovir (50 mg/kg twice daily) was effective in reducing mortality rates, although less so than pritelivir, at a level comparable to 1 mg/kg pritelivir.

shedding, and recurrence [62]. Ten animals in each experimental group were intravaginally infected with HSV-2 and inspected daily for herpetic lesions, from which a disease score was derived.

Because patients typically seek medical intervention when lesions are manifest, indicating the presence of symptoms, initiation of medication often occurs later in the course of the disease. Therefore, in this experimental design, initiation of therapy was deliberately delayed until day 4 postinfection, coinciding with lesion progression, and was maintained until day 14 postinfection. In pritelivir-treated animals (20 mg/kg bid), redness and swelling resolved within 24 h of treatment initiation (Figure 2.7a). Notably, a smaller proportion of animals (2 of 10) developed new lesions after initiation of pritelivir therapy compared to 9 of 10 animals treated with valacyclovir (150 mg/kg bid). The mean cumulative increase in symptom severity score within the first 24 h of therapy was 0.4 in pritelivir-treated animals compared to 1.4 in valacyclovir-treated animals. The effect on severity score was statistically significant for pritelivir compared to placebo ($P < 0.05$). In addition to the reduced incidence of new lesions after treatment initiation, pritelivir-treated animals demonstrated the onset of healing within 1 day. Furthermore, the mean time to healing was significantly shorter with pritelivir (7.1 ± 0.7 days) compared to the control groups (12.1 ± 1.2 on valacyclovir, 14.3 ± 1.6 on placebo). Thus, pritelivir effectively reduced the healing time by 50%, whereas valacyclovir achieved only a 15% reduction (2 days) compared to the placebo control group.

2.5.2.2 Mouse Model for HSE

A murine lethal challenge model of HSE that has been widely applied to study the potential protective effects of antiviral agents including pritelivir [63] was employed to evaluate the *in vivo* activity of orally-administered pritelivir in the treatment of acute HSV-1 and HSV-2 infections including acyclovir-resistant strains [66]. Unlike the aforementioned pritelivir study using this model, however, treatment was delayed 72 h as opposed to 6 h post intranasal viral inoculation, as this is thought to more closely mimic the clinical situation characterized by the development of symptoms prior to seeking medical attention. Following lethal infection, twice-daily treatment with either oral pritelivir or acyclovir, both alone and in combination at various dosage combinations, was initiated and continued for 7 days in 15 female BALB/c mice per treatment group. Compared with vehicle-treated mice, single-agent treatment with pritelivir at dosages from 0.3 to 30 mg/kg significantly reduced mortality within 21 days post virus inoculation (primary endpoint) against HSV-1 strain E-377 ($P < 0.001$), and significantly increased survival ($P < 0.005$) of mice infected with an acyclovir-resistant HSV-1 strain 11360 at dosages of 1 and 3 mg/kg. For mice infected with the HSV-2 MS strain, all pritelivir dosages higher than 0.3 mg/kg were found to be significantly effective versus vehicle ($P < 0.005$) (Figure 2.7b), and significantly improved survival ($P < 0.0001$) for mice infected with the acyclovir-resistant HSV-2 strain (12247) at dosages of 1–3 mg/kg. Additionally, combination therapies of pritelivir (0.1 or 0.3 mg/kg/dose) with acyclovir (10 mg/kg/dose) were found to be protective against HSV-2, strain MS ($P < 0.0001$) when compared with vehicle-treated mice. Although unconfirmed via direct experimentation, the observation of a significant increase in mean days to death ($P < 0.05$) associated with the use of combination therapies may support the potential for synergy between the two antiviral agents [66].

2.5.2.3 Pritelivir in Immunocompromised Mouse Model

Because immunocompromised patients with HSV are at increased risk of reactivation, the *in vivo* efficacy of pritelivir was assessed using a murine infection model in which 12 nude (athymic) female BALB/c mice were immunocompromised via scarification of a neck skin site with HSV-1, SC16 cl-2 [64]. The inoculated mice showed visible clinical signs of infection within 2 days postinfection and received a single daily dose of either 1.5 or 15 mg/kg of pritelivir intraperitoneally either once only on study day 3 postinfection or once on study days 3 and 4 postinfection ($n = 2$ per treatment group). Although efficacy results from such small sample sizes are not subject to statistical analysis, and are therefore of limited utility with regard to drawing solid conclusions, it should be noted that the two immunocompromised mice treated with 15 mg/kg pritelivir once daily for 2 days showed substantial reductions in clinical lesions on postinfection day 6. Additionally, reduced infectious viruses at the secondary sites of infection (ear and brainstem) were also observed in these immunocompromised animals on postinfection day 6 [64].

2.6 Clinical Data

2.6.1 Phase 1 Program

The pharmacology of pritelivir has extensively been studied in healthy volunteers. Initial Phase 1 trials were conducted to evaluate the PK and safety of pritelivir to support further clinical development [69]. As is typical in the early stages of development when there is only limited information on developmental toxicity available, the

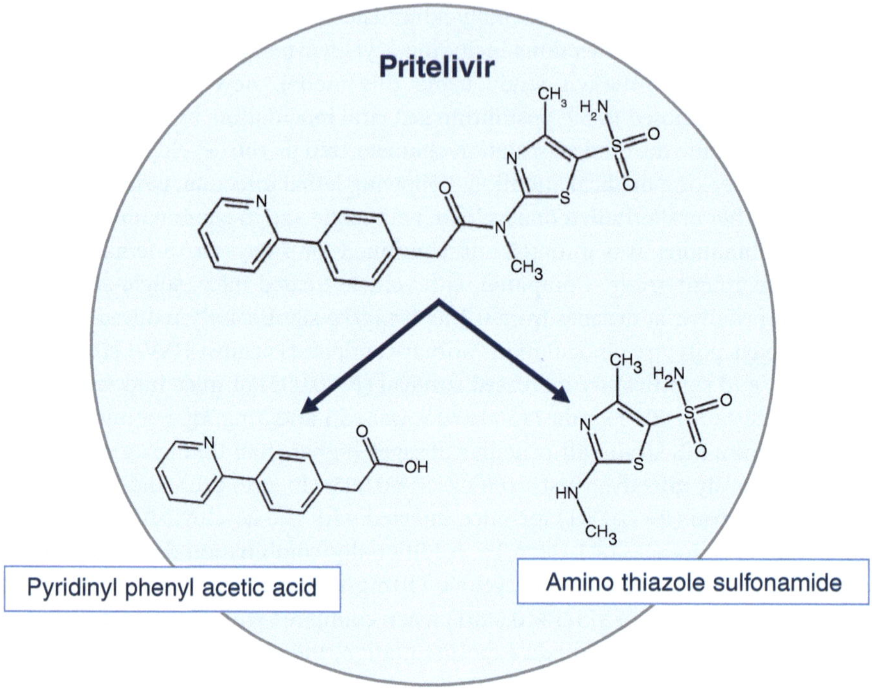

Figure 2.8 In a Phase 1 trial in healthy human volunteers, pritelivir was shown to be metabolized primarily by amide hydrolysis to pyridinyl phenyl acetic acid and aminothiazole sulfonamide.

majority of volunteers in these studies were male; however, one cohort of healthy female subjects was included in the single-ascending-dose trial portion of the program. Briefly, single- and multiple ascending-dose study results demonstrated linear PK up to 480 and 400 mg, respectively, and results from additional studies showed a half-life range of 52–83 h, achievement of steady state between 8 and 13 days, the absolute bioavailability of 72% under fasted conditions, a median time to maximum concentration of 3.5 h after dosing, and a low impact of food. Following single- and multiple ascending-doses, pritelivir was observed to be safe and well tolerated up to 600 and 200 mg, respectively.

Healthy male volunteers were also studied in two open-label Phase 1 mass balance trials aimed to identify pritelivir metabolites after single and multiple oral doses of pritelivir (100 mg), as well as to describe the PK of pritelivir and its metabolites [70]. Results indicated that pritelivir was mainly metabolized through amide hydrolysis (Figure 2.8) with subsequent glucuronidation and N-demethylation, and the metabolites primarily excreted by the kidneys.

2.6.2 Genital HSV

Initial clinical efficacy data with pritelivir were generated in two Phase 2 clinical trials in otherwise healthy (immunocompetent) individuals with genital HSV-2 [71, 72]. For both trials, viral shedding, i.e. the presence of virus at mucosal surfaces, was chosen as the primary efficacy outcome not only due to its relevance regarding the likelihood of disease transmission, but also its objective, reproducible, and precise nature [73]. These characteristics allow for the detection of differences between doses and treatments with a reasonable sample size. This is achieved by exploiting the biological property of the virus that >80% of HSV-2 seropositive individuals reactivate the virus in the genital tract, even without apparent lesions, which can be detected by adequate methods such as quantitative PCR [74].

The first Phase 2 trial was a proof-of-concept study to evaluate the efficacy and safety of oral pritelivir at doses of 5, 25, or 75 mg daily, following appropriate loading doses on day 1 to achieve immediate steady-state exposure, compared to placebo on mucocutaneous viral shedding for 28 days in 156 participants [71]. Due to the long half-life of the drug in the blood, a 400 mg once-weekly dose group was also included. Results revealed that the rates of HSV shedding among placebo-treated patients and those treated with pritelivir 5 mg daily were similar (16.6% and 18.2%, respectively) (Figure 2.9a). However, a dose-related effect of pritelivir treatment was detected, with response rates of 9.3% and 2.1% of days for those in the 25 and 75 mg daily treatment groups, respectively, and 5.3% of days for those in the 400 mg weekly treatment group. Additionally, the relative risk of viral shedding significantly differed between patients in both the 75 mg daily and 400 mg weekly pritelivir treatment groups as compared with the placebo-treated group (0.13 and 0.32, respectively; both $P < 0.001$). Findings also showed that patients treated with pritelivir 75 mg/kg daily and 400 mg weekly had significantly lower risks of genital lesions compared with placebo-treated patients (0.13 for each pritelivir group; both $P < 0.02$). Similarly, these two pritelivir-treatment groups demonstrated significant reductions in the percentage of days with genital lesions as compared with the placebo group (1.2% for each pritelivir group vs. 9.0% in the placebo group. Safety data indicated that the rate of adverse events was similar in all groups [71].

The second of these trials was a randomized, double-blind, crossover trial in which pritelivir 100 mg following a loading dose of 400 mg on day 1 was compared with valacyclovir 500 mg for suppression of genital HSV shedding and lesions [72]. For the crossover, patients randomized to their first drug received it for 28 days prior to a 28-day washout period, followed by 28 days of treatment with the second

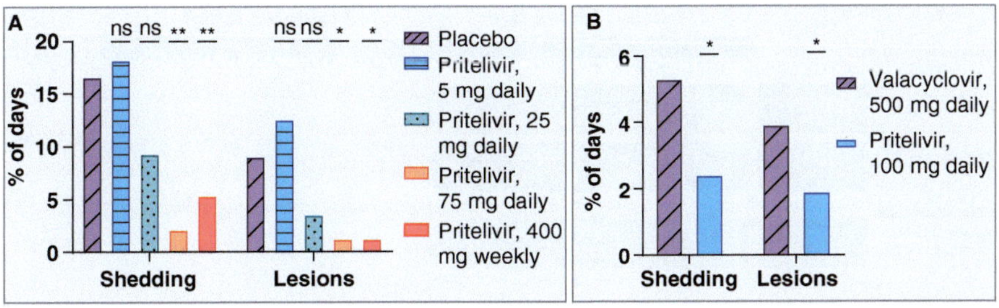

Figure 2.9 In a proof-of-concept trial, pritelivir at doses of 75 mg daily and 400 mg weekly significantly reduced both viral shedding and genital lesions in this patient population (Figure 2.9a). Results from an additional trial comparing pritelivir with valacyclovir showed that treatment with pritelivir 100 mg daily was associated with significantly lower rates of viral shedding and genital lesions compared to valacyclovir 500 mg daily (Figure 2.9b).

drug. During both treatment periods, patients collected genital swabs four times daily for HSV testing via quantitative PCR. During the trial, the FDA imposed a clinical hold based on hematological and skin-related findings in a concurrent non-human primate chronic toxicity study, leading to premature termination of the trial by the sponsor. At the time of termination, 56 of the 91 randomized patients had completed both treatment periods. Results revealed that the relative risk for HSV shedding significantly differed between pritelivir and valacyclovir treatment (0.42; P = 0.01) (Figure 2.9b). The risk score was based on the detection of HSV shedding in 173 of 7276 (2.4%) swabs during the patients' pritelivir treatment period and 392 of 7453 (5.3%) during the valacyclovir treatment period. In addition, the presence of genital lesions also significantly differed between patients during the pritelivir and valacyclovir treatment periods (1.9% vs. 3.9% of days, respectively; P = 0.04). Regarding safety, adverse events occurred in 62.3% and 69.2% of patients during the pritelivir and valacyclovir treatment periods, respectively [72]. Of note, the hold was subsequently lifted based on clinical safety data and following an additional chronic toxicity study where no comparable findings were observed [54].

Based on favorable safety, tolerability, and PK profiles in healthy volunteers, as well as experience from Phase 2 trials in immunocompetent persons with genital HSV-2 infection, 100 mg once-daily with a loading dose of 400 mg on day 1 was selected as the therapeutic dose to be used in subsequent Phase 2 and 3 trials in immunocompromised patients with resistant infection [69].

2.6.3 Resistant Infections

Encouraged by the positive results of both trials in otherwise healthy individuals with genital HSV-2 infection, further development focused on the patient population with the greatest medical need and limited treatment options: immunocompromised patients with resistant HSV infections. The Phase 2/3 PRIOH-1 ("Pritelivir Immunocompromised Herpes") trial was designed to compare the efficacy and safety of pritelivir versus foscarnet for the treatment of acyclovir-resistant mucocutaneous HSV infections in immunocompromised subjects (NCT03073967) [75]. The trial was initiated as a Phase 2 study (Figure 2.10). Based on key learnings

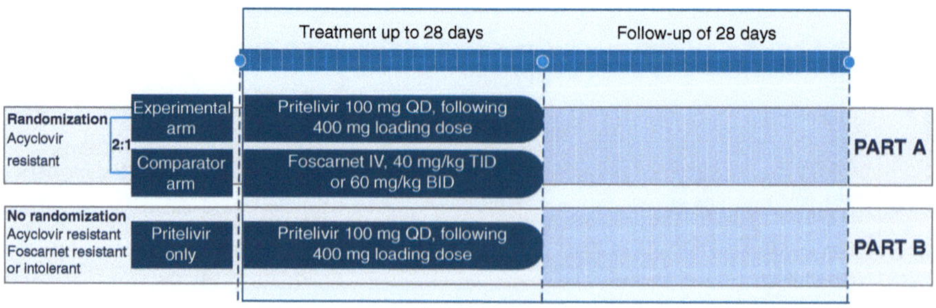

Figure 2.10 In Part A of the Phase 2 portion of the Phase 2/3 PRIOH-1 trial, acyclovir-resistant patients were randomized to receive either pritelivir or foscarnet for up to 28 days with 28 days of follow-up. Patients in Part B were acyclovir-resistant and foscarnet-resistant/intolerant and were treated with pritelivir only at the same dosage for the same duration of treatment and follow-up as patients in Part A.

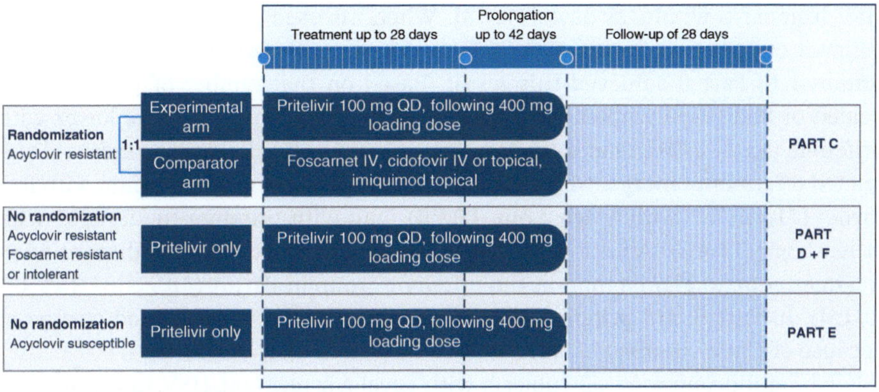

Figure 2.11 The Phase 3 portion of the Phase 2/3 PRIOH-1 trial consists of four parts. In Part C, acyclovir-resistant patients are randomized to receive either pritelivir or Investigator's Choice, i.e. foscarnet IV, cidofovir (IV or topical), or imiquimod topical for up to 28 days with the option of subsequent treatment extension up to 42 days and 28 days of follow-up, as well as two "Recurrence Calls" 60 and 90 days after the end of treatment (not shown). Parts D and F, both fully enrolled, were nonrandomized and evaluated the same dose of pritelivir with the same treatment duration, extension, and follow-up as in Part C in acyclovir-resistant and foscarnet-resistant/intolerant patients. Part E is similar to Parts D and F, except that patients are enrolled who are susceptible to acyclovir (and foscarnet).

and promising results with pritelivir in Phase 2, the Phase 2 portion was stopped early and rolled over into the Phase 3 portion, which also includes design adjustments such as a different primary endpoint, investigator's choice (foscarnet IV, cidofovir IV or topical, imiquimod topical) instead of foscarnet only as a comparator, and an option to prolong treatment for up to 42 days if needed (Figure 2.11). At the time of writing, Phase 3 is still ongoing. Therefore, only top-line data from the Phase 2 portion are briefly summarized here, pending the final results of the Phase 3 portion.

The randomized, open-label, multicenter, comparator-controlled Phase 2 portion of the PRIOH-1 trial had two arms, referred to as Parts A and B (Figure 2.10) [76]. The aim of Part A was to compare the efficacy and safety of pritelivir administered as 100 mg orally once daily following a loading dose of 400 mg with foscarnet administrated intravenously (40 mg/kg every 8 h or 60 mg/kg every 12 h) in male or female immunocompromised patients ≥18 years of age with acyclovir-resistant mucocutaneous HSV infection. In Part B, patients who developed foscarnet resistance/intolerance before entering the study or during treatment in Part A were treated with pritelivir as in Part A for up to 28 days. Of the 23 patients treated with pritelivir in both parts, 11 had HIV infection and 12 had malignancy, transplant, or an autoimmune disease as underlying causes of immune suppression [76]. Of the seven patients treated with foscarnet in Part A, two had HIV infection and five had transplant as an underlying cause.

Top-level results from Parts A and B of the Phase 2 portion of PRIOH-1 showed that 83% (19/23) of patients on pritelivir, demonstrated full resolution of their

HSV lesionsNo within 28 days [77, 78]. When assessed per part, 93% (14/15) of patients on pritelivir versus 57% (4/7) on foscarnet in Part A and 63% (5/8) on pritelivir in Part B achieved this result. Based on the number of patients who healed or had post-treatment resistance testing, rates appeared to be lower with pritelivir (4.8%, 1/21) than with foscarnet (20.0%, 1/5). Regarding safety, drug-related treatment-emergent adverse events (TEAEs) were more common with foscarnet (71.4%, 5/7 with at least one TEAE) than with pritelivir (8.7%, 2/23 with at least one TEAE). Additionally, more foscarnet-treated patients discontinued treatment due to TEAEs than in the pritelivir group in Part A (42.9% [3/7] vs. 0% [0/15]). In Part B, 1/8 pritelivir-treated patients (12.5%) discontinued treatment because of CMV viremia [77, 78].

For immunocompromised subjects with acyclovir-resistant HSV infections and foscarnet resistance or intolerance, who are unable to participate in the PRIOH-1 trial, an EAP has been initiated by the manufacturer (NCT05844436) [79]. As of the data cut-off date of 4 November, 2022, 122 requests for pritelivir were made for this patient population [77, 78]. Of these, 79 led to treatment in a total of 53 patients with up to six treatment cycles. The majority of treatments (43%) were in the United States. followed by the United Kingdom (14%), France (11%), Canada (9%), and additional countries involved in the program (23%). It is important to acknowledge that data captured in an EAP may not be directly comparable to that of a controlled clinical trial, given the differing nature and intent of these programs. Nevertheless, of the 79 treatments, 69 were completed and evaluable at the data cut-off date and were subjected to an interim analysis. Of the 69 treatments, 54% (37/69) and 61% (42/69) led to healing within 28 and 42 days, respectively. As for safety, 30% (16/53) of patients were reported to have at least one TEAE, with 7.5% (4/53) considered drug-related [77, 78]. To date, 97 individual patients have been treated with up to 10 treatment cycles.

Results from use of the pritelivir EAP are also making their way into the literature in the form of real-world case studies (Table 2.1). For example, Cannon et al. published a case involving an immunosuppressed (HIV-negative) allogenic peripheral blood stem cell transplant recipient [80]. The patient, a 75-year-old woman with acyclovir-resistant HSV-2, was refractory to second-line anti-HSV therapy. Through the EAP, the patient received pritelivir and showed both symptomatic and virological control throughout the course of its administration. In another case study, two patients with acyclovir-resistant HSV-2 who had received hematopoietic stem cell transplant (one for sickle-cell disease and one for acute myeloid leukemia) showed sustained improvement when pritelivir was requested and obtained through the EAP [81]. An additional case study of two hematopoietic cell transplantation recipients treated with pritelivir obtained through the EAP for acyclovir-resistant HSV oral lesions showed partial response to outpatient treatment by week 1 of treatment and complete response by week 4 in both patients [82]. No adverse events were noted in either patient. Pritelivir obtained through the EAP has also been used to successfully treat a patient with clinically refractory and progressive HSV pneumonitis while on treatment with foscarnet and acyclovir after their discontinuation [83].

Table 2.1 Selected details from pritelivir early access program cases.

References	HSV infection	Reason for immunosuppression	Treatment	Outcome
[80]	Acyclovir resistant genital HSV-2	Allogenic peripheral blood stem cell transplant	Pritelivir 100 mg for 4 weeks	Marked lesion improvement, HSV PCR negative; patient died of hospital-acquired pneumonia
[81]	Acyclovir resistant genital HSV-2	HSCT	3 treatment cycles with pritelivir 100 mg for 42 days each, other treatments in between	Complete or almost complete healing after each cycle
	Acyclovir resistant genital HSV-2		2 treatment cycles with pritelivir 100 mg (28 days and ongoing)	Complete healing after each treatment
[82]	Acyclovir resistant oral HSV-1	Allogenic HCT	Pritelivir 100 mg for 34 days	Partial response after 7 days and complete healing by day 34
			Pritelivir 100 mg for 30 days	Clinical response within 1 week, infection completely resolved after 30 days
[83]	Acyclovir and foscarnet-resistant HSV-1 pneumonitis	Allogenic HCT	Pritelivir 100 mg for 4 weeks	HSV undetectable in BAL after 4 weeks of treatment; patient developed an idiopathic lung syndrome and multiorgan failure and died while on pritelivir treatment

BAL, bronchoalveolar lavage; HCT, hematopoietic cell transplantation; HSCT, hematopoietic stem cell transplantation; HSV, herpes simplex virus.

2.7 Pritelivir Resistance

Numerous nonclinical studies have been conducted to investigate resistance to pritelivir. For example, results from selecting resistant HSV mutants *in vitro* indicated that the frequency of pritelivir resistance ranged from 0.5 to 4.5×10^{-6} [53]. By comparison, the frequency of acyclovir resistance is at least one order of magnitude higher, ranging from 10^{-3} to 10^{-4} [43]. When large numbers of plaque-forming units were screened in a different experimental setup, single pritelivir-resistant virions could be identified in crude laboratory extracts and a few clinical isolates at a frequency of 10^{-4} to 10^{-5}. It should be noted that all of these strains and isolates were fully sensitive to pritelivir in the standard plaque assays [84, 85]. Using a specific PCR technique, it was shown that the identified mutants were preexisting and not induced via treatment with pritelivir [86].

In addition, results from *in vivo* studies in animals infected with helicase–primase inhibitor-resistant strains demonstrated either reduced pathogenicity or the potential for effective treatment with pritelivir by increasing the dose and duration of treatment [67]. Furthermore, a study conducted by Sulka et al. showed that low levels of preexisting resistant virus did not change any positive effects of pritelivir treatment when administered using mutant to wild-type mixtures of up to 1 : 50 [87].

Regarding clinical studies, no evidence was found that pritelivir induced known resistance-mediating mutations or amino acid variations at other loci in swabs from the first Phase 2 trial in immunocompetent individuals with genital herpes [71, 88]. In line with this, no changes to the reference sequence were detected in UL5 or UL52 in swabs from the second Phase 2 trial in genital HSV-2 patients [72].

The first case of pritelivir resistance emerging in the clinic was detected during the Phase 2 portion of the PRIOH-1 trial in immunocompromised patients with acyclovir-resistant infections [89]. In Part A, 1 out of 15 patients included in the resistance analysis showed resistance to pritelivir (6.7%) compared to one out of five patients (20%) with resistance to foscarnet. Combining the populations for Parts A and B, only 1 out of 21 patients (4.8%) included in the resistance analysis was resistant to pritelivir. However, based on the limited number of immunocompromised patients treated with pritelivir to date, no conclusions can be drawn about incidence or impact in the clinic [89].

2.8 Conclusion and Outlook

There remains a need for safe and effective therapies to treat HSV, especially in immunocompromised patients. Currently, available treatments for these vulnerable patients are either toxic or lack proven efficacy. Ideally, alternative treatment options should not only be highly effective but also have favorable pharmacological properties and good oral bioavailability. Pritelivir has demonstrated both of these characteristics. Based on its novel and resistance-breaking mode of action, pritelivir may offer an effective alternative for acyclovir-resistant mucocutaneous HSV infections. The benefits of this approach may include rapid resolution of severe mucocutaneous lesions, fewer side effects compared to foscarnet, and, due to oral administration, no need for hospitalization.

Studies of pritelivir have demonstrated that it is highly active against HSV-1 and HSV-2, selective and specific, and resistance-breaking due to a different mode of action than nucleoside analogs. Pritelivir administration was shown to reduce viral shedding and result in fewer days with lesions, superior to valacyclovir. Pritelivir showed a safety profile similar to that of placebo and valacyclovir.

In immunocompromised subjects with acyclovir-resistant mucocutaneous HSV infection, pritelivir treatment resulted in a considerably higher healing rate compared to foscarnet. Furthermore, there were fewer treatment discontinuations due to TEAEs. Pritelivir also offers potential benefits, such as a favorable pharmacokinetic profile for daily dosing and an oral route of administration, which eliminates the need for IV administration.

No treatment-emergent resistance has been observed in nonclinical studies or under suppressive treatment in immunocompetent individuals. In immunocompromised subjects, pritelivir administration appears to be associated with lower apparent rates of treatment-emergent resistance than foscarnet.

Results from the ongoing Phase 3 portion of PRIOH-1 are eagerly awaited and should provide further information on the efficacy, safety, and tolerability of pritelivir for the treatment of acyclovir-resistant mucocutaneous HSV infections in immunocompromised individuals. Based on the overall positive clinical profile, additional clinical investigations in patients with other HSV infections, such as neonatal HSV or encephalitis, may be warranted.

References

1 Roizman B, Knipe DM, Whitley RJ. Herpes simplex viruses. In: Knipe DM, Howley PM, editors. *Fields virology*. 5th ed. New York: Lippincott Williams & Wilkins; 2006. 2502–601.

2 Whitley R, Baines J. Clinical management of herpes simplex virus infections: past, present, and future. F1000Res. 2018;7:1726.

3 Looker KJ, Magaret AS, May MT, Turner KM, Vickerman P, Gottlieb SL, et al. Global and regional estimates of prevalent and incident herpes simplex virus type 1 infections in 2012. PLoS Onc. 2015;10:e0140765.

4 Looker KJ, Magaret AS, Turner KM, Vickerman P, Gottlieb SL, Newman LM. Global estimates of prevalent and incident herpes simplex virus type 2 infections in 2012. PLoS One. 2015;10:e114989.

5 Kennedy PG, Rovnak J, Badani H, Cohrs RJ. A comparison of herpes simplex virus type 1 and varicella-zoster virus latency and reactivation. *J Gen Virol.* 2015;96:1581–602.

6 Preston CM, Efstathiou S. Molecular basis of HSV latency and reactivation. Cambridge: Cambridge University Press; 2007.

7 Mindel A, Marks C. Psychological symptoms associated with genital herpes virus infections: epidemiology and approaches to management. *CNS Drugs.* 2005;19:303–12.

8 Ross K, Johnston C, Wald A. Herpes simplex virus type 2 serological testing and psychosocial harm: a systematic review. *Sex Transm Infect.* 2011; 87:594–600.

9 Rowe AM, St Leger AJ, Jeon S, Dhaliwal DK, Knickelbein JE, Hendricks RL. Herpes keratitis. *Prog Retin Eye Res.* 2013;32:88–101.

10 Mark KE, Wald A, Magaret AS, Selke S, Olin L, Huang ML, et al. Rapidly cleared episodes of herpes simplex virus reactivation in immunocompetent adults. *J Infect Dis.* 2008;198:1141–49.

11 Sacks SL, Griffiths PD, Corey L, Cohen C, Cunningham A, Dusheiko GM, et al. HSV shedding. *Antiviral Res.* 2004;63(Suppl 1):S19–26.

12 Gupta R, Warren T, Wald A. Genital herpes. *Lancet.* 2007;370:2127–37.

13 Sacks SL, Griffiths PD, Corey L, Cohen C, Cunningham A, Dusheiko GM, et al. HSV-2 transmission. *Antiviral Res.* 2004;63(Suppl 1):S27–35.

14 Mertz GJ. Asymptomatic shedding of herpes simplex virus 1 and 2: implications for prevention of transmission. *J Infect Dis.* 2008;198:1098–100.

15 Koelle DM, Posavad CM, Barnum GR, Johnson ML, Frank JM, Corey L. Clearance of HSV-2 from recurrent genital lesions correlates with infiltration of HSV-specific cytotoxic T lymphocytes. *J Clin Invest.* 1998;101:1500–8.

16 Zhu J, Koelle DM, Cao J, Vazquez J, Huang ML, Hladik F, et al. Virus-specific CD8+ T cells accumulate near sensory nerve endings in genital skin during subclinical HSV-2 reactivation. *J Exp Med.* 2007;204:595–03.

17 Jordi C, Miguel M, Pilar P-R. Infections caused by herpes viruses other than cytomegalovirus in solid organ transplant recipients. *Enfermedades Infecciosas y Microbiología Clínica.* 2012;30:63–9.

18 Ramaswamy M, Geretti AM. Interactions and management issues in HSV and HIV coinfection. *Expert Rev Anti-infec Ther.* 2007;5:231–43.

19 Styczynski J. Management of herpesvirus infections in hematopoietic cell transplant recipients. *Transplantology.* 2021;2:8–21.

20 Chang W-J, Wang H-Y, Huang Y-C, Lin C-Y, Leu S-W, Hsieh M-J, et al. Outcomes of herpes simplex virus pneumonitis in critically Ill patients. *Viruses.* 2022;14:205.

21 Riediger C, Sauer P, Matevossian E, Müller MW, Büchler P, Friess H. Herpes simplex virus sepsis and acute liver failure. *Clin Transplant.* 2009;23:37–1.

22 Cunningham AL, Dwyer DE. The pathogenesis underlying the interaction of HIV and herpes simplex virus after co-infection. *J HIV Ther.* 2004;9:9–13.

23 James SH, Kimberlin DW, Whitley RJ. Antiviral therapy for herpesvirus central nervous system infections: neonatal herpes simplex virus infection, herpes simplex encephalitis, and congenital cytomegalovirus infection. *Antiviral Res.* 2009;83:207–13.

24 Stoopler ET, Kuperstein AS, Sollecito TP. How do I manage a patient with recurrent herpes simplex? J *Can Dent Assoc.* 2012;78:c154.

25 Hodge RAV, Field HJ. Antiviral agents for herpes simplex virus. *Adv pharmacol.* 2013;67:1–38.

26 O'Brien JJ, Campoli-Richards DM. Acyclovir: an updated review of its antiviral activity, pharmacokinetic properties and therapeutic efficacy. *Drugs.* 1989;37:233–09.

27 Corey L, Wald A, Patel R, Sacks SL, Tyring SK, Warren T, et al. Once-daily valacyclovir to reduce the risk of transmission of genital herpes. *N Engl J Med.* 2004;350:11–20.

28 Hammer K, Dietz J, Lo TS, Johnson EM. A systematic review on the efficacy of topical acyclovir, penciclovir, and docosanol for the treatment of herpes simplex labialis. *Dermatology*. 2018;6:118–23.

29 James HJK, D.W. Herpes Simplex Virus in Transplant Recipients. http://www.antimicrobe.org/t02.asp. Accessed 29Feb2024.

30 Meyers JD, Flournoy N, Thomas ED. Infection with herpes simplex virus and cell-mediated immunity after marrow transplant. *J Infect Dis.* 1980;142:338–46.

31 Reitano M, Tyring S, Lang W, Thoming C, Worm AM, Borelli S, et al. Valaciclovir for the suppression of recurrent genital herpes simplex virus infection: a large-scale dose range-finding study. International valaciclovir HSV study group. *J Infect Dis.* 1998;178:603–10.

32 Piret J, Boivin G. Resistance of herpes simplex viruses to nucleoside analogues: mechanisms, prevalence, and management. *Antimicrob Agents Chemother.* 2011;55:459–72.

33 CDC. 2021. Sexually Transmitted Infections Treatment Guidelines; Genital Herpes. https://www.cdc.gov/std/treatment-guidelines/herpes.htm. Accessed 29Feb2024.

34 Bernstein DI, Spruance SL, Arora SS, Schroeder JL, Meng TC. Evaluation of imiquimod 5% cream to modify the natural history of herpes labialis: a pilot study. *Clin Infect Dis.* 2005;41:808–14.

35 Perkins N, Nisbet M, Thomas M. Topical imiquimod treatment of aciclovir-resistant herpes simplex disease: case series and literature review. *Sex Transm Infect.* 2011;87:292–5.

36 Schacker TW, Conant M, Thoming C, Stanczak T, Wang Z, Smith M. Imiquimod 5-percent cream does not alter the natural history of recurrent herpes genitalis: a phase II, randomized, double-blind, placebo-controlled study. *Antimicrob Agents Chemother.* 2002;46:3243–8.

37 Spruance S, Aoki FY, Tyring S, Stanberry L, Whitley R, Hamed K. Short-course therapy for recurrent genital herpes and herpes labialis. *J Fam Pract.* 2007;56:30–6.

38 Schalkwijk HH, Snoeck R, Andrei G. Acyclovir resistance in herpes simplex viruses: prevalence and therapeutic alternatives. *Biochem Pharmacol.* 2022;206:115322.

39 Anton-Vazquez V, Mehra V, Mbisa JL, Bradshaw D, Basu TN, Daly ML, et al. Challenges of aciclovir-resistant HSV infection in allogeneic bone marrow transplant recipients. *J Clin Virol.* 2020;128:104421.

40 Frobert E, Burrel S, Ducastelle-Lepretre S, Billaud G, Ader F, Casalegno JS, et al. Resistance of herpes simplex viruses to acyclovir: an update from a ten-year survey in France. *Antiviral Res.* 2014;111:36–41.

41 Kakiuchi S, Tsuji M, Nishimura H, Yoshikawa T, Wang L, Takayama-Ito M, et al. Association of the emergence of acyclovir-resistant herpes simplex virus type 1 with prognosis in hematopoietic stem cell transplantation patients. *J Infect Dis.* 2017;215:865–73.

42 Piret J, Boivin G. Antiviral resistance in herpes simplex virus and varicella-zoster virus infections: diagnosis and management. *Curr Opin Infect Dis.* 2016; 29:654–62.

43 Bacon TH, Levin MJ, Leary JJ, Sarisky RT, Sutton D. Herpes simplex virus resistance to acyclovir and penciclovir after two decades of antiviral therapy. *Clin Microbiol Rev.* 2003;16:114–28.

44 Larder BA, Derse D, Cheng YC, Darby G. Properties of purified enzymes induced by pathogenic drug-resistant mutants of herpes simplex virus. Evidence for virus variants expressing normal DNA polymerase and altered thymidine kinase. *J Biol Chem.* 1983;258:2027–33.

45 Schmit I, Boivin G. Characterization of the DNA polymerase and thymidine kinase genesof herpes simplex virus isolates from AIDS patients in whom acyclovirand foscarnet therapy sequentially failed. *J Infect Dis.* 1999;180:487–90.

46 Piret J, Boivin G. Antiviral drug resistance in herpesviruses other than cytomegalovirus. *Rev Med Virol.* 2014;24:186–18.

47 Engel JP, Englund JA, Fletcher CV, Hill EL. Treatment of resistant herpes simplex virus with continuous-infusion acyclovir. *JAMA.* 1990;263:1662–4.

48 Drugs@FDA. Foscavir. https://www.accessdata.fda.gov/scripts/cder/daf/index.cfm?event=overview.process&ApplNo=020068. Accessed 29Feb2024.

49 Safrin S, Crumpacker C, Chatis P, Davis R, Hafner R, Rush J, et al. A controlled trial comparing foscarnet with vidarabine for acyclovir-resistant mucocutaneous herpes simplex in the acquired immunodeficiency syndrome. The AIDS Clinical Trials Group. *N Engl J Med.* 1991;325:551–5.

50 Jacobson MA. Review of the toxicities of foscarnet. J Acquir Immune Defic Syndr. (1988). 1991;5 (Suppl 1):S11–7.

51 Danve-Szatanek C, Aymard M, Thouvenot D, Morfin F, Agius G, Bertin I, et al. Surveillance network for herpes simplex virus resistance to antiviral drugs: 3-year follow-up. *J Clin Microbiol.* 2004;42:242–9.

52 Kleymann G. Discovery, SAR and medicinal chemistry of herpesvirus helicase primase inhibitors. *Curr Med Chem-Anti-Infect Agents.* 2004;3:69–3.

53 Kleymann G, Fischer R, Betz UA, Hendrix M, Bender W, Schneider U, et al. New helicase-primase inhibitors as drug candidates for the treatment of herpes simplex disease. *Nat Med.* 2002;8:392–8.

54 Birkmann A, Bonsmann S, Kropeit D, Pfaff T, Rangaraju M, Sumner M, et al. Discovery, chemistry, and preclinical development of pritelivir, a novel treatment option for acyclovir-resistant herpes simplex virus infections. *J Med Chem.* 2022;65:13614–28.

55 Weller SK, Kuchta RD. The DNA helicase-primase complex as a target for herpes viral infection. *Expert Opin Ther Targets.* 2013;17:1119–32.

56 Chono K, Katsumata K, Kontani T, Kobayashi M, Sudo K, Yokota T, et al. ASP2151, a novel helicase-primase inhibitor, possesses antiviral activity against varicella-zoster virus and herpes simplex virus types 1 and 2. *J Antimicrob Chemother.* 2010;65:1733–41.

57 Crute JJ, Grygon CA, Hargrave KD, Simoneau B, Faucher AM, Bolger G, et al. Herpes simplex virus helicase-primase inhibitors are active in animal models of human disease. *Nat Med.* 2002;8:386–91.

58 Gege C, Bravo FJ, Uhlig N, Hagmaier T, Schmachtenberg R, Elis J, et al. A helicase-primase drug candidate with sufficient target tissue exposure affects latent neural herpes simplex virus infections. *Sci Transl Med.* 2021;13:eabf8668.

59 Hou J, Gao H, Fan Y, Wang Y, Qin M, Di Y, et al. Pharmacokinetics and safety study of HN0037, a novel anti-human herpes simplex virus inhibitor, in healthy volunteers. *Clin Pharmacol Drug Dev.* 2022;11(12):1467–73. doi:10.1002/cpdd.1138.

60 Shoji N, Tanese K, Sasaki A, Horiuchi T, Utsuno Y, Fukuda K, et al. Pharmaceuticals and medical device agency approval summary: amenamevir for the treatment of herpes zoster. *J Dermatol.* 2020;47:683–88.

61 Field HJ, Huang ML, Lay EM, Mickleburgh I, Zimmermann H, Birkmann A. Baseline sensitivity of HSV-1 and HSV-2 clinical isolates and defined acyclovir-resistant strains to the helicase-primase inhibitor pritelivir. *Antiviral Res.* 2013;100:297–99.

62 Baumeister J, Fischer R, Eckenberg P, Henninger K, Ruebsamen-Waigmann H, Kleymann G. Superior efficacy of helicase-primase inhibitor BAY 57-1293 for herpes infection and latency in the guinea pig model of human genital herpes disease. *Antivir Chem Chemother.* 2007;18:35–48.

63 Betz UA, Fischer R, Kleymann G, Hendrix M, Rubsamen-Waigmann H. Potent in vivo antiviral activity of the herpes simplex virus primase-helicase inhibitor BAY 57-1293. *Antimicrob Agents Chemother.* 2002;46:1766–72.

64 Biswas S, Jennens L, Field HJ. The helicase primase inhibitor, BAY 57-1293 shows potent therapeutic antiviral activity superior to famciclovir in BALB/c mice infected with herpes simplex virus type 1. *Antiviral Res.* 2007;75:30–35.

65 Kaufman HE, Varnell ED, Gebhardt BM, Thompson HW, Atwal E, Rubsamen-Waigmann H, et al. Efficacy of a helicase-primase inhibitor in animal models of ocular herpes simplex virus type 1 infection. *J Ocul Pharmacol Ther.* 2008;24(1):34–42.

66 Quenelle DC, Birkmann A, Goldner T, Pfaff T, Zimmermann H, Bonsmann S, et al. Efficacy of pritelivir and acyclovir in the treatment of herpes simplex virus infections in a mouse model of herpes simplex encephalitis. *Antiviral Res.* 2017;149:1–6.

67 Biswas S, Sukla S, Goldner T, Field HJ, Kropeit D, Paulsen D, et al. Pharmacokinetics-pharmacodynamics of the helicase-primase inhibitor pritelivir following treatment of wild-type or pritelivir-resistant virus infection in a murine herpes simplex virus 1 infection model. *Antimicrob Agents Chemother.* 2014;58:3843–52.

68 Bernstein DI. Use of the Guinea pig model of genital herpes to evaluate vaccines and antivirals: review. *Antiviral Res.* 2020;180:104821.

69 Kropeit D, Bonsmann S, von Richter O, McCormick D, Pausch J, Sumner M, et al. First-in-human, single- and multiple-ascending-dose, food-effect, and absolute bioavailability trials to assess the pharmacokinetics, safety, and tolerability of pritelivir, a nonnucleoside helicase-primase inhibitor against herpes simplex virus in healthy subjects. *Clin Pharmacol Drug Dev.* 2023;12(7):749–60 doi: 10.1002/cpdd.1241.

70 Bonsmann S, McCormick D, Pausch J, de Vries M, Sumner M, Birkmann A, et al. Mass balance and metabolite profile after single and multiple oral doses of pritelivir in healthy subjects. *Clin Pharmacol Drug Dev*. 2024;13:389–403. doi: 10.1002/cpdd.1358.

71 Wald A, Corey L, Timmler B, Magaret A, Warren T, Tyring S, et al. Helicase-primase inhibitor pritelivir for HSV-2 infection. *N Engl J Med*. 2014;370:201–10.

72 Wald A, Timmler B, Magaret A, Warren T, Tyring S, Johnston C, et al. Effect of pritelivir compared with valacyclovir on genital HSV-2 shedding in patients with frequent Recurrences: a randomized clinical trial. *JAMA*. 2016;316:2495–503.

73 Agyemang E, Magaret AS, Selke S, Johnston C, Corey L, Wald A. Herpes simplex virus shedding rate: surrogate outcome for genital herpes recurrence frequency and lesion rates, and phase 2 clinical trials end point for evaluating efficacy of antivirals. *J Infect Dis*. 2018;218:1691–99.

74 Wald A, Zeh J, Selke S, Warren T, Ryncarz AJ, Ashley R, et al. Reactivation of genital herpes simplex virus type 2 infection in asymptomatic seropositive persons. *N Engl J Med*. 2000;342:844–50.

75 ClinicalTrials.gov. Trial on Efficacy and Safety of Pritelivir Tablets for Treatment of Acyclovir-resistant Mucocutaneous HSV (Herpes Simplex Virus) Infections in Immunocompromised Subjects (PRIOH-1). https://www.clinicaltrials.gov/study/NCT03073967 (accessed 1 March 2024).

76 Workowski K, Albrecht J, Avery R, Chandrasekar P, Chemaly R, Issa N, et al. Pritelivir in immunocompromised patients with mucocutaneous acyclovir-resistant herpes simplex virus infections - first case series, abstr IDWeek, Virtual, Abstract No. 1069990; 2021.

77 Birkmann A, Marini A, Mueller A, Polak TB, Rangaraju M, Sumner M, et al. Pritelivir for the treatment of resistant HSV infections in immunocompromised patients: update on an ongoing Phase 3 trial and Early Access Program, abstr STI & HIV 2023 World Congress, Chicago, IL, USA, Session HIV PrEP, Viral STIs; 2023.

78 Birkmann A. Helicase-primase inhibitor pritelivir for the treatment of resistant HSV infections in immunocompromised patients, abstr IDWeek, Boston, MA, USA, Pipeline Session; 2023.

79 ClinicalTrials.gov. Expanded Access Intermediate Size Treatment Protocol: Pritelivir for Immunocompromised Subjects With Treatment Resistant Herpes Simplex Virus Type 1 or 2. https://www.clinicaltrials.gov/study/NCT05844436 (accessed 1 March 2024).

80 Cannon L, Tholouli E, Ward C, Farooq H, Kingston M. Use of pritelivir in refractory aciclovir-resistant herpes simplex virus type 2. *Int J STD AIDS*. 2021;32(10):978–80. doi: 10.1177/09564624211006568:9564624211006568.

81 Serris A, Pouvaret A, Loiseau C, Abid H, Burrel S, Fourgeaud J, et al. Pritelivir for recurrent aciclovir-resistant herpes simplex virus 2 infections in immunocompromised patients. *J Antimicrob Chemother*. 2022;77:2303–05.

82 Bosetti D, Bernardi C, Maulini M, Giannotti F, Mamez AC, Masouridi-Levrat S, et al. Salvage treatment of refractory HSV oral lesions with pritelivir in allogeneic hematopoietic cell transplant recipients. *Antimicrob Agents Chemother*. 2023;67:e0173222.

83 Huntjens DW, Dijkstra JA, Verwiel LN, Slijkhuis M, Elbers P, Welkers MRA, et al. Optimizing antiviral dosing for HSV and CMV treatment in immunocompromised patients. *Pharmaceutics*. 2023;15(1):163.

84 Biswas S, Jennens L, Field HJ. Single amino acid substitutions in the HSV-1 helicase protein that confer resistance to the helicase-primase inhibitor BAY 57-1293 are associated with increased or decreased virus growth characteristics in tissue culture. *Arch Virol*. 2007;152:1489–500.

85 Biswas S, Smith C, Field HJ. Detection of HSV-1 variants highly resistant to the helicase-primase inhibitor BAY 57-1293 at high frequency in 2 of 10 recent clinical isolates of HSV-1. *J Antimicrob Chemother*. 2007;60:274–79.

86 Sukla S, Biswas S, Birkmann A, Lischka P, Zimmermann H, Field HJ. Mismatch primer-based PCR reveals that helicase-primase inhibitor resistance mutations pre-exist in herpes simplex virus type 1 clinical isolates and are not induced during incubation with the inhibitor. *J Antimicrob Chemother*. 2010;65:1347–52.

87 Sukla S, Biswas S, Birkmann A, Lischka P, Ruebsamen-Schaeff H, Zimmermann H, et al. Effects of therapy using a helicase-primase inhibitor (HPI) in mice infected with deliberate mixtures of wild-type HSV-1 and an HPI-resistant UL5 mutant. *Antiviral Res*. 2010;87:67–3.

88 Edlefsen PT, Birkmann A, Huang ML, Magaret CA, Kee JJ, Diem K, et al. No evidence of pritelivir resistance among herpes simplex virus type 2 isolates after 4 weeks of daily therapy. *J Infect Dis*. 2016;214:258–64.

89 Birkmann A, Greninger A, Huang ML, Rangaraju M, Selke S, Timmler B, et al. Resistance analysis in a phase 2 clinical trial with the helicase-primase inhibitor pritelivir in immunocompromised adults with acyclovir resistant herpes simplex virus (HSV) Infection, abstr ICAR, Seattle, WA, USA, Abstract #53; 2022.

Part II

Immunodeficiency Virus

3

The Discovery and Early Development of the HIV-1 Integrase Strand Transfer Inhibitor Dolutegravir (S/GSK1349572)

Sherene Min[1], Soong Hoon Kim[2], Mark Underwood[1], Brian Wynne[1], William Spreen[1]

[1] *ViiV Healthcare, 406 Blackwell Street, Suite 300, Durham, NC, USA*
[2] *ViiV Healthcare, 36 E Industrial Road, Branford, CT, USA*

3.1 Introduction

Human immunodeficiency virus (HIV) continues to be a global health threat [1]. Over 40 million people have died from complications related to HIV and acquired immunodeficiency syndrome (AIDS) since the emergence of this disease in the early 1980s [1, 2]. At the end of 2022, an estimated 39 million people were living with HIV globally, with ~1.3 million new HIV diagnoses and ~630 000 HIV-related deaths recorded that year [1]. No cure currently exists for HIV, but highly effective antiretroviral therapy (ART) has turned it into a chronic health condition for many individuals living with HIV [3]. One of the most significant advances in the treatment of HIV-1 in the past 15 years has been the introduction of integrase strand transfer inhibitors (INSTIs) [2, 4].

INSTIs are a class of antiretroviral drugs that work by blocking the HIV integrase enzyme [4]. Integrase is essential for HIV replication, allowing the virus to insert its genetic material into the DNA of human cells [4, 5]. By blocking integrase, INSTIs prevent HIV from infecting new cells [2]. The first marketed INSTI was raltegravir (RAL; approved for use in the United States in 2007) [6] and the second was elvitegravir (EVG; approved for use in the United States in 2012) [7]. However, improvements in dosing frequency, coadministration without the need for pharmacokinetic (PK) boosting agents, increased barrier to resistance, and coverage of INSTI-resistant isolates were needed [4, 5].

Trends in Antiviral Drug Development, First Edition. Edited by Michael J. Sofia and Zhengqiang Wang.

Dolutegravir (DTG) arose from a joint venture between GSK and Shionogi that began in 2001. The joint venture was established to develop, manufacture, and commercialize HIV and central nervous system therapeutics. An HIV-specific chemistry working group was formed in 2002 to catalyze the discovery and advancement of novel INSTI compounds. The joint venture led to the discoveries of the second-generation INSTIs S/GSK1349572 (DTG) and S/GSK1265744 (cabotegravir).

DTG, a second-generation INSTI, has become a mainstay of HIV-1 treatment since approval of the first product, Tivicay (DTG) [8], for use in the United States in 2013. DTG preferentially blocks the strand transfer step of integration of the viral genome into host cell DNA and is designed to retain activity against virus resistant to other INSTIs (e.g. RAL and EVG) [4, 9, 10]. The proposed mechanism of inhibition involves the drug molecule chelating to two Mg^{2+} ions in the integrase DD(35)E catalytic active site [5]. DTG was selected for clinical trials because results from nonclinical studies demonstrated a favorable safety profile, PK supporting once-daily dosing, and the potential for activity against INSTI-resistant virus [4, 9]. Additionally, the drug interaction profile was predicted to be favorable, with glucuronidation as the primary route of metabolism and no demonstration of induction or inhibition of cytochrome P450 (CYP) enzymes *in vitro* [11]. DTG was shown to have potent *in vitro* activity, with a half-maximal inhibitory concentration (IC_{50}) against HIV in peripheral blood mononuclear cells of 0.50 nM and a protein-adjusted 90% inhibitory concentration (PA-IC_{90}) of 64 ng/mL [4]. Throughout the course of clinical development, DTG, in combination with other antiretroviral agents, demonstrated robust potency, durability, a high barrier to resistance, and a favorable safety profile, as well as improved convenience with once-daily dosing with no need for a PK boosting agent [12–16].

Globally, three additional antiretroviral combination products that include DTG have been approved for the treatment of HIV-1: Triumeq (DTG/abacavir/lamivudine [DTG/ABC/3TC], approved for use in adults and children aged ≥3 months and weighing ≥6 kg who are ART-naive or ART-experienced) [17], Juluca (DTG/rilpivirine, approved for use in adults who are virologically suppressed on a stable ART regimen) [18], and Dovato (DTG/3TC, approved for use in adults who are ART-naive or virologically suppressed on a stable ART regimen) [19]. Tivicay is indicated in combination with other antiretroviral agents for the treatment of HIV-1 in adults and children aged ≥4 weeks and weighing ≥3 kg [8, 20]. Altogether, DTG-containing products have indications in ART-naive and ART-experienced adults, adolescents, and children [17–19]. Current treatment guidelines from the US Department of Health and Human Services and the European AIDS Clinical Society recommend DTG-based regimens for first-line therapy in people living with HIV-1 and for those who have responded to therapy but wish to switch regimens [21, 22]. Through a unique licensing agreement with the Medicines Patent Pool, generic forms of DTG are available for adults in at least 95 countries and for children in at least 123 countries [23]. This chapter will describe the discovery and early clinical development of DTG.

3.2 Medicinal Chemistry

3.2.1 Discovery of S/GSK1349572 (DTG)

3.2.1.1 Evolution of the Chemical Structure Leading to DTG

HIV integrase transfers viral DNA to the host chromosome in a two-step process [5]. In the first step, integrase catalyzes the hydrolysis and removal of a dinucleotide from each end of the viral DNA (i.e. 3′-processing), generating two 3′-hydroxyl groups. In the second step, integrase mediates the cleavage of the host DNA and the incorporation of viral DNA via the transesterification of the two 3′-hydroxyl groups, with cell-repair mechanisms completing the strand transfer. The integrase active site contains two magnesium ions complexed to a triad of conserved DD(35)E carboxylate residues (Asp64, Asp116, and Glu152) that mediate the 3′-processing and strand transfer. Small molecules designed to simultaneously coordinate two magnesium ions (two-metal-binding pharmacophore; Figure 3.1a) disrupt this catalytic process [4], thus leading to the successful discovery of the first-generation INSTIs RAL and EVG (Figure 3.1b) [5]. Although these drugs were critical components

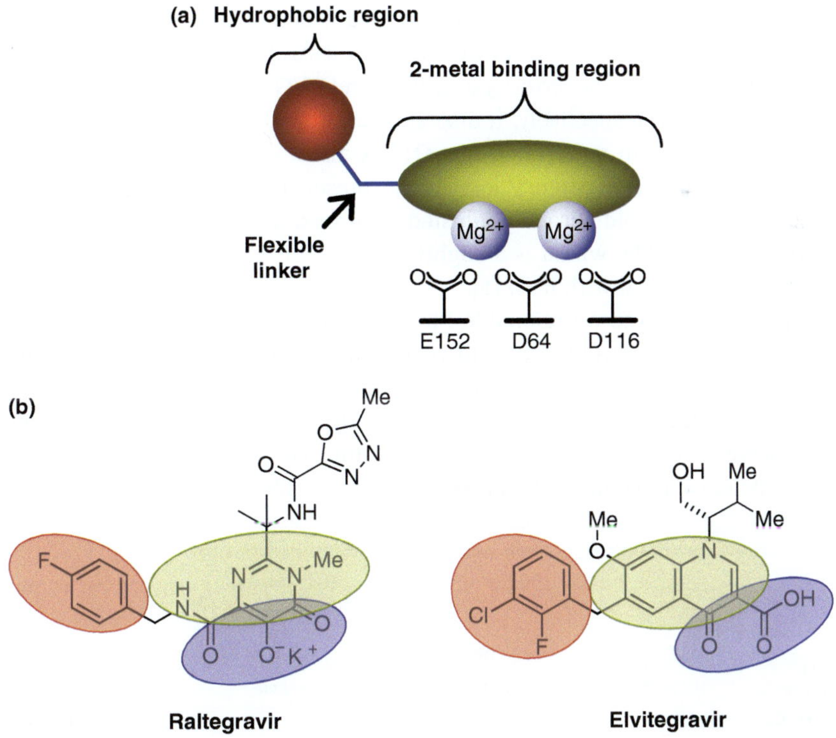

Figure 3.1 (a) Two-metal-binding pharmacophore model for HIV-1 integrase strand transfer inhibitors. *Source:* Reproduced with [4] permission from American Chemical Society © 2013. (b) First-generation HIV-1 integrase strand transfer inhibitors. *Source:* Adapted with [5] permission from American Chemical Society © 2012.

of ART, there was a need for improved INSTIs that addressed the limitations of RAL and EVG. For example, both compounds have poor PK exposure: RAL requires twice-daily dosing, and EVG requires coadministration of a PK booster, which raises concerns about drug–drug interactions. For second-generation INSTIs, there was a strong interest in improving PK exposure during coadministration of other antiretrovirals in an ART regimen. Moreover, second-generation INSTIs needed to combat emergent RAL- and EVG-resistant mutations and have a higher barrier to overall resistance [4].

The general design principle embedded in the two-metal-binding pharmacophore model was likewise adopted for second-generation INSTIs [5]. Most chemical series align within this narrow structural framework: a conserved two-metal-binding region flanked by a hydrophobic tail through a flexible linker [4]. Hydroxynaphthpyridinone (**1**), hydroxyquinoline (**2**), and hydroxynaphthpyridine (**3**) were among the early chemical series examined (Figure 3.2) [5]. These structures share the triad of Lewis basic groups that coordinate the magnesium ions and exhibited modest-to-good strand transfer inhibition [5].

A key design insight was the replacement of the nitrogen of **3** with an oxygen in **4** (Figure 3.2). As a more electronegative element, oxygen was expected to enhance coordination with magnesium according to the hard and soft acids and bases theory [5]. Consistent with this prediction, compound **4** demonstrated improved strand transfer inhibition versus **3** and showed potent antiviral activity [5].

While **4** demonstrated superiority over RAL and EVG against integrase mutations Y143R and Q148K, respectively, it also showed limitations. Specifically, **4** exhibited a ~200-fold loss in potency against Q148K and a ~21-fold loss against N155H (Table 3.1) [5, 26]. The potency loss in the presence of human serum albumin (20 mg/mL) was 12-fold (protein-adjusted half-maximal inhibitory concentration [$^{\text{MT-4}}$PA-IC_{50}], 120 nM), which is a slightly larger shift compared with RAL but a smaller shift compared with EVG [5].

Compound **4** showed a reasonable PK profile across preclinical studies in rats, dogs, and monkeys, with moderate-to-good oral bioavailability [5]. Encouragingly, the concentration of **4** (5 mg/kg orally) at 24 h post-dose in rats, dogs, and monkeys was above the targeted $^{\text{MT-4}}$PA-IC_{50}. Additionally, no off-target issues of concern were observed. Thus, **4** served as an anchor point to test whether further improvements in potency and compound properties were possible.

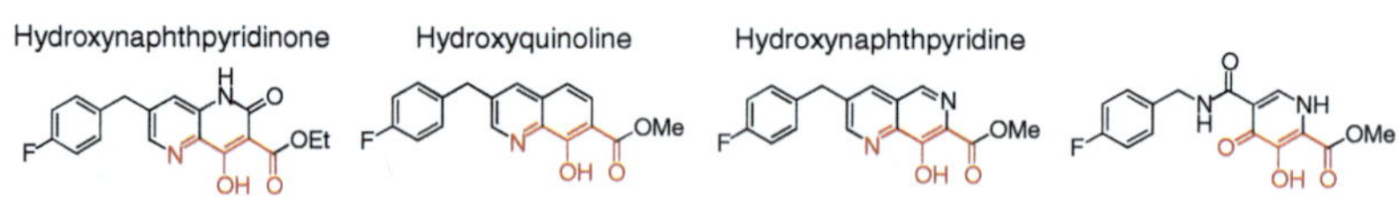

	1	2	3	4
Strand transfer IC_{50}[a]	13 nM[25,b]	354 nM[24]	417 nM[24]	18 nM[5,c]
Antiviral IC_{50}[d]	1200 nM[25,b]	—	—	10 nM[5,d]

Figure 3.2 Enzyme and inhibition activity of early second-generation INSTIs [5, 24, 25]. IC_{50}, half-maximal inhibitory concentration. [a]Strand transfer inhibitory activities; assay conditions differed for **2** and **3** compared with **1** and **4**. [b]Values are from one experiment. [c]Values are the mean from three independent experiments. [d]Anti-HIV activities using MT-4 cells.

Table 3.1 Antiviral potency against integrase mutations for RAL, EVG, and **4** [5, 26].

	RAL (SD)	EVG (SD)	4 (SD)
MT-4/MTT WT IC_{50}, nM[a]	6.2 (2.2)	1.3 (0.3)	10.0 (2.9)
Q148K, FC	83.0 (6.6)	>1700.0	200.0 (25.0)
N155H, FC	8.4 (1.8)	25.0 (7.8)	21.0 (8.9)
Y143R, FC	16.0 (3.9)	1.8 (0.2)	1.5 (0.6)
$^{MT\text{-}4}$PA-IC_{50}, nM[b]	29.0	29.0	120.0
HSA potency shift[c]	4.7	22.0	12.0

EVG, elvitegravir; FC, fold change; HSA, human serum albumin; IC_{50}, half-maximal inhibitory concentration; MTT, tetrazolium-based colorimetric assay; PA-IC_{50}, protein-adjusted half-maximal inhibitory concentration; RAL, raltegravir; WT, wild type.

a. Anti-HIV activities using MT-4 cells.
b. Based on MT-4/MTT IC_{50}.
c. Potency shift in the presence of human serum albumin.

Source: Adapted from [5] with permission from American Chemical Society © 2012.

With its ester functionality, **4** was expected to be unstable *in vivo* and undergo enzyme-mediated hydrolysis to corresponding acid **5**, which showed poor activity (Table 3.2); therefore, it was hypothesized that amides may resolve the issue of instability. Compared with **4** (Figure 3.3a), the loss of activity observed with amide **6** may be attributed to the disruption of two-metal-binding conformation (Figure 3.3b) [27]. Therefore, acyclic amide **6** (MT-4/MTT WT IC_{50}, 110 nM; Table 3.2) was rebuilt as constrained cyclic **7** (MT-4/MTT WT IC_{50}, 2 nM; unsaturated ring) and **8** (MT-4/MTT WT IC_{50}, 6 nM; saturated ring) [27]. Both bicyclic amide rings showed significant improvement in activity against WT HIV-1 in MT-4 cells compared with **6**, which supports the hypothesis that the two-metal-binding model requires specific orientation of coordinating groups. Only modest differences in potency between the saturated and unsaturated series were found in WT HIV-1 in MT-4 cells; however, saturated rings were generally more potent against the resistance-associated Q148K mutation. Compound **9**, one of the earlier compounds examined that contained a substitution on the ring and not on the amide, was configurationally unstable and was isolated as a mixture [27]. Compound **9** demonstrated potent antiviral activity against both WT HIV-1 (MT-4/MTT WT IC_{50}, 4 nM) and the Q148K mutation (2.1-fold change), with a low (3.7-fold) potency shift in the presence of human serum albumin (Table 3.2), supporting the hypothesis that cyclic amides with a locked two-metal-binding orientation could restore potency while replacing the ester functionality found in **4**.

Further stabilization of **9** through ring annulation into a tricyclic ring system afforded **10** (five-membered ring) and **11** (six-membered ring) [27]. Both compounds were configurationally stable and equally potent in the MT-4/MTT WT viral assay, but **11** showed improved activity against Q148K and a smaller potency shift in the presence of human serum albumin (Table 3.3) [27]. Separation of **11** into its pure enantiomers **11S** and **11R** revealed a dependence of the antiviral activity ($^{\text{pseudo-HIV [pHIV]}}IC_{50}$, ^{pHIV}PA-$IC_{50}$, and Q148K fold change) on the stereochemical

Table 3.2 Antiviral profile of acyclic and bicyclic compounds [27].

5

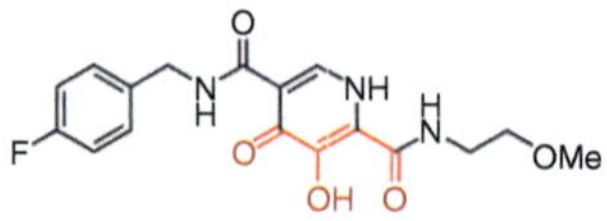

6

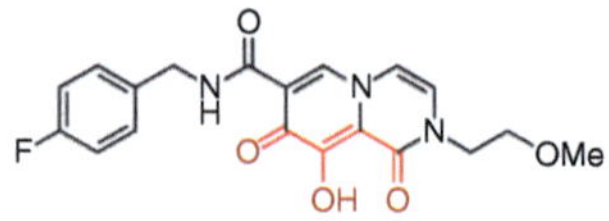

7

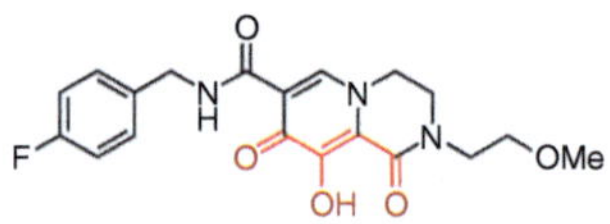

8

9

Compound	MT-4/MTT WT IC_{50}, nM[a,b]	MT-4/MTT CC_{50}, nM[a,c]	Q148K, FC[a]	$^{MT\text{-}4}$PA-IC_{50}, nM[d]	Potency shift[e]
4	10.0 (2.9)	19 000 (850)	200.0 (25.0)	120	12.0
5	1400.0 (400.0)	>200 000	NA	NA	NA
6	110.0 (42.0)	16 000 (1400)	NA	NA	NA
7	1.6 (1.1)	>200 000	35.0 (6.8)	27	17.0
8	5.9 (0.0)	52 000 (8 300)	18.0 (2.5)	18	3.1
9	3.8 (0.5)	83 000 (22 000)	2.1 (0.5)	14	3.7

CC_{50}, half-maximal cytotoxic concentration; FC, fold change; IC_{50}, half-maximal inhibitory concentration; MTT, tetrazolium-based colorimetric assay; NA, not available; PA-IC_{50}, protein-adjusted half-maximal inhibitory concentration; WT, wild type.

a. Values are the mean (SD) from three independent experiments.
b. Anti-HIV activities using MT-4 cells.
c. Cytotoxicity using MT-4 cells.
d. Based on MT-4/MTT IC_{50}.
e. Potency shift in the presence of human serum albumin.

Source: Adapted from [27] with permission from American Chemical Society © 2013.

disposition of the compounds [4]. Both **11S** and **11R** were found to be configurationally stable, establishing the practical importance of their different profiles.

As the final optimization step, the effect of chiral substitution on the tricyclic ring was examined (Table 3.4) [4]. The C-3 methyl–substituted enantiomeric pair **12** and **13** showed large differences in Q148K viral potency and shifts in the pHIV luciferase assay [4]. The addition of a fluorine group (**14**) in the aryl ring offered a further modest improvement in Q148K potency, and this modification was incorporated into subsequent analogs. The C-2 methyl diastereomers **15** and **16** led to small but incremental potency changes in Q148K as well as in ^{pHIV}PA-IC_{50}. A larger substituent (**17**, isopropyl for methyl) or a polar group (**18**, carbomethoxy for methyl) did not significantly alter the Q148K viral and shift profile. The introduction of a C-4 R-methyl group (**19**) did not substantially change the overall profile, whereas **20**, the enantiomer of **19**, demonstrated the best potency against the resistant Q148K mutation. Due to its PK profile and potency against the Q148K mutation, **20**, which is now known as DTG, was progressed to early clinical development.

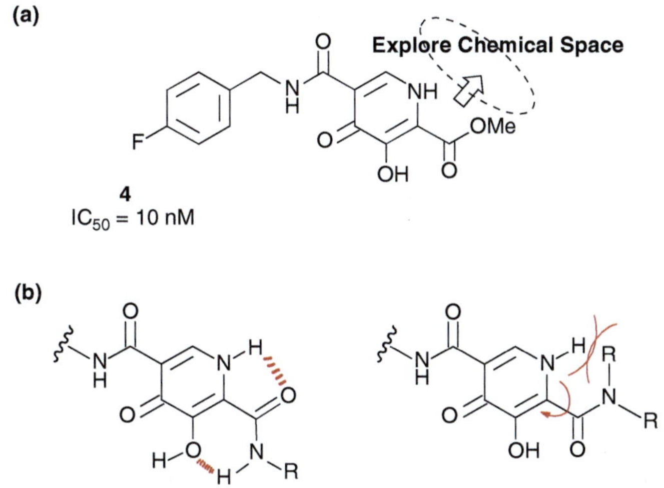

Figure 3.3 (a) Compound **4** (monocyclic carbamoyl pyridone lead inhibitor 1); (b) unfavorable intramolecular hydrogen bonding and steric hindrance potentially disrupt the coplanarity of the metal chelation motif by amide modifications. *Source:* Adapted from [27] with permission from American Chemical Society © 2013.

Table 3.3 Initial structure–activity relationship of tricyclic rings [27].

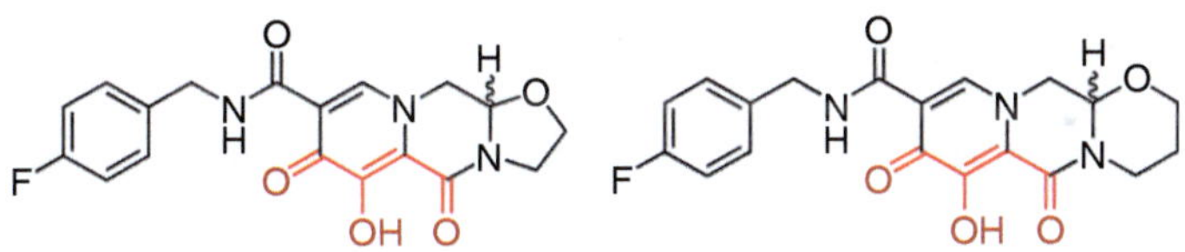

10 **11**

Compound	MT-4/MTT WT IC_{50}, nM[a,b]	MT-4/MTT CC_{50}, nM[a,c]	Q148K, FC[a]	MT-4PA-IC_{50}, nM[d]	Potency shift[e]
9	3.8 (0.5)	83 000 (22 000)	2.1 (0.5)	14.1	3.7
10	2.7 (0.6)	72 000 (24 000)	19.0 (5.1)	140.4	52.0
11	3.4 (1.1)	18 000 (1600)	2.8 (0.3)	26.9	7.9

CC_{50}, half-maximal cytotoxic concentration; FC, fold change; IC_{50}, half-maximal inhibitory concentration; MTT, tetrazolium-based colorimetric assay; PA-IC_{50}, protein-adjusted half-maximal inhibitory concentration; WT, wild type.

a. Values are the mean (SD) from three independent experiments.
b. Anti-HIV activities using MT-4 cells.
c. Cytotoxicity using MT-4 cells.
d. Based on MT-4/MTT IC_{50}.
e. Potency shift in the presence of human serum albumin.

Source: Adapted from [27] with permission from American Chemical Society © 2013.

Table 3.4 Effects of substitution on six-membered ring tricyclic carbamoyl pyridones [4].[a]

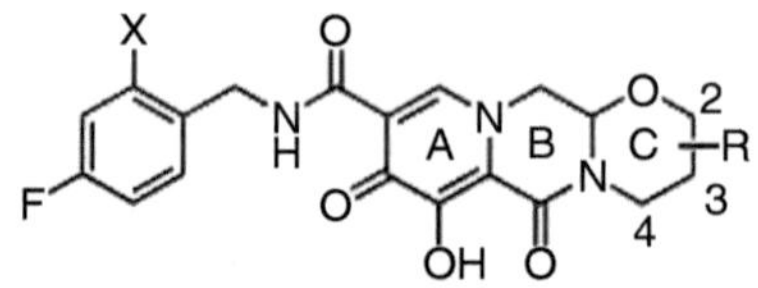

Compound	X	Ring C	$^{pHIV}IC_{50}$ (95% CI), nM	$^{pHIV}PA\text{-}IC_{50}$ (95% CI), nM[b]	Q148K, FC[c]
11S	H	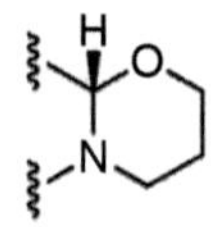	1.9 (0.4, 9.6)	81 (12, 540)	2.4
11R	H	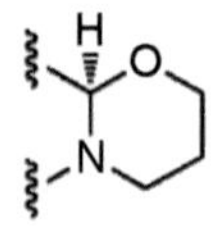	2.2 (1.4, 3.5)	9 (4, 25)	11.0
12	H	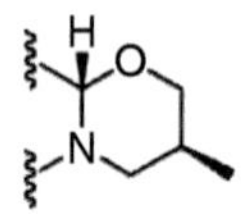	0.8 (0.4, 1.6)	210 (120, 350)	36.0
13	H	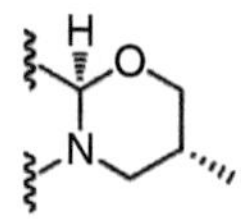	0.7 (0.5, 0.9)	6 (4, 7)	9.9
14	F	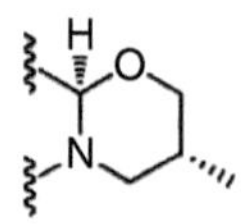	0.5 (0.4, 0.5)	5 (3, 10)	5.7
15	F	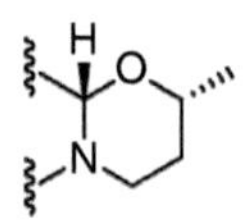	0.9 (0.7, 1.1)	7 (2, 23)	2.1
16	F	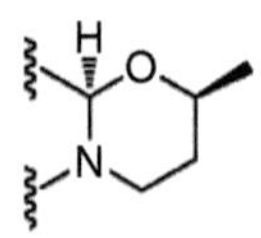	0.5 (0.4, 0.7)	2 (1, 3)	2.8
17	F	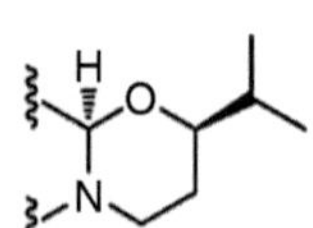	0.7 (0.6, 0.8)	4 (3, 6)	1.9
18	F	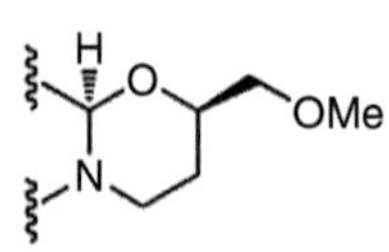	0.4 (0.3, 0.5)	1 (0, 2)	3.5

Table 3.4 (Continued)

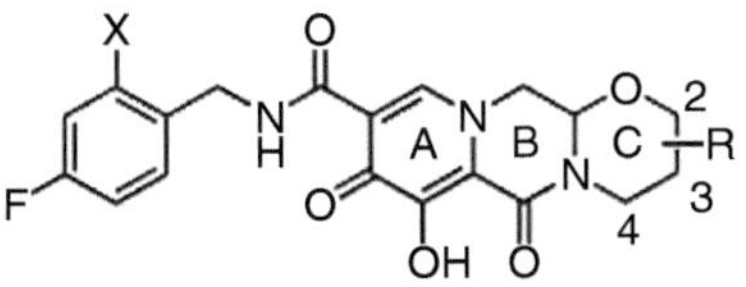

Compound	X	Ring C	$^{pHIV}IC_{50}$ (95% CI), nM	$^{pHIV}PA\text{-}IC_{50}$ (95% CI), nM[b]	Q148K, FC[c]
19	F	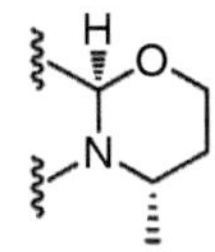	0.5 (0.5, 0.6)	4 (2, 8)	2.2
20 (DTG)	F	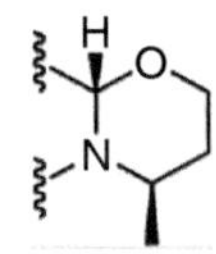	1.7 (1.3, 2.2)	22 (15, 33)	0.4

DTG, dolutegravir; FC, fold change; IC_{50}, half-maximal inhibitory concentration; PA-IC_{50}, protein-adjusted half-maximal inhibitory concentration; pHIV, pseudo-HIV.

a. Data represent the mean of at least three replicates.

b. In the presence of human serum albumin.

c. Fold change against the resistant variant in the pHIV assay.

Source: Adapted from [4] with permission from American Chemical Society © 2013.

3.3 Virology

3.3.1 *In Vitro* Studies Indicated Robust Efficacy and High Barrier to Resistance of DTG

In vitro virology data supported the progression of DTG into clinical development. Studies investigated DTG antiviral potency in several cell lines, protein-adjusted half-maximal effective concentration (PA-EC_{50}) by extrapolation of titrated human serum, estimation of 90% effective concentration (EC_{90}), *in vitro* strand transfer assay, resistance selection passage, activity versus HIV-1 with resistance to other classes of antiretrovirals, activity versus wild-type HIV-2, and cellular assays supportive of mechanism of action [26, 28]. In multiple antiviral assays, DTG inhibited HIV replication with low-to-sub nanomolar potency and had a selectivity index of 9400 (Table 3.5) [4]. Measurements of the fold shift from EC_{50} to EC_{90} ranged from threefold to fourfold, and an EC_{90} fourfold higher than the EC_{50} was selected as a conservative measure [26]. Titration of human serum into the MT-4 assay enabled extrapolation of the potency shift of 100% human serum as a PA-EC_{50} of 38 nM. These data allowed the estimation of a clinical PA-IC_{90} of 64 ng/mL for DTG [4].

DTG demonstrated a considerable improvement in resistance profile compared with RAL and EVG *in vitro*, providing the foundation for its status as a second-

Table 3.5 Virologic efficacy profile of DTG [4, 26].

	DTG
INST, mean (95% CI), nM	2.7 (1.7, 4.1)[a]
$^{MT\text{-}4}IC_{50}$, mean (95% CI), nM	2.0 (0.8, 5.1)[a]
$^{PBMC}IC_{50}$, mean (95% CI), nM	0.5 (0.3, 0.8)[a]
PA-EC_{50}, nM	38
PA-EC_{90}, nM	152
100% human serum fold shift	75[a]
PA-IC_{90}, ng/mL	64[a]

DTG, dolutegravir; IC_{50}, half-maximal inhibitory concentration; INST, integrase strand transfer (assay); PA-EC_{50}, protein-adjusted half-maximal effective concentration; PA-EC_{90}, protein-adjusted 90% effective concentration; PA-IC_{90}, protein-adjusted 90% inhibitory concentration; PBMC, peripheral blood mononuclear cell.

a. Data shown are the mean of at least three replicates.

generation INSTI (Figure 3.4) [9, 26]. Notably, DTG showed activity against site-directed molecular clones with the RAL-resistant signature mutations Y143R, Q148K, N155H, and G140S/Q148H [26]. When tested against a panel of 50 first-generation INSTI-resistant site-directed mutants, only two mutant viruses (E138K/Q148K and Q148R/N155H) led to a >10-fold change in EC_{50} for DTG, whereas 23 and 32 mutant viruses led to a >10-fold change in EC_{50} for RAL and EVG, respectively.

The high barrier to resistance of DTG is consistent with its potent activity against many singly and multiply mutated viruses [26, 29], evaluations that could be targeted for clinically relevant mutations that emerge with first-generation INSTIs.

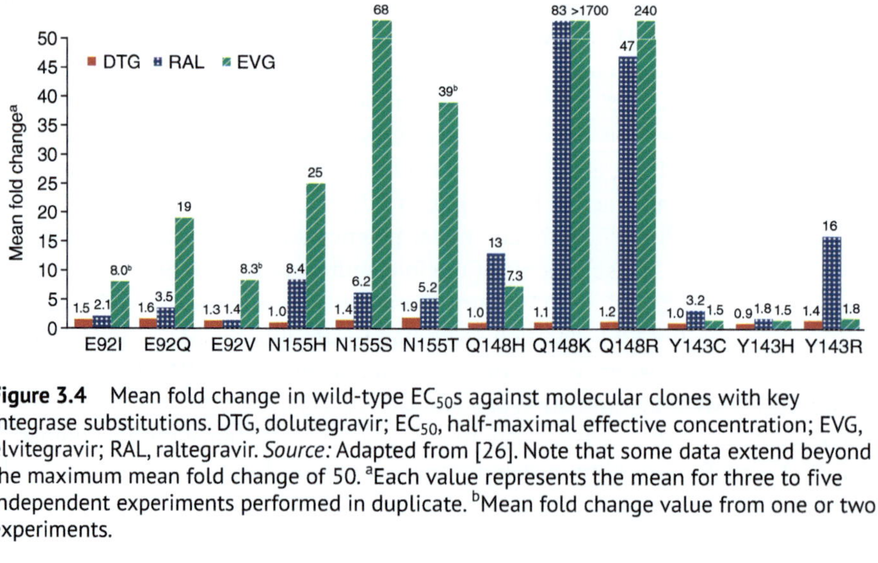

Figure 3.4 Mean fold change in wild-type EC_{50}s against molecular clones with key integrase substitutions. DTG, dolutegravir; EC_{50}, half-maximal effective concentration; EVG, elvitegravir; RAL, raltegravir. *Source:* Adapted from [26]. Note that some data extend beyond the maximum mean fold change of 50. [a]Each value represents the mean for three to five independent experiments performed in duplicate. [b]Mean fold change value from one or two experiments.

Additional studies evaluated dissociation rates from integrase–DNA complexes with integrase mutations to further understand the mechanism underlying the barrier to resistance of DTG. Compared with RAL and EVG, DTG dissociated considerably more slowly from a wild-type integrase–DNA complex, and longer binding was observed for DTG versus RAL or EVG in integrase–DNA complexes containing key substitutions [9]. Notably, there was an overall inverse relationship between dissociative half-life and fold change in EC_{50}. Dissociation rates of INSTIs measured using additional clinically relevant singly and multiply mutated integrase proteins [30] confirmed and extended DTG slower dissociation rates (Figure 3.5) [31]. Importantly, rapid dissociation of DTG (half-life <1 h) was only observed for integrase with multiple RAL signature substitutions or Q148 + ≥2 RAL secondary substitutions, indicating that accumulation of certain RAL-associated substitutions can confer resistance to DTG. Either additive or synergistic effects were observed when DTG was tested in combination with approved representative HIV antiretrovirals, or with antivirals for hepatitis B and C virus that may be used for patients coinfected with HIV; no antagonistic effects were seen [26].

Resistance passage experiments were performed to identify mutations that are selected under increasing drug pressure [32], and findings supported the high barrier to resistance of DTG [9]. HIV-1 strain IIIB passage versus DTG in cultures containing MT-2 cells through 112 days resulted in only four viruses with mutations; no mutations reduced susceptibility >5-fold to DTG. In contrast, mutations with >5-fold reduced susceptibility to RAL were observed on day 28, and highly resistant

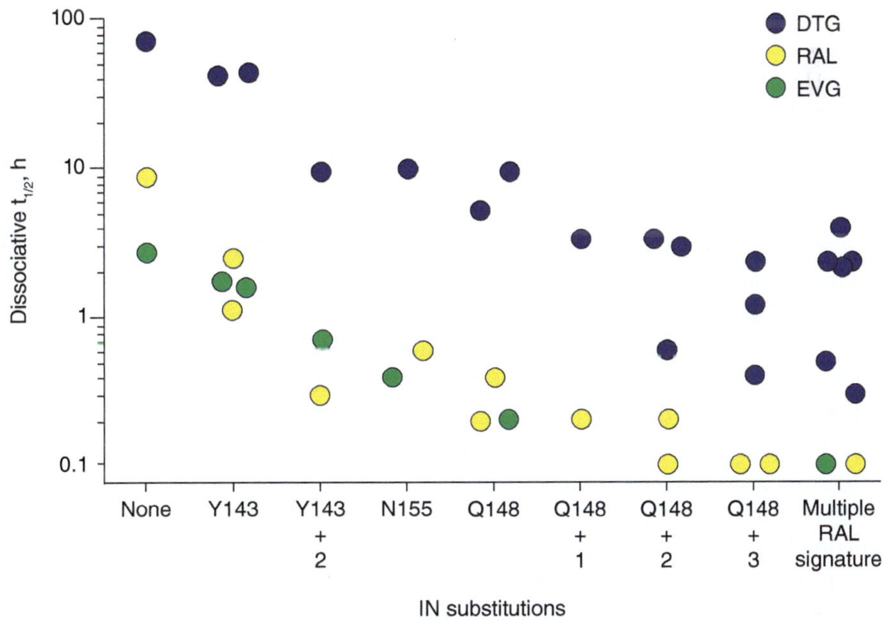

Figure 3.5 Dissociation of INSTIs from integrase proteins with or without integrase substitutions using population genotypes from participants in the pilot VIKING study ViiV Healthcare / Jules Levin. IN, integrase; INSTI, integrase strand transfer inhibitor; $t_{1/2}$, half-life.

mutations (fold change in EC_{50} >100) were observed with RAL. Furthermore, in dose-escalation *in vitro* passage experiments, the maximum concentration of INSTI allowing for viral replication was considerably lower for DTG compared with RAL and EVG (6.4 vs. 800 and 160 nM, respectively).

Studies on crystal structure provided insight into the molecular mechanisms underlying the high barrier to resistance of DTG [33–36]. Hare and colleagues presented the crystal structure of the prototype foamy virus (PFV) intasome composed of a tetramer of PFV integrase protein plus viral DNA ends (a surrogate for the HIV-1 intasome) [33, 34]. Their important work provided the overarching architecture and revealed the viral DNA and positioning for the catalytic pocket. They reported that DTG occupies the intasome active site, similar to first-generation INSTIs [34]. Subsequently, HIV-1 intasome models were used to better understand the docking of DTG, RAL, and EVG based on insights from the PFV intasome [36]. It was posited that the streamlined architecture of the metal-chelating scaffold of DTG is a factor that renders DTG more independent of mutations that deform the catalytic pocket. Furthermore, DTG has a larger and more aromatic metal-chelating scaffold that may allow it to delocalize the catalytic Mg^{2+} positive charge and form stronger coordinate bonds.

In vitro studies identified that the R263K and G118R integrase substitutions conferred low-level resistance to DTG [35, 37, 38]. In serial passage experiments of DTG in healthy donor primary human umbilical cord blood mononuclear cells infected with subtype B, C, and A/G viruses, the R263K mutation was observed in 5/5 subtype B selections and 1/2 A/G selections [38]. The G118R mutation was also observed in subtype C and A/G viruses. The R263K mutation confers low-level resistance to DTG (fold change of 2.3) [35]. The G118R substitution led to a 3.1-fold change in susceptibility to DTG and similar low-level resistance to RAL and EVG [37]. However, R263K and G118R significantly impaired HIV-1 viral replication capacity and strand transfer activity, thus impairing the fitness of the virus [35, 37, 38].

Findings from clinical studies of DTG confirmed the improved resistance profile of DTG and provided further understanding of INSTI mutations that could confer DTG resistance. The Phase 3 SAILING study assessed DTG versus RAL in ART-experienced, INSTI-naive adults with resistance to ≥2 antiretroviral classes [12], and the Phase 3b DAWNING study evaluated DTG in adults who had prior virologic failure on a first-line ART regimen [39]. Findings from these studies revealed that DTG-selected resistance was distinct from the first-generation INSTIs RAL and EVG [40, 41]. Substitutions commonly included G118R and/or R263K, which were rarely observed with RAL or EVG treatment and were identified *in vitro* with DTG passage studies, as well as a constellation of other resistance-associated mutations that were present in multiply mutated viruses and included N155H, H51H/Y, E138E/K, G140S, Q148H/R, T66T/I, and L74I [35, 41, 42]. Further analysis revealed that DTG remained bound *in vitro* to integrase–DNA complexes with integrase G118R or R263K mutations for a prolonged time compared with RAL and EVG (Table 3.6).

To better understand resistance dynamics and impact on viral fitness at the molecular level, a cryogenic electronic microscopy intasome structure was used to develop HIV-1 integrase homology models [41, 43]. HIV-1 integrase structural analyses supported that three dominant mechanisms may collectively lead to INSTI

Table 3.6 Dissociation of INSTIs from integrase–DNA complexes with INSTI resistance-associated substitutions.

Integrase-linked substitution(s)	$t_{1/2}$, h[a]		
	DTG	RAL	EVG
Wild type	92.0	9.2	2.6
R263K	52.0	9.6	2.2
G118R	10.7	2.7	0.9
G118R, E138K	10.1	3.2	1.1

DTG, dolutegravir; EVG, elvitegravir; INSTI, integrase strand transfer inhibitor; RAL, raltegravir; $t_{1/2}$, half-life.

a. For reference, $t_{1/2}$ values for DTG, RAL, and EVG with a wild-type integrase–DNA complex were previously measured to be 71, 8.8, and 2.7 h, respectively [9].

Source: Adapted from Ref. [41].

resistance when the G118R substitution is selected by DTG [41]. The steric occupation of G118R in the catalytic site can both prevent INSTI binding and host target DNA binding, a possible explanation for both resistance to DTG and reduced viral replication capacity. Also, G118R may promote strand transfer and integration of viral DNA through interactions with viral DNA 5′ phosphate of the catalytic adenosine and stabilize the product host target DNA [41]. The R263K substitution in integrase may eliminate multiple hydrogen bonds with the substrate and/or viral DNA catalytic loop, allowing for more flexibility and ultimately impacting the binding kinetics such that the viral DNA ends that are joined with the host genome ends during integration have a higher binding preference than INSTIs. R263K may also disrupt the usual interactions between HIV-1 integrase and viral DNA termini, thereby decreasing viral replication capacity.

3.4 Clinical Development

Throughout its clinical development, DTG has demonstrated potent antiviral activity and durability, a high barrier to resistance, a favorable safety and tolerability profile, and improved convenience with once-daily dosing and no requirement for a PK-boosting agent.

3.4.1 PK Studies Supported Once-daily 50 mg Dosing of DTG

The preclinical PK profile of DTG dosed as 5 mg/kg as a suspension in rats, dogs, and primates had low clearance values across species [4]. Oral bioavailability (76–87% in fasted rats and primates) was good and consistent across species; no significant CYP3A4 inhibition was observed. DTG was predicted to achieve the target trough concentration coverage of PA-IC_{90} in humans with low, once-daily, oral doses (<100 mg/day).

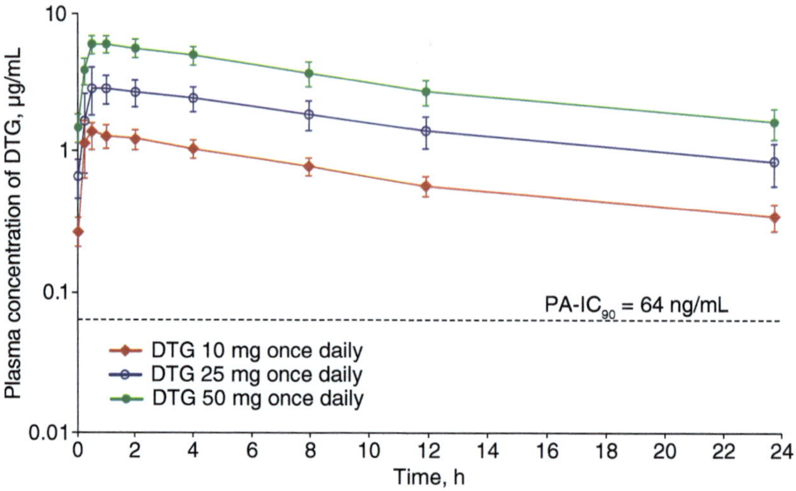

Figure 3.6 Mean (SD) plasma concentration–time profiles for DTG oral suspension on day 10 after repeat-dose administration. Dashed horizontal line represents the *in vitro* PA-IC_{90} of 64 ng/mL. DTG, dolutegravir; PA-IC_{90}, protein-adjusted 90% inhibitory concentration.

Subsequent Phase 1 PK studies of DTG used an oral suspension formulation for evaluation in healthy participants while the tablet formulation was under development [13]. In the initial single-dose study, suspension doses of 2–100 mg were administered, and in the multiple-dose study, doses of 10, 25, and 50 mg once daily were administered for 10 days. The PK profile suggested that once-daily, low-milligram doses would achieve therapeutic concentrations, with a half-life of ~15 h and linear PK across the dose range studied. After 10 days of 50 mg dose administration, steady-state concentration at the end of the dosing interval was 1.6 µg/mL, ~25-fold higher than the PA-IC_{90} (64 ng/mL) for wild-type HIV-1 (Figure 3.6).

Solid tablet formulations of 10, 25, and 50 mg were developed for administration in subsequent clinical studies. Coadministration of DTG 50 mg under fed conditions increased plasma DTG exposure [44]. However, exposures in the fed or fasted state were deemed adequate for antiviral activity, and thus DTG 50 mg once daily could be coadministered with or without food [44]. DTG tablets demonstrated a solubility-limited absorption [30]. The increase in DTG exposure with the tablet formulation appeared dose proportional from 25 to 50 mg but less than dose proportional from 50 to 100 mg [45, 46].

3.4.2 Clinical Studies in People Living with HIV-1

Phase 2 and 3 studies of DTG in people living with HIV-1 demonstrated durable efficacy, a favorable safety profile, and a high barrier to HIV-1 resistance. Phase 3 studies evaluated DTG in ART-naive participants with comparisons to the first-generation INSTI RAL, as well as efavirenz (EFV)-based and ritonavir-boosted darunavir (DRV/r)-based ART (Table 3.7). Additional Phase 3 studies evaluated DTG in participants with prior treatment failure.

Table 3.7 Efficacy of DTG in Phase 3/3b studies of people living with HIV-1.

Study name	Phase	Population	DTG dose	Comparator ART	Time point, week	Plasma HIV-1 RNA <50 copies/mL, n/N (%)
SPRING-2 [50]	3	ART-naive	50 mg once daily[a]	RAL 400 mg twice daily[a]	48	DTG: 361/411 (88) RAL: 351/411 (85)
					96	DTG: 332/411 (81) RAL: 314/411 (76)
SINGLE [15, 51]	3	ART-naive	50 mg once daily[b]	EFV 600 mg/ TDF 300 mg/ FTC 200 mg once daily	48	DTG: 364/414 (88) EFV/TDF/FTC: 338/419 (81)
					96	DTG: 80[c] EFV/TDF/FTC: 72[c]
					144	DTG: 71[c] EFV/TDF/FTC: 63[c]
FLAMINGO [52, 53]	3b	ART-naive	50 mg once daily[a]	DRV/r 800 mg/100 mg once daily[a]	48	DTG: 217/242 (90) DRV/r: 200/242 (83)
					96	DTG: 194/242 (80) DRV/r: 164/242 (68)
SAILING [12]	3	ART-experienced, INSTI-naive with resistance to ≥2 ARV classes	50 mg once daily[d]	RAL 400 mg twice daily[d]	24	DTG: 281/354 (79) RAL: 252/361 (70)
					48	DTG: 251/354 (71) RAL: 230/361 (64)
VIKING-3 [29, 56]	3	ART-experienced with prior INSTI resistance and limited treatment options[e]	50 mg twice daily	None	Day 8[f]	Mean change from baseline in HIV-1 RNA, $-1.43\ \log_{10}$ copies/mL
					Week 24[d]	126/183 (69)
					Week 48[d]	116/183 (63)

ABC, abacavir; ART, antiretroviral therapy; ARV, antiretroviral; DRV/r, ritonavir-boosted darunavir; DTG, dolutegravir; EFV, efavirenz; EVG, elvitegravir; FTC, emtricitabine; RAL, raltegravir; 3TC, lamivudine; TDF, tenofovir disoproxil fumarate.

a. Plus TDF/FTC or ABC/3TC.
b. Plus ABC/3TC.
c. n/N not reported.
d. Plus optimized background therapy.
e. Resistance to RAL and/or EVG and to ≥2 other ARV classes with ≥1 fully active ARV at screening.
f. Functional monotherapy period.

Given the promising PK and safety results of DTG in healthy participants, a randomized, placebo-controlled, Phase 2a study in ART-naive and ART-experienced (but INSTI-naive) people living with HIV-1 was conducted [10]. Monotherapy with DTG was administered to understand the inherent antiviral effect of DTG and demonstrated potent antiviral effect and a clear dose-response relationship. Mean decline in HIV-1 RNA (a validated surrogate marker for treatment response) [47] on day 11 from baseline was 1.51, 2.03, and 2.46 $\log_{10}$ copies/mL at doses of 2, 10, and 50 mg, respectively. The observed antiviral response with DTG 50 mg was greater than that seen in similar studies with first-generation INSTI monotherapy, including RAL 400 mg twice daily (−1.70 $\log_{10}$ copies/mL) and ritonavir-boosted EVG 50 mg once daily (−2.03 $\log_{10}$ copies/mL) [48]. Together with a favorable safety profile, these results supported the further evaluation of DTG in larger and longer-term Phase 2b studies of people living with HIV-1.

The randomized, dose-ranging, Phase 2b SPRING-1 study was conducted to select a once-daily dose of DTG for evaluation in Phase 3 trials. People living with HIV-1 naive to ART received DTG 10, 25, or 50 mg plus open-label background nucleoside reverse transcriptase inhibitors (NRTIs) [49]. A rapid antiviral response was observed, with 93% (144/155) of participants who received DTG achieving HIV-1 RNA <50 copies/mL (the recommended endpoint for determining treatment efficacy by the US Food and Drug Administration [47]) by 16 weeks, and this was sustained for up to 48 weeks (91% [48/53], 88% [45/51], and 90% [46/51] achieving HIV-1 RNA <50 copies/mL in the DTG 10, 25, and 50 mg groups, respectively). All DTG doses were well tolerated, and because similar efficacy, safety, and tolerability were seen with all three doses at week 16, with confirmation at week 24, the 50 mg dose was selected for Phase 3 evaluation.

For the Phase 3 evaluation, due to favorable efficacy, PK, safety, and virologic profile, DTG had the possibility of replacing standard-of-care therapy at the time, so multiple studies in ART-naive adults living with HIV-1 were conducted with different standard-of-care comparators. In the multicenter, Phase 3 SPRING-2 study, ART-naive adults living with HIV-1 were randomly assigned to receive either DTG 50 mg once daily or RAL 400 mg twice daily, plus background NRTIs [50]. At week 48, 88% and 85% in the DTG and RAL groups, respectively, achieved HIV-1 RNA <50 copies/mL, which was sustained through week 96. Increases in CD4+ cell count from baseline to week 96 were similar between groups, and only 10 participants (2%) in each group discontinued due to adverse events (AEs). Of those with protocol-defined virologic failure (PDVF; DTG, n = 22; RAL, n = 29), no DTG-treated participants had treatment-emergent INSTI or NRTI resistance compared with five participants in the RAL group.

The randomized, Phase 3 SINGLE study was designed to compare once-daily DTG 50 mg plus a fixed-dose combination of ABC/3TC with fixed-dose EFV/TDF/emtricitabine (FTC), which was the only single-tablet regimen recommended by US treatment guidelines at the time [51]. At week 48, DTG + ABC/3TC was statistically superior to EFV/TDF/FTC, with 88% and 81% of participants achieving HIV-1 RNA <50 copies/mL, respectively ($P = 0.003$). While high rates of virologic suppression were achieved with both regimens, median time to suppression was faster with DTG + ABC/3TC compared with EFV/TDF/FTC (28 vs. 84 days), and DTG + ABC/3TC was associated with fewer discontinuations due to AEs (2% vs. 10%). These results were maintained at week 144, with a higher proportion of participants with virologic suppression in the DTG + ABC/3TC group compared with EFV/TDF/FTC (71% vs. 63%; P = 0.01). These findings were unique among Phase 3 studies, demonstrating the superiority of a new antiretroviral agent (DTG) over established ART. Similar to results from SPRING-2, no resistance mutations were observed in the DTG + ABC/3TC group, whereas seven participants in the EFV/TDF/FTC group developed resistance mutations [15]. Findings from the pivotal SINGLE study supported approval of the three-drug INSTI-based regimen DTG/ABC/3TC as a novel single-tablet formulation.

Altogether, the durable virologic response, favorable safety and tolerability, and absence of treatment-related resistance observed in the pivotal SPRING-2 and SINGLE studies, as well as the Phase 3b FLAMINGO study, which demonstrated the favorable efficacy and safety of DTG-based versus DRV/r-based ART [52, 53] (Table 3.7), supported DTG as an additional first-line treatment option for ART-naive people living with HIV-1.

3.4.3 Clinical Studies in Individuals with Prior Treatment Failure

Due to the increasing need for safe and effective second-line ART options, the SAILING study evaluated the efficacy and safety of DTG compared with RAL in ART-experienced individuals with prior drug resistance and limited treatment options who could potentially benefit from a second-generation INSTI [12]. SAILING was a randomized, noninferiority, Phase 3 study that enrolled eligible participants with two consecutive plasma HIV-1 RNA measurements of ≥400 copies/mL (unless they had >1000 copies/mL at screening), resistance to ≥2 antiretroviral classes, and one or two fully active agents available to create a background therapy. Participants were randomly assigned to DTG 50 mg once daily or RAL 400 mg twice daily, plus investigator-selected background therapy. Results from SAILING demonstrated that DTG was superior to RAL, with 71% versus 64% of participants in the DTG and RAL groups, respectively, achieving HIV-1 RNA <50 copies/mL at week 48 (P = 0.030). Furthermore, significantly fewer participants in the DTG versus RAL group had evidence of treatment-emergent INSTI resistance (1% [4/354] vs. 5% [17/361]; P = 0.003), and AE rates were similar across groups.

To further assess the efficacy and safety of DTG in individuals with limited treatment options, including those with prior treatment failure with RAL, the open-label, pilot, Phase 2b VIKING study was conducted [30]. In the initial cohort of participants (Cohort 1), DTG 50 mg once daily demonstrated promising antiviral activity [30, 54]. Subsequent PK/pharmacodynamic modeling predicted that DTG 50 mg twice daily would appreciably increase the proportion with virologic response on day 11 in this population, yet clinical pharmacology data indicated a less than dose-proportional increase in DTG exposure from 50 to 100 mg [54, 55]. Therefore, a dose of 50 mg twice daily was assessed in Cohort 2 of VIKING [30]. With this higher dose, virologic response was more pronounced in Cohort 2 than Cohort 1 at week 24 (proportion of participants with HIV-1 RNA <50 copies/mL, 75% [18/24] vs. 41% [11/27]). DTG was well tolerated, and there was no clinically significant difference in the safety profile between doses. Results from this pilot study supported Phase 3 evaluation of DTG 50 mg twice-daily dosing in heavily treatment-experienced individuals living with HIV-1.

The open-label, Phase 3 VIKING-3 study investigated the efficacy and safety of DTG 50 mg twice daily in ART-experienced adults with resistance to RAL and/or EVG while continuing their failing regimen (without RAL or EVG) through day 7 (DTG functional monotherapy), after which the background regimen was optimized [29]. Participants had advanced disease and had used a median of 14 prior antiretrovirals. A robust antiviral response was demonstrated, as the mean change from baseline in plasma HIV-1 RNA at day 8 was −1.43 $\log_{10}$ copies/mL (95% CI, −1.52, −1.34) and 69% of participants achieved plasma HIV-1 RNA <50 copies/mL at week 24, with a sustained virologic response at week 48. While this is a lower rate of virologic suppression compared with the other Phase 3 DTG studies described in Sections 3.4.2 and 3.4.3 [12, 50, 51], DTG demonstrated a high rate of virologic response in a patient population with few to no viable regimens available. Multivariate analyses demonstrated a strong association between baseline DTG susceptibility and treatment response [29]; virologic response at week 48 decreased with increasing baseline resistance [56]. DTG 50 mg twice daily was generally well tolerated despite advanced disease status, with only 3% (5/183) of participants discontinuing due to AEs [56], similar to discontinuation rates in the other Phase 3 studies described in Sections 3.4.2 and 3.4.3 [12, 50, 51]. Results from VIKING-3 supported DTG 50 mg twice daily as an efficacious treatment option for those who are INSTI-experienced with demonstrated or suspected INSTI-associated resistance substitutions that confer INSTI resistance [29].

Through the clinical development that supported global marketing authorizations, DTG exhibited key advantages over first-generation INSTIs, including once-daily administration without a PK booster for individuals naive to INSTIs, robust virologic potency, a high barrier to resistance, durable efficacy, and a favorable safety profile. Importantly, superiority over existing HIV-1 therapy was demonstrated in several Phase 3 studies, including among individuals with very limited treatment options. DTG proved to be an advancement in the field of HIV-1 therapy and quickly became the guideline-recommended INSTI of choice [21, 22, 57].

Acknowledgments

This work was funded by ViiV Healthcare. The authors would like to acknowledge the contributions of Eric Gillis to this work. Editorial assistance was provided under the direction of the authors by Seth Hurley, PhD, and Jennifer Rossi, MA, ELS, MedThink SciCom, and funded by ViiV Healthcare. All authors are employees of ViiV Healthcare and own stock in GSK.

References

1 World Health Organization. HIV and AIDS. 2023. https://www.who.int/news-room/fact-sheets/detail/hiv-aids (accessed 4 December 2023).

2 Métifiot M, Marchand C, Pommier Y. HIV integrase inhibitors: 20-year landmark and challenges. *Adv Pharmacol.* 2013;67:75–105. doi: 10.1016/B978-0-12-405880-4.00003-2.

3 Deeks SG, Lewin SR, Havlir DV. The end of AIDS: HIV infection as a chronic disease. *Lancet.* 2013;382(9903):1525–33. doi: 10.1016/S0140-6736(13)61809-7.

4 Johns BA, Kawasuji T, Weatherhead JG, Taishi T, Temelkoff DP, Yoshida H, et al. Carbamoyl pyridone HIV-1 integrase inhibitors 3. A diastereomeric approach to chiral nonracemic tricyclic ring systems and the discovery of dolutegravir (S/GSK1349572) and (S/GSK1265744). *J Med Chem.* 2013;56(14):5901–16. doi: 10.1021/jm400645w.

5 Kawasuji T, Johns BA, Yoshida H, Taishi T, Taoda Y, Murai H, et al. Carbamoyl pyridone HIV-1 integrase inhibitors. 1. Molecular design and establishment of an advanced two-metal binding pharmacophore. *J Med Chem.* 2012;55(20):8735–44. doi: 10.1021/jm3010459.

6 Isentress [prescribing information]. Rahway, NJ: Merck Sharp & Dohme LLC; 2022.

7 Vitekta [prescribing information]. Foster City, CA: Gilead Sciences, Inc; 2014.

8 Tivicay [prescribing information]. Durham, NC: ViiV Healthcare; 2022.

9 Hightower KE, Wang R, Deanda F, Johns BA, Weaver K, Shen Y, et al. Dolutegravir (S/GSK1349572) exhibits significantly slower dissociation than raltegravir and elvitegravir from wild-type and integrase inhibitor-resistant HIV-1 integrase-DNA complexes. *Antimicrob Agents Chemother.* 2011;55(10):4552–59. doi: 10.1128/AAC.00157-11.

10 Min S, Sloan L, DeJesus E, Hawkins T, McCurdy L, Song I, et al. Antiviral activity, safety, and pharmacokinetics/pharmacodynamics of dolutegravir as 10-day monotherapy in HIV-1-infected adults. *AIDS.* 2011;25(14):1737–45. doi: 10.1097/QAD.0b013e32834a1dd9.

11 Reese MJ, Savina PM, Generaux GT, Tracey H, Humphreys JE, Kanaoka E, et al. In vitro investigations into the roles of drug transporters and metabolizing enzymes in the disposition and drug interactions of dolutegravir, a HIV integrase inhibitor. *Drug Metab Dispos.* 2013;41(2):353–61. doi: 10.1124/dmd.112.048918.

12 Cahn P, Pozniak AL, Mingrone H, Shuldyakov A, Brites C, Andrade-Villanueva JF, et al. Dolutegravir versus raltegravir in antiretroviral-experienced, integrase-inhibitor-naive adults with HIV: week 48 results from the randomised, double-blind, non-inferiority SAILING study. *Lancet.* 2013;382(9893):700–8. doi: 10.1016/S0140-6736(13)61221-0.

13 Min S, Song I, Borland J, Chen S, Lou Y, Fujiwara T, et al. Pharmacokinetics and safety of S/GSK1349572, a next-generation HIV integrase inhibitor, in healthy volunteers. *Antimicrob Agents Chemother.* 2010;54(1):254–58. doi: 10.1128/AAC.00842-09.

14 Stellbrink HJ, Reynes J, Lazzarin A, Voronin E, Pulido F, Felizarta F, et al. Dolutegravir in antiretroviral-naive adults with HIV-1: 96-week results from a randomized dose-ranging study. *AIDS.* 2013;27(11):1771–78. doi: 10.1097/QAD.0b013e3283612419.

15 Walmsley S, Baumgarten A, Berenguer J, Felizarta F, Florence E, Khuong-Josses MA, et al. Brief report: dolutegravir plus abacavir/lamivudine for the treatment

of HIV-1 infection in antiretroviral therapy-naive patients: week 96 and week 144 results from the SINGLE randomized clinical trial. *J Acquir Immune Defic Syndr.* 2015;70(5):515–19. doi: 10.1097/QAI.0000000000000790.

16 Patel DA, Snedecor SJ, Tang WY, Sudharshan L, Lim JW, Cuffe R, et al. 48-week efficacy and safety of dolutegravir relative to commonly used third agents in treatment-naive HIV-1-infected patients: a systematic review and network meta-analysis. *PLoS One.* 2014;9(9):e105653. doi: 10.1371/journal.pone.0105653.

17 Triumeq [prescribing information]. Durham, NC: ViiV Healthcare; 2023.

18 Juluca [prescribing information]. Durham, NC: ViiV Healthcare; 2022.

19 Dovato [prescribing information]. Durham, NC: ViiV Healthcare; 2023.

20 Panel on Antiretroviral Therapy and Medical Management of Children Living with HIV. Guidelines for the use of antiretroviral agents in pediatric HIV infection. 2023. https://clinicalinfo.hiv.gov/en/guidelines/pediatric-arv (accessed 4 December 2023).

21 European AIDS Clinical Society. Guidelines version 12.0. 2023. https://www.eacsociety.org/media/guidelines-12.0.pdf (accessed 4 December 2023).

22 Panel on Antiretroviral Guidelines for Adults and Adolescents. Guidelines for the use of antiretroviral agents in adults and adolescents with HIV. 2023. https://clinicalinfo.hiv.gov/en/guidelines/adult-and-adolescent-arv (accessed 4 December 2023).

23 Medicines Patent Pool. Dolutegravir – Adult (DTG). 2014. https://medicinespatentpool.org/licence-post/dolutegravir-adult-dtg (accessed 4 December 2023).

24 Murai H, Endo T, Kurose N, Taishi T, Yoshida H, inventors; Shionogi and Co Ltd. Heterocyclic compounds having inhibitory activity against HIV integrase. US Patent US 2006/0128669 A1, 2006; filed 11 August 2003 and issued 15 June 2006.

25 Boros EE, Edwards CE, Foster SA, Fuji M, Fujiwara T, Garvey EP, et al. Synthesis and antiviral activity of 7-benzyl-4-hydroxy-1,5-naphthyridin-2(1H)-one HIV integrase inhibitors. *J Med Chem.* 2009;52(9):2754–61. doi: 10.1021/jm801404b.

26 Kobayashi M, Yoshinaga T, Seki T, Wakasa-Morimoto C, Brown KW, Ferris R, et al. In vitro antiretroviral properties of S/GSK1349572, a next-generation HIV integrase inhibitor. *Antimicrob Agents Chemother.* 2011;55(2):813–21. doi: 10.1128/AAC.01209-10.

27 Kawasuji T, Johns BA, Yoshida H, Weatherhead JG, Akiyama T, Taishi T, et al. Carbamoyl pyridone HIV-1 integrase inhibitors. 2. Bi- and tricyclic derivatives result in superior antiviral and pharmacokinetic profiles. *J Med Chem.* 2013;56(3):1124–35. doi: 10.1021/jm301550c.

28 Charpentier C, Larrouy L, Collin G, Damond F, Matheron S, Chêne G, et al. In-vitro phenotypic susceptibility of HIV-2 clinical isolates to the integrase inhibitor S/GSK1349572. *AIDS.* 2010;24(17):2753–55. doi: 10.1097/QAD.0b013e32833f9e36.

29 Castagna A, Maggiolo F, Penco G, Wright D, Mills A, Grossberg R, et al. Dolutegravir in antiretroviral-experienced patients with raltegravir- and/or elvitegravir-resistant HIV-1: 24-week results of the phase III VIKING-3 study. *J Infect Dis.* 2014;210(3):354–62. doi: 10.1093/infdis/jiu051.

30 Eron JJ, Clotet B, Durant J, Katlama C, Kumar P, Lazzarin A, et al. Safety and efficacy of dolutegravir in treatment-experienced subjects with

raltegravir-resistant HIV type 1 infection: 24-week results of the VIKING study. *J Infect Dis.* 2013;207(5):740–48. doi: 10.1093/infdis/jis750.

31 Hightower K, DeAnda F, Wang R, Vavro C, Underwood M. Combinations of primary and secondary integrase mutations in the VIKING pilot study: effects on and rationale for dolutegravir dissociation. Slides presented at International Workshop on HIV and Hepatitis Virus Drug Resistance and Curative Strategies, Sitges; 2012 (5–9 June 2012).

32 Oliveira M, Brenner BG, Wainberg MA. Isolation of drug-resistant mutant HIV variants using tissue culture drug selection. *Methods Mol Biol.* 2009;485:427–33. doi: 10.1007/978-1-59745-170-3_29.

33 Hare S, Gupta SS, Valkov E, Engelman A, Cherepanov P. Retroviral intasome assembly and inhibition of DNA strand transfer. *Nature.* 2010;464(7286):232–36. doi: 10.1038/nature08784.

34 Hare S, Smith SJ, Métifiot M, Jaxa-Chamiec A, Pommier Y, Hughes SH, et al. Structural and functional analyses of the second-generation integrase strand transfer inhibitor dolutegravir (S/GSK1349572). *Mol Pharmacol.* 2011;80(4):565–72. doi: 10.1124/mol.111.073189.

35 Wainberg MA, Han YS. Will drug resistance against dolutegravir in initial therapy ever occur? *Front Pharmacol.* 2015;6:90. doi: 10.3389/fphar.2015.00090.

36 DeAnda F, Hightower KE, Nolte RT, Hattori K, Yoshinaga T, Kawasuji T, et al. Dolutegravir interactions with HIV-1 integrase-DNA: structural rationale for drug resistance and dissociation kinetics. *PLoS One.* 2013;8(10):e77448. doi: 10.1371/journal.pone.0077448.

37 Quashie PK, Mesplède T, Han YS, Veres T, Osman N, Hassounah S, et al. Biochemical analysis of the role of G118R-linked dolutegravir drug resistance substitutions in HIV-1 integrase. *Antimicrob Agents Chemother.* 2013;57(12):6223–35. doi: 10.1128/AAC.01835-13.

38 Quashie PK, Mesplède T, Han YS, Oliveira M, Singhroy DN, Fujiwara T, et al. Characterization of the R263K mutation in HIV-1 integrase that confers low-level resistance to the second-generation integrase strand transfer inhibitor dolutegravir. *J Virol.* 2012;86(5):2696–705. doi: 10.1128/JVI.06591-11.

39 Aboud M, Kaplan R, Lombaard J, Zhang F, Hidalgo JA, Mamedova E, et al. Dolutegravir versus ritonavir-boosted lopinavir both with dual nucleoside reverse transcriptase inhibitor therapy in adults with HIV-1 infection in whom first-line therapy has failed (DAWNING): an open-label, non-inferiority, phase 3b trial. *Lancet Infect Dis.* 2019;19(3):253–64. doi: 10.1016/S1473-3099(19)30036-2.

40 Underwood MR, Dudas K, Horton J, Wang R, Deanda F, Griffith S, et al. Analysis and characterization of treatment-emergent resistance in ART-experienced, integrase inhibitor–naive subjects with dolutegravir (DTG) vs raltegravir (RAL) in SAILING (ING111762). Poster presented at the International Workshop on HIV and Hepatitis Virus Drug Resistance and Curative Strategies, Toronto; 2013 (4–8 June 2013).

41 Underwood M, Horton J, Nangle K, Hopking J, Smith K, Aboud M, et al. Integrase inhibitor resistance mechanisms and structural characteristics in antiretroviral therapy-experienced, integrase inhibitor-naive adults with HIV-1 infection treated with dolutegravir plus two nucleoside reverse transcriptase inhibitors in the

DAWNING study. *Antimicrob Agents Chemother*. 2022;66(1):e0164321. doi: 10.1128/AAC.01643-21.

42 Underwood MR, Horton J, DeAnda F. Raltegravir (RAL)-resistant isolates without primary RAL integrase (IN) mutations are cross-resistant to elvitegravir (EVG) but not dolutegravir (DTG). Poster presented at the International Workshop on Antiviral Drug Resistance, Berlin; 2014 (3–7 June 2014).

43 Passos DO, Li M, Yang R, Rebensburg SV, Ghirlando R, Jeon Y, et al. Cryo-EM structures and atomic model of the HIV-1 strand transfer complex intasome. *Science*. 2017;355(6320):89–92. doi: 10.1126/science.aah5163.

44 Song I, Borland J, Chen S, Patel P, Wajima T, Peppercorn A, et al. Effect of food on the pharmacokinetics of the integrase inhibitor dolutegravir. *Antimicrob Agents Chemother*. 2012;56(3):1627–29. doi: 10.1128/AAC.05739-11.

45 Lovern M, Underwood M, Nichols G, Song I. PK/PD modeling supports dose escalation decision in VIKING. Poster presented at the 10th International Congress on Drug Therapy in HIV Infection, Glasgow; 2010 (7–11 November 2010).

46 Zhang J, Hayes S, Sadler BM, Minto I, Brandt J, Piscitelli S, et al. Population pharmacokinetics of dolutegravir in HIV-infected treatment-naive patients. *Br J Clin Pharmacol*. 2015;80(3):502–14. doi: 10.1111/bcp.12639.

47 US Food and Drug Administration. Human immunodeficiency virus-1 infection: developing antiretroviral drugs for treatment. Guidance for industry. 2015. https://www.fda.gov/files/drugs/published/Human-Immunodeficiency-Virus-1-Infection--Developing-Antiretroviral-Drugs-for-Treatment.pdf (accessed 18 January 2023).

48 Lalezari J, Sloan L, DeJesus E, Hawkins T, McCurdy L, Song I, et al. Potent antiviral activity of S/GSK1349572, a next generation integrase inhibitor (INI), in INI-naive HIV-1-infected patients: ING111521 protocol. Slides presented at the 5th IAS Conference on HIV Pathogenesis, Treatment and Prevention, Cape Town; 2009 (19–22 July 2009).

49 van Lunzen J, Maggiolo F, Arribas JR, Rakhmanova A, Yeni P, Young B, et al. Once daily dolutegravir (S/GSK1349572) in combination therapy in antiretroviral-naive adults with HIV: planned interim 48 week results from SPRING-1, a dose-ranging, randomised, phase 2b trial. *Lancet Infect Dis*. 2012;12(2):111–18. doi: 10.1016/S1473-3099(11)70290-0.

50 Raffi F, Rachlis A, Stellbrink HJ, Hardy WD, Torti C, Orkin C, et al. Once-daily dolutegravir versus raltegravir in antiretroviral-naive adults with HIV-1 infection: 48 week results from the randomised, double-blind, non-inferiority SPRING-2 study. *Lancet*. 2013;381(9868):735–43. doi: 10.1016/S0140-6736(12)61853-4.

51 Walmsley SL, Antela A, Clumeck N, Duiculescu D, Eberhard A, Gutiérrez F, et al. Dolutegravir plus abacavir-lamivudine for the treatment of HIV-1 infection. *N Engl J Med*. 2013;369(19):1807–18. doi: 10.1056/NEJMoa1215541.

52 Molina JM, Clotet B, van Lunzen J, Lazzarin A, Cavassini M, Henry K, et al. Once-daily dolutegravir versus darunavir plus ritonavir for treatment-naive adults with HIV-1 infection (FLAMINGO): 96 week results from a randomised, open-label, phase 3b study. *Lancet HIV*. 2015;2(4):e127–36. doi: 10.1016/S2352-3018(15)00027-2.

53 Clotet B, Feinberg J, van Lunzen J, Khuong-Josses MA, Antinori A, Dumitru I, et al. Once-daily dolutegravir versus darunavir plus ritonavir in antiretroviral-naive adults with HIV-1 infection (FLAMINGO): 48 week results from the randomised open-label phase 3b study. *Lancet.* 2014;383(9936):2222–31. doi: 10.1016/S0140-6736(14)60084-2.

54 Lovern M, Underwood M, Song I, Nichols G. PK/PD modeling supports the dose-escalation decision in VIKING [abstract]. *J Int AIDS Soc.* 2010;13(suppl 4):P182.

55 Wajima T, Kubota R. Pharmacokinetic/Pharmacodynamic modeling and long-term simulation of dolutegravir (DTG, S/GSK1349572) in integrase resistant patients with a simple viral dynamic model. Poster presented at the 20th Population Approach Group in Europe, Athens; 2011 (7–10 June 2011).

56 Vavro CL, Huang J, Avatapally C, Min S, Ait-Khaled M. Durable efficacy and limited integrase resistance evolution in subjects receiving dolutegravir after failing prior integrase inhibitor (INI) regimens: week 48 results from VIKING-3. Poster presented at the 12th European Meeting on HIV and Hepatitis Treatment Strategies and Antiviral Drug Resistance, Barcelona; 2014 (26–28 March 2014).

57 Gandhi RT, Bedimo R, Hoy JF, Landovitz RJ, Smith DM, Eaton EF, et al. Antiretroviral drugs for treatment and prevention of HIV infection in adults: 2022 recommendations of the International Antiviral Society-USA Panel. *JAMA.* 2023;329(1):63–84. doi: 10.1001/jama.2022.22246.

4

The Discovery and Development of Islatravir (4′-Ethynyl-2-fluoro-2′-deoxyadenosine [EFdA], MK-8591)

Alexa A. Snyder[1], Xin Wen[1], Eleftherios Michailidis[1], Karen A. Kirby[1], Hiroaki Mitsuya[2,3], Stefan G. Sarafianos[1]

[1] *Department of Pediatrics, Laboratory of Biochemical Pharmacology, Center for ViroScience and Cure, Emory University School of Medicine and Children's Healthcare of Atlanta, 1760 Haygood Drive NE, Atlanta, GA 30322, USA*
[2] *National Center for Global Health & Medicine Research Institute, 1-21-1 Toyama, Shinjuku, Tokyo 162-8655, Tokyo, Japan*
[3] *Center for Cancer Research, National Cancer Institute, NIH, 9000 Rockville Pike, Bethesda, MD 20852, USA*

4.1 Introduction

Since the start of the human immunodeficiency virus (HIV) epidemic in the early 1980s, there have been roughly 85.6 million individuals infected by the virus [1]. Currently, there are over 39 million people living with HIV, yet there is still no simple cure [2]. Nonetheless, antiretroviral therapy (ART) has been highly successful at reducing viral loads to below detectable levels, thus reducing transmission, and increasing life expectancy close to near-normal levels. Left untreated, HIV infection leads to acquired immunodeficiency syndrome (AIDS) [3, 4]. There are many FDA-approved drug classes to treat HIV infection that belong to the following classes: nucleoside reverse transcriptase inhibitors (NRTIs), non-nucleoside reverse transcriptase inhibitors (NNRTIs), protease inhibitors (PIs), fusion inhibitors, CCR5 antagonists, integrase strand transfer inhibitors (INSTIs), attachment inhibitors, post-attachment inhibitors, and capsid-targeting antivirals [5]. These antivirals are used in ARTs that usually include multiple drugs from at least two different classes targeting multiple steps in the HIV replication cycle [6]. Most HIV treatments are typically given as once daily oral medication regimens. Failure to adhere to these strict dosing regimens can lead to HIV resistance and viral rebound. Therefore, there is a need for novel antivirals that have the potential to be long-acting.

Trends in Antiviral Drug Development, First Edition. Edited by Michael J. Sofia and Zhengqiang Wang.

Islatravir (ISL), also known as EFdA (4′-ethynyl-2-fluoro-2′-deoxyadenosine), is a first-in-class nucleoside reverse transcriptase translocation inhibitor (NRTTI). Here, together with some general background, we provide an overview of the development of this NRTTI, key chemical features, biochemical and structural mechanisms of inhibition, resistance and hypersusceptibility, combination therapies, animal studies, synthetic advances, clinical trials, and potential delivery methods of EFdA for long-acting treatment of HIV. The studies highlighted in this review illustrate both the past and future significance of EFdA in the development of NRTTIs as a class of antivirals for HIV treatment.

4.2 HIV Replication Cycle

HIV infects CD4+T cells that play an essential role in the immune response [7]. Following fusion of the virion to the cell plasma membrane [8, 9], the capsid core is released into the cytoplasm. This capsid core contains two positive single-stranded RNA (+ssRNA) strands and essential viral proteins like reverse transcriptase (RT), integrase, protease, and nucleocapsid. The core exploits the cellular microtubule system and associated motor proteins to transport to the nucleus and subsequently traverses the nuclear pore complex [10, 11]. During capsid transportation, RT converts the viral +ssRNA into double-stranded DNA (dsDNA), initiating reverse transcription in the cytoplasm and completing it in the nucleus [12–14]. The capsid core is thought to uncoat in the nucleus, releasing the newly synthesized dsDNA and associated proteins [15]. The viral DNA is then integrated preferentially into the euchromatin of the host genome by HIV integrase [16]. This integrated proviral DNA is used for viral RNA transcription, hijacking the host transcriptional machinery. Viral RNAs are transported to the cytoplasm and translated into the Gag polyprotein precursor ($Pr55^{Gag}$) containing the matrix (MA), capsid (CA), nucleocapsid (NC), and P6 structural proteins and the GagPol polyprotein precursor ($Pr160^{Gag\text{-}Pol}$) which also contains protease, RT, and integrase. Other viral proteins include the viral envelope glycoproteins and the viral accessory and regulatory proteins (Vif, Vpr, Vpu, and Nef) [17–19]. These proteins and the two copies of the full-length +ssRNA viral genome are then trafficked to the plasma membrane, packaged, and assembled, while encapsidating the viral RNA. Budding then occurs with the aid of host factors such as the endosomal sorting complex required for transport (ESCRT) machinery [20]. Once the virion is released, maturation occurs by the viral protease cleaving the gag precursors into mature proteins, leading to a morphological rearrangement of the virion [19]. To stop HIV infection, current ARTs target multiple steps of this replication cycle.

4.3 Structure and Function of HIV RT

HIV reverse transcription is an essential step of the HIV replication cycle; therefore, RT is highly targeted for antiretroviral treatment. RT is an asymmetric heterodimer composed of a p66 and p51 subunit (Figure 4.1). The p66 subunit is the catalytically active subunit, having both DNA polymerase and RNase H activity. This subunit is

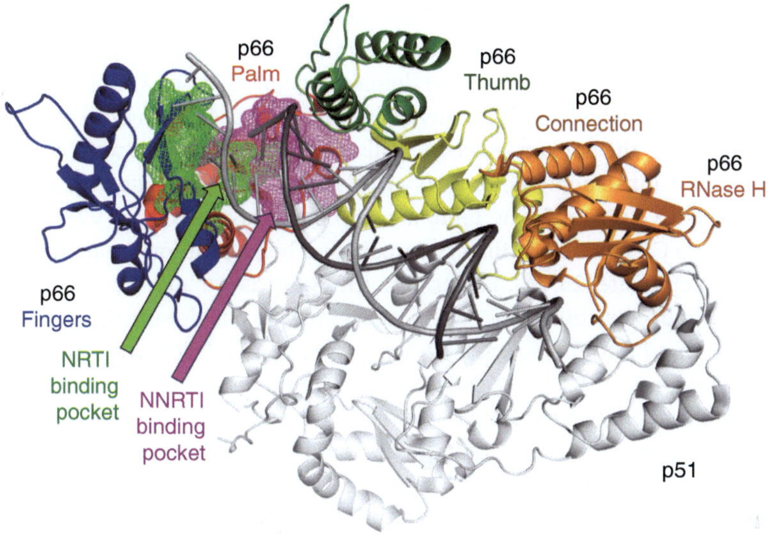

Figure 4.1 Structure of HIV RT in complex with dsDNA. HIV-1 RT in complex with double-strand DNA (DNA shown in gray). The p51 is shown as light gray and the p66 subunit is shown as multicolor: fingers (blue), palm (red), thumb (dark green), connection (yellow) subdomains, and RNase H domain (orange). The nucleoside RT inhibitor (NRTI) binding pocket at the polymerase active site is shown in bright green and non-nucleoside RT inhibitor binding pocket (NNIBP) is shown in pink. *Source*: Adapted from [29].

often described as a right hand, comprising five subdomains: the fingers (residues 1–84, 119–154), the palm (residues 85–118, 155–241), the thumb (residues 242–313), the linker/connection (residues 314–426), and the RNase H domain (residues 427–560). The sequence of the p51 subunit is derived from the p66 subunit, but it lacks the RNase H domain. The p51 subunit has no enzymatic function and folds differently than the p66 counterpart [3, 21–23]. The nucleic acid binding cleft of RT is formed by the fingers, palm, thumb, connection, and RNase H subdomains of the p66 subunit, and connection and thumb subdomains of the p51 subunit. This cleft can hold ~17 base pairs of DNA/DNA substrate and ~18 base pairs of RNA/DNA substrate [24]. When RT is bound to a DNA/DNA substrate, the DNA is in A-form at the polymerase active site and the B-form past the polymerase active site [25]. Alternatively, when RT is bound to RNA/DNA substrate, this substrate takes on an H-form conformation [26]. RT alters its conformation when a nucleic acid substrate is bound or unbound in the fingers, thumb, and palm subdomains [21, 26–28].

HIV RT has several important structural elements that contribute to its function. The polymerase active site of RT is composed of three catalytic aspartic acids (D110, D185, and D186) that bind two divalent cations (Mg^{2+} and/or Mn^{2+}) [22, 30] and are essential for dNTP incorporation and DNA synthesis. The two metals promote the nucleophilic attack of the primer's 3′-hydroxy on the incoming deoxynucleotide triphosphates (dNTPs), stabilizing the transition state and facilitating the release of the β- and γ-phosphate groups. Once the deoxynucleotide monophosphate (dNMP) is incorporated into the primer end, positioned at the N-site or pretranslocation site, RT can translocate. Upon translocation, RT moves the dNMP into the P-site or post-translocation site, and freeing the N-site for binding of the next incoming dNTP. The RNase H domain is responsible for cleaving the RNA genome in the RNA/DNA substrate nonspecifically. The catalytic core of the RNase H domain is composed of a DEDD motif (D443, E478, D498, D549), which binds two divalent cations; Mg^{2+} being most physiologically relevant, although the domain is known to also use Mn^{2+} or Co^{2+} as well [31].

Other important structural elements of RT include the template grip, the primer grip, and the YXDD motif. The template grip consists of residues V75, R78, N81, E89, P157, and G93 in the p66 fingers domain and helps position the template at the polymerase active site [25, 26]. The primer grip comprises a highly conserved region of RT, the β12–β13 hairpins. This region is essential for positioning the 3′-hydroxy of the primer strand into the polymerase active site. The YXDD motif is present in all retroviral RTs. In HIV RT, the X is M184. This motif, elucidated by structural studies, is thought to change conformation between pre- and post-translocation states, acting as a springboard to move the primer terminus forward after dNTP incorporation [32, 33]. Other key residues include R72 [34] and K65 [28] which bind β- and γ-phosphates of the incoming dNTP, essential for proper dNTP binding and incorporation. Y115 also plays a key role in differentiating between deoxy- and ribonucleotide substrates. Due to this function, it is often termed the "steric gate" [35–37].

4.4 DNA Synthesis by RT

Utilizing these key features described above, HIV RT can convert the +ssRNA genome into DNA through a multistep process. The DNA polymerase domain synthesizes DNA using either a DNA or RNA template, and the RNase H domain cleaves RNA that is part of an RNA/DNA complex [38]. Reverse transcription starts with the host tRNALys3 binding to the primer-binding site (PBS) of the viral genome and synthesizing through the U5 site, and ending at the R region of the viral RNA 5′-end, forming the minus-strand strong stop product. Meanwhile, RNase H digests the RNA portion of the RNA/DNA complex, exposing the R region on the 3′-terminus of minus-strand cDNA, which hybridizes with the R region at the 3′-end of the +ssRNA genome (intra- or intermolecularly), forming the minus-strand transfer product. Minus-strand cDNA synthesis continues using the +RNA genome as the template. When RT passes the polypurine tract (PPT), the RNase H domain cleaves its borders and generates a unique plus-strand RNA primer. Plus-strand DNA synthesis proceeds using the PPT primer and the minus-strand cDNA template back to the U5 region and stops after copying the PBS sequence on the tRNA, generating a plus-strand DNA of the PBS. The tRNA is then removed by the RNase H domain and the complementary PBS sequences at both ends hybridize, forming the plus-strand transfer product. Plus-strand cDNA synthesis continues toward the 3′-end. The last steps of HIV reverse transcription involve strand displacement synthesis to the PBS and PPT ends, which, along with host repair and ligation mechanisms, are essential to open the circular intermediate to a linear double-strand viral DNA with a long terminal repeat (LTR) at each end (Figure 4.2). This dsDNA is then ready for host-genome integration. Any unintegrated products are ligated into 1-LTR or 2-LTR circles by host factors [22].

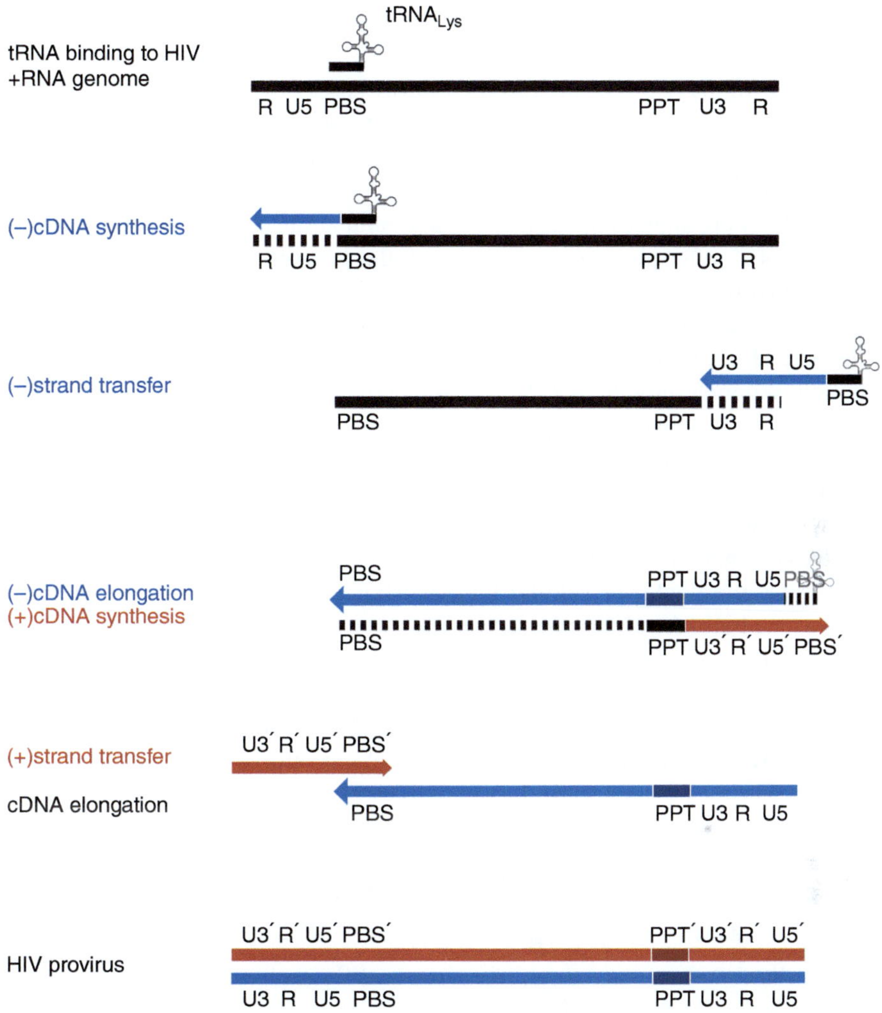

Figure 4.2 HIV-1 reverse transcription process. The process of HIV-1 reverse transcription begins with the binding of host tRNALys to the PBS located near the 3′-end of the viral genome (two copies per virion). Synthesis of (−)cDNA initiates toward the 5′-end concurrently with the degradation of the viral RNA genome by RNase H, which exposes the R sequence on the (−)cDNA that complements the R sequence on the 5′-end of the RNA genome, facilitating the first "jump." (−)cDNA synthesis proceeds toward the 5′-end, with RNase H degrading the RNA genome. When RT passes through the poly-purine tract (PPT), RNase H cleaves only its borders, leaving the intact PPT to serve as a primer for (+) cDNA synthesis. (+)cDNA synthesis progresses and RNase H degrades the tRNALys, exposing the PBS on the (+)cDNA, which complements the PBS on the (−)cDNA, thus facilitating the second "jump." Both strands of cDNA elongation proceed with RNase H ultimately degrading the entire RNA genome, resulting in the formation of the double-stranded viral DNA (HIV provirus).

Black: viral RNA genome.
Blue: minus-strand cDNA.
Red: plus-strand cDNA.

Source: Created with BioRender.

4.5 RT Inhibitor Classes

There are currently two FDA-approved HIV drug classes that target HIV RT: NRTIs and NNRTIs. Another class of antivirals is being developed with a novel mechanism of action called NRTTIs. EFdA is the first NRTTI to make it to Phase 3 clinical trials for the treatment of HIV.

NRTIs: They are analogs of nucleosides and bind at the polymerase active site of RT (Figure 4.1), after phosphorylation by cellular kinases. All current HIV NRTIs lack a 3′-hydroxy, thus, when they are incorporated into the elongating nucleic acid substrate, another dNTP cannot be added. This mechanism of action is termed immediate chain termination (ICT) or is often referred to as obligate chain termination. When developing NRTIs, efficient activation from the nucleoside to the nucleotide form (active triphosphate [TP] form) [39–41], the stability and metabolic profile of the NRTI, and a high barrier to resistance are important considerations. For most NRTIs, the rate-limiting step is the addition of the α-phosphate. Tenofovir (TFV) (Figure 4.3), belongs to the class of acyclic nucleoside phosphonates (ANPs), which contain a phosphonate, and thus only needs the addition of the β- and γ-phosphates by cellular kinases. High stability of NRTIs is needed to ensure a long half-life, which correlates with lower and less frequent dosing. Lastly, NRTIs that have a high barrier to resistance are necessary to combat the extensive prevalence of drug-resistant viruses. Resistance mutations have occurred at amino acid positions 41, 65, 67, 68, 69, 70, 74, 115, 151, 184, 210, and 215 of RT with NRTI treatment [42]. NRTIs are critical components of current ART treatments, yet due

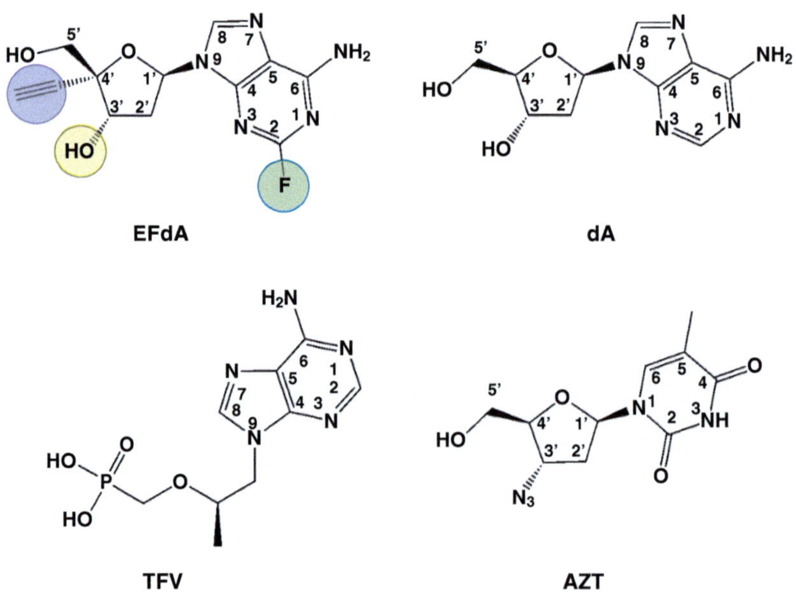

Figure 4.3 Chemical structures of EFdA, dA, TFV, and AZT. The 4′-ethynyl, 2-fluoro, and 3′-hydroxy groups are shown in blue, green, and yellow, respectively.

to drug resistance, new nucleoside analogs with novel mechanisms of action are needed [3, 21, 22, 43].

NNRTIs: They are allosteric inhibitors of HIV RT, binding at a distinct hydrophobic pocket located ~10 Å away from the polymerase active site at the base of the thumb (residues L100, K101, K103, V106, T107, V108, V179, Y181, Y188, V189, G190, F227, W229, L234, and Y318 of p66 and E138 of p51) (Figure 4.1) [21]. This pocket only exists in the presence of NNRTIs. When an NNRTI is present, RT forms an entrance, tunnel, and groove for binding. Binding in this pocket causes conformational changes in residues Y181 and Y188, induces an extended primer grip [44], overextends the thumb restricting its mobility [45], and reduces the flexibility of the YMDD motif [28]. Together, these changes inhibit the ability of RT to synthesize DNA. A major consideration in the development of current NNRTIs is ensuring a high barrier to resistance. The NNRTI binding pocket is not essential for DNA synthesis, therefore, NNRTI resistance mutations emerge rather easily. Recent NNRTIs have higher flexibility, allowing "wiggling and jiggling" in the pocket, reducing the frequency and impact of resistance mutations [3, 21, 22, 43, 46].

EFdA, a Novel NRTTI: EFdA, an analog of 2′-deoxyadenosine (dA) (Figure 4.3), is classified as an NRTTI owing to its impact on RT translocation. It possesses several critical functional groups – 4′-ethynyl, 2-fluoro, and 3′-hydroxy – that influence its mechanism of action and favorable pharmaceutical profile (Figure 4.3). EFdA is a highly potent nucleoside analog, with low toxicity and high stability, making it a strong candidate for long-acting antiviral treatment.

4.6 EFdA Development

Nucleoside and nucleotide analogs have been used to treat viral infections for several decades. The chemical makeup of canonical nucleosides and nucleotides includes a ribose sugar and an aromatic heterocyclic base (adenine, guanine, cytosine, thymine, and uracil). In addition to the ribose ring and base, nucleotides have at least one phosphate group and are the building blocks for all genomic material (RNA/DNA). Over the years, several modifications to the ribose ring, bases, glycosidic bond, and phosphate groups have been explored. Scientists have taken advantage of these modifications to target various viral enzymes. Modifications to the scaffold of these nucleosides/nucleotides can have a large impact on bond angles and lengths, chirality of atoms, charge, polarity, potential bonding interactions, and can affect their structural conformation, including their sugar puckering [47–53]. Most current nucleoside/nucleotide analogs benefit from several changes to their scaffold that improve the therapeutic properties of these inhibitors. One such inhibitor that exploits several of these modifications is EFdA. EFdA is an adenosine analog that exploits a 4′-ethynyl modification on the ribose ring, a fluoro substitution at the 2-position on the adenine base, and a 3′-hydroxy group resembling the canonical nucleoside structure. Although many substitutions to the nucleoside scaffold have been explored [54–61], here we focus on the effects of modifications at the 4′-, 2-, and 3′-positions, which are important for the development of EFdA.

4.6.1 4′-Ethynyl Ribose Modifications

Modifying the ribose ring of nucleosides has long been a strategy for making inhibitors that target various viruses. These modifications include functional group additions, atom substitutions, carbocyclic ring variations, size modifications, and atom removal, leading to acyclic nucleosides [47, 62]. The 4′-position of the ribose ring was one of the late positions to be explored. The first 4′-modified ribose ring was discovered in the late 1950s. Nucleocidin was isolated from *Streptomyces calvus* in 1957 [63], but its chemical structure was not elucidated until later, revealing a 4′-fluoro modification [64–67]. After the discovery of nucleocidin, other ribonucleosides with a 4′-fluoro were synthesized [64, 68]. Some of these were explored as potential *S*-adenosyl-L-homocysteine hydrolase (SAHase) inhibitors [69] or anticancer agents [70]. Several 4′-substituted-2′-deoxynucleosides (4′-SdNs) have been explored since, including: 4′-fluoro, 4′-methoxy, 4′-azido, 4′-cyano, 4′-methyl, or 4′-ethynyl for antiviral activity [71, 72]. Methyl substitutions were synthesized at the 4′-position [73, 74], inducing a preferential south (C2′-endo/C3′-exo) conformation of the ribose ring [74]. Several ribonucleosides substituted with 4′-methoxy [68, 75, 76] and azido groups [77–79] were also synthesized. Due to the antiviral success of zidovudine (AZT) (Figure 4.3), which contains a 3′-azido group, 4′-azido substitutions were tested in combination with 4′-methoxy modifications. These compounds exhibited potent anti-HIV activity, even against AZT-resistant strains [80–82]. Interestingly, the 4′-azido modification induced a north (C2′-exo/C3′-endo) conformation of the ribose ring [83]. It was later revealed that RT prefers substrates in the north conformation [84, 85]. 4′-Cyano [86–88] and 4′-methyl [87, 89, 90] modifications were also tested on ribo- and deoxyribonucleotide scaffolds. Almost all these synthesized compounds showed anti-HIV activity, although many of them had cytotoxicity. 4′-C-methylnucleosides, 4′-methylBVDU 9 and 4′-methylBVaraU 10 compounds also inhibited varicella-zoster virus (VZV) activity *in vitro* [91].

4′-ethynyl substitutions began to be explored in the late 1990s. Several 4′-ethynyl substituted deoxynucleosides were synthesized and had moderate HIV [92], herpes simplex virus (HSV)-1, and HSV-2 inhibition [93, 94], as well as low cytotoxicity [92, 94]. Other 4′-ethynyl compounds, 4′-C-ethynyl-β-D-ribo-pentofuranosylpyrimidines [95] and the L-enantiomer of 4′-C-ethynyl-2′-deoxycytidine [96], did not have antiviral activity against the viruses tested. More 4′-ethynyl-purine nucleosides were synthesized with 4′-ethynyl deoxyadenosine exhibiting the highest anti-HIV activity with low toxicity [91]. Kodama et al. expanded these 4′-ethynyl modifications to deoxycytidine, deoxyribofuranosyl-2,6-diaminopurine, deoxyguanosine, and lastly deoxyadenosine (the base of EFdA). These molecules showed potent EC_{50} values, ranging from 0.0003 to 0.01 μM, against HIV-1, HIV-2, and clinical isolates [97]. Kohgo et al. also showed that 4′-*C*-ethynyl-2′-deoxy purine nucleosides exhibited a higher potency against HIV activity compared to methyl, ethyl, and cyano modifications [98, 99].

Later, methyl, vinyl, and ethynyl groups were added at the 4′-position of 2′-, 3′-didehydro-3′-deoxythymidine (D4T, Stavudine) as well. D4T is an FDA-approved HIV antiviral; it is a thymidine analog that has a didehydro ring [100, 101].

The 4′-ethynyl substitution had the greatest potency against HIV and lowest toxicity in cells [102]. 4′-Ethynyl D4T was phosphorylated 4-fold more efficiently by cellular kinases than D4T. This compound was shown to act favorably with other HIV antivirals [103]. 2′-,3′-Didehydro-3′-deoxy-4′-ethynyl thymidine (4′-Ed4T) was then synthesized and showed even higher potency than 4′-D4T analogs. This compound had lower toxicity against mitochondrial DNA synthesis and acted synergistically with 3TC and FTC [104]. Ultimately, Kohgo et al. synthesized 2′-deoxy-4′-C-ethynyl-2-fluoro-adenosine (4′Ed2FA or EFdA). This compound showed highly potent anti-HIV activity (but moderately potent anti-HBV activity), low cytotoxicity, and was stable in blood plasma [105, 106]. Another 4′-ethynyl adenosine monophosphate analog, EdAP, exhibited antiviral activity against influenza A virus and hepatitis B virus [107, 108]. Altogether, 4′-substitutions have been explored since the 1950s, and to date, 4′-ethynyl modifications have shown the highest antiviral activity against HIV and several other viruses.

4.6.2 2-Fluoro Adenosine Modifications

Several modifications have been introduced to nucleoside bases to test and improve the antiviral potential of nucleoside analogs. These include, but are not limited to, changes in the position of the nitrogens, the size of nucleobases, and addition of functional groups to various positions of the bases [47, 48]. Other changes include modifications in the glycosidic linkage between the base and the ribose ring. There were also attempts to increase or decrease the size of the base through incorporation of additional rings, long hydrocarbon chains, or a perylene group. Commonly, several positions of purines and pyrimidines have been modified. The most frequent positions modified include the 5-position of pyrimidines and the 6-position of purines [47, 109]. Here, we discuss modifications that lead up to the 2-fluoro modification of adenosine on EFdA, including -NH_3, -OH, -SMe, -Cl, and -F. One of the first tested modifications was the addition of a 2-amino to a guanosine base [110]. Abacavir (ABC), an FDA-approved antiviral since 1998, has a 2-NH_2 substitution on the purine base of a carbocyclic nucleoside [111]. Hydroxy modifications at the 2-position of the adenine base in nucleotides with a phosphonate were reported in two patents, but the antiviral activity was not published [112, 113]. SMe modifications at the 2-position of adenosine have also been tested. Although these nucleotide di- and triphosphate analogs were not tested for antiviral activity, these compounds were fully resistant to alkaline phosphatase hydrolysis [114]. Lastly, halogen substitutions to the 2-position of adenine derivatives have been used since the 1990s. 2-Chloro groups were first tested on adenosine monophosphates. These molecules showed anticancer activity, but no viral inhibition [115–118]. 2-Fluoro modifications to a dideoxyadenosine derivatives were shown to inhibit HIV-1 [119]. Fluoro, chloro-, and hydroxy- modifications at this position were also tested on adenine N-[2-(phosphonomethoxy)ethyl] derivatives. These compounds showed inhibition toward both RNA and DNA viruses [120]. Modifications at the 2-position of adenine have been studied extensively and displayed various activities, including reduced degradation, anticancer effects, and antiviral activity, halogenated substitutions being the most potent.

4.6.3 3′-Hydroxy Ribose Modification

The 3′-position of the ribose ring has been extensively modified. Endogenous dNTPs include a 3′-hydroxy, which is essential for phosphorylation and subsequent nucleotide additions by cellular or other DNA polymerases. Current FDA-approved NRTIs for the treatment of HIV lack a 3′-hydroxy. Without a 3′-hydroxy, NRTIs block nucleic acid elongation through obligate (or immediate) chain termination. This mechanism of action has been used effectively to inhibit HIV-1 RT for several decades. Some of the first 3′-ribose modifications were dideoxycytidine (ddC) and dideoxyinosine (ddI) [60, 121, 122]. These nucleoside analogs lack a 3′-hydroxy. Other substitutions to the 3′-position have also been explored, AZT being the most famous example. This antiviral was the first NRTI approved to treat HIV and has a 3′-azido group. Other notorious examples of NRTIs lacking a 3′-hydroxy include lamivudine (3TC) and emtricitabine ((-)FTC) which are categorized as oxathiolanes due to the sulfur at the 3′-position [123, 124]. A 3′-carbon to oxygen substitution was also explored, revealing (+)-(2′R,4′R)-dioxolane cytidine and the corresponding (−)-(2′R,4′R)-dioxolane guanosine, which also had potent antiviral activity [125–127].

Furthermore, alterations of the ribose ring to an acyclic variation or a carbocyclic ring have been investigated. One of the first acyclic nucleoside analogs discovered was acyclovir which inhibits HSV types 1 and 2, VZV, Epstein–Barr virus, and, to a lesser degree cytomegalovirus [128–130] Researchers hypothesized that adding a phosphate group to these acyclic analogs would circumvent the rate limiting step of nucleoside phosphorylation to a dNMP. However, problems arose with membrane permeability and an increase in susceptibility to phosphatases. This led to the development of phosphonate antivirals, including tenofovir (TFV). This inhibitor is an acyclic nucleotide that contains a phosphonate and is one of the most widely used antivirals against HIV and HBV (Figure 4.3) [131]. TFV was originally formulated as tenofovir disoproxil fumarate (TDF), but now more commonly formulated as tenofovir alafenamide (TAF), which increased oral bioavailability by almost 10-fold [132]. Another example of ribose ring substitutions is the use of carbocyclic analogs. Here, we focus on 5 member rings with modifications to the "equivalent of the 3′-position." Carbocylic analogs were first shown to be some of the most potent inhibitors against S-adenosyl-L-homocysteine hydrolase (SAHase), but interestingly also showed potent inhibition of viral polymerases [48]. (-)-carbovir (carbocyclic 2′-,3′-didehydro-2′-3′-dideoxyguanosine) was one of the carbocyclic nucleosides that showed an antiviral effect against HIV [133, 134]. However, it had low oral availability [135]. This limitation led to the development of ABC. This inhibitor has an aminocyclopropyl group at the 6-position of the base that is intracellularly metabolized to the active carbovir-TP. ABC was shown to have better bioavailability than carbovir [136–139]. An interesting characteristic of both inhibitors is their utilization of an unsaturated 5-member ring, which increases rigidity and decreases the fragility of the glycosidic bond, thus making them potent inhibitors.

While most chain-terminating antivirals lack a 3′-hydroxy, some of the most potent nucleoside analog inhibitors retain this group. Some examples of such inhibitors include entecavir and sofosbuvir which target hepatitis B and C viruses, respectively [140–145]. Both antivirals act as nonobligate chain terminators, whereby, once incorporated in the elongating primer they inhibit further nucleic acid synthesis [146]. Similarly, remdesivir, most commonly known to inhibit SARS-CoV-2, blocks the viral RNA-dependent RNA polymerase [147]. Other antivirals containing the 3′-hydroxy group act through delayed chain termination, nonobligate chain termination, and mutagenic functionality [62]. Although obligate chain terminators have shown remarkable success at inhibiting several viral polymerases, there has been increasing emphasis on developing antivirals that contain a 3′-hydroxy group due to their ability to have multiple mechanisms of action and increased phosphorylation efficiency by cellular kinases compared to 3′-hydroxy-deficient antivirals [62].

4.7 EFdA: A Compound with 4′-Ethynyl, 2-Fluoro, and 3′-Hydroxy Modifications

All the previous studies led to the development of EFdA, which combines 4′-, 2-, and 3′-modifications to adenosine [88, 105, 106], attributing to its high binding affinity, a high barrier to resistance, efficient phosphorylation by cellular kinases into its active triphosphate form, low toxicity, and an extended half-life of triphosphorylated EFdA in cells [141]. Kawamoto et al. systematically showed how the presence of 4′-ethynyl, 2-fluoro, and 3′-hydroxy groups improved the potency and drug profile of these adenosine analogs (Table 4.1). EFdA was highly potent (EC_{50} ~0.07 nM) against HIV-1 wild-type (WT) virus and also effectively inhibited several drug-resistant strains; it exhibited low cytotoxicity (therapeutic index ~134 000) [88], consistent with its inefficient recognition by mitochondrial DNA polymerase γ [148, 149]. Of note, 2′-deoxy-4′-C-ethynyl-2-chloroadenosine (ECldA), also had an excellent therapeutic index of ~339 000. EFdA has been reported to inhibit hepatitis B virus, albeit not as potently (EC_{50} ~160 nM) compared to HIV [141]. The compound is efficiently phosphorylated by cellular 2′-deoxycytidine kinase (dCK) [88], leading to a triphosphorylated active form with a half-life of approximately 17 h [149]. Furthermore, EFdA demonstrates high stability in blood plasma [149]. The favorable antiviral properties of EFdA are a result of specific modifications to the adenosine base and ribose ring, and are described below.

4.7.1 Effects of the 4′-Ethynyl Modification on EFdA

The 4′-ethynyl of EFdA alters the conformation of the ribose ring, increases incorporation efficiency, and decreases deamination by adenosine deaminase (ADA). Deoxynucleosides, in solution, are in a dynamic equilibrium between the north (C2′-exo/C3′-endo) and south (C2′-endo/C3′-exo) conformations, with natural nucleosides favoring the south conformation. In contrast, Kirby et al. showed that the 4′-ethynyl of EFdA causes its ribose ring to favor the north conformation (C2′-exo/C3′-endo) (Figure 4.4). At the HIV RT active site, the ribose ring must be in the north conformation for proper positioning of the 3′-hydroxy group for nucleophilic attack to the next incoming dNTP. Using molecular modeling, it was also observed that if EFdA-monophosphate (MP) was in the south conformation once incorporated, the 3′-hydroxy group would sterically clash with Y115 of RT [150]. The ribose rings of

Table 4.1 Adenosine analogs: potency and selectivity.

Compound abbreviation	Chemical structure	EC_{50} (μM)		Selectivity	
		HIV-1_{IIIB}	HIV-2_{EHO}	CC_{50} (μM)	Index (CC_{50}/EC_{50} HIV-1_{IIIB})
2′-Deoxy-4′-*C*-ethynyl-adenosine (EdA)	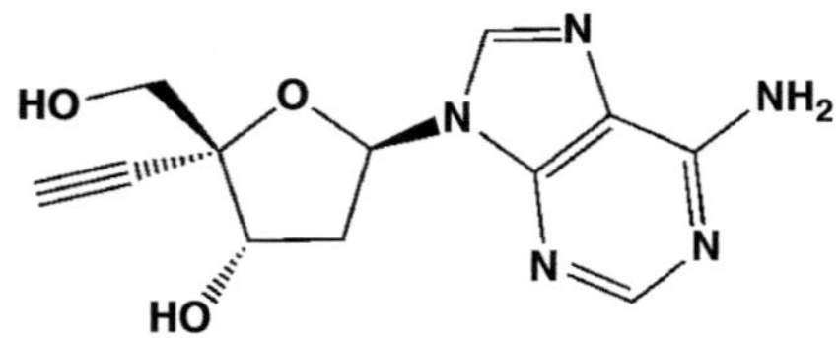	0.0095 ± 0.0027	0.006 ± 0.0015	104 ± 6.2	11 000
2′-Deoxy-4′-*C*-ethynyl-2-fluoro-adenosine (EFdA)	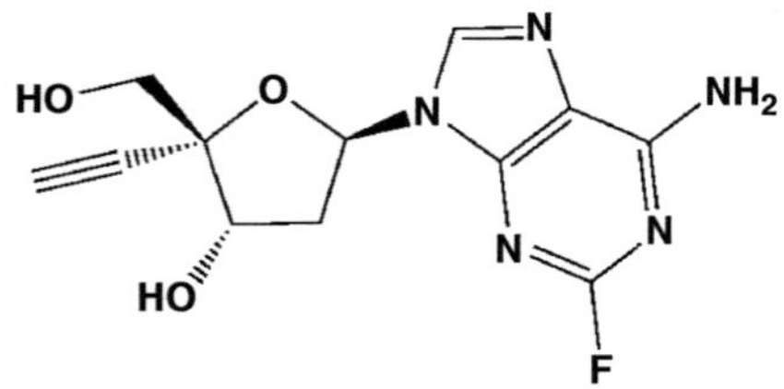	0.00007 ± 0.000017	0.00009 ± 0.000022	9.8 ± 3.4	134 000
2′,3′-Dideoxy-4′-*C*-ethynyl-2-fluoro-adenosine (EFddA)	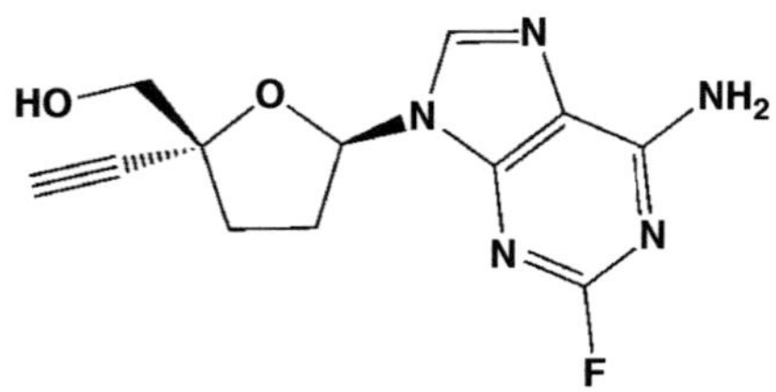	1.17 ± 0.29	1.07 ± 0.23	230 ± 33	196

2′,3′-Didehydro-3′-deoxy-4′-*C*-ethynyl-2-fluoro-adenosine (EFd4A)	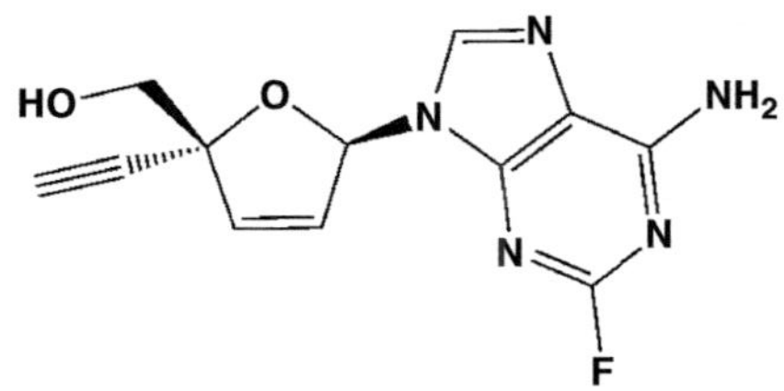	0.11 ± 0.033	0.089 ± 0.0007	98 ± 26	899
2′-Deoxy-4′-*C*-cyano-2-fluoro-adenosine (CNFdA)	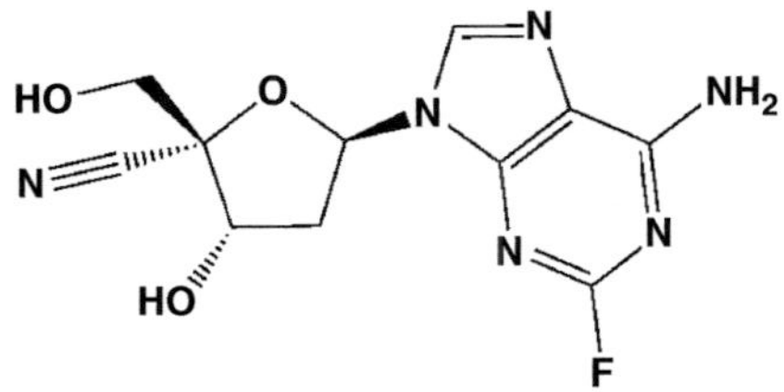	0.1 ± 0.034	0.09 ± 0.0087	>340	>3 300
2′-Deoxy-4′-*C*-ethynyl-2-chloro-adenosine (ECldA)	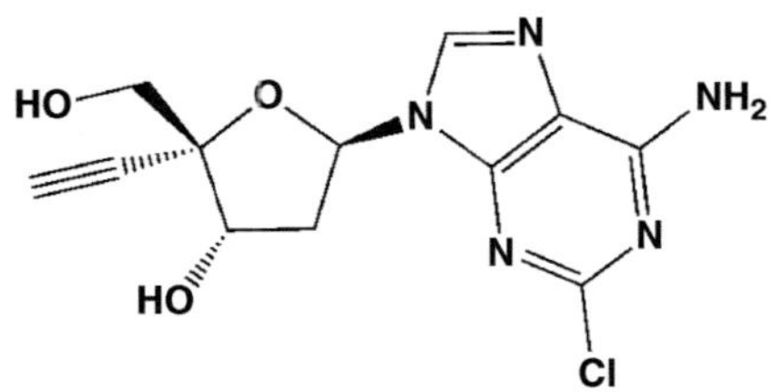	0.00069 ± 0.00018	0.0006 ± 0.0000028	230 ± 16	339 000
2′,3′-Dideoxyinosine (ddI)	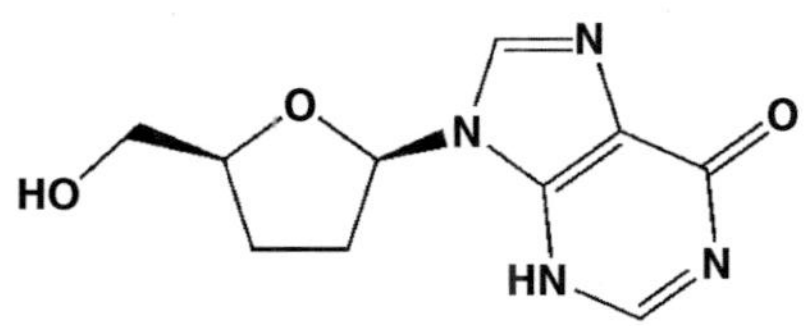	27 ± 12	24 ± 4.4	>100	>4

Source: Adapted from [88].

4′-cyano analogs were also shown to favor the north conformation, while 4′-methyl and 4′-ethyl substituents favored the north conformation less often [151]. The preference of EFdA to be in the north conformation can partially be explained by favorable molecular orbital interactions. The lone pairs of the O4 and O3′ interact with the antibonding orbitals of the 4′-ethynyl group, leading to increased stabilization in the north conformation. Moreover, analogs containing 4′-substitutions are more efficiently incorporated into a template/primer substrate by RT than the endogenous substrate, dA, with 4′-ethynyl analogs having the highest incorporation efficiency [151]. The 4′-ethynyl also decreases the ability of ADA to degrade EFdA, due to a lower K_m. This is likely due to steric clashes between the ethynyl group of EFdA and the sulfur and ε-carbon of M155. Overall, the 4′-ethynyl group of EFdA contributes strongly to the sugar ring puckering, which impacts the interactions EFdA makes with RT, ADA, and potentially dCK. Specifically, these studies showed that this modification increased binding affinity to RT and decreased susceptibility to ADA.

4.7.2 Effects of the 2′-Fluoro Modification on EFdA

The addition of halogen groups to lead compounds has long been a technique used by medicinal chemists to increase the half-life of pharmaceuticals [152–154]. The 2-fluoro group of EFdA is a prime example of this. Fluoride molecules act as a nearly isosteric replacement for hydrogen, although they will often have a profound effect on the metabolic properties of the molecule to which they are attached due to their high electronegativity. The natural substrate, dA, is metabolized by the cellular enzyme ADA, which converts adenosine to inosine [155–158]. The presence of 2-fluoro group significantly inhibits this metabolic step [88, 151]. Addition of the fluoro group changes the pK_a of the N1 atom of adenosine, thus pulling electrons away from the C2 atom and increasing its positive charge [159]. This change in charge distribution increases the energy barrier of the N1 protonation, which is the first step of the deamination cycle. Thus, this fluoro group does not act sterically, as the 4′-ethynyl does at the ADA active site, but rather affects the charge on

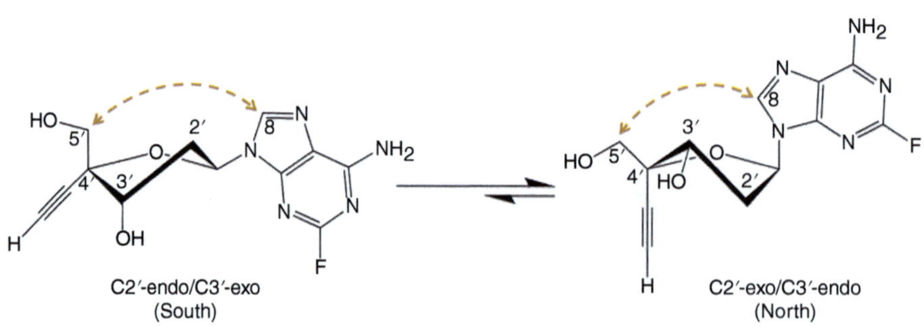

Figure 4.4 Equilibrium between the north (left) and south (right) conformations of the ribose sugar ring in EFdA. The 4′-ethynyl causes the north conformation to be favored more often (75% vs. 25%). These different conformations affect the position of the 3′-hydroxy, as well as the distance between C5′ of the ribose and C8 of the adenine base (orange dotted arrows), thus affecting the major and minor groove spacing of EFdA-containing double-stranded DNA (dsDNA) sequences. *Source*: Modified from Kirby et al. [151] / American Society for Microbiology.

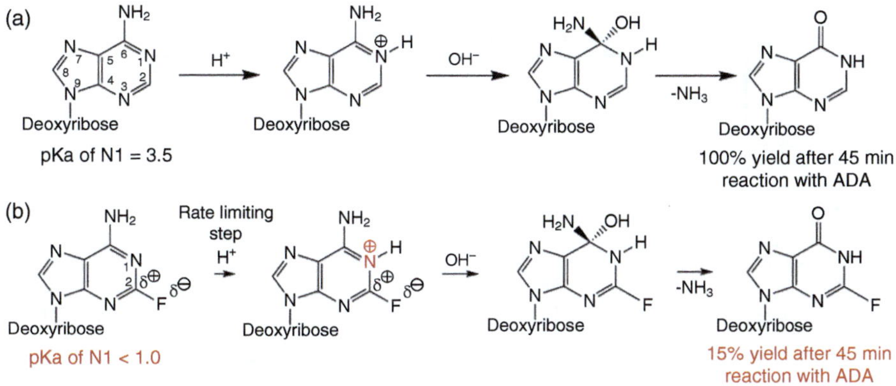

Figure 4.5 Reaction mechanism of the hydrolytic deamination of dA (a) and 2-FdA (b) by ADA. The electronegative 2-fluoro group causes a significant decrease in pK_a at the N1 position, resulting in a high-energy barrier for the N1 protonation (first) step of the reaction (red N^+ in (b)). *Source*: Modified from Kirby et al. [151] / American Society for Microbiology.

the base to significantly slow the deamination process (Figure 4.5). Also, the effects of 2-fluoro on the ribose conformation were negligible, suggesting that it primarily impacts the metabolic stability of EFdA [151].

4.7.3 Effects of the 3′-Hydroxy Modification on EFdA

Kawamoto et al. showed that dCK is the cellular kinase primarily responsible for the conversion of EFdA into its active form, EFdA-TP. In these studies, they showed that knocking down dCK resulted in a reduction of EFdA activity, while reintroduction of dCK restored EFdA activity in cell-based assays. Of note, there remained some moderate activity of EFdA in the dCK knockdown cells, suggesting EFdA may be phosphorylated by other kinases, but to a lesser degree [88]. The presence of a 3′-hydroxy group plays a crucial role in the phosphorylation of nucleosides by cellular kinases. To better understand this effect, structural studies were conducted to characterize the interactions between natural nucleosides and dCK. These studies revealed that the 3′-hydroxy makes hydrogen bond interactions with residues G197 and Y86 at the dCK active site [160]. Although the structure of EFdA in complex with dCK has not been solved, it is expected that it would make similar interactions as the natural nucleoside. Of note, the 3′-hydroxy group also contributes to the multiple mechanisms of RT inhibition by EFdA (discussed below).

4.7.3.1 Mechanisms of Inhibition: EFdA

4.7.3.1.1 Biochemical Mechanisms of Inhibition

EFdA differs from other HIV NRTIs because of its 3′-hydroxy group. This crucial distinction enables EFdA to operate through three mechanisms of inhibition: (i) immediate chain termination, (ii) delayed chain termination, and (iii) efficient mismatched incorporation which leads to RT inhibition and decreased unblocking (Figure 4.6) [161–163]. Once EFdA-MP is incorporated at the end of the primer,

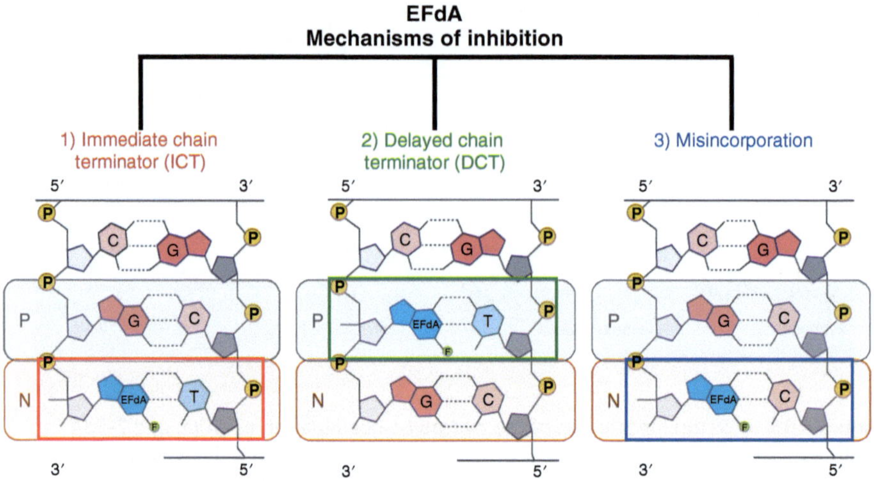

Figure 4.6 EFdA mechanisms of inhibition. EFdA can inhibit as an ICT, DCT, and can be misincorporated, leading to translocation inhibition. *Source*: Created with BioRender.

DNA polymerization can either stall (immediate chain termination) or continue by addition of only a single nucleotide (delayed chain termination). Michailidis et al revealed that EFdA can act as an immediate or delayed chain terminator (ICT or DCT) in a sequence-dependent manner, although the majority of EFdA-MP was incorporated at the N-site [161]. Hence, EFdA was initially termed a translocation-defective reverse transcriptase inhibitor (TDRTI). Interestingly, RT misincorporated EFdA-MP more efficiently than canonical nucleotides, opposite a noncognate base. Thus, once EFdA is misincorporated, RT cannot further support DNA synthesis. This misincorporation also decreased the unblocking efficiency of RT [162]. In conclusion, biochemical studies have shown that EFdA may act through three unique mechanisms of action. These mechanisms related to translocation inhibition are distinct from traditional HIV NRTIs, thus this inhibitor is now categorized as an NRTTI. Furthermore, these studies reported increased incorporation efficiency [161] and tighter binding of EFdA-TP to RT/DNA compared to dATP [162]. The multiple mechanisms of action against RT and these favorable binding properties of EFdA may contribute to its high potency and high barrier to resistance.

4.7.3.1.2 Structural Mechanisms of Inhibition

To further understand the inhibition mechanism of EFdA and the key interactions of EFdA-TP with RT, Salie et al. solved crystal structures of EFdA-TP and EFdA-MP intermediates bound to RT in complex with double-stranded DNA (dsDNA). These structures include an RT/DNA_{ddGMP}/EFdA-TP ternary complex, an RT/DNA/EFdA-MP complex with EFdA-MP in both the N- and P-sites, and an RT/DNA/EFdA-MP mismatch complex where EFdA-MP is mismatched at the N-site. The first intermediate solved was a 2.4 Å structure of EFdA-TP bound to a dideoxyguanosine monophosphate (ddGMP)-terminated DNA/RT complex to position EFdA-TP at the N-site without incorporation into the 3′-end of the primer. This structure revealed that the 4′-ethynyl group exploits a conserved hydrophobic pocket at the polymerase active

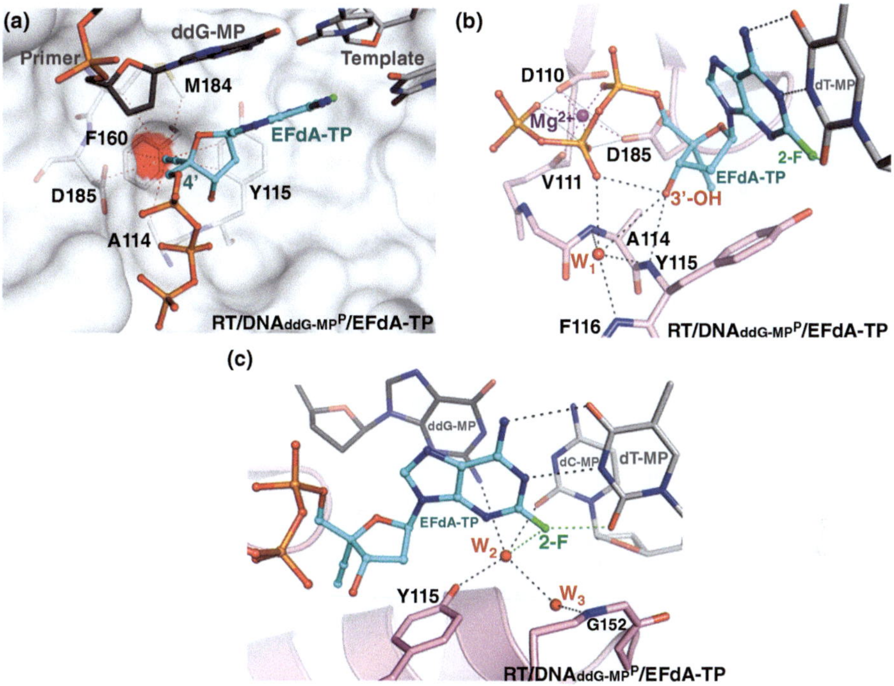

Figure 4.7 EFdA-TP interactions at the polymerase active site of RT/DNA$_{ddGMP}$/EFdA-TP. (a) Hydrophobic interactions (atoms within 4 Å; red dotted lines) of the 4′-ethynyl at a hydrophobic pocket (red surface) near the polymerase active site; (b) Polar interactions between the EFdA-TP 3′-OH, the β-phosphate, ordered water W_1, and peptide backbone atoms are shown as dotted lines (H-bonds, black; metal binding, purple; halogen bond, green); (c) Interactions of water W_2 with Y115, 2-F, the P-site base pair, and water W_3, which is also bound to the main-chain N-H of G152. *Source*: Adapted from [29] / National Academy of Sciences.

site (residues: A114, Y115, F160, and M184, and the aliphatic chain of D185) (Figure 4.7a) [29]. These interactions explain the increased incorporation efficiency and tighter binding of EFdA-TP to the RT/DNA complex compared to dA [161, 162]. This structure also revealed other key polar interactions, including a network of ordered water molecules that stabilize the binding of EFdA-TP and the associated nucleic acid substrate to RT (Figure 4.7b,c). New interactions were identified between K220 and the γ-phosphate of EFdA-TP. These studies provide structural insight into the mechanism of EFdA-TP inhibition. These interactions that stabilize EFdA-TP at the polymerase active site may explain why EFdA acts primarily as a *de facto* immediate chain terminator. If RT can translocate and move EFdA-MP into the P-site, there is steric crowding of EFdA-MP between the 4′-ethynyl group and the backbone of Y183. Primer distortion also occurs at the P-site when EFdA-MP is present there, affecting the distance between neighboring phosphates in the primer near the active site and blocking translocation (Figure 4.8a–d). EFdA-TP can also be incorporated as a mismatched pairing, creating a Hoogsteen base pairing, which has been shown to inhibit RT transcription. These findings elucidated the structural mechanisms of EFdA inhibition of HIV-1 RT; hindering DNA synthesis through translocation defects [29].

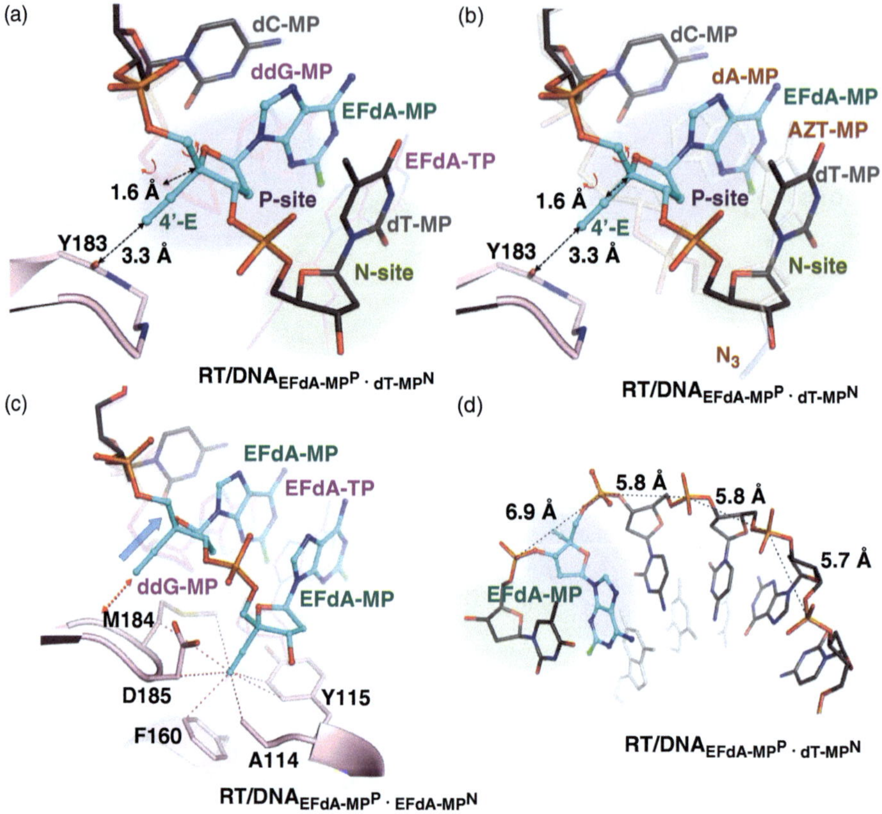

Figure 4.8 Structural changes in EFdA-containing complexes. (a) Superposition of N-complex RT/DNA$_{EFdA\text{-}MP}{}^{P}{\bullet}_{dT\text{-}MP}{}^{N}$ and ternary complex RT/DNA$_{ddG\text{-}MP}{}^{P}$/EFdA-TP (magenta DNA). (b) Superposition of two N complexes of RT, in the presence or absence of EFdA-MP at the P-site [RT/DNA$_{EFdA\text{-}MP}{}^{P}{\bullet}_{dT\text{-}MP}{}^{N}$ and RT/DNA$_{dA\text{-}MP}{}^{P}{\bullet}_{AZT\text{-}MP}{}^{N}$ (PDB ID 1N6Q in light brown) [32]. In both (a) and (b), without the localized 1.6-Å primer shift (marked dotted arrow) of EFdA-MP^{P}, the currently observed 3.3-Å distance between the 4′-ethynyl (4′-E) and main-chain carbonyl of Y183 would have been highly unfavorable at ~1.7 Å. This shift is accomplished through a change in the γ-torsion angle about the C4′−C5′ (a) vs. (b). (c) Primer distortion due to interactions of P-site EFdA-MP of RT/DNA$_{EFdA\text{-}MP}{}^{P}{\bullet}_{EFdA\text{-}MP}{}^{N}$ with Y183 (red arrow) compared with RT/DNA$_{ddG\text{-}MP}{}^{P}$/EFdA-TP (magenta DNA); hydrophobic interactions of the 4′-E at the active-site hydrophobic pocket (red dotted lines connect atoms within 4 Å). (d) The change in γ results in increased distance between neighboring phosphates near the polymerase active site. *Source*: Adapted from [29] / National Academy of Sciences.

4.8 Mechanism of Resistance/Hypersusceptibility to EFdA

4.8.1 Mechanisms of Viral Resistance to EFdA

There are three potential mechanisms of resistance that RT could develop to counter EFdA. However, these mechanisms are rare due to EFdA's overall high potency and the reduced viral fitness of resistant strains. Potentially, a mutation could cause decreased incorporation efficiency, increased excision/unblocking ability, and/or have

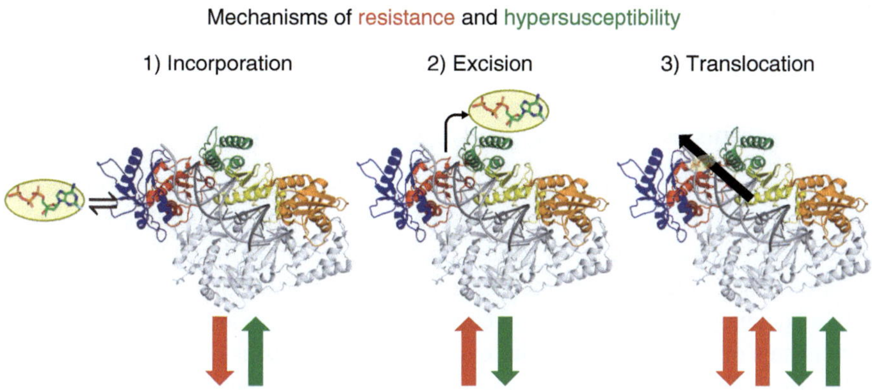

Figure 4.9 Mechanism of RT resistance or hypersusceptibility to EFdA-TP. Red arrows indicate mechanism of RT resistance (decreased incorporation, increased excision, and increased/decreased translocation) and green arrows indicated mechanisms of RT hypersusceptibility (increased incorporation, decreased excision, and increased/decreased translocation) to EFdA.

an effect on the translocation efficiency of RT that may lead to resistance (Figure 4.9). A mutation that causes resistance to EFdA could cause steric hindrance or a loss of interaction that decreases EFdA-TP affinity. It may also affect the incorporation efficiency of RT. These mutations are called NRTI discriminatory mutations. Some common discriminatory resistance mutations include: M184V/I affecting 3TC, FTC, and ABC, K65R affecting tenofovir, 3TC, FTC, and ABC, K70E affecting ABC and tenofovir, L74VI affecting ABC, and Y115F affecting ABC and tenofovir susceptibility [164]. A resistance mutation could also increase the ability of RT to unblock or excise an incorporated nucleotide utilizing ATP. These mutations are called excision enhancement mutations (EEMs), or thymidine-associated mutations (TAMs) because they were historically known to enhance D4T and AZT excision [165, 166]. These mutations include M41L, D67N, K70R, T215FY, and K291QE [164]. TAMs, specifically T215FY, have been shown to increase π-π interactions between the aromatic ring of the amino acid residue and the adenine base [166–169]. Lastly, a resistance mutation could affect the translocation efficiency of RT on an EFdA-MP terminated primer.

The first mutations reported to cause EFdA resistance were I142V/T/165R/M184V and were found through passaging experiments [88]. These mutations exhibited a moderate resistance to EFdA (22-fold increase in EC_{50}). Individually, M184V showed 7.5-fold, I142V 0.9-fold, and T165R 1.5-fold resistance to EFdA. Although this triple mutant virus showed impaired viral fitness, E138K/M184V/I was later reported to emerge from passaging experiments. E138K was shown to be a compensatory mutation, remedying the viral replication deficiency caused by M184V/I [170]. There was an initial concern that these mutations would appear in patients, but they have not. Other mutations that caused increased resistance to EFdA were M41L/M184V/A114S and M184V/A114S increasing the EC_{50} value of EFdA by 24- to 65-fold. Interestingly, this A114S/M184V double mutation is almost 50 times more susceptible to TFV [171, 172]. Resistance mutations M184V/V106I (22.8-fold) and M184V/I42V (25.2-fold) also decrease the susceptibility of EFdA [173].

These studies concluded that EFdA has a high barrier to resistance, as shown by *in vitro* and *in vivo* studies.

4.8.2 Mechanisms of Viral Hypersusceptibility to EFdA

Because there are three potential mechanisms of resistance to EFdA, there are also three potential mechanisms of hypersusceptibility. Hypersusceptibility is the phenomenon in which a mutation causes increased susceptibility of the virus to an inhibitor. This is often determined as a decrease in EC_{50} value or an increase in antiviral potency to the mutant virus compared to WT. There are three hypothesized mechanisms of resistance, thus there are three mechanisms of HIV hypersusceptibility to EFdA. These mechanisms include increased incorporation efficiency, decreased excision/unblocking ability, and/or effects on the translocation efficiency of RT that may impact RT susceptibility to inhibitors (Figure 4.9). The effects of one or more hypersusceptibility mutations could decrease the translocation ability of RT, thus increasing the translocation inhibition by EFdA-TP. Alternatively, mutations could increase the translocation efficiency of RT, moving EFdA-MP into the P-site. This would decrease the excision ability of RT as EFdA-MP can only be excised from the N-position, thereby increasing susceptibility. There have been several mutations reported in the literature that show resistance to another RT inhibitor and hypersusceptibility to EFdA. The K65R substitution in RT is the major resistance mutation selected for in TFV treated individuals. This mutation causes threefold resistance to TFV but is 2.5-fold more sensitive to EFdA. This hypersusceptibility was due to an excision deficiency in RT_{K65R}, or in other words, a decrease in the ability of this mutant RT to unblock/excise EFdA-MP from the primer. This mutation had minor impact on the incorporation and translocation efficiency of RT [174]. Another reported hypersusceptibility mutation is F227C. The F227C mutation is a resistant mutation to several NNRTIs such as DOR. F227C causes ~70-fold decrease in DOR potency compared to WT [175]. This mutation often arises with several other mutations, including: V106A/I, A98G, H221Y, Y118L, P225H, and Y318F. HIV-1 F227C is 10–20-fold more susceptible to EFdA compared to WT [171]. Understanding these resistance and hypersusceptibility mutations to EFdA are critical for therapeutic use and highlight the potential benefits of using EFdA in combination regimens as discussed in the next paragraph.

4.8.3 Combination Therapies with EFdA

All HIV antiviral treatment is given as a combination therapy. Targeting multiple stages of the HIV lifecycle ensures maximum suppression, reducing the risk of drug-resistant mutations. The combination of FTC and TFV is one of the most widely used HIV treatment pairings and has been shown to have synergistic antiviral activity together [176, 177]. There may be an optimal partner for HIV treatment with EFdA. This candidate would have a synergistic or additive pairing, the potential to be long-acting, or favorable cross-resistance profile to another antiviral. Some pairings have been evaluated. One study assessed EFdA in combination with NRTIs: AZT, 3TC, FTC, TDF, and NNRTIs: efavirenz (EFV), etravirine (ETR), and rilpivirine

(RPV) against HIV-1_{NL4-3} infection. Synergy was analyzed using MacSynergy II and CalcuSyn. Most pairings were determined to be additive, although EFdA with RPV showed minor synergy and EFdA with 3TC showed minor antagonism. The mechanisms of synergy between EFdA and RPV are unknown, although it is hypothesized that it may be involved in the reduction of phosphorolytic cleavage [178]. EFdA was also evaluated with lenacapavir (LEN) as a potential long-acting antiviral partner. LEN can be used as a once every six-month injectable, yet does not have a long-acting partner. Cilento et al. sought out to determine synergistic, additive, or antagonistic pairings between LEN and various long-acting inhibitors, including EFdA [179]. EFdA and LEN were analyzed with SynergyFinder plus, a program that uses ZIP, LOEWE, HSA, and BLISS models (Figure 4.10) [180]. These studies showed the EFdA and LEN made an additive pairing using the ZIP, LOEWE, and BLISS model, and made a synergistic pairing using the HSA model. The lack of antagonism makes these two antivirals a promising combination for long-acting therapy. Furthermore, complementary resistance profiles may be a strategy used for combination pairings. This strategy involves pairing two antivirals where a set of mutations causes resistance to one drug and hypersusceptibility to another. As mentioned previously, EFdA is resistant to M184V/I, M184V/A114S, but is hypersusceptible to F227C, V106A/F227C, and K65R. Interestingly, the mutations that cause resistance to EFdA are hypersusceptible to TFV. Although EFdA and TFV are both dA analogs, they do not

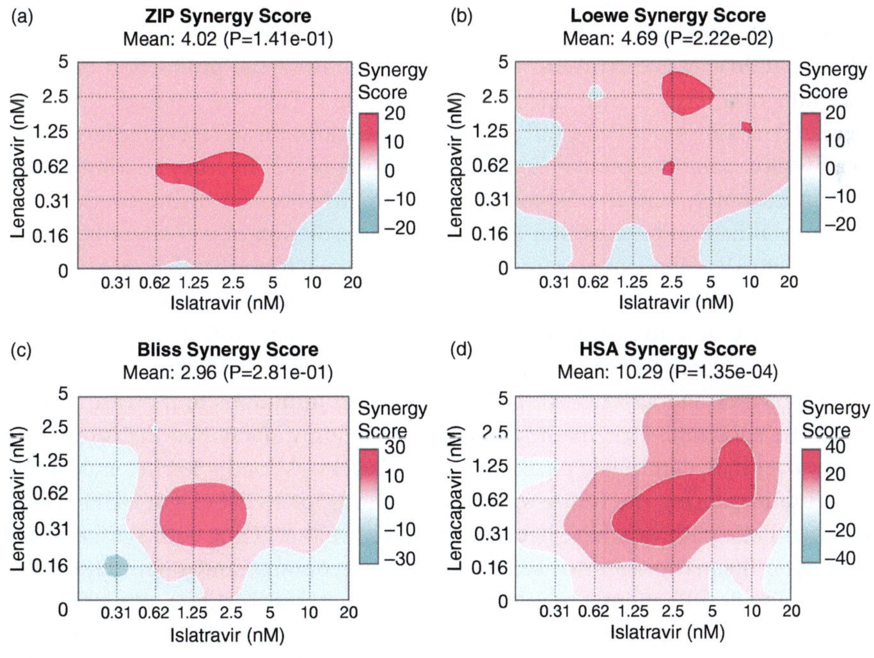

Figure 4.10 Synergy matrices and scores of EFdA paired with Lenacapavir. (a–d) Synergy matrices are representative of the synergy score of this antiviral pairing for each dose-response concentration. Cumulative synergy scores are shown above. Synergy scores are calculated using ZIP, LOEWE, BLISS, and HSA models. *Source*: Adapted from [179] / CC BY 4.0 / MDPI.

seem to be antagonistic. It was shown that EFdA and TFV actually have differential incorporation preferences on a nucleic acid substrate by RT [162]. This may also be due to no competition of α phosphorylation, the rate limiting-step of EFdA conversion to EFdA-TP, by dCK. This is because TFV has a phosphonate group present. Furthermore, DOR and EFdA have complementary resistance profiles. RT_{F227C} is hypersusceptibile to EFdA and resistant to DOR. These antivirals are currently in clinical trials together due to their high barrier of resistance together [171]. There are several strategies to optimize combination therapies with EFdA including the ones described above, yet more can be explored.

4.8.4 Synthesis Advances

EFdA can be synthesized through many routes ranging from 2 to 18 steps. EFdA was first synthesized in 18 steps using an expensive starting material, 2-amino-2′-deoxyadenosine, and the yield was low at only 2.5% [181]. The number of steps to synthesize EFdA was then reduced in 2011, to 12-steps. Kageyama et al. were able to reduce the overall number of steps using the starting material (*R*)-glyceraldehyde acetonide and increased the yield to 18%. The disadvantage of this synthetic strategy was the need for careful chromatographic separation of intermediates throughout the synthesis scheme [182]. To alleviate these challenges, Fukuyama et al. created a 14-step synthesis that reduced the number of chromatographic purifications to four and instituted exclusively diastereoselective constructions of the asymmetric C4′ [183]. Soon after, scientists from Merck & Co., Inc. developed an asymmetric enantioselective synthesis of EFdA that allowed the production of EFdA at the kilogram level at high purity and a 90% yield. This was through three key aspects: (i) establishment of a fully-substituted 4′-carbinol stereocenter leveraging biocatalysts capabilities (enzymes), (ii) the addition of a 3′-hydroxy bearing carbon using an extremely selective Noyori-type asymmetric transfer hydrogenation, and (iii) a crystallization process that allowed for the isolation of a *N*-silyl nucleoside intermediate allowing isolation of the correct anomer from the glycosylation step [184]. In 2019, EFdA was synthesized using *in vitro* biocatalytic cascade synthesis. Scientists were able to synthesize the stereospecific EFdA using five genetically-engineered enzymes and four auxiliary enzymes at a 51% yield. Some of the engineered enzymes had an astonishing 34 amino acid mutations to increase activity toward the substrate. This method, the atom economy, decreased the number of steps by over half, eliminated purification steps, and was conducted in a mild aqueous solution, altering the future synthesis methods for EFdA (Figure 4.11) [185]. Other synthetic pathways have been developed as well utilizing both synthetic and biocatalytic means to improve selectivity, atom economy, and safety [186–191]. Recently, the development and pilot plant-scale synthesis of EFdA was published with the aim to convert this process to produce >100 kg of EFdA per synthesis; 20 kg of EFdA was synthesized using an aerobic oxidation process to produce (*R*)-ethynylglyceraldehyde-3-phosphate using an engineered galactose oxidase enzyme to get a 90% yield of highly pure EFdA. This synthesis made large advances in developing a green and sustainable synthesis of EFdA using aerobic oxidation at an industrial scale [192].

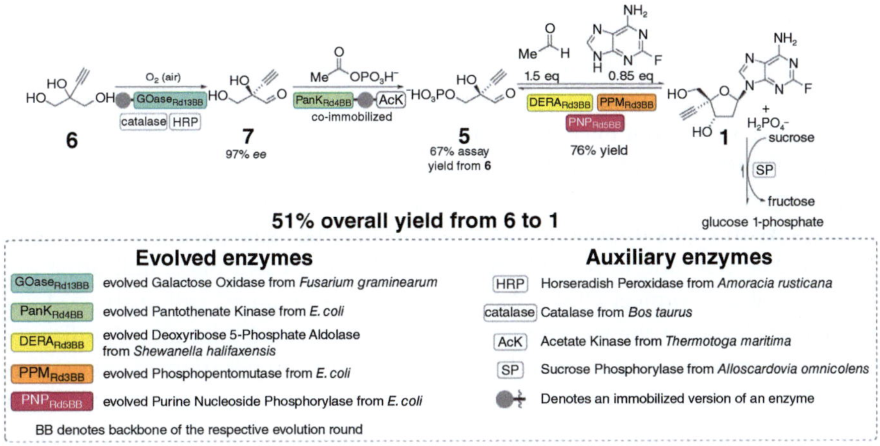

Figure 4.11 Synthesis scheme of EFdA utilizing a biocatalytic pathway. Evolved enzymes shown in colored boxes and wild-type enzymes are shown in white boxes. *Source*: Adapted from [185] / American Association for the Advancement of Science - AAAS.

4.9 Animal Studies

It is essential to test novel antivirals in established animal models to determine *in vivo* antiviral effects and potential toxicity before clinical trials. Over the years, numerous humanized mouse and nonhuman primate models have been developed for HIV studies [193]. Utilizing these tools, multiple preclinical studies of EFdA were conducted with different dosing regimens and routes of viral inoculation to evaluate its pharmacokinetics, antiviral effects, and toxicity in humanized rodents and macaque models (Table 4.2).

4.9.1 EFdA Pharmacokinetic Studies in Rodents

Several studies were conducted to determine the pharmacokinetics (PK) and *in vivo* toxicity of EFdA using BALB/c mice [194] and NSG-humanized Thy/Liv mice [195]. 20 mg/mL of EFdA was administered intraperitoneally to BALB/c mice [194], and the maximal plasma concentration was reached within the first 30 min but rapidly decreased. High doses (50 mg/mL) of EFdA were administered to mice twice daily for 14 consecutive days, and no apparent adverse effect beyond ruffled fur was observed in all treated mice. A similar experiment was done by Stoddart et al. in humanized mice [195]. EFdA was administered orally to NSG-hu Thy/Liv mice at a single dose of 5 mg/kg. EFdA concentration reached a peak plasma level of 3.2 μM 30 min post-administration, consistent with the 20 mg/kg dose Hattori et al. did in BALB/c mice [194]. Here, EFdA had an elimination half-life of 3.8 h, and the plasma level remained at least 10 times the IC_{90} for multiple HIV-1 strains 12 h after dosing. More importantly, no systemic or local toxicity was observed in treated mice. These studies demonstrated that EFdA is a promising drug candidate for HIV-1 therapy.

4.9.2 EFdA Antiviral Activity in Humanized Mice

To determine the effect that EFdA has on HIV suppression, several animal studies were conducted. To evaluate the *in vivo* anti-HIV activity of EFdA, NOD/SCID/Janus kinase 3 knockout mice transplanted with human peripheral blood mononuclear cell (hPBMC) were inoculated with HIV-1_{JR-FL} intraperitoneally [194]. EFdA (0.5 mg/kg) was administered intraperitoneally to mice twice daily for 14 consecutive days. With EFdA treatment, the HIV-infected mice showed a >10-fold higher CD4+/CD8+ cell ratio in the spleen compared to untreated mice, indicating that EFdA was able to protect CD4+ T cell against HIV-1-induced cell death. p24 and HIV-1 RNA levels in the spleen and plasma of the treated mice also significantly reduced, suggesting EFdA potently suppresses HIV-1 replication in mice transplanted with hPBMC. Antiviral accessibility and adherence are important for individuals living with HIV to maintain viral suppression and prevent viral rebound. Therefore, oral treatment of EFdA was also assessed in humanized mice in more recent years [195]. In this study, NSG-BLT mice (NSG mice transplanted with human bone marrow, liver, and thymus tissues) were inoculated with HIV-1_{JR-CSF} intraperitoneally. Orally, EFdA was given twice daily at various dosages (0.1–10 mg/kg/day) to mice before HIV-1 inoculation. All mice dosed with 1 and 10 mg/kg/day of EFdA treatment remained nonviremic for up to 5 weeks post-treatment, while 50% of mice with 0.1 mg/kg/day treatment remained nonviremic, indicating oral EFdA treatment can suppress viral infection in humanized mice. Additionally, 10 mg/kg/day of EFdA was orally administered to HIV-infected mice 6 weeks postinfection, and similar antiviral activity was observed. However, five out of six treated mice experienced viral rebound 2 weeks post-treatment-cessation, and these rebound viruses harbored the M184V mutation in the viral reverse transcriptase. Nonetheless, orally administered EFdA was able to suppress viral infection and replication, even for variant viruses, in humanized mice. To further evaluate the *in vivo* impact of EFdA in HIV-1 infected humanized mice, its systematic antiviral efficiency was assessed, with focus on the two major sites of HIV transmission, the female reproductive tract (FRT) and gastrointestinal (GI) tract [196]. Here, HIV-1_{JR-CSF} was injected to mice intravenously and 10 mg/kg of EFdA was administered orally. HIV-RNA levels in peripheral blood (PB) and cervicovaginal lavage decreased to below detectable level. HIV-RNA levels in the lymph nodes (LN), liver, lung, and spleen of EFdA-treated mice reduced by at least 1-log, and those in the GI tract and FRT reduced by more than 2-log. Additionally, EFdA treatment significantly prevented the depletion of CD4+ T cells in the PB, mucosal tissues, and lymphoid tissues. Although EFdA treatments have been successful in suppressing viral replications in humanized mice models, potential reservoirs in lymphoid tissues is still a major concern for viral rebound. To evaluate the earliest steps of HIV spread from lymphoid tissues, Maidji et al. [197] quantified the distribution of HIV in the lymphoid tissues of HIV-infected humanized mice with and without EFdA-suppression and after drug cessation. They discovered that even under EFdA suppression, lymphoid tissues in these mice contained rare HIV reservoirs capable of initiating viral rebound. A sizable portion of the HIV reservoir was found in CD3+ T cells, but a smaller subset of CD163+ macrophages also contained HIV RNA or DNA. Nevertheless, these studies demonstrated that EFdA could systematically inhibit HIV infection and prevent infection-related CD4+ T cell death in humanized mice, supporting its strong potential for further development as an oral antiretroviral.

HIV transmission through the female reproductive tract, male genital tract, and oral and rectal tissues are the most common routes of transmission [198]. Therefore, it is important to evaluate the antiviral activity of EFdA for HIV inoculated specific in these regions. To this end, Kovarova et al. [199] evaluated the effect of EFdA as a pre-exposure prophylaxis (PrEP) agent for both vaginal and oral HIV infection in NSG-BLT mice. With once daily oral EFdA (10 mg/kg) administration followed by three consecutive high-dose HIV challenges, significant reduction of HIV-RNA in the mice serum, cervicovaginal secretions, and saliva was observed. Additionally, EFdA PrEP reduced HIV-RNA and -DNA levels in PB and all tissues tested without significant CD4+ cell depletion. Using the NSG-BLT mice, Kovarova et al. later developed a novel *in vivo* model that allows detailed analysis of HIV infection throughout the male genital tract (MGT) [200]. Male NSG-BLT mice were treated with EFdA (1.8 mg/kg/day) orally and exposed to HIV-1_{CH040} via the penis. Treatment with EFdA reduced HIV replication in the MGT by 2- to 3-logs, leading to the restoration of CD4+ T cell levels. Furthermore, EFdA PrEP was also found to be effective in preventing penile HIV acquisition. These studies further supported the high efficiency of EFdA in suppressing HIV infection via different routes of transmission [200].

4.9.3 EFdA Antiviral Activity in Rhesus Macaques

Further evaluations of EFdA in nonrodent models were done by multiple groups to demonstrate its high potency and efficiency *in vivo*. Murphy-Corb et al. [201] assessed the effect of EFdA on SIV in rhesus macaques (RMs), a common animal model used for novel antiretroviral development. Here, two SIV-infected RMs with advanced AIDS were administered 0.4 mg/kg of EFdA subcutaneously twice daily. Within 1-week of EFdA treatment, both macaques showed 3- to 4-log reduction in plasma viral load, and eventually reduced to below detectable level. Furthermore, clinical symptoms of AIDS, which include diarrhea, weight loss, and poor activity were resolved within the first month of EFdA treatment. More importantly, the macaques did not show any observable signs of clinical or pathological toxicity within the 6 months of continuous treatment. EFdA was able to sustain viral suppression for the entire treatment period, although the virus level rebounded after treatment cessation. The rebound virus harbored the M184V/I mutation in the viral reverse transcriptase, yet EFdA was still about to maintain viral suppression during the treatment period, which is consistent with a mice study [195]. PK of EFdA in male RMs was evaluated by Stoddart et al. [195]; 5 mg/kg of EFdA was administered orally and was rapidly absorbed. The peak plasma level of 1.6 μM was reached 90 min post ingestion and remained at or above the IC_{90} level 24 h post the single dose. The elimination half-life of EFdA in rhesus macaque was 1.9 h. Though no SIV challenge was done *in vivo*, PBMCs were isolated from the rhesus macaque 24 h post EFdA administration and shown to be protected against SIV infection without further addition of the drug. This indicated that the active EFdA-TP remained high enough in the cells to protect against SIV infection. Additionally, Stoddart et al. showed that a significant amount of EFdA was detected in the cerebrospinal fluid (CSF) 8 h post administration, indicating that EFdA can cross the blood–brain barrier. These data provide support for further development of EFdA as an oral drug for HIV [202].

Implants containing 40–50% EFdA in PLA, PCL, and EVA polymers developed by Barrett et al. from Merck & Co., Inc. [203] were also tested in RMs with subcutaneous insertion. The release rates from these implants were consistent with those observed in the rat study, showing low variability among subjects. A predictive model was developed to estimate human PK of the EFdA-implants based on this nonhuman primate study, suggesting that the projected plasma levels of EFdA from the implants would exceed those achieved by a 2 mg oral weekly dose and be comparable to a 10 mg oral weekly dose. Another *in vivo* study was done with EFdA and intrarectal SHIV challenge in RM by Markowitz et al. from Merck & Co., Inc. [204]. Here, EFdA (0.1–3.9 mg/kg) was administered orally at three different dose regimens before and after the intrarectal SHIV challenge. While most control macaques became infected after one to four rectal SHIV challenges, all animals treated with high and medium doses of EFdA (3.9 and 1.3 mg/kg) remained uninfected after 12 challenges. This was confirmed by the absence of plasma viremia, undetectable proviral DNA in circulating PBMCs, and no seroconversion up to week 24 of the study. The efficacy of EFdA in preventing infection was significantly high, with a 41.5-fold lower risk of infection compared to control animals. The mean EFdA plasma concentration was 2.8 nM while the mean intracellular concentration of the active triphosphate moiety was 4.1 μM. However, at a lower dose of 0.1 mg/kg, the protection was less effective. Six of eight animals were protected at this dose, indicating a 7.2-fold lower risk of infection compared to controls. EFdA also exhibited prevention efficacy in RM, and therefore could be used as a PrEP agent. These studies further support the future development of EFdA as a PrEP agent and antiretroviral drug.

4.9.4 EFdA Long-acting Activity in Various Animals

The unique PK properties, particularly its high potency and prolonged duration of action, predicted EFdA as a suitable candidate for long-acting therapy. To evaluate EFdA as a long-acting agent *in vivo*, drug-eluting device implants were designed and studied in rodent models [203, 205]. Slow-dosing implant devices made from three different materials, poly (lactic acid) (PLA), poly(caprolactone) (PCL), and poly (ethylene vinyl acetate) (EVA) were evaluated by Barrett et al. from Merck & Co., Inc. [203]. Implants containing multiple EFdA loads in PLA, PCL, and EVA polymers were subcutaneously implanted in Wistar Han rats. For all implants evaluated, a small and transient burst release of EFdA was observed but reached a pseudo-steady-state plasma concentration within a few days. A consistent mass balance of the released drug was also observed, with 70–80% of the antiviral accounted for when comparing PK data and residual drug content in excised implants. After a one-time administration, the EFdA implants demonstrated sustained release and effective drug concentrations in the rat plasma for over 6 months with a 1.6-log decrease in viral load. More importantly, the implants were well-tolerated with no significant toxicity observed, and all animals survived until the scheduled study termination. To evaluate EFdA as a candidate for anti-HIV combination therapy as a long-acting agent, Beloor et al. developed removable implant containing a preclinical NNRTI, compound I, and EFdA, which had previously been shown synergistic effect [205, 206]. Here, implants loaded with both compound I and EFdA were surgically inserted to humanized BTL mice prior to HIV-1_{JRCSF} infection, resulting in an initial burst in drug release with serum concentrations well above the effective concentration needed for viral suppression. These implants maintained high drug levels for at least 56 days, significantly reducing plasma viral loads and protecting CD4+ T cells in treated mice compared to controls. Other studies have been conducted for the development of a subcutaneous reservoir-style, poly(ε-caprolactone) EFdA-eluting implant [207] that was tested in Wistar rats at 0.1, 0.3, and 1 mg/kg [208]. The mean absorption rate and cumulative mass from 154 days were roughly 0.072 mg/day and 8.67 mg, respectively. This absorption was described as zero-order (constant), rather than first-order (dose-dependent), which may be advantageous for controlling release for future formulations. This implant is expected to get clinical PK above EFdA-TP's PrEP efficacy threshold [208]. These results support the potential of EFdA for further development as a long-acting antiretroviral [205, 208].

Formulations with EFdA that involve female-controlled vaginal insertions and films had been developed and studied *in vitro* and *in vivo* [209–211]. Recently, a 3D-printed intravaginal ring (IVR) containing EFdA had shown long-acting potential in sheep and macaque models [211]. In this study, EFdA was absorbed into silicone poly(urethane) IVRs in two different ways: by itself (EFdA IVR) and together with two birth-control agents, etonogestrel (ENG) and ethinyl estradiol (EE), as a multiple prevention technology (MPT IVR). *In vitro* evaluations of these IVRs showed that drug release is sustainable for 150 days for the MPT IVRs and over 90 days for EFdA IVRs. Human-size MPT IVRs (containing 14 mg ENG, 3 mg EE, and 300 mg EFdA) were subsequently inserted vaginally in female Merino sheep for 14 days for PK and safety evaluations. EFdA concentration in plasma peaked at approximately 10 ng/mL and maintained at approximately 1 ng/mL for 14 days. On day 14, MPT IVRs were removed from the sheep, and approximately 0.1 ng/mL EFdA was still detected in plasma on day 21. Similar PK profile was observed in sheep vaginal tissues and fluids where EFdA concentrations peaked at 14 days and maintained at approximately 1000 ng per unit tested. Peak EFdA concentrations observed in rectal fluids were significantly lower at 10 ng/swab. No systematic toxicity was observed; however, mild vaginal epithelial hyperplasia, parakeratosis, submucosal and mucosal eosinophil infiltration, and neutrophile infiltrates were identified in all MPT IVR-treated sheep. The other formulation of macaque-sized IVR was loaded with EFdA (62 mg) only and vaginally inserted in three female pig-tailed macaques. Plasma concentration of EFdA maintained above 10 ng/mL for 28 days by rapidly decreased upon IVR removal. More importantly, plasma concentration of EFdA-TP, the active gradient of EFdA, maintained above 2-log higher. Similar PK trend was observed in vaginal tissue, vaginal fluids, and rectal fluids, with EFdA concentrations of 10 000 ng/g, 100–10 000 ng/swab, and 10–100 ng/swab, respectively, but rapidly decreased upon IVR removal. Furthermore, no adverse effect was observed in all treated macaques, and CD3, CD4, and CD8 T-cells frequencies in the macaque PBMCs maintained stable for at least 30 days. This study demonstrated the 3D-printed IVRs loaded with EFdA (with or without birth-control agents) achieved their target PK and safety in two animal models while maintaining EFdA's efficacy concentration, supporting its use as an MPT, although neither viral challenge nor antiviral activity studies were conducted.

Another independent research team conducted a similar study with the objective of developing subcutaneous implant devices for both anti-HIV and birth control applications [212]. Here, biodegradable ε-polycaprolactone implants loaded with low (62 mg) and high (98 mg) concentrations of EFdA along with implants loaded with ENG (33 mg) were developed. For their *in vivo* study, low- and high-dose EFdA implants were subcutaneously inserted in the same pigtailed macaques (low-dose implant on day 0, then high-dose implant on day 36) for PK evaluation. Another group of macaques was inserted with high-dose EFdA implant along with the ENG implant. Plasma concentration of EFdA was maintained at 1–10 nM for at least 90 days for all treated groups. More importantly, EFdA-TP levels in PBMCs of all treated macaques reached 10–100 fmol/10^6 cells. EFdA and EFdA-TP levels in plasma and PBMCs (respectively) were maintained above the lower limit of quantification (LLOQ) for up to 10 days post implant removal. EFdA penetration in rectal and vaginal tissues of all treated macaques at different dosages was confirmed on day 21 (Table 4.2). Furthermore, Daly et al. demonstrated that the presence of ENG in macaques did not impact the concentrations of EFdA-TP in blood or tissues [212]. Although previous studies indicate that birth-control agents might affect nucleotide metabolism [213, 214], the presence of ENG enhanced the EFdA-TP:dATP ratios in the vaginal tissues of macaques treated with a high dose of EFdA. However, it had no effect on PBMCs and rectal tissues. This suggests that ENG is unlikely to adversely affect the viral inhibitory capacity of EFdA. Two high-dose EFdA implants broke after implantation and caused mild redness and swelling at implant sites. All intact implants were well-tolerated, and no systematic toxicity was observed in any of the animals. This study further supports the long-acting potential of EFdA and the use of MTP.

Recently, a refillable nanofluidic implant device loaded with EFdA was developed and evaluated in rectally and vaginally SHIV-infected rhesus macaques [215]. In this study, EFdA is loaded on nanochannels harbored by a silicon membrane inside the titanium implant device. EFdA is loaded transcutaneously after implantation. One week after SHIV infection, implants were inserted into the dorsum of 8 macaques. Plasma concentration of EFdA was maintained at a median of 6.38 nM, and EFdA-TP concentrations in PBMC were maintained at a median of 0.33 pmol/10^6 cells for 4 weeks. Furthermore, EFdA-TP concentrations across sites of HIV replication or reservoir (rectum, cervix, adipose tissues, and inguinal lymph nodes) were maintained at medians of 5.41–18.02 fmol/mg. Plasma viral RNA (vRNA) levels were monitored during the 1-month test period in all treated macaques. Averaged plasma vRNA reduction was $-2.29 \log_{10}$ copies/mL at week 4, although two macaques showed viral rebound compared to week 3 due to the emergence of the M184V mutation in RT. Nevertheless, strong negative correlations were observed in all treated macaques between EFdA plasma concentration and plasma viral loads, as well as between

Table 4.2 Summary of animal studies with EFdA (ISL).

Animal model	Administration/ formulation	Dosage[a]	PK	Antiviral effect	References
Mice (PBMC-transplanted NOD/SCID/ Janus kinase 3 K/O, BALB/c)	**Injection** (Intraperitoneal, EFdA dissolved in PBS)	Twice daily (5–50 mg/kg) for 14 consecutive days. PK study: single dose (20 mg/kg)	Cmax = 13.6 μM at 30 min Area under PK curve: 17 ± 5.7 μM/h	10-fold higher CD4+/CD8+ cell ratio in the spleen 10-fold reduction in p24-expressing cells in the spleen and HIV-1 RNA and p24 levels in plasma	[194]
Rhesus macaques	**Injection** (Subcutaneous, EFdA dissolved in PBS)	Twice-daily (0.4 mg/kg) of EFdA to advanced AIDS macaques	N/A	3- to 4-log reduction in plasma viral load within 1 week of treatment Clinical symptoms of AIDS significantly resolved within 1 week of treatment Sustained suppression of virus harboring the M184V/I mutation in the viral RT enzyme for up to 4–6 months	[201]
Mice (NSG-hu Thy/Liv)	**Oral** (EFdA dissolved in PBS)	Twice-daily dose of EFdA (0.1–10 mg/kg) PK study: single dose (5 mg/kg)	Cmax: 13.6 μM at 30 min Half-life: 3.8 h Area under PK curve: 9.5 μM/h	EFdA-treated mice remained nonviremic for up to 5 weeks EFdA treatment in mice suppressed HIV replication, even for M184V virus	[195]
Rhesus macaques	**Oral** (EFdA dissolved in grape juice)	Single dose at 5 mg/kg	Cmax: 1.6 μM at 1.5 h Half-life: 1.9 h Area under PK curve: 6.5 μM/h	PBMCs from EFdA-treated macaque were protected against *ex vivo* SIV infection without further drug addition	[195]
Mice (NSG-BLT humanized mice)	**Oral** (EFdA dissolved in PBS)	Once-daily dose at 10 mg/kg for 3 weeks	N/A	HIV-RNA levels in PB and CVL of mice reduced to below detectable level HIV-RNA levels in the LN, liver, lung, and spleen of mice reduced by at least 1-log HIV-RNA levels in the GI tract and FRT reduced by more than 2-log	[196]

(Continued)

Table 4.2 (Continued)

Animal model	Administration/ formulation	Dosage[a]	PK	Antiviral effect	References
Mice (NSG-BLT humanized mice)	**Oral** (EFdA dissolved in PBS)	Once-daily dose at 10 mg/kg	N/A	Protected against three consecutive high-dose HIV challenges via vaginal or oral infection Reduced HIV-RNA and DNA levels in PB and other tissues as PrEP agent	[199]
Rats (Wistar Han rat)	**Implant** (Subcutaneous, EFdA loaded PLA, PCL, EVA polymers)	100–400 mg/kg for 40–80 wt% EFdA implants	Initial burst of EFdA (<5% of total EFdA mass) observed. EFdA release rate from PCL polymers maintained above 10 μg/day for over 500 days. Consistent mass balance of the released EFdA observed	EFdA implants demonstrated sustained release and effective drug concentrations in rat plasma for over 6 months Viral loads reduced by 1.6-log compared to no treatment controls	[203]
Rhesus macaques	**Implant** (Subcutaneous, EFdA loaded PLA, PCL, EVA polymers)	40–50 wt% EFdA in PLA, 50 wt% EFdA in PCL, and 50 wt% EFdA in EVA	Similar to PK in rats: Initial burst of EFdA observed. EFdA release rate from polymers maintained above 100 μg/day for at least 70 days	A predictive model was developed to estimate human PK of the EFdA-implants based on this rhesus macaque study, and results demonstrated the projected plasma levels of EFdA from the implants would exceed those achieved by a 2 mg oral weekly dose and be comparable to a 10 mg oral weekly dose	[203]
Mice (NSG-BLT humanized mice)	**Oral** (EFdA dissolved in PBS)	Once-daily dose at 10 mg/kg	N/A	EFdA systematically inhibits HIV infection and prevents infection-related CD4+ T cell-death in mice However, even under EFdA suppression, lymphoid tissues in these mice contained very rare HIV reservoirs capable of initiating viral rebound	[197]

Rhesus macaques	**Oral** (EFdA dissolved in 5 mL/kg of 10% Tween 80)	Weekly dose at 0.1–3.9 mg/kg	Not shown	Undetectable proviral DNA levels in plasma and PBMCs of macaques after 12 intrarectal SHIV challenges with high and medium doses of EFdA (3.9 mg/kg and 1.3 mg/kg) and maintained for up to week 24. 41.5-fold lower risk of infection compared to control animals. With low EFdA dosage (0.1 mg/kg), 75% macaques were protected against SHIV challenges with a 7.2-fold lower risk of infection compared to controls	[204]
Mice (NSG-BLT humanized mice)	**Injection** (Intraperitoneal, EFdA dissolved in PBS) **Implant** (Subcutaneous)	Compound I (190 mg/kg) and EFdA (10 mg/kg); Compound I (120 mg), EFdA (120 mg), polymer (180 mg)	Area under PK curve: Compound I – NP: 3874 ± 243.1 µg × h/mL. EFdA – NP: 6978 ± 518 µg × h/mL. Compound I – implant: $10\,058 \pm 2074$ µg × h/mL. EFdA – implant: 9129 ± 1323 µg × h/mL. Clearance: Compound I – NP: 0.8 mL/min/kg EFdA–NP: 0.02 mL/min/kg Compound I – implant: 0.6 mL/min/kg EFdA – implant: 1.3 mL/min/kg	Serum drug concentrations maintained well above the effective concentration needed for viral suppression after initial burst from implants. These implants maintained high drug levels for at least 56 days with significant reduction of plasma viral loads and CD4+ T cell death	[205]

(Continued)

Table 4.2 (Continued)

Animal model	Administration/ formulation	Dosage[a]	PK	Antiviral effect	References
Mice (NSG-BLT humanized mice)	**Oral** (EFdA dissolved in PBS)	Daily dose at 1.8 mg/kg	N/A	2- to 3-logs reduction in HIV replication in the MGT and increased restoration of CD4+ T cell levels after HIV-1 exposed via the penis. Effectively prevent penile HIV acquisition as a PrEP agent	[200]
Merino sheep (female)	**IVR** (Silicone poly(urethane))	Etonogestrel (14 mg), ethinyl estradiol (3 mg), and EFdA (300 mg)	EFdA plasma concentration peaked at 10 ng/mL on day 1 and maintained above 1 ng/mL for up to 14 days	None performed	[211]
Pigtailed macaques (female)	**IVR** (Silicone poly(urethane))	EFdA (62 mg)	Plasma concentration of EFdA maintained at approximately 10 ng/mL for 28 days. EFdA-TP concentration in plasma maintained at approximately 1000 ng/mL for 28 days	None performed	[211]
Pigtailed macaques	**Implant** (Subcutaneous, biodegradable ε-polycaprolactone)	EFdA (33, 62, and 92 mg) 62 mg-EFdA loaded implant + 92 mg-EFdA loaded implant were inserted in the same macaques 36 days apart. 92 mg-EFdA loaded implant and 33 mg ENG-loaded implant were inserted in the same macaques 36 days apart	low (62 mg), mid (98 mg), and high (62 mg + 98 mg) doses of EFdA (averaged levels): Plasma Cmax = 1.3–2.8 nM Plama Tmax = 26–43 days PBMC Cmax = 36–164 fmol/10^{-6} cells PBMC Tmax = 21–67 days PBMC EFdA-TP: 20.5–63.6 fmol/10^{-6} cells Vaginal EFdA-TP = 9.6–30.5 pmol/g Rectal EFdA-TP = 18.5–36.9 pmol/g	None performed	[212]

Rhesus macaques	**Implant** (Subdermal implant, titanium refillable nanofluidic implant)	EFdA-loaded (31–41 mg) EFdA loaded transcutaneously immediately after implantation	Mean EFdA delivery dosage of 0.21 ± 0.07 mg/kg/day. Peak plasma EFdA concentration of 9.46 nM, and maintain a median concentration of 6.38 nM for 4-week study. Peak EFdA-TP PBMC concentration = 0.41 pmol/10^6 cells, and maintained a median concentration of 0.33 pmol/10^6 cells for 4-week study	Mean viral load reduction of $-2.30 \pm 0.53 \log_{10}$ copies/mL by week 2 of treatment	[215]
Rats (Wistar rat)	**Implant** (poly (ε-caprolactone))	EFdA loaded (0.1, 0.3, and 1 mg/kg) Subcutaneous implant	Mean absorption rate from 0 to 154 days: 0.072 ± 0.024 mg/day. Cumulative mass at 154 days: 8.67 ± 3.22 mg C_{max}, $C_{avg}t_{1/2}$, and AUC_{last} were: 22.6, 44.7, 201.3 ng/mL, 0.675; 2.283, 9.566 ng/mL, 11.8, 27.1, 17.8/h, and 16.2, 54.8, 229.6 h ng/mL for dose levels 0.1, 0.3, 1.0 mg/kg, respectively	None performed	[208]

a. No systematic toxicity observed in all animal studies.

EFdA-TP concentration in PBMC and plasma viral loads. Moreover, no systematic toxicity was observed in all treated macaques. However, histological examination revealed moderate reaction in tissues around insertion sites compared to tissues in PBS-loaded macaques, potentially due to inflammatory responses to the drug. Still, this refillable implant formulation had shown higher potency than daily or weekly oral EFdA dosage, supporting its potential as a long-acting agent.

4.10 Clinical Trials

EFdA, referred to as islatravir (ISL), was licensed by Merck & Co., Inc. in 2012 from the Yamasa Corporation [216] and clinical trials began in 2015. To date, there have been 22 clinical trials conducted involving EFdA (Table 4.3) [217–238]. Phase 1 trials began in the Fall of 2015 and were completed in 2017 [224]. These initial trials showed that EFdA was well tolerated, EFdA-TP levels were dose-proportional, and the half-life was determined to be between 78.5 and 128 h. Phase 2 clinical trials began in 2017 to determine the efficacy and safety of EFdA-based regimens. During these trials, only two participants taking a high dose of EFdA (25 mg) discontinued due to adverse events [222]. Phase 2 clinical trials with a monthly dose were conducted in 2019 for pre-exposure prophylaxis (PrEP) [235]. These studies found that safety, tolerability, and PK achieved the threshold needed for this treatment, thus clinical trials could progress. Phase 3 clinical trials began in 2020 [226, 227, 230, 232]. Because of its initial clinical success as a long-acting agent Gilead Sciences and Merck & Co., Inc. agreed in 2021 to codevelop and commercialize EFdA and lenacapavir (LEN), a long-acting capsid-targeting antiviral [239]. These joint plans and clinical trials were halted because CD4+T cells and white blood cells were decreasing in individuals taking EFdA (both HIV positive and negative). By the end of 2021, all EFdA-related clinical trials were put on full or partial hold [240]. In 2022, working with the FDA, Merck & Co., Inc. lowered the dosage of EFdA (<2 mg/dose) to circumvent the lowered T cell counts observed, while still suppressing the virus. This lowered dosage was paired with doravirine (DOR) and LEN and administered as either once per day or weekly [221]. Currently, EFdA monthly PrEP trials have been discontinued, but there are still two clinical trials actively recruiting. These recruiting studies will be testing EFdA, at the lowered dose, with DOR [217, 229]. There was a study planned to begin in 2024 to evaluate the safety, tolerability and pharmacokinetics of once-yearly implants at various EFdA doses; however, this clinical trial was withdrawn [241]. Despite initial setbacks, the clinical development of EFdA is progressing.

4.11 Long-acting Methods of Delivery for EFdA Treatment

ART treatment has been moving toward long-acting therapies to improve patient convenience and adherence. Patient nonadherence to strict dosing regimens can lead to viral resistance, thereby treatment failure. Due to the long half-life and favorable selectivity index of EFdA, it may have the potential to be used as a long-acting

Table 4.3 Clinical trials of EFdA (ISL) sorted by date.

NCT number	Study title	Administration	Interventions	Phase	Dates	Study status	References
NCT02217904	A study of islatravir (MK-8591) in ART-naive, human immunodeficiency virus-1 infected participants (MK-8591-003)	**Oral** Single dose	Drug: 1 mg ISL|Drug: 2 mg ISL|Drug: 10 mg ISL|Drug: 30 mg ISL|Drug: 0.5 mg ISL|Drug: 0.25 mg ISL	1	09/2015–05/2017	Completed	[224]
NCT03272347	Islatravir (MK-8591) with doravirine and lamivudine in participants infected with human immunodeficiency virus type 1 (MK-8591-011)	**Oral** Daily	Drug: ISL|Drug: Placebo to ISL|Drug: DOR|Drug: Placebo to DOR|Drug: 3TC|Drug: Placebo to 3TC|Drug: DOR/3TC/TDF|Drug: Placebo to DOR/3TC/TDF|Drug: DOR/ISL	2	11/2017–03/2022	Completed	[222]
NCT04003103	Safety and pharmacokinetics of oral islatravir (MK-8591) once monthly in participants at low risk of human immunodeficiency virus 1 (HIV-1) infection (MK-8591-016)	**Oral** Monthly	Drug: ISL|Drug: Placebo	2	09/2019–11/2022	Completed	[235]
NCT04223791	Switch to doravirine/islatravir (DOR/ISL) in human immunodeficiency virus 1 (HIV-1) participants treated with bictegravir/emtricitabine/tenofovir alafenamide (BIC/FTC/TAF) (MK-8591A-018)	**Oral** Daily	Drug: DOR/ISL|Drug: BIC/FTC/TAF|Drug: Placebo to BIC/FTC/TAF|Drug: Placebo to FDC DOR/ISL	3	02/2020–05/2024	Active, not recruiting	[227]
NCT04223778	Safety and efficacy of a switch to doravirine/islatravir in participants with HIV-1 (MK-8591A-017)	**Oral** Daily	Drug: DOR/ISL|Drug: ART	3	02/2020–03/2024	Active, not recruiting	[230]

(Continued)

Table 4.3 (Continued)

NCT number	Study title	Administration	Interventions	Phase	Dates	Study status	References
NCT04233879	Study of doravirine/islatravir (DOR/ISL 100 mg/0.75 mg) to evaluate the antiretroviral activity, safety, and tolerability in treatment-naïve participants with human immunodeficiency virus type 1 (HIV-1) infection (MK-8591A-020)	**Oral** Daily	Drug: DOR/ISL|Drug: BIC/FTC/TAF|Drug: Placebo to BIC/FTC/TAF|Drug: Placebo to DOR/ISL	3	02/2020–03/2025	Active, not recruiting	[226]
NCT04233216	Doravirine/islatravir (DOR/ISL) in heavily treatment-experienced (HTE) participants for human immunodeficiency virus type 1 (HIV-1) infection (MK-8591A-019)	**Oral** Daily	Drug: ISL|Drug: DOR|Drug: DOR/ISL|Drug: Placebo to ISL|Drug: Placebo to DOR	3	03/2020–11/2023	Active, not recruiting	[232]
NCT04303156	Pharmacokinetics of islatravir in participants with severe renal impairment (MK-8591-026)	**Oral** Single dose	Drug: ISL	1	06/2020–10/2020	Completed	[237]
NCT04568603	Islatravir and methadone pharmacokinetics (MK-8591-029)	**Oral** Daily	Drug: ISL|Drug: Methadone	1	10/2020–07/2021	Completed	[233]
NCT04515641	Single-dose islatravir in moderate hepatic impairment (MK-8591-030)	**Oral** Single dose	Drug: ISL	1	11/2020–09/2021	Completed	[236]

NCT04295772	Doravirine/islatravir (DOR/ISL) in pediatric participants with human immunodeficiency virus type 1 (HIV-1) who are <18 years of age and weigh ≥35 kg (MK-8591A-028)	**Oral** Daily	Drug: DOR/ISL	2	11/2020–01/2023	Completed	[231]
NCT04644029	Oral ISL QM as PrEP in cisgender women at high risk for HIV-1 infection (MK-8591-022)	**Oral** Monthly	Drug: ISL|Drug: Placebo to FTC/TDF|Drug: FTC/TDF|Drug: Placebo to ISL	3	02/2021–07/2024	Active, not recruiting	[219]
NCT04564547	Dose-ranging, switch study of islatravir (ISL) and ulonivirine (MK-8507) once-weekly in virologically suppressed adults with human immunodeficiency virus type 1 (HIV-1) [MK-8591-013]	**Oral** Weekly	Drug: ISL|Drug: Ulonivirine|Drug: BIC/FTC/TAF|Drug: Placebo to ISL|Drug: Placebo to Ulonivirine|Drug: Placebo to BIC/FTC/TAF	2	03/2021–12/2023	Active, not recruiting	[225]
NCT04652700	Oral islatravir (MK-8591) once-monthly as pre-exposure prophylaxis (PrEP) in men and transgender women who are at high risk for HIV-1 infection (MK-8591-024)	**Oral** Monthly	Drug: ISL|Drug: FTC/TDF|Drug: FTC/TAF|Drug: Placebo to ISL|Drug: Placebo to FTC/TDF|Drug: Placebo to FTC/TAF	3	03/2021–08/2023	Completed	[223]
NCT04776252	Open-label, follow-up of doravirine/islatravir (DOR/ISL 100 mg/0.75 mg) for participants with human immunodeficiency virus-1 (HIV-1) infection (MK-8591A-033)	**Oral** Daily	Drug: MK-8591A	3	09/2021–10/2025	Active, not recruiting	[234]

(Continued)

Table 4.3 (Continued)

<table>
<tr><th>NCT number</th><th>Study title</th><th>Administration</th><th>Interventions</th><th>Phase</th><th>Dates</th><th>Study status</th><th>References</th></tr>
<tr><td>NCT05052996</td><td>Study evaluating the safety and efficacy of islatravir in combination with lenacapavir in virologically suppressed people with HIV</td><td>Oral
Weekly</td><td>Drug: ISL|Drug: LEN|Drug: B/F/TAF</td><td>2</td><td>10/2021–11/2027</td><td>Active, not recruiting</td><td>[221]</td></tr>
<tr><td>NCT05130086</td><td>A study of islatravir (MK-8591) in trans and gender diverse participants (MK-8591-035)</td><td>Oral
Monthly</td><td>Drug: ISL</td><td>2</td><td>10/2022–03/2024</td><td>Withdrawn</td><td>[238]</td></tr>
<tr><td>NCT05630755</td><td>A switch to doravirine/islatravir (DOR/ISL) in participants with human immunodeficiency virus type 1 (HIV-1) who are virologically suppressed on bictegravir/emtricitabine/tenofovir alafenamide (BIC/FTC/TAF) (MK-8591A-052)</td><td>Oral
Daily</td><td>Drug: DOR/ISL|Drug: BIC/FTC/TAF|Drug: Placebo to BIC/FTC/TAF|Drug: Placebo to DOR/ISL</td><td>3</td><td>02/2023–10/2025</td><td>Active, not recruiting</td><td>[218]</td></tr>
<tr><td>NCT05631093</td><td>A switch to doravirine/islatravir (DOR/ISL) in participants with human immunodeficiency virus type 1 (HIV-1) who are virologically suppressed on ART (MK-8591A-051)</td><td>Oral
Daily</td><td>Drug: ART|Drug: DOR/ISL</td><td>3</td><td>02/2023–10/2025</td><td>Active, not recruiting</td><td>[228]</td></tr>
</table>

NCT05705349	DOR/ISL in HIV-1 antiretroviral treatment-naïve participants (MK-8591A-053)	**Oral** Daily	Drug: DOR/ISL|Drug: BIC/FTC/TAF|Drug: Placebo to DOR/ISL|Drug: Placebo to BIC/FTC/TAF	3	03/2023–11/2026	Recruiting	[217]
NCT05766501	A study of doravirine/islatravir (DOR/ISL, MK-8591A) for the treatment of human immunodeficiency virus 1 (HIV-1) infection in participants who previously received DOR/ISL (MK-8591A-054)	**Oral** Daily	Drug: DOR/ISL	3	03/2023–01/2026	Recruiting	[229]
NCT05115838	Radiopaque matrix MK-8591 implant in participants at low-risk for human immunodeficiency virus type 1 (HIV-1) infection (MK-8591-043)	**Implant**	Drug: ISL|Drug: Placebo	2	01/2024–10/2025	Withdrawn	[220]

antiviral with the proper formulation and dosing. There are several long-acting dosing methods that could potentially be used to administer EFdA. These methods include subdermal implants, intravaginal rings (IVRs), microneedle array patches, subdermal injections, long-acting hydrogels, and oral regimens that are dosed less often [242]. EFdA treatment with several of these methods has been successful in animal trials. Although oral dosing with lower frequency has been less successful in clinical trials thus far, EFdA may be tested as a subdermal implant in future clinical trials and is predicted to be more effective with this method compared to oral dosing [243]. Another potential for the administration of EFdA is microneedle array patches. These patches are a newer technology that utilize microneedles to administer a treatment to a targeted area without pain, the need to be administered by a trained medical staff, and they are removable [244, 245]. One of the challenges of long-acting therapies is discontinuation of treatment if there is an adverse reaction. This is particularly a problem with long-acting injectables and oral medications. These microneedle array patches and implants have the advantage of being removable. One disadvantage of these patches is that they can only hold a finite amount of medication. Hence, the high potency of EFdA makes it a viable candidate for this type of regimen. In summary, while each of these long-acting methods has its set of advantages and disadvantages, progress in all these methods is poised to enhance long-acting treatment options for individuals living with HIV.

4.12 NRTTIs as a Drug Class

EFdA (ISL) is the first-in-class NRTTI targeting HIV to progress into clinical trials. Distinguished by its high potency, extended half-life, and minimal cytotoxicity *in vitro*, EFdA operates as a translocation inhibitor, stopping reverse transcription through multiple mechanisms of action. NRTTIs are now considered their own antiviral class. Building on the lessons from both successes and setbacks with EFdA, Merck & Co., Inc. is developing other NRTTIs such as MK-8527. This compound is a 7-deaza-deoxyadenosine analog with a chloro at the 2-position of the adenosine base. It demonstrates high potency, EC_{50} 3.4 nM in MT4 cells and 0.2 nM in PBMCs, and has low to moderate clearance with good oral absorption in animal studies. MK-8527 is now in Phase 1/2 clinical trials [246–249]. Notably, this inhibitor has the potential to be used as a long-acting antiviral and is being tested as a once/month regimen in a clinical trial [249].

References

1 World Health Organization. HIV 2023 [Available from: https://www.who.int/data/gho/data/themes/hiv-aids#:~:text=Since%20the%20beginning%20of%20the,at%20the%20end%20of%202022.

2 UNAIDS. Global HIV & AIDS statistics—Fact sheet: UNAIDS 2022 [Available from: https://www.unaids.org/en/resources/fact-sheet.

3 Cilento ME, Kirby KA, Sarafianos SG. Avoiding drug resistance in HIV reverse transcriptase. *Chem Rev*. 2021;121(6):3271–96.

4 Lucas S, Nelson AM. HIV and the spectrum of human disease. *J Pathol*. 2015;235(2):229–41.

5 FDA-Approved HIV Medicines 2023 [Available from: https://hivinfo.nih.gov/understanding-hiv/fact-sheets/fda-approved-hiv-medicines.

6 HIVinfo.NIH.gov. HIV Overview 2023 [Available from: https://hivinfo.nih.gov/understanding-hiv/fact-sheets/fda-approved-hiv-medicines#:~:text=People%20on%20ART%20take%20a,two%20different%20HIV%20drug%20classes.

7 Luckheeram RV, Zhou R, Verma AD, Xia B. CD4+ T cells: differentiation and functions. *Clin Dev Immunol*. 2012;2012(1):925135.

8 Sharma M, Marin M, Wu H, Prikryl D, Melikyan GB. Human immunodeficiency virus 1 preferentially fuses with pH-neutral endocytic vesicles in cell lines and human primary CD4+ T-cells. *ACS Nano*. 2023;17(17):17436–50.

9 Chen B. Molecular mechanism of HIV-1 entry. *Trends Microbiol*. 2019;27(10):878–91.

10 Naghavi MH. HIV-1 capsid exploitation of the host microtubule cytoskeleton during early infection. *Retrovirology*. 2021;18(1):1–11.

11 Shen Q, Wu C, Freniere C, Tripler TN, Xiong Y. Nuclear import of HIV-1. *Viruses*. 2021;13(11):2242.

12 Dharan A, Bachmann N, Talley S, Zwikelmaier V, Campbell EM. Nuclear pore blockade reveals that HIV-1 completes reverse transcription and uncoating in the nucleus. *Nat Microbiol*. 2020;5(9):1088–95.

13 Francis AC, Marin M, Prellberg MJ, Palermino-Rowland K, Melikyan GB. HIV-1 uncoating and nuclear import precede the completion of reverse transcription in cell lines and in primary macrophages. *Viruses*. 2020;12(11):1234.

14 Burdick RC, Li C, Munshi M, Rawson JM, Nagashima K, Hu W-S, et al. HIV-1 uncoats in the nucleus near sites of integration. *Proc Natl Acad Sci*. 2020;117(10):5486–93.

15 Müller TG, Zila V, Müller B, Kräusslich H-G. Nuclear capsid uncoating and reverse transcription of HIV-1. *Annu Rev Virol*. 2022;9(1):261–84.

16 Lusic M, Siliciano RF. Nuclear landscape of HIV-1 infection and integration. *Nat Rev Microbiol*. 2017;15(2):69–82.

17 Dutilleul A, Rodari A, Van Lint C. Depicting HIV-1 transcriptional mechanisms: a summary of what we know. *Viruses*. 2020;12(12):1385.

18 Behrens RT, Sherer NM. Retroviral hijacking of host transport pathways for genome nuclear export. *mBio*. 2023;14(6):e00070-23.

19 Freed EO. HIV-1 assembly, release and maturation. *Nat Rev Microbiol*. 2015;13(8):484–96.

20 Morita E, Sandrin V, McCullough J, Katsuyama A, Hamilton IB, Sundquist WI. ESCRT-III protein requirements for HIV-1 budding. *Cell Host Microbe*. 2011;9(3):235–42.

21 Singh K, Marchand B, Kirby KA, Michailidis E, Sarafianos SG. Structural aspects of drug resistance and inhibition of HIV-1 reverse transcriptase. *Viruses*. 2010;2(2):606–38.

22 Sarafianos SG, Marchand B, Das K, Himmel DM, Parniak MA, Hughes SH, et al. Structure and function of HIV-1 reverse transcriptase: molecular mechanisms of polymerization and inhibition. *J Mol Biol*. 2009;385(3):693–713.

23 Harrison J, Passos DO, Bruhn JF, Bauman JD, Tuberty L, DeStefano JJ, et al. Cryo-EM structure of the HIV-1 Pol polyprotein provides insights into virion maturation. *Sci Adv*. 2022;8(27):eabn9874.

24 Götte M, Maier G, Gross HJ, Heumann H. Localization of the active site of HIV-1 reverse transcriptase-associated RNase H domain on a DNA template using site-specific generated hydroxyl radicals. *J Biol Chem*. 1998;273(17):10139–46.

25 Ding J, Das K, Hsiou Y, Sarafianos SG, Clark Jr AD, Jacobo-Molina A, et al. Structure and functional implications of the polymerase active site region in a complex of HIV-1 RT with a double-stranded DNA template-primer and an antibody Fab fragment at 2.8 Å resolution. *J Mol Biol*. 1998;284(4):1095–111.

26 Sarafianos SG, Das K, Tantillo C, Clark AD, Ding J, Whitcomb JM, et al. Crystal structure of HIV-1 reverse transcriptase in complex with a polypurine tract RNA: DNA. *EMBO J*. 2001;20(6):1449–61.

27 Rodgers D, Gamblin S, Harris B, Ray S, Culp J, Hellmig B, et al. The structure of unliganded reverse transcriptase from the human immunodeficiency virus type 1. *Proc Natl Acad Sci*. 1995;92(4):1222–6.

28 Huang H, Chopra R, Verdine GL, Harrison SC. Structure of a covalently trapped catalytic complex of HIV-1 reverse transcriptase: implications for drug resistance. *Science*. 1998;282(5394):1669–75.

29 Salie ZL, Kirby KA, Michailidis E, Marchand B, Singh K, Rohan LC, et al. Structural basis of HIV inhibition by translocation-defective RT inhibitor 4′-ethynyl-2-fluoro-2′-deoxyadenosine (EFdA). *Proc Natl Acad Sci*. 2016;113(33):9274–9.

30 Khan N, Chen X, Geiger JD. Role of divalent cations in HIV-1 replication and pathogenicity. *Viruses*. 2020;12(4):471.

31 Beilhartz GL, Götte M. HIV-1 ribonuclease H: structure, catalytic mechanism and inhibitors. *Viruses*. 2010;2(4):900–26.

32 Sarafianos SG, Clark AD, Das K, Tuske S, Birktoft JJ, Ilankumaran P, et al. Structures of HIV-1 reverse transcriptase with pre-and post-translocation AZTMP-terminated DNA. *EMBO J*. 2002;21(23):6614–24.

33 Tuske S, Sarafianos SG, Clark Jr AD, Ding J, Naeger LK, White KL, et al. Structures of HIV-1 RT–DNA complexes before and after incorporation of the anti-AIDS drug tenofovir. *Nat Struct Mol Biol*. 2004;11(5):469–74.

34 Sarafianos SG, Pandey VN, Kaushik N, Modak MJ. Site-directed mutagenesis of arginine 72 of HIV-1 reverse transcriptase: catalytic role and inhibitor sensitivity(∗). *J Biol Chem*. 1995;270(34):19729–35.

35 Martín-Hernández AM, Domingo E, Menéndez-Arias L. Human immunodeficiency virus type 1 reverse transcriptase: role of Tyr115 in deoxynucleotide binding and misinsertion fidelity of DNA synthesis. *EMBO J*. 1996;15(16):4434–42.

36 Gao G, Orlova M, Georgiadis MM, Hendrickson WA, Goff SP. Conferring RNA polymerase activity to a DNA polymerase: a single residue in reverse transcriptase controls substrate selection. *Proc Natl Acad Sci*. 1997;94(2):407–11.

37 Boyer PL, Sarafianos SG, Arnold E, Hughes SH. Analysis of mutations at positions 115 and 116 in the dNTP binding site of HIV-1 reverse transcriptase. *Proc Natl Acad Sci.* 2000;97(7):3056–61.

38 Krupkin M, Jackson LN, Ha B, Puglisi EV. Advances in understanding the initiation of HIV-1 reverse transcription. *Curr Opin Struct Biol.* 2020;65:175–83.

39 De Clercq E. Antiviral drugs in current clinical use. *J Clin Virol.* 2004;30(2):115–33.

40 Perno C-F, Yarchoan R, Cooney DA, Hartman N, Gartner S, Popovic M, et al. Inhibition of human immunodeficiency virus (HIV-1/HTLV-IIIBa-L) replication in fresh and cultured human peripheral blood monocytes/macrophages by azidothymidine and related 2′, 3′-dideoxynucleosides. *J Exp Med.* 1988;168(3):1111–25.

41 Parniak MA, Sluis-Cremer N. Inhibitors of HIV-I reverse transcriptase. *Adv Pharmacol.* 2000;49:67–109.

42 Database SUDR. Position Summaries [Available from: https://hivdb.stanford.edu/cgi-bin/PositionPhenoSummary.cgi.

43 Ruiz FX, Arnold E. Evolving understanding of HIV-1 reverse transcriptase structure, function, inhibition, and resistance. *Curr Opin Struct Biol.* 2020;61:113–23.

44 Tantillo C, Ding J, Jacobo-Molina A, Nanni RG, Boyer PL, Hughes SH, et al. Locations of anti-AIDS drug binding sites and resistance mutations in the three-dimensional structure of HIV-1 reverse transcriptase: implications for mechanisms of drug inhibition and resistance. *J Mol Biol.* 1994;243(3):369–87.

45 Kohlstaedt L, Wang J, Friedman J, Rice P, Steitz T. Crystal structure at 3.5 Å resolution of HIV-1 reverse transcriptase complexed with an inhibitor. *Science.* 1992;256(5065):1783–90.

46 Das K, Bauman JD, Clark Jr AD, Frenkel YV, Lewi PJ, Shatkin AJ, et al. High-resolution structures of HIV-1 reverse transcriptase/TMC278 complexes: strategic flexibility explains potency against resistance mutations. *Proc Natl Acad Sci.* 2008;105(5):1466–71.

47 Seley-Radtke KL, Yates MK. The evolution of nucleoside analogue antivirals: a review for chemists and non-chemists. Part 1: Early structural modifications to the nucleoside scaffold. *Antivir Res.* 2018;154:66–86.

48 Yates MK, Seley-Radtke KL. The evolution of antiviral nucleoside analogues: a review for chemists and non-chemists. Part II: Complex modifications to the nucleoside scaffold. *Antivir Res.* 2019;162:5–21.

49 Cano-Soldado P, Pastor-Anglada M. Transporters that translocate nucleosides and structural similar drugs: structural requirements for substrate recognition. *Med Res Rev.* 2012;32(2):428–57.

50 Roy B, Depaix A, Périgaud C, Peyrottes S. Recent trends in nucleotide synthesis. *Chem Rev.* 2016;116(14):7854–97.

51 Balzarini J. Metabolism and mechanism of antiretroviral action of purine and pyrimidine derivatives. *Pharm World Sci.* 1994;16:113–26.

52 Wang C, Song Z, Yu H, Liu K, Ma X. Adenine: an important drug scaffold for the design of antiviral agents. *Acta Pharm Sin B.* 2015;5(5):431–41.

53 Kataev VE, Garifullin BF. Antiviral nucleoside analogs. *Chem Heterocycl Compd (N Y)*. 2021;57(4):326–41.

54 Mitsuya H, Yarchoan R, Broder S. Molecular targets for AIDS therapy. *Science*. 1990;249(4976):1533–44.

55 Mitsuya H, Broder S. Inhibition of infectivity and replication of HIV-2 and SIV in helper T-cells by 2′, 3′-dideoxynucleosides in vitro. *AIDS Res Hum Retrovir*. 1988;4(2):107–13.

56 Klecker Jr RW, Collins JM, Yarchoan R, Thomas R, Jenkins JF, Broder S, et al. Plasma and cerebrospinal fluid pharmacokinetics of 3′-azido-3′-deoxythymidine: a novel pyrimidine analog with potential application for the treatment of patients with AIDS and related diseases. *Clin Pharmacol Ther*. 1987;41(4):407–12.

57 Mitsuya H, Jarrett RF, Matsukura M, di Marzo Veronese F, DeVico AL, Sarngadharan M, et al. Long-term inhibition of human T-lymphotropic virus type III/lymphadenopathy-associated virus (human immunodeficiency virus) DNA synthesis and RNA expression in T cells protected by 2′, 3′-dideoxynucleosides in vitro. *Proc Natl Acad Sci*. 1987;84(7):2033–7.

58 Mitsuya H, Broder S. Strategies for antiviral therapy in AIDS. *Nature*. 1987;325(6107):773–8.

59 Yarchoan R, Broder S. Development of antiretroviral therapy for the acquired immunodeficiency syndrome and related disorders. *N Engl J Med*. 1987;316(9):557–64.

60 Mitsuya H, Broder S. Inhibition of the in vitro infectivity and cytopathic effect of human T-lymphotrophic virus type III/lymphadenopathy-associated virus (HTLV-III/LAV) by 2′, 3′-dideoxynucleosides. *Proc Natl Acad Sci*. 1986;83(6):1911–5.

61 Mitsuya H, Weinhold KJ, Furman PA, St Clair MH, Lehrman SN, Gallo RC, et al. 3′-Azido-3′-deoxythymidine (BW A509U): an antiviral agent that inhibits the infectivity and cytopathic effect of human T-lymphotropic virus type III/lymphadenopathy-associated virus in vitro. *Proc Natl Acad Sci*. 1985;82(20):7096–100.

62 Deval J. Antimicrobial strategies: inhibition of viral polymerases by 3′-hydroxyl nucleosides. *Drugs*. 2009;69:151–66.

63 Thomas S, Singleton V, Lowery J, Sharpe R, Pruess L, Porter J, et al. Nucleocidin, a new antibiotic with activity against trypanosomes. *Antibiot Annu*. 1957;716–21.

64 Owen GR, Verheyden JP, Moffatt JG. 4′-Substituted nucleosides. 3. Synthesis of some 4′-fluorouridine derivatives. *J Org Chem*. 1976;41(18):3010–7.

65 Morton GO, Lancaster JE, Van Lear GE, Fulmor W, Meyer WE. Structure of nucleocidin. III. Revised structure. *J Am Chem Soc*. 1969;91(6):1535–7.

66 Jenkins ID, Verheyden JP, Moffatt JG. 4′-Substituted nucleosides. 2. Synthesis of the nucleoside antibiotic nucleocidin. *J Am Chem Soc*. 1976;98(11):3346–57.

67 Moffatt J, Jenkins I, Verheyden J. Synthesis of the nucleoside antiobiotic nucleocidin. *J Am Chem Soc*. 1971;93(17):4323–4.

68 Verheyden J, Jenkins ID, Owen GR, Dimitrijevich SD, Richards CM, Srivastava PC, et al. Synthesis of certain 4′-substituted nucleosides. *Ann N Y Acad Sci*. 1975;255:151–65.

69 Guillerm D, Muzard M, Allart B, Guillerm G. Synthesis of 4′-fluoroadenosine as an inhibitor of S-adenosyl-L-homocysteine hydrolase. *Bioorg Med Chem Lett.* 1995;5(14):1455–60.

70 Ajmera S, Bapat AR, Danenberg PV, Stephanian E. Synthesis and interaction with uridine phosphorylase of 5′-deoxy-4′, 5–difluorouridine, a new prodrug of 5-fluorouracil. *J Med Chem.* 1988;31(6):1094–8.

71 Ohrui H, Mitsuya H. 4′-C-substituted-2′-deoxynucleosides: a family of antiretroviral agents which are potent against drug-resistant HIV variants. *Curr Drug Targets Infect Disord.* 2001;1(1):1–10.

72 Hayakawa H, Kohgo S, Kitano K, Ashida N, Kodama E, Mitsuya H, et al. Potential of 4′-C-substituted nucleosides for the treatment of HIV-1. *Antivir Chem Chemother.* 2004;15(4):169–87.

73 Ohrui H, Nishizaki T, Waga T, Meguro H. Synthetic study on 4C′-methyl nucleosides: Part I, Synthesis of 4C-methyl-D-ribofuranose derivative. *Nucleic Acids Symp Ser.* 1991(25):1–2.

74 Waga T, Nishizaki T, Miyakawa I, Ohrui H, Meguro H. Synthesis of 4′-C-methylnucleosides. *Biosci Biotechnol Biochem.* 1993;57(9):1433–8.

75 Verheyden JP, Moffatt JG. 4′-Substituted nucleosides. I. Synthesis of 4′-methoxyuridine and related compounds. *J Am Chem Soc.* 1975;97(15):4386–95.

76 Sasaki T, Minamoto K, Hattori K. Facile cyclization-decyclization reaction of a 4′, 5′-didehydroadenosine derivative. New route to the chemical modification of adenosine. *J Am Chem Soc.* 1973;95(4):1350–1.

77 Moffatt J. Chemical transformations of the sugar moiety of nucleosides. Nucleoside Analogues: Chemistry, Biology, and Medical Applications. New York: Springer; 1979. p. 71–164.

78 Klumpp K, Lévêque V, Le Pogam S, Ma H, Jiang W-R, Kang H, et al. The novel nucleoside analog R1479 (4′-azidocytidine) is a potent inhibitor of NS5B-dependent RNA synthesis and hepatitis C virus replication in cell culture. *J Biol Chem.* 2006;281(7):3793–9.

79 Perrone P, Daverio F, Valente R, Rajyaguru S, Martin JA, Lévêque V, et al. First example of phosphoramidate approach applied to a 4 ′-substituted purine nucleoside (4′-azidoadenosine): conversion of an inactive nucleoside to a submicromolar compound versus hepatitis C virus. *J Med Chem.* 2007;50(22):5463–70.

80 Maag H, Rydzewski RM, McRoberts MJ, Crawford-Ruth D, Verheyden JP, Prisbe EJ, et al. Synthesis and anti-HIV activity of 4′-azido-and 4′-methoxynucleosides. *J Med Chem.* 1992;35(8):1440–51.

81 Chen MS, Suttmann RT, Papp E, Cannon PD, McRoberts MJ, Bach C, et al. Selective action of 4′-azidothymidine triphosphate on reverse transcriptase of human immunodeficiency virus type 1 and human DNA polymerases. *Biochemistry.* 1993;32(23):6002–10.

82 Chen MS, Suttmann R, Wu J, Prisbe E. Metabolism of 4′-azidothymidine. A compound with potent and selective activity against the human immunodeficiency virus. *J Biol Chem.* 1992;267(1):257–60.

83 Maag H, Nelson JT, Steiner JLR, Prisbe EJ. Solid-state and solution conformations of the potent HIV inhibitor, 4′-azidothymidine. *J Med Chem.* 1994;37(4):431–8.

84 Marquez VE, Ezzitouni A, Russ P, Siddiqui MA, Ford H, Feldman RJ, et al. HIV-1 reverse transcriptase can discriminate between two conformationally locked carbocyclic AZT triphosphate analogues. *J Am Chem Soc.* 1998;120(12):2780–9.

85 Marquez VE, Ezzitouni A, Russ P, Siddiqui MA, Ford Jr H, Feldman RJ, et al. Lessons from the pseudorotational cycle: conformationally rigid AZT carbocyclic nucleosides and their interaction with reverse transcriptase. *Nucleosides Nucleotides Nucleic Acids.* 1998;17(9–11):1881–4.

86 Counde O, Wu HY, Fraser-Smith EB, Walker KA. Synthesis of 4′-cyanothymidine and analogs as potent inhibitors of HIV. *Tetrahedron Lett.* 1992;33(1):37–40.

87 Nomura M, Shuto S, Tanaka M, Sasaki T, Mori S, Shigeta S, et al. Nucleosides and nucleotides. 185. Synthesis and biological activities of 4′α-C-branched-chain sugar pyrimidine nucleosides. *J Med Chem.* 1999;42(15):2901–8.

88 Kawamoto A, Kodama E, Sarafianos SG, Sakagami Y, Kohgo S, Kitano K, et al. 2′-Deoxy-4′-C-ethynyl-2-halo-adenosines active against drug-resistant human immunodeficiency virus type 1 variants. *Int J Biochem Cell Biol.* 2008;40(11):2410–20.

89 Griffon J-F, Dukhan D, Pierra C, Benzaria S, Loi A, La Colla P, et al. 4′-C-methyl-β-D-ribofuranosyl purine and pyrimidine nucleosides revisited. *Nucleosides Nucleotides Nucleic Acids.* 2003;22 (5–8):707–9.

90 Waga T, Ohrui H, Meguro H. Synthesis and biological evaluation of 4′-C-methyl nucleosides. *Nucleosides Nucleotides.* 1996;15(1–3):287–304.

91 Kitano K, Machida H, Miura S, Ohrui H. Synthesis of novel 4′-C-methyl-pyrimidine nucleosides and their biological activities. *Bioorg Med Chem Lett.* 1999;9(6):827–30.

92 Sugimoto I, Shuto S, Mori S, Shigeta S, Matsuda A. Nucleosides and nucleotides. 183. Synthesis of 4′ α-branched thymidines as a new type of antiviral agent. *Bioorg Med Chem Lett.* 1999;9(3):385–8.

93 Kohgo S, Horie H, Ohrui H. Synthesis of 4′-C-ethynyl-beta-D-arabino- and 4′-C-ethynyl-2′-deoxy-beta-D-ribo-pentofuranosyl pyrimidines, and their biological evaluation. *Biosci Biotechnol Biochem.* 1999;63(6):1146–9.

94 Kohgo S, Kodama E, Shigeta S, Saneyoshi M, Machida H, Ohrui H. Synthesis of 4′-substituted nucleosides and their biological evaluation. *Nucleic Acids Symp Ser.* 1999(42):127–8.

95 Yamaguchi R, Imanishi T, Kohgo S, Horie H, Ohrui H. Synthesis of 4′-C-ethynyl-β-D-ribo-pentofuranosyl pyrimidines. *Biosci Biotechnol Biochem.* 1999;63(4):736–42.

96 Kohgo S, Mitsuya H, Ohrui H. Synthesis of the L-enantiomer of 4′-C-ethynyl-2′-deoxycytidine. *Biosci Biotechnol Biochem.* 2001;65(8):1879–82.

97 Kodama E-I, Kohgo S, Kitano K, Machida H, Gatanaga H, Shigeta S, et al. 4′-Ethynyl nucleoside analogs: potent inhibitors of multidrug-resistant human immunodeficiency virus variants in vitro. *Antimicrob Agents Chemother.* 2001;45(5):1539–46.

98 Kohgo S, Yamada K, Kitano K, Sakata S, Hayakawa H, Nameki D, et al. Synthesis of 4′-C-ethynyl and 4′-C-cyano purine nucleosides from natural nucleosides and their anti-HIV activity. *Nucleosides Nucleotides Nucleic Acids.* 2003;22(5–8):887–9.

99 Kohgo S, Yamada K, Kitano K, Iwai Y, Sakata S, Ashida N, et al. Design, efficient synthesis, and anti-HIV activity of 4′-C-cyano- and 4′-C-ethynyl-2′-deoxy purine nucleosides. *Nucleosides Nucleotides Nucleic Acids*. 2004;23(4):671–90.

100 Hitchcock M. 2′, 3′-Didehydro-2′, 3′-dideoxythymidine (D4T), an anti-HIV agent. *Antivir Chem Chemother*. 1991;2(3):125–32.

101 Horwitz JP, Chua J, Rooge MAD, Noel M, Klundt IL. Nucleosides. IX. The formation of 2′, 3′-unsaturated pyrimidine nucleosides via a novel β-elimination reaction1, 2. *J Org Chem*. 1966;31(1):205–11.

102 Haraguchi K, Takeda S, Tanaka H, Nitanda T, Baba M, Dutschman G, et al. Synthesis of a highly active new anti-HIV agent 2′, 3′-didehydro-3′-deoxy-4′-ethynylthymidine. *Bioorg Med Chem Lett*. 2003;13(21):3775–7.

103 Dutschman GE, Grill SP, Gullen EA, Haraguchi K, Takeda S, Tanaka H, et al. Novel 4′-substituted stavudine analog with improved anti-human immunodeficiency virus activity and decreased cytotoxicity. *Antimicrob Agents Chemother*. 2004;48(5):1640–6.

104 Nitanda T, Wang X, Kumamoto H, Haraguchi K, Tanaka H, Cheng Y-C, et al. Anti-human immunodeficiency virus type 1 activity and resistance profile of 2′, 3′-didehydro-3′-deoxy-4′-ethynylthymidine in vitro. *Antimicrob Agents Chemother*. 2005;49(8):3355–60.

105 Ohrui H, Kohgo S, Hayakawa H, Kodama E, Matsuoka M, Nakata T, et al. 2′-Deoxy-4′-C-ethynyl-2-fluoroadenosine: a nucleoside reverse transcriptase inhibitor with highly potent activity against all HIV-1 strains, favorable toxic profiles and stability in plasma. *Nucleic Acids Symp Ser*. 2006;50(1):1–2.

106 Ohrui H, Kohgo S, Hayakawa H, Kodama E, Matsuoka M, Nakata T, et al. 2′-Deoxy-4′-C-ethynyl-2-fluoroadenosine: a nucleoside reverse transcriptase inhibitor with highly potent activity against wide spectrum of HIV-1 strains, favorable toxic profiles, and stability in plasma. *Nucleosides Nucleotides Nucleic Acids*. 2007;26(10–12):1543–6.

107 Takeuchi T, Sriwilaijaroen N, Sakuraba A, Hayashi E, Kamisuki S, Suzuki Y, et al. Design, synthesis, and biological evaluation of EdAP, a 4′-ethynyl-2′-deoxyadenosine 5′-monophosphate analog, as a potent influenza A inhibitor. *Molecules*. 2019;24(14):2603.

108 Kamata M, Takeuchi T, Hayashi E, Nishioka K, Oshima M, Iwamoto M, et al. Synthesis of nucleotide analogues, EFdA, EdA and EdAP, and the effect of EdAP on hepatitis B virus replication. *Biosci Biotechnol Biochem*. 2020;84(2):217–27.

109 Kamzeeva PN, Aralov AV, Alferova VA, Korshun VA. Recent advances in molecular mechanisms of nucleoside antivirals. *Curr Issues Mol Biol*. 2023;45(8):6851–79.

110 Dvořáková H, Holý A, Rosenberg I. Synthesis of some 2′-C-alkyl derivatives of 9-(2-phosphonomethoxyethyl) adenine and related compounds. *Collect Czechoslov Chem Commun*. 1994;59(9):2069–94.

111 Daluge SM, Good SS, Faletto MB, Miller WH, St Clair MH, Boone LR, et al. 1592U89, a novel carbocyclic nucleoside analog with potent, selective anti-human immunodeficiency virus activity. *Antimicrob Agents Chemother*. 1997;41(5):1082–93.

112 Krawczyk SH. Inventor. Nucleobase phosphonate analogs for antiviral treatment. US patent WO2005012324A2. 2004. https://patentimages.storage.googleapis.com/b6/67/ed/eca670aa1bc8ac/WO2005012324A2.pdf

113 Averett DR, Cho J-R. Inventors. Nucleoside phosphonate derivatives useful in the treatment of HIV infections. US patent WO/2005/079812. 2005.

114 Eliahu S, Martín-Gil A, Perez de Lara MJ, Pintor J, Camden J, Weisman GA, et al. 2-MeS-β, γ-CCl2-ATP is a potent agent for reducing intraocular pressure. *J Med Chem.* 2010;53(8):3305–19.

115 Robak T, Błasińska-Morawiec M, Błasiński J, Hellmann A, Hałaburda K, Konopka L, et al. 2-Chlorodeoxyadenosine (cladribine) in the treatment of hairy cell leukemia and hairy cell leukemia variant: 7-year experience in Poland. *Eur J Haematol.* 1999;62(1):49–56.

116 Levy SG, Wasson DB, Carson DA, Cottam HB. Synthesis of 2-Chloro-2′-5′-dideoxy-5′-difluoromethylphosphinyladenosine: a nonhydrolyzable isosteric, isopolar analog of 2-chlorodeoxyadenosine monophosphate. *Synthesis.* 1996;1996(07):843–6.

117 Saven A, Carrera C, Carson D, Beutler E, Piro L. 2-Chlorodeoxyadenosine treatment of refractory chronic lymphocytic leukemia. *Leuk Lymphoma.* 1991;5(sup1):133–8.

118 Tallman MS, Peterson L, Hakimian D, Gillis S, Polliack A, editors. Treatment of hairy-cell leukemia: current views. *Semin Hematol.* 1999;36(2):155–63.

119 Tanaka M, Srinivas RV, Ueno T, Kavlick MF, Hui FK, Fridland A, et al. In vitro induction of human immunodeficiency virus type 1 variants resistant to 2′-beta-Fluoro-2′, 3′-dideoxyadenosine. *Antimicrob Agents Chemother.* 1997;41(6):1313–8.

120 Holý A, Günter J, Dvořáková H, Masojidkova M, Andrei G, Snoeck R, et al. Structure– antiviral activity relationship in the series of pyrimidine and purine N-[2-(2-phosphonomethoxy) ethyl] nucleotide analogues. 1. Derivatives substituted at the carbon atoms of the base. *J Med Chem.* 1999;42(12):2064–86.

121 de Clercq E. Milestones in the discovery of antiviral agents: nucleosides and nucleotides. *Acta Pharm Sin B.* 2012;2(6):535–48.

122 Horwitz JP, Chua J, Noel M, Donatti JT. Nucleosides. XI. 2′, 3′-dideoxycytidine. *J Org Chem.* 1967;32(3):817–8.

123 Schinazi RF, McMillan A, Cannon D, Mathis R, Lloyd RM, Peck A, et al. Selective inhibition of human immunodeficiency viruses by racemates and enantiomers of cis-5-fluoro-1-[2-(hydroxymethyl)-1, 3-oxathiolan-5-yl] cytosine. *Antimicrob Agents Chemother.* 1992;36(11):2423–31.

124 Soudeyns H, Yao X, Gao Q, Belleau B, Kraus J-L, Nguyen-Ba N, et al. Anti-human immunodeficiency virus type 1 activity and in vitro toxicity of 2′-deoxy-3′-thiacytidine (BCH-189), a novel heterocyclic nucleoside analog. *Antimicrob Agents Chemother.* 1991;35(7):1386–90.

125 Kim HO, Schinazi RF, Nampalli S, Shanmuganathan K, Cannon DL, Alves AJ, et al. 1, 3-Dioxolanylpurine nucleosides (2R, 4R) and (2R, 4S) with selective anti-HIV-1 activity in human lymphocytes. *J Med Chem.* 1993;36(1):30–7.

126 Kim HO, Shanmuganathan K, Alves AJ, Jeong LS, Beach JW, Schinazi RF, et al. Potent anti-HIV and anti-HBV activities of (−)-L-β-dioxolane-C and (+)-L-β-dioxolane-T and their asymmetric syntheses. *Tetrahedron Lett.* 1992;33(46):6899–902.
127 Kim HO, Ahn SK, Alves AJ, Beach JW, Jeong LS, Choi BG, et al. Asymmetric synthesis of 1, 3-dioxolane-pyrimidine nucleosides and their anti-HIV activity. *J Med Chem.* 1992;35(11):1987–95.
128 Elion G. Acyclovir: discovery, mechanism of action, and selectivity. *J Med Virol.* 1993;Suppl 1:2–6.
129 King DH. History, pharmacokinetics, and pharmacology of acyclovir. *J Am Acad Dermatol.* 1988;18(1):176–9.
130 O'Brien JJ, Campoli-Richards DM. Acyclovir: an updated review of its antiviral activity, pharmacokinetic properties and therapeutic efficacy. *Drugs.* 1989;37:233–309.
131 Holý A, Rosenberg I. Synthesis of isomeric and enantiomeric O-phosphonylmethyl derivatives of 9-(2, 3-dihydroxypropyl) adenine. *Collect Czech Chem Commun.* 1987;52(11):2775–91.
132 Ray AS, Fordyce MW, Hitchcock MJ. Tenofovir alafenamide: a novel prodrug of tenofovir for the treatment of human immunodeficiency virus. *Antivir Res.* 2016;125:63–70.
133 Carter SG, Kessler JA, Rankin CD. Activities of (-)-carbovir and 3′-azido-3′-deoxythymidine against human immunodeficiency virus in vitro. *Antimicrob Agents Chemother.* 1990;34(6):1297–300.
134 Vince R, Hua M, Brownell J, Daluge S, Lee F, Shannon WM, et al. Potent and selective activity of a new carbocyclic nucleoside analog (carbovir: NSC 614846) against human immunodeficiency virus in vitro. *Biochem Biophys Res Commun.* 1988;156(2):104653.
135 Yeom Y, Remmel R, Huang S, Hua M, Vince R, Zimmerman C, et al. Pharmacokinetics and bioavailability of carbovir, a carbocyclic nucleoside active against human immunodeficiency virus, in rats. *Antimicrob Agents Chemother.* 1989;33(2):171–5.
136 Foster RH, Faulds D. Abacavir. *Drugs.* 1998;55(5):729–8.
137 Daluge SM, Good SS, Faletto MB, Miller WH, St Clair MH, Boone LR, et al. 1592U89, a novel carbocyclic nucleoside analog with potent, selective anti-human immunodeficiency virus activity. *Antimicrob Agents Chemother.* 1997;41(5):1082–93.
138 Yuen GJ, Weller S, Pakes GE. A review of the pharmacokinetics of abacavir. *Clin Pharmacokinet.* 2008;47:351–71.
139 Faletto MB, Miller WH, Garvey EP, St Clair MH, Daluge SM, Good SS, et al. Unique intracellular activation of the potent anti-human immunodeficiency virus agent 1592U89. *Antimicrob Agents Chemother.* 1997;41(5):1099–107.
140 Gentile I, Borgia F, R Buonomo A, Castaldo G, Borgia G. A novel promising therapeutic option against hepatitis C virus: an oral nucleotide NS5B polymerase inhibitor sofosbuvir. *Curr Med Chem.* 2013;20(30):3733–42.

141 Takamatsu Y, Tanaka Y, Kohgo S, Murakami S, Singh K, Das D, et al. 4′-modified nucleoside analogs: potent inhibitors active against entecavir-resistant hepatitis B virus. *Hepatology*. 2015;62(4):1024–36.

142 Lawitz E, Mangia A, Wyles D, Rodriguez-Torres M, Hassanein T, Gordon SC, et al. Sofosbuvir for previously untreated chronic hepatitis C infection. *N Engl J Med*. 2013;368(20):1878–87.

143 Michailidis E, Kirby KA, Hachiya A, Yoo W, Hong SP, Kim S-O, et al. Antiviral therapies: focus on hepatitis B reverse transcriptase. *Int J Biochem Cell Biol*. 2012;44(7):1060–71.

144 Woo G, Tomlinson G, Nishikawa Y, Kowgier M, Sherman M, Wong DK, et al. Tenofovir and entecavir are the most effective antiviral agents for chronic hepatitis B: a systematic review and Bayesian meta-analyses. *Gastroenterology*. 2010;139(4):1218–29.e5.

145 Bisacchi G, Chao S, Bachard C, Daris J, Innaimo S, Jacobs G, et al. BMS-200475, a novel carbocyclic 2′-deoxyguanosine analog with potent and selective anti-hepatitis B virus activity in vitro. *Bioorg Med Chem Lett*. 1997;7(2):127–32.

146 Langley DR, Walsh AW, Baldick CJ, Eggers BJ, Rose RE, Levine SM, et al. Inhibition of hepatitis B virus polymerase by entecavir. *J Virol*. 2007;81(8):3992–4001.

147 Yin W, Mao C, Luan X, Shen D-D, Shen Q, Su H, et al. Structural basis for inhibition of the RNA-dependent RNA polymerase from SARS-CoV-2 by remdesivir. *Science*. 2020;368(6498):1499–504.

148 Sohl CD, Singh K, Kasiviswanathan R, Copeland WC, Mitsuya H, Sarafianos SG, et al. Mechanism of interaction of human mitochondrial DNA polymerase γ with the novel nucleoside reverse transcriptase inhibitor 4′-ethynyl-2-fluoro-2′-deoxyadenosine indicates a low potential for host toxicity. *Antimicrob Agents Chemother*. 2012;56(3):1630–4.

149 Nakata H, Amano M, Koh Y, Kodama E, Yang G, Bailey CM, et al. Activity against human immunodeficiency virus type 1, intracellular metabolism, and effects on human DNA polymerases of 4′-ethynyl-2-fluoro-2′-deoxyadenosine. *Antimicrob Agents Chemother*. 2007;51(8):2701–8.

150 Kirby K, Singh K, Michailidis E, Marchand B, Kodama E, Ashida N, et al. The sugar ring conformation of 4′-ethynyl-2-fluoro-2′-deoxyadenosine and its recognition by the polymerase active site of hiv reverse transcriptase. *Cell Mol Biol*. 2011;57(1):40.

151 Kirby KA, Michailidis E, Fetterly TL, Steinbach MA, Singh K, Marchand B, et al. Effects of substitutions at the 4′ and 2 positions on the bioactivity of 4′-ethynyl-2-fluoro-2′-deoxyadenosine. *Antimicrob Agents Chemother*. 2013;57(12):6254–64.

152 Shah P, Westwell AD. The role of fluorine in medicinal chemistry. *J Enzyme Inhib Med Chem*. 2007;22(5):527–40.

153 Hagmann WK. The many roles for fluorine in medicinal chemistry. *J Med Chem*. 2008;51(15):4359–69.

154 Böhm HJ, Banner D, Bendels S, Kansy M, Kuhn B, Müller K, et al. Fluorine in medicinal chemistry. *ChemBioChem*. 2004;5(5):637–43.

155 Orozco M, Canela EI, Franco R. Theoretical study of the hydroxyl nucleophilic attack on the 6-aminopyrimidine molecule: functional implications in the reaction mechanism of nucleoside deaminative enzymes. *J Org Chem*. 1990;55(9):2630–7.

156 Orozco M, Lluis C, Mallol J, Canela EI, Franco R. Theoretical approximation to the reaction mechanism of adenosine deaminase. *Quant Struct-Activ Relat.* 1989;8(2):109–14.

157 Frick L, Wolfenden R, Smal E, Baker DC. Transition-state stabilization by adenosine deaminase: structural studies of its inhibitory complex with deoxycoformycin. *Biochemistry.* 1986;25(7):1616–21.

158 Wolfenden R, Frick L. Mechanisms of enzyme action and inhibition: transition state analogues for acid-base catalysis. *J Protein Chem.* 1986;5(2):147–55.

159 Suydam IT, Strobel SA. Fluorine substituted adenosines as probes of nucleobase protonation in functional RNAs. *J Am Chem Soc.* 2008;130(41):13639–48.

160 Sabini E, Ort S, Monnerjahn C, Konrad M, Lavie A. Structure of human dCK suggests strategies to improve anticancer and antiviral therapy. *Nat Struct Biol.* 2003;10(7):513–9.

161 Michailidis E, Marchand B, Kodama EN, Singh K, Matsuoka M, Kirby KA, et al. Mechanism of inhibition of HIV-1 reverse transcriptase by 4′-ethynyl-2-fluoro-2′-deoxyadenosine triphosphate, a translocation-defective reverse transcriptase inhibitor. *J Biol Chem.* 2009;284(51):35681–91.

162 Michailidis E, Huber AD, Ryan EM, Ong YT, Leslie MD, Matzek KB, et al. 4′-Ethynyl-2-fluoro-2′-deoxyadenosine (EFdA) inhibits HIV-1 reverse transcriptase with multiple mechanisms. *J Biol Chem.* 2014;289(35):24533–48.

163 Muftuoglu Y, Sohl CD, Mislak AC, Mitsuya H, Sarafianos SG, Anderson KS, et al. Probing the molecular mechanism of action of the HIV-1 reverse transcriptase inhibitor 4′-ethynyl-2-fluoro-2′-deoxyadenosine (EFdA) using pre-steady-state kinetics. *Antivir Res.* 2014;106:1–4.

164 Database SUDR. NRTI Resistance Notes.

165 Larder BA, Kemp SD. Multiple mutations in HIV-1 reverse transcriptase confer high-level resistance to zidovudine (AZT). *Science.* 1989;246(4934):1155–8.

166 Boyer PL, Sarafianos SG, Arnold E, Hughes SH. Selective excision of AZTMP by drug-resistant human immunodeficiency virus reverse transcriptase. *J Virol.* 2001;75(10):4832–42.

167 Acosta-Hoyos AJ, Scott WA. The role of nucleotide excision by reverse transcriptase in HIV drug resistance. *Viruses.* 2010;2(2):372–94.

168 Chamberlain P, Ren J, Nichols C, Douglas L, Lennerstrand J, Larder B, et al. Crystal structures of zidovudine-or lamivudine-resistant human immunodeficiency virus type 1 reverse transcriptases containing mutations at codons 41, 184, and 215. *J Virol.* 2002;76(19):10015–9.

169 Boyer PL, Sarafianos SG, Arnold E, Hughes SH. Nucleoside analog resistance caused by insertions in the fingers of human immunodeficiency virus type 1 reverse transcriptase involves ATP-mediated excision. *J Virol.* 2002;76(18):9143–51.

170 Oliveira M, Brenner BG, Xu H, Ibanescu R-I, Mesplède T, Wainberg MA, et al. M184I/V substitutions and E138K/M184I/V double substitutions in HIV reverse transcriptase do not significantly affect the antiviral activity of EFdA. *J Antimicrob Chemother.* 2017;72(11):3008–11.

171 Diamond TL, Ngo W, Xu M, Goh SL, Rodriguez S, Lai M-T, et al. Islatravir has a high barrier to resistance and exhibits a differentiated resistance profile from approved nucleoside reverse transcriptase inhibitors (NRTIs). *Antimicrob Agents Chemother*. 2022;66(6):e00133-22.

172 Cilento ME, Reeve AB, Michailidis E, Ilina TV, Nagy E, Mitsuya H, et al. Development of human immunodeficiency virus type 1 resistance to 4′-ethynyl-2-fluoro-2′-deoxyadenosine starting with wild-type or nucleoside reverse transcriptase inhibitor-resistant strains. *Antimicrob Agents Chemother*. 2021;65(12):10.1128/aac.01167-21.

173 Aulicino PC, Sharma S, Truong K, Kukunoor V, Ghei K, Arazi-Caillaud S, et al. Variable antiviral activity of islatravir against M184I/V mutant HIV-1 selected during antiretroviral therapy. *J Antimicrob Chemother*. 2023;79(2):370–4.

174 Michailidis E, Ryan EM, Hachiya A, Kirby KA, Marchand B, Leslie MD, et al. Hypersusceptibility mechanism of Tenofovir-resistant HIV to EFdA. *Retrovirology*. 2013;10:1–12.

175 Lai M-T, Feng M, Xu M, Ngo W, Diamond TL, Hwang C, et al. Doravirine and islatravir have complementary resistance profiles and create a combination with a high barrier to resistance. *Antimicrob Agents Chemother*. 2022;66(5):e02223-21.

176 Feng JY, Ly JK, Myrick F, Goodman D, White KL, Svarovskaia ES, et al. The triple combination of tenofovir, emtricitabine and efavirenz shows synergistic anti-HIV-1 activity in vitro: a mechanism of action study. *Retrovirology*. 2009;6(1):1–16.

177 Borroto-Esoda K, Vela JE, Myrick F, Ray AS, Miller MD. In vitro evaluation of the anti-HIV activity and metabolic interactions of tenofovir and emtricitabine. *Antivir Ther*. 2006;11(3):377–84.

178 Hachiya A, Reeve AB, Marchand B, Michailidis E, Ong YT, Kirby KA, et al. Evaluation of combinations of 4′-ethynyl-2-fluoro-2′-deoxyadenosine with clinically used antiretroviral drugs. *Antimicrob Agents Chemother*. 2013;57(9):4554–8.

179 Cilento ME, Ong YT, Tedbury PR, Sarafianos SG. Drug interactions in lenacapavir-based long-acting antiviral combinations. *Viruses*. 2022;14(6):1202.

180 Zheng S, Wang W, Aldahdooh J, Malyutina A, Shadbahr T, Tanoli Z, et al. SynergyFinder plus: toward better interpretation and annotation of drug combination screening datasets. *Genom Proteom Bioinform*. 2022;20(3):587–96.

181 Kageyama M, Miyagi T, Yoshida M, Nagasawa T, Ohrui H, Kuwahara S, et al. Concise synthesis of the anti-HIV nucleoside EFdA. *Biosci Biotechnol Biochem*. 2012;76(6):1219–25.

182 Kageyama M, Nagasawa T, Yoshida M, Ohrui H, Kuwahara S. Enantioselective total synthesis of the potent anti-HIV nucleoside EFdA. *Org Lett*. 2011;13(19):5264–6.

183 Fukuyama K, Ohrui H, Kuwahara S. Synthesis of EFdA via a diastereoselective aldol reaction of a protected 3-keto furanose. *Org Lett*. 2015;17(4):828–31.

184 McLaughlin M, Kong J, Belyk KM, Chen B, Gibson AW, Keen SP, et al. Enantioselective synthesis of 4′-ethynyl-2-fluoro-2′-deoxyadenosine (EFdA) via enzymatic desymmetrization. *Org Lett*. 2017;19(4):926–9.

185 Huffman MA, Fryszkowska A, Alvizo O, Borra-Garske M, Campos KR, Canada KA, et al. Design of an in vitro biocatalytic cascade for the manufacture of islatravir. *Science*. 2019;366(6470):1255–9.

186 Patel NR, Nawrat CC, McLaughlin M, Xu Y, Huffman MA, Yang H, et al. Synthesis of islatravir enabled by a catalytic, enantioselective alkynylation of a ketone. *Org Lett.* 2020;22(12):4659–64.

187 Nawrat CC, Whittaker AM, Huffman MA, McLaughlin M, Cohen RD, Andreani T, et al. Nine-step stereoselective synthesis of islatravir from deoxyribose. *Org Lett.* 2020;22(6):2167–72.

188 Hong CM, Xu Y, Chung JY, Schultz DM, Weisel M, Varsolona RJ, et al. Development of a commercial manufacturing route to 2-fluoroadenine, the key unnatural nucleobase of islatravir. *Org Process Res Dev.* 2020;25(3):395–404.

189 Patel NR, Huffman MA, Wang X, Ding B, McLaughlin M, Newman JA, et al. Five-step enantioselective synthesis of islatravir via asymmetric ketone alkynylation and an ozonolysis cascade. *Chem Eur J.* 2020;26(62):14118–23.

190 Johnson HC, Zhang S, Fryszkowska A, Ruccolo S, Robaire SA, Klapars A, et al. Biocatalytic oxidation of alcohols using galactose oxidase and a manganese (III) activator for the synthesis of islatravir. *Org Biomol Chem.* 2021;19(7):1620–5.

191 Chen Y, Jo J, Hernandez E, Wang H, Bernardoni F. Biocatalytic cascade process of islatravir: analytical and regulatory control strategy of minor enantiomer. *J Pharm Biomed Anal.* 2023;234:115536.

192 Shaw MH, Fryszkowska A, Alvizo O, Attadgie I, Borra-Garske M, Devine PN, et al. Development of a biocatalytic aerobic oxidation for the manufacturing route to islatravir. *Org Process Res Dev.* 2024;28(9):3545–59.

193 Hatziioannou T, Evans DT. Animal models for HIV/AIDS research. *Nat Rev Microbiol.* 2012;10(12):852–67.

194 Hattori S, Ide K, Nakata H, Harada H, Suzu S, Ashida N, et al. Potent activity of a nucleoside reverse transcriptase inhibitor, 4′-ethynyl-2-fluoro-2′-deoxyadenosine, against human immunodeficiency virus type 1 infection in a model using human peripheral blood mononuclear cell-transplanted NOD/SCID Janus kinase 3 knockout mice. *Antimicrob Agents Chemother.* 2009;53(9):3887–93.

195 Stoddart CA, Galkina SA, Joshi P, Kosikova G, Moreno ME, Rivera JM, et al. Oral administration of the nucleoside EFdA (4′-ethynyl-2-fluoro-2′-deoxyadenosine) provides rapid suppression of HIV viremia in humanized mice and favorable pharmacokinetic properties in mice and the rhesus macaque. *Antimicrob Agents Chemother.* 2015;59(7):4190–8.

196 Shanmugasundaram U, Kovarova M, Ho PT, Schramm N, Wahl A, Parniak MA, et al. Efficient inhibition of HIV replication in the gastrointestinal and female reproductive tracts of humanized BLT mice by EFdA. *PLoS One.* 2016;11(7):e0159517.

197 Maidji E, Moreno ME, Rivera JM, Joshi P, Galkina SA, Kosikova G, et al. Cellular HIV reservoirs and viral rebound from the lymphoid compartments of 4′-ethynyl-2-fluoro-2′-deoxyadenosine (EFdA)-suppressed humanized mice. *Viruses.* 2019;11(3):256.

198 Tebit DM, Ndembi N, Weinberg A, Quiñones-Mateu ME. Mucosal transmission of human immunodeficiency virus. *Curr HIV Res.* 2012;10(1):3–8.

199 Kovarova M, Shanmugasundaram U, Baker CE, Spagnuolo RA, De C, Nixon CC, et al. HIV pre-exposure prophylaxis for women and infants prevents vaginal and oral HIV transmission in a preclinical model of HIV infection. *J Antimicrob Chemother.* 2016;71(11):3185–94.

200 Kovarova M, Wessel SE, Johnson CE, Anderson SV, Cottrell ML, Sykes C, et al. EFdA efficiently suppresses HIV replication in the male genital tract and prevents penile HIV acquisition. *mBio*. 2023;14(4):e02224-22.

201 Murphey-Corb M, Rajakumar P, Michael H, Nyaundi J, Didier PJ, Reeve AB, et al. Response of simian immunodeficiency virus to the novel nucleoside reverse transcriptase inhibitor 4′-ethynyl-2-fluoro-2′-deoxyadenosine in vitro and in vivo. *Antimicrob Agents Chemother*. 2012;56(9):4707–12.

202 Stoddart CA, Galkina SA, Joshi P, Kosikova G, Moreno ME, Rivera JM, et al. Oral administration of the nucleoside EFdA (4′-ethynyl-2-fluoro-2′-deoxyadenosine) provides rapid suppression of HIV viremia in humanized mice and favorable pharmacokinetic properties in mice and the rhesus macaque. *Antimicrob Agents Chemother*. 2015;59(7):4190–8.

203 Barrett SE, Teller RS, Forster SP, Li L, Mackey MA, Skomski D, et al. Extended-duration MK-8591-eluting implant as a candidate for HIV treatment and prevention. *Antimicrob Agents Chemother*. 2018;62(10):e01058-18.

204 Markowitz M, Gettie A, Bernard LS, Andrews CD, Mohri H, Horowitz A, et al. Once-weekly oral dosing of MK-8591 protects male rhesus macaques from intrarectal challenge with SHIV109CP3. *J Infect Dis*. 2019;221(9):1398–406.

205 Beloor J, Kudalkar SN, Buzzelli G, Yang F, Mandl HK, Rajashekar JK, et al. Long-acting and extended-release implant and nanoformulations with a synergistic antiretroviral two-drug combination controls HIV-1 infection in a humanized mouse model. *Bioeng Transl Med*. 2022;7(1):e10237.

206 Kudalkar SN, Beloor J, Quijano E, Spasov KA, Lee W-G, Cisneros JA, et al. From in silico hit to long-acting late-stage preclinical candidate to combat HIV-1 infection. *Proc Natl Acad Sci*. 2018;115(4):E802–11.

207 Li L, Gatto GJ, Brand RM, Krovi SA, Cottrell ML, Norton C, et al. Long-acting biodegradable implant for sustained delivery of antiretrovirals (ARVs) and hormones. *J Control Release*. 2021;340:188–99.

208 Kinsale TS, Cottrell ML, Li L, Brand R, Gatto G, Luecke E, et al. Pharmacokinetic modeling to guide preclinical development of an islatravir-eluting reservoir-style biodegradable implant for long-acting HIV PrEP. *Pharmaceutics*. 2024;16(2):201.

209 Zhang W, Parniak MA, Sarafianos SG, Cost MR, Rohan LC. Development of a vaginal delivery film containing EFdA, a novel anti-HIV nucleoside reverse transcriptase inhibitor. *Int J Pharm*. 2014;461(1–2):203–13.

210 Zhang W, Hu M, Shi Y, Gong T, Dezzutti CS, Moncla B, et al. Vaginal microbicide film combinations of two reverse transcriptase inhibitors, EFdA and CSIC, for the prevention of HIV-1 sexual transmission. *Pharm Res*. 2015;32(9):2960–72.

211 Young IC, Srinivasan P, Shrivastava R, Janusziewicz R, Thorson A, Cottrell ML, et al. Next generation 3D-printed intravaginal ring for prevention of HIV and unintended pregnancy. *Biomaterials*. 2023;301:122260.

212 Daly MB, Wong-Sam A, Li L, Krovi A, Gatto GJ, Norton C, et al. Pharmacokinetic study of islatravir and etonogestrel implants in macaques. *Pharmaceutics*. 2023;15(12):2676.

213 Shen Z, Rodriguez-Garcia M, Patel MV, Bodwell J, Kashuba ADM, Wira CR. Hormonal contraceptives differentially suppress TFV and TAF inhibition of

HIV infection and TFV-DP in blood and genital tract CD4+ T cells. *Sci Rep.* 2017;7(1):17697.

214 Cottrell ML, Prince HMA, Schauer AP, Sykes C, Maffuid K, Poliseno A, et al. Decreased tenofovir diphosphate concentrations in a transgender female cohort: implications for human immunodeficiency virus preexposure prophylaxis. *Clin Infect Dis.* 2019;69(12):2201–4.

215 Pons-Faudoa FP, Trani ND, Capuani S, Facchi I, Wood AM, Nehete B, et al. Antiviral potency of long-acting islatravir subdermal implant in SHIV-infected macaques. *J Control Release.* 2024;366:18–27.

216 Merck Advances Phase 3 Trial to Evaluate Investigational Islatravir as Once-Monthly Oral PrEP for Women at High Risk for Acquiring HIV-1 [press release]. 2020.

217 DOR/ISL in HIV-1 Antiretroviral Treatment-naïve Participants (MK-8591A-053) [Available from: https://classic.clinicaltrials.gov/show/NCT05705349.

218 A Switch to Doravirine/Islatravir (DOR/ISL) in Participants With Human Immunodeficiency Virus Type 1 (HIV-1) Who Are Virologically Suppressed on Bictegravir/Emtricitabine/Tenofovir Alafenamide (BIC/FTC/TAF) (MK-8591A-052) [Available from: https://classic.clinicaltrials.gov/show/NCT05630755.

219 Oral ISL QM as PrEP in Cisgender Women at High Risk for HIV-1 Infection (MK-8591-022) [Available from: https://classic.clinicaltrials.gov/show/NCT04644029.

220 Radiopaque Matrix MK-8591 Implant in Participants at Low-Risk for Human Immunodeficiency Virus Type 1 (HIV-1) Infection (MK-8591-043) [Available from: https://classic.clinicaltrials.gov/show/NCT05115838.

221 Study Evaluating the Safety and Efficacy of Islatravir in Combination With Lenacapavir in Virologically Suppressed People With HIV [Available from: https://classic.clinicaltrials.gov/show/NCT05052996.

222 Islatravir (MK-8591) With Doravirine and Lamivudine in Participants Infected With Human Immunodeficiency Virus Type 1 (MK-8591-011) [Available from: https://classic.clinicaltrials.gov/show/NCT03272347.

223 Oral Islatravir (MK-8591) Once-Monthly as Preexposure Prophylaxis (PrEP) in Men and Transgender Women Who Are at High Risk for HIV-1 Infection (MK-8591-024) [Available from: https://classic.clinicaltrials.gov/show/NCT04652700.

224 A Study of Islatravir (MK-8591) in Anti-Retroviral Therapy-Naive, Human Immunodeficiency Virus-1 Infected Participants (MK-8591-003) [Available from: https://classic.clinicaltrials.gov/show/NCT02217904.

225 Dose Ranging, Switch Study of Islatravir (ISL) and Ulonivirine (MK-8507) Once-Weekly in Virologically-Suppressed Adults With Human Immunodeficiency Virus Type 1 (HIV-1) MK-8591-013 [Available from: https://classic.clinicaltrials.gov/show/NCT04564547.

226 Study of Doravirine/Islatravir (DOR/ISL 100 mg/0.75 mg) to Evaluate the Antiretroviral Activity, Safety, and Tolerability in Treatment-Naïve Participants With Human Immunodeficiency Virus Type 1 (HIV-1) Infection (MK-8591A-020) [Available from: https://classic.clinicaltrials.gov/show/NCT04233879.

227 Switch to Doravirine/Islatravir (DOR/ISL) in Human Immunodeficiency Virus 1 (HIV-1) Participants Treated With Bictegravir/Emtricitabine/Tenofovir Alafenamide (BIC/FTC/TAF) (MK-8591A-018) [Available from: https://classic.clinicaltrials.gov/show/NCT04223791.

228 A Switch to Doravirine/Islatravir (DOR/ISL) in Participants With Human Immunodeficiency Virus Type 1 (HIV-1) Who Are Virologically Suppressed on Antiretroviral Therapy (ART) (MK-8591A-051) [Available from: https://classic.clinicaltrials.gov/show/NCT05631093.

229 A Study of Doravirine/Islatravir (DOR/ISL, MK-8591A) for the Treatment of Human Immunodeficiency Virus 1 (HIV-1) Infection in Participants Who Previously Received DOR/ISL (MK-8591A-054) [Available from: https://classic.clinicaltrials.gov/show/NCT05766501.

230 Safety and Efficacy of a Switch to Doravirine/Islatravir in Participants With HIV-1 (MK-8591A-017) [Available from: https://classic.clinicaltrials.gov/show/NCT04223778.

231 Doravirine/Islatravir (DOR/ISL) in Pediatric Participants With Human Immunodeficiency Virus Type 1 (HIV-1) Who Are <18 Years of Age and Weigh ≥35 kg (MK-8591A-028) [Available from: https://classic.clinicaltrials.gov/show/NCT04295772.

232 Doravirine/Islatravir (DOR/ISL) in Heavily Treatment-Experienced (HTE) Participants for Human Immunodeficiency Virus Type 1 (HIV-1) Infection (MK-8591A-019) [Available from: https://classic.clinicaltrials.gov/show/NCT04233216.

233 Islatravir and Methadone Pharmacokinetics (MK-8591-029) [Available from: https://classic.clinicaltrials.gov/show/NCT04568603.

234 Open-label, Follow-up of Doravirine/Islatravir for Participants With Human Immunodeficiency Virus-1 (HIV-1) Infection (MK-8591A-033) [Available from: https://classic.clinicaltrials.gov/show/NCT04776252.

235 Safety and Pharmacokinetics of Oral Islatravir (MK-8591) Once Monthly in Participants at Low Risk of Human Immunodeficiency Virus 1 (HIV-1) Infection (MK-8591-016) [Available from: https://classic.clinicaltrials.gov/show/NCT04003103.

236 Single-Dose Islatravir in Moderate Hepatic Impairment (MK-8591-030) [Available from: https://classic.clinicaltrials.gov/show/NCT04515641.

237 Pharmacokinetics of Islatravir in Participants With Severe Renal Impairment (MK-8591-026) [Available from: https://classic.clinicaltrials.gov/show/NCT04303156.

238 A Study of Islatravir (MK-8591) in Trans and Gender Diverse Participants (MK-8591-035) [Available from: https://classic.clinicaltrials.gov/show/NCT05130086.

239 Jointly Develop and Commercialize Long-Acting, Investigational Treatment Combinations of Lenacapavir and Islatravir in HIV [press release]. 2021.

240 Merck Co. I. Merck Announces Clinical Holds on Studies Evaluating Islatravir for the Treatment and Prevention of HIV-1 Infection. 2021.

241 Matthews RP, Patel M, Barrett SE, Haspeslagh L, Reynders T, Zhang S, et al. Safety and pharmacokinetics of islatravir subdermal implant for HIV-1 pre-exposure prophylaxis: a randomized, placebo-controlled phase 1 trial. *Nat Med.* 2021;27(10):1712–7.

242 Shi Y, Lu A, Wang X, Belhadj Z, Wang J, Zhang Q. A review of existing strategies for designing long-acting parenteral formulations: focus on underlying mechanisms, and future perspectives. *Acta Pharm Sin B.* 2021;11(8):2396–415.

243 Kim H-y, Zhang L, Hendrix CW, Haberer JE, von Kleist M. Modeling of HIV-1 prophylactic efficacy and toxicity with islatravir shows non-superiority for oral dosing, but promise as a subdermal implant. bioRxiv. 2024:10.1101/2024.04.15.589489.

244 Kinvig H, Cottura N, Lloyd A, Frivold C, Mistilis J, Jarrahian C, et al. Evaluating islatravir administered via microneedle array patch for long-acting HIV pre-exposure prophylaxis using physiologically based pharmacokinetic modelling. *Eur J Drug Metab Pharmacokinet.* 2022;47(6):855–68.

245 Donnelly RF, Prausnitz MR. The promise of microneedle technologies for drug delivery. *Drug Deliv Transl Res.* 2024;14:573–80.

246 Merck to Initiate New Phase 3 Clinical Program with Lower Dose of Daily Oral Islatravir in Combination with Doravirine for Treatment of People with HIV-1 Infection [press release]. 2022.

247 A Study of MK-8527 in Human Immunodeficiency Type 1 Virus (HIV-1) Infected Participants (MK-8527002) [Available from: https://classic.clinicaltrials.gov/show/NCT03615183.

248 MK-8527 Single-Dose Trial in HIV-1 Infected Participants (MK-8527-004) [Available from: https://classic.clinicaltrials.gov/show/NCT05494736.

249 Safety and Pharmacokinetic Study of Oral MK-8527 QM in Participants at Low-Risk for HIV-1 Infection (MK-8527-007) [Available from: https://classic.clinicaltrials.gov/show/NCT06045507.

Part III

Hepatitis Viruses

5

Discovery of the RNA Interference Therapeutic Imdusiran, a GalNAc-conjugated siRNA

Emily Thi

Arbutus Biopharma Inc., 701 Veterans Circle Warminster, PA, USA

5.1 Introduction

Short interfering RNAs (siRNAs) have been a transformative therapeutic modality over the past decade, enabling selective reduction in human proteins across various diseases [1, 2]. Effective application of siRNAs *in vivo* has required the development of delivery platforms needed to ensure stability of siRNAs as well as delivery to the intended tissue. Successful siRNA delivery platforms utilized previously involve lipid nanoparticle (LNP)-based systems as well as *N*-acetylgalactosamine-conjugated siRNAs with or without the use of agents to facilitate the endosomal escape of siRNA that is required to enable engagement with the RNA-induced silencing complex (RISC) present in the host cell cytoplasm (Figure 5.1) [1, 3–5]. siRNA delivered via LNP has been clinically validated against liver targets with the approval of Onpattro (patisiran), an LNP-siRNA formulation approved for the treatment of polyneuropathy caused by hereditary transthyretin-mediated amyloidosis in 2018 [6]. In the infectious diseases field, LNP-siRNA has also been used effectively against viral targets, with proof-of-concept preclinical antiviral effects demonstrated against acute viral infections such as Ebola and other hemorrhagic fever viruses [7–9]. Against chronic viral infections such as hepatitis B virus (HBV) and hepatitis delta virus (HDV), LNP-siRNA have demonstrated potent antiviral effects in preclinical models [10, 11]; however, clinical efficacy and applicability against viral infections has been hampered by the dose-limiting risk of intravenous infusion reactions associated with the LNP and the need for concomitant steroid use and close monitoring for management of this risk [12]. In contrast, subcutaneously administered GalNAc-conjugated siRNA have seen clinical success against endogenous targets, such as the recently approved Amvuttra (vutrisiran) [13]. To date, a GalNAc-siRNA antiviral has yet to be approved [1].

Trends in Antiviral Drug Development, First Edition. Edited by Michael J. Sofia and Zhengqiang Wang.

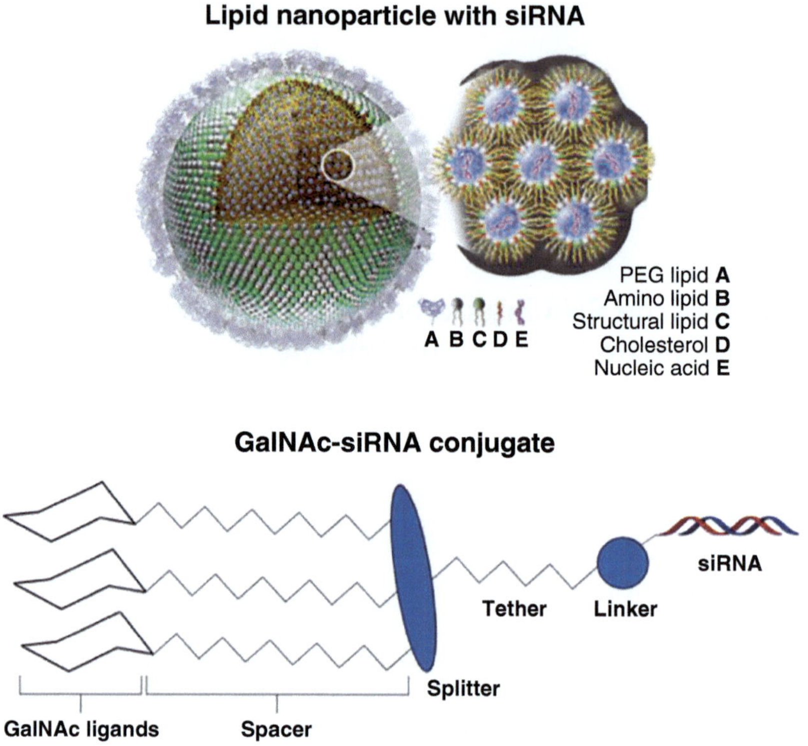

Figure 5.1 Structural components of typical LNP-siRNA and GalNAc-siRNA conjugates.

While both systems enable effective delivery of siRNA *in vivo*, LNP and GalNAc conjugate systems fundamentally differ in their compositions as well as mechanisms for delivery to the liver, the target organ for HBV. These differences also affect the depth and durability of antiviral activity, and these should be considered for selecting the appropriate delivery platform against the viral target. In this chapter, we describe the discovery and preclinical characterization of imdusiran, an siRNA therapeutic against HBV, beginning with a comparison of HBV antiviral efficacy between LNP-mediated siRNA delivery and GalNAc conjugate systems in preclinical *in vivo* models. Subsequent selection of the GalNAc conjugate platform entailed optimization of the GalNAc ligand to enable greater efficacy through modification of valency, and siRNA trigger selection against different HBV targets. Optimization of these elements resulted in imdusiran, a potent, highly efficacious GalNAc-siRNA against HBV with demonstrated clinical activity in patients chronically infected with HBV [14, 15].

5.2 Comparison of Preclinical Anti-HBV Efficacy Between Lipid Nanoparticle-encapsulated and GalNAc-conjugated siRNA

Lipid nanoparticles used to mediate siRNA delivery typically consist of a phospholipid, cholesterol, a poly(ethylene glycol) (PEG)-conjugated lipid, and an ionizable lipid that plays a key role in facilitating particle formation and endosomal release of the siRNA trigger payload [16, 17]. LNP for siRNA delivery is administered intravenously, and delivery to the liver is mediated by the association of LNP with apolipoproteins in the blood and uptake into hepatocytes via low-density lipoprotein receptors [5, 16, 17]. Upon uptake into endosomes, the eventual acidification during endosomal maturation triggers protonation of the ionizable lipid present in the LNP and fusion of the LNP with the endosomal membrane, resulting in a burst of siRNA payload release into the cytoplasm where engagement with the RISC complex occurs to enable silencing of the target RNA [5, 16, 17]. The effectiveness of LNP-mediated delivery of siRNA against HBV has been demonstrated in preclinical mouse models of HBV, where greater than 3 $\log_{10}$ reduction in HBsAg was observed at doses ≥0.3 mg/kg in adeno-associated virus (AAV)-HBV mice and up to 1.9 $\log_{10}$ decline in HBsAg was observed in HBV-infected humanized chimeric mice with biweekly treatment [10]. This effect on HBsAg, which is required for HDV virion production, was also found to have a concomitant impact on HDV viremia, reducing serum HDV RNA by 1.6 $\log_{10}$ with biweekly treatment in a HBV/HDV humanized liver chimeric mouse model [11].

In contrast, GalNAc-conjugated siRNAs are subcutaneously administered, and delivery to the liver is mediated by the interaction of the GalNAc ligand component with the asialoglycoprotein receptor (ASGPR) expressed predominantly on hepatocytes. Uptake of GalNAc-siRNA occurs via endosomes; however, the mechanism by which GalNAc-conjugated siRNAs are released from endosomes is not clear. Less than 1% of GalNAc-conjugated siRNA is thought to immediately escape from endosomes [4], with the remaining residing in endosomal structures that act as depots within hepatocytes that serve to drive prolonged durability of the siRNA silencing activity [18]. Thus, GalNAc-siRNAs do not need to be dosed as frequently as LNP-siRNA to maintain a similar silencing effect. A clear example of this advantage in dosing frequency is illustrated by a comparison of the approved siRNA therapeutics for the treatment of hereditary transthyretin amyloidosis: while the LNP-delivered patisiran requires dosing every 3 weeks, the GalNAc-siRNA vutrisiran is dosed every 3 months [19].

These fundamental differences between the mechanisms of siRNA delivery with LNPs and GalNAc conjugates also impact the timing of the depth and durability of antiviral efficacy with siRNA delivered with these systems. Acute viral infections, such as with hemorrhagic fever viruses, may benefit from the rapid time to onset of RNA silencing activity enabled by LNP siRNA delivery, which provides effective antiviral activity during the short acute phase of the viral infection, at which time the viral load is exponentially increasing prior to the development of host immunity which ultimately clears the infection [7–9, 20]. For chronic viral infections such as HBV, however, the durability of antiviral efficacy may be more impactful. In a preclinical AAV mouse model of HBV (AAV-HBV), the same siRNA trigger encapsulated and delivered in a LNP formulation achieves similar depth of serum HBsAg reduction as the GalNAc conjugate at a 10-fold lower dose, and the nadir of silencing is reached earlier for LNP-siRNA (7 days post-dose) compared to the GalNAc-siRNA (14 days post-dose). However, LNP-siRNA is less durable in maintaining this effect, likely due to the rapid release of siRNA mediated by the LNP and lack of depot establishment that otherwise occurs with the GalNAc conjugate which facilitates continued engagement of the RISC complex and HBV silencing

(Figure 5. 2). These preclinical data suggest that for HBV targeting, a GalNAc conjugate approach would be most beneficial for enabling durable antiviral activity against this chronic viral infection.

5.3 Comparison of Tetravalent Versus Trivalent GalNAc Ligands

Having determined that a GalNAc conjugate system would be most appropriate for the development of an HBV siRNA antiviral, efforts were undertaken to improve the activity through optimization of the GalNAc ligand structure. Amongst other areas of focus, which included the ligand structure [21], the contribution of the valency of the GalNAc ligands to activity was also investigated. GalNAc ligands to date generally have a trivalent structure (Figure 5.1) in keeping with what is thought to be the predominant trimer subunit presentation of the ASGPR; however, tetra-, bi- and monoantennary ligands have been previously described [22], and *in vitro* comparison of binding affinity to ASGPR has been reported with a hierarchy of tetra- > tri- > bi- > monoantennary galactosides [22]. A comparison of binding affinity in HepG2 cells utilizing fluorescently-labeled tetravalent and trivalent GalNAc ligands also demonstrated better binding affinity for the tetravalent GalNAc ligand compared to the trivalent ligand (Figure 5.3). As GalNAc moieties are highly labile, the greater binding affinity observed for tetravalent compared to trivalent ligands is thought to be due to the retention of the trivalent ligand structure should there be a loss of one of the GalNAc moieties in the tetravalent ligand. Given this data, a tetravalent GalNAc ligand was selected as the backbone for imdusiran.

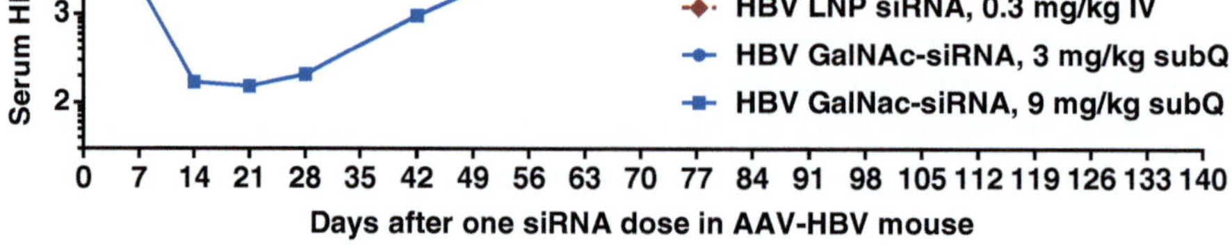

Figure 5.2 siRNA administered via LNP mediates a shorter duration of efficacy compared to the same siRNA administered as a GalNAc conjugate. Serum HBsAg in AAV-HBV mice administered a single dose of LNP-siRNA or GalNAc-siRNA targeting HBV on day 0. Serum HBsAg quantitated from group pooled serum collected at the indicated time points from N=5 mice/group. IV, intravenous; subQ, subcutaneous administration.

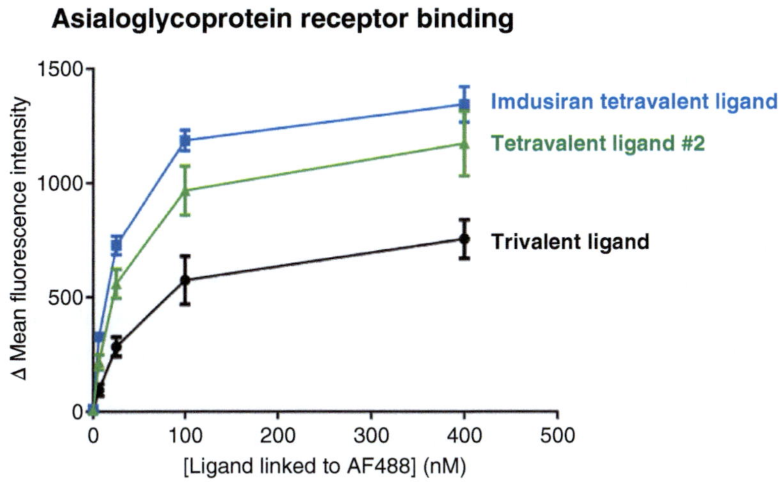

Figure 5.3 Tetravalent GalNAc ligands have increased binding to ASGPR compared to trivalent ligand. HepG2 cells were incubated with Alexafluor 488-labeled GalNAc ligand, and the uptake of fluorescence was measured by flow cytometry.

5.4 siRNA Selection: Comparison of siRNA Activity Against HBsAg and HBx Targets

Due to the presence of overlapping open reading frames in the HBV genome, silencing of multiple HBV RNA species and HBV antigens is possible with the use of a single siRNA trigger targeting a site located on all HBV transcripts. As HBsAg expression is of particular interest due to this HBV antigen's role in mediating host immune tolerance [23–25], as well as the ability of HBsAg to be produced from integrated HBV DNA, it was of interest to compare the efficacy obtained using an siRNA directly targeting the coding region of HBsAg with that using an siRNA targeting a site common to all HBV RNAs, located in the HBx region, but upstream of the HBV DNA integration breakpoint hotspot located around HBV nucleotide position 1800 [26, 27]. GalNAc-conjugated siRNAs were administered to AAV-HBV mice and knockdown of HBsAg was compared. Despite showing similar potencies (0.09 nM for HBsAg-targeting siRNA vs. 0.2 nM for HBx targeting siRNA) in a dual luciferase reporter assay system which allows for assessment of siRNA activity against the RNA target in the absence of viral infection, the GalNAc-siRNA targeting the HBsAg coding sequence was found to be less durable compared to the GalNAc-siRNA targeting HBx *in vivo* after a single dose (Figure 5.4). An siRNA strategy enabling silencing of HBx (which would not be achieved with a targeting strategy against HBsAg-coding region alone) may also yield additional benefits as this viral protein has been implicated as a negative regulator of type I interferon signaling [28–30], and silencing of HBx may also aid in the reappearance of the chromosome 5/6 complex (SMC5/6) which is a host factor that suppresses cccDNA transcription and maintains epigenetic HBV silencing [31]. Given these potential advantages, the

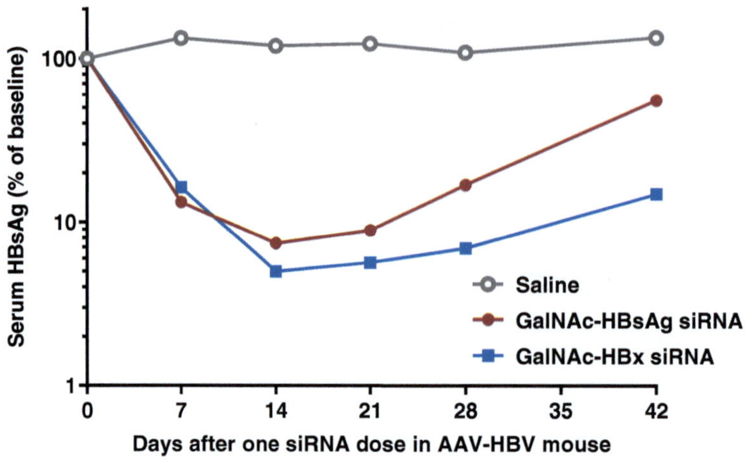

Figure 5.4 siRNA targeting the HBsAg coding region has a shorter duration of effect compared to siRNA targeting HBx. Serum HBsAg in AAV-HBV mice administered a single subcutaneous dose of GalNAc-siRNA targeting HBV at 3 mg/kg on day 0. Serum HBsAg quantitated from group pooled serum collected at the indicated time points from N = 5 mice/group.

siRNA targeting HBx was selected for further development as the siRNA trigger component in imdusiran.

5.5 Imdusiran Design and Preclinical Characterization

5.5.1 Abrogation of siRNA Immunostimulatory Potential

Nucleic acids, including siRNA and antisense oligonucleotides, are known to have the potential to activate the human immune system through a variety of nucleic acid receptor-mediated pathways [3, 32–35]. Immune responses to RNA oligonucleotides are primarily triggered by the activation of TLRs, in particular TLR7, TLR8, and TLR3, which are constitutively expressed within the endosomal compartment of human dendritic cells, monocytes and macrophages or on the surface of myeloid dendritic cells and epithelial cells, respectively. Activation of these pathways leads to the rapid, high-level production of cytokines such as IFN-α, and chemokines such as IP-10. High levels of cytokine responses can contribute to undesirable side effects such as toxicity and off-target drug activities. The inherent immunostimulatory characteristics of nucleic acid-based therapeutics are governed by the nucleotide sequence and chemistries used. Chemical modification of nucleotides is intended to minimize recognition of the nucleic acid by the mammalian immune system and has been demonstrated to abrogate innate immunostimulatory activity of siRNA as well as improve resistance against nuclease-mediated degradation [3, 32–36]. Although chemical modification is beneficial in eliminating the immunostimulatory properties of siRNA, this is often at the cost of reduced siRNA potency [3, 32–35]. With these in mind, a minimalistic approach was undertaken in regards to the chemical modification of the HBx-targeting siRNA of imdusiran; 2′-fluoro-modified ribonucleotides were sparingly employed and positioning of the 2′-*O*-methyl-modified ribonucleotides was empirically tested in AAV-HBV mouse studies to abrogate immunostimulatory properties of the siRNA while retaining siRNA efficacy. As a consequence, the siRNA of imdusiran is composed of 82% of 2′-*O*-methyl- (32 of 39 nucleotides) and 15% of 2′-fluoro-modified ribonucleotides (6 of 39 nucleotides) which improve resistance against nuclease-mediated degradation and abrogate the immunostimulatory potential inherent in nonhost nucleic acid polymers (Figure 5.5) but maintains efficacy. Phosphorothioate linkages were also added to stabilize against degradation by nucleases. In contrast, other GalNAc-siRNA in development for HBV have been described as containing 23% of 2′-fluoro-modified ribonucleotides (10/44 nucleotides) [5, 37].

In a human whole blood cytokine release assay, imdusiran treatment did not result in the production of proinflammatory cytokines, including IFN-α, IL-6, IL-1β, IL-1RA, IP-10, and TNF-α, which were previously identified as resulting from innate immune stimulation with exogenous nucleic acids [34, 35] (Figure 5.6). These results suggest that the minimal chemical modification approach employed in developing imdusiran siRNA was effective in eliminating innate immune responses.

5.5.2 Imdusiran siRNA Target Site Sequence

As imdusiran contains a single siRNA trigger, it was essential that the siRNA target site be highly conserved across HBV genotypes and HBV viral sequences described to date in HBV sequence databases. An HBx-targeting site was eventually selected that was present on all HBV RNA derived from either covalently closed circular DNA (cccDNA) or integrated HBV DNA, and was highly conserved (≥97.1%) among the 9098 HBV genotype A–H sequences available in the HBVdb database [38] (Table 5.1). This siRNA sequence also had demonstrated activity against representative clones of HBV genotypes A–H in a transient transfection human hepatoma HepG2 cell system (Table 5.2).

The core antisense strand sequence, including the seed region, was also selected to minimize off-target activity based on low sequence matches to the human transcriptome, the rhesus monkey transcriptome, and the microRNA database miRbase [33]. The decreased potential for off-target activity was confirmed in primary human hepatocytes (PHH). No evidence of off-targeting by imdusiran siRNA

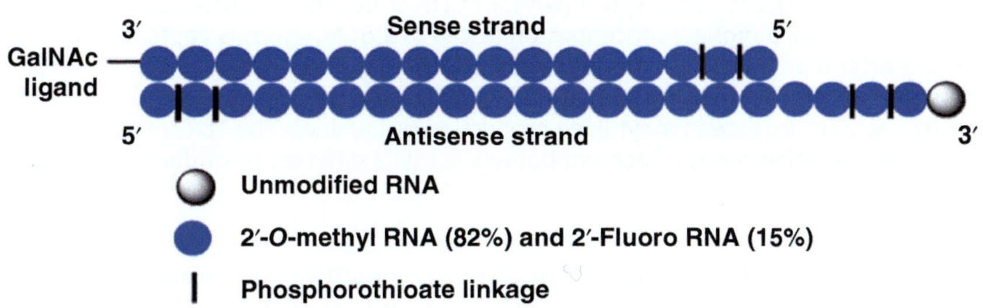

Figure 5.5 Imdusiran siRNA structure.

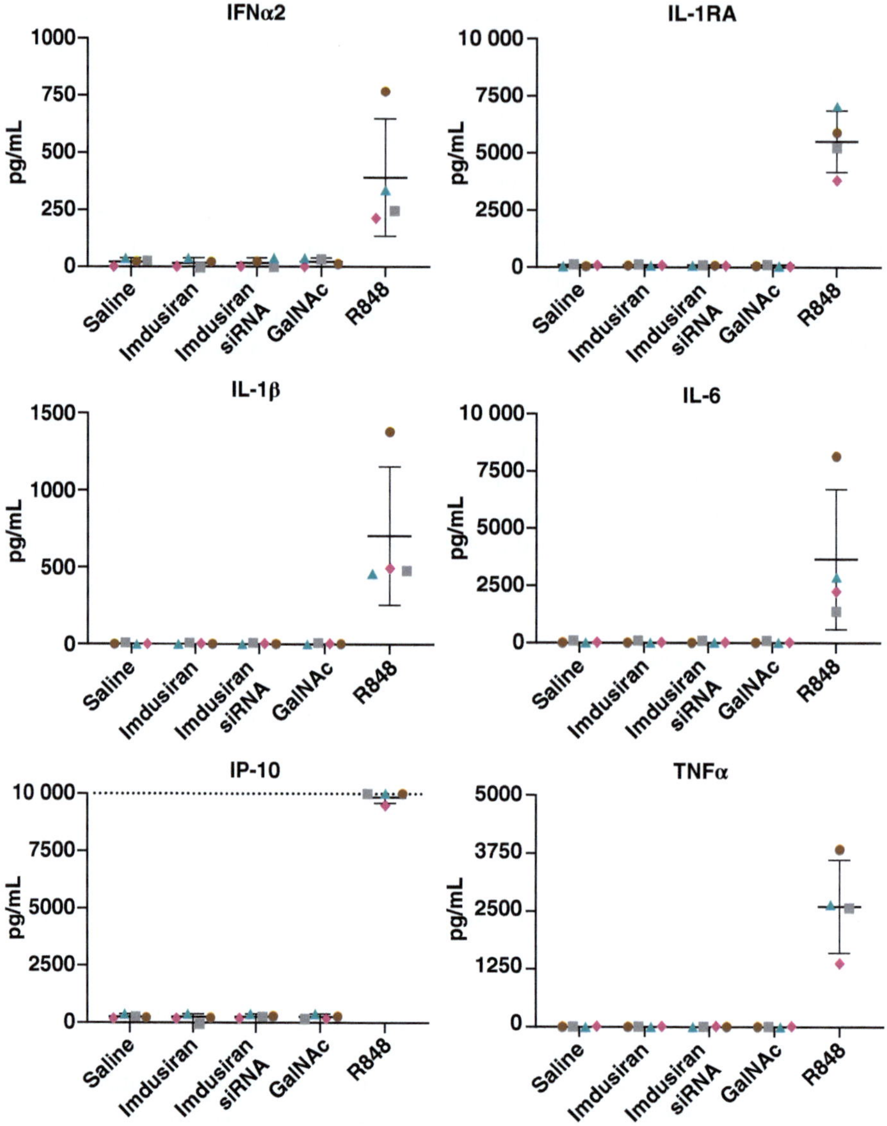

Figure 5.6 Imdusiran does not intrinsically induce cytokine production. Human whole blood from healthy donors was incubated for 24 h with imdusiran (GalNAc-ligand conjugated to siRNA), imdusiran siRNA (without GalNAc ligand), or GalNAc ligand individual components up to the highest concentration tested of 1.6 μM, which is approximately twofold greater than the C_{max} achieved from a 100 mg/kg imdusiran dose in preclinical models (at least 100-fold higher than the 60 mg clinical dose of imdusiran). Culture supernatants were assessed for cytokine production by Luminex. The TLR7/8 agonist R848 was used as a positive control. Each symbol represents a different individual healthy donor.

was detected in PHH under these conditions; in contrast, a previously described GalNAc-siRNA with off-targeting activity [39, 40] elicited changes in host gene expression (Figure 5.7).

Table 5.1 Imdusiran target site conservation in HBV genotypes A–H.

Genotype	A	B	C	D	E	F	G	H
No. of sequences analyzed[a]	942	2371	3524	1518	325	369	82	34
% Target site conservation[b]	98.9	99.8	99.6	98.1	99.2	99.2	96.4	97.1

a. 9098 HBV genotype A–H sequences available in HBVdb database accessed on 15 May 2019.
b. Complete complementarity to imdusiran antisense nucleotides 2–18.

In addition to confirming minimal off-targeting potential by imdusiran siRNA, it was of interest to determine the degree of imdusiran activity retained in situations where the HBV target site sequence contained mismatches. As an siRNA, imdusiran targets HBV RNA for RNAi-mediated degradation based on complementarity between a distinct HBV RNA sequence and the imdusiran siRNA antisense strand (guide) sequence. Given this mechanism of action, changes to the nucleotide sequence within the imdusiran target site may potentially impact imdusiran activity. Sequence changes within an siRNA target site may have differential effects on RNAi activity, ranging from enhancement to complete inhibition of siRNA activity, and are influenced by the identity and position of the nucleotide change [41, 42].

The effect of nucleotide changes within the imdusiran target site on imdusiran activity was examined by first identifying HBV sequences available in the public database of HBV sequences (HBVdb, [38]) which contained mismatches within the imdusiran target site. The most commonly occurring polymorphisms identified within the imdusiran target site consisted of 10 single nucleotide polymorphisms,

Table 5.2 Activity of imdusiran against representatives of HBV genotype A–H.

GenBank accession no.[a]	HBV genotype	EC_{50}[b] ± SD (μM)		
		Intracellular HBV DNA	Secreted HBsAg	Cytotoxicity CC_{50}
HE97437	A	1.24 ± 0.71	0.17 ± 0.14	>6.53
JN406371	B	0.75 ± 0.40	0.31 ± 0.08	>6.53
AB24634	C	1.43 ± 0.63	0.59 ± 0.15	>6.53
U95551	D	2.20 ± 1.14	0.38 ± 0.23	>6.53
HE97438	E	1.68 ± 0.43	0.32 ± 0.08	>6.53
HE97436	F	0.72 ± 0.32	0.27 ± 0.16	>6.53
AP00726	G	2.11 ± 0.11	0.38 ± 0.20	>6.53
AB06431	H	1.05 ± 0.01	0.21 ± 0.06	>6.53
AB17974	H	1.79 ± 1.08	0.89 ± 0.44	>6.53

a. Each of these variants has a fully conserved imdusiran target site.
b. Mean of n = 2–4 biological replicates ± standard deviation, 7 days after imdusiran application.

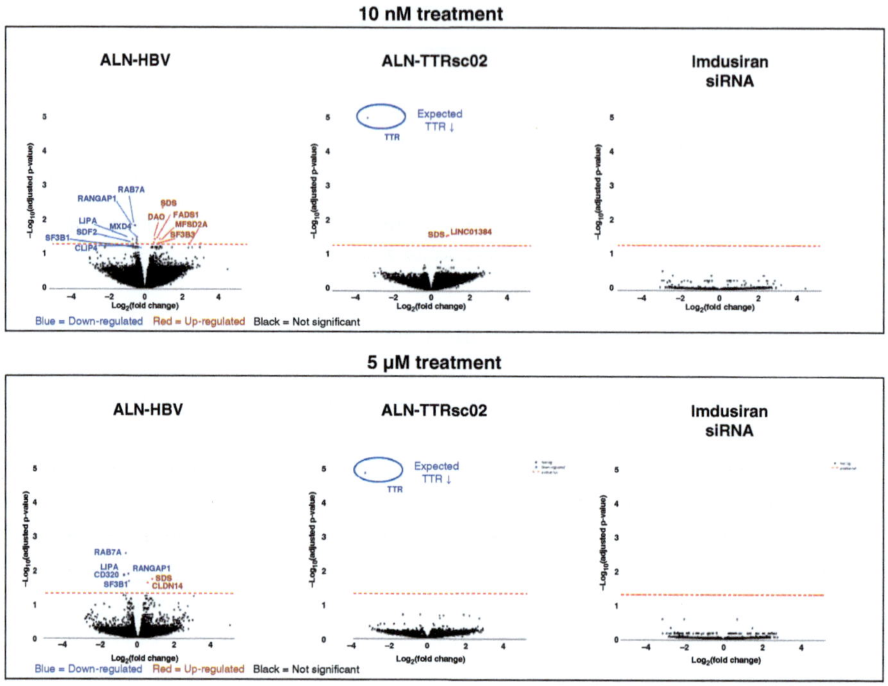

Figure 5.7 Off-targeting analysis in PHH.

Primary human hepatocytes from three different donors were treated with imdusiran siRNA, and total RNA changes of the pooled isolated RNA were assessed by RNAseq. A comparator GalNAc-siRNA previously described as possessing off-target activity and a clinical candidate GalNAc-siRNA against the endogenous liver target transthyretin were included as references. Total RNA changes at 3 days after treatment with imdusiran siRNA at a pharmacological concentration (10 nM) and a high concentration (5 μM) that represents a concentration that is 9-fold higher than the liver concentration of the predominant imdusiran antisense strand active metabolite in nonhuman primates dosed with 30 mg/kg (30-fold higher than the 60 mg clinical dose). Changes in RNA were assessed at 3 days post-treatment to align with the timing of the nadir of HBV RNA silencing by imdusiran. Volcano plots indicate the differentially expressed genes compared to saline treatment control that met a significance threshold of false discovery rate-adjusted $P < 0.05$. No differentially expressed genes were identified that met this criteria when PHH were treated with imdusiran siRNA. Red dashed line indicates P-value cut-off.

which were present at ≥0.5% prevalence in any one HBV genotype, or at ≥0.3% prevalence in at least two genotypes within 9098 HBV sequences available in the HBVdb [38] (Table 5.3). These nucleotide changes were then generated in a reference genotype D HBV sequence and imdusiran activity was assessed using HepG2 cells transiently transfected with HBV plasmids containing either wild-type or imdusiran target site variants. Imdusiran was found to retain activity against these HBV variants (Table 5.3). Additional assessments of imdusiran target site polymorphisms identified in CHB subjects in imdusiran clinical studies are ongoing and it will be of interest to compare any target site polymorphisms observed in the clinical population with those present in the HBVdb database.

Table 5.3 Activity of imdusiran against HBV variants with polymorphisms at target site.

HBV genome variant	Secreted HBsAg EC_{50}[a] ± SD (nM)	Cytotoxicity CC_{50} (nM)	Prevalence in HBVdb[b] (%)
T to A at nucleotide 1580	1.07±0.54	>65.30	Gt E=0.5%
C to G at nucleotide 1582	1.45±0.53	>65.30	Gt E=0.8%
C to T at nucleotide 1587	1.01±0.25	>65.30	Gt B=0.2% Gt C, D=0.1% Gt G=3.6%
G to A at nucleotide 1588	1.41±0.80	>65.30	Gt A, F=0.3% Gt D=0.1%
G to C at nucleotide 1588	1.75±0.70	>65.30	Gt A=0.1% Gt E=0.5%
G to T at nucleotide 1588	1.39±0.78	>65.30	Gt D=0.1% Gt F=0.5%
C to A at nucleotide 1589	1.88 ± 0.61	>65.30	Gt A=0.9% Gt B, C, D=0.1%
C to G at nucleotide 1589	2.53±1.34	>65.30	Gt D=0.1% Gt H=2.9%
T to G at nucleotide 1590	2.23±1.23	>65.30	Gt A, E=0.3% Gt D=1.2%
T to C at nucleotide 1591	2.43±1.39	>65.30	Gt B, C=0.1% Gt G=1.2%
Wild-type	1.46±0.42	>65.30	Conserved in ≥97.5% across Gt sequences

Gt, genotype.

a. Mean of n=3 biological replicates, 4 days after imdusiran application.

b. Prevalence=number of sequences with variant/total number of sequences for that HBV genotype per HBVdb accessed in May 2019. Number of sequences per HBV genotype: Genotype A=959; Genotype B=2460; Genotype C=3295; Genotype D=1527; Genotype E=371; Genotype F=369; Genotype G=83; Genotype H=34. Prevalence=0% for that variant in any genotype not otherwise noted.

5.6 Imdusiran Antiviral Activity

5.6.1 Imdusiran Antiviral Activity in Primary Mouse and Human Hepatocyte HBV Models

Imdusiran exerted HBV antiviral effects in primary hepatocyte HBV models: mouse hepatocytes (PMHs) isolated from an AAV-HBV mouse, as well as in PHHs infected with HBV. These cell systems enabled the assessment of imdusiran activity in which imdusiran uptake into hepatocytes is mediated by ASGPR. As PMHs and PHHs typically lose ASGPR expression within 24–48 h of *in vitro* culture [43], cells were treated with imdusiran within 3–4 h from the start of *in vitro* culture in order to ensure imdusiran uptake via the ASGPR. This enabled retention of the natural imdusiran uptake pathway in the same cell systems where the anti-HBV effects of the compound were characterized.

In PMHs isolated from AAV-HBV infected mice, imdusiran exerted an inhibitory effect on HBV RNA, HBsAg, and HBeAg (Figure 5.8). Reduction of HBV subgenomic RNA coincided with decreases in secreted HBsAg (Figure 5.8). These inhibitory effects were reduced in the presence of asialofetuin, a glycoprotein that competes for ASGPR binding [22, 43], confirming the mechanism of imdusiran cellular uptake through ASGPR (Figure 5.8).

Imdusiran antiviral effects were also confirmed in PHHs from a donor that was previously characterized to be permissive to HBV infection. Imdusiran showed potent inhibition of HBV RNA, resulting in suppression of viral antigen synthesis and rcDNA synthesis (Figure 5.9). Uptake of imdusiran by ASGPR was again

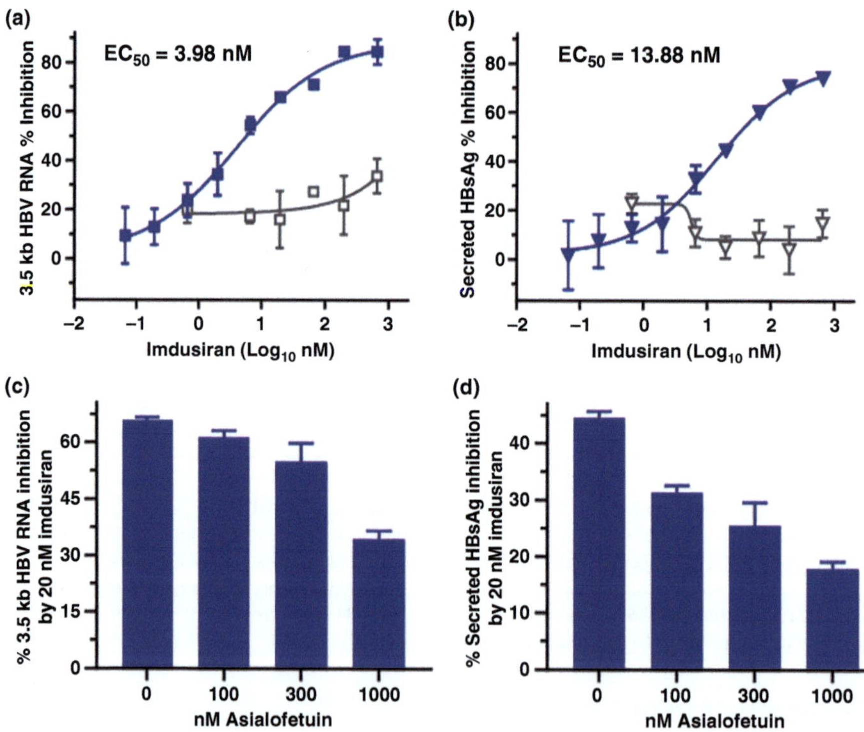

Figure 5.8 ASGPR-mediated uptake and anti-HBV activities in HBV-infected primary mouse hepatocytes.

Primary mouse hepatocytes were isolated from AAV-HBV mice and treated with imdusiran (filled symbols) or non-GalNAc-ligand conjugated imdusiran siRNA (open symbols) as a negative control for ASGPR-mediated uptake at 3 h after culturing freshly isolated hepatocytes. (a) Levels of 3.5 kb HBV RNA (pgRNA); and (b) levels of secreted HBsAg at 48 h post-treatment. Application of the ASGPR ligand asialofetuin competitively inhibits imdusiran cell uptake and anti-HBV activity against: (c) 3.5 kb HBV RNA; and (d) secreted HBsAg. Each data point represents the mean of technical triplicates with SD error bars and results are expressed relative to PBS negative control treatment.

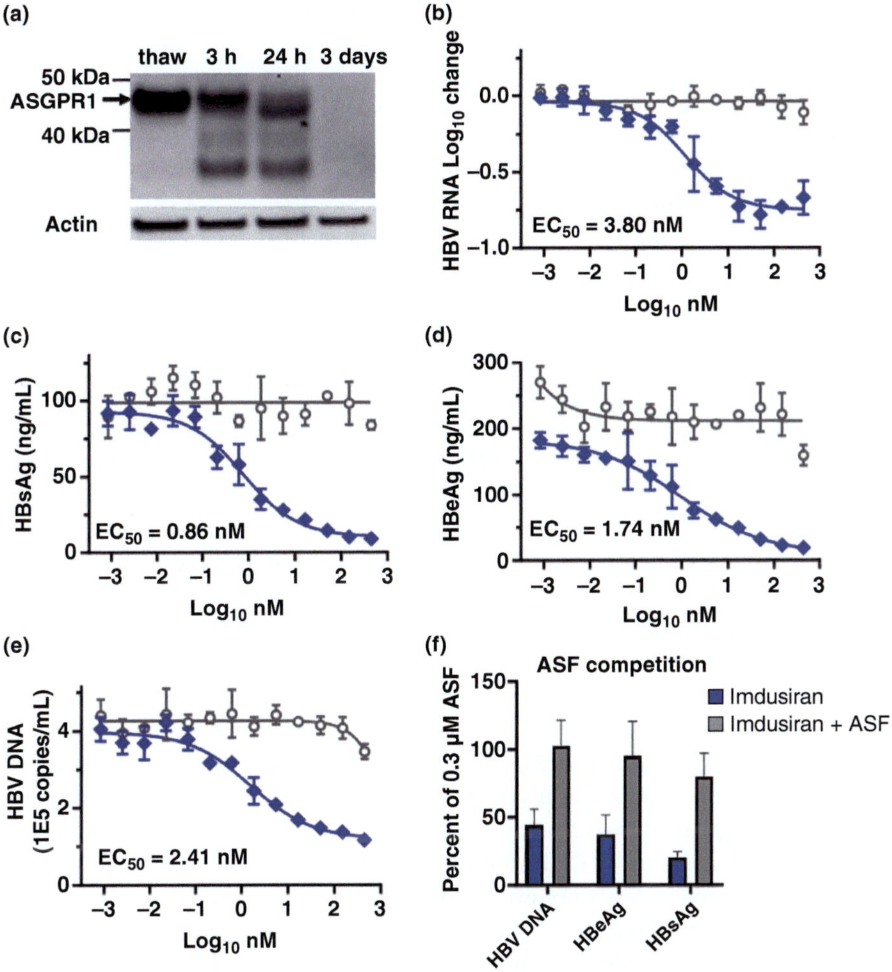

Figure 5.9 ASGPR-mediated uptake and anti-HBV activities in infected primary human hepatocytes. (a) ASGPR protein expression reduced with time after thaw of cryopreserved PHH; (b) HBV RNA; (c) HBsAg; (d) HBeAg; and (e) HBV DNA after treatment with imdusiran (blue diamonds) or non-GalNAc-ligand conjugated imdusiran siRNA (open circles) as a negative control for ASGPR-mediated uptake. (f) Treatment with the glycoprotein asialofetuin (ASF), which competes for ASGPR uptake, results in inhibition of imdusiran activity. *Source*: Lam et al., (2024) / American Chemical Society

confirmed by direct competition with asialofetuin: blocking ASGPR-mediated uptake with asialofetuin abrogated the observed imdusiran antiviral activity in this PHH HBV infection system (Figure 5.9). As previously described for other GalNAc-siRNA [44], imdusiran activity was retained in the presence of 40% human serum (Table 5.4).

5.6.2 Imdusiran *In Vivo* Antiviral Activity

Imdusiran anti-HBV activity was evaluated after single and repeat dosing in an AAV mouse model to verify drug efficacy in a whole body system. Imdusiran treatment

Table 5.4 Activity of imdusiran in the presence of 40% human serum in HBV-infected PHHs.

Viral endpoint	No human serum EC_{50} ± SD (nM)[a]	40% human serum EC_{50} ± SD (nM)[a]	EC_{50} fold shift[b]
rcDNA	2.23 ± 0.74	1.57 ± 1.00	0.7×
HBV RNA	8.01 ± 3.17	7.55 ± 0.75	1.1×
HBsAg	1.36 ± 0.32	1.00 ± 0.30	0.7×
HBeAg	2.51 ± 0.60	1.88 ± 0.70	0.7×
Cytotoxicity	>450	>450	1.0×
GAPDH RNA	>450	>450	1.0×

a. Mean and standard deviation of four biological replicates.
b. Fold shift calculated as mean of individual fold shift values obtained from four biological replicates comparing EC_{50} values obtained in the presence or absence of 40% human serum in the same experiment.

resulted in up to 91% reduction of all HBV RNAs, which correlated with concomitant dose-dependent reductions in all assayed viral biomarkers in serum and liver (Figure 5.10). Reductions in serum HBsAg were dose-dependent for animals treated at 1–9 mg/kg, with mean 1.3 $\log_{10}$ and 2.8 $\log_{10}$ decreases observed for the 1 mg/kg and 3 mg/kg dose levels at 14 days post-dosing, respectively. A mean maximal 3.7 $\log_{10}$ decrease in serum HBsAg was observed at day 14 relative to day 0 baseline levels in animals administered a single dose of imdusiran at 9 mg/kg (Figure 5.10). Serum HBsAg suppression was durable, with up to 2.2 $\log_{10}$ inhibition still present at 10 weeks after single-dose administration. This durability in silencing activity was also observed in CHB subjects treated with imdusiran and supported a clinical dosing frequency of once every 8 weeks [14, 15]. In AAV-HBV mice, repeat dose administration of imdusiran resulted in additional serum HBsAg reductions (Figure 5.10), with the cumulative nadir of serum HBsAg remaining unchanged 4 weeks later at week 14, suggesting a delay in viral rebound after repeat-dosing with imdusiran, likely due to the formation of siRNA endosomal depots in hepatocytes that sustain RNA silencing activity [18].

These effects on HBV viral markers coincided with the degree of imdusiran antisense strand loading in the RISC complex in the liver, further confirming imdusiran's direct antiviral activity through RNAi (Figure 5.11).

5.7 Imdusiran Combination Treatment

5.7.1 Imdusiran Combination with Standard of Care and Investigational Agents

It is likely that the achievement of a functional cure for CHB requires combination treatment with modalities targeting different elements of the viral lifecycle and host immune response. Imdusiran *in vitro* combination with standard-of-care agents tenofovir alafenamide (TAF) or PEG-IFN-α2a, or with investigational agents

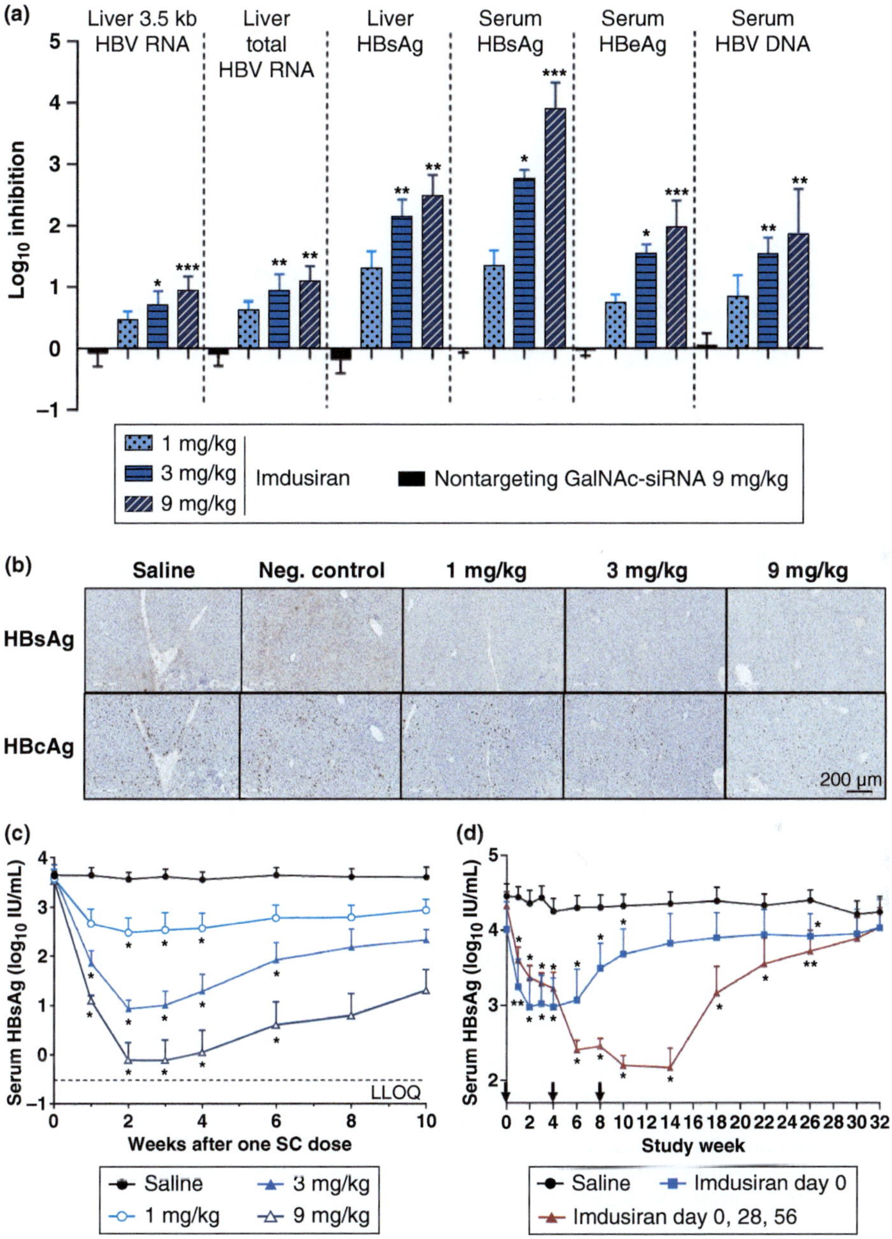

Figure 5.10 Anti-HBV activity in AAV-HBV mouse model. (a) 3.5 kb HBV RNA (pgRNA), total HBV RNA, HBsAg in liver and serum, serum HBeAg and serum HBV DNA on day 14 after imdusiran or nontargeting GalNAc-siRNA (negative control) single subcutaneous dose treatment on day 0. (b) Immunohistochemistry analysis staining for HBsAg and HBcAg of liver from AAV-HBV mice on day 14 after imdusiran or nontargeting GalNAc-siRNA (negative control) single subcutaneous dose treatment on day 0. (c) Serum HBsAg in AAV-HBV mice administered single subcutaneous dose of imdusiran on day 0. (d) Serum HBsAg in AAV-HBV mice administered either single subcutaneous dose of imdusiran on day 0 (blue squares) or repeat subcutaneous doses of imdusiran (red triangles) at 1 mg/kg at day 0, week 4, and week 8 as indicated by black arrows. *$P < 0.05$, **$P < 0.01$, ***$P < 0.01$ by two-way ANOVA with Geisser–Greenhouse correction and Dunnett's multiple comparisons test.

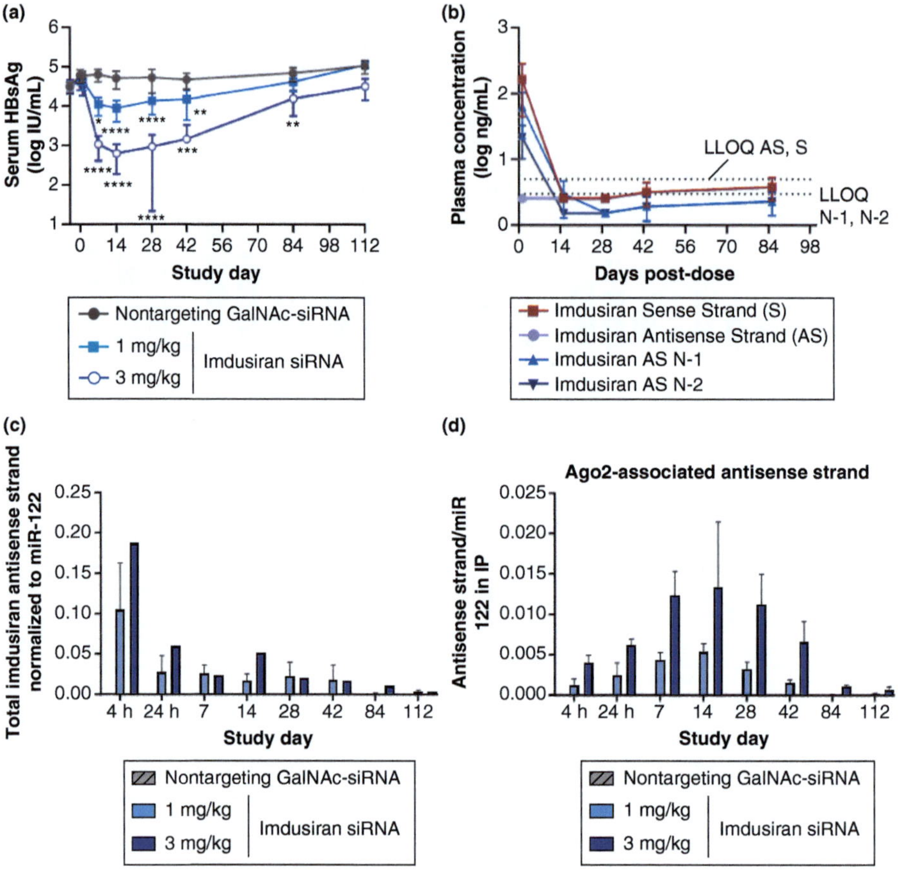

Figure 5.11 Kinetics of HBsAg reduction and imdusiran concentrations in plasma, liver, and on RISC complexes. (a) Serum HBsAg levels after a single subcutaneous dose of imdusiran siRNA or nontargeting GalNAc siRNA negative control treatment on day 0. (b) Plasma concentrations of imdusiran sense and antisense strands as well as the two active antisense metabolites N-1 and N-2. (c) Liver levels of imdusiran antisense strand as detected by stem-loop RT-PCR. Data are normalized to miR-122. (d) Liver levels of imdusiran antisense strand associated with the RISC complex, the multiprotein complex that binds to and specifically cleaves target RNAs. Detection was by stem-loop RT-PCR of Ago2 immunoprecipitated RNA normalized to levels of miR-122. LLOQ, lower limit of quantitation. Data are means ± standard deviation. *$P < 0.05$; **$P < 0.01$; ***$P < 0.001$; ****$P < 0.0001$ by mixed-effects analysis with Dunnett's multiple comparisons test against nontargeting GalNAc-siRNA control.

such as the capsid assembly modulator (CAM) AB-506, or the HBV RNA destabilizer AB-452, resulted in additive to synergistic effects for inhibition of both viral replication (HBV DNA) and viral antigen (HBsAg), with no statistically significant antagonism (Table 5.5) in PHH infected with HBV. These results suggest that imdusiran can be combined with standard-of-care agents such as nucleos(t)ide analogs or the immunomodulator PEG-IFN-α2a with no antagonistic effects on these agents' activities, and that combination treatments with other investigational agents may be considered.

Table 5.5 *In vitro* dual combination studies of imdusiran with standard-of-care and investigational agents in HBV-infected PHHs.

HBV biomarker	Inhibitor A	Inhibitor A EC_{50} N = 1	Imdusiran EC_{50} (nM) N = 1	Synergy volume (%)[a] N = 2	Antagonism volume (%)[a] N = 2	Conclusion
HBV DNA	TAF (nM)	0.083	<8.032	88.42, 2.46	−1.10, −2.33	Additive to moderate synergy
HBsAg		4.119	<8.032	0, 0	−2.46, −1.74	Additive
HBV DNA	PEG-IFN-α2a (IU/mL)	1.192	<8.032	0, 8.59	−15.53, −0.19	Additive
HBsAg		12.910	<8.032	0, 0.02	0, −2.74	Additive
HBV DNA	AB-506 (μM)	0.076	<8.032	106.50, 17.24	0, 0	Additive to strong synergy
HBsAg		>4.0	<8.032	2.33, 0	0, 0	Additive
HBV DNA	AB-452 (μM)	0.006	<8.032	45.01, 4.55	0, 0	Additive to minor synergy
HBsAg		0.011	<8.032	0, 0	−22.78, −16.16	Additive

a. Synergy/antagonism values were interpreted according to MacSynergy II guidelines [45] at 99.9% confidence interval as follows: <25, insignificant synergism/antagonism, 25–50, minor but significant synergism/antagonism, 50–100, moderate synergism/antagonism – may be important *in vivo*, >100, strong synergism/antagonism – probably important *in vivo*, ~1000, possible errors – check data and repeat experiment.

In addition, imdusiran was not found to be susceptible to HBV variants resistant to nucleos(t)ide analogs therapy [46], and imdusiran activity was retained against HBV variants resistant to CAM treatment [47–49] in transiently transfected Huh7 or HepG2 systems (Tables 5.6 and 5.7). These data suggest that imdusiran is cross-resistant against these HBV variants and should be able to maintain antiviral activity should HBV resistance arise against nucleos(t)ide analogs or CAMs during combination treatment. Taken together, the data are supportive of the amenability of imdusiran combination treatment with nucleos(t)ide analogs therapy and other investigational modalities.

5.8 Perspectives

Current treatments for CHB require life-long therapy as the prospects for a cure are low – only an estimated 1–12% of patients treated with NAs or PEG-IFN-α achieve HBsAg loss and sustained suppression of HBV DNA after cessation of treatment, the current definition of a functional cure [50]. As such, new approaches, featuring new modalities, are needed to overcome gaps in the current toolbox of standard-of-care agents. Reduction of viral antigens, including HBsAg expression, would provide potential benefits in reducing the immunosuppressive environment needed to

Table 5.6 Activity of imdusiran against nucleos(t)ide analog-resistant variants of HBV.

HBV polymerase variant[a]	Intracellular HBV DNA EC_{50}[b] ± SD (μM)			HBsAg EC_{50}[b] ± SD (μM)			Cytotoxicity CC_{50} (μM)		
	Imdusiran	ETV	3TC	Imdusiran	ETV	3TC	Imdusiran	ETV	3TC
rtM204I	2.24 ± 0.22	ND	>100	0.61 ± 0.61	ND	>100	>6.53	ND	>100
rtM204I + V173L	2.10 ± 0.73	ND	>100	0.24 ± 0.16	ND	>100	>6.53	ND	>100
rtM204I + S202G	1.75 ± 0.65	37.18 ± 12.89	ND	0.56 ± 0.54	>1	ND	>6.53	>1	ND
rtM204V + L180M	1.78 ± 1.08	ND	>100	0.23 ± 0.11	ND	>100	>6.53	ND	>100
rtM204I + S202G + M250V	1.65 ± 0.71	10.06 ± 13.59	ND	0.51 ± 0.59	>1	ND	>6.53	>1	ND
Wild-type	1.94 ± 0.23	0.02 ± 0.01	1.47 ± 1.40	0.37 ± 0.30	>0.1	>10	>6.53	>0.1	>10

ND: not determined.

a. Variants were constructed on wild-type genotype D HBV (GenBank accession no. U95551). Each of these variants has a fully conserved imdusiran target site.

b. Mean of n = 2 biological replicates, 7 days after imdusiran application.

Table 5.7 Activity of imdusiran against HBV variants resistant to capsid assembly modulator.

HBV core protein variant[a]	HBsAg or HBV DNA[c] EC_{50}[b] ± SD (nM)	
	Imdusiran	AB-506
T33N	31.8±20.2	15210±1160
L30F	24.6±4.0	880±90
L37Q	30.0±17.8	520±90
I105T	30.6±7.3	1530±450
T109M	31.6±21.2	60±30
Y38F+T109S	33.3±3.8	240±170
Wild-type	29.5±13.7	50±10

a. Variants were constructed on wild-type genotype D HBV (GenBank accession no. U95551). Each of these variants has a fully conserved imdusiran target site.
b. Concentration at which 50% inhibition (vs. the DPBS control condition) is achieved.
c. Calculated from three biological replicates. EC_{50} against HBsAg for imdusiran and against HBV DNA for AB-506.

maintain chronic HBV infection and providing a more conducive environment for sustained anti-HBV immune responses needed for functional cure. Given that the HBV genome features overlapping open reading frames to express viral antigens, targeting multiple aspects of the HBV lifecycle with a single HBV siRNA trigger is possible and supports the use of RNAi against this viral target.

Although success has been previously observed using LNP-delivered siRNA against multiple viral targets, these involve acute infections for which rapid onset of antiviral activity is most beneficial [7–9, 20], and resolution of infection occurs upon induction of adaptive immune responses after 5–7 days; as such, the durability of the antiviral response mediated by the siRNA beyond this period is of limited value. These elements do not apply to chronic viral infections such as HBV, where patients may be infected for decades, and the antiviral adaptive immune response is deficient in clearing the viral infection. In preclinical AAV-HBV mouse studies, the shorter durability of the antiviral effect of LNP-delivered siRNA is in stark contrast to the extended durability conferred by GalNAc conjugation. This limitation was also observed in the clinic, where LNP-siRNA targeting HBV was generally unsuccessful in mediating durable anti-HBV effects, and significant rebound in viral markers was observed between dosing intervals [51].

Building upon these experiences led to the discovery of imdusiran, a potent, pan-genotypic GalNAc-conjugated siRNA that effectively inhibits HBV viral antigen expression as well as viral replication through the silencing of HBV RNA in the liver. Antiviral effects mediated by imdusiran were durable, as demonstrated in an AAV-HBV mouse model, with up to 2.2 $\log_{10}$ serum HBsAg inhibition still present at 10 weeks after a single dose, and the durability of antiviral effect was also similarly observed in a Phase 1b study of imdusiran [14, 15]. These data validate the use of GalNAc conjugation as an effective RNAi strategy against CHB.

Notably, in the ongoing clinical studies with imdusiran in CHB patients, sustained HBsAg reduction has been observed up to 72 weeks after the last dose of imdusiran, suggesting potential restoration of antiviral immune control beyond the viral suppression conferred by imdusiran direct RNAi activity [14, 15]. As such, imdusiran treatment may confer multiple benefits in not only mediating a direct antiviral effect, but also potentially resetting the antiviral immune status of CHB patients toward a greater level of immune control. Combinations of imdusiran with immunomodulators may be able to build upon this new foundation of immune reawakening to ultimately bring about HBV functional cure. Such strategies are currently being examined in Phase 2 clinical studies.

References

1 Corydon IJ, Fabian-Jessing BK, Jakobsen TS, Jorgensen AC, Jensen EG, Askou AL, et al. 25 years of maturation: a systematic review of RNAi in the clinic. *Mol Ther Nucleic Acids*. 2023;33:469–82.

2 Friedrich M, Aigner A. Therapeutic siRNA: State-of-the-Art and Future Perspectives. *BioDrugs*. 2022;36(5):549–71.

3 Ali Zaidi SS, Fatima F, Ali Zaidi SA, Zhou D, Deng W, Liu S. Engineering siRNA therapeutics: challenges and strategies. *J Nanobiotechnology*. 2023;21(1):381.

4 Guo S, Zhang M, Huang Y. Three 'E' challenges for siRNA drug development. *Trends Mol Med*. 2024;30(1):13–24.

5 Hu B, Zhong L, Weng Y, Peng L, Huang Y, Zhao Y, et al. Therapeutic siRNA: state of the art. *Signal Transduct Target Ther*. 2020;5(1):101.

6 Administration US FDA. FDA approves first-of-its kind targeted RNA-based therapy to treat a rare disease 2018 [updated August 10, 2018. Available from: https://www.fda.gov/news-events/press-announcements/fda-approves-first-its-kind-targeted-rna-based-therapy-treat-rare-disease.

7 Thi EP, Lee AC, Geisbert JB, Ursic-Bedoya R, Agans KN, Robbins M, et al. Rescue of non-human primates from advanced Sudan ebolavirus infection with lipid encapsulated siRNA. *Nat Microbiol*. 2016;1(10):16142.

8 Thi EP, Mire CE, Lee AC, Geisbert JB, Ursic-Bedoya R, Agans KN, et al. siRNA rescues nonhuman primates from advanced Marburg and Ravn virus disease. *J Clin Invest*. 2017;127(12):4437–48.

9 Thi EP, Mire CE, Lee AC, Geisbert JB, Zhou JZ, Agans KN, et al. Lipid nanoparticle siRNA treatment of Ebola-virus-Makona-infected nonhuman primates. Nature. 2015;521(7552):362–5.

10 Thi EP, Dhillon AP, Ardzinski A, Bidirici-Ertekin L, Cobarrubias KD, Cuconati A, et al. ARB-1740, a RNA Interference Therapeutic for Chronic Hepatitis B Infection. *ACS Infect Dis*. 2019;5(5):725–37.

11 Ye X, Tateno C, Thi EP, Kakuni M, Snead NM, Ishida Y, et al. Hepatitis B virus therapeutic agent ARB-1740 has inhibitory effect on hepatitis delta virus in a new dually-infected humanized mouse model. *ACS Infect Dis*. 2019;5(5):738–49.

12 Dunning J, Sahr F, Rojek A, Gannon F, Carson G, Idriss B, et al. Experimental treatment of ebola virus disease with TKM-130803: a single-arm phase 2 clinical trial. *PLoS Med*. 2016;13(4):e1001997.

13 Keam SJ. Vutrisiran: first approval. *Drugs*. 2022;82(13):1419–25.

14 Gane E, Berliba E, Kim YJ, Holmes J, Lim YS, Strasser S, et al. AB-729, a GalNac-siRNA, results in robust reductions of HBV DNA and HBsAg in subjects with chronic hepatitis B infection. Asian Pacific Association for the Study of the Liver 30th Annual Conference 2021, Bangkok, Thailand.

15 Yuen MF, Berliba E, Kim YJ, Holmes JA, Lim YS, Strasser SI, et al. Safety and pharmacodynamics of the GalNAc-siRNA AB-729 in subjects with chronic hepatitis B infection. *Hepatology*. 2020;72:62A.

16 Lam K, Leung A, Martin A, Wood M, Schreiner P, Palmer L, et al. Unsaturated, Trialkyl ionizable lipids are versatile lipid-nanoparticle components for therapeutic and vaccine applications. *Adv Mater*. 2023;35(15):e2209624.

17 Lam K, Schreiner P, Leung A, Stainton P, Reid S, Yaworski E, et al. Optimizing lipid nanoparticles for delivery in primates. Adv Mater. 2023;35(26):e2211420.

18 Brown CR, Gupta S, Qin J, Racie T, He G, Lentini S, et al. Investigating the pharmacodynamic durability of GalNAc-siRNA conjugates. *Nucleic Acids Res*. 2020;48(21):11827–44.

19 Plante-Bordeneuve V, Perrain V. Vutrisiran: a new drug in the treatment landscape of hereditary transthyretin amyloid polyneuropathy. Expert Opin Drug Discov. 2024;19(4):393–402.

20 Thi EP, Mire CE, Ursic-Bedoya R, Geisbert JB, Lee ACH, Agans KN, et al. Marburg virus infection in nonhuman primates: therapeutic treatment by lipid-encapsulated siRNA. *Sci Transl Med*. 2014;6(250):250ra116.

21 Holland RJ, Lam K, Ye X, Martin AD, Wood MC, Palmer L, et al. Ligand conjugate SAR and enhanced delivery in NHP. *Mol Ther*. 2021;29(10):2910–9.

22 Westerlind U, Westman J, Tornquist E, Smith CI, Oscarson S, Lahmann M, et al. Ligands of the asialoglycoprotein receptor for targeted gene delivery, part 1: synthesis of and binding studies with biotinylated cluster glycosides containing N-acetylgalactosamine. *Glycoconj J*. 2004;21(5):227–41.

23 Ferrari C. HBV and the immune response. Liver Int. 2015;35 Suppl 1:121–8.

24 Khanam A, Chua JV, Kottilil S. Immunopathology of chronic hepatitis B infection: role of innate and adaptive immune response in disease progression. *Int J Mol Sci*. 2021;22(11):5497–519.

25 Kuipery A, Gehring AJ, Isogawa M. Mechanisms of HBV immune evasion. *Antiviral Res*. 2020;179:104816.

26 Tu T, Budzinska MA, Vondran FWR, Shackel NA, Urban S. Hepatitis B virus DNA integration occurs early in the viral life cycle in an In Vitro infection model via sodium taurocholate cotransporting polypeptide-dependent uptake of enveloped virus particles. *J Virol*. 2018;92(11):e02007-17.

27 Yang Z, Zeng J, Chen Y, Wang M, Luo H, Huang AL, et al. Detection of HBV DNA integration in plasma cell-free DNA of different HBV diseases utilizing DNA capture strategy. *Virol Sin*. 2024;39(4):655–66.

28 Chen H, Jiang L, Chen S, Hu Q, Huang Y, Wu Y, et al. HBx inhibits DNA sensing signaling pathway via ubiquitination and autophagy of cGAS. *Virol J*. 2022;19(1):55.

29 Cho IR, Oh M, Koh SS, Malilas W, Srisuttee R, Jhun BH, et al. Hepatitis B virus X protein inhibits extracellular IFN-alpha-mediated signal transduction by downregulation of type I IFN receptor. *Int J Mol Med*. 2012;29(4):581–6.

30 Kumar M, Jung SY, Hodgson AJ, Madden CR, Qin J, Slagle BL. Hepatitis B virus regulatory HBx protein binds to adaptor protein IPS-1 and inhibits the activation of beta interferon. *J Virol*. 2011;85(2):987–95.

31 Allweiss L, Giersch K, Pirosu A, Volz T, Muench RC, Beran RK, et al. Therapeutic shutdown of HBV transcripts promotes reappearance of the SMC5/6 complex and silencing of the viral genome in vivo. *Gut*. 2022;71(2):372–81.

32 Behlke MA. Chemical modification of siRNAs for in vivo use. *Oligonucleotides*. 2008;18(4):305–19.

33 Broering R, Real CI, John MJ, Jahn-Hofmann K, Ickenstein LM, Kleinehr K, et al. Chemical modifications on siRNAs avoid Toll-like-receptor-mediated activation of the hepatic immune system in vivo and in vitro. *Int Immunol*. 2014;26(1):35–46.

34 Judge AD, Bola G, Lee AC, MacLachlan I. Design of noninflammatory synthetic siRNA mediating potent gene silencing in vivo. *Mol Ther*. 2006;13(3):494–505.

35 Judge AD, Sood V, Shaw JR, Fang D, McClintock K, MacLachlan I. Sequence-dependent stimulation of the mammalian innate immune response by synthetic siRNA. *Nat Biotechnol*. 2005;23(4):457–62.

36 Lee Y, Urban JH, Xu L, Sullenger BA, Lee J. 2′Fluoro modification differentially modulates the ability of RNAs to activate pattern recognition receptors. *Nucleic Acid Ther*. 2016;26(3):173–82.

37 Janas MM, Zlatev I, Liu J, Jiang Y, Barros SA, Sutherland JE, et al. Safety evaluation of 2′-deoxy-2′-fluoro nucleotides in GalNAc-siRNA conjugates. *Nucleic Acids Res*. 2019;47(7):3306–20.

38 Hayer J, Jadeau F, Deleage G, Kay A, Zoulim F, Combet C. HBVdb: a knowledge database for Hepatitis B Virus. *Nucleic Acids Res*. 2013;41(Database issue):D566–70.

39 Gane E, Lim YS, Kim JB, Jadhav V, Shen L, Bakardjiev AI, et al. Evaluation of RNAi therapeutics VIR-2218 and ALN-HBV for chronic hepatitis B: results from randomized clinical trials. *J Hepatol*. 2023;79(4):924–32.

40 Schlegel MK, Janas MM, Jiang Y, Barry JD, Davis W, Agarwal S, et al. From bench to bedside: Improving the clinical safety of GalNAc-siRNA conjugates using seed-pairing destabilization. *Nucleic Acids Res*. 2022;50(12):6656–70.

41 Dahlgren C, Zhang HY, Du Q, Grahn M, Norstedt G, Wahlestedt C, et al. Analysis of siRNA specificity on targets with double-nucleotide mismatches. Nucleic Acids Res. 2008;36(9):e53.

42 Du Q, Thonberg H, Wang J, Wahlestedt C, Liang Z. A systematic analysis of the silencing effects of an active siRNA at all single-nucleotide mismatched target sites. *Nucleic Acids Res*. 2005;33(5):1671–7.

43 Severgnini M, Sherman J, Sehgal A, Jayaprakash NK, Aubin J, Wang G, et al. A rapid two-step method for isolation of functional primary mouse hepatocytes: cell characterization and asialoglycoprotein receptor based assay development. *Cytotechnology*. 2012;64(2):187–95.

44 Agarwal S, Allard R, Darcy J, Chigas S, Gu Y, Nguyen T, et al. Impact of serum proteins on the uptake and RNAi activity of GalNAc-conjugated siRNAs. *Nucleic Acid Ther*. 2021;31(4):309–15.

45 Prichard MN, Shipman C, Jr. A three-dimensional model to analyze drug-drug interactions. *Antiviral Res.* 1990;14 (4–5):181–205.

46 Choi YM, Lee SY, Kim BJ. Naturally occurring hepatitis B virus reverse transcriptase mutations related to potential antiviral drug resistance and liver disease progression. *World J Gastroenterol.* 2018;24(16):1708–24.

47 Janssen HLA, Hou J, Asselah T, Chan HLY, Zoulim F, Tanaka Y, et al. Randomised phase 2 study (JADE) of the HBV capsid assembly modulator JNJ-56136379 with or without a nucleos(t)ide analogue in patients with chronic hepatitis B infection. *Gut.* 2023;72(7):1385–98.

48 Mani N, Cole AG, Phelps JR, Ardzinski A, Burns R, Chiu T, et al. Preclinical characterization of AB-506, an inhibitor of HBV replication targeting the viral core protein. *Antiviral Res.* 2022;197:105211.

49 Verbinnen T, Talloen W, Janssen HLA, Zoulim F, Shukla U, Vandenbossche JJ, et al. Viral sequence analysis of chronic hepatitis B patients treated with the capsid assembly modulator JNJ-56136379 in the JADE phase 2a study. *Antiviral Res.* 2023;216:105660.

50 European Association for the Study of the Liver. European Association for the Study of the Liver. EASL 2017 clinical practice guidelines on the management of hepatitis B virus infection. *J Hepatol.* 2017;67(2):370–98.

51 Streinu-Cercel A, Gane E, Cheng W, Sievert W, Roberts S, Ahn SH, et al. SAT-155 - A phase 2a study evaluating the multi-dose activity of ARB-1467 in HBeAg positive and negative virally suppressed subjects with hepatitis B. *J Hepatol.* 2017;66(1):S688–S9.

6

Discovery and Development of ARO-HBV/JNJ-3989

Christine I. Wooddell[1], Oliver Lenz[2], Thomas Schluep[1], Man-Fung Yuen[3], Michael Biermer[2]

[1] *Arrowhead Pharmaceuticals, Pasadena, CA, USA*
[2] *Janssen Pharmaceuticals NV, Beerse, Belgium*
[3] *Department of Medicine, The University of Hong Kong, Hong Kong, China*

6.1 Introduction

6.1.1 Chronic Hepatitis B Virus Infection

Globally, more than 300 million people are chronically HBV infected (CHB), and 0.8 million a year die of HBV-related complications [1]. In the absence of treatment, 15–40% of chronically infected individuals may develop serious sequelae of liver disease, including hepatocellular carcinoma (HCC), cirrhosis of the liver, and liver failure [2]. HBV is highly infectious and transmitted through body fluids. CHB is defined as being serologically positive for HBV surface antigen (HBsAg) at two time points that are at least six months apart. Individuals who spontaneously or induced by therapy achieve HBV DNA and HBsAg serum clearance off all treatment, representing immunological control of the infection, are considered "functionally cured" and have a greatly decreased probability of developing HCC and other sequelae of liver disease [3]. However, once a chronic infection is established functional cures are rarely obtained. Nucleos(t)ide analogs (NA), such as entecavir and tenofovir, are considered standard of care for treating CHB. These NAs effectively suppress HBV replication, reducing risks of developing HCC and liver decompensation, but have limited effect on HBsAg and typically must be taken for the rest of the patient's life to prevent reactivation of the virus, which usually happens if the patient stops treatment [4]. Only 1–4% of patients treated with such antivirals achieve functional cure annually [5–11]. The stigma and discrimination of living with HBV greatly impact patients' lives and hinder diagnosis and treatment [12]. For all these reasons, a finite treatment regimen that can induce functional cure in more CHB patients is urgently needed.

Trends in Antiviral Drug Development, First Edition. Edited by Michael J. Sofia and Zhengqiang Wang.

6.1.1.1 HBV Structure and Associated Molecular and Cellular Biology

HBV is a lipid-enveloped virus with a 3.2 kilobase (kb) double-stranded relaxed circular DNA (rcDNA) genome. It enters hepatocytes via the sodium taurocholate cotransporting polypeptide (NTCP) receptor, followed by the release of the rcDNA and the formation of a covalently closed circular DNA (cccDNA) in the nucleus. The cccDNA functions as a minichromosome transcribed by the host RNA polymerase II to produce the viral RNAs [13]. The major HBV transcripts are the 3.5 kb pregenomic RNA (pgRNA), the 3.5 kb pre-core messenger RNA (mRNA) that encodes the e antigen (HBeAg), the HBsAg mRNAs that encode the large HBsAg (L-HBsAg, 2.4 kb) or the middle (M) and small (S) HBsAg (2.1 kb), and the X protein (HBx) mRNA (Figure 6.1). M- and S-HBsAg are transcribed utilizing two distinct translation start codons from the 2.1 kb mRNA. HBsAg refers to all three forms of the S protein collectively, though S-HBsAg is the most abundant of these, while L-HBsAg with the pre-s1 domain is needed for binding to NTCP. All transcripts from cccDNA utilize the same HBV polyadenylation signal (polyA) and become polyadenylated by the endogenous mRNA polyadenylation factors. The pgRNA and pre-core mRNA are somewhat longer than one genome length due to the transcription of a few hundred bases more than one round of the circular template. pgRNA is the template for translation of core and HBV polymerase (pol), in addition to being the substrate for reverse transcription by HBV polymerase. The transcription start site of the pre-core (HBeAg) is upstream of that for pgRNA and the translation start codon for the pre-core is also upstream of the start codon for core. By this mechanism, HBeAg includes an upstream signal peptide for secretion that the core does not have, even though the open reading frame (ORF) of the core protein that forms the capsid has the same sequence as most of the ORF for HBeAg, but the two proteins have distinctly different structures.

HBV polymerase is covalently attached to the pgRNA following its translation prior to the pgRNA/pol complex being encapsidated by typically 240 subunits of the core protein. Within the capsid, HBV polymerase reverse transcribes pgRNA to form the first DNA strand in minus orientation. Pol then produces a partial plus strand to result in a partially double-stranded, relaxed, circular DNA (rcDNA) with a 7–9 base terminal redundancy. The capsid is enveloped by a lipid membrane containing HBsAg and secreted as the infectious virus (Dane particle). HBV spreads from cell to cell when the Dane particle is taken up by another hepatocyte after the pre-S1 peptide in L-HBsAg of the envelope binds to NTCP receptor expressed on hepatocytes. After rcDNA is transported into the nucleus it is converted into the fully double-stranded cccDNA by endogenous factors. Alternatively, rcDNA may be transported to the nucleus of the parent hepatocyte to produce another cccDNA.

Replication to produce rcDNA is a process that requires pol to switch strands three times utilizing the DR1 and DR2 elements that are shown in Figure 6.1. A double-stranded linear DNA (dslDNA) forms as an alternate product of the replication process when the primed DNA minus strand fails to switch strands, resulting in the linear DNA with a DR1 element on each end [14, 15]. These alternate replication products are also encapsidated and either secreted to be taken up by another hepatocyte or cycled into the nucleus of the parent cell. This linear DNA can integrate into the host chromosomes, typically losing part of the sequence interior to the DR1

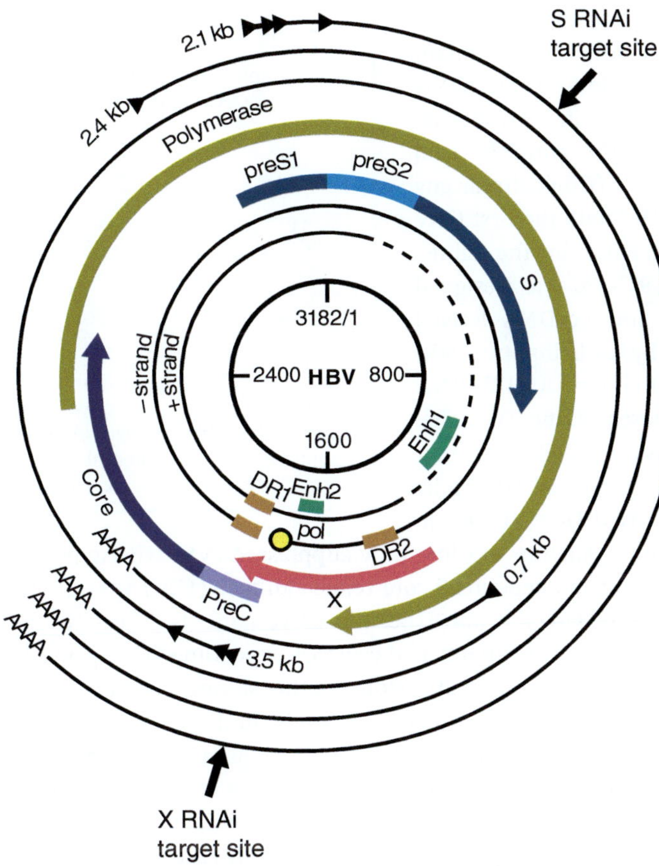

Figure 6.1 Schematic representation of HBV transcripts.

The major HBV transcripts are ~3.5 kb for pre-core (HBeAg) and pgRNA (encodes core and HBV polymerase); 2.4 kb for pre-S1 (encodes L-HBsAg); 2.1 kb for pre-S2 (encodes M-HBsAg) and S (encodes S-HBsAg); and 0.7 kb for HBx. Positions of enhancers Enh1 and Enh2 (green); direct repeats DR1 and DR2 (brown); and positions of the S siRNA and X siRNA target sites (arrows) are indicated. *Source*: Adapted from [86] / with permission of Elsevier.

elements at either end. Although dslDNA does not replicate the virus because it lacks the ends, the integrated HBV (iDNA) can disrupt host genes, produce HBV-host fusion proteins, change expression levels, cause gene rearrangements, and lead to HCC [16–18]. The pre-core/core promoter is missing from the iDNA, thus preventing expression of HBeAg or core from iDNA, but all the elements required to transcribe HBsAg are retained: promoter, enhancers, and complete ORF. The only missing element for mRNA expression is the polyadenylation signal that is downstream of DR1. Following the integration of the iDNA, however, the polyA signal can be provided by the host DNA into which the iDNA integrates [19].

Integrated HBV plays a major role in CHB because much of the abundant HBsAg produced in patients that are HBeAg negative and in many that are HBeAg positive following long-term viral suppression is from iDNA rather than cccDNA [19–21]. HBV integrates even very early in HBV infection, but the amount of iDNA increases

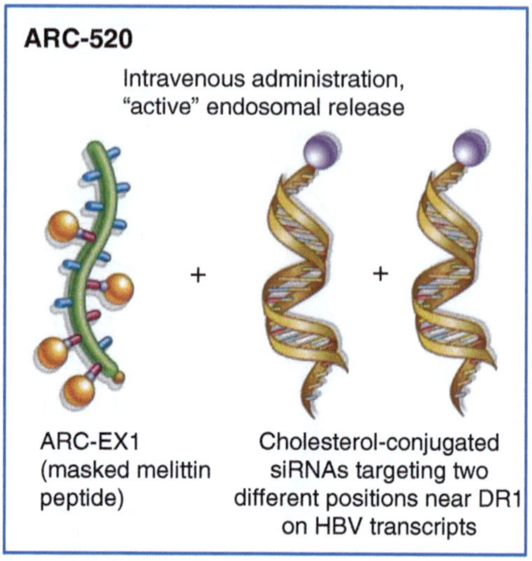

with the progression of liver disease [20, 22, 23]. This abundant HBsAg expression from iDNA was discovered in Arrowhead Pharmaceuticals' clinical trial Heparc-2001 (Clinicaltrials.gov NCT 02065336) with the first HBV RNAi therapeutic in the clinic, ARC-520. ARC-520 was designed to silence all cccDNA-derived transcripts by utilizing two siRNAs that both bound to HBV RNA within 40 bases of the DR1 site. A single dose of ARC-520 (3 mg/kg or 4 mg/kg) reduced HBsAg by a mean of only 0.3 $\log_{10}$ IU/mL in HBeAg-negative patients and in virally suppressed (VS) HBeAg-positive patients. However, HBeAg and hepatitis B core-related antigen (HBcrAg) were both reduced by approximately 1 $\log_{10}$ in the HBeAg-positive patients. These results suggested that the HBeAg-positive patients had cccDNA expressing HBeAg and HBcrAg, but much of the HBsAg was originating from another source. Then a cohort of NA-naïve HBeAg positive patients was added to the study. In these patients, HBsAg, HBeAg, and HBcrAg were all reduced by 1.3–1.5 $\log_{10}$, suggesting that the HBsAg in these patients was primarily derived from cccDNA, along with the HBeAg and HBcrAg. Proof of HBsAg expression from iDNA was demonstrated by Iso-seq, next-generation full-length mRNA sequencing of liver RNA in HBeAg-negative chimpanzees that were chronically HBV infected. Most of their HBsAg transcripts were fused to chimpanzee sequence and utilized host sequences as polyA signals. HBeAg-positive chimpanzees, on the other hand, had HBsAg transcripts that terminated following the HBV polyA signal, confirming that the vast majority of their HBsAg was produced from cccDNA. Utilizing an siRNA that targeted within the S ORF, Arrowhead then demonstrated deep reduction of HBsAg up to 3 $\log_{10}$ following three injections of this siRNA, confirming that the HBV transcripts could be reduced in the HBeAg-negative chimpanzees when the siRNA target sequence was present in the transcripts. Grudda et al. subsequently demonstrated that iDNA maintains HBsAg expression during NA antiviral treatment of CHB patients [20].

As can be expected of an ancient virus that has evolved to produce life-long infection, every HBV transcript and its products serve to maintain HBV infection by not only directly producing virus but also by altering cell physiology and suppressing immune control [24, 25]. HBsAg impacts many types of immune cells, including Kupffer cells, monocytes, macrophages, dendritic cells, natural killer cells, T cells, and B cells to maintain infection and prevent the development of anti-HBsAg antibodies [26]. Decades of exposing HBV-specific T cells and B cells to high levels of HBsAg results in impaired function and loss of such cells. Quantitative HBsAg levels strongly correlated with reduced IL-4-secreting CD4+ T cells in CHB patients [27]. HBsAg also enhances the activity of T regulatory cells and increases the differentiation and expansion of monocytes into myeloid-derived suppressor cells, both of which serve to abrogate an anti-HBV immune response [28, 29]. The abundant HBsAg inhibits both innate and adaptive immune responses to maintain the chronicity of infection [24, 25, 30–33]. HBsAg impairs dendritic cell maturation and function, reducing IL-12 production through disruption of the MAPK pathway. HBsAg can impair the activation of all toll-like receptor (TLR) pathways and reduce the cytokine production associated with each [26, 34–40]. High amounts of HBsAg expression that result in the "ground glass" phenotype cause ER stress that decreases the health of hepatocytes [41]. Oxidative stress and mitochondrial injury are also features of CHB infection [42]. In addition to producing energy for the cell, mitochondria play a key role in mediating the innate immune response and controlling apoptosis. The multifunctional X protein has been implicated in gene activation, enhancement of replication, involvement in cell cycle and apoptosis regulation, calcium signaling, ROS regulation, and altered mitochondrial function [42–47]. HBx is required to initiate and maintain HBV replication [48], which it does by inhibiting the host restriction complex SMC5/6 to allow persistent transcription of cccDNA [49–51]. Pol replicates the virus in addition to suppressing the immune response to infection [52–54]. Core interacts with Kupffer cells to suppress TLR-2 activation and promote CD8+ T-cell exhaustion [55]. HBeAg suppresses the maturation of Kupffer cells and enhances PD-L1 expression in them [56].

6.1.1.1.1 Clinical Course of HBV Infection

HBV-infected patients who do not clear/control the virus during acute infection and enter the chronic state generally evolve through distinct phases of infection defined by HBV HBeAg status, serum levels of HBV DNA, and alanine aminotransferase (ALT). The first phase of infection is the "high replicative, low inflammatory phase" (or chronic HBeAg positive infection, formerly "immune tolerant") characterized by HBeAg positivity, very high amounts of HBV DNA, high amounts of HBsAg and normal ALT indicating absence of necro-inflammatory activity. These patients tend to be younger and have greater potential for an immune response [57, 58]. CHB progresses to an HBeAg-positive "immune active" phase (chronic HBeAg-positive hepatitis) exhibiting periodic elevations of liver enzymes and generally high, but fluctuating viral markers (e.g. HBV DNA, HBsAg). Immune pressure frequently results in patients becoming HBeAg negative and anti-HBeAg antibody positive, bringing patients into a 3rd phase named "low replication, low inflammatory phase" (or HBeAg-negative infection, formerly: "inactive carrier status") with low levels of viral proteins and normal or slightly elevated liver enzymes. This can be followed by a reactivation phase of viral replication and necro-inflammatory activity (HBeAg-negative chronic hepatitis). The phases with ongoing necro-inflammatory activity especially drive the development of liver cirrhosis, decompensation, and HCC. The greatest need for treatment is in patients who have high replication with liver inflammation, but it can be argued that functional cure of younger patients with a greater capacity for immune response would prevent these individuals from progressing to more severe disease and reduce integration of HBV DNA into the host chromosome and clonal hepatocyte expansion, which have been associated with development of HCC [59–61].

6.2 Development of RNAi Therapeutic ARO-HBV/JNJ-73763989 (JNJ-3989)

6.2.1 RNAi as a Therapeutic Modality

RNAi is a conserved naturally occurring catalytic mechanism during which a double-stranded small interfering RNA (siRNA) or fully synthetic RNA-like molecule herein also named "siRNA" is utilized by the RNA-induced silencing complex (RISC) to cleave complementary mRNA [62]. Therapeutic siRNAs typically have ribose modified with 2′ fluoro or 2′-*O*-methyl in addition to other modifications to increase stability and abrogate the innate immune response to double-stranded RNA. The Argonaut2 protein of RISC separates the strands of the siRNA, retaining the antisense guide strand to form the catalytic RISC that cleaves the target RNA. Cleaved RNA is subsequently degraded by nucleases with the net result that the encoded protein is not produced. Proteins that were previously considered undruggable can now be targeted by RNAi with siRNA complementary to the target mRNA [63]. The first FDA-approved RNAi drug was patisiran for hereditary transthyretin amyloidosis (hATTR) in 2018. Givosiran for acute hepatic porphyria and lumasiran for primary hyperoxaluria type 1 followed in 2019 and 2020, respectively. All three of these siRNA drugs target genes in hepatocytes, but patisiran was delivered intravenously in a lipid nanoparticle, whereas givosiran and lumasiran were delivered subcutaneously and targeted to hepatocytes in the liver via an *N*-acetylgalactosamine (GalNAc) ligand to be taken up and internalized via asialoglycoprotein receptor (ASGPR), which is highly abundant on hepatocytes. Dozens of other RNAi therapeutics are in clinical trials.

6.2.2 Use of siRNA to Treat HBV Infection

RNAi can be harnessed to reduce the viral products that allow for chronicity of infection and hereby to be part of the strategy to develop a functional cure for patients. Hepatocytes are efficiently targeted with a GalNAc ligand on the siRNA. To function as a compact genome with overlapping reading frames, HBV has sequences that are conserved between all HBV genotypes. All ccc-DNA-derived transcripts have a common 3-prime end and utilize the same polyA signal. siRNA targeting all cccDNA and iDNA-derived viral RNA is not expected to directly reduce cccDNA, but it can prevent the formation of new cccDNA molecules by inhibiting viral replication via the direct antiviral effect of cleaving pgRNA, reducing components of the Dane particles, and by reducing expression of cccDNA via reduction of the HBV X protein (HBx) expression to mediate transcriptional silencing [64, 65]. Other siRNA agents targeting HBV that are currently in clinical development are AB-729, RG6346 (RO7445482), and VIR-2218.

While siRNAs targeting host proteins are generally designed to reduce the target protein by 50–95%, serum HBV proteins, especially HBsAg, are abundantly produced in patients, and multi-$\log_{10}$ of reduction would be required to clear these products. HBsAg levels below 100 IU/mL are known to lead to an improved prognosis for patients, but whether this improvement mediated by direct-acting antivirals like siRNA would have the same benefit is unknown. Although there are conserved sequences between different genotypes, individual genomes and quasi-species present further variation in sequences that could potentially impact binding of the siRNA.

6.2.2.1 JNJ-3989 siRNAs Have Broad Cross-reactivity to HBV Genotypes and HBV Transcripts

The developers of ARO-HBV/JNJ-3989 aimed to maximally reduce HBsAg derived from both cccDNA and iDNA while also reducing all other cccDNA-derived transcripts in CHB patients of all genotypes [66]. HBV genomes that differ from one another by ≥8% of their sequence are defined as separate genotypes (A–J). The first step to develop an RNAi therapeutic to target HBV was determining all highly conserved 17-base sequences in the HBV genome, as described for the first-generation HBV RNAi therapeutic ARC-520 [67]. Bases 2–18 are the most important for target sequence selection, though bases added to either end can affect the overall activity of an siRNA. These conserved sequences were then evaluated for suitable sequence-based characteristics to have the capacity for RNAi activity. Any sequences deemed to have potential off-target concerns, such as sequences complementary to mRNA of human genes, were deselected. An siRNA typically has reduced activity with even one mismatch to an mRNA target site, but some can retain activity with a mismatch, depending on the position of the mismatch. Any potential off-target genes were characterized and empirically evaluated for knockdown of gene expression in a human hepatocyte cell line (Hep3B). GalNAc-targeted siRNAs are productively taken up only in hepatocytes; thus, expression of off-target genes in hepatocytes is most relevant.

The conserved sequences selected for evaluation of RNAi activity were synthesized with GalNAc ligand in a triantennary configuration and chemical modifications to enhance the siRNA's activity and were tested in a mouse model of chronic HBV infection. As previously described, an overlength HBV genome (HBV1.3) minicircle derived from a plasmid expressing a genotype D HBV was hydrodynamically injected (HDI) into immunodeficient NOD-SCID mice to generate the HDI HBV model [67]. The selection of siRNAs was performed by measuring HBsAg and HBeAg in the serum of these mice. Suppression of viral proteins was normalized to both the pre-dose expression level and the expression level in control mice injected with saline rather than siRNA. This HDI HBV model allowed for characterization of the siRNAs' magnitude of effect and duration of activity in hepatocytes *in vivo*.

ARO-HBV/JNJ-3989

Subcutaneous administration, "passive" endosomal release

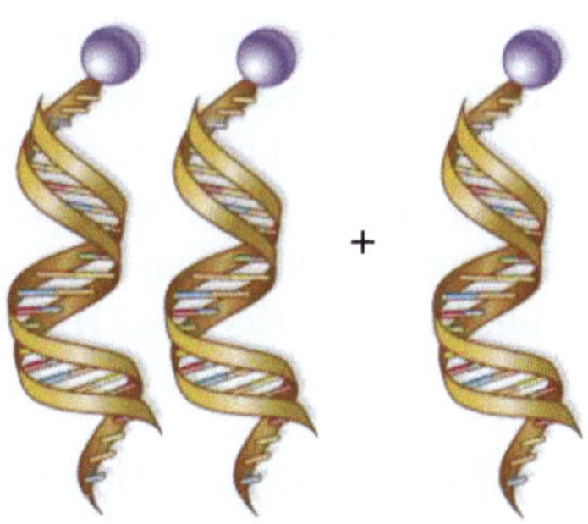

N-acetyl-galactosamine-conjugated siRNAs 2:1 targeting within the S and X gene ORFs

Two siRNAs in combination were selected to become ARO-HBV/JNJ-3989 to target transcripts of both cccDNA and iDNA and have greater genotype coverage [66]. The selected HBV S gene siRNA (JNJ-73763976)

is within the S gene ORF and is expected to target not only the L, M, and S-HBsAg transcripts but also pgRNA that would be cleaved in the middle of the pol ORF and pre-core transcripts for which the target site is downstream of its ORF (Figure 6.1). Targeting directly in the S gene ORF will prevent the ORF from being translated to produce HBsAg, whether it was transcribed from cccDNA or iDNA. The antisense strand sequence at positions 2–18 was perfectly complementary to 95.32% of complete, well-characterized genotype A-H HBV genomes in NCBI and complementary to 99.73% of genomes that were either perfectly matched or had one mismatch. The selected X gene siRNA (JNJ-73763924) targets within the X gene and is expected to target all ccc-DNA-derived transcripts near the 3-prime end of each. The X siRNA sequence is perfectly complementary to 95.65% of the HBV genomes and complementary to 99.17% of genomes that are either perfectly matched or have one mismatch. Both siRNAs together are expected to match 99.73% of genomes with perfect complementarity and >99.99% of genomes when allowing up to one mismatch. Addition of the X siRNA to the S siRNA increases coverage of HBV genotypes and genomes as well as targeting additional HBV transcripts. Specifically, the X siRNA is expected to target X transcripts, which are not targeted by the S-siRNA, and spliced transcripts in which the S target site is spliced out.

6.2.2.1.1 Reduction of Serum HBsAg, HBeAg, and HBV DNA in Mouse Model of Chronic HBV Infection

Combining the X siRNA with the S siRNA enhanced HBeAg reduction in the HDI mouse model of HBV infection (Figure 6.2a [66]). Mice given three doses of 3.2 mg/kg S siRNA alone once every 3 weeks (Q3W) and measured on day 57 had a 3.47 $\log_{10}$ reduction of HBsAg, normalized to pretreatment and controls. HBsAg was reduced nearly the same amount by 3.2 mg/kg S siRNA plus 0.8 mg/kg X siRNA (3.41 $\log_{10}$). HBeAg, however, was reduced 0.75 $\log_{10}$ from 3.2 mg/kg S siRNA, 0.47 $\log_{10}$ from 0.8 mg/kg X siRNA, but 2.07 $\log_{10}$ from the combined 3.2 mg/kg S siRNA plus 0.8 mg/kg X siRNA. Thus, the combined siRNAs increased the knockdown of the pre-core transcript more than either one alone.

ARO-HBV/JNJ-3989, comprised of S siRNA and X siRNA at a 2:1 ratio, durably reduced all HBV transcripts [66]. A single dose of 4 mg/kg ARO-HBV reduced transcripts measured by a probe in the X region (expected to include all transcripts) by 95% and those measured in the core region (pgRNA and pre-core) by 90% at the nadir 3 weeks post-dose [66]. Three repeat doses of 4 mg/kg ARO-HBV given Q3W in HDI HBV model mice demonstrated that each dose yielded a further reduction in HBsAg, HBeAg, and HBV DNA in serum (Figure 6.2b). HBsAg and serum HBV DNA decreased more than 3 $\log_{10}$ (multiple data points to below the lower level of quantitation (< LLOQ)) and HBeAg more than 2 $\log_{10}$ to < LLOQ.

6.3 Pharmacokinetics and Safety of JNJ-3989 in Humans

Pharmacokinetics and safety of JNJ-3989 were evaluated in healthy adults (18–65 years) that were non-Japanese or Japanese in one study (ClinicalTrials.gov NCT04002752 [68]) or of Chinese origin in another [69] and in CHB patients that were Asian,

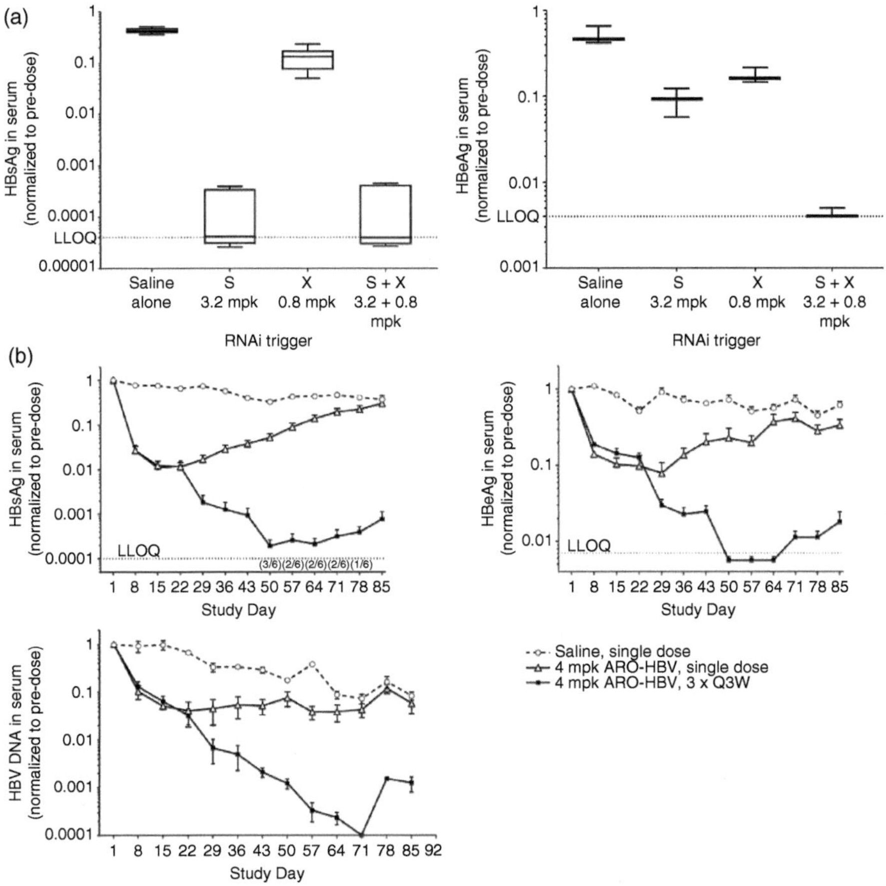

Figure 6.2 Reduction of HBsAg, HBeAg, and HBV DNA in HDI HBV mouse model serum. (a) HDI HBV model mice were given a single subcutaneous injection of saline as control, 3.2 mg/kg of S siRNA, 0.8 mg/kg of X siRNA or the combined 3.2 mg/kg S siRNA + 0.8 mg/kg X siRNA. Serum HBsAg and HBeAg were measured on day 56 and are shown normalized to pretreatment levels with standard error; (b) HDI HBV mice were given a single 4 mg/kg dose of ARO-HBV or three subcutaneous injections 3 weeks apart and HBsAg, HBeAg, and serum HBV DNA were measured at the indicated times and normalized to pretreatment levels. Numbers in parentheses indicate the number of mice out of 6 total that had expression below the lower limit of quantitation (LLOQ) at indicated time points. *Source*: Adapted from [66] / with permission of Elsevier.

non-Japanese (NCT03365947 [68]). Participants were given a single subcutaneous (SC) injection of 25–400 mg JNJ-3989 and plasma concentrations were measured over 48 h. The median time to maximum concentration as the siRNAs were released from the subcutaneous space into the plasma ranged from 6–10 h. The mean plasma half-life was 4–9 h with the siRNAs expected to be either taken up into the liver by association of the GalNAc-siRNA with ASGPR or eliminated through the kidneys. The mean elimination plasma half-life was approximately 5 h. These parameters all point to release from the subcutaneous space followed by uptake into the liver via ASGPR occurring relatively quickly, as the plasma half-life for the siRNAs was short. Exposure increased in a dose-proportional manner and was not notably different between healthy volunteers and CHB patients.

Safety and tolerability were assessed through day 28 for healthy volunteers given a single dose of JNJ-3989 and through day 116 for CHB patients given an injection of 200 mg JNJ-3898 every 4 weeks (Q4W) along with daily NA plus daily oral JNJ-56136379 (JNJ-6379), a capsid assembly modulator of type E (CAM-E) in one cohort. JNJ-3989 was generally well tolerated with no difference between healthy volunteers and CHB patients. In each of these studies all adverse events were mild/moderate.

Preclinical pharmacology and toxicity with doses up to 300 mg/kg had been explored in mice, rats, and monkeys [68]. The no adverse effect level (NOAEL) for single and multiple doses in monkeys was 300 mg/kg.

6.4 Clinical Studies in Chronically HBV-infected Patients

More than 500 CHB patients have been treated with JNJ-3989 in multiple clinical trials through December 2023 [68, 70, 71]. In a Phase 2a first-in-patient study (AROHBV1001, NCT03365947), HBeAg positive and negative, treatment naïve and VS patients received three doses of 25–400 mg JNJ-3989 by subcutaneous injection at weekly, biweekly or every 4-week frequency (Q1W, Q2W, or Q4W) in combination with daily oral NA to assess the effect of the timing of the three JNJ-3989 doses and duration of response up to day 392. In one cohort, patients in addition to 200 mg JNJ-3989 + NA received daily 250 mg oral CAM-E JNJ-6379 to assess the tolerability of this combination. REEF-1 (NCT03982186) was a Phase 2b study that evaluated the antiviral activity and safety of JNJ-3989 given for 48 weeks of Q4W dosing in combination with oral NA, with or without CAM-E JNJ-6379, with an option to stop NA in those patients who met the primary NA stopping criteria at week 48 or during the 48 week follow up period. REEF-2 (NCT04129554) was a placebo-controlled Phase 2b study investigating the effect of controlled NA withdrawal in long-term VS HBeAg-negative patients at the end of 48 weeks of JNJ-3989-based treatment compared to NA control. PENGUIN (NCT04667104) and REEF-IT (NCT04439539) are ongoing Phase 2 studies, at the time of writing, exploring the addition of the immune-modulator pegIFN-alpha2a (pegIFN) to JNJ-3989-based treatments in VS (PENGUIN) or in not currently treated (NCT) HBeAg-positive patients (REEF-IT). REEF-D (NCT04535544) is an ongoing Phase 2 study in patients with chronic hepatitis B/D coinfection. Antiviral activity in each study was assessed by measuring serum HBsAg, HBeAg, HBcrAg, HBV RNA, and HBV DNA. All studies were conducted in accordance with the ethical principles of the Declaration of Helsinki and were consistent with Good Clinical Practice.

6.4.1 Direct Antiviral Treatment Approach with JNJ-3989-based Regimens

6.4.1.1 Pharmacological Response to JNJ-3989 with Short-term Treatment

Phase 2a study AROHBV1001 assessed reductions in serum HBV parameters in HBeAg-positive and HBeAg-negative patients who received 25, 50, 100, 200, 300, or 400 mg JNJ-3989 given three times in combination with daily oral NA [70].

Prior to the study, some patients were NA-naïve and others were VS. Treatment with JNJ-3989 + NA resulted in HBsAg reduction from day 1 to day 112 in every group, indicating that JNJ-3989 targets cccDNA and iDNA-derived transcripts (Figure 6.3a). Smaller mean HBsAg reductions were observed in 25 mg JNJ-3989

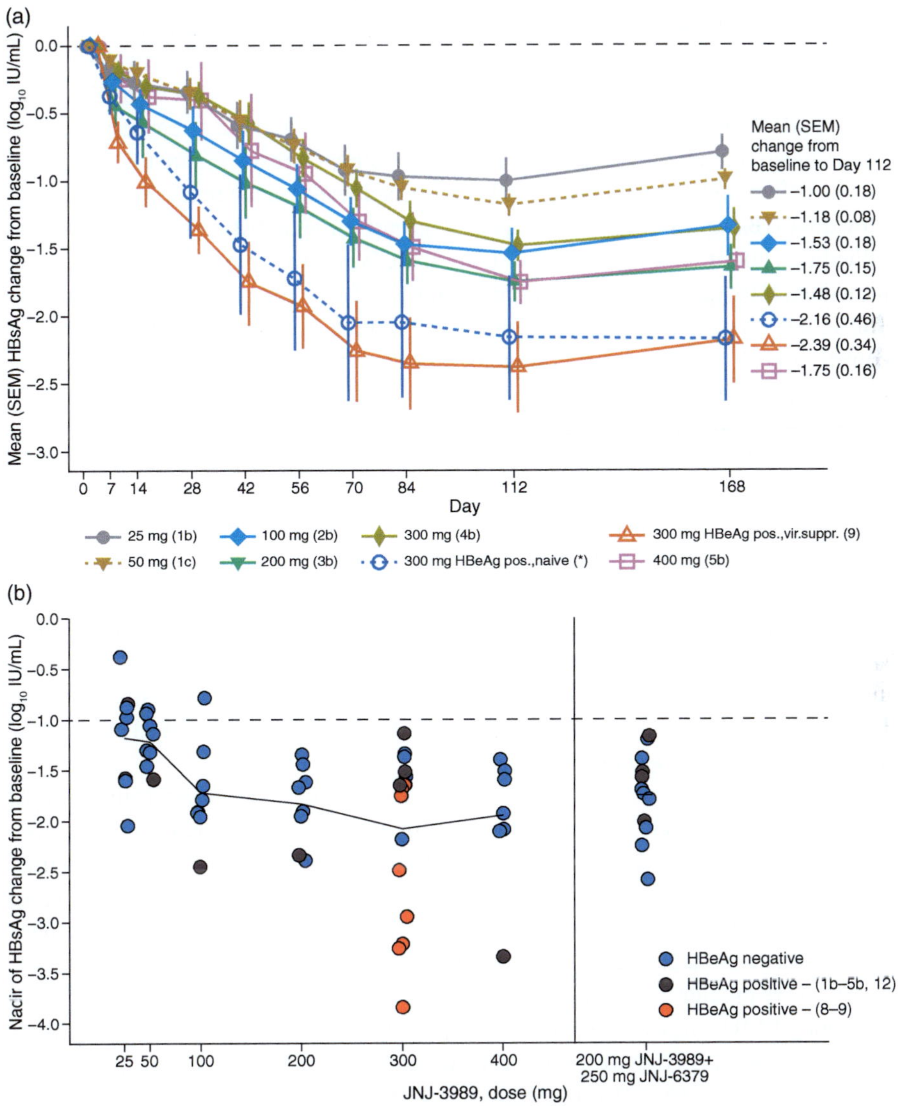

Figure 6.3 Mean and individual changes in HBsAg following short-term treatment of three doses Q4W JNJ-3989 in combination with NA in Phase 2a study AROHBV1001. (a) Mean change in HBsAg from day 0 to day 168 in patients that received JNJ-3989 at the indicated dosages plus NA. (b) Individual nadir of HBsAg change from day 0 to end of follow-up (FU) day 392 in the JNJ-3989 Q4W dose cohorts on continued NA and in the cohort that received 200 mg JNJ-3989 Q4W and 250 mg JNJ-6379 daily for 12 weeks with continued NA. Solid black line in Figure 6.3b connects the mean nadir of each dose. *Source*: Adapted from [70] / Elsevier / CC BY 4.0.

(1.00 $\log_{10}$ IU/mL) and 50 mg JNJ-3989 (1.18 $\log_{10}$ IU/mL) arms than with 100–400 mg JNJ-3989 (1.48–2.39 $\log_{10}$ IU/mL) among the Q4W groups. Most patients received three doses of JNJ-3989 every 4 weeks, but patients in three cohorts were dosed three times weekly with 100, 200, or 300 mg, and in another cohort, three times biweekly with 100 mg JNJ-3989. There was no notable benefit in terms of mean HBsAg decline with shorter dosing intervals [70]. The HBsAg means across all arms reached nadir at day 112, i.e. 56 days after the last Q4W dose. From cohort to cohort, the HBsAg decline was somewhat dose-responsive, with lower responses mainly in the 25 and 50 mg cohort, but Cohorts 8 and 9 with HBeAg positive patients that were NA-naïve (2.16 $\log_{10}$ IU/mL reduction) or VS (2.39 $\log_{10}$ IU/mL reduction), respectively, had greater mean reductions of HBsAg than those in other cohorts (Figure 6.3). While all but one patient in the 100–400 mg Q4W cohorts achieved >1 log reduction in HBsAg, individual HBsAg declines within the dose group varied substantially, as shown by individual patient HBsAg nadirs in Figure 6.3b. While the majority of patients had immediate HBsAg declines upon administration of JNJ-3989, a subgroup of mainly HBeAg-negative patients experienced HBsAg reduction only from week 4 or week 8 onwards [70]. This observation was not dose- or dosing frequency-dependent. The underlying mechanism for this delayed response is not understood.

Consistent with greater reduction of HBsAg than HBeAg in the mouse model with ARO-HBV/JNJ-3989, mean HBsAg reduction was also more pronounced than HBeAg in patients with baseline HBeAg >1.11 $\log_{10}$ PEIU/mL with a mean reduction at day 112 of 1.47 $\log_{10}$ PEIU/mL to nadir for HBeAg compared to 2.11 $\log_{10}$ IU/mL for HBsAg. Similarly, HBcrAg in patients that had >10 kU/mL was reduced with a mean of 1.20 $\log_{10}$ kU/mL to nadir and 1.97 $\log_{10}$ IU/mL for HBsAg in the same patients. Serum HBV RNA in patients that had baseline levels >2.65 $\log_{10}$ U/mL was reduced similarly to HBsAg with mean HBV RNA reduction of 1.93 $\log_{10}$ U/mL and HBsAg reduction in the same patients of 1.86 $\log_{10}$ IU/mL. Interestingly, while these were the observed patterns, some patients had more pronounced declines of HBcrAg and/or HBeAg than HBsAg.

For 39/40 patients that received 100–400 mg JNJ-3989, FU data were available until day 392, which was 336 days after the last treatment with JNJ-3989 (Figure 6.4). These patients remained on NA treatment. Fifteen of 39 (38%) patients maintained ≥1 $\log_{10}$ IU/mL HBsAg decrease to day 392 and were defined as "sustained HBsAg responders." The mean (standard error, SE) sustained HBsAg decrease in these patients was 1.96 (0.20) log IU/mL. Mean (SE) reductions of all viral parameters at day 392 were greater for "sustained" than "non-sustained HBsAg responders": 1.71 (0.45) $\log_{10}$ PEIU/mL versus 0.77 (0.22) $\log_{10}$ PEIU/mL HBeAg; 2.20 (0.43) $\log_{10}$ kU/mL versus 0.43 (0.15) $\log_{10}$ kU/mL HBcrAg; and 2.36 (0.46) $\log_{10}$ U/mL versus 0.78 (0.20) $\log_{10}$ U/mL HBV RNA, respectively. There were no safety concerns for treatment with JNJ-3989 + NA, with or without CAM-E JNJ-6379. ALT elevations were all relatively minor and not considered clinically relevant. A subset of patients not only sustained the HBsAg response but had decreasing viral markers during the follow-up period well after RNAi treatment. Further study is required to determine if these profound post-RNAi reductions are due to sustained host control. Interestingly, fewer ALT elevations occurred in "sustained HBsAg responders."

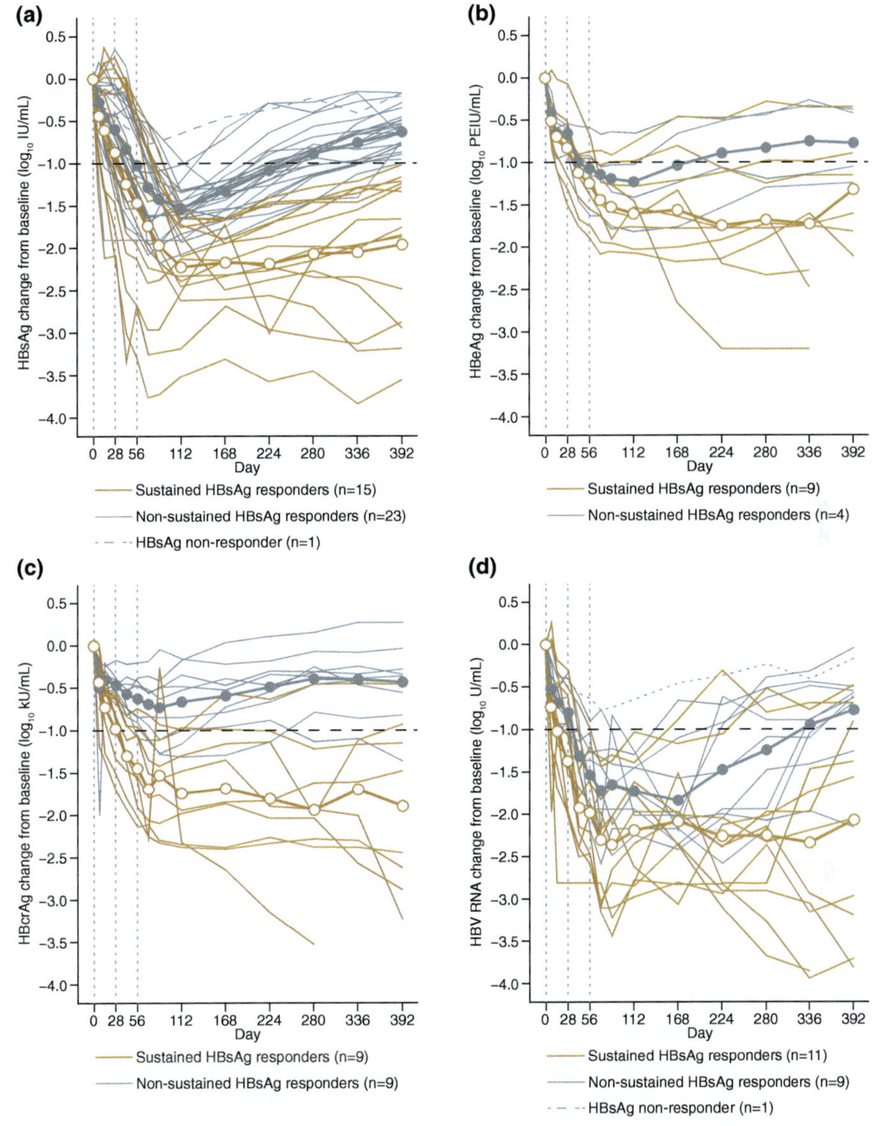

Figure 6.4 Mean and individual changes in serological markers of HBV infection from day 0 to day 382 by HBsAg responder status in Q4W 100–400 mg dose cohorts. (a) HBsAg; (b) HBeAg; (c) HBcrAg; and (d) HBV RNA. Sustained HBsAg responders defined as patients with ⩾1 log_{10} HBsAg reduction from day 0 to day 392. Only patients with HBeAg, HBcrAg, and HBV RNA >1 log_{10} of lower limit of quantitation were included in panels (b), (c), and (d), respectively. Circles represent mean. *Source*: Adapted from [70] / Elsevier / CC BY 4.0.

The promising results from short-term treatment with JNJ-3989 in AROHBV1001 triggered the start of two larger clinical studies with the aim to assess the safety and efficacy of JNJ-3989, to better characterize dose response, treatment duration, combinations and off-treatment response. They also guided the study design of the Phase 2b double-blinded studies REEF-1 (73763989HPB2001) and REEF-2 (73763989PAHPB2002):

a. Dose: The observation of similar mean HBsAg reductions with three monthly doses of 100, 200, 300, and 400 mg was based on small cohorts of 4–12 patients that were not balanced for HBeAg status, treatment history, or HBsAg levels at study start. To confirm the optimal dose of JNJ-3989 with a longer treatment duration and larger sample size, REEF-1 enrolled 90 patients per cohort and stratified randomization for HBeAg status (positive vs. negative) and treatment history at screening (NCT vs. VS) with a minimum HBsAg level >100 IU/mL at screening. Three dose levels, 40, 100, and 200 mg of JNJ-3989, were tested and the statistical methodology MCP-Mod was used to allow for interpolation of results if required [72].

b. Duration: Assess whether HBsAg kinetics under siRNA treatment for longer than 12 weeks would follow a constant and linear decline or if the reduction of HBsAg would level off despite continued treatment.

c. Durability of response: After short-term treatment with JNJ-3989 in study AROHBV1001, most patients had increasing serum levels of HBsAg during FU. However, 336 days after the last dose, 38% had maintained HBsAg levels that were >1 $\log_{10}$ lower than baseline, suggesting an improved viral control by the host immune system. An important question was whether more patients would show a sustained response if levels of HBsAg could be reduced further – ideally to seroclearance – or if suppressed HBsAg levels were maintained for a longer duration with extended treatment.

d. Combination of antiviral approaches: REEF-1 and REEF-2 explored the combination of JNJ-3989 with the CAM-E JNJ-6379 that had shown profound reductions of HBV RNA and HBV DNA in a Phase 1b and the Phase 2 JADE study [73, 74] and the inhibition of cccDNA replenishment via a secondary mode of action in preclinical studies [75].

e. The role of NA during treatment and FU: To minimize the risk of viral breakthrough with the development of resistance and to maximize viral suppression, JNJ-3989 was combined with NA in all treatment arms either as continued treatment in VS patients or as newly started NA at baseline in NCT patients. To assess the proportion of patients achieving functional cures, the studies needed to allow for the discontinuation of NA at the end of treatment (EOT). In REEF-1, a key secondary endpoint, and in REEF-2, the primary endpoint was the proportion of patients with HBsAg seroclearance (<0.05 IU/mL) 24 weeks after the end of all interventions (including NA). REEF-1 and REEF-2 followed two different approaches to safely discontinue NA at the end of investigational treatment. In REEF-1 NA treatment could be withdrawn at the EOT or during FU if predefined stopping criteria based on HBsAg, HBV DNA, HBeAg, and ALT were met. REEF-2 focused on non-cirrhotic, HBeAg-negative patients with long-term NA treatment and viral suppression, a population for which the controlled withdrawal of NA treatment is supported by EASL treatment guidelines [11]. Another important consideration was that the controlled withdrawal of NA had been associated with higher rates of off-treatment HBsAg seroclearance and functional cure in patients with low HBsAg, compared to cohorts with continued NA treatment [76, 77]. REEF-2 was specifically designed to assess the effect that stopping of NA had on off-treatment response in patients with JNJ-3989-induced low HBsAg levels.

6.4.1.2 REEF-1 Clinical Study

REEF-1 was a Phase 2b double-blind, active-controlled, randomized study in CHB patients (18–65 years) conducted at 108 hospitals and outpatient centers across 19 countries in Asia, Europe, and North and South America [71]. Randomization was stratified by HBeAg status (positive or negative) and by NA treatment at screening (NCT or VS), a total of 470 patients. As shown in Figure 6.5, all patients including the placebo group (n = 45) received daily oral NA (entecavir, tenofovir disoproxil fumarate, or tenofovir alafenamide). Dual groups received NA plus subcutaneous siRNA JNJ-3989 (Q4W) of 40 mg (n = 93), 100 mg (n = 93) or 200 mg JNJ-3989 (n = 96) Q4W or NA plus 250 mg daily oral CAM-E JNJ-6379 (n = 48). Triple therapy comprised NA plus 100 mg JNJ-3989 (Q4W) plus daily 250 mg JNJ-6379 (n = 95). Three hundred ten (66%) of 470 patients were male, 244 (52%) were White, 190 (40%) were Asian, and the mean age was 43 years (standard deviation [SD] 10.7). Two hundred ninety-eight (63%) of patients were virologically suppressed and 328 (70%) were HBeAg negative at screening. Treatment was for 48 weeks and the FU phase was an additional 48 weeks or longer. Results from interim FU analysis at 24 weeks are available.

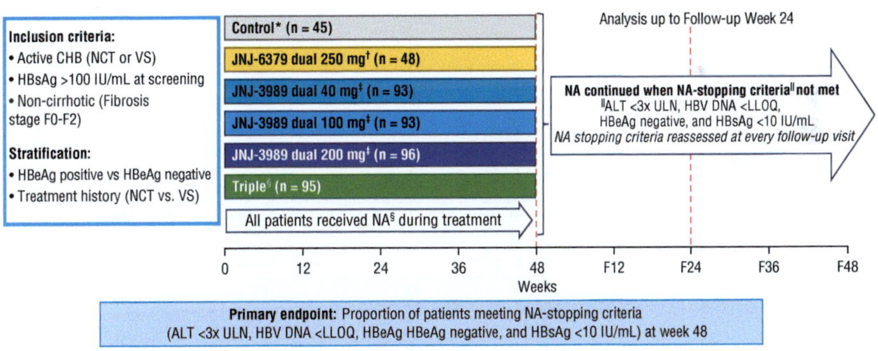

Figure 6.5 REEF-1 study design. ALT, alanine aminotransferase; HBV, hepatitis B virus; JNJ-3989, JNJ-73763989; JNJ-6379, JNJ-56136379.

* Once per day entecavir (0.5 mg), tenofovir disoproxil fumarate (300 mg), or tenofovir alafenamide (25 mg).

† Oral JNJ-6379 (250 mg) once per day plus oral nucleos(t)ide analogs once per day plus JNJ-3989 placebo every 4 weeks.

‡ Subcutaneous injection of JNJ-3989 every 4 weeks plus oral nucleos(t)ide analogs once per day plus JNJ-6379 placebo once per day.

§ Subcutaneous injection of JNJ-3989 (100 mg) every 4 weeks plus oral JNJ-6379 (250 mg) once per day plus oral nucleos(t)ide analogs once per day. Nucleos(t)ide analog stopping criteria were ALT less than three times the ULN, HBV DNA below the lower limit of quantitation, HBeAg negativity, and HBsAg less than 10 IU/mL.

Source: Adapted from [71] / with permission of Elsevier.

The primary efficacy endpoint was patients meeting the NA stopping criteria at week 48 (Figure 6.6). The secondary endpoints were HBsAg seroclearance at 24 weeks off all study treatments; meeting the NA stopping criteria during the FU phase; and change in HBsAg, HBeAg, HBV DNA, HBcrAg, and HBV RNA. At the end of the 48-week treatment, 47 (10%) of 470 patients met the primary endpoint for stopping NA: ALT <3× upper limit of normal (ULN), HBV DNA

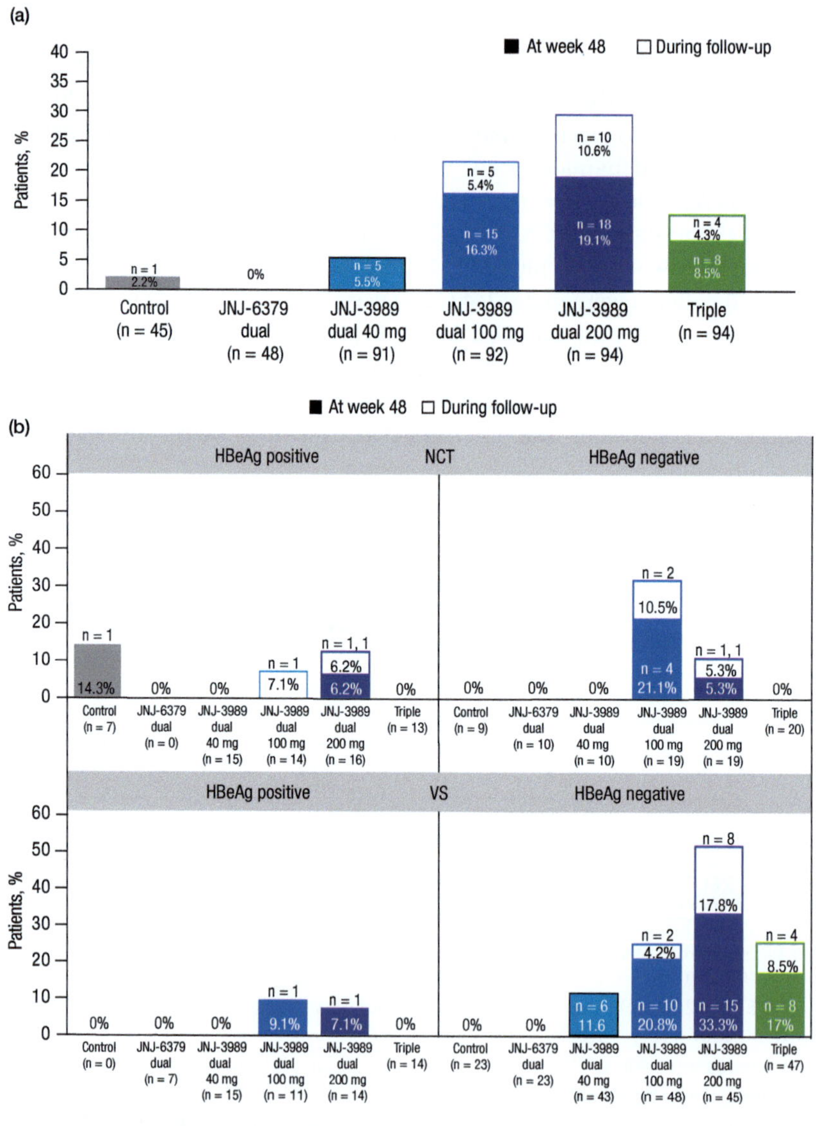

Figure 6.6 Percentages of patients meeting NA-stopping criteria at week 48 and during the 24-week follow up by (a) overall and (b) stratification factors. Nucleos(t)ide analog-stopping criteria are ALT less than three times the ULN, hepatitis B virus DNA below the lower limit of quantitation, HBeAg negative, and HBsAg less than 10 IU/mL. *Source*: Adapted from [71] / with permission of Elsevier.

below the lower level of quantitation, HBeAg negative, and HBsAg <10 IU/mL. An additional 19 (4%) met these criteria during the FU phase. The group that received 200 mg JNJ-3989 with NA had the highest proportion that met these criteria with 18 (19%) of 94 patients at week 48 and 10 (11%) during FU (Figure 6.6). Among the patients who received 200 mg JNJ-3989 and met NA-stopping criteria, 67% remained off NA with HBV DNA <LLOQ and HBsAg <100 IU/mL at the 24-week FU time point.

Thirty-eight (83%) of the JNJ-3989-treated patients who met the NA-stopping criteria at week 48 were virologically suppressed and HBeAg negative. Of the virologically suppressed, HBeAg-negative patients who received 200 mg JNJ-3989 with NA, 15 (33%) met the NA stopping criteria during the treatment phase, and an additional 8 (18%) met these criteria by the week 24 FU. The CAM-E JNJ-6379 provided no additional benefit in these studies.

By week 48 (EOT) all groups that received JNJ-3989 had substantial reductions of HBsAg with a dose-dependent mean reduction in HBsAg (Figure 6.7). At week 48, 98% of patients in the 200 mg JNJ-3989 group had 1 $\log_{10}$ or greater decline in HBsAg and 74% of patients in this group with available data at week 48 had a 2 $\log_{10}$ decline or greater with a mean reduction of 2.6 $\log_{10}$ IU/mL at week 48 that remained 1.9 $\log_{10}$ IU/mL reduced at week 24 FU. The week 48 mean reduction of HBsAg from the 100 mg JNJ-3989 dose was 2.1 $\log_{10}$ IU/mL that remained reduced by 1.5 $\log_{10}$ IU/mL at week 24 FU. The greatest mean decreases in HBsAg were observed in the subgroup of HBeAg positive and NCT patients for whom in the 200 mg JNJ-3989 + NA arm, the mean (SE) HBsAg reduction at week 48 was 3.56 (0.35) $\log_{10}$ IU/mL and at 24-week FU was 2.63 $\log_{10}$ IU/mL (Figure 6.8). In all JNJ-3989-containing arms, the mean HBsAg reduction kinetics showed a steep decline during the first 16–24 weeks of treatment followed by a continued shallower decline until treatment week 48 (Figures 6.7 and 6.8).

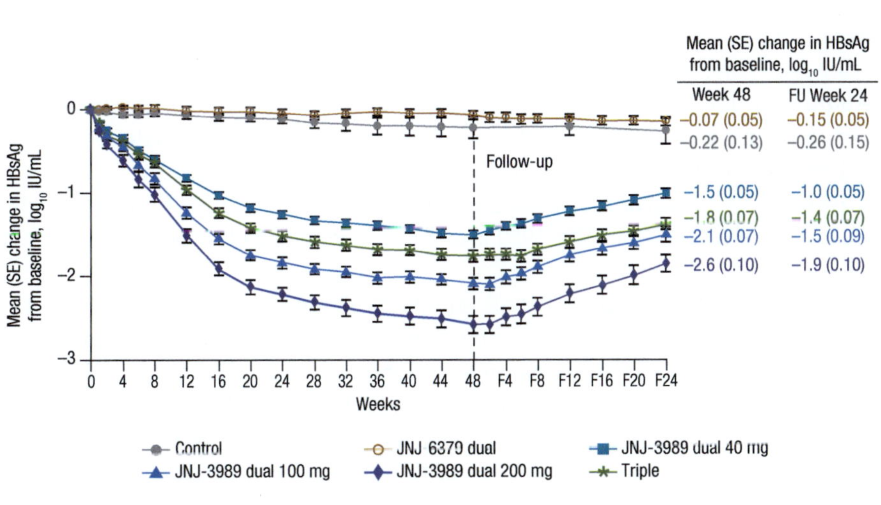

Figure 6.7 Mean change in HBsAg in REEF-1 study patients. Error bars show SE. *Source:* Adapted from [71] / with permission of Elsevier.

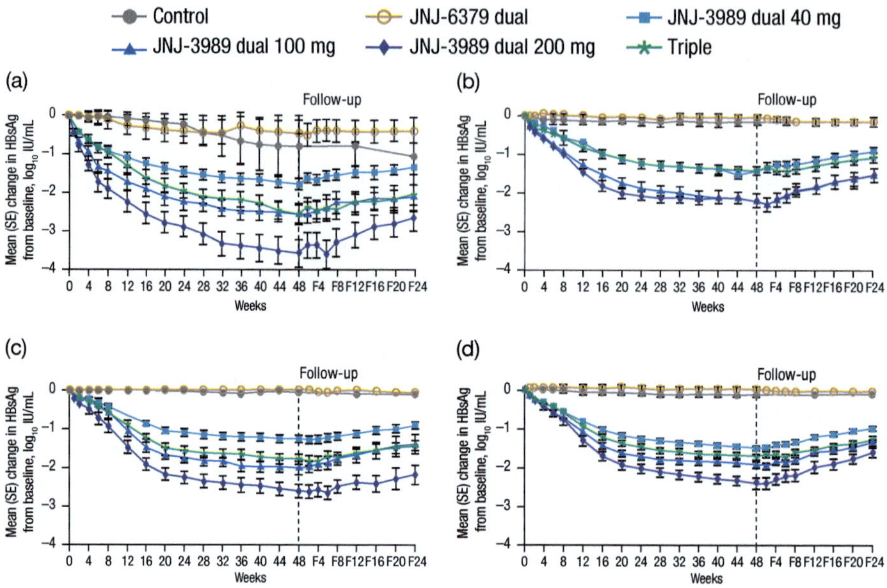

Figure 6.8 Mean (SE) change in HBsAg over time in (a) NCT HBeAg-positive; (b) NCT HBeAg-negative; (c) VS HBeAg-positive; and (d) VS HBeAg-negative patients. F, follow-up; NCT, not currently treated; SE, standard error; VS, virologically suppressed. *Source*: Adapted from [71] / with permission of Elsevier.

Interestingly, as observed previously in the ARO-HBV-1001 study, a large variability in HBsAg reduction was observed in REEF-1 ranging from 0.9 to $6.1\,\log_{10}$ IU/mL at week 48 in the 200 mg JNJ-3989 group. HBsAg seroclearance was rare with eight JNJ-3989 treated patients achieving at least transient HBsAg seroclearance and one HBsAg seronegative off-treatment (i.e. functional cure). HBsAg <100 IU/mL at week 48 was achieved in 68 (75%) of patients treated with 200 mg JNJ-3989.

Addition of the CAM-E JNJ-6379 to JNJ-3989 did not provide a benefit. In the group combining 100 mg JNJ-3989 with JNJ-6379, mean HBsAg reduction from baseline at week 48 was $1.8\,\log_{10}(0.07)$ IU/mL and $1.4\,\log_{10}(0.07)$ IU/mL at 24-week FU. In terms of proportion of patients reaching specific cut-offs of HBsAg levels, the addition of JNJ-6379 to JNJ-3989 also did not provide a clear benefit over JNJ-3989 + NA alone.

The mean HBsAg reduction for all groups receiving JNJ-3989 was $\geq 1\,\log_{10}$ IU/mL at 24 weeks of FU, demonstrating that meaningful HBsAg reductions persisted beyond the end of RNAi treatment. As observed in the short-term treatment, patients in each JNJ-3989 group had stable ($\pm 0.2\,\log_{10}$) or decreased ($>0.2\,\log_{10}$) HBsAg from the EOT to FU week 24, 28 (17.2%) of the 163 patients who received 100 or 200 mg JNJ-3989 and 27 (33.3%) of the 81 patients who received the triple therapy (Figure 6.9).

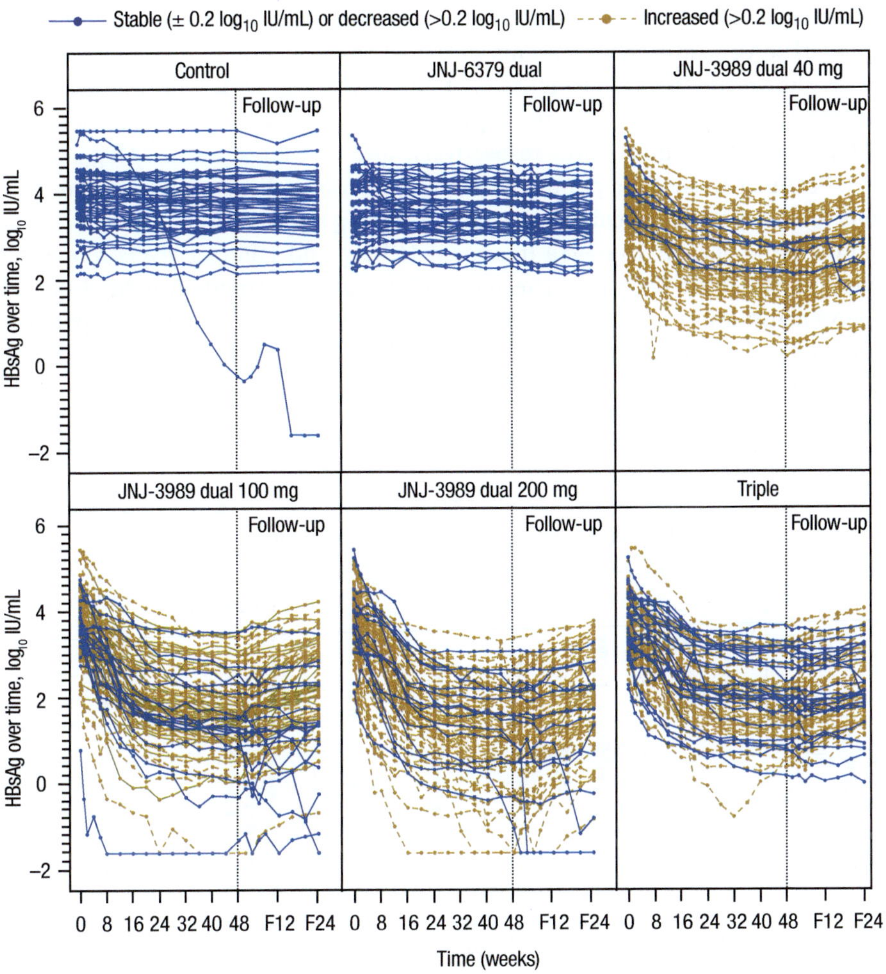

Figure 6.9 REEF-1: Individual patient HBsAg (log_{10} IU/mL) over time for patients who had stable or decreasing, or increasing HBsAg from week 48 to FU week 24. *Source*: Adapted from [71] / with permission of Elsevier.

HBsAg category, n (%)	JNJ-3989 (40 mg) + NA (n = 81)	JNJ-3989 (100 mg) + NA (n = 81)	JNJ-3989 (200 mg) + NA (n = 82)	JNJ-3989 (100 mg) + 250 mg JNJ-6379 + NA ("Triple") (n = 81)
Stable (±0.2 log_{10}) or decreased (>0.2 log_{10}) HBsAg from EOT	7 (8.6)	15 (18.5)	13 (15.9)	27 (33.3)
Increased >0.2 log_{10} HBsAg from EOT	74 (91.4)	66 (81.5)	68 (82.9)	53 (65.4)

Pronounced HBcrAg and HBeAg reductions were observed mainly in JNJ-3989 treated HBeAg positive NCT patients with high baseline levels of these viral markers, as shown for HBeAg in Figure 6.10. Mean HBeAg reduction from baseline at week 48 and FU week 24 with 200 mg JNJ-3989 + NA in this population was 2.2 $\log_{10}$ and 2.52 $\log_{10}$ IU/mL, respectively, and 2.56 $\log_{10}$ and 2.74 $\log_{10}$ U/mL for HBcrAg at these time points, respectively. Despite these pronounced reductions, the proportion of patients becoming HBeAg and/or HBcrAg negative was limited. However, off JNJ-3989 treatment HBeAg and HBcrAg levels remained generally stable or declined. The mean (SD) baseline (pretreatment) level of HBeAg across study arms was 2.31 (1.16) $\log_{10}$ IU/mL in the NCT HBeAg positive patients, and much lower in VS HBeAg positive patients with a mean (SD) of 0.27 (0.79) $\log_{10}$ IU/mL.

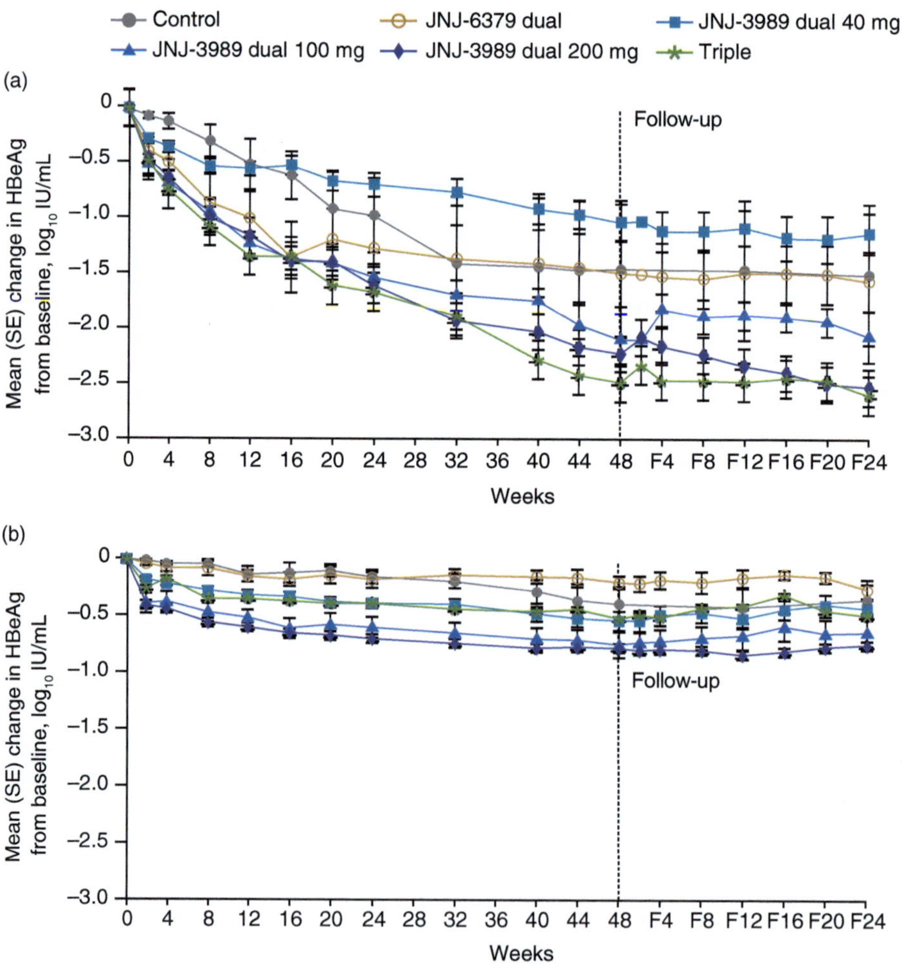

Figure 6.10 REEF-1: Mean (SE) change in HBeAg from baseline over time in HBeAg-positive patients. (a) NCT or; (b) VS ($\log_{10}$ IU/mL). *Source*: Adapted from [71] / with permission of Elsevier.

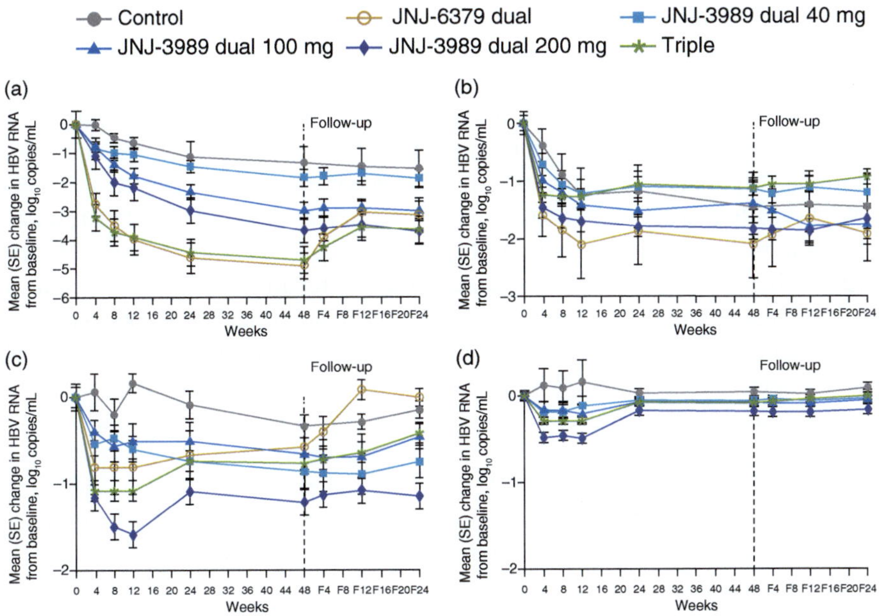

Figure 6.11 REEF-1: Mean (SE) change in HBV RNA over time in (a) NCT HBeAg-positive; (b) NCT HBeAg-negative; (c) VS HBeAg-positive; and (d)* VS HBeAg-negative patients.

Note: the y-axis scale is not the same for all panels. *There were too few VS HBeAg-negative patients in the JNJ-6379 dual arm to include data in d. The baseline mean (SD) HBV RNA level in NCT HBeAg positive was 6.94 (1.18) $\log_{10}$ copies/mL with all having HBV RNA >LOD; in NCT, HBeAg negative was 3.66 (1.40) $\log_{10}$ copies/mL with 31.9% having HBV RNA <LOD; in VS, HBeAg positive was 3.32 (1.21) $\log_{10}$ copies/mL with 34.8% having HBV RNA <LOD; and in VS, HBeAg negative was 2.29 (0.32) $\log_{10}$ copies/mL with 89.0% having HBV RNA <LOD.

Source: Adapted from [71] / with permission of Elsevier.

As with HBeAg, the greatest reductions in HBV RNA were in the NCT HBeAg-positive patients (Figure 6.11). This is the only subgroup of patients with mean RNA levels greater than the LLOQ of the assay used, which is 4.04 $\log_{10}$ copies/mL. NCT HBeAg-positive patients who received 200 mg JNJ-3989 + NA had mean (SE) reductions from baseline of 3.66 (0.30) $\log_{10}$ copies/mL and 3.70 (0.31) $\log_{10}$ copies/mL at the EOT week 48 and FU week 24, respectively, with 28.6% and 38.5% of these patients reaching HBV RNA below the limit of detection (<LOD) at week 48 and FU week 24, respectively. In contrast, 89% of the VS HBeAg-negative patients had HBV RNA <LOD at baseline, and the mean HBV RNA levels in NCT HBeAg negative and VS HBeAg positive patients at baseline were <LLOQ.

Viral sequence analyses were performed to assess the presence of nucleotide polymorphisms in the target regions of the S and X siRNAs. Baseline nucleotide polymorphisms in the JNJ-3989 S- and X-siRNA target regions were present in 11% and 14% of NCT patients, respectively. The majority of patients with polymorphisms carried single nucleotide polymorphisms. Importantly, no apparent impact of these baseline polymorphisms was observed on the JNJ-3989–induced HBsAg and HBeAg declines [78].

6.4.1.3 REEF-2 Clinical Study

REEF-2 was a Phase 2b double-blind, active-controlled, randomized study in non-cirrhotic, HBeAg negative CHB patients (18–65 years) who were virologically suppressed with NA for at least 2 years and had HBsAg >100 IU/mL at study entry. The study was conducted at 33 hospitals and outpatient centers across seven countries in Europe [79]. Eighty-five patients in the active arm received 200 mg SC JNJ-3989 Q4W plus daily oral JNJ-6379 (250 mg) and daily oral NA (ETV, TF, or TAF) and 45 patients in the control arm received SC placebo of JNJ-3989, oral placebo of JNJ-6379, and daily oral NA. At treatment week 48, NA treatment was to be withdrawn in all patients and the follow-up (FU) phase was an additional 48 weeks, with primary endpoint analysis at FU week 24. The mean age was 46 years (SD 10.2), 66% of patients were male, 66% were White, and 19% were Asian. At screening, 78.5% had HBsAg >1000 IU/mL.

The primary efficacy endpoint, which was the proportion of patients achieving HBsAg seroclearance (<0.05 IU/mL) 24 weeks after all study interventions including NA stopped and was not met by any patient in either arm. At week 48 (end of treatment) mean reductions of HBsAg from baseline were 1.89 $\log_{10}$ and 0.06 $\log_{10}$ IU/mL in the active and control arm, respectively. After 48 weeks of FU, mean HBsAg levels were 1.46 $\log_{10}$ and 0.49 $\log_{10}$ IU/mL lower than baseline, for the active and control arm, respectively. A total of 71.1% of patients in the active arm achieved HBsAg <100 IU/mL at week 48; during FU proportions were reduced to 67.1% at FU week 24 and to 46.9% at the end of study (FU week 48). At FU week 48, 31.6% of patients in the active arm had declining or stable ($\pm$ 0.2 $\log_{10}$ change) HBsAg levels, compared to treatment week 48, and 81.5% had maintained an HBsAg >1 $\log_{10}$ lower than the baseline level (Figure 6.12).

Treatment with JNJ-3989 + JNJ-6379 and NA was well tolerated and safe. The discontinuation of NA treatment at week 48 led to significant virologic and biochemical flares, predominantly in the control arm: 26.8% had a flare with HBV DNA >100 000 IU/mL and 24.3% had an ALT peak of >10× ULN. In the active arm, numbers were significantly smaller with 2.6% and 1.3% that had HBV DNA >100 000 IU/mL and ALT > 10× ULN, respectively. One patient in the control arm had severe HBV reactivation that despite timely initiation of NA re-treatment required liver transplantation [80].

In this study, functional cure (HBsAg seronegativity after stopping all treatment) was not achieved at FU week 24 nor FU week 48, which may be explained by the requirement of HBsAg levels >100 IU/mL at screening (>78% had HBsAg >1000 IU/mL). Multiple studies with NA withdrawal have suggested that low HBsAg values (preferentially <100 IU/mL) are associated with a higher likelihood of achieving HBsAg seroclearance after NA cessation [81]. All but one patient in the control arm still had HBsAg >100 IU/mL at time of NA cessation. However, it is important to note that 71.1% in the active arm had achieved values below 100 IU/mL at time of NA withdrawal and 46.9% maintained HBsAg <100 IU/mL after 48 weeks off all treatment (Figure 6.12).

After NA discontinuation, HBV DNA was at least transiently detectable in all patients. NA treatment was reinitiated in 7.8% of the active arm and 27.3% of the

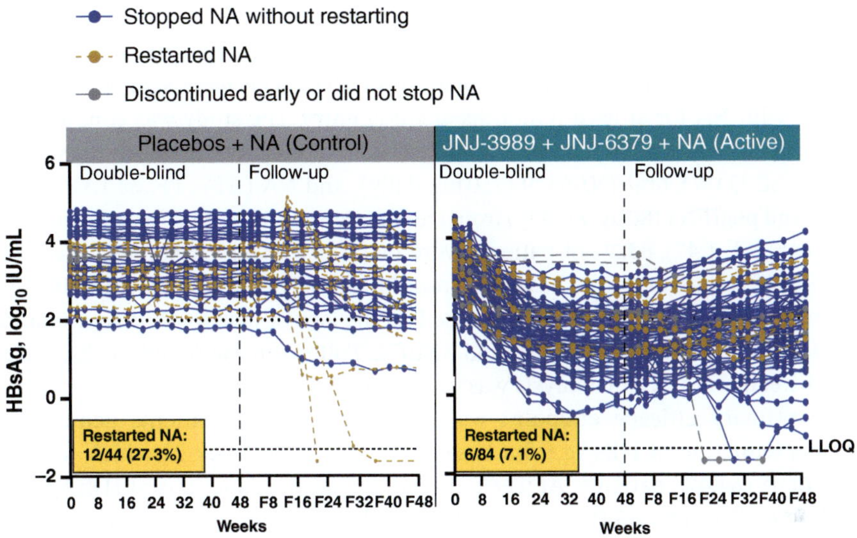

Figure 6.12 REEF-2: Individual patient HBsAg (log_{10} IU/mL) over time for patients in the control arm (NA only for 48 weeks) and treatment arm (200 mg JNJ-3989 + 250 mg CAM-E JNJ-6379 + NA for 48 weeks). NA stopped at week 48 and was not re-started (blue lines) or restarted (yellow lines) during FU. Gray lines indicate patients that discontinued early or did not stop NA.

Source: Adapted from [79] / American Association for the Study of Liver Diseases.

control arm. Among all patients who remained off NA at FU week 48, HBV DNA values were <2000 IU/mL and <LLOQ in 46.3% and 2.4%, respectively, in the control arm and 80.5% and 22.1%, respectively, in the active arm. The combination of HBV DNA <LLOQ and HBsAg <100 IU/mL, referred to as partial cure, was reported in 15.6% of the active arm versus 0% in the control arm.

6.4.2 Combination Approaches of JNJ-3989 with Immune Modulation

While profound reduction of viral markers, including sustained off-treatment response in some patients was achieved with JNJ-3989-based treatment in the absence of immune modulators, the rate of sustained HBsAg seroclearance or functional cure remained low. Thus, studies in combination with immune modulators were started to assess if JNJ-3989-reduced HBsAg levels would result in improved efficacy once immune modulators are added.

The addition of pegIFN-alpha2a (pegIFN) to a JNJ-3989-based regimen once HBsAg levels were reduced by JNJ-3989 treatment was tested in two Phase 2 studies that focused on two distinct patient populations. PENGUIN explored a treatment duration of 24 weeks in VS CHB patients. REEF-IT was in HBeAg-positive CHB patients who were not currently treated and treatment was for 48 weeks or longer. In both studies, interferon was added during the last 12 weeks of JNJ-3989 treatment.

6.4.2.1 PENGUIN Clinical Study

PENGUIN is a Phase 2b open-label, single-arm study in non-cirrhotic patients (18–65 years) with HBeAg negative or positive CHB who were virologically suppressed with NA for at least 6 months at study entry. The study was conducted in New Zealand, Taiwan, Japan, and Poland. A total of 48 patients received JNJ-3989 (200 mg SC Q4W), oral JNJ-6379 (250 mg daily), and NA (ETV, TF, or TAF) for 24 weeks and pegIFN (180 μg weekly) from treatment week 12 to 24. The mean age was 44.6 years (SE 1.47), 83.3% of patients were male, 70.8% were Asian, and the mean duration of NA treatment at baseline was 4.51 (SE 0.5) years. Treatment with JNJ-6379 was discontinued in PENGUIN based on the lack of benefit of its combination with JNJ-3989 observed in REEF-1 and REEF-2. Follow-up was for 48 weeks, and an end-of-treatment interim analysis was presented [82].

The primary efficacy endpoint was the proportion of patients achieving a $\geq 2\log_{10}$ reduction of HBsAg from baseline at treatment week 24 which was met by 31 of 48 (64.4%) patients. At weeks 12 and 24, mean reductions of HBsAg from baseline (SE) were 1.43 (0.07) and 2.18 (0.08) $\log_{10}$ IU/mL, respectively. At Week 24, 91.7% and 47.9% of patients had HBsAg <100 and <10 IU/mL, respectively. One patient had achieved HBsAg seroclearance at week 24. Between week 12 and week 24, the kinetics of HBsAg did not show a clear improved decline due to the addition of pegIFN.

6.4.2.2 REEF-IT Clinical Study

REEF-IT is a Phase 2b open-label, single-arm study in non-cirrhotic patients (18–55 years) with HBeAg positive CHB who were not currently treated with NA for at least 6 months prior to study entry. The study was conducted at 30 centers in Asia, Europe, and North America. A total of 54 patients were enrolled and received JNJ-3989 (200 mg SC Q4W), oral JNJ-6379 (250 mg daily) and NA (TDF or TAF) for 48 weeks or longer and pegIFN (180 μg weekly) during the last 12 weeks of treatment. The period of JNJ-3989 dosing prior to addition of pegIFN is the "induction phase" and the end of pegIFN treatment is the EOT; thus, pegIFN was received between the end of induction (EOI) and EOT. The mean age was 33.6 years, 51.9% of patients were male, and 55.6% were Asian. The mean (SD) level at baseline for HBsAg was 4.65 (0.54) $\log_{10}$ IU/mL, for HBeAg was 2.49 (1.08) $\log_{10}$ IU/mL, for HBV DNA was 7.93 (1.04) $\log_{10}$ IU/mL, and 68.5% of participants met the definition of immune-tolerance with HBV DNA $>10^7$ IU/mL and normal ALT. Treatment with JNJ-6379 was discontinued in REEF-IT based on the lack of benefit of its combination with JNJ-3989 observed in REEF-1 and REEF-2. Follow-up was for 48 weeks, and the primary analysis when all patients had reached the FU week 24 time point was presented [71].

The primary endpoint of functional cure (HBsAg <0.05 IU/mL at FU week 24 after stopping NA) was not achieved by any patient in the study as only one participant met protocol-defined NA stopping criteria at the end of treatment. This patient had achieved HBsAg and HBeAg seroclearance at the end of treatment but had detectable HBsAg 24 weeks after stopping NA. Overall, HBsAg seroclearance (HBsAg <0.05 IU/mL) at ≥1 time point was achieved by 11/54 (20%) patients. At

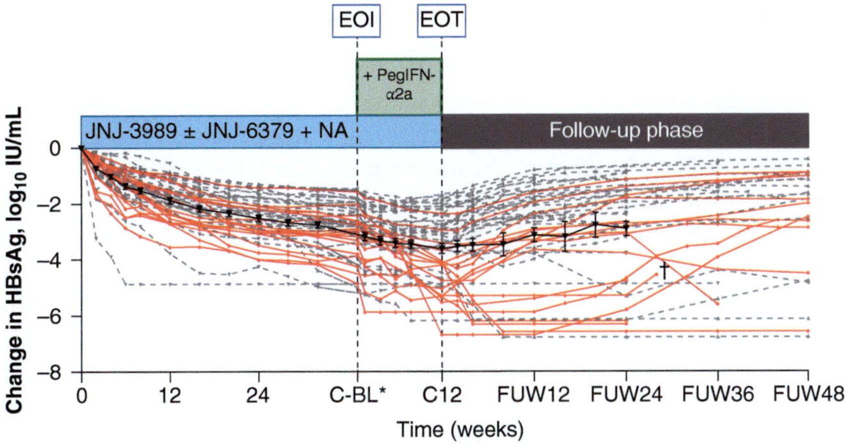

Figure 6.13 REEF-IT: Mean (SE) and individual HBsAg change from baseline kinetics during treatment with JNJ-3989±JNJ-6379+NA with the addition of pegIFN-alpha2a starting at the EOI and FU starting at end of treatment (EOT). Black line indicates the mean (SE). Gray lines represent individual patients. Red lines represent individual patients with an additional >0.5 $\log_{10}$ reduction from the start of the consolidation phase to C12 compared with predicted HBsAg decline based on linear extrapolation for this exploratory analysis.

* Data between weeks 36 and C-BL are excluded as they were not available for all patients.

† One patient met the NA stopping criteria at EOT but did not achieve functional cure.

Source: Adapted from [87] / with permission of Elsevier.

the last available time point, 6 of these 11 patients were still HBsAg negative, and two were HBeAg and HBsAg negative. Mean (SE) HBsAg changes from baseline were −2.85 (0.13), −3.61 (0.18), and −2.90 (0.26) $\log_{10}$ IU/mL at EOI, EOT, and FU week 24, respectively (Figure 6.13). Based on exploratory analyses, 17/49 (35%) patients had an additional >0.5 $\log_{10}$ IU/mL reduction in HBsAg, compared to predicted HBsAg decline when pegIFN was added to JNJ-3989 ± JNJ-3679 + NA treatment; and some of these patients maintained the reduction during the FU phase (Figure 6.13). At the start of pegIFN addition to the regimen (EOI) 48.9% and 23.4% of patients had reached an HBsAg cut-off of <100 and <10 IU/mL, respectively, and 80.9% and 34.0% had achieved a >2 or >3 $\log_{10}$ reduction from baseline.

The different response patterns to the addition of pegIFN as an immune-modulator observed in the PENGUIN and REEF-IT studies are remarkable and may be explained by a less exhausted HBV-specific T-cell status in the NCT, HBeAg-positive population with younger age in REEF-IT. While at the time of pegIFN addition reductions of HBsAg from baseline were greater in REEF-IT than in PENGUIN, the proportion of patients with HBsAg <100 IU/mL was greater in PENGUIN likely due to lower baseline HBsAg levels. Another potential contribution to better responsiveness to pegIFN addition is the longer duration of HBsAg suppression in REEF-IT where siRNA-based treatment was given for ≥36 weeks prior to pegIFN addition versus 12 weeks in PENGUIN.

6.4.3 HBsAg Targeting siRNA JNJ-3989 in Chronic Hepatitis D

HDV/HBV coinfection is considered the viral hepatitis with the fastest disease progression and the worst clinical outcome leading to cirrhosis, liver decompensation, and HCC. HDV is a satellite virus that requires the HBsAg from HBV to form infectious viral particles for its lifecycle. Treatment options for HDV/HBV coinfection are limited and are off-label use of pegIFN with limited response rate or the once daily SC viral entry inhibitor bulevirtide (myrcludex), which has recently been approved in parts of the world.

Given the encouraging HBsAg reduction seen under JNJ-3989 treatment in HBV-infected patients and the crucial role of HBsAg in the lifecycle of HDV we postulated that a JNJ-3989-induced HBsAg reduction should also result in HDV inhibition. This hypothesis was further supported by data in chimeric HBV/HDV infected mice showing HDV RNA reduction after administration of an HBsAg lowering siRNA [83] as well as the findings from the bulevirtide clinical studies showing a reduction in HDV replication in the presence of bulevirtide, an entry inhibitor of the HBsAg–NTCP interaction.

6.4.3.1 REEF-D Clinical Study

REEF-D, a Phase 2, multicenter, randomized, double-blind, placebo-controlled study in adult patients with chronic hepatitis D (CHD) was designed with two parts, a proof-of-concept Part 1 followed by a confirmatory, larger Part 2. In both parts, patients were randomized 4 : 1 to receive JNJ-3989 (100 mg SC every 4 weeks) + NA for 144 weeks (active arm) or placebo + NA for 52 weeks followed by JNJ-3989 + NA for 96 weeks (deferred treatment arm) with 48 weeks of off-treatment FU. Changes in serum viral markers (HBsAg and HDV RNA) and safety (ALT levels) were assessed. The primary composite endpoint was the proportion of patients with HDV RNA decline from baseline of $\geq 2\log_{10}$ IU/mL (or undetectable) and normal ALT at week 48. Antiviral activity in Part 1 was used to expand the study (Part 2). Data from Part 1, week 48 interim analysis were reported at the International Liver Conference [84].

Of 22 patients enrolled in REEF-D Part 1, 17 and 5 were randomized to the active and deferred treatment arms, respectively. Treatment with JNJ-3989 led to robust reductions in HBsAg and, subsequently, HDV RNA (Figure 6.14). Twelve of 17 (70.6%) patients in the active arm had treatment-emergent ALT elevations (confirmed $\geq 3\times$ ULN and $\geq 2\times$ nadir) between weeks 8 and 20, which were associated with HDV RNA rebound, limited transient HBV DNA increase in most patients, but no clear impact on HBsAg decline, and resulted in early treatment discontinuation. Grade 3/4 ALT elevations were observed with no cases of decompensation. In general, ALT levels returned to baseline values after treatment discontinuation in those patients. Among the nine JNJ-3989 treated patients still on treatment at week 48, mean (SE) HBsAg change from baseline was $-1.75\,(0.29)\log_{10}$ IU/mL in the active arm versus $-0.07\,(0.04)\log_{10}$ IU/mL for the five patients in the deferred treatment arm. Mean (SE) change from baseline in HDV RNA was $-1.52\,(0.38)\log_{10}$ IU/mL and $-0.23\,(0.15)\log_{10}$ IU/mL in the active and deferred treatment arms, respectively. Four of 17 (23.5%) patients in the active arm achieved the primary composite endpoint versus none from the deferred treatment arm. Among the five patients without

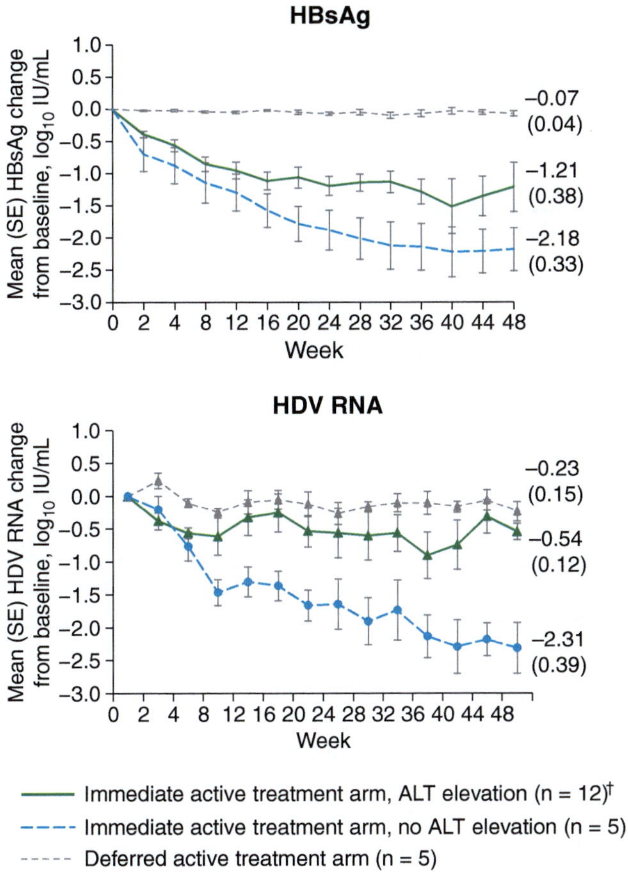

Figure 6.14 REEF-D: Proof of concept showing reduction of HDV RNA serum levels under treatment with HBsAg targeting JNJ-3989. HBsAg and HDV RNA over time by ALT elevation status. †Data in this group are available for 12, 9, 6, and 4 patients at Weeks 12, 24, 36, and 48, respectively. *Source*: Adapted from [84] / American Association for the Study of Liver Diseases.

ALT elevations, four (80.0%) achieved the primary composite endpoint, two (40.0%) of whom had HDV RNA <LLOQ of the assay. Aside from ALT elevations, safety outcomes were unremarkable. Predefined antiviral criteria to initiate Part 2 of the study were met and the observation of ALT flares predominantly in patients with high HBsAg and HDV RNA levels at baseline led to the exclusion of patients with cirrhosis or HBsAg >10 000 IU/mL and HDV RNA >100 000 IU/mL.

This is the first proof-of-concept study in humans demonstrating that an HBsAg-targeting agent leads to simultaneous reduction of HBsAg and HDV RNA serum levels in patients with CHD. The unexpected pattern of frequent and clinically relevant ALT elevations was not observed in JNJ-3989 studies with HBV mono-infection, suggesting an interaction of intracellular HBsAg lowering with hepatitis D viral products that leads to disruption of hepatocytes. The mechanism of this finding is not yet understood, and the sample size of Part 1 was too small for firm conclusions. However, ALT elevations leading to treatment discontinuation were not observed in participants with HBsAg values <10 000 IU/mL in Part 1 of REEF-D. Patients without ALT elevations generally had a continuous reduction of HDV RNA, 4/5 of whom met the primary efficacy endpoint. At the time of writing, the confirmatory Part-2 of REEF-D is ongoing.

6.5 Discussion and Perspectives

Achieving a functional cure for CHB has a high medical need and current treatment options are insufficient for this widely accepted optimal treatment outcome. Despite multiple therapeutic approaches having entered the clinic, treatments with high functional cure rates in broad patient populations, which require restoration of the host immune system, remain elusive. The ability of JNJ-3989 siRNAs to reduce levels of all viral proteins, and especially HBsAg, which is the key culprit for the inefficient anti-HBV immune response in CHB, may lead/contribute to a restoration of the immune response and ultimately to functional cure [85].

JNJ-3989 is a promising agent for the treatment of HBV because the siRNAs are efficiently targeted to hepatocytes and given the organization of the HBV genome can degrade all viral RNAs produced by cccDNA, including the pgRNA that is the precursor of infectious rcDNA, and HBsAg produced from iDNA. JNJ-3989 has the direct antiviral effect of reducing each component of infectious viral particles and the HBx protein that is involved in multiple steps of the viral lifecycle, including blocking the SMC5/6 complex that would otherwise inhibit expression of cccDNA [64]. In the absence of HBx, therefore, cccDNA transcription and all resulting products are expected to be reduced. Multiple Phase 2 studies with JNJ-3989-containing treatment regimens present a robust database of patients with CHB, and confirmed the profound reductions of viral parameters including HBsAg in patients of all CHB subgroups, HBeAg positive or negative, not currently treated or virologically suppressed, and those at baseline positive for HBeAg, HBcrAg and HBV RNA. HBsAg and serum HBV RNA reductions from baseline were usually greater than the reduction of HBeAg and HBcrAg and were greatest in NCT, HBeAg-positive patients. In some patients with HBsAg loss, this led to the unusual combination of persisting HBeAg and/or HBcrAg positivity despite HBsAg negativity.

In all subgroups, the decline of HBsAg was most profound during the first 16–20 weeks of treatment with a continued but more subtle further reduction beyond 24 weeks of treatment. The 200 mg dose of JNJ-3989 subcutaneously injected every 4 weeks for up to 48 weeks showed the highest reductions from baseline of HBsAg and other viral parameters and led to the highest rate of patients reaching an HBsAg level of <100 or <10 IU/mL.

Despite the profound reduction of HBsAg during JNJ-3989 + NA treatment with essentially all patients reaching HBsAg <1 000 IU/mL and many achieving HBsAg <100 IU/mL, the rate of functional cure remained low. A proportion of patients had sustained off-treatment HBsAg suppression and HBV DNA suppression, suggesting a new steady state between the virus and the host immune system. This observation indicates that siRNA-induced reductions in HBsAg levels with a treatment period of less than 1 year, despite resetting levels of viral expression, are not sufficient to reach a state of functional cure. Longer treatment and thereby longer HBsAg suppression might ultimately provide the opportunity for sufficient restoration but this avenue was not further explored, given that toward the end of 48 weeks of treatment further HBsAg decline was rather minimal. Low EOT HBsAg levels (<100 IU/mL) were described to be associated with functional cure in HBeAg negative long-term NA-treated patients after NA withdrawal, but this approach in JNJ-3989-treated patients (REEF-2) did not lead to functional cure during the study even though 71% of patients had reached HBsAg levels <100 IU/mL at end of all treatment. However, patients who were treated with JNJ-3989 in the active arm experienced fewer and less severe off-treatment flares, and a proportion of patients had sustained HBV DNA and HBsAg suppression. Functional cure can occur years after NA stop and/or pegIFN treatment; it remains to be seen if these JNJ-3989-treated patients will develop functional cure at a higher rate than patients who did not receive this treatment.

To achieve higher rates of functional cure in a wide range of CHB populations, JNJ-3989 may need to be combined with immune-modulating agents. The robust and consistent siRNA reduction of HBsAg and other viral products may increase responsiveness to immune stimulation. The addition of pegIFN in patients with JNJ-3989-induced low HBsAg levels showed proof of concept in a relevant proportion of NCT HBeAg-positive patients (REEF-IT) but had limited effect in VS HBeAg-negative patients in the PENGUIN study. In REEF-IT, 11 (20%) patients at least transiently achieved HBsAg seroclearance and 11 (20%) were HBeAg negative at the last observed study timepoint. Importantly, 67% had HBV DNA <LLOQ and nearly 100% had HBV DNA <100 IU/mL in a population reported to respond poorly to NA or PegIFN. Other combinations of JNJ-3989 currently under investigation are therapeutic vaccination and/or low-dose PD-1 inhibition.

In patients with CHD, the direct reduction of HBsAg by JNJ-3989 has led to profound on-treatment declines of HDV RNA. The unexpected finding of ALT elevations leading to treatment discontinuation is specific to HDV coinfection.

Overall, based on a large clinical database, JNJ-3989 has delivered robust and consistent reduction of HBsAg levels and other viral parameters across various subgroups of patients with CHB. While sustained off-treatment reductions of HBsAg in some patients suggest that a new steady state between the virus and host immune-control was achieved, this rarely led to functional cure as the desired outcome of finite treatment approaches. Thus, combination of JNJ-3989 with compounds of other modes of action, including immune modulation is likely needed for achieving this goal.

References

1 Collaborators GBDHB. Global, regional, and national burden of hepatitis B, 1990–2019: a systematic analysis for the Global Burden of Disease Study 2019. *Lancet Gastroenterol Hepatol*. 2022;7(9): 796–829.

2 Lok AS, McMahon BJ. Chronic hepatitis B: update 2009. *Hepatology*. 2009;50(3): 661–2.

3 Anderson RT, Choi HSJ, Lenz O, Peters MG, Janssen HLA, Mishra P, et al. Association between seroclearance of hepatitis B surface antigen and long-term clinical outcomes of patients with chronic hepatitis B virus infection: systematic review and meta-analysis. *Clin Gastroenterol Hepatol.* 2021;19(3): 463–72.
4 Nassal M. HBV cccDNA: viral persistence reservoir and key obstacle for a cure of chronic hepatitis B. *Gut.* 2015;64(12): 1972–84.
5 Hara T, Suzuki F, Kawamura Y, Sezaki H, Hosaka T, Akuta N, et al. Long-term entecavir therapy results in falls in serum hepatitis B surface antigen levels and seroclearance in nucleos(t)ide-naive chronic hepatitis B patients. *J Viral Hepat.* 2014;21(11): 802–8.
6 Ko KL, To WP, Mak LY, Seto WK, Ning Q, Fung J, et al. A large real-world cohort study examining the effects of long-term entecavir on hepatocellular carcinoma and HBsAg seroclearance. *J Viral Hepat.* 2020;27(4): 397–406.
7 Lam YF, Seto WK, Wong D, Cheung KS, Fung J, Mak LY, et al. Seven-year treatment outcome of entecavir in a real-world cohort: effects on clinical parameters, HBsAg and HBcrAg levels. *Clin Transl Gastroenterol.* 2017;8(10): e125.
8 Buti M, Tsai N, Petersen J, Flisiak R, Gurel S, Krastev Z, et al. Seven-year efficacy and safety of treatment with tenofovir disoproxil fumarate for chronic hepatitis B virus infection. *Dig Dis Sci.* 2015;60(5): 1457–64.
9 Chan HL, Fung S, Seto WK, Chuang WL, Chen CY, Kim HJ, et al. Tenofovir alafenamide versus tenofovir disoproxil fumarate for the treatment of HBeAg-positive chronic hepatitis B virus infection: a randomised, double-blind, phase 3, non-inferiority trial. *Lancet Gastroenterol Hepatol.* 2016;1(3): 185–95.
10 Buti M, Gane E, Seto WK, Chan HL, Chuang WL, Stepanova T, et al. Tenofovir alafenamide versus tenofovir disoproxil fumarate for the treatment of patients with HBeAg-negative chronic hepatitis B virus infection: a randomised, double-blind, phase 3, non-inferiority trial. *Lancet Gastroenterol Hepatol.* 2016;1(3): 196–206.
11 European Association for the Study of the Liver. Electronic address, European Association for the Study of the Liver. EASL 2017 Clinical Practice Guidelines on the management of hepatitis B virus infection. *J Hepatol.* 2017;67(2): 370–98.
12 Matthews PC, Jack K, Wang S, Abbott J, Bryce K, Cheng B, et al. A call for advocacy and patient voice to eliminate hepatitis B virus infection. *Lancet Gastroenterol Hepatol.* 2022;7(4): 282–5.
13 Bock CT, Schwinn S, Locarnini S, Fyfe J, Manns MP, Trautwein C, et al. Structural organization of the hepatitis B virus minichromosome. *J Mol Biol.* 2001;307(1): 183–96.
14 Staprans S, Loeb DD, Ganem D. Mutations affecting hepadnavirus plus-strand DNA synthesis dissociate primer cleavage from translocation and reveal the origin of linear viral DNA. *J Virol.* 1991;65(3): 1255–62.
15 Tu T, Budzinska MA, Shackel NA, Urban S. HBV DNA integration: molecular mechanisms and clinical implications. *Viruses.* 2017;9(4): 75.
16 Jiang Z, Jhunjhunwala S, Liu J, Haverty PM, Kennemer MI, Guan Y, et al. The effects of hepatitis B virus integration into the genomes of hepatocellular carcinoma patients. *Genome Res.* 2012;22(4): 593–601.

17 Schluter V, Meyer M, Hofschneider PH, Koshy R, Caselmann WH. Integrated hepatitis B virus X and 3′ truncated preS/S sequences derived from human hepatomas encode functionally active transactivators. *Oncogene*. 1994;9(11): 3335–44.

18 Ramirez R, van Buuren N, Gamelin L, Soulette C, May L, Han D, et al. Targeted long-read sequencing reveals comprehensive architecture, burden, and transcriptional signatures from hepatitis B virus-associated integrations and translocations in hepatocellular carcinoma cell lines. *J Virol*. 2021;95(19): e0029921.

19 Wooddell CI, Yuen MF, Chan HL, Gish RG, Locarnini SA, Chavez D, et al. RNAi-based treatment of chronically infected patients and chimpanzees reveals that integrated hepatitis B virus DNA is a source of HBsAg. *Sci Transl Med*. 2017;9(409): eaan0241.

20 Grudda T, Hwang HS, Taddese M, Quinn J, Sulkowski MS, Sterling RK, et al. Integrated hepatitis B virus DNA maintains surface antigen production during antiviral treatment. *J Clin Invest*. 2022;132(18): e161818.

21 Meier MA, Calabrese D, Suslov A, Terracciano LM, Heim MH, Wieland S. Ubiquitous expression of HBsAg from integrated HBV DNA in patients with low viral load. *J Hepatol*. 2021;75(4): 840–7.

22 Tu T, Budzinska MA, Vondran FWR, Shackel NA, Urban S. Hepatitis B virus DNA integration occurs early in the viral life cycle in an in vitro infection model via sodium taurocholate cotransporting polypeptide-dependent uptake of enveloped virus particles. *J Virol*. 2018;92(11): e02007–17.

23 Zhao XL, Yang JR, Lin SZ, Ma H, Guo F, Yang RF, et al. Serum viral duplex-linear DNA proportion increases with the progression of liver disease in patients infected with HBV. *Gut*. 2016;65(3): 502–11.

24 Zhao HJ, Hu YF, Han QJ, Zhang J. Innate and adaptive immune escape mechanisms of hepatitis B virus. *World J Gastroenterol*. 2022;28(9): 881–96.

25 Bertoletti A, Ferrari C. Innate and adaptive immune responses in chronic hepatitis B virus infections: towards restoration of immune control of viral infection. *Gut*. 2012;61(12): 1754–64.

26 Tout I, Loureiro D, Mansouri A, Soumelis V, Boyer N, Asselah T. Hepatitis B surface antigen seroclearance: immune mechanisms, clinical impact, importance for drug development. *J Hepatol*. 2020;73(2): 409–22.

27 Gu Y, Lian Y, Gu L, Chen L, Li X, Zhou L, et al. Correlations between cytokines produced by T cells and clinical-virological characteristics in untreated chronic hepatitis B patients. *BMC Infect Dis*. 2019;19(1): 216.

28 Kondo Y, Ninomiya M, Kakazu E, Kimura O, Shimosegawa T. Hepatitis B surface antigen could contribute to the immunopathogenesis of hepatitis B virus infection. *ISRN Gastroenterol*. 2013; 2013: 935295.

29 Fang Z, Li J, Yu X, Zhang D, Ren G, Shi B, et al. Polarization of monocytic myeloid-derived suppressor cells by hepatitis B surface antigen is mediated via ERK/IL-6/STAT3 signaling feedback and restrains the activation of T cells in chronic hepatitis B virus infection. *J Immunol*. 2015;195(10): 4873–83.

30 Pollicino T, Caminiti G. HBV-integration studies in the clinic: role in the natural history of infection. *Viruses*. 2021;13(3): 368.

31 Utzschneider DT, Alfei F, Roelli P, Barras D, Chennupati V, Darbre S, et al. High antigen levels induce an exhausted phenotype in a chronic infection without impairing T cell expansion and survival. *J Exp Med.* 2016;213(9): 1819–34.

32 Yuen MF, Chen DS, Dusheiko GM, Janssen HLA, Lau DTY, Locarnini SA, et al. Hepatitis B virus infection. *Nat Rev Dis Primers.* 2018;4: 18035.

33 Guidotti LG, Isogawa M, Chisari FV. Host-virus interactions in hepatitis B virus infection. *Curr Opin Immunol.* 2015;36: 61–6.

34 Op den Brouw ML, Binda RS, van Roosmalen MH, Protzer U, Janssen HL, van der Molen RG, et al. Hepatitis B virus surface antigen impairs myeloid dendritic cell function: a possible immune escape mechanism of hepatitis B virus. *Immunology.* 2009;126(2): 280–9.

35 Buschow SI, Biesta PJ, Groothuismin, ZMA, Erler NS, Vanwolleghem T, Ho E, et al. TLR7 polymorphism, sex and chronic HBV infection influence plasmacytoid DC maturation by TLR7 ligands. *Antivir Res.* 2018;157: 27–37.

36 Ouaguia L, Leroy V, Dufeu-Duchesne T, Durantel D, Decaens T, Hubert M, et al. Circulating and hepatic BDCA1+, BDCA2+, and BDCA3+ dendritic cells are differentially subverted in patients with chronic HBV infection. *Front Immunol.* 2019;10: 112.

37 Xu Y, Hu Y, Shi B, Zhang X, Wang J, Zhang Z, et al. HBsAg inhibits TLR9-mediated activation and IFN-alpha production in plasmacytoid dendritic cells. *Mol Immunol.* 2009;46(13): 2640–6.

38 Wang S, Chen Z, Hu C, Qian F, Cheng Y, Wu M, et al. Hepatitis B virus surface antigen selectively inhibits TLR2 ligand-induced IL-12 production in monocytes/macrophages by interfering with JNK activation. *J Immunol.* 2013;190(10): 5142–51.

39 Li X, Zhou L, Gu Y, Chen L, Gu L, Huang Y. Quantative HBsAg level correlates dendritic cells maturation in chronic hepatitis B patients. *Clin Res Hepatol Gastroenterol.* 2020;44(3): 321–8.

40 Jiang M, Broering R, Trippler M, Poggenpohl L, Fiedler M, Gerken G, et al. Toll-like receptor-mediated immune responses are attenuated in the presence of high levels of hepatitis B virus surface antigen. *J Viral Hepat.* 2014;21(12): 860–72.

41 Li Y, Xia Y, Cheng X, Kleiner DE, Hewitt SM, Sproch J, et al. Hepatitis B surface antigen activates unfolded protein response in forming ground glass hepatocytes of chronic hepatitis B. *Viruses.* 2019;11(4): 386.

42 Lee YI, Hwang JM, Im JH, Lee YI, Kim NS, Kim DG, et al. Human hepatitis B virus-X protein alters mitochondrial function and physiology in human liver cells. *J Biol Chem.* 2004;279(15): 15460–71.

43 Bouchard MJ, Schneider RJ. The enigmatic X gene of hepatitis B virus. *J Virol.* 2004;78(23): 12725–34.

44 Rawat S, Clippinger AJ, Bouchard MJ. Modulation of apoptotic signaling by the hepatitis B virus X protein. *Viruses.* 2012;4(11): 2945–72.

45 Casciano JC, Bouchard MJ. Hepatitis B virus X protein modulates cytosolic Ca(2+) signaling in primary human hepatocytes. *Virus Res.* 2018;246: 23–7.

46 Waris G, Huh KW, Siddiqui A. Mitochondrially associated hepatitis B virus X protein constitutively activates transcription factors STAT-3 and NF-kappa B via oxidative stress. *Mol Cell Biol.* 2001;21(22): 7721–30.

47 Kim SJ, Khan M, Quan J, Till A, Subramani S, Siddiqui A. Hepatitis B virus disrupts mitochondrial dynamics: induces fission and mitophagy to attenuate apoptosis. *PLoS Pathog*. 2013;9(12): e1003722.

48 Lucifora J, Arzberger S, Durantel D, Belloni L, Strubin M, Levrero M, et al. Hepatitis B virus X protein is essential to initiate and maintain virus replication after infection. *J Hepatol*. 2011;55(5): 996–1003.

49 Decorsiere A, Mueller H, van Breugel PC, Abdul F, Gerossier L, Beran RK, et al. Hepatitis B virus X protein identifies the Smc5/6 complex as a host restriction factor. *Nature*. 2016;531(7594): 386–9.

50 Livingston CM, Ramakrishnan D, Strubin M, Fletcher SP, Beran RK. Identifying and characterizing interplay between hepatitis B virus X protein and Smc5/6. *Viruses*. 2017;9(4): 69.

51 Niu C, Livingston CM, Li L, Beran RK, Daffis S, Ramakrishnan D, et al. The Smc5/6 complex restricts HBV when localized to ND10 without inducing an innate immune response and is counteracted by the HBV X protein shortly after infection. *PLoS One*. 2017;12(1): e0169648.

52 Yu S, Chen J, Wu M, Chen H, Kato N, Yuan Z. Hepatitis B virus polymerase inhibits RIG-I- and Toll-like receptor 3-mediated beta interferon induction in human hepatocytes through interference with interferon regulatory factor 3 activation and dampening of the interaction between TBK1/IKKepsilon and DDX3. *J Gen Virol*. 2010;91(Pt 8): 2080–90.

53 Lei Q, Li T, Kong L, Li L, Ding X, Wang X, et al. HBV-Pol is crucial for HBV-mediated inhibition of inflammasome activation and IL-1beta production. *Liver Int*. 2019;39(12): 2273–84.

54 Liu Y, Li J, Chen J, Li Y, Wang W, Du X, et al. Hepatitis B virus polymerase disrupts K63-linked ubiquitination of STING to block innate cytosolic DNA-sensing pathways. *J Virol*. 2015;89(4): 2287–300.

55 Li M, Sun R, Xu L, Yin W, Chen Y, Zheng X, et al. Kupffer cells support hepatitis B virus-mediated CD8+ T cell exhaustion via hepatitis B core antigen-TLR2 interactions in mice. *J Immunol*. 2015;195(7): 3100–9.

56 Yu X, Lan P, Hou X, Han Q, Lu N, Li T, et al. HBV inhibits LPS-induced NLRP3 inflammasome activation and IL-1beta production via suppressing the NF-kappaB pathway and ROS production. *J Hepatol*. 2017;66(4): 693–702.

57 Kennedy PT, Sandalova E, Jo J, Gill U, Ushiro-Lumb I, Tan AT, et al. Preserved T-cell function in children and young adults with immune-tolerant chronic hepatitis B. *Gastroenterology*. 2012;143(3): 637–45.

58 Wang FS, Li J, Zhang C. Why is the functional cure rate of young children with chronic hepatitis B receiving antiviral therapy considerably high? *Hepatol Int*. 2024;18(1): 296-298.

59 Zhu S, Dong Y, Wang L, Liu W, Zhao P. Early initiation of antiviral therapy contributes to a rapid and significant loss of serum HBsAg in infantile-onset hepatitis B. *J Hepatol*. 2019;71(5): 871–5.

60 Mason WS, Gill US, Litwin S, Zhou Y, Peri S, Pop O, et al. HBV DNA integration and clonal hepatocyte expansion in chronic hepatitis B patients considered immune tolerant. *Gastroenterology*. 2016;151(5): 986– 98.e4.

61 Wong DK, Cheng SCY, Mak LL, To EW, Lo RC, Cheung TT, et al. Among patients with undetectable hepatitis B surface antigen and hepatocellular carcinoma, a high proportion has integration of HBV DNA into hepatocyte DNA and no cirrhosis. *Clin Gastroenterol Hepatol.* 2020;18(2): 449–56.

62 Ameres SL, Martinez J, Schroeder R. Molecular basis for target RNA recognition and cleavage by human RISC. *Cell.* 2007;130(1): 101–12.

63 Zhu Y, Zhu L, Wang X, Jin H. RNA-based therapeutics: an overview and prospectus. *Cell Death Dis.* 2022;13(7): 644.

64 Allweiss L, Giersch K, Pirosu A, Volz T, Muench RC, Beran RK, et al. Therapeutic shutdown of HBV transcripts promotes reappearance of the SMC5/6 complex and silencing of the viral genome in vivo. *Gut.* 2022;71(2): 372–81.

65 Balagopal A, Grudda T, Ribeiro RM, Saad YS, Hwang HS, Quinn J, et al. Single hepatocytes show persistence and transcriptional inactivity of hepatitis B. *JCI Insight.* 2020;5(19): e140584.

66 Wooddell C, Zui R, Hamilton H, Chu Q, Sternard H, Schumacher J, et al. Development of subcutaneously administered RNAi therapeutic ARO-HBV for chronic hepatitis B virus infection. *J Hepatol.* 2018;68: S18–9.

67 Wooddell CI, Rozema DB, Hossbach M, John M, Hamilton HL, Chu Q, et al. Hepatocyte-targeted RNAi therapeutics for the treatment of chronic hepatitis B virus infection. *Mol Ther.* 2013;21(5): 973–85.

68 Gane E, Yuen MF, Kakuda TN, Ogawa T, Takahashi Y, Goeyvaerts N, et al. JNJ-73763989 pharmacokinetics and safety: liver-targeted siRNAs against hepatitis B virus, in Japanese and non-Japanese healthy adults, and combined with JNJ-56136379 and a nucleos(t)ide analogue in patients with chronic hepatitis B. *Antivir Ther.* 2022;27(3): 13596535221093856.

69 Li H, Niu X, Zhang Y, Zhang D, Zhang Y, Wang L, et al. Pharmacokinetics, safety, and tolerability of the siRNA JNJ-73763989 in healthy Chinese adult participants. *Clin Pharmacol Drug Dev.* 2023;12(2): 175–80.

70 Yuen MF, Locarnini S, Lim TH, Strasser SI, Sievert W, Cheng W, et al. Combination treatments including the small-interfering RNA JNJ-3989 induce rapid and sometimes prolonged viral responses in patients with CHB. *J Hepatol.* 2022;77(5): 1287–98.

71 Yuen MF, Asselah T, Jacobson IM, Brunetto MR, Janssen HLA, Takehara T, et al. Efficacy and safety of the siRNA JNJ-73763989 and the capsid assembly modulator JNJ-56136379 (bersacapavir) with nucleos(t)ide analogues for the treatment of chronic hepatitis B virus infection (REEF-1): a multicentre, double-blind, active-controlled, randomised, phase 2b trial. *Lancet Gastroenterol Hepatol.* 2023;8(9): 790–802.

72 Pinheiro J, Bornkamp B, Glimm E, Bretz F. Model-based dose finding under model uncertainty using general parametric models. *Stat Med.* 2014;33(10): 1646–61.

73 Janssen HLA, Hou J, Asselah T, Chan HLY, Zoulim F, Tanaka Y, et al. Randomised phase 2 study (JADE) of the HBV capsid assembly modulator JNJ-56136379 with or without a nucleos(t)ide analogue in patients with chronic hepatitis B infection. *Gut.* 2023;72(7): 1385–98.

74 Zoulim F, Lenz O, Vandenbossche JJ, Talloen W, Verbinnen T, Moscalu I, et al. JNJ-56136379, an HBV capsid assembly modulator, is well-tolerated and has antiviral activity in a phase 1 study of patients with chronic infection. *Gastroenterology*. 2020;159(2): 521–33.e9.

75 Berke JM, Dehertogh P, Vergauwen K, Mostmans W, Vandyck K, Raboisson P, et al. Antiviral properties and mechanism of action studies of the hepatitis B virus capsid assembly modulator JNJ-56136379. *Antimicrob Agents Chemother*. 2020;64(5): e02439–19.

76 Berg T, Simon KG, Mauss S, Schott E, Heyne R, Klass DM, et al. Long-term response after stopping tenofovir disoproxil fumarate in non-cirrhotic HBeAg-negative patients - FINITE study. *J Hepatol*. 2017;67(5): 918–24.

77 Berg T, Lampertico P. The times they are a-changing - a refined proposal for finite HBV nucleos(t)ide analogue therapy. *J Hepatol*. 2021;75(2): 474–80.

78 Verbinnen T, Jezorwski J, Biermer M, Shukla U, Augustyns I, Jacobson IM, et al. Viral sequence analysis of not-currently treated chronic hepatitis B patients enrolled in the REEF-1 study and impact of baseline nucleotide polymorphisms in the small interfering RNA JNJ-3989 trigger target regions on viral antigen declines. AASLD The Liver Meeting; Washington, D.C., 2022.

79 Agarwal K, Buti M, van Bommel F, Lampertico P, Janczewska E, Bourliere M, et al. Efficacy and safety of combination treatment with siRNA JNJ-73763989 and capsid assembly modulator JNJ-56136379 (bersacapavir) in HBeAg negative virologically suppressed chronic hepatitis B patients: follow-up week 48 end of study results from REEF-2. AASLD The Liver Meeting; Washington, D.C., 2022.

80 Agarwal K, Lok J, Carey I, Shivkar Y, Biermer M, Berg T, et al. A case of HBV-induced liver failure in the REEF-2 phase II trial: implications for finite treatment strategies in HBV 'cure'. *J Hepatol*. 2022;77(1): 245–8.

81 Hirode G, Choi HSJ, Chen CH, Su TH, Seto WK, Van Hees S, et al. Off-Therapy response after Nucleos(t)ide analogue withdrawal in patients with chronic hepatitis B: an international, multicenter, multiethnic cohort (RETRACT-B Study). *Gastroenterology*. 2022;162(3): 757–71.e4.

82 Gane E, Janczewska E, Takehara T, Chuang WL, Peng CY, Hlebowicz M, et al. Efficacy and safety of siRNA JNJ-73763989, capsid assembly modulator JNJ-56136379, nucleos(t)ide analog, and pegylated interferon alpha-2a for the treatment of chronic hepatitis B: week 24 results from the phase 2 PENGUIN study. AASLD The Liver Meeting; Washington, D.C.,2022.

83 Thi EP, Dhillon AP, Ardzinski A, Bidirici-Ertekin L, Cobarrubias KD, Cuconati A, et al. ARB-1740, a RNA interference therapeutic for chronic hepatitis B infection. *ACS Infect Dis*. 2019;5(5): 725–37.

84 Wedemeyer H, Gane E, Agarwal K, Tabak OF, Forns X, Akarca US, et al. Treatment with siRNA JNJ-73763989 plus Nucleos(t)ide Analogue (NA) decreases HBsAg and HDV RNA levels in patients with Chronic Hepatitis D (CHD): part 1 of the REEF-D study. EASL The International Liver Conference, 2023.

85 Paratala BS, Park JJ, Ganchua SC, Gane E, Yuen MF, Lee ACH, et al. Inhibition of hepatitis B surface antigen in chronic hepatitis B subjects by RNA interference therapeutic AB-729 is accompanied by upregulation of HBV-specific T-cell activation markers. EASL The International Liver Conference, 2021.

86 Gish RG, Given BD, Lai CL, Locarnini SA, Lau JY, Lewis DL, et al. Chronic hepatitis B: virology, natural history, current management and a glimpse at future opportunities. *Antivir Res.* 2015;121: 47–58.

87 Kennedy PT, Fung S, Buti M, Yilmaz G, Chuang WL, Asselah T, et al. Efficacy and safety of siRNA JNJ-73763989, capsid assembly modulator-E JNJ-56136379, and Nucleos(t)ide Analogs with pegylated interferon alpha-2a added in immune-tolerant patients with chronic hepatitis B virus infection: interim results from the phase 2 REEF-IT study. AASLD The Liver Meeting, 2023.

7

The Discovery and Development of Sofosbuvir as the Backbone of HCV Curative Therapies

Michael J. Sofia

OnCore Consulting LLC

7.1 Introduction

The hepatitis C virus (HCV) is a liver tropic virus that, on chronic infection, leads to liver cirrhosis and liver cancer. Approximately 75 million individuals globally are chronically infected with HCV, and individuals who inject illicit drugs are at particular risk for infection [1]. Early therapy, introduced in 1991, was based on interferon (IFN) and over time with the introduction of pegylated interferon (PEG-IFN) and the addition of ribavirin (RBV)-sustained virological response (SVR = cure) was achieved in approximately 40% of patients; however, severe side effects over the 48-week treatment regimen and limited genotype coverage hampered the use and uptake of this therapy [2]. Consequently, there arose a need for tolerable, safe, and short-duration oral therapies that could achieve significantly higher SVR rates across a wide patient population. On 6 December 2013, sofosbuvir was approved by the US FDA as the backbone of the first IFN-free combination therapy for the treatment and cure of chronic hepatitis C infection (cHCV). This event marked a turning point in the treatment of HCV and set the stage for future efforts to eliminate HCV from the infected population. Sofosbuvir went on to become the backbone of treatment for most major HCV curative combination regimens that include the fixed dose regimens Harvoni®, Epclusa®, and Vosevi®. Sofosbuvir was also designated by the World Health Organization as an essential medicine in 2015 and combination regimens with sofosbuvir as the backbone have contributed to the cure of millions of individuals worldwide and have had an important impact on organ transplant practices [3, 4].

The discovery of sofosbuvir began with the objective to identify a drug that would embody certain key characteristics desired for an HCV curative therapy. These desirable characteristics included good oral bioavailability, pan-genotypic

Trends in Antiviral Drug Development, First Edition. Edited by Michael J. Sofia and Zhengqiang Wang.

character, a high barrier to the development of resistant virus, the ability to be combined with other direct-acting antivirals (DAA) to produce highly potent regimens, a treatment duration of less than 48 weeks, minimal side effects, elimination of the need for IFN, and ideally, liver targeting to increase potency and reduce systemic adverse events. To achieve these objectives, the HCV NS5B RNA-dependent RNA polymerase (RdRp) was chosen as the molecular target. The RdRp is one of seven viral nonstructural proteins that are involved in supporting the virus lifecycle (Figure 7.1) [5]. This viral polymerase exhibits a highly conserved active site across all genotypes, which makes the possibility of developing a pan-genotypic agent possible and the critical nature of its active-site residues meant that development of a substrate-based inhibitor (i.e. nucleos[t]ide) might provide a high barrier to the development of resistant virus [6, 7]. This is in stark contrast to the other highly explored nonstructural protein targets, NS3/4a protease and NS5A replication complex protein, which exhibited significant sequence variability at the key drug binding sites, resulting in a very low barrier to the development of resistance and coverage of only genotype 1 viral variants [7–13].

7.2 Identification of the 2′-α-F, 2′-β-*C*-Methyl Cytidine Nucleoside PSI-6130

The search for molecules with the desired characteristics began with the synthesis and screening of novel nucleoside analogs based on known characteristics of viral polymerase inhibitors. To accomplish the screening effort, the HCV replicon assay

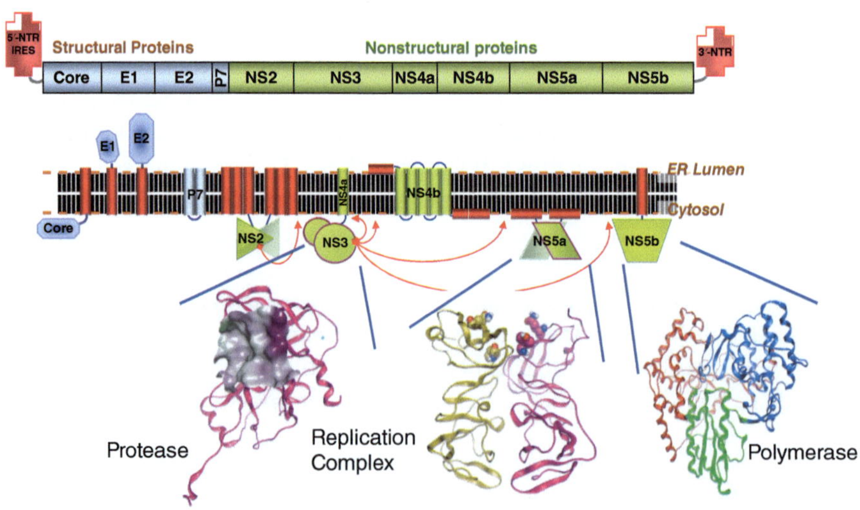

Figure 7.1 HCV genome showing the structural and nonstructural viral proteins. Shown are the PDB structures of the three key drug targets: the NS3/4a protease (PDB code: 3RC6), the NS5A replication complex protein (PDB code 3FQM), and the NS5B RNA-dependent RNA polymerase (PDB code 1C2P).

was employed. This whole cell assay used a noninfectious clone that lacks the three key structural proteins needed by the virus to produce active virions.[14] The use of a cell-based system was critical to identifying substrate-based RdRp inhibitors because the nucleoside triphosphate metabolite is the active inhibitor, and thus the presence of three cellular kinases to successively phosphorylate the nucleoside and produce the active triphosphate need to be present to enable the phosphorylation events.

Extensive analog synthesis eventually led to the identification of the unique 2′-α-F, 2′-β-*C*-methyl cytidine nucleoside PSI-6130 (Figure 7.2) [15, 16]. The configuration of the 2′-F and 2′-methyl substituents delivered a unique profile for this molecule. Other isomeric combinations of F and methyl substituents at the 2′-position of the nucleoside were detrimental to either potency or cytotoxicity profile making the combination of the 2′-α-F, 2′-β-*C*-methyl substitution pattern ideal [17]. Even when compared to other 2′-substituted nucleosides being developed at the time, specifically nucleosides with the 2′-α-hydroxy-2′-β-methyl substitution pattern: NM283, MK-0608 (Figure 7.2), the 2′-α-F, 2′-β-*C*-methyl substituted nucleoside PSI-6130 was superior in both resistance and overall safety profile [18, 19].

Further exploration of the SAR around the 2′-α-F, 2′-β-*C*-methyl nucleoside platform proved fruitless. Any variations to the nucleobase or the furanose sugar portions of the nucleoside led to inactive inhibitors of HCV replication [20]. Consequently, it appeared that 2′-α-F, 2′-β-*C*-methyl cytidine nucleoside, PSI-6130, was singular in its HCV inhibitory properties. In addition, PSI-6130 was shown to have pan-genotypic character when evaluated against a series of HCV polymerase

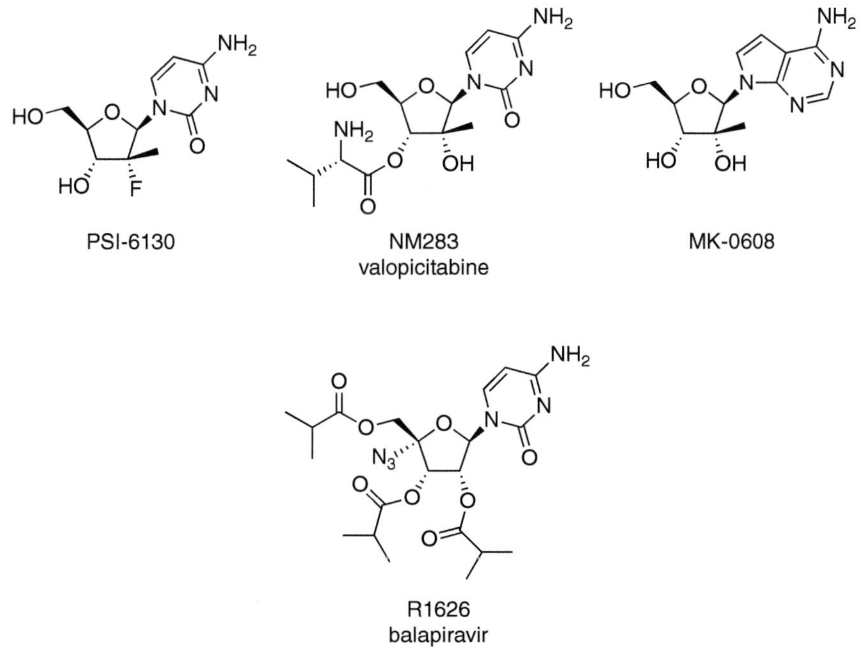

Figure 7.2 Structures of early HCV NS5B nucleoside polymerase inhibitors.

enzymes from genotype 1–4 viruses [20, 21]. Although the S282T amino acid substitution in the RdRp did show susceptibility to PSI-6130, the fold reduction in inhibitory potency was nonconsequential, and this amino acid substitution produced a virus with a low level of fitness [13]. It was subsequently shown that the S282T RdRp viral variant was not a preexisting variant observed in clinical samples from HCV patients [22, 23].

An attempt was made to explore the PSI-6130 as a clinical development candidate. Preclinical pharmacokinetic studies showed that PSI-6130 demonstrated modest bioavailability and was plagued by the production of a uridine metabolite that proved inactive as an inhibitor of HCV. No adverse toxicological findings were observed for PSI-6130 in a preclinical GLP 28-day study in two species. As was observed preclinically, in Phase 1 clinical studies PSI-6130 again showed significant metabolism to the inactive uridine metabolite and only modest bioavailability (%F = 24) [17]. Consequently, PSI-6130 was terminated as a clinical development candidate.

7.3 Identification of the 3′,5′-Diisobutyrate Ester Prodrug RG7128 and Clinical Proof-of-concept

Attempts to block production of the uridine metabolite, which was thought to result from gut metabolism, led to the preparation of simple prodrugs to enhance the absorption. This work produced the 3′,5′-diisobutyrate ester prodrug RG7128 (Figure 7.3), which retained the potency of PSI-6130 and showed improved preclinical pharmacokinetic properties (Table 7.1) [17]. RG7128 was progressed into clinical development and at doses of 1000 and 1500 mg BID in a 14-day monotherapy clinical study successfully demonstrated a 2.7 log IU/mL decline of HCV RNA levels in genotype 1 chronically-infected patients [24, 25]. The combination of RG7128 with PEG-IFN + RBV also showed a 90% rapid virological response rate (RVR = undetectable HCV RNA levels at end of 28 days of therapy) in genotype 2 and 3 HCV patients, demonstrating for the first time the potential for pan-genotypic

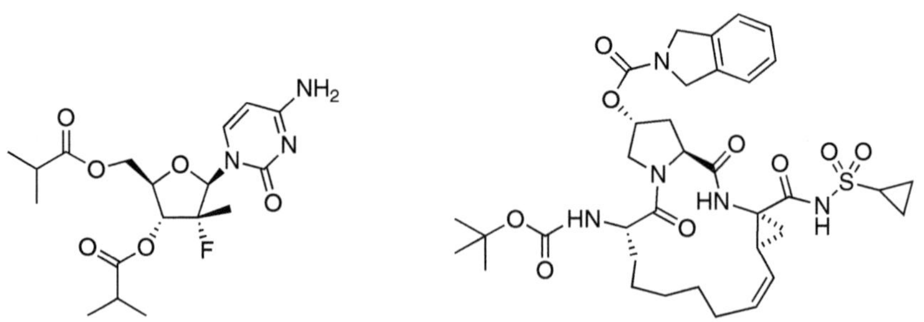

RG7128
mericitabine

ITMN/RG7227
danoprevir

Figure 7.3 Chemical structures of the nucleoside prodrug NS5B polymerase inhibitor RG7128 (mericitabine) and the NS3/4a protease inhibitor RG7227 (danoprevir).

Table 7.1 *In vitro* potency and rat pharmacokinetic study of PSI-6130 and RG7128.

Compound	EC90 (μM) Clone A	CC_{50} (μM)	Caco2 Papp (×10−6 cm/s)	Rat PK 10 mg/kg AUC_{0-24} (mg h/mL) C_{max} (mg/mL)
PSI-6130	3.03	>100	0.21	$AUC_{(parent)} = 2.97$ $C_{max} = 0.6$
RG7128	2.5	>100	6.4	$AUC_{(parent)} = 16.14$ $C_{max} = 1.86$

character in a direct-acting antiviral [24, 25]. The groundbreaking study involving the combination of RG7128 with the NS3/4a protease inhibitor danoprevir (ITNM-191/RG7227, Figure 7.3) for the first time tested whether the combination of two DAAs in the absence of PEG-IFN would produce similar viral load kinetic declines and similar RVRs compared to PEG-IFN-containing combinations. This study known as INFORM-1 did indeed show that a DAA combination was comparable to PEG-IFN-containing combinations and thus set the stage for future IFN-free combination studies (Table 7.2) [26].

7.4 Development of the 5′-Phosphoramidate Uridine Nucleotide Prodrug for Liver Targeting and Clinical Proof-of-concept

Even with positive clinical results observed with RG7128 either as monotherapy or in combination with other agents, RG7128 presented some limitations as a development candidate in the fast-moving competitive field of HCV drug development.

Table 7.2 INFORM-1 study results combining RG7128 (mericitabine) and RG7227 (danoprevir) in genotype 1 chronically infected HCV patients for 28 days.

RG7128 + mericitabine (mg, dose frequency)	RG7227 + denoprevir (mg, dose frequency)	Number of patients	Patient population	HCV RNA medium change from baseline ($\log_{10}$ IU/mL)
500, BID	100, TID	8	Naive	−3.9
500, BID	200, TID	8	Naive	−5.2
1000, BID	100, TID	7	Naive	−4.8
1000, BID	200, TID	8	Naive	−4.8
1000, BID	600, BID	8	TF (non-null)	−4.0
1000, BID	900, BID	8	TF (null)	−4.9
1000, BID	900, BID	8	Naive	−5.1

TF, previously treated failures.

RG7128 required BID dosing, a high drug load and still was plagued by significant metabolism to the inactive uridine metabolite. Consequently, an effort was undertaken to improve on these characteristics. To do so, a radical approach was pursued based on a developing understanding of PSI-6130 metabolism. Studies showed that PSI-6130 had two routes of metabolism (Figure 7.4) [20, 27–29]. One involved sequential phosphorylation producing the corresponding nucleotide triphosphate, which had a half-life in hepatocytes of approximately 5 h and was the alternate substrate for the HCV RdRp. This triphosphate becomes incorporated into the growing RNA chain, subsequently leading to a chain termination event. The other metabolic pathway produced the uridine nucleoside, PSI-6206, which could not be phosphorylated because it was determined not to be a substrate for any cellular kinases. Incompatibility with the first phosphorylation step of the phosphorylation cascade is a common problem for non-natural nucleoside analogs. However, further studies revealed that the cytidine 5′-monophosphate was partially converted to the uridine (PSI-6206) 5′-monophosphate derivative in hepatocytes and that this uridine monophosphate was further converted to the 5′-triphosphate, which acted as a potent inhibitor of the HCV RdRp. In fact, this uridine triphosphate had a long half-life (>24 h) in hepatocytes [28]. These results indicated that if the uridine monophosphate could be delivered to the liver and converted to the triphosphate, the long hepatocyte residence time might afford a drug that required less frequent dosing and possibly lower drug requirements because it would accumulate in hepatocytes. However, the challenge was how to deliver a chemically and metabolically unstable nucleotide 5′-monophosphate, which also would not penetrate biological membranes because of its highly charged nature.

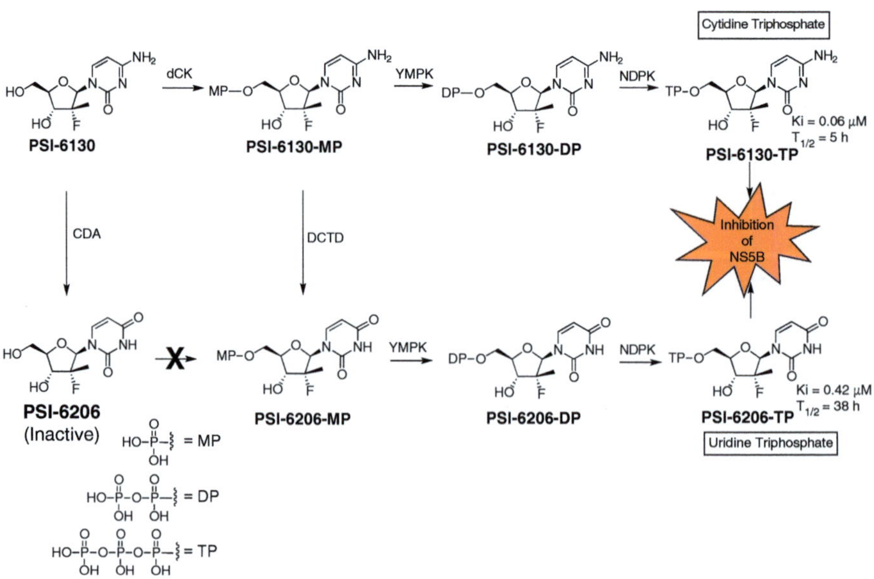

Figure 7.4 Metabolic profile of PSI-6130.

The answer to this challenge focused on the development of a 5′-monophosphate prodrug strategy that would not only protect the 5′-phosphate from degradation but would also deliver the nucleotide monophosphate directly to the liver – a liver targeting strategy [30]. Several nucleotide monophosphate prodrug approaches had been developed to bypass the nonproductive first nucleoside phosphorylation step common to nonnatural nucleoside analogs, but none demonstrated clinical *in vivo* proof-of-concept, and none had been shown to be liver targeting. However, we had surmised that the phosphoramidate prodrug construct might provide the right set of functional characteristics that could be used to develop a liver targeting strategy. The concept was to leverage liver first-pass metabolism in which the terminal ester group of the prodrug moiety would be cleaved by liver esterases, thus launching a series of chemical and enzymatic events in the hepatocyte, revealing the naked 5′-monophosphate, and because the monophosphate was highly negatively charged it would be trapped in the hepatocyte (Figure 7.5) [20, 31]. To test this hypothesis, the simple phosphoramidate methyl ester of the uridine nucleoside 5′-monophosphate (PSI-7672, Figure 7.5) was prepared and when tested in the HCV replicon whole cell assay, it proved to be more potent than either PSI-6130 or RG7128 ($EC_{90} = 0.86$) [31]. The next step was to optimize the substituent combination of the uridine nucleotide

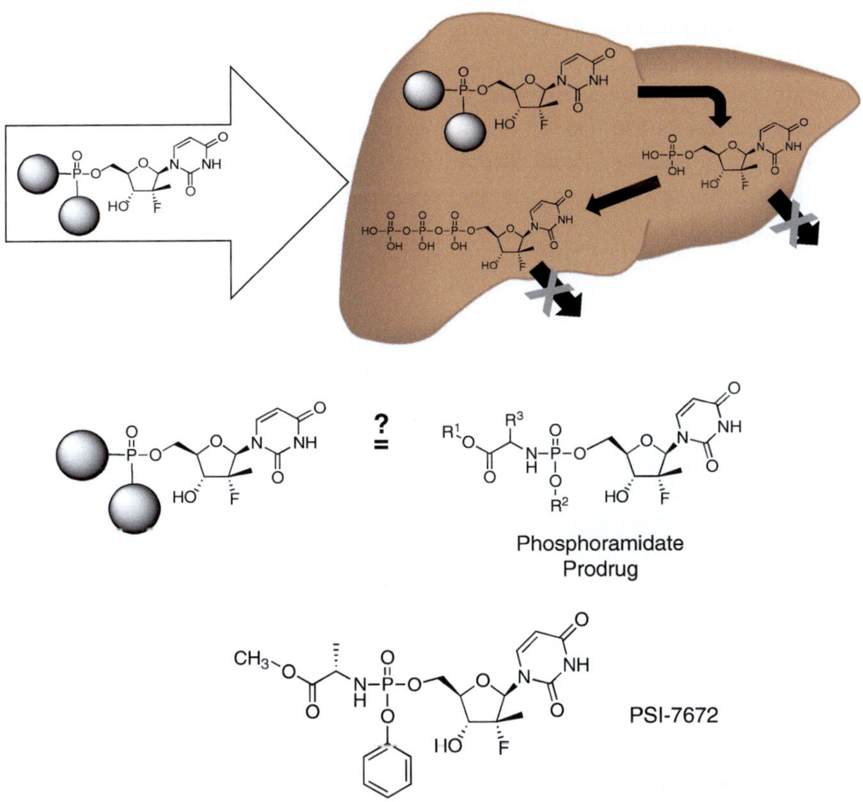

Figure 7.5 Concept of liver targeting a nucleotide prodrug using first-pass metabolism and trapping of the triphosphate metabolite in hepatocytes.

phosphoramidate prodrug moiety that would support *in vivo* studies and subsequent drug development. To accomplish this, criteria were developed that would support oral delivery and selective 5′-monophosphate release in hepatocytes. These criteria included intrinsic cellular activity, gastrointestinal stability, good oral absorption, plasma stability, and efficient liver metabolism of the phosphoramidate moiety to the naked 5′-monophosphate eventually leading to triphosphate formation. A screening cascade was established to interrogate each of the criteria, including important safety assessments relevant to nucleos(t)ide drugs – broad cellular cytotoxicity, mitochondrial toxicity, inhibition of host DNA and RNA polymerases, and bone marrow toxicity (Figure 7.6). Although the uridine nucleoside PSI-6206 itself demonstrated an excellent cellular safety profile, it became clear that the nature of the substituents on the corresponding 5′-phosphoramidate played an important role in modulating cytotoxicity. Several phosphoramidate prodrugs showed exceptional HCV replicon activity but unfortunately showed unacceptable cellular toxicity. The toxicity of some analogs was generally ascribed to the nature of the prodrug substituents which were released into the cell during the process of prodrug disassembly to generate the naked 5′-phosphate. A selected diverse set of compounds that met potency, *in vitro* safety, and gastrointestinal stability criteria, were progressed into *in vivo* assessment and metabolism studies in primary hepatocytes from human, rat, dog, and monkey [31].

In vivo assessment was achieved using surrogate animal model systems. Direct *in vivo* animal model efficacy was impractical because HCV only infects chimpanzees, which are difficult to access, handle, and expensive, especially since this program required the need for multiple compound *in vivo* assessments. Central to the development of our surrogate animal model was the key assumption that liver levels of the uridine triphosphate were directly correlated to the efficacy of the molecule.

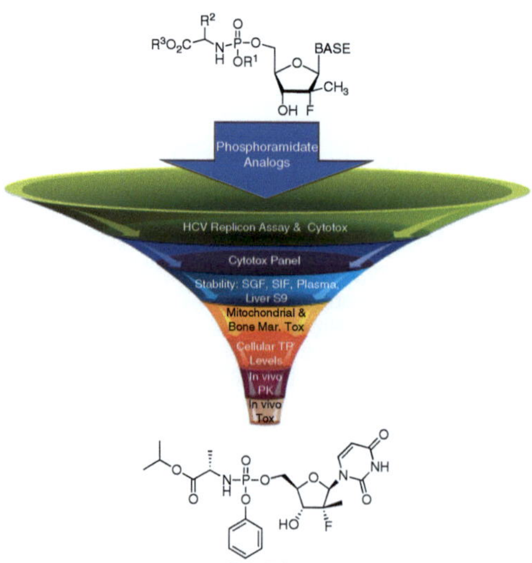

PSI-7851

Characteristic	PSI-7851
EC_{90} (μM)	0.48
Genotype Coverage	1a, 1b, 2a, 3a, 4a
HCV Drug Combinations	IFN, RBV, Nucs & NonNucs Polymerase, Protease, NS4A
Cytotoxicity (μM)	>100
Mitochondrial Toxicity (μM)	>100
Bone Marrow Tox (erythroid, myeloid) IC50 (mM)	>50
Metabolic Stability	Stable in SGF, SIF, and plasma
T½ (h) of Triphosphate in Primary Human Hepatocytes	38
Liver Triphosphate Levels: (ng/g) Rat Dog Monkey	1935 10560 57
Liver:Plasma Ratio	13-26
Resistance	S282T

SGF = simulated gastric fluid
SIF = simulated intestinal fluid

Figure 7.6 Design of the compound assessment cascade leading to identification of the prodrug clinical candidate PSI-7851.

Table 7.3 Phase 1b clinical study results of PSI-7851 in HCV genotype 1 patients for 3 days.

PSI-7851 dose	Number of subjects	Mean HCV RNA reduction after 3 days ($\log_{10}$ IU/mL)	Subjects with >1 $\log_{10}$ HCV RNA reduction during treatment
50 mg QD	8	−0.49	3
100 mg QD	8	−0.61	2
200 mg QD	8	−1.01	6
400 mg QD	8	−1.95	7
Placebo	8	−0.04	0

Consequently, a comprehensive bioanalytical method was developed for rat, dog, and cynomolgus monkey measuring plasma and liver levels of the parent prodrug, of each of the 5′-mono-, di-, and triphosphate analogs and of the parent uridine nucleoside. The results of these *in vivo* studies and of the primary hepatocyte metabolism studies showed that the isopropyl ester phosphoramidate prodrug PSI-7851 was superior to the other analogs tested and therefore, it became the clinical development candidate (Figure 7.6) [31].

PSI-7851 was shown to be a potent inhibitor of HCV replication ($EC_{90}=0.31\mu M$), with a high barrier to resistance and pan-genotypic character. In the replicon cellular assay, the S282T HCV RdRp variant was shown to deliver 14-fold reduced sensitivity to PSI-7851, but as mentioned earlier, this variant also exhibited a low level of fitness [31, 32]. PSI-7851 did not demonstrate any ADMET liabilities, showed a clean nonclinical animal toxicology profile, and was AMES and genotox negative. In addition, its triphosphate was shown not to be a substrate for any host DNA or RNA polymerases [31–33].

The prodrug PSI-7851 was first assessed in a standard single and multiple ascending dose clinical study in healthy volunteers with no observed adverse events. Subsequent evaluation in chronically infected HCV genotype 1 patients evaluated the drug for 3 days at doses of 50, 100, 200 and 400 mg once daily (QD). A dose-dependent decline in HCV RNA levels of −0.49 to −1.95 $\log_{10}$ IU/mL was seen with no adverse events nor any drug resistance noted (Table 7.3) [34]. These results were a proof-of-concept for the liver-targeted prodrug strategy.

7.5 Sofosbuvir: The Backbone of HCV Curative Therapies

When deciding to move PSI-7851 into clinical development, it was noted that the molecule was a mixture of diastereomers at the phosphorus center of the prodrug moiety. However, attempts to separate these isomers were not undertaken until clinical proof-of-concept of the prodrug strategy was achieved because, at the time, no chemical technology existed to selectively make each of the isomers for testing, and new technology would have to be developed. With proof-of-concept in hand, a crystallization and subsequent diastereoselective synthesis of the more intrinsically active Sp isomer, PSI-7977 (Sofosbuvir), was achieved (Figure 7.7) [31, 35, 36]. Subsequently,

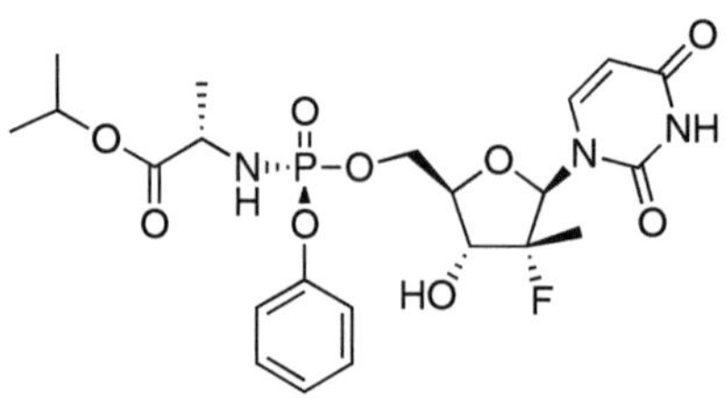

PSI-7977
Sofosbuvir (SOF)

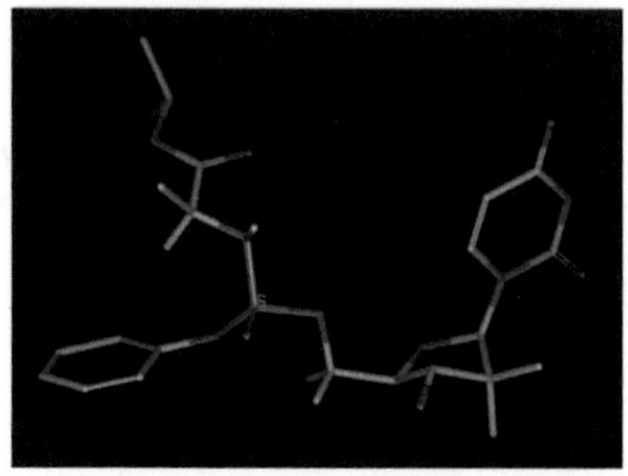

Sofosbuvir X-ray crystal structure
Sp stereochemistry at phosphorus
CSD Entry: CUZROG

Compound	Clone A replicon EC_{90} (μM)	ET replicon EC_{90} (μM)
PSI-7851	0.48	0.22
PSI-7977 (Sp Isomer)	0.28	0.17
PSI-7976 (Rp Isomer)	2.96	1.85

Figure 7.7 The structure of sofosbuvir (SOF, PSI-7977), the intrinsic potency of the individual diastereomers relative to the PSI-7851, and the X-ray crystal structure of the more active isomer sofosbuvir.

in a Phase 2b clinical study, genotype 1 patients were dosed with PSI-7977 at either 100, 200, or 400 mg QD for 24 days in combination with PEG-IFN and RBV and demonstrated a rapid virological response (RVR) of 88–94% with a mean reduction in HCV RNA of 5.1–5.3 $\log_{10}$ IU/mL (Table 7.4) [37]. It was also shown that PSI-7977 alone was more effective at reducing HCV RNA levels by 1.8 $\log_{10}$ IU/mL compared to PSI-7851 monotherapy, showing that the single isomer was more efficacious than the isomeric mixture. Subsequent clinical pharmacology studies showed no food effect or drug–drug interaction concerns with this prodrug [38–40].

Table 7.4 Sofosbuvir plus PEG-IFN and RBV dosed for 28 days in a Phase 2 clinical trial.

Sofosbuvir PSI-7977 dose (QD)	Number of patients (Pts)	Mean HCV RNA Δ from baseline (BL)	HCV RNA <LOD[a] N (%)
100 mg + SOC[b]	16	−5.3	14/16 (88%)
200 mg + SOC[b]	18	−5.1	17/18 (94%)
400 mg + SOC[b]	15	−5.3	14/15 (93%)
Placebo + SOC[b]	14	−2.8	3/14 (21%)

a. LOD means limit of detection (<15 IU/mL) by Roche Taqman Assay.
b. SOC = PEG-IFN and RBV.

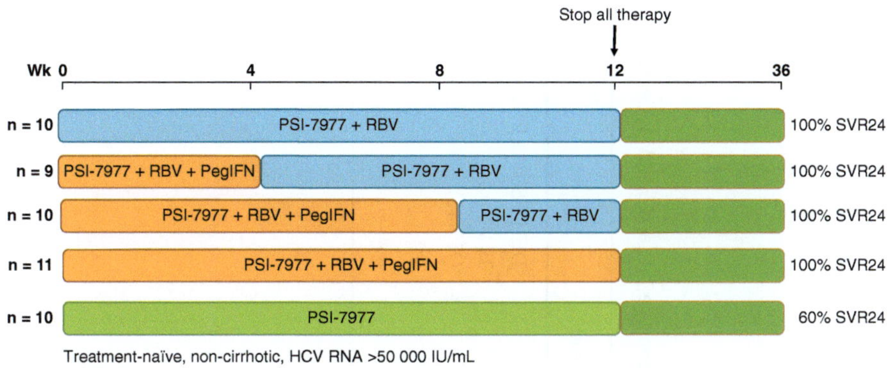

Figure 7.8 Electron study design evaluating sofosbuvir (PSI-7977) 400 mg QD with or without PEG-IFN and RBV for 12 weeks in genotype 2 and 3 HCV patients.

Even though sofosbuvir showed a best-in-class RVR in combination with PEG-IFN and RBV, the aspirational objective was to assess whether PEG-IFN could be eliminated from the combination and whether these patients could ultimately achieve a cure, i.e. SVR12 or SVR24 (no detectable virus or sustained virological response at 12- or 24-weeks following cessation of therapy). The exploratory study known as Electron set out to answer this question (Figure 7.8). This multi-arm study was designed to administer sofosbuvir to genotypes 2 and 3 chronically infected HCV patients, either alone or in combination with PEG-IFN + RBV, or in combination with RBV alone for 12 weeks, and then assess the patients' viral load 24 weeks after the end of treatment, i.e. SVR24 [41]. The results of this study would ultimately change the landscape of HCV drug development forever. All patients in sofosbuvir + RBV and sofosbuvir + PEG-IFN + RBV arms of the Electron study cohorts achieved SVR24 and were cured of their HCV infection (Figure 7.8) [41].

Early studies in genotype 1 patients showed that the IFN-free sofosbuvir regimen, either alone or in combination with the nucleoside RBV, was not quite as effective (84% SVR24 for patients with no prior therapy and 10% SVR24 for treatment-experienced patients after 12 weeks of therapy) at achieving an HCV cure as was shown in the genotypes 2 and 3 patient population Electron study. However, subsequent clinical studies combining sofosbuvir with the NS5A inhibitor, daclatasvir (DCV), or the NS3/4a protease inhibitor simeprevir (SIM) with or without RBV did produce high SVR rates (80–100%) (Tables 7.5 and 7.6) [42–44]. Although DAAs like daclatasvir and simeprevir were very effective at reducing HCV RNA levels, they lacked pan-genotypic character and generally had a low barrier to the development of resistance [45]. Consequently, the combination with sofosbuvir protected the patient from emerging resistance observed with these other DAAs and elicited a pan-genotypic profile, making it the backbone of HCV curative DAA combinations [7].

Final validation of the liver targeting character of sofosbuvir was achieved when, in a liver transplant study, patients were pre-dosed with sofosbuvir and levels of drug and metabolites were measured in the liver and plasma. High levels of both the sofosbuvir-derived monophosphate and triphosphate were observed in the livers, and little to no sofosbuvir was noted in the plasma of these patients, confirming what was predicted in the preclinical studies [46, 47].

Table 7.5 Sofosbuvir (400 mg, QD) plus daclatasvir (60 mg, QD) in a Phase 2 combination clinical study SVR results in genotype 1, 2, and 3 HCV patients.

	Genotype 2 or 3 (naive)			**Genotype 1 (naive)**	
	DCV + SOF 24 weeks (N = 14)	DCV + SOF + RBV 24 weeks (N = 14)	DCV + SOF + RBV 24 weeks (N = 15)	DCV + SOF 12 weeks (N = 41)	DCV + SOF + RBV 12 weeks (N = 41)
%SVR12	93	86[a]	100	100	95[c]
%SVR24	100	93	100	95[b]	93[d]

a. HCV RNA levels missing for one patient.
b. HCV RNA levels missing for two patients who had SVR at 36 weeks.
c. HCV RNA levels missing for one patient who had SVR at 24 weeks and one lost to follow-up.
d. HCV RNA levels missing for two patients who had SVR at 36 weeks and one lost to follow-up.

Table 7.6 Sofosbuvir (400 mg, QD) plus simeprevir (150 mg, QD) with or without RBV combination Phase 2 clinical study SVR results in HCV genotype 1 patients for 12 or 24 weeks of therapy.

	SIM (150 mg QD) + SOF (400 mg QD) SVR12 (week on therapy)	SIM (150 mg QD) + SOF (400 mg QD) + RBV (WB) SVR (week on therapy)
Cohort 1		
81 patients	93% (24 weeks)	79% (24 weeks)
GT 1 nonresponders	93% (12 weeks)	96% (12 weeks)
Metavir fibrosis score 0–2		
Cohort 2		
87 patients	100% (24 weeks)	93% (24 weeks)
GT 1 nonresponders or naive	93% (12 weeks)	93% (12 weeks)
Metavir fibrosis scores 3–4		

WB, weight-based dosing.

Phase 3 registration clinical trials evaluated sofosbuvir combinations in HCV chronically infected patients across genotypes 1 through 6 (Table 7.7) [48, 49]. For genotype 2 patients who were either naive to treatment, previous PEG-IFN treatment failures or were unable or unwilling to take PEG-IFN, sofosbuvir (400 mg QD) was combined with RBV and given for 12 or 16 weeks. Cure rates ranged from 91

Table 7.7 Sofosbuvir Phase 3 studies in genotypes 1, 2, 3, 4, 5, and 6 HCV patients[a].

Genotype	Neutrino GT 1, 4, 5, 6 SOF + PEG-IFN + RBV 12 weeks (Treatment naive) (N = 327) SVR12	Fission[b] GT 2, 3 SOF + RBV 12 weeks (Treatment naive) (N = 499) SVR12	Positron GT 2, 3 SOF + RBV 12 weeks (IFN intolerant, ineligible or unwilling) (N = 278) SVR12	Fusion GT 2, 3 SOF + RBV (IFN non-responders and relapsers) (N = 201) SVR12	Valence GT 2, 3 SOF + RBV (Naive and experienced) (N = 419) SVR12
GT 1	90%	-	-	-	-
GT 2	-	97%	93%	86% (12 weeks) 94% (26 weeks)	93% (12 weeks)
GT 3	-	56%	61%	30% (12 weeks) 62% (16 weeks)	84% (24 weeks)
GT 4	96%	-	-	-	-
GT 5	100%	-	-	-	-
GT 6	100%	-	-	-	-

a. In each study arm SOF given as a 400 mg QD dose, RBV was dosed on a weight-based basis, PEG-IFN dosed weekly 180 mg SQ.
b. Noninferiority study versus PEG-IFN + RBV arm.

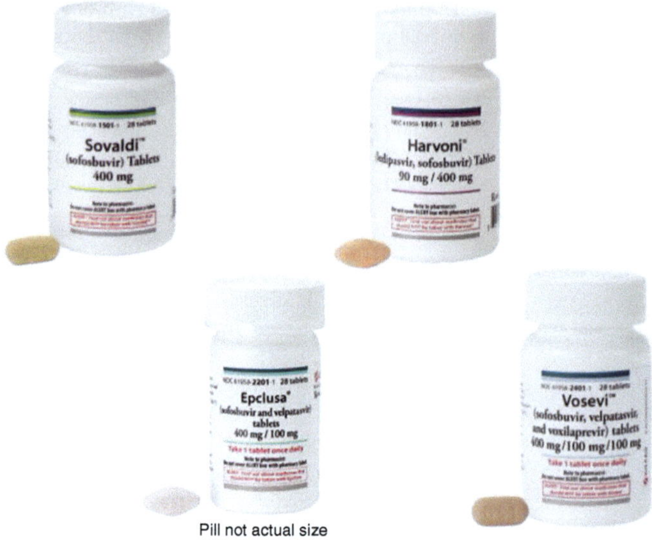

Figure 7.9 Sofosbuvir based regimens include Sovaldi, Harvoni, Epclusa, and Vosevi.

to 94% and were not dependent on their state of liver cirrhosis. Genotype 3 patients were in general less responsive to short 12-week sofosbuvir + RBV combination treatment and required longer treatment durations to achieve higher SVR12 rates, and the state of liver cirrhosis in these patients also appeared to have a negative impact on SVR rates. Genotype 3 patients who were intolerant or unwilling to take PEG-IFN demonstrated an SVR12 of only 61% after 12 weeks of therapy which rose to 84% if treatment was extended to 24 weeks. For genotype 1 and 4 patients, sofosbuvir + PEG-IFN + RBV was the most effective combination, and after 12 weeks of therapy, it gave SVR12 cure rates of 89% and 96%, respectively. Genotype 5 and 6 patients were quite responsive to sofosbuvir + PEG-IFN + RBV and after 12 weeks of therapy achieved SVR12 rates of 100%. In the Phase 3 studies clinical resistance was not a concern and adverse events were not shown to be drug related [48, 49].

On 6 December 6 2013, sofosbuvir, in combination with the generic drug RBV, was approved by the US FDA as the first IFN-free cure for patients with genotype 2 and 3 HCV infection and for genotype 1 IFN-ineligible patients. It was also approved in combination with PEG-IFN and RBV for genotype 1 and 4 patients. Sofosbuvir was sold by Gilead Sciences under the trade name Sovaldi®. In January, 2014, sofosbuvir was approved by the EMA for the treatment and cure of HCV.

DAA HCV curative combinations of sofosbuvir with daclatasvir and simeprevir were eventually approved to treat genotype 1 patients without the use of PEG-IFN; however, the important breakthrough occurred in October 2014 with the approval of the drug Harvoni, which was the first fixed-dose IFN-free combination of sofosbuvir and the NS5A inhibitor ledipasvir [50–52]. This afforded a one pill once-a-day 12-week drug therapy to cure genotype 1 HCV infected patients with a cure rate of 93–99%. Further development of sofosbuvir-based HCV curative regimens led to the fixed-dose pan-genotypic regimens Epclusa (sofosbuvir + velpatasvir) and Vosevi (sofosbuvir + velpatavir + voxilaprevir) (Figure 7.9) [53, 54].

7.6 Conclusion

The discovery and development of sofosbuvir was a landmark achievement in public health and ushered in a new era of short-duration, IFN-free cures for chronically infected HCV patients. It has been estimated that since sofosbuvir's launch in 2013, sofosbuvir-based regimens have cured over 10 million individuals chronically infected with HCV [55, 56]. It has also been the backbone of regimens used to eliminate HCV from high endemic populations in Egypt, Georgia, and other countries [57, 58].

References

1 Grebely J, Dore GJ, Morin S, Rockstroh JK, Klein MB. Elimination of HCV as a public health concern among people who inject drugs by 2030—What will it take to get there? *J Int AIDS Soc*. 2017;20(1): 22146.

2 Alqahtani SA, Sulkowski M. The role of interferon for the treatment of chronic hepatitis C virus infection. In: Sofia MJ, editor. HCV: The Journey from Discovery to a Cure. Topic in Medicinal Chemistry. 1. Switzerland: Springer Nature; 2019. 98–113.

3 Organization WH. Hepatitis C Fact Sheet 2024 [Available from: https://www.who.int/news-room/fact-sheets/detail/hepatitis-c.

4 Zahid MN. Transplantation of organs from hepatitis C virus-positive donors under direct-acting antiviral regimens. *J Clin Med*. 2022;11(3): 770.

5 Saeed M, Billerbeck E, Rice CM. HCV molecular virology and animal models. In: Sofia MJ, editor. HCV: The Journey from Discovery to a Cure. Topics in Medicinal Chemistry. 1. Switzerland: Springer Nature; 2019. 30–68.

6 Lohmann V, Korner F, Herian U, Bartenschlager R. Biochemical properties of hepatitis C virus NS5B RNA-dependent RNA polymerase and identification of amino acid sequence motifs essential for enzymatic activity. *J Virol*. 1997;71(11): 8416–28.

7 Martinez MA, Franco S. Therapy implications of hepatitis C virus genetic diversity. *Viruses*. 2020;13(1): 41.

8 Romano KP, Ali A, Royer WE, Schiffer CA. Drug resistance against HCV NS3/4A inhibitors is defined by the balance of substrate recognition versus inhibitor binding. *Proc Natl Acad Sci U S A*. 2010;107(49): 20986–91.

9 Sarrazin C. The importance of resistance to direct antiviral drugs in HCV infection in clinical practice. *J Hepatol*. 2016;64(2): 486–504.

10 Karino Y, Toyota J, Ikeda K, Suzuki F, Chayama K, Kawakami Y, et al. Characterization of virologic escape in hepatitis C virus genotype 1b patients treated with the direct-acting antivirals daclatasvir and asunaprevir. *J Hepatol*. 2013;58(4): 646–54.

11 Nakamoto S, Kanda T, Wu S, Shirasawa H, Yokosuka O. Hepatitis C virus NS5A inhibitors and drug resistance mutations. *World J Gastroenterol*. 2014;20(11): 2902–12.

12 Lohmann V, Korner F, Koch J, Herian U, Theilmann L, Bartenschlager R. Replication of subgenomic hepatitis C virus RNAs in a hepatoma cell line. *Science*. 1999;285(5424): 110–3.

13 Najera I. Resistance to HCV nucleoside analogue inhibitors of hepatitis C virus RNA-dependent RNA polymerase. *Curr Opin Virol.* 2013;3(5): 508–13.

14 Lohmann V, Bartenschlager R. On the history of hepatitis C virus cell culture systems. *J Med Chem.* 2014;57(5): 1627–42.

15 Clark JL, Hollecker L, Mason JC, Stuyver LJ, Tharnish PM, Lostia S, et al. Design, synthesis, and antiviral activity of 2′-Deoxy-2′-fluoro-2′-C-methyl-cytidine, a potent inhibitor of hepatitis C virus replication. *J Med Chem.* 2005;48(17): 5504–8.

16 Stuyver LJ, McBrayer TR, Tharnish PM, Clark J, Hollecker L, Lostia S, et al. Inhibition of hepatitis C replicon RNA synthesis by beta-D-2′-deoxy-2′-fluoro-2′-C-methylcytidine: a specific inhibitor of hepatitis C virus replication. *Antivir Chem Chemother.* 2006;17(2): 79–87.

17 Sofia MJ, Furman PA, Symonds WT. 2′-F-2′-C-methyl nucleosides and nucleotides for the treatment of hepatitis C virus: from discovery to the clinic. In: Barrish JC, Carter PH, Cheng PTW, Zahler R, editors. RSC Drug Discovery Ser Accounts in Drug Discovery. 4. UK: Royal Society of Chemistry; 2010. 238–66.

18 Sofia MJ, Chang W, Furman PA, Mosley RT, Ross BS. Nucleoside, nucleotide and non-nucleside inhibitors of hepatitis C virus NS5B RNA-dependent RNA polymerase. *J Med Chem.* 2012;55: 2481–531.

19 Migliaccio G, Tomassini JE, Carroll SS, Tomei L, Altamura S, Bhat B, et al. Characterization of resistance to non-obligate chain-terminating ribonucleoside analogs that inhibit hepatitis C virus replication in vitro. *J Biol Chem.* 2003;278(49): 49164–70.

20 Sofia MJ, Furman PA. The discovery of sofosbuvir: a liver-targeted nucleotide prodrug for the treatment and cure of HCV. In: Sofia MJ, editor. HCV: The Journey from Discovery to a Cure. Topics in Medicinal Chemistry. 1. Switzerland: Springer-Nature; 2019. 141–70.

21 Stuyver LJ, McBrayer TR, Tharnish PM, Clark J, Hollecker L, Lostia S, et al. Inhibition of hepatitis C replicon RNA synthesis by beta-D-2′-deoxy-2′-fluoro-2′-C-methylcytidine: a specific inhibitor of hepatitis C virus replication. *Antivir Chem Chemothe.* 2006;17(2): 79–87.

22 Le Pogam S, Seshaadri A, Kosaka A, Chiu S, Kang H, Hu S, et al. Existence of hepatitis C virus NS5B variants naturally resistant to non-nucleoside, but not to nucleoside, polymerase inhibitors among untreated patients. *J Antimicrob Chemother.* 2008;61(6): 1205–16.

23 Le Pogam S, Yan JM, Kosaka A, Ji Y, Chin DJ, Gonzaludo N, et al. No evidence of drug resistance or baseline S282T resistance mutation among GT1 and GT4 HCV infected patients on nucleoside polymerase Inhibitor RG7128 and PEG-INF/RBV combination treatment for up to 12 weeks: interim analysis from the propel study. 61st Annual Meeting of the American Association for the Study of Liver Diseases; October 29–November 2; Boston, MA, 2010.

24 Brown NA. Progress towards improving antiviral therapy for hepatitis C with hepatitis C virus polymerase inhibitors. Part I: Nucleoside analogues. *Expert Opin Investig Drugs.* 2009;18(6): 709–25.

25 Burton JR. HCV NS5B polymerase inhibitors. *Clin Liver Dis.* 2009;13: 12.

26 Gane EJ, Roberts SK, Stedman CA, Angus PW, Ritchie B, Elston R, et al. Oral combination therapy with a nucleoside polymerase inhibitor (RG7128) and danoprevir for chronic hepatitis C genotype 1 infection (INFORM-1): a randomised, double-blind, placebo-controlled, dose-escalation trial. *Lancet.* 2010;376(9751): 1467–75.

27 Murakami E, Bao H, Ramesh M, McBrayer TR, Whitaker T, Micolochick SHM, et al. Mechanism of activation of b-D-2′-deoxy-2′-fluoro-2′-C-methylcytidine and inhibition of hepatitis C virus NS5B RNA polymerase. *Antimicrob Agents Chemother.* 2007;51(2): 503–9.

28 Ma H, Jiang WR, Robledo N, Leveque V, Ali S, Lara-Jaime T, et al. Characterization of the metabolic activation of hepatitis C virus nucleoside inhibitor beta-D-2′-Deoxy-2′-fluoro-2′-C-methylcytidine (PSI-6130) and identification of a novel active 5′-triphosphate species. *J Biol Chem.* 2007;282(41): 29812–20.

29 Murakami E, Niu C, Bao H, Micolochick SHM, Whitaker T, Nachman T, et al. The mechanism of action of beta-D-2′-deoxy-2′-fluoro-2′-C-methylcytidine involves a second metabolic pathway leading to beta-D-2′-deoxy-2′-fluoro-2′-C-methyluridine 5′-triphosphate, a potent inhibitor of the hepatitis C virus RNA-dependent RNA polymerase. *Antimicrob Agents Chemother.* 2008;52(2): 458–64.

30 Hecker SJ, Erion MD. Prodrugs of phosphates and phosphonates. *J Med Chem.* 2008;51(8): 2328–45.

31 Sofia MJ, Bao D, Chang W, Du J, Nagarathnam D, Rachakonda S, et al. Discovery of a beta-*D*-2′-Deoxy-2′-α-fluoro-2′-b-*C*-methyluridine nucleotide prodrug (PSI-7977) for the treatment of hepatitis C virus. *J Med Chem.* 2010;53(19): 7202–18.

32 Lam AM, Murakami E, Espiritu C, Steuer HM, Niu C, Keilman M, et al. PSI-7851, a pronucleotide of beta-*D*-2′-deoxy-2′-fluoro-2′-*C*-methyluridine monophosphate, is a potent and pan-genotype inhibitor of hepatitis C virus replication. *Antimicrob Agents Chemother.* 2010;54(8): 3187–96.

33 Arnold JJ, Sharma SD, Feng JY, Ray AS, Smidansky ED, Kireeva ML, et al. Sensitivity of mitochondrial transcription and resistance of RNA polymerase II dependent nuclear transcription to antiviral ribonucleosides. *PLoS Pathog.* 2012;8(11): e1003030.

34 Lawitz E, Rodriguez-Torres M, Denning J, Albanis E, Compropst M, Berrey MM, et al. Pharmacokinetics, pharmacodynamics, and tolerability of GS-7851 a nucleotide analog polymerase inhibitor, following multiple ascending doses in patients with chronic hepatits C infection. *Antimicrob Agents Chemother.* 2013;57(3): 1209–17.

35 Ross BS, Reddy GP, Zhang H-R, Rachakonda S, Sofia MJ. Synthesis of diastereomerically pure nucleotide phosphoramidates. *J Org Chem.* 2011;76(20): 8311–9.

36 Murakami E, Tolstykh T, Bao H, Niu C, Steuer HMM, Bao D, et al. Mechanism of activation of PSI-7851 and Its diastereoisomer PSI-7977. *J Biol Chem.* 2010;285(45): 34337–47.

37 Rodriguez-Torres M, Lawitz E, Kowdley KV, Nelson DR, Dejesus E, McHutchison JG, et al. Sofosbuvir (GS-7977) plus peginterferon/ribavirin in treatment-naive patients with HCV genotype 1: a randomized, 28-day, dose-ranging trial. *J Hepatol.* 2013;58(4): 663–8.

38 Mathias A, Cornpropst MT, Clemons D, Denning J, Symonds W. No Clinically significant pharmacokinetic drug-drug interactions between sofosbuvir (GS-7977) and the Immunosuppressants cyclosporine A or tacrolimus in healthy volunteers. 63rd Annual Meeting of the American Association for the Study of Liver Diseases; November 9–13; Boston, MA, 2012.

39 Karageorgopoulos DE, El-Sherif O, Bhagani S, Khoo SH. Drug interactions between antivirals and new or emerging direct-acting antivirals in HIV/hepatitis C virus coinfection. *Curr Opin Infect Dis*. 2014;27(1): 36–45.

40 Koff RS. Review article: the efficacy and safety of sofosbuvir, a novel, oral nucleotide NS5B polymerase inhibitor, in the treatment of chronic hepatitis C virus infection. *Aliment Pharmacol Ther*. 2014;39(5): 478–87.

41 Gane EJ, Stedman CA, Hyland R, Ding X, Svarovskaia E, Symonds WT, et al. Nucleotide polymerase inhibitor sofosbuvir plus ribavirin for heptitis C. *New Eng J Med*. 2013;368: 34–44.

42 Sulkowski MS, Gardiner DF, Rodriguez-Torres M, Reddy KR, Hassanein T, Jacobson I, et al. Daclatasvir plus sofosbuvir for previously treated or untreated chronic HCV infection. *New Eng J Med*. 2014;370(3): 211–21.

43 Childs-Kean LM, Hand EO. Simeprevir and sofosbuvir for treatment of chronic hepatitis C infection. *Clin Ther*. 2015;37(2): 243–67.

44 Lawitz E, Sulkowski MS, Ghalib R, Rodriguez-Torres M, Younossi ZM, Corregidor A, et al. Simeprevir plus sofosbuvir, with or without ribavirin, to treat chronic infection with hepatitis C virus genotype 1 in non-responders to pegylated interferon and ribavirin and treatment-naive patients: the COSMOS randomised study. *Lancet*. 2014;384(9956): 1756–65.

45 Najera I. Resistance to HCV nucleoside analogue inhibitors of hepatitis C virus RNA-dependent RNA polymerase. *Curr Opin Virol*. 2013;3(5): 508–13.

46 Babusis D, Curry MP, Denning J, Park Y, Murakami E, Afdhal N, et al. Nucleotide analog levels in liver explants form HCV infected subjects undergoing liver transplantation after up to 24 weeks sofosbuvir (GS-7977) with ribavirin treatment. *Hepatology*. 2013;58(Suppl.): 737A.

47 Martel-Laferriere V, Dieterich DT. GS-7977: a promising nucleotide analog ns5b polymerase inhibitor of HCV. *Future Virol*. 2012;7(6): 537–46.

48 Jacobson IM, Gordon SC, Kowdley KV, Yoshida EM, Rodriguez-Torres M, Sulkowski MS, et al. Sofosbuvir for hepatitis C genotype 2 or 3 in patients without treatment options. *New Eng J Med*. 2013;368(20): 1867–77.

49 Lawitz E, Mangia A, Wyles D, Rodriguez-Torres M, Hassanein T, Gordon SC, et al. Sofosbuvir for previously untreated chronic hepatitis C infection. *New Eng J Med*. 2013;368(20): 1878–87.

50 Afdhal N, Reddy KR, Nelson DR, Lawitz E, Gordon SC, Schiff E, et al. Ledipasvir and sofosbuvir for previously treated HCV genotype 1 infection. *New Eng J Med*. 2014;370(16): 1483–93.

51 Afdhal N, Zeuzem S, Kwo P, Chojkier M, Gitlin N, Puoti M, et al. Ledipasvir and sofosbuvir for untreated HCV genotype 1 infection. *New Eng J Med*. 2014;370(20): 1889–98.

52 McQuaid T, Savini C, Seyedkazemi, S. Sofosbuvir. A significant paradigm change in HCV treatment. *J Clin Transl Hepatol*. 2015;3: 27–35.

53 Everson G, Towner WJ, Davis MN, Wyles D, Nahass R, Thuluvath PJ, et al. Sofosbuvir with valopatasvir in treatment-naive noncirrhotic patients with genotype 1 to 6 hepatitis C virus infection: a randomized trial. *Ann Intern Med*. 2015;163(11): 818–26.

54 Bourleire M, Gordon SC, Flamm SL, Cooper CL, Ramji A, Tong M, et al. Sofosbuvir, velpatasvir, and voxilaprevir for previously treated HCV infection. *New Eng J Med*. 2017;376(22): 2134–46.

55 Voeller A, Razavi-Shearer D, Blach S, Gamkrelidze I, Razavi-Shearer K, Razavi H. Total HCV patients treated with direct acting antivirals since 2014. American Association for the Study of Liver Disease; November 10, 2023; Boston, MA,2023.

56 Mullard A. A decade of HCV drugs. *Nat Rev Drug Discov*. 2024;23(1): 8.

57 Hassanin A, Kamel S, Waked I, Fort M. Egypt's ambitious strategy to eliminate hepatitis C virus: a case study. *Glob Health Sci Pract*. 2021;9(1): 187–200.

58 Averhoff F, Shadaker S, Gamkrelidze A, Kuchuloria T, Gvinjilia L, Getia V, et al. Progress and challenges of a pioneering hepatitis C elimination program in the country of Georgia. *J Hepatol*. 2020;72(4): 680–87.

8

The Discovery and Development of Harvoni® and Epclusa®: Ending the Interferon Era; Curing All Hepatitis C Genotypes

John O. Link

Gilead Sciences (Ret.), john.link@yahoo.com

8.1 Cure

Hepatitis C virus (HCV) infection is one of the few widespread, life-threatening chronic diseases where high cure rates have been achieved. Based on the chronicity and high mutability of HCV, and with therapy historically requiring powerful immune activation via injected interferon (IFN), a high cure rate, safe, well-tolerated, all direct-acting antiviral (DAA) therapy for HCV was nearly unimaginable just over a decade ago. In the year of the 10th anniversary of its approval, we detail the discovery and development of ledipasvir/sofosbuvir (Harvoni®), the first DAA therapy for curing Genotype 1 (GT1) HCV infection, and the combination drug sofosbuvir/velpatasvir (Epclusa®), which subsequently expanded cure to all six major HCV genotypes (Figure 8.1).

8.2 Introduction

A single-stranded RNA virus in the Flaviviridae family, the hepatitis C virus (HCV) was originally dubbed non-A non-B hepatitis. Harvey J. Alter, Michael Houghton, and Charles M. Rice were awarded the 2020 Nobel Prize in Medicine for their work leading to the 1989 discovery of HCV. HCV is a genetically diverse, highly mutable virus with six major genotypes (GT 7 and 8 are known but rare) and 86 subtypes, with GT distributions varying worldwide [1]. GT1 is the most prevalent (estimated at 49% of total infected individuals) [2], and unfortunately proved the most difficult to treat with IFN-based therapies that were the standard of care (SOC) from 1968 to 2014 [3]. HCV is a major global health problem, with 58 million individuals chronically

Trends in Antiviral Drug Development, First Edition. Edited by Michael J. Sofia and Zhengqiang Wang.

	Structure	Harvoni®	Epclusa®	Vosevi®
sofosbuvir SOF GS-7977	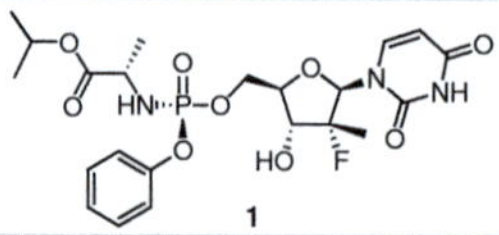 1	400 mg	400 mg	400 mg
ledipasvir LDV GS-5885	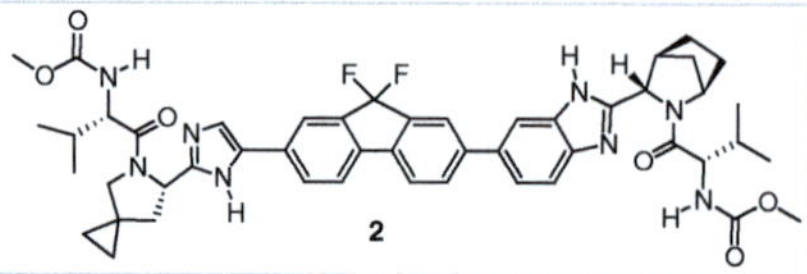 2	90 mg		
velpatasvir VEL GS-5816	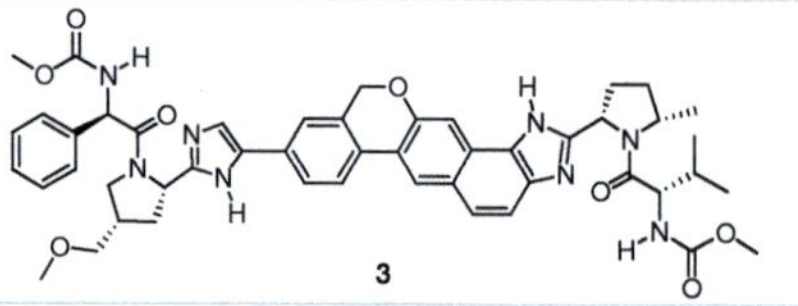 3		100 mg	100 mg
voxilaprevir VOX GS-9857	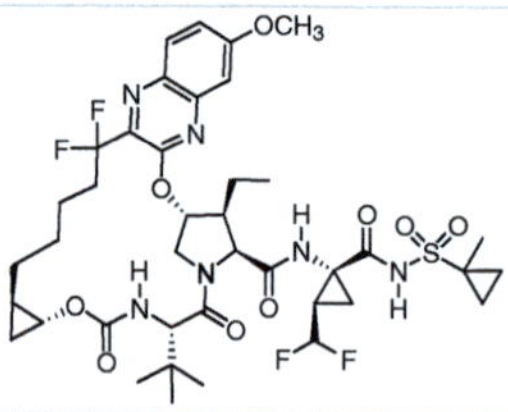			100 mg
	Duration, weeks	8[a], 12[b]	12[d]	12[e]
	GTs treated	1, 4–6[c]	1–6	1–6
	Includes PI?	No	No	Yes
	1st FDA Approval	Oct 10, 2014	Jun 28, 2016	Jul 18, 2017

[a]Considered in GT1 treatment-naïve patients without cirrhosis if pretreatment HCV RNA is <6 million IU/mL.
[b]May include RBV, or extend to 24 weeks depending on GT, treatment experience, and level of cirrhosis.
[c]Approved for GT4-6 treatment 12 November 2015.
[d]Include RBV for treating PLWHCV and decompensated cirrhosis.
[e]DAA treatment experienced. Vosevi® is included here as a sofosbuvir-based STR but is not otherwise discussed.

Figure 8.1 Components and attributes of the SOF-containing single-tablet regimens (STRs) Harvoni®, Epclusa®, and Vosevi®. Lower combination dosage forms are available for pediatric treatment. Generic daclatasvir with licensed generic sofosbuvir is the additional sofosbuvir-based regimen in use.

infected, and is a leading cause of cirrhosis, liver failure, and hepatocellular carcinoma (HCC) and resulted in 400 000 deaths yearly as of 2019 [4]. Transmission is through blood contact and individuals can be silently infected for many years, and the opioid epidemic is proving to be a major avenue of transmission. In 2013 viral hepatitis (HBV and HCV) was the seventh leading cause of death worldwide, outpacing infectious diseases such as HIV, tuberculosis, and malaria [5]. In the United States, HCV accounted for more deaths than all other CDC-notifiable infections combined [6]. There is no vaccine for HCV prevention, attributed in part to the high mutability of HCV [7]. Ironically, highly efficacious curative HCV drugs have been achieved despite the diversity and mutability of the virus; in the face of such viral mutability, tremendous resources and efforts over more than 25 years have resulted in safe, short-duration, and effective cures for over 95% of infected individuals.

Although IFN-based therapies (eventually with the addition of ribavirin, RBV) represented an early breakthrough and became the SOC from 1991 to 2014, IFN proved extremely difficult for those with HCV infection due to low efficacy and debilitating side-effects. Side effects of IFN regimens are often described as causing "flu-like" symptoms such as fever, nausea, headache, fatigue, depression, and muscle ache. RBV problematically has overlapping toxicity with IFN and significantly adds anemia and additional fatigue. Likening the toxicities to "flu-like" symptoms belies the true difficulties individuals endure when facing 24- or 48-week IFN/RBV-based therapy. The severe intolerability was only widely revealed through "real world" studies. In clinical trial settings, cure rates of IFN-based therapies improved from 6% (1969, 6 months of IFN) to 54–6% (2002, 12 months of pegylated IFN [PEG] and RBV, in a mixed GT1-3 population) [8]. But the reality for people living with HCV (PLWHCV) treated outside controlled clinical trials ("real-world" settings) was dire. A compilation of 25 independent US real-world studies, including 13 000 HCV-infected individuals from 2002 to 2012, revealed the true barrier to cure, with only 3% of all individuals achieving a sustained viral response (SVR, or cure, is confirmed when HCV RNA is below the lower limit of quantitation in an individual's blood 12 weeks after completion of treatment; Figure 8.2) [9]. The steepest barrier to achieving SVR was the start of therapy mainly based on physical or psychological ineligibility, with multiple subsequent barriers. Taking a representative set of 100 individuals from the study, 61 were contraindicated for therapy. Of the 39 eligible for therapy, over half refused to initiate therapy based on their concerns with side effects or the desire to defer treatment hoping for approval of improved therapies. Of the 19 who initiated treatment, six discontinued due to intolerability or futility criteria based on insufficient viral-load reduction at specified timepoints (significant safety and efficacy monitoring are required during IFN treatment). Of the 13 who endured the complex and challenging 24- or 48-week treatment, only 3 achieved SVR. Thus, in the intent-to-treat (ITT) calculation, the SVR in this real-world study would be 16% (based on the number of people achieving durable viral response divided by the number starting therapy), but importantly this calculation does not

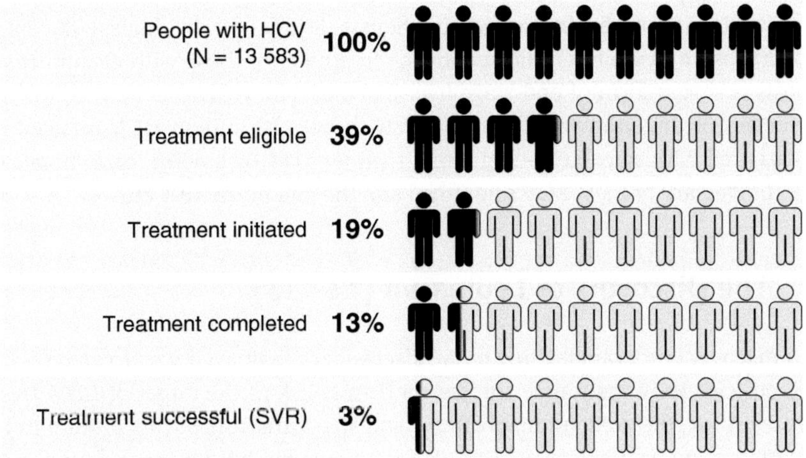

Figure 8.2 Real-world IFN/ribavirin SVR rates are much lower than in clinical trials. From 25 independent studies (2002–2012).

reflect the large number of individuals unable or unwilling to start therapy. The practical reality for PLWHCV can truly be seen when accounting for those who do not initiate therapy; by such analysis only 3% of all individuals in these IFN-based studies were successfully cured. These real-world studies underscored the desperate need for improved therapies.

The prospect for PLWHCV changed rapidly in 2014 when ledipasvir/sofosbuvir (SOF) (Harvoni) marked the end of the IFN era as the first-approved all-DAA drug for the cure of GT1 HCV infection [10]. Following in 2016, SOF/VEL (Epclusa) expanded and further simplified therapy as the first pan-genotypic DAA approved for the treatment of all six major genotypes of HCV, including individuals with decompensated cirrhosis. Herein we describe the discovery and development of Harvoni and Epclusa (Figure 8.1). Harvoni, as the first approved all DAA STR for the cure of HCV, was initially approved for the treatment of GT1 HCV – the most prevalent GT worldwide – with expansion in 2015 to include treatment of individuals with GT 4, 5, and 6, or HIV/HCV coinfection, and remained for years as the only drug combination to accomplish high cure rates in as little as 8 weeks. Harvoni is an STR, a once-daily complete regimen in a single pill, comprised of 400 mg sofosbuvir (SOF), a pangenotypic drug and the only approved nucleotide nonstructural protein 5B (NS5B) inhibitor, and long half-life, first FDA approved HCV nonstructural protein 5A (NS5A) inhibitor ledipasvir. The discovery of sofosbuvir is detailed in Chapter 7 herein. Epclusa, approved in 2016, is the only pangenotypic STR and is a combination of 400 mg SOF and 100 mg of pangenotypic, panfibrotic NS5A inhibitor velpatasvir, and is indicated for all common HCV GTs [1–6]. The cure rates over 95%, safety, simplicity, and broad inclusion criteria of these regimens opened treatment for millions of individuals ineligible for IFN therapy or who had deferred treatment (Figure 8.1). For example, with Harvoni or Epclusa HIV/HCV coinfected individuals experience the same cure rates as HCV monoinfected individuals (*vide infra*), while their treatment was contraindicated for IFN therapy due to exceedingly low HCV cure rates. Further, these therapies do not contain an HCV protease inhibitor (PI), which makes them important for treating individuals with compensated or decompensated cirrhosis. PIs carry a warning concerning hepatic injury in individuals with compensated cirrhosis and are not indicated for the treatment of those with decompensated cirrhosis (a severe form of end-stage liver disease). The year 2023 marked the tenth anniversary of the approval of sofosbuvir with over 10 million HCV-infected individuals cured with sofosbuvir-containing regimens [11, 12], while 2024 heralds the tenth anniversary of LDV/SOF's approval and the end of the IFN era.

8.3 The Discovery of Ledipasvir [13–15]

In our discovery program leading to the discovery of ledipasvir we targeted the HCV NS5A protein. HCV resistance studies confirm NS5A as the target of LDV. Despite extensive study, the detailed mechanistic and structural role of this protein in HCV replication remains a mystery. NS5A is not an enzyme but has been found to bind to a range of viral and host proteins and possibly RNA [16]. There are no purified NS5A-based biochemical assays, and antiviral potency measurements are achieved by assessing the rate of replication in replicon cellular assays in the presence of varying inhibitor concentrations. As such, small-molecule approaches to targeting NS5A have been empirical and have not been structure-enabled. Several chemotypes targeting NS5A had been discovered through phenotypic screens and pursued as optimization leads [17], and daclatasvir was the first of these to achieve clinical proof-of-concept [18]. NS5A was one of many host and viral targets undertaken in search to find a cure for hepatitis C infection.

Genetic diversity is a fundamental basis for HCV chronicity. The HCV family is comprised of eight genotypes and 86 subtypes. Further, there are also a multitude of quasispecies present *within* each HCV-infected individual's viral population. These fluxional quasispecies, sometimes termed a "viral swarm," are based on the error rate of HCV polymerase and the size of the viral genome; it has been calculated that all possible single, double, and most triple mutants are produced daily in a PLWHCV's viral pool [19]. This fluid "viral swarm" may form a basis for the ability of HCV to evade the host immune system. Mutability is also a basis for the rapid generation of drug resistance in HCV and a key area of focus in drug design. Most agents (sofosbuvir being largely an exception) when dosed in monotherapy rapidly result in the selection of multiple viral quasispecies called resistance-associated substitutions (RAS). GT and subtype diversity along with viral mutability posed significant challenges to HCV drug discovery.

With HCV mutability as a backdrop, our approach was based on learnings and expertise from the discovery and development of HIV STRs. STRs provide multiple benefits, particularly in addressing viral variants, along with adherence and therefore efficacy. While individual drugs can be designed to inhibit a range of viral variants, concomitant use of multiple drugs of differing mechanism serves to inhibit a broader range of variants. The simplicity of taking a single pill once daily supports adherence. Adherence to a multidrug cocktail supports a consistently high drug pressure which is paramount to avoid virologic breakthrough and resistance selection which can occur during missed doses. Resistance generation can also occur if only a subset of drugs are taken in a multi-pill regimen, and co-formulation in an STR obviates this concern [20]. The target profile for an HCV STR required (i) oral bioavailability sufficient for oral administration, (ii) a low daily dose (potency and pharmacokinetics [PK]) so that all components could be combined into a practical single-pill size, (iii) appropriate physicochemical properties for coformulation compatibility, (iv) a low potential for drug–drug interactions (DDI) as a perpetrator with the co-formulated drugs and with coadministered medications in the broad HCV-infected population, (v) low potential as a DDI victim with co-formulated drugs and particularly with coadministered drugs since dose cannot be adjusted within the STR, (vi) PK half-life consistent with once-daily dosing, (vii) sufficiently high resistance barrier to minimize the number of agents required to allow a single pill, (viii) a good safety profile acceptable for treatment of the broad population of individuals infected with HCV. The stringent requirements (i–vi) for an STR are tightly connected to PK properties and oral bioavailability. The NS5A inhibitors ledipasvir and velpatasvir occupy nontraditional drug space. Since the mid-1990s, a plethora of "rules of oral bioavailability" and "druglikeness" have been promulgated and to varying degrees adopted by drug-discovers and their organizations. LDV and VEL

Table 8.1 Ledipasvir and velpatasvir are orally bioavailable drugs achieving the stringent properties requirements of STRs while breaking most "rules" of bioavailability.

	Metric	Rule value	LDV	VEL
Lipinski rule of 5 [21]	Molecular weight	≤500 Da	**889**	**883**
	CLogP[a]	≤5	**6.7**	**5.7**
	H-bond donors	≤5	**4**	**4**
	H-bond acceptors[b]	≤10	**14**	**16**
Veber rules [22]	Rotatable bonds[c]	≤10	**12**	**13**
	Polar surface area[c]	<140 A^2	**174**	**193**
Ring rule [23]	No. of rings	≤5 is 95th P_i	**10**	**9**
Aromatic ring rule [24]	No. of aromatic rings	≤3	**5**	**6**

Red box and text indicate parameter has broken the rule.

a. Calculated with CheBioDraw 14.0, CamBridgesoft Corp.

b. As originally defined by Lipinski.

c. Calculated using Pipeline Pilot.

handily break most of these rules Table 8.1 [21–24] and are not only bioavailable drugs, but they achieve the stringent PK requirements of an STR. In the discovery of LDV and VEL, we did not follow a rule-based approach and instead pursued free-thinking hypothesis generation and testing without limitations to the creativity that is at the core of drug discovery [25].

Many questions remained. A cure would require multiple drugs, but their number, wild-type (WT, parent viral genome sequence without substitutions) antiviral potency, and activity against the multitude of viral species were unknown. It was further unknown if an all-DAA combination could be sufficient for cure, or if host immune activation would also be required; some researchers in the field questioned if IFN could be replaced. As early HCV-directed programs progressed, the mutability and viral diversity of HCV made inhibitor series with potential for broad GT potency rare. Attaining sufficient potency against multiple subtypes and resistant mutants (viral variants with RAS) within even a single GT posed a significant challenge. When we initiated our NS5A program, the majority of our ongoing 20 HCV programs targeting various mechanisms were focused on GT1 infection as the highest unmet need (representing 50% of HCV-infected individuals worldwide and suffering the lowest IFN/RBV SVR) [2]. Targeting GT1 HCV and achieving the criteria for an STR was our focus in the discovery of ledipasvir. In our NS5A discovery program we sought a high level of antiviral potency, combined with a long PK half-life to suppress GT1 viral replication and avoid virologic breakthrough, resistance, and treatment failure.

Throughout our approach we discovered a wide diversity of core structures and sought to understand and improve their contribution to potency and PK properties in elaborated inhibitors [26]. A series of inhibitors with symmetric central "cores" incorporating bis-benzimidazoles and capped with methoxymethyl-carbamate

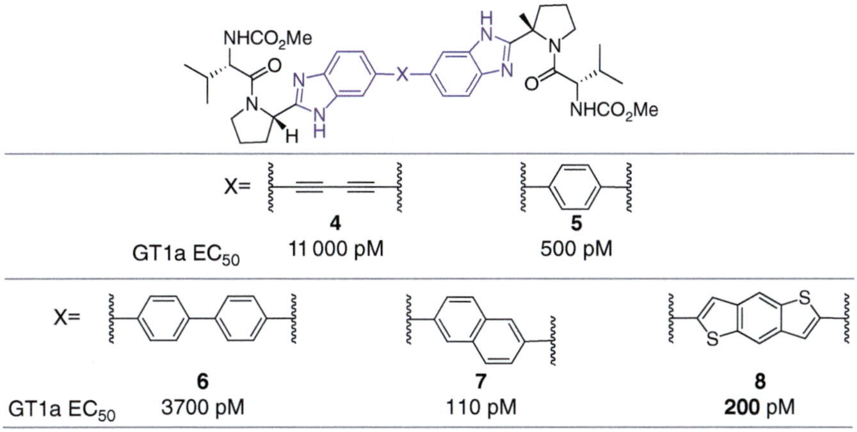

Figure 8.3 Symmetric core bis-benzimidazole inhibitor potency. "Core" highlighted in purple. Here and hereafter the GT1a strain is H77.

(Moc) valine-amides of pyrrolidines are detailed in Figure 8.3. Throughout this chapter, "cores" are defined as the central portion of the molecule spanning between the pyrrolidine or modified pyrrolidine groups (indicated as the purple-colored portion of the molecule in Figure 8.3). In all examples, GT1a and GT1b potency was measured but for each example GT1b potency was superior to GT1a so is generally not reported here. In the general structure shown in Figure 8.3, connection of the benzimidazoles by a bond or an alkyne (structures not shown) did not achieve a 50% effective inhibitory concentration (EC_{50}) at a 44 000 picomolar (pM) inhibitor concentration (the highest concentration screened) in a GT1a replicon cellular assay. Increasing the length with bis-alkyne **4** afforded measurable potency to 11 000 pM. A phenyl ring connection (**5**), which is intermediate in length between one and two alkynes, afforded 500 pM potency. Clearly, the central portion of the molecule is not simply a rigid spacer; presumably, an aryl ring contacting the NS5A target provides distinct improvements in potency over alkyne counterparts. Despite having potential for more hydrophobic contact, increasing length to biphenyl **6** afforded less potency (3700 pM) than the monoaryl, while fused aryls, including naphthyl **7** (110 pM) and benzdithiophene **8** (200 pM) were the most potent in the series. With the span of distance and varying motifs represented by compounds **4–8**, the data were consistent with fused aryl rings, such as naphthyl **7** at 110 pM as preferred groups.

Crystal structures of NS5A without the N-terminal helix domain have been achieved; two structures for each of GT1b and GT1a have been crystallized as dimers, with all four dimers in differing orientations [27–29]. A number of groups have pursued dimeric, symmetric NS5A inhibitors. In a departure from this approach, we were intrigued by the potential to diversify our structures by assessing unsymmetric cores (here we use unsymmetric to describe structures that do not "read" as symmetric from one end to the other end of the "core"). Figure 8.4 shows a series of unsymmetric-core imidazole-X-benzimidazole inhibitors. With a phenyl "X" group the molecule didn't achieve 50% inhibition at the highest concentration (compound not shown), but phenyl-alkyne **9** afforded inhibition at 2500 pM. It was

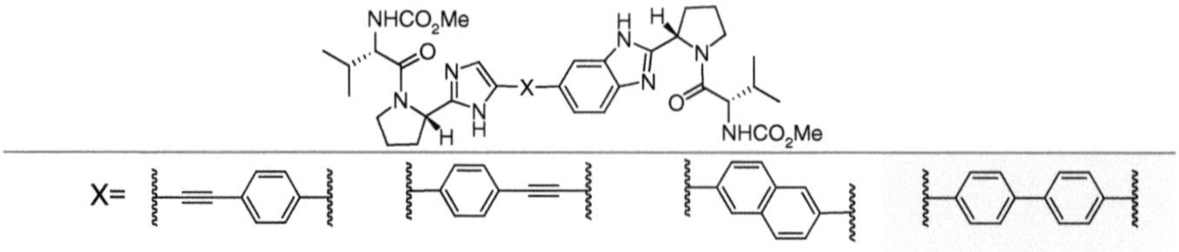

	9	10	11	12
GT1a EC_{50}	2500 pM	380 pM	71 pM	170 pM

Compound 12:

	Rat	Dog
PK half-life (h)	1.6	2.3
F%	11.5	-

Figure 8.4 Unsymmetric core imidazole-X-benzimidazole inhibitor potency. Fused ring core **11** is the most potent.

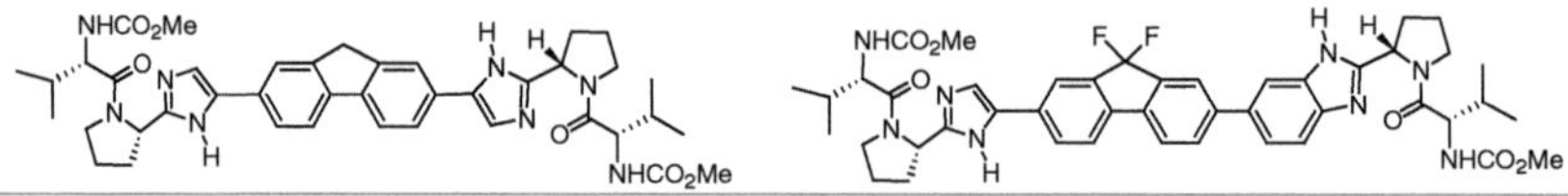

	13	14
GT1a EC_{50}	94 pM	40 pM

Compound 14: HLM Pred E_h < 0.16%

	Rat	Dog
PK half-life (h)	1.8	2.6
F%	36.7	-

Figure 8.5 Fluorene-based core provides potency in **13** but is unstable to oxidation. Unsymmetric difluorofluorene-based core affords improved potency and hepatic extraction at the lower limit of the HLM assay.

intriguing to find that reversing the orientation of phenyl-alkyne **10** resulted in a differentiated molecule that would not have pertained if the core's end-groups were identical, and afforded a striking improvement in potency to 380 pM. There is a distinct SAR to unsymmetric end-groups coupled to unsymmetric "X" groups within the core. In this imidazole-X-benzimidazole series, the naphthyl-based core **11** was highly potent at 71 pM, with biphenyl **12** also very potent at 170 pM. The half-life of **12** in rat and dog proved promising at 1.6 and 2.3 h, respectively, although the rat oral-bioavailability (F) was low at 11.5%. In these studies, the unsymmetric series produced inhibitors with potency improvements over the symmetric benzimidazole-X-benzimidazole series, and this is clearly seen for "X" groups such as biphenyl or naphthyl that were incorporated in each series. We further noted that in each series fused aryls as "X" groups were preferred, with naphthyl affording the most potent member in each series. Using learnings from the potency of the biphenyl and the potency benefits of fused-ring systems was a basis for pursuing fused tricyclic ring "X" groups.

Fused tricyclic "X" groups were assessed in the symmetric-core series of imidazole-X-imidazole (Figure 8.5). Assessment of tricyclic X groups was facilitated in the symmetric series based on synthetic tractability relative to the unsymmetric series. Fluorene **13** afforded an encouraging level of potency at 94 pM, but the propensity of the central methylene toward oxidation (even air oxidation upon standing) led us to work to block this reactive position. The fluorenone oxidation product afforded a loss in potency to 300 pM (not shown) suggesting to us that either polarity in this region is not preferable, or there is some steric requirement in the vector of the carbonyl. Blocking oxidation of **13** by substitution of the fluorene methylene with two methyl groups afforded a significant loss of potency to 7000 pM (compound not shown).

We still sought to block the oxidation of the fluorene methylene and considered that a difluoro-substitution of the methylene might provide an optimal blocking group through the hypothesis that the out-of-plane bulk in the dimethyl-substituted methylene was the basis for its potency loss. Because fused-aryls had performed better in the unsymmetric imidazole-X-benzimidazole series, and based on the challenging synthesis of the difluorofluorene we did not incorporate it in the symmetric series but went directly to more important unsymmetric series. The resulting unsymmetric imidazole-difluorofluorene-benzimidazole inhibitor **14** proved to be the most potent inhibitor described to this point at 40 pM. Relative to biaryl compound **12**, PK half-lives for **14** in rat and dog were improved and the oral bioavailability in rat was significantly and unexpectedly improved (from 11.5% to 36.7%). Importantly, **14** exhibited metabolic stability at the experimental lower limit of our human liver microsome (HLM) assay with a predicted hepatic extraction (E_h) of <13%. The lower limit of stability that can be measured in the HLM assay is based on the length of time that microsomes are metabolically active and by the limits of quantitation of disappearance of the parent molecule.

Many pyrrolidine mimetics were synthesized. The rigid 2.2.1 azabicyclic ring inhibitor **15** (Figure 8.6) was more potent at 210 pM than the molecule utilizing a piperidine in the same position (450 pM, not shown). Importantly, **15** had a long PK half-life of 5.3 h in dog. We sought to determine if the beneficial PK half-life of the bicyclic bridged ring system might translate to an unsymmetric series.

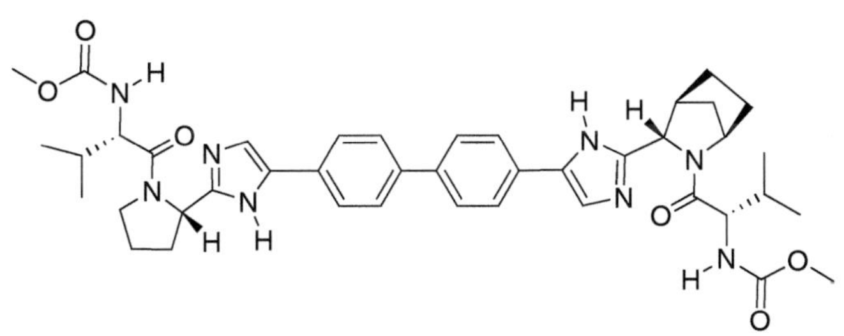

15

GT1a EC_{50} 210 pM

HLM Pred. CL < 0.16 L/h/kg

	Rat	*Dog*
PK half-life (h)	1.5	5.3

Figure 8.6 The 2.2.1 azabicyclo ring inhibitor **15** has a long half-life in dog.

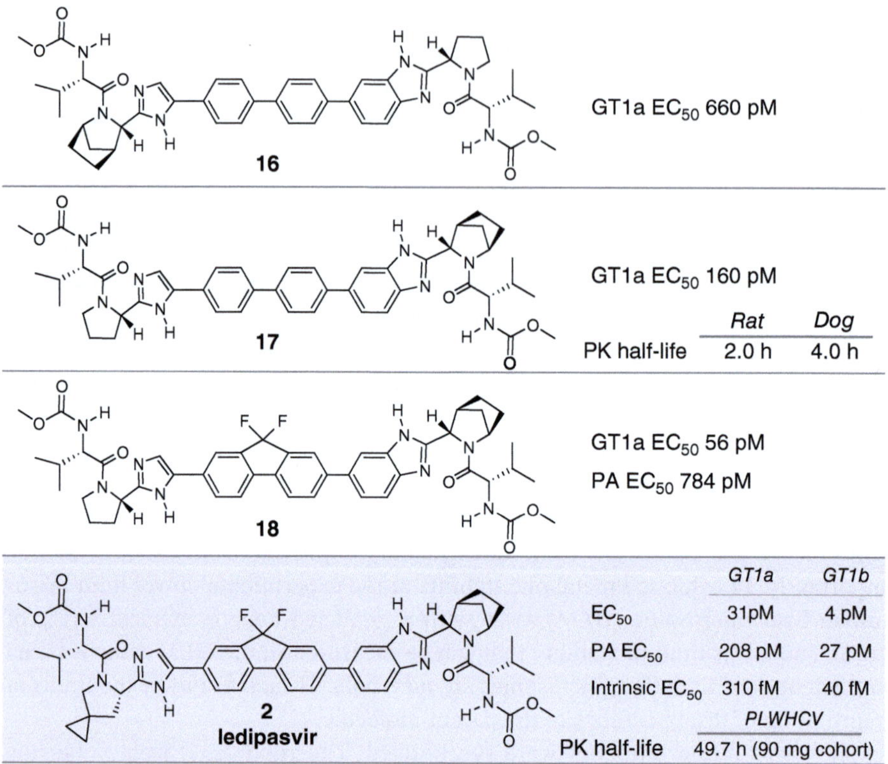

Figure 8.7 Differential end substitution with an unsymmetric core leads to potency synergy in **17**. Adding difluorofluorene further improves potency. Spirocyclopropyl-pyrrolidine improves potency and protein-adjusted potency to afford LDV.

Assessing the 2.2.1 azabicyclic ring on either side of the unsymmetric imidazole-biphenyl-benzimidazole core provided striking and divergent potency results that underscored the importance of working in the unsymmetric series (Figure 8.7). It was initially discouraging to find that placing the bicyclic pyrrolidine on the imidazole side of the core in compound **16** resulted in weaker potency at 660 pM than obtained in the symmetric compound **15** (210 pM). It was then exciting to discover that with the 2.2.1 ring system connected to the benzimidazole side, compound **17** proved more potent at 160 pM than when the 2.2.1 azabicyclic ring was placed on the imidazole side (**16**), or when utilized in the symmetric series (**15**). As we had hoped, the half-life in rat and dog were both prolonged in compound **17** (2.0 and 4.0 h, respectively) relative to pyrrolidine **12**. These results demonstrated the importance of differentiating sides on an unsymmetric core – a wider range of chemical space could be pursued with the potential for significant improvements in the attributes of the resulting molecules.

Although a challenging target for synthesis, the potency-enhancing imidazole-difluorofluorene-benzimidazole core structure was synthesized with the 2.2.1 azabicyclic ring on the preferred benzimidazole side (**18**) and gratifyingly drove the potency to 56 pM, producing the most potent compound described to this point.

To project the potency of this important compound *in vivo*, the protein-adjusted (PA) EC_{50} based on human serum binding was found to be 784 pM [30]. To further investigate the beneficial properties of multi-ring pyrrolidines, modifications were made to replace the remaining pyrrolidine. Elaboration to a spirocyclopropyl-pyrrolidine produced ledipasvir **2**, an exceptionally potent GT1a inhibitor at 31 pM (Figure 8.7). Significantly, the addition of the spirocyclopropyl group differentially drove the PA EC_{50} affording a very high potency of 208 pM (100% human serum). When accounting for protein-binding in the replicon assay by dialysis against the cell-culture medium (bovine serum albumin is present in the assay), ledipasvir was found to have a GT1 intrinsic EC_{50} of 310 femtomolar (fM).

Ledipasvir exhibited the most favorable PK profile among the compounds described here, with long half-lives (4.67, 7.41, and 10.3 h) and good bioavailability (32.5, 53, and 41.1%) in rat, dog, and monkey respectively and low metabolism with a 7–10% E_h in those species (Figure 8.8). Like **14**, ledipasvir was stable at lower limit of measurement in an HLM assay. To quantify the predicted clearance (CL), tritiated ledipasvir was synthesized and demonstrated exceptionally low metabolite formation in human hepatocytes resulting in a PredCL = 0.012 L/h/kg equivalent to a 1% hepatic extraction [13]. Ledipasvir was projected to have a human half-life sufficient to support once-daily dosing based on volumes of distribution (V_{ss}) higher than total body water in preclinical species (ranging from 1.19 to 2.66 L/h/kg), along with an exceptionally low predicted CL based on human hepatocytes.

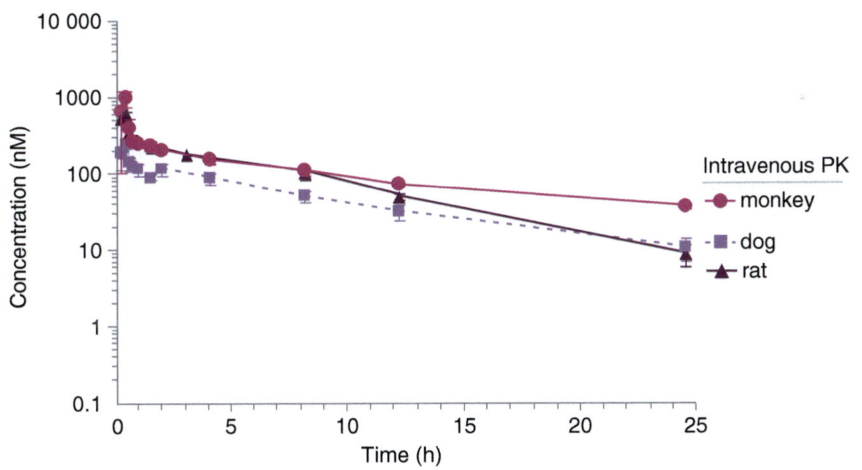

Species	Dose (mpk)	$t_{1/2}$ (h)	E_h	Vss (L/kg)	%F[a]
Rat	1	4.67	10%	2.66	32.5
Dog	0.2	7.41	7%	1.19	53.0
Monkey	0.5	10.3	10%	2.15	41.1

[a]All parameters except F% are from IV dosing.

Figure 8.8 Preclinical LDV PK curves. Ledipasvir preclinical PK curves and parameters: long half-lives, low hepatic extraction, and good bioavailability in rat, dog, and monkey.

Prior to the discovery of ledipasvir, it had become apparent that the NS5A inhibitor program had the potential to deliver a clinical candidate, and that various advanced inhibitor series were posing increasingly challenging chemical syntheses. In a forward-looking approach, process chemists joined the research team prior to candidate identification to study and develop more robust chemistry in the chemical series being pursued to accelerate timelines upon candidate molecule identification; this early collaboration proved both enjoyable for the expanded team and fruitful for the program. Once ledipasvir was identified, process and medicinal chemists continued to collaborate to improve the chemistry to enable toxicology and other investigational new drug application (IND) enabling studies to commence rapidly. Concurrently, the virology group, while engaging in IND-enabling studies, initiated efforts to generate a broader range of replicon assays to prepare for the transition to a pangenotype NS5A program. As they developed replicons beyond GT1 and GT1 RAS, the virology and high-throughput biology groups screened the structurally diverse existing inhibitors from the ledipasvir discovery program to identify molecules with broader genotype activity, so that when the medicinal chemists completed their early LDV synthesis development supporting efforts, starting-points would already be identified for the pangenotype program leading to velpatasvir.

8.4 Early Development of Ledipasvir

With a favorable potency and PK profile, and no significant findings in 14-day safety studies in rat and dog, ledipasvir entered clinical trials. The efforts to drive a strong preclinical PK profile (including an extremely low human PredCL) translated clinically [13]. Ledipasvir's 49.7-h half-life at the prescribed 90 mg dose in PLWHCV stands as the longest half-life of any marketed NS5A drug (Figure 8.9) [31]. Ledipasvir concentrations increased in both HCV-uninfected individuals (3–100 mg) and in PLWHCV (1–90 mg) over the ranges tested [31]. Twenty-four hours after the third 90 mg dose (C_{tau}), the LDV concentration was 610-fold above its PA EC_{50} (Figure 8.9). Based on its low potential for cytochrome P450 metabolism, LDV can be coadministered in the presence of potent cytochrome P450 (CYP) inhibitors, such as ritonavir or cobicistat, and therefore does not require dose adjustment when used in the treatment of HIV/HCV coinfected individuals who are prescribed HIV-PI regimens which require PK enhancement through use of these CYP inhibitors. This is an important result since (i) HIV/HCV coinfected individuals are a significant population and were contraindicated for therapy in the IFN era and (ii) dose adjustments are not compatible with formulation as an STR. Additionally, sofosbuvir is not subject to CYP metabolism, so no dose adjustment is needed for Harvoni® in the presence of CYP inhibitors or inducers, thereby broadening and simplifying treatment. Overall, these PK results for LDV demonstrated potential for once-daily dosing and use in an STR.

In people infected with GT1a HCV, rapid viral load reduction (VLR) was observed with three once-daily LDV doses from 1 to 90 mg [31]. Similarly, rapid VLR was

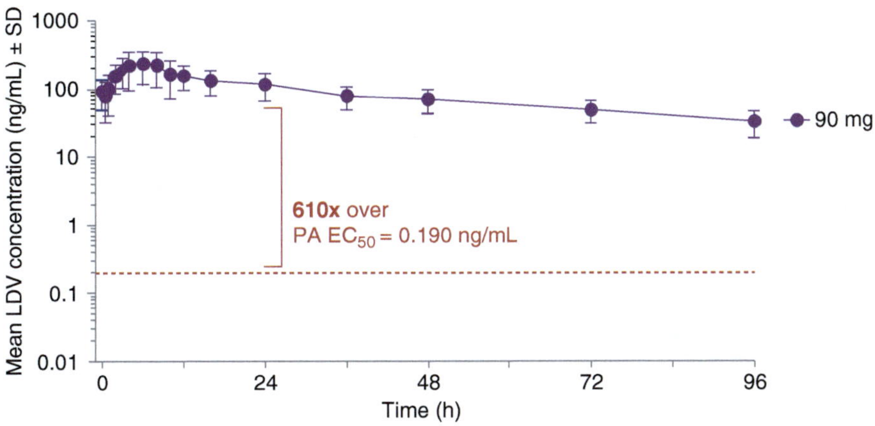

C_{max} (nM)	$t_{1/2}$ (h)	C_{tau} (ng/mL)	C_{tau} (nM)
247.7 (45.4)	49.7 (37.8, 54.3)	115.9 (42.6)	130.4 (42.6)

Data included are mean (CV%), except for $t_{1/2}$ which are median (Q1, Q3).

Figure 8.9 LDV PK curve and parameters in 10 PLWHCV post the third (final) 90 mg dose are consistent with once-daily dosing. LDV has a long 49.7 h half-life with a concentration 610-fold over the protein-adjusted potency 24 h after the third dose. Day 3 dose was administered at time zero. Error bars are standard deviations around the mean.

observed for GT1b infected individuals at the 10 mg dose (Figure 8.10, three doses shown for clarity). HCV NS5A inhibitors display the most rapid VLR of all HCV antiviral drug modalities [32]. Also shown in Figure 8.10, viral rebound (increase in viral load after dosing) initiated slowly at the end of the dosing period, with GT1b-infected subjects showing a slower rebound, reflecting the higher potency of LDV both against GT1b WT and RAS, along with the less diverse landscape of GT1b RAS (GT1b 10 mg dose mean maximal VLR = 3.3 $\log_{10}$). The potency of LDV against GT1 RAS can be found in Table 8.8 (*vide infra*); in the pangenotypic NS5A inhibitor velpatasvir discovery program we achieved improved activity against RAS. The median maximal VLR for GT1a cohorts was (dose, VLR $\log_{10}$ HCV RNA IU/mL): 1 mg, 2.3; 3 mg, 3.1; 10 mg, 3.2; 30 mg, 3.3; 90 mg 3.1. By happenstance, the absence of GT1a RAS in the 1 and 3 mg cohorts along with the presence of RAS of decreasing susceptibility in the higher dose groups (from 30 to 90 mg, Figure 8.10) somewhat obscured the LDV dose response (the GT1a doses at 3 and 90 mg doses shown in Figure 8.10) [33]. This understanding, along with early clinical data demonstrating superior efficacy of 90 over 30 mg LDV doses (combination studies with GS-9190 HCV polymerase, GS-9256 HCV PI, and RBV) along with good safety results in the higher-dose cohorts prompted the selection of the 90 mg LDV dose for subsequent combination with the 400 mg SOF dose in Phase 3 studies [34]. Indeed the properties of both LDV and SOF met the stringent criteria required for incorporation into an STR [20]. All LDV/SOF studies were accomplished with an SOF 400 mg/LDV 90 mg STR.

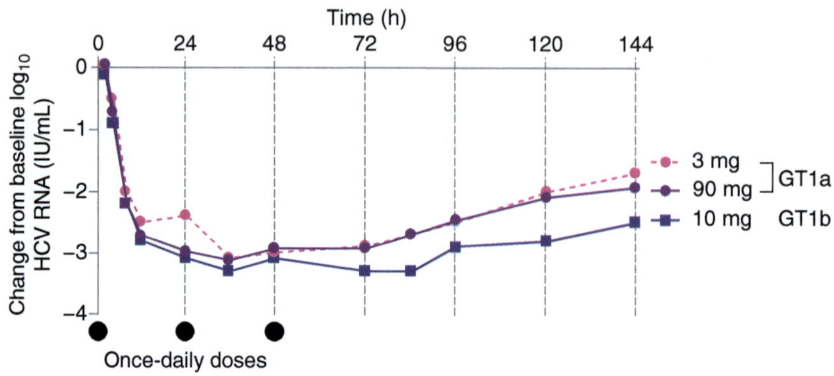

	GT1a			GT1b
LDV dose[a]	10 mg	30 mg	90 mg	10 mg
# of subjects with baseline RAS	1	2	1	2
RAS composition[b]	12% Y93C	Q30E/Q K24K/R	L31M	L31M 1.3% Y93H
Mean maximal VLR ($\log_{10}$)	3.2	3.3	3.1	3.3

[a]Ten participants in each cohort.
[b]Where percentage composition is given, measurement is by deep sequencing; other results by population sequencing (detectable when > 25% of total population). EC_{50} values for GT1a and GT1b RAS provided *vide infra* **Table 8**.

Figure 8.10 Ledipasvir VLR curves for 3 and 90 mg (GT1a) and 10 mg (GT1b) doses. Rapid and persistent antiviral activity is observed. Parameters including mean maximum VLR and RAS composition for individuals with baseline RAS in 3-day LDV dosing in monotherapy, GT1a (10, 30, 90 mg) and GT1b (10 mg).

8.5 Ledipasvir/Sofosbuvir STR Clinical Trial Results and Real-world Effectiveness

The ION registrational clinical trials in HCV GT1-infected individuals were undertaken (Table 8.2) with once-daily LDV (90 mg)/SOF (400 mg), formulated as an STR, and included some arms with twice-daily, weight-based dosages of RBV. ION-1 enrolled a treatment naive (TN) population with 16% of subjects having compensated cirrhosis treated for 12 or 24 weeks with or without RBV. SVR was high at 97–99% with the 12-week/no RBV arm affording a 99% SVR and demonstrating no benefit for longer treatment or inclusion of RBV in this population. Eight hundred sixty-five PLWHCV were randomized and treated, with 3.65% discontinuing treatment and only three participants experiencing virologic failure (0.35%, one subject with breakthrough and two with relapse). In this ITT protocol, discontinuations and those lost to follow-up (LTFU) are counted as treatment failures in the calculation of SVR%. ION-2 enrolled a treatment-experienced (TE) group (IFN/RBV ± PI), including 20% with compensated cirrhosis, and had a 12/24 week ±RBV design analogous to ION-1. SVR was high across all groups, and adding RBV or extension to 24 weeks did not significantly enhance SVR. ION-3 assessed shortened therapy in a non-cirrhotic population, with 8-week LDV/SOF ± RBV compared to 12-week LDV/SOF.

Table 8.2 Real-world results for SOF/LDV are similar to controlled clinical trials and stand in contrast to IFN-based therapy for people with GT1 HCV (see Figure 8.2).

	Study/regimen	Study participant characteristics	SVR%	(n/N)
Clinical trial efficacy	**Gilead-sponsored Phase 3**			
	ION-1; SOF/LDV 12 weeks	100% TN, 16% cirrhosis	**99**	(211/214)
	ION-2; SOF/LDV 12 weeks	100% TE, 20% cirrhosis	**94**	(102/109)
	ION-2; SOF/LDV + RBV 12 weeks	100% TE, 20% cirrhosis	**96**	(107/111)
	ION-2; SOF/LDV + 24 weeks	100% TE, 20% cirrhosis	**99**	(108/109)
	ION-3; SOF/LDV 8 weeks	100% TN	**94**	(202/215)
	ION-4; SOF/LDV 12 weeks	100% HIV/HCV coinfected; 55% TE, 20% cirrhosis	**96**	(322/335)[a]
Real-world effectiveness	**HCV-Target (US)**			
	SOF/LDV 8 weeks	HCV RNA < 6MM IU/mL, 4% TE,1% cirrhosis[b]	**96**	(271/282)
	SOF/LDV 12 weeks	36% TE, 27% cirrhosis	**97**	(881/910)
	SOF/LDV 24 weeks	90% TE, 76% cirrhosis	**95**	(486/510)
	SOF/LDV + RBV 12 weeks	59% TE, 52% cirrhosis	**97**	(206/212)
	SOF/LDV + RBV 24 weeks	91% TE, 74% cirrhosis	**95**	(77/81)
	Deutsches Hepatitis C Register			
	SOF/LDV 8 weeks	Overall (2.5% cirrhosis[b])	**98.3**	(821/835)
	SOF/LDV 12 weeks	Overall	**98.1**	(1208/1231)
		Cirrhosis	**94.3**	(165/175)
	Korean real-world study[c]	Overall	**98.1**	(202/206)
		HCV RNA < 6MM IU/mL, 8 or 12 weeks	**100**	(55/55)
		Compensated cirrhosis	**96.4**	(53/55)
		Decompensated cirrhosis	**100**	(6/6)
		Post liver transplant	**88.9**	(8/9)

TE, treatment experienced (typically PEG/RBV + PI); TN, treatment naive.

a. Included eight individuals with GT4 infection.

b. Not recommended per protocol.

c. Therapy included RBV in groups, such as treatment-experienced with compensated cirrhosis, decompensated cirrhosis, or post liver transplant.

SVR was high from 93% to 95% with LDV/SOF + RBV having the numerically lowest SVR. Accordingly, shortened therapy to 8 weeks was not found to be inferior to longer therapy or the addition of RBV.

Ending the IFN era, LDV/SOF STR (Harvoni®) heralded a medical revolution as the first approved all-DAA cure for GT1 PLWHCV, the most prevalent GT worldwide. Harvoni® was approved on 10 October 2014, in the United States and the European Union the following month, based on the ION studies. With therapy in as few as 8 weeks there is still no HCV therapy shorter than LDV/SOF. It provided a safe, simple, and effective cure with exceedingly high SVR. ION 1–3 participants from populations that had historically experienced lower SVR in IFN-based therapies were not significantly different from the general population of enrollees. Subsequently, ION-4 enrolled HIV/HCV coinfected people, 55% TE and 20% with cirrhosis and resulted in a 96% SVR, showing that individuals previously contraindicated for IFN-based therapy due to poor cure rates now attained results similar to other groups. The restrictive eligibility requirements and hesitance to initiate therapy that posed significant barriers to start IFN-based therapy (Figure 8.2) are not present for LDV/SOF. Also, notably absent were the complex monitoring and stopping criteria of IFN-based regimens. LDV/SOF was safe and well-tolerated in these studies, although there was no placebo-comparator arm. Headache, nausea, and fatigue were the most common adverse events and were more common in the RBV-containing arms. Less than 5% of participants reported grade-3 or −4 events. Please see publications detailing the ELECTRON clinical trials referenced here for information regarding the safety results for LDV/SOF ± RBV. To quantify the differences between an IFN regimen and LDV/SOF, a functional assessment of chronic illness therapy-fatigue score (FACET-F) and fatigue subscale (FACET-FS) were developed. For PEG + SOF + RBV 12 weeks, individuals feel worse during therapy, and then return nearly to baseline 4 weeks after the end of treatment (EOT), and only show the improvements of cure with a score above baseline 12 weeks after EOT. On LDV/SOF, individuals feel measurably better from the first weeks of treatment and improvement continues through EOT and weeks 4 and 12 after therapy [35].

Real-world effectiveness databases have reported results subsequent to the LDV/SOF approval. HCV-Target is a large real-world US cohort. Results in Table 8.2 show that 8 weeks of LDV/SOF in a largely cirrhosis-free population and meeting the US guidance of baseline HCV RNA <6MM IU/mL achieved 96% SVR. Other cohorts with varying percentages of TE and individuals with cirrhosis were treated with or without RBV for 12 or 24 weeks and achieved 95–97% SVR [36]. Further in Table 8.2, the German DHC-R [37] provides results from a large registry, with 98% of individuals achieving SVR with either 8 or 12 weeks of LDV/SOF therapy. Assessing a subgroup of 175 individuals with cirrhosis afforded 94.3% SVR. While these results, like most real-world studies, read out data with per-protocol analysis (computing SVR only including individuals who completed treatment), the ITT analysis also affords high SVR (94% ITT for 8 weeks, 94.9 ITT for 12 weeks). A recent report of a retrospective analysis of real-world results from Korea is also shown in Table 8.2 with breakouts of some of the populations that historically proved difficult to treat, such as individuals with decompensated cirrhosis (100% SVR, six individuals) or who are post-liver transplant (88.9% SVR, nine individuals) [38]. Adverse events (AEs) were considered mild and occurred in 12% of the population (including some attributed to RBV), with only two individuals (0.7%) discontinuing due to AEs and neither discontinuation was attributed to LDV/SOF. The comparable SVR between real-world and clinical trial results, along with the safety and simplicity of LDV/SOF therapy, characterizes the DAA revolution for PLWHCV and stands in clear contrast to the experience of those seeking treatment in the IFN era (Figure 8.2).

8.6 Initiation of a Pangenotypic NS5A Inhibitor Program

Directly after the discovery of ledipasvir and early work to progress it to clinical studies, research efforts were initiated toward the discovery of a pan-GT NS5A inhibitor to provide a cure for PLWHCV regardless of genotype. As we transitioned from targeting GT1 to an expanded GT set, we developed replicon systems to expand our GT, RAS, and subtype set. Utilizing public and in-house patient isolate data we analyzed sequences for variants and their prevalence. Based on these analyses, over 60 viable replicons were investigated although some variants did not show significant resistance shifts to our NS5A inhibitors. As a model of the HCV diversity in human infection, these replicons proved useful to help guide our inhibitor structure–activity relationship (SAR) discoveries. As our chemical matter evolved, so did our understanding of the breadth of potency our molecules possessed against the range of HCV variants in this evolving replicon panel. We had early cores, favored for their contributions to good PK profiles and GT1 potency. As new replicons became available for screening, these favored cores were abandoned as we discovered that they did not have sufficient potential for potent inhibition of some GT2 variants. Accordingly, combining potent pangenotype activity with PK properties viable for an STR presented significant challenges, which we ultimately overcame through the discovery of velpatasvir.

Positions in the NS5A sequence where RASs are common are 28, 30, 31, 32, 58, 92, and 93 among others. In GT2a and 2b as in GT1a, L31 is WT, with L31M being a RAS, but in GT2a and GT2b L31M is nearly as frequent as WT. As an aside, GT4 has M31 as WT and proved less difficult to inhibit than when this residue was present in GT1 or GT2. This GT2 M31 eliminated our early cores. Although it was difficult to abandon early cores associated with favorable PK properties, the focus on pangenotypic potency was critical for success. The importance of this approach was further validated, after our discovery of velpatasvir; another group disclosed that the prevalence of GT2 M31 RAS in their clinical trial population rendered half of the GT2 cohort highly resistant to their NS5A pangenotypic candidate that otherwise demonstrated good GT1, 3, and 4 VLR [39].

We sought cores that provided a strong base of potent activity across genotypes and RAS which we could build upon, along with contributions to the metabolic stability and PK properties that we ultimately desired in a pan-GT NS5A molecule for an STR. As we discovered molecules that had broader genotype and subtype activity, we discovered parallel improvements against RAS in various genotypes. There are varying HCV genotype distributions worldwide, as well as variations in resources and infrastructure. Prior therapies had varying levels of treatment population selection criteria, requiring HCV genotyping, baseline RAS assessment, information on treatment history, and fibrosis staging, resulting in contraindications, additions of agents, or adjustment to the lengths of treatment. We sought simplicity in the pangenotypic STR, where dose and duration of therapy were constant across genotypes and baseline characteristics, with SVR sufficiently high that only a simpler qualitative HCV RNA test could be utilized without the need for genotyping before treatment and regardless of fibrosis stage, along with an excellent safety profile and minimal drug interactions to place few restrictions on who could be cured and to allow for minimal testing before and minimal monitoring during therapy.

8.7 Discovery of the Pangenotypic NS5A Inhibitor Velpatasvir [40–42]

The theme of fused-ring system cores pursued in discovering LDV continued to prove important in the discovery of pangenotype NS5A inhibitors. During our NS5A programs, we designed and synthesized numerous cores; many of these novel ring systems possessed multiple fused-ring systems in varying configurations and their chemical syntheses became increasingly challenging. Detailed work compared focused changes on these ring systems and Table 8.3 provides examples of a series of fused tetracycle-based core inhibitors. We standardized on pyrrolidine-valine-methoxycarbonyl end-groups to establish the contribution of the core to broad genotype activity. Genotype 2a J6 (derived from a patient isolate with baseline M31), and a GT2b patient isolate strain (also M31) initially afforded some of the more challenging replicons to inhibit [40]. As direct patient isolates and based on the prevalence of methionine at position 31 in GT2, we took significant guidance from results in these replicons. The overall trend for broad genotype activity improved in the series from tetracyclic inhibitors **19–21**, where oxygen atoms replaced methylenes and isomers were assessed. The subtlety of this approach is apparent by comparing **20** and **21**, where a positional isomer of one ether oxygen affords improved potency against GT1a WT, and 2a L31, and M31. With **21** having improved potency, we studied a wide range of Moc-amino-acid capping groups. This approach afforded varying potency patterns across genotypes relative to Moc-valine. The benefits of Moc-D-phenylglycine capping both sides in inhibitor **22** are shown in Table 8.3. While affording some loss in potency against GT1 activity, utilizing D-phenylglycine (D-Phg) afforded broad improvements in GT2 potency over the valine-based inhibitor **21**. Based on this and other studies, we found that the total molecule is limited by the contribution of the core, and capping groups can add further benefit. Unfortunately, for the tetracyclic series, the oxidative metabolic stability could not be improved with inhibitors typically in the range of 0.18–0.39 L/h/kg. Based on these Pred E_h values and on their moderate Vss, this series did not provide sufficient potential for once-daily dosing.

Table 8.3 Fused cores can improve broad genotype activity. D-phenylglycine in **22** also affords improvement across GTs.

	GT EC_{50} (pM)										$PredE_h$
	1a	2a JFH1 (L31)	2a JFH1 (L31M)	2a J6 (M31)	2b (M31)	3a	4a	5a	6a	6e	HLM
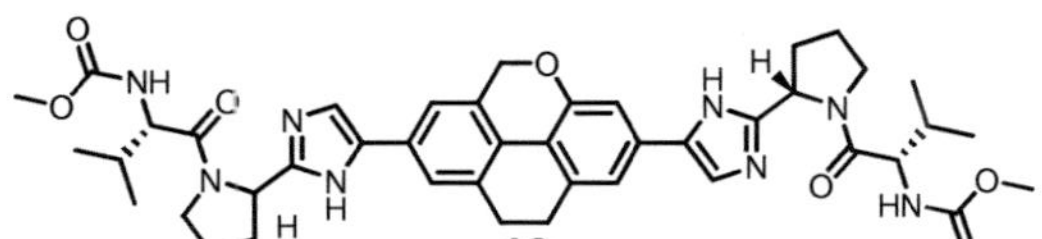 **19**	65	16	-	23,000	-	-	-	-	-	-	33%
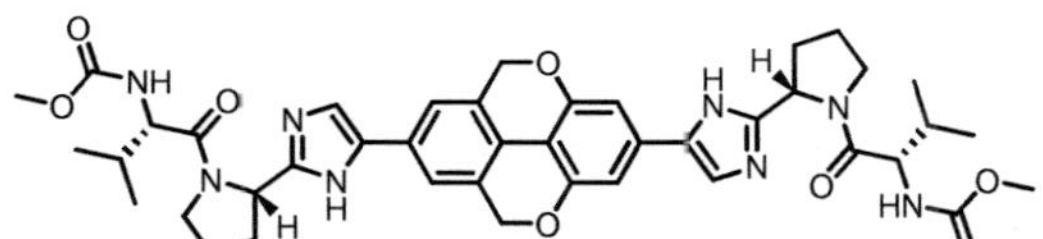 **20**	87	19	9500	-	-	-	-	-	-	-	-
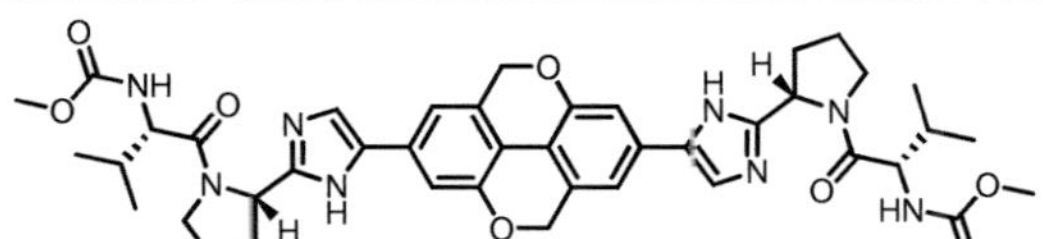 **21**	29	7	5900	8810	1770	35	19	56	16	1160	15%
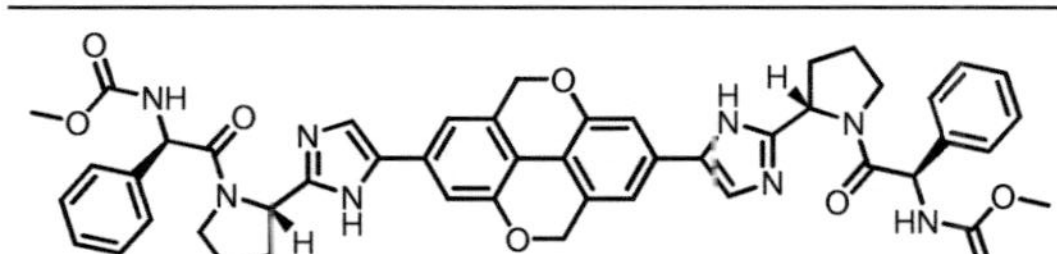 **22**	93	7	145	240	22	-	-	-	-	-	16%

The replicons here and afterwards, unless otherwise noted are GT1a (strain H77); GT1b Con-1; GT2a JFH1; GT2a JFH1 (L31M) subgenomic transient replicon cells with L31M mutation (RAS); GT2a J6 (M31) (note this patient isolate has M as baseline sequence at position 31); GT2b (M31) MD2b-1 NS5A (note, this patient isolate has M as baseline sequence at position 31); GT3a S52 NS5A transient chimeric replicon based on GT1b Renilla luciferase (Rluc) backbone; GT4a ED43 NS5A transient chimeric replicons based on GT1b Rluc backbone; GT5a SA13 NS5A (9-184) transient chimeric replicons based on GT1b Rluc backbone; GT6a HK6 stable subgenomic HCV replicon; GT6e D88 NS5A (9-184) transient chimeric replicons based on GT1b Rluc backbone.
GT1a, GT1b, and GT2a JFH1 are stable subgenomic replicon cells.
Adapted from based on GT2a JFH1 Rluc backbone. For these replicons see: GS-5816, a Second-Generation HCV NS5A Inhibitor With Potent Antiviral Activity, Broad Genotypic Coverage, and a High Resistance Barrier, Cheng, G.; Yu, M.; Peng, B.; Lee, Y.-J.; Trejo-Martin, A.; Gong, R.; Bush, C.; Worth, A.; Nash, M.; Chan, K.; Yang, H.; Beran, R.; Tian, Y.; Perry, J.; Taylor, J.; Yang, C.; Paulson, M.; Delaney, W.; Link, J. O. J. Hepatol. 2013, 58(suppl):S484. https://www.natap.org/2013/EASL/EASL_34.htm (Last accessed February 8, 2024)
For these replicons, see Ref. [48].

Table 8.4 Striking improvement in broad genotype potency and metabolic stability with the discovery of the pentacyclic core in **24**, and further improvements from methoxymethylpyrrolidine **25**, including GT1a RAS (footnote below).

	GT EC_{50} (pM)						$PredE_h$	Fraction abs. (F_A%)		
	1a	2a	2a J6 (M31)	2b (M31)	3a	4a	HLM	Rat	Dog	Cyno
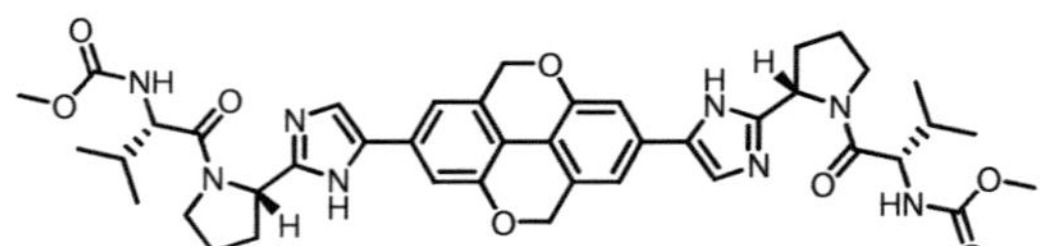 **21**	29	7	8810	1770	35	19	15%	18	45	11
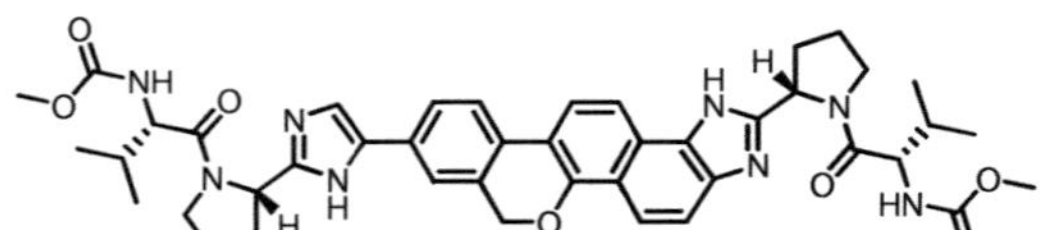 **23**	41	16	20 861	32 349	12	130	18%	31	54	-
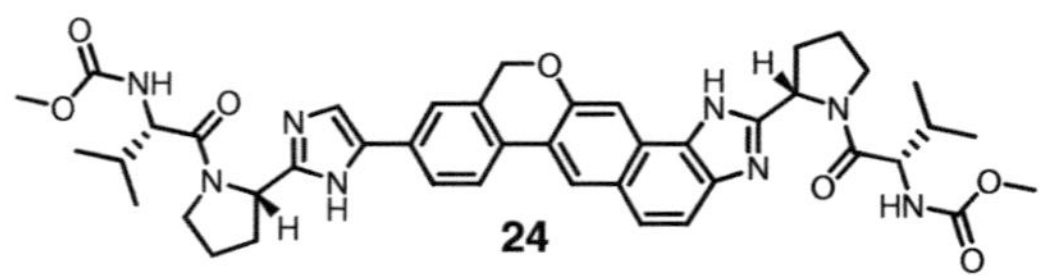 **24**	16	4	1155	162	4	8	<13%	15	50	-
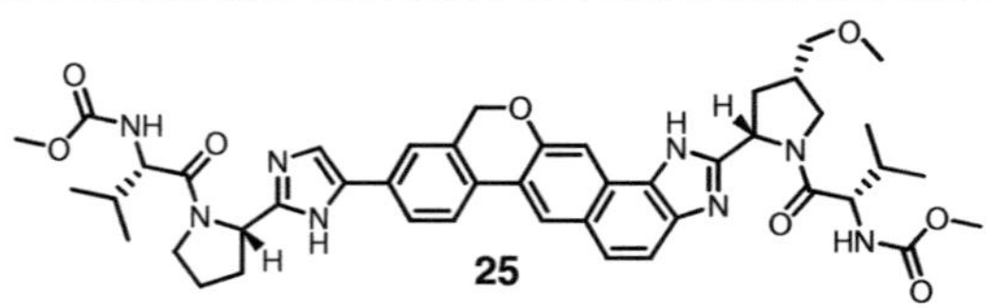 **25**	[a]21	8	280	135	24	14		16	56	-

a. Compound **25** has low to subnanomolar potency against GT1a RAS (GT1a RAS [EC50 nM]): M28T (0.19), Q30H (0.15), Q30R (0.22), L31M (0.37), Y83C (0.22), Q30E (10.4), Y93H (4.7). Improvements are over LDV GT1a RAS activity as shown in Table 8.8 *vide infra*.

Efforts on further ring fusions continued, and Table 8.4 details a fundamental breakthrough for the program. Two fused pentacyclic-ring cores were conceptualized and their complex syntheses were undertaken. Efforts to synthesize the pentacycle in **23** were completed first, and while hopes for this scaffold were high, it underperformed relative to tetracycle **21** against key GT2 strains. The synthesis of **24** was still underway and there was some question if the significant synthetic challenge was worth the time investment based on results from **23**. Fortunately, the chemist synthesizing **24** was undaunted and prevailed. The pentacycle in inhibitor **24** contributed the best pangenotype activity among over 100 novel cores that we had synthesized [43, 44]. Here again an unsymmetric core proved to possess beneficial attributes, and these benefits became further apparent as we differentially substituted the ends of this core. The basis for the striking difference in activity between potent inhibitor **24** and its isomer **23** remains unexplained. In addition, **24** demonstrated metabolic stability at the lower limit of the HLM assay (<13% E_h) affording a further breakthrough relative to the tetracyclic series where metabolic stability consistent with QD (once-daily dosing) had not been achieved. The decreased fraction absorbed (F_a) in rat and dog for **24** relative to **21** (Table 8.4) was a portent of the forthcoming challenges in the pentacyclic core series. The narrowed chemical space that we discovered to provide access to broader genotypic potency presented PK challenges, e.g. insufficient metabolic stability in the tetracycle series and low solubility/permeability in the pentacycles.

In further work, extensive studies on pyrrolidine modifications had led to the discovery that the unique 4-methoxymethyl pyrrolidine substituent in concert with the pentacyclic core (**25**) afforded further broadened genotype potency (Table 8.4) and afforded low to subnanomolar potency against GT1a RAS (footnote Table 8.4). Utilizing this unique pyrrolidine with the pentacyclic core (**24**) afforded good potency across GT1-4 while maintaining stability in HLM. The contribution of methoxymethyl-pyrrolidine to PK properties will be discussed *vide infra*.

As we progressed to polycyclic fused-ring cores, the *in vitro* behavior of the compounds proved challenging. Low solubility in buffer media along with adherence to surfaces made it difficult to attain viable measurements in Caco2 assays, based on the high surface area in the assay wells. *In vitro* permeability assays were therefore not viable. We utilized *in vivo* PK studies to gauge permeability and used the percent fraction absorbed (F_a%) in preclinical species (particularly rat and cyno) to guide us to compounds with viable permeability in humans. Note that reported here is F_a%, where the contribution of *in vivo* CL has been removed to focus on permeability (rather than relying on F%, which is confounding as it incorporates both permeability and CL). The F_a% calculation also assumes the absence of gut metabolism [45]. Dosing was from a non-precipitating formulation to remove the contribution of solubility from the measurement. Regarding solubility, we could rely on the expertise of our formulations group to innovate, as with ledipasvir, where they discovered a spray-dried amorphous suspension formulation to assist with drug dissolution [46].

While synthesizing more pentacyclic core for further investigation, we studied the attributes of the D-Phg capping group in the context of the tetracyclic core. In the series **21–26** to **22** with sequential replacement of each valine by D-Phg, the GT2a J6, and GT2b activity each significantly improved (Tables 8.4 and 8.5). It was

Table 8.5 D-phenylglycine improves pan-genotype activity and shows improved potencies with pentacyclic core or 4-methylpyrrolidine but decreases F_a.

	GT EC_{50} (pM)						PredE[h]	Fraction abs. (F_A%)		
	1a	2a	2a J6 (M31)	2b (M31)	3a	4a	HLM	Rat	Dog	Cyno
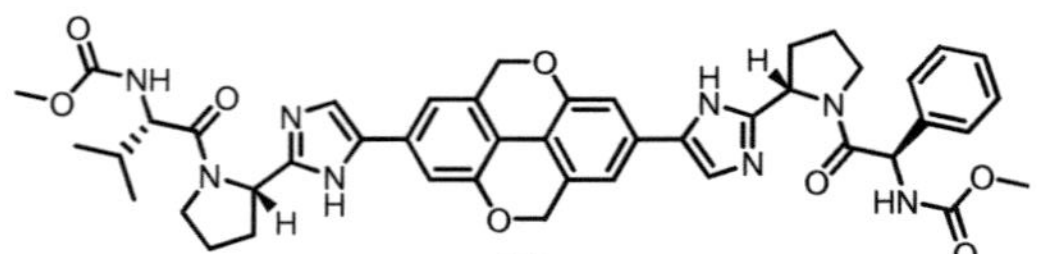 26	50	4	1820	76	-	-	20%	-	-	-
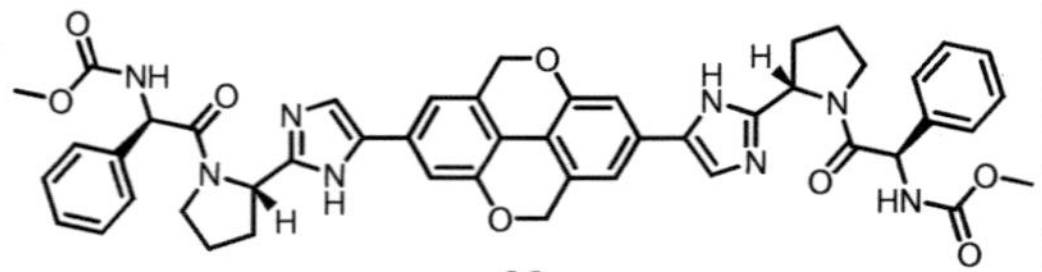 22	93	7	145	22	-	-	16%	2	34	2
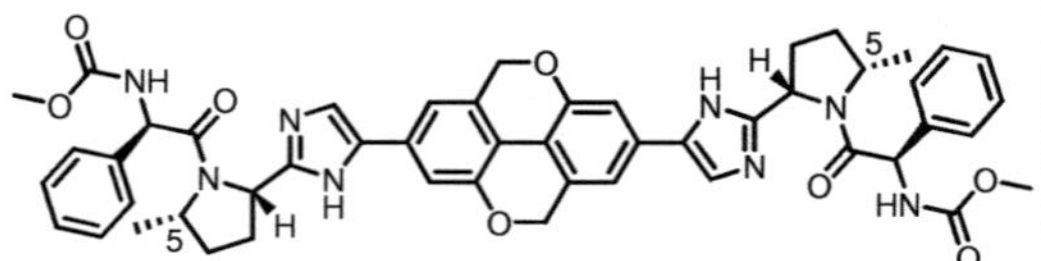 27	150	8	1835	94	4	15	<13%	3	43	-
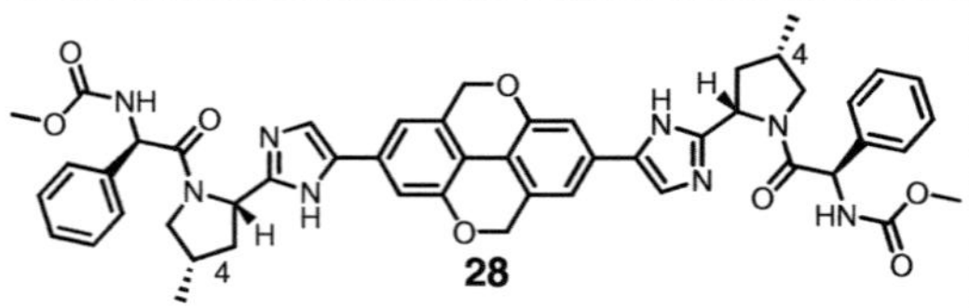 28	31	5	58	26	6	13		2	30	

	GT EC_{50} (pM)						PredE[h]	Fraction abs. (F_A%)		
	1a	2a	2a J6 (M31)	2b (M31)	3a	4a	HLM	Rat	Dog	Cyno
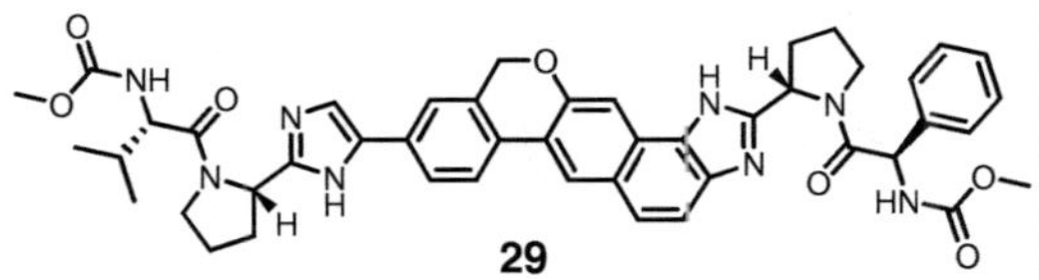 **29**	31	21	134	12	5	8		13	56	16
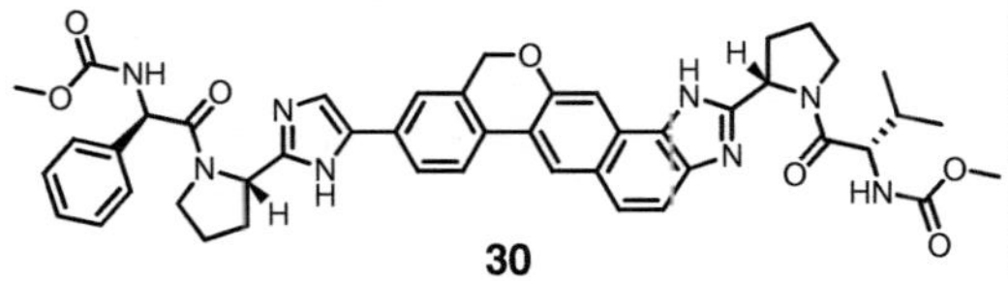 **30**	32	33	6	73	16	6	<13%	-	-	-
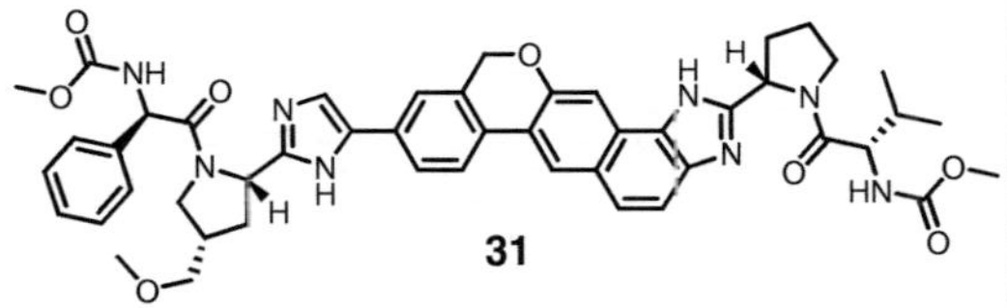 **31**	11	6	13	9	17	7[a]		12		3

a. For compound 31: GT4a ED43 stable subgenomic replicon.

nonetheless concerning that with two D-Phg groups in **22**, the F_a in both rat and cyno had dropped to 2% suggesting to us that incorporation of two D-Phg groups in a candidate molecule would be untenable. When studying a range of pyrrolidine substitutions and variants, the 5- and 4-methyl pyrrolidines had afforded potency and/or PK benefits in some cases. We sought to compare the potency attributes of the 4- and 5-methyl pyrrolidine in combination with D-Phg. Combination of the methylated pyrrolidines with D-Phg capping showed a "mismatch" with the 5-methyl pyrrolidine (**27** generally less potent than unsubstituted pyrrolidine **22**) and "match" with the 4-methyl (**28** generally more potent than **22**). The superiority of 4-methyl substitution served to inform pyrrolidine/D-Phg combination choices moving forward. Pentacyclic inhibitors with D-Phg (**29** and **30**; Table 8.5) exhibited improvements in GT2a J6 and GT2b relative to valine (**24**; Table 8.4). Importantly, the benefit of an unsymmetric core was recapitulated, where D-Phg on the imidazole side in inhibitor **30** showed a favorable potency pattern (relative to the opposing side D-Phg inhibitor **29**), with all GT1-4 replicons testing below 100 pM. While substitution with D-Phg improved the pangenotype activity, it again proved clearly detrimental to permeability. As noted, we considered compounds with two D-Phg groups untenable (**22**, **27**, **28** each with rat and cyno F_a at or below 3% when tested). Single D-Phg substitution in the pentacycle was also problematic, with **29** showing 13% and 16% Eh in rat and cyno, respectively. The challenges that D-Phg presented with absorption continued, with multiple avenues being attempted for improvement.

We continued to utilize our learnings on RAS and pan-GT activity to drive pan-GT potency further. The information that 4-methyl substitution with D-Phg is a "match" (**28**), and with the beneficial RAS and broad-genotype attributes that the 4-methoxymethyl pyrrolidine in the pentacyclic series demonstrated in compound **25**, we sought to combine the best pan-GT and RAS potency-driving elements. Since the D-Phg performed well on the imidazole side of the core in compound **30**, its combination with 4-methoxymethyl in **31** afforded a potent pangenotypic profile with all GT1-4 replicons tested below 15 pM EC_{50} (Table 8.5). Importantly, as with other pentacycles, inhibitor **31** had an E_h at the lower limit of the assay. While these attributes had achieved our potency and metabolic stability goals, the low F_a in rat and cyno (cynomolgus monkey) (12% and 3% F_a) was deeply concerning. With methoxymethyl-pyrrolidine pentacycle **31** we had an exceptional pangenotype activity profile and good metabolic stability, but in a compound that had low odds of acceptable clinical oral absorption.

D-Phg-pentacycle **31**, with such a low permeability, was too high-risk for progression to clinical studies; a path to drive good permeability in a compound with the excellent pan-GT potency and metabolic stability of **31** needed to be discovered. Multiple approaches were taken, and the fraction-absorbed values for various pyrrolidine modifications are shown in Table 8.6. Incorporation of 4-methylpyrrolidine on the pentacycle-side in **32** (Table 8.6) maintained excellent potency (not shown) but the F_a was low at 9% and 4% in rat and cyno, respectively. Employing fused cyclopropylpyrrolidine in **33**, which can be thought of as intermediate in structure between 4- and 5-methyl substitution, again had a low F in rat of 11%. In a striking result, modification with the 5-methyl pyrrolidine on the pentacycle side both provided the overall best potency profile and dramatically improved F_a to 36% in rat

Table 8.6 Solely velpatasvir (**3**) combines excellent pan-GT activity and good absorption in rat/dog/cyno.

	Fraction abs. (FA%)		
	Rat	**Dog**	**Cyno**
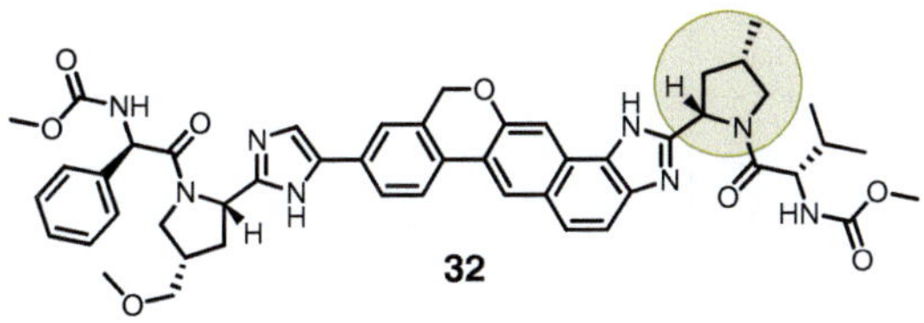 **32**	9	-	4
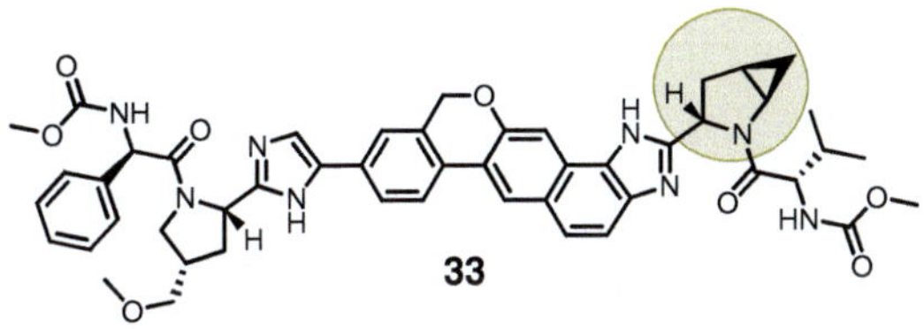 **33**	11	13	
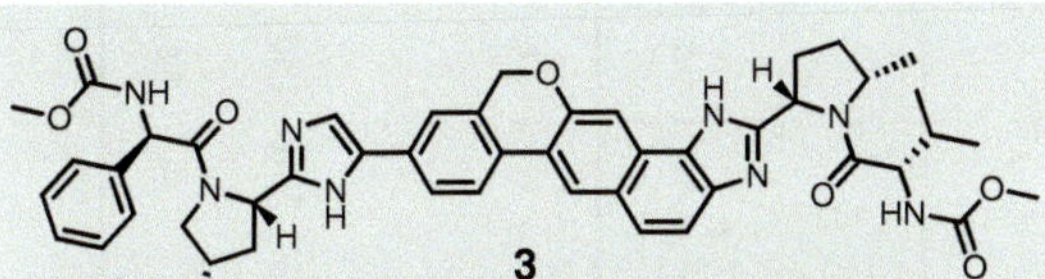 **3** **velpatasvir**	36	29	37
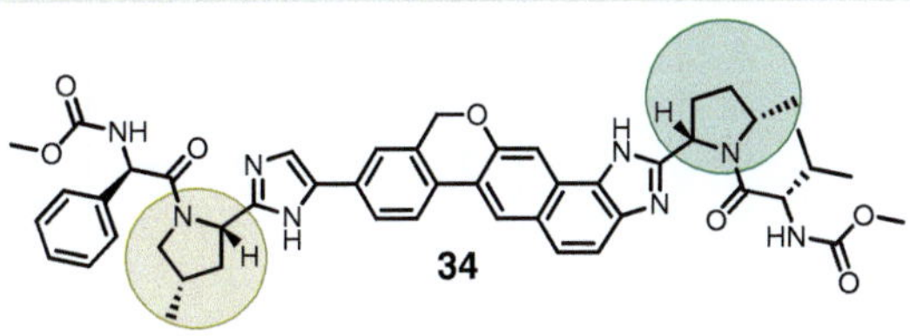 **34**	4	-	-
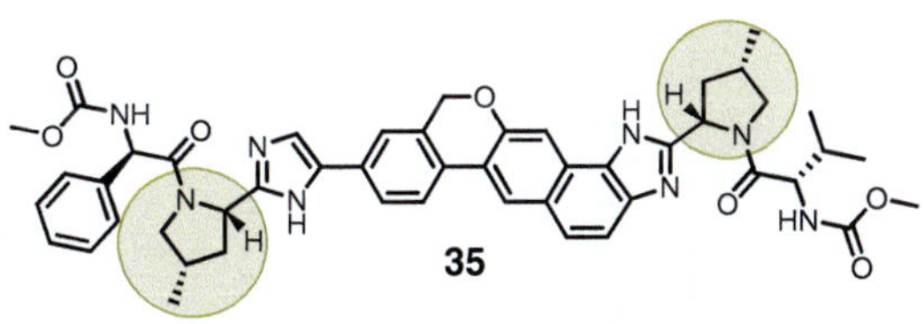 **35**	4	-	-

All compounds in this table are stable at the lower limit of the HLM assay (< 13% Eh). Compounds **32**, **33**, **3** all have GT1a, 2a (M31), 2b, 3a potency <30 pM.

and 37% in cyno (potency follows in Table 8.7). The combination of low picomolar pangenotype potency, good F_a%, and HLM stability of <0.16 L/h/kg afforded an outstanding profile and **3**, velpatasvir, was selected to undergo further characterization for progression to clinical candidacy.

To further understand the unique combination of the elements in VEL, we left the beneficial 5-methyl in place and swapped the 4-methoxymethyl for methyl

Table 8.7 Tetracyclic core with VEL end-groups (**36**) or interchanging the end-groups of VEL on pentacyclic core (**37**) afford inferior compounds. LDV and VEL potency across genotypes.

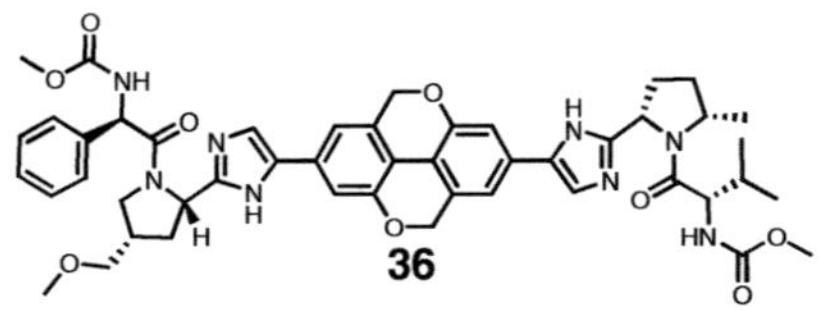

36

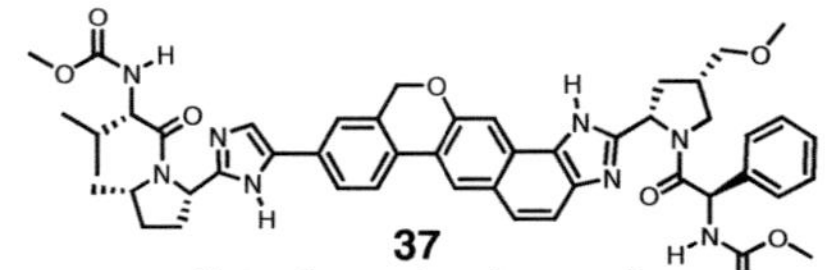

37
"interchanged end-groups"
cyno F_a = 13%

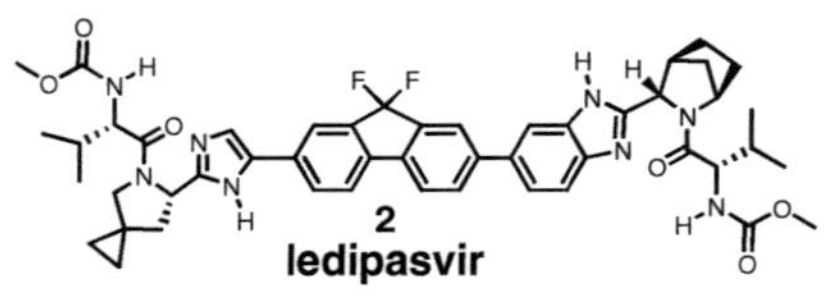

2
ledipasvir

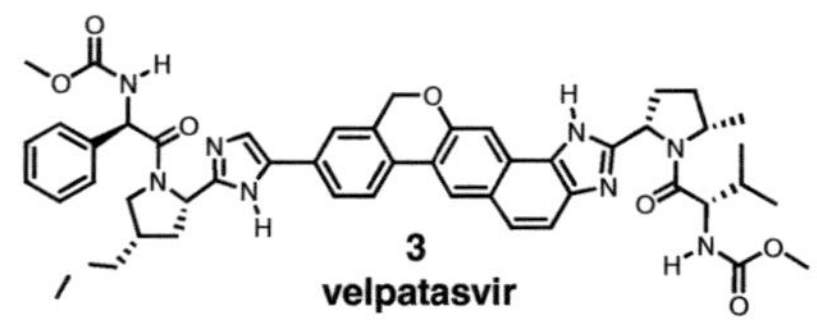

3
velpatasvir

	GT EC_{50} (pM)										PredCL (L/h/Kg)	E_h	Cyno PK
	1a	1b	2a (L31)	2a J6 (31 M)	2b (31 M)	3a	4a	5a	6a	6e	HLM or Hhep	HLM or Hhep	F_a%
36	16	16	-	480	290	120	13	81	22	110	0.20[a]	17%	26
37	-	-	-	-	-	-	-	-	-	-	-	-	13
ledipasvir (**2**)	31	4	21,000	249,000	530,000	168,000	390	150	1,100	264,000	0.012[b]	1%	47
velpatasvir (**3**)	14	16	8	16	6	4	9	54	6	130	0.060[b]	5%	37

GT3a S52 transiently transfected subgenomic HCV replicon.
GT4a ED43 stable subgenomic HCV replicon.
a. Measured in human liver microsomes (HLM).
b. Measured with tritiated LDV or VEL in human hepatocytes (Hhep).

(compound **34**), which eroded permeability to 4% F_a in rat. Further, the substitution of both pyrrolidines with 4-methyl again resulted in a F_a of 4% (compound **35**). To understand the role of the core, the pentacyclic core of VEL was replaced by the tetracycle-based core while retaining the end-groups of VEL (compound **36**, Table 8.7). Replacing VEL's core eroded potency against all genotypes except GT1b and GT6e, along with lower F_a% in cyno. "Flipping" the core of VEL (alternatively imagined as interchanging the end-groups of VEL) resulted in a low 13% F_a in cyno (compound **37**), again showcasing the benefits of differential end-group substitution on an unsymmetric core. No metric or rule-based approach to bioavailability could predict that addition, deletion, or positional changes of methyl or methoxy groups, or end-group swapping on a complex ~900 MW molecule would lead to a very specific and favorable combination as seen in velpatasvir and as witnessed through the related structures in Tables 8.5–8.7. These results demonstrate that the creativity to identify differentiating elements within a complex chemical structure, and the persistence to investigate small structural changes in rulebreaker drug space can lead to beneficial PK results.

The potency across genotypes 1–6 and various subtypes for VEL and LDV (Table 8.7) exemplifies the significant advancements made during our pangenotype NS5A efforts. The results in Table 8.8 contrasting VEL and LDV show how our efforts to improve pangenotype and subtype potency in VEL translated to parallel improvements against resistance in GT1a, as represented by the general picomolar potency of VEL against GT1a RAS [47, 48]. VEL also has very favorable picomolar to low nanomolar potency against common GT2 and GT3 RAS (Table 8.8).

8.8 The Development of Velpatasvir

Preclinical and clinical PK properties for VEL are shown in Figure 8.11. The PredE_h of VEL is low across species, with humans the lowest at 5% E_h. Such a low PredEh is not typically measured within a research setting. Pushing the limits of incubation times and higher concentrations of human hepatocytes yielded a lower limit of the assay to 13% E_h as seen herein. As with LDV, to measure the low PredE_h, tritiated VEL was incubated with human hepatocytes followed by highly accurate quantitation of metabolites. In preclinical species, there was a good *in vitro*/*in vivo* concordance for predicted and measured E_h. With volumes of distribution higher than body water (1.43–1.61 L/h/kg), good *in vivo* half-lives (from 2.25 to 5.2 h), moderate bioavailability (25–30%), and excellent metabolic stability with a PredE_h of 5% from HLM, VEL was predicted to have attributes consistent with once-daily dosing and viability for combination in an STR. VEL had no findings of concern in nonclinical safety studies and proceeded into Phase 1 clinical trials.

After such intensive SAR and PK work, it was exhilarating to find that VEL had PK consistent with once-daily dosing. VEL exhibited a long 15.2–16.2-h half-life (100 and 150 mg doses) and concentrations 24 h after dosing of 53 nM (single 100 mg dose) and 72 nM (150 mg, end of 14 doses) which were significantly above its average protein adjusted EC_{50} of 1 nM (average GT1-6). The PK curve for the 150 mg dose is shown in Figure 8.11 [49]. Oral bioavailability was 50% calculated based on the human oral CL/F and human predicted E_h from preclinical studies,

Table 8.8 Velpatasvir and ledipasvir *in vitro* potency against GT1-3 RAS.

	GT1a[a]							GT1b[a]	GT2a[b]			GT2b[b]		GT3a[c]	
	M28T	**Q30H**	**Q30R**	**L31M**	**Y93C**	**Q30E**	**Y93H**	**Y93H**	**K44R**	**N62V**	**N62S**	**R44K**	**P58S**	**A30K**	**P58S**
VEL EC_{50} (nM)	0.105	0.032	0.031	0.22	0.053	0.25	8.5	0.011	0.005	0.01	0.003	0.011	0.01	0.21	3.2
LDV EC_{50} (nM)	1.9	5.7	19.6	17	49.6	169	52.0	7.2							

a. GT1 RAS are transiently transfected in the GT1a or 1b subgenomic replicon.
b. GT2a and 2b RAS replicons encode full-length NS5A from GT2a clinical isolates.
c. GT3a RAS are from site-directed mutagenesis in the WT replicon.

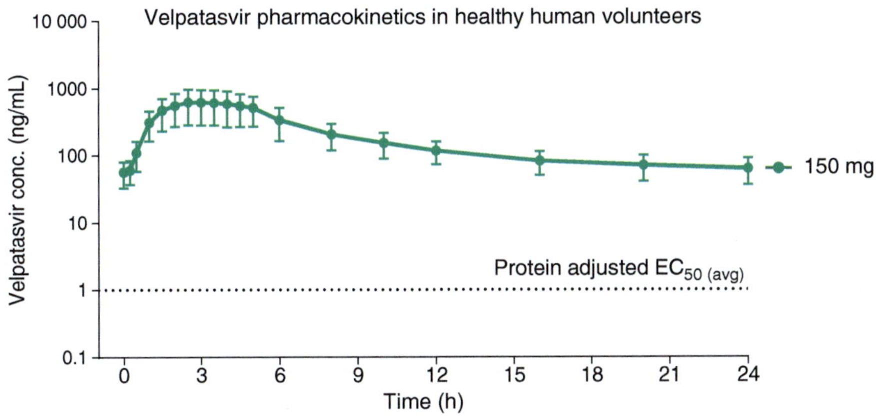

	in vitro	in vivo						
	PredE$_h$ LM	Dose/Route (mg/kg) or [b](mg)	E$_h$	Vss (L/kg)	$t_{1/2}$ (h)	F%	C_{24h} one dose (nM)	C_{tau} 14 doses (nM)
rat	18.5%	2 (IV); 2 (PO)	23.5%	1.61	2.25	28	-	-
dog	20%	0.25 (IV); 0.5 (PO)	13.9%	1.43	5.20	25	-	-
monkey	<10%	0.5 (IV); 1 (PO)	11.8%	1.60	5.03	30	-	-
human	5%[a]	100 mg (PO)	-	-	15.7	~50[e]	53	nd
human		150 mg (PO)	-	-	16.2,[c] 15.2[d]	-	48	72

[a] Measured using ^{1}H-VEL.
[b] Human-dose in milligrams.
[c] Half-life after first dose.
[d] Half-life after fourteenth dose.
[e] Calculated from human oral CL/F and human PredCL from preclinical data.

Figure 8.11 Velpatasvir fourteenth dose PK curve in healthy human subjects (150 mg once-daily). Velpatasvir *in vitro* and *in vivo* PK parameters in rat, dog, cyno monkey, and healthy human subjects. Velpatasvir is orally bioavailable and has PK consistent with once-daily dosing and inclusion in an STR.

which validated our preclinical approach whereby we relied on F_a values in rat and cyno to guide us for candidate selection.

VEL was the first HCV NS5A inhibitor tested in a Phase 1 clinical design to investigate VLR in individual cohorts of GT1a, 1b, 2, 3, and 4 infected individuals. VEL demonstrated rapid, potent, and persistent pangenotypic VLR in GT1-4-infected subjects with three once-daily monotherapy doses (Figure 8.12). These results validated our discovery program pan-GT potency testing and optimization paradigm in the discovery of VEL (VLR curves for GT2-4 shown for clarity, Figure 8.12). The mean maximum VLRs for VEL at the 150 mg dose ranged from 3.3 to 4.4 $\log_{10}$ (Figure 8.12). Vel's enhanced GT1a VLR relative to LDV (4.0 vs. 3.3 $\log_{10}$ IU/mL, 150 vs. 90 mg dose, respectively) is consistent with VEL's increased potency against GT1a RAS and also validated the use of RAS replicon testing to guide our discovery of VEL [50]. Figure 8.12 also shows the frequency and composition of pretreatment baseline NS5A RAS in these cohorts, with 12–100% of the subjects in the 150 mg cohorts having baseline RASs. Importantly, potent VLR for volunteers with RAS is shown in Figure 8.12. The presence of GT1a L31M, along with one individual

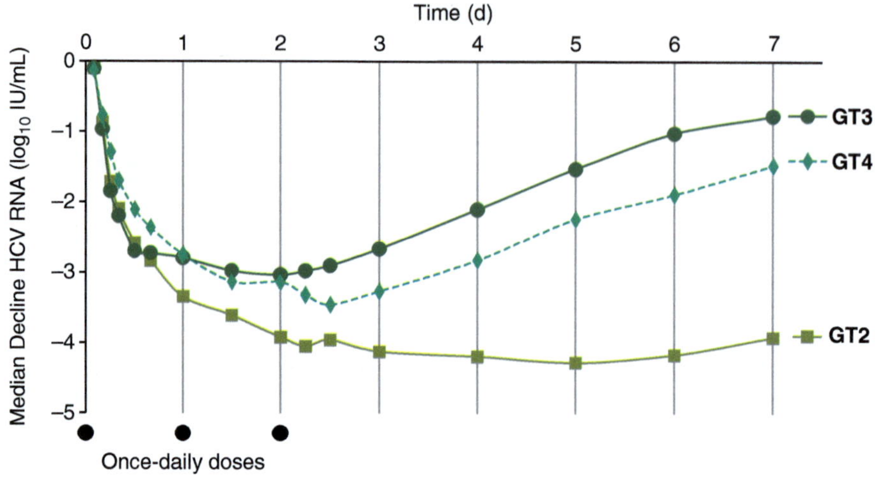

	GT1a		GT1b	GT2b				GT3		GT4	
Mean Max VLR for all volunteers, VEL 150 mg dose	4.0 (n = 7)		4.0 (n = 8)	4.4 (n = 8)				3.3 (n = 6)		3.5 (n = 2)	
Volunteers with baseline RAS(%), n = 11[a]	L31L/M	M28T(48) L31M(16) H58D(17)	Y93H (14)	L31M (12)	L31M (>99)	L31M (30)	L31M (>99)	Y93H (17)	A30K (>99)	L30H(97) L30S(2)	L30H(46) L30R(54)
Mean max, VLR for volunteers with RAS $\log_{10}$	2.9		4.5	4.1				2.7	2.9	3.5	
Mean max, VLR $\log_{10}$ in volunteers without RAS	4.4		4.4	4.6				3.5		NA	

[a]Some individuals have multiple RAS.

Figure 8.12 Viral load reduction curves with three once-daily 150 mg doses of velpatasvir monotherapy in GT2-4-infected volunteers. Mean maximum VLR ($\log_{10}$) with 150 mg VEL in volunteers with RAS at baseline comparing VLR for all volunteers, those with RAS, and those without RAS. Potent VLR is seen with or without RAS.

having three RAS making up the majority of their pool afforded potent VLR. GT2b L31M had little effect with a VLR of 4.1 $\log_{10}$ in four individuals, and high potency was seen in GT3 volunteers harboring Y93H (17%) or A30K (>99%) [51]. While the median viral load decline was rapid in all cohorts, the rate of rebound after dosing completion varied. As seen in Figure 8.12, the 150 mg GT2b cohort VL continued to decline through day 5 and slowly started to rebound to $-4 \log_{10}$ on day 7. In our discovery program we had access to multiple GT2 subtype and RAS replicons, and VEL very potently inhibits the virus in GT2-infected individuals. The GT3 cohort contrasts with a potent yet less pronounced VLR and earlier rebound phase; it is possible that there is a greater prevalence of viral resistance variants in this genotype. Nonetheless, it is worth noting that the GT3 VLR observed here during therapy with VEL is comparable to the VLR produced through LDV treatment of GT1a-infected volunteers. Finally, when assessing individuals without identified RAS, the mean maximum VLR ranged from 4.4 to 4.6 $\log_{10}$ IU/mL for GT1, 2, and 4 while GT3 = 3.5 $\log_{10}$ IU/mL. Overall, the long half-life consistent with once-daily dosing and the potent pangenotypic antiviral activity in PLWHCV validated our preclinical discovery approach and made VEL our candidate-of-choice for combination with sofosbuvir in a pangenotypic STR.

8.9 Clinical Trial Efficacy and Real-world Effectiveness with the Pangenotypic Sofosbuvir/Velpatasvir STR

A 12-week therapy with the SOF/VEL (400 mg/100 mg) STR was initiated in GT1-6 PLWHCV in the multicenter placebo-controlled ASTRAL 1–5 trials as outlined in Table 8.9 [52–56]. Broad enrollment criteria in ASTRAL 1–3 and 5 included treatment naive or experienced and individuals with or without cirrhosis. ASTRAL-4 enrolled individuals with decompensated cirrhosis while ASTRAL-5 enrolled GT1-4 HIV/HCV coinfected individuals. Both groups were contraindicated for therapy in the IFN era and those with decompensated cirrhosis had transplant as their only option. In the DAA era, those with compensated cirrhosis have limited options as HCV PI-containing regimens have a warning for use in those with advanced liver disease due to the risk of hepatotoxicity and those with decompensated cirrhosis are contraindicated for PI use. ASTRAL-1 enrolled GT1, 2, 4–6 while ASTRAL-2 and ASTRAL-3 focused on GT2 and GT3 respectively. As seen in Table 8.8 SVR rates were high with GT1, 2, 4–6, achieving 99% SVR with GT3 and HIV/HCV coinfected GT1-4 subjects each achieving 95% SVR. Those with decompensated cirrhosis achieved 83% SVR; with added RBV the SVR rose to 94% [55]. Baseline RAS were observed by deep sequencing. In ASTRAL-1, 276/616 (42%) had RAS and 257/259 achieved SVR. In ASTRAL-2, 60% of the 134 had RAS, with 52% as L31M and all achieved SVR. In ASTRAL-3, 43/274 (16%) had RAS (A30K, L31M, Y93H); 38/43 achieved SVR (88%) and 21/25 with Y93H had SVR (88%); of the 231 without RAS, 97% had SVR. In ASTRAL-4 72/255 (28%) had RAS with SVR achieved in 64/72 (89%) versus SVR in 169/183 (92%) with no RAS. A total of 100% of GT1-infected individuals treated with SOF/VEL + RBV had SVR while 98% with no RAS had SVR. For SOF/VEL 12 weeks 80% achieved SVR with RAS versus 96% SVR without RAS. With 24 weeks SOF/VEL 90% achieved SVR with RAS versus 98% without RAS.

The results from the ASTRAL trials significantly simplified therapy. SOF/VEL was generally well-tolerated – and further – ASTRAL-1 had a placebo group (n = 116). Please see the publications detailing the ASTRAL clinical trials for information regarding the safety results for SOF/VEL ± RBV. Patient-reported outcome (PRO) questionnaires showed a clear divergence between the treated and placebo group [57]. For the treated group, PROs revealed improved general health and emotional well-being starting at week 4 of treatment as their HCV fatigue abated during therapy and the gains continued for those with SVR 12 and 24 weeks after cure, while the placebo group remained unchanged. Improvements during therapy stand in contrast to the patient experience during the IFN era, where those treated had difficulty completing therapy due to the adverse effects precipitated by IFN/RBV. SOF/VEL 12-week therapy received FDA approval as Epclusa on 28 June 2016, for treatment naive or experienced individuals with GT1-6 HCV infection without cirrhosis or with compensated cirrhosis. For those with decompensated cirrhosis, RBV is added. Subsequently after the ASTRAL-5 studies, SOF/VEL was also approved for HIV/HCV coinfected individuals. SOF/VEL is a pangenotypic STR.

The challenge prior to the introduction to LDV/SOF had been achieving high SVR rates in real-world circumstances as shown in Figure 8.2. Taking real-world cures to all genotypes remained an important step. SOF/VEL has high SVR rates, a good

Table 8.9 Clinical trial efficacy or real-world effectiveness for 12-week SOF/VEL therapy.

	Study	Study participant characteristics		SVR%	(n/N)
Clinical trial efficacy	**ASTRAL-1**	32% TE, 19% cirrhosis	GT1,2,4,5,6 overall	**99**	(618/624)
	ASTRAL-2	19% TE, 19% cirrhosis	GT2	**99**	(133/134)
	ASTRAL-3	26% TE, 29% cirrhosis	GT3	**95**	(264/277)
	ASTRAL-4	64% TE, 100% decompensated cirrhosis	GT1–4	**83**	(75/90)
	ASTRAL-4 (+ RBV)	54% TE, 100% decompensated cirrhosis	GT1–4	**94**	(82/87)
	ASTRAL-5	29% TE, 18% cirrhosis, 100% HIV/HCV-coinfected	GT1–4	**95**	(101/106)
Real-world effectiveness	**GECCO**	24% TE (including DAA), 26% cirrhosis, 9% HIV/HCV-coinfected, 19% + RBV	GT3	**99.5**	(213/214)
	"Global"	TN/TE ± compensated cirrhosis	Overall	**98.9**	(5141/5196)
			GT1	**99.1**	(1599/1613)
			GT2	**99.3**	(1535/1546)
			GT3	**99.3**	(1649/1677)
			GT4	**99.6**	(238/239)
			GT5–6	**98.5**	(67/68)
			Compensated cirrhosis	**97.9**	(1055/1078)
			Treatment experienced	**98.4**	(632/642)
			Unknown GT	**100**	(42/42)
			No cirrhosis, F stage unknown	**97.4**	(480/493)
			F stage unknown	**100**	(82/82)
			Treatment history unknown	**100**	(33/33)

Experienced defined as previous treatment with interferon, ribavirin, telaprevir, boceprevir, and/or simeprevir.

TE, treatment experienced; TN, treatment naïve.

safety profile, and broadly inclusive treatment criteria that include those with compensated or decompensated cirrhosis in a once-daily 12-week pangenotypic STR; these are attributes for good translation in all-important real-world settings. Table 8.9 outlines results from real-world effectiveness studies for SOF/VEL. The real-world German Hepatitis C Cohort (GECCO) reported a high 99.5% cure rate for GT3 infection with 213/214 achieving SVR in a group with a wide range of baseline characteristics, including treatment experience (including DAAs), cirrhosis including decompensated cirrhosis, RAS, and HIV/HCV coinfection [58]. A global real-world study ("Global") focused on the simplicity of therapy with SOF/VEL to provide data supporting minimal monitoring in elimination efforts. This was a study of SOF/VEL for 12 weeks without RBV with the treatment of GT1-6 PLWHCV ± cirrhosis (no decompensated cirrhosis), ± treatment experienced (no all-DAA treatment-experienced) in a large cohort of 5552 PLWHCV [59]. The results reported SVR based on virological reasons (i.e. breakthrough or relapse) are shown in Table 8.9, where failures for non-virological reasons were excluded from the SVR results (non-virological reasons include those lost to follow-up, early treatment discontinuation, etc.). The overall SVR was 98.9% (5141/5196) and the SVR for each GT1-6 ranged from 98.3% to 99.6%. GT3 at 98.3% achieved a similar SVR rate to the other genotypes and to the ASTRAL-3 results; this Global study is the largest real-world GT3 cohort to date. In early 2017, a few years after the introduction of LDV/SOF and other DAAs, GT3 infection had been dubbed "the last stand of HCV" due to challenges in curing this population, along with the increased rate of GT3 disease progression to cirrhosis and higher rates of HCC [60, 61]. The clinical trial and real-world results in Table 8.9 demonstrate high effectiveness in individuals with GT3 infection. The only subgroup identified with a statistically significant increased risk of failure were those with stage F4 fibrosis with a 97.9% SVR. The authors noted that in the "unknown" groups (unknown GT, cirrhosis, fibrosis stage, and treatment history) the SVR rates were all high and in-line with the overall results. They also noted high overall SVR without the use of RBV in the study. Based on each of these factors, the authors suggested that a streamlined 12-week SOF/VEL regimen without RBV and without extensive pre- or on-treatment monitoring could simplify efforts toward HCV elimination. A "minimal monitoring" (MinMon) Phase 4 clinical trial further supports this thesis (Figure 8.13) [62]. Inclusion criteria were simplified: treatment naive, ±compensated cirrhosis, ±HIV coinfection. The minimal monitoring approach included no genotyping; the

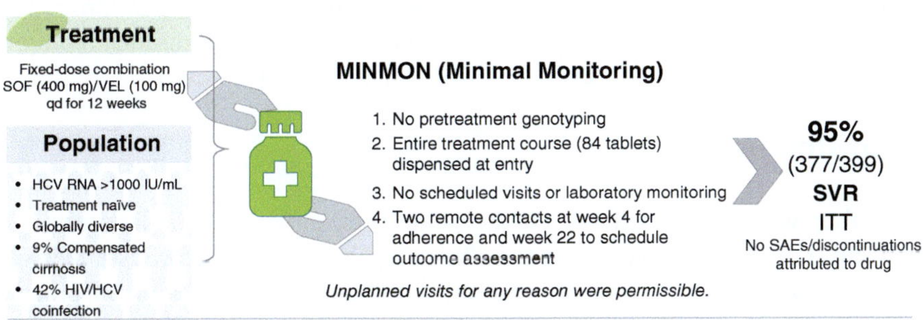

Figure 8.13 MinMon trial ACTG 5360, a Phase 4, open-label, single-arm (38 sites worldwide) study to support HCV elimination efforts.

full 12-week course of 84 SOF/VEL pills were dispensed (no RBV), and there were no scheduled visits and no laboratory monitoring although unplanned visits were permitted. This was an ITT study (missing results count as failure in SVR) and with 399 PLWHCV starting treatment the SVR was 95% (377/399).

8.10 Conclusion

Year 2023 marked the tenth anniversary of the approval of sofosbuvir (Sovaldi®), with over 10 million HCV-infected people cured worldwide with sofosbuvir-based regimens. The year 2024 celebrates the tenth anniversary of the approval of ledipasvir/sofosbuvir (Harvoni®) and the end of the IFN era as the first all-DAA therapy. The subsequent approval of SOF/VEL made a single DAA therapy available to individuals infected by any major HCV genotype at any stage of fibrosis. These transformative therapies provide cures to individuals who have a chronic, debilitating, life-threatening illness. The discovery and development of sofosbuvir (see previous chapter), ledipasvir/sofosbuvir, and sofosbuvir/velpatasvir are the culmination of intensive, creative, and exhaustive efforts within discovery research, pharmaceutical development and manufacturing, and clinical research groups, and the dedication of the participants in the clinical trials. Sofosbuvir is unique as the only approved HCV NS5B nucleotide drug; sofosbuvir's exceptionally high barrier to resistance is well suited to combination with the extremely potent and rapid viral-load reducing NS5A inhibitors ledipasvir or pangenotypic velpatasvir. These active pharmaceutical ingredients meet the stringent potency, drug interaction, safety, and PK criteria for inclusion in STRs. The safety, simplicity, and effectiveness of these once-daily pills broadly provide high cure rates typically in 8–12 weeks. Despite the high mutability of HCV, it is striking that just two direct-acting antiviral agents in combination, without an immunomodulator, are so highly effective at HCV cure. Cure has deep implications for PLWHCV. The "number needed to treat" (NNT) is a metric utilized in evidence-based medicine to define the number of individuals needed to undergo treatment to afford a defined favorable outcome for a single individual.* LDV/SOF has an NNT = 3 for avoidance of HCV-related death over a range of baseline characteristics including ±cirrhosis and/or ±treatment experience [63]. HCV cure not only leads to an improved quality of life but also significantly reduces HCV-related mortality. A testament to their safety, simplicity, and efficacy, these STRs have demonstrated cure rates in real-world settings that are comparable to those observed in controlled clinical trials, and panfibrotic and pangenotypic SOF/VEL possess the attributes favorable for use in minimal monitoring approaches to support worldwide HCV elimination efforts [64, 65]

Acknowledgment

The author would like to thank all the volunteers, patients, their families, and study-site staff who participated in the clinical trials involving sofosbuvir, ledipasvir, and velpatasvir. The author would also like to thank all the researchers, process

* For example, for statins the NNT is 55–114 in order for a patient in a high-risk population to avoid a single cardiovascular event during a 5-year period.

development and manufacturing scientists, and clinical researchers dedicated to the discovery and development of LDV/SOF and SOF/VEL.

Compliance with Ethical Standards

Conflict of Interest: John O. Link is a former employee of Gilead Sciences.

Ethical Approval: All procedures performed in the studies involving human participants were in accordance with the ethical standards of the institutional and/or national research committee and with the 1964 Helsinki Declaration and its later amendments or comparable ethical standards.

Informed Consent was obtained from all individual participants included in the study.

References

1 Hedskog C, Parhy B, Chang S, Zeuzem S, Moreno C, Shafran SD, et al. Identification of 19 novel hepatitis C virus subtypes-Further expanding HCV classification. *Open Forum Infect Dis.* 2019; 6(3):ofz076. doi: 10.1093/ofid/ofz076.

2 Petruzziello A, Marigliano S, Loquercio G, Cozzolino A, Cacciapuoti C. Global epidemiology of hepatitis C virus infection: An up-date of the distribution and circulation of hepatitis C virus genotypes. *World J Gastroenterol.* 2016;22(34): 7824–7840.

3 Moon AM, Green PK, Berry K, Ioannou GN. Transformation of hepatitis C antiviral treatment in a national healthcare system following the introduction of direct antiviral agents. *Aliment Pharmacol Ther.* 2017;45(9): 1201–1212.

4 World Health Organization. (2022). WHO publishes updated guidance on hepatitis C infection – with new recommendations on treatment of adolescents and children, simplified service delivery and diagnostics. Retrieved March 2, 2024, from https://www.who.int/news/item/24-06-2022-WHO-publishes-updated-guidance-on-hepatitis-C-infection.

5 Stanaway JD, Flaxman AD, Naghavi M, Fitzmaurice C, Vos T, Abubakar I, et al. The global burden of viral hepatitis from 1990 to 2013: Findings from the Global Burden of Disease Study 2013. *The Lancet.* 2016;388(10049): 1081–1088.

6 Ly KN, Hughes EM, Jiles RB, Holmberg SD. Rising mortality associated with hepatitis C virus in the United States, 2003–2013. *Clin Infect Dis.* 2016;62(10): 1287–1288.

7 Tzarum N, Giang E, Kong L, He L, Prentoe J, Augestad E, et al. Genetic and structural insights into broad neutralization of hepatitis C virus by human VH1-69 antibodies. *Sci Adv.* 2019;5(1): eaav1882. doi: 10.1126/sciadv.aav1882.

8 Strader DB, Seeff LB. A brief history of the treatment of viral hepatitis C. *Clin Liver Dis.* 2012;1(1): 6–11.

9 North CS, Hong BA, Adewuyi SA, Pollio DE, Jain MK, Devereaux R, et al. Hepatitis C treatment and SVR: The gap between clinical trials and real-world treatment aspirations. *General Hospital Psychiatry.* 2012;35(2): 122–128.

10 Pockros PJ. The dawn of a new era in the treatment of HCV infection: A view from the USA. *Clin Liver Dis.* 2014;4(5): 107–108.

11 Mascolini M. (2023). Total HCV patients treated with direct acting antivirals since 2014: Generics of latest DAAs are key to continuing global expansion of HCV therapy. Conference Reports for NATAP. Retrieved February 19, 2024, from https://www.natap.org/2023/AASLD/AASLD_130.htm.

12 Mullard AA. decade of HCV drugs. *Nat Rev Drug Discov.* 2024;23(6): 8–9.

13 Link JO, Taylor JG, Xu L, Mitchell M, Guo H, Liu H, et al. Discovery of ledipasvir (GS-5885):A potent, once-daily oral NS5A inhibitor for the treatment of hepatitis C virus infection. *J Med Chem.* 2014;57(5): 2033–2046.

14 Sofia MJ, and Link JO. Harvoni: A Combination Therapy for Curing HCV. In: *Comprehensive Medicinal Chemistry III* (ed. S. Chackalamannil, D. Rotella, and S. Ward), 2017; 558–582. Amsterdam: Elsevier Ltd.

15 Link JO. The discovery of ledipasvir (GS-5885): The potent once-daily oral HCV NS5A inhibitor in the single-tablet regimen Harvoni®. In: *HCV: The Journey from Discovery to a Cure Volume II Topics in Medicinal Chemistry* (ed. M. J. Sofia), 2019; 57–80. New York: Springer International Publishing.

16 Macdonald A, Harris M, Hepatitis C. virus NS5A: Enigmatic but still promiscuous 10 years on! *J Gen Virol.* 2015;96(Pt 4): 727–738.

17 Schmitz U, Tan SL. NS5A-from obscurity to new target for HCV therapy. Recent Patents on Anti-Infective Drug Discovery. 2008;3(2): 77–92.

18 Gao M, Nettles RE, Belema M, Snyder LB, Nguyen VN, Fridell RA et al. Chemical genetics strategy identifies an HCV NS5A inhibitor with a potent clinical effect. *Nature.* 2010;465(7294): 96–100.

19 Nguyen T, Guedj J. HCV kinetic models and their implications in drug development. CPT: Pharmacometrics Systems Pharmacology. 2015;4(4): 231–242.

20 Choudhary MC, Mellors JW. The transformation of HIV therapy: One pill once a day. Antiviral Therapy. 2022;27(2). doi: 10.1177/13596535211062396.

21 Lipinski CA, Lombardo F, Dominy BW, Feeney PJ. Experimental and computational approaches to estimate solubility and permeability in drug discovery and development settings. *Adv Drug Deliv Rev.* 1997;23 (1–3): 3–25.

22 Veber DF, Johnson SR, Cheng HY, Smith BR, Ward KW, Kopple KD, et al. Molecular properties that influence the oral bioavailability of drug candidates. *J Med Chem.* 2022;45(12): 2615–2623.

23 Ritchie TJ, Simon JF. The impact of aromatic ring count on compound developability--are too many aromatic rings a liability in drug design? *Drug Discov Today.* 2009;14 (21–22): 1011–1020.

24 Taylor RD, MacCoss M, Lawson, ADG. Rings in drugs. *J Med Chem.* 2014;57(14): 5845–5859.

25 Halford, B. Wrestling with Lipinski's rule of 5. *Chem & Eng News.* 2023;101(8): 16–19.

26 Guo H, Kato D, Kirschberg, TA et al. (2010). Patent Application. WO 2010/132601 A1.

27 TellinghuisenTL, Marcotrigiano J, Rice CM. Structure of the zinc-binding domain of an essential component of the hepatitis C virus replicase. *Nature.* 2005;435(7040): 374–379.

28 Love RA, Brodsky O, Hickey MJ, Wells PA, Cronin CN. Crystal structure of a novel dimeric form of NS5A domain I protein from hepatitis C virus. *J Virol.* 2009;83(9): 4395–4403.

29 Lambert SM, Langley DR, Garnett JA, Angell R, Hedgethorne K, Meanwell NA et al. The crystal structure of NS5A domain 1 from genotype 1a reveals new clues to the mechanism of action for dimeric HCV inhibitors. *Protein Science.* 2014;23(6): 723–734.

30 Mo H, Yang C, Wang K, Huang M, Murray B, Qi X, et al. Estimation of inhibitory quotient using a comparative equilibrium dialysis assay for prediction of viral response to hepatitis C virus inhibitors. *J Viral Hepat.* 2011;18(5): 338–348.

31 Lawitz EJ, Gruener D, Hill JM, Marbury T, Moorehead L, Mathias A, et al. A phase 1, randomized, placebo-controlled, 3-day, dose-ranging study of GS-5885, an NS5A inhibitor, in patients with genotype 1 hepatitis C. *J Hepatol.* 2012;57(1): 24–31.

32 Chatterjee A, Smith PF, Perelson AS, Hepatitis C. viral kinetics: The past, present, and future. *Clin Liver Dis.* 2013;17(1): 13–26.

33 Wong KA, Worth A, Martin R, Svarovskaia E, Brainard DM, Lawitz E, et al. Characterization of hepatitis C virus resistance from a multiple-dose clinical trial of the novel NS5A inhibitor GS-5885. *Antimicrob Agents Chemother.* 2013;57(12): 6333–6340.

34 Osinusi A, McHutchison, JG. The clinical development of ledipasvir/sofosbuvir (LDV/SOF, Harvoni®). *In HCV: The Journey from Discovery to a Cure Volume II Topics in Medicinal Chemistry* (ed. M. J. Sofia), 2019; 237–280. New York: Springer International Publishing.

35 Younossi ZM, Stepanova M, Sulkowski M, Naggie S, Henry L, Hunt S, et al. Sofosbuvir and ledipasvir improve patient-reported outcomes in patients co-infected with hepatitis C and human immunodeficiency virus. *J Viral Hepat.* 2016;23(11): 857–865.

36 Terrault NA, Zeuzem S, Di Bisceglie AM, Lim JK, Pockros PJ, Frazier LM, et al. Combination in patients with hepatitis C virus infection and factors associated with sustained virologic response. *Gastroenterology.* 2016;151(6): 1131–1140.

37 Buggisch P, Vermehren J, Mauss S, Günther R, Schott E, Pathil, A, et al. Real-world effectiveness of 8-week treatment with ledipasvir/sofosbuvir in chronic hepatitis C. *J Hepatol.* 2018;68(4): 663–671.

38 Lee SK, Lee SW, Lee HL, Kim HY, Kim CW, Im Chang U, et al. Real-life experience of ledipasvir and sofosbuvir for HCV infected Korean patients: A multicenter cohort study. *Korean J Intern Med.* 2022;37(6): 1167–1175.

39 Levin J. (2013). IDX719, HCV NS5A Inhibitor, Demonstrates Pan-Genotypic Activity after Three Days of Monotherapy in Genotype 1, 2, 3 or 4 HCV-Infected Subjects. Conference Reports for NATAP. Retrieved February 2, 2024, from https://www.natap.org/2013/HCV/013113_03.htm

40 Link JO, Taylor JG, Trejo-Martin A, Kato D, Katana AA, Krygowski ES et al. Discovery of velpatasvir (GS-5816):A potent pan-genotypic HCV NS5A inhibitor in the single-tablet regimens Vosevi® and Epclusa®. *Bioorg Med Chem Lett.* 2019;29(16): 2415–2427.

41 Link JO. (2022). The discovery of velpatasvir (GS-5816):The potent pan-genotypic once-daily oral HCV NS5A inhibitor in the single-tablet regimens Epclusa and Vosevi. In:*HCV*: the *Journey from Discovery to a Cure Volume II Topics in Medicinal Chemistry* (ed. M. J. Sofia), 81–110. New York: Springer International Publishing.

42 Link JO. (2018). Case History: Curing genotype 1–6 hepatitis C: Velpatasvir, the NS5A inhibitor in Epclusa® and Vosevi®. In: *Medicinal Chemistry Reviews Volume* 53(ed. J. J. Bronson), 541–564. Washington, DC: Medicinal Chemistry Division of the American Chemical Society.

43 Bacon EM, Cottell JJ, Katana AA, et al. (2012). Patent Application. WO 2012/068234 A2.

44 Bacon EM, Cottell JJ, Katana AA, et al. (2013). Patent Application. WO/2013/075029 A1.

45 Rowland M, and Tozer TN. (2011). *Clinical Pharmacokinetics and Pharmacodynamics: Concepts and Applications* (4th ed.) Philadelphia: Wolters Kluwer Health/Lippincott William & Wilkins.

46 Yang T, Oliyai R, Kent KM. The making of the one pill—Developing single tablet regimens for HIV and for HCV. *Antiviral Therapy*. 2022;27: 1–9. (2). https://doi.org/10.1177/13596535211067606.

47 Cheng G, Tian Y, Doehle B, Corsa A, Lee, YJ, Gong, R, et al. Inhibitor ledipasvir. *Antimicrobial Agents and Chemotherapy*. 2016;60(3): 1847–1853.

48 Cheng G, Tian Y, Yu M, Trejo-Martin A, Gong R, Bush, C, et al. (2013). GS-5816, a second-generation HCV NS5A inhibitor with potent antiviral activity, broad genotypic coverage, and a high resistance barrier. EASL 48th Annual Meeting, Amsterdam, The Netherlands, 24 to 28 April 2013. Retrieved from http://www.natap.org/2013/EASL/EASL_34.htm.

49 Mogalian E, German P, Kearney, BP, Yang, CY, Brainard D, Link, J, et al. (2017). Preclinical pharmacokinetics and first-in-human pharmacokinetics, safety, and tolerability of velpatasvir, a pangenotypic hepatitis C virus NS5A inhibitor, in healthy subjects. Antimicrob Agents Chemother. 61, e02084–16. https://doi.org/10.1128/AAC.02084-16.

50 Lawitz EJ, Freilich B, Link JO, German P, Mo H, Han L, et al. A phase 1, randomized, dose-ranging study of GS-5816, a once-daily NS5A inhibitor, in patients with genotype 1–4 hepatitis C virus. *J Viral Hepat*. 2015. 22(12): 1011–1019.

51 Lawitz EJ, Dvory-Sobol H, Doehle BP, Worth AS, McNally J, Brainard DM, et al. Clinical resistance to velpatasvir (GS-5816), a novel pan-genotypic inhibitor of the hepatitis C virus NS5A protein. *Antimicrob Agents Chemother*. 2016;60(9): 5368–5378.

52 Feld JJ, Jacobson IM, Hezode C, Asselah T, Ruane PJ, Gruener N, et al. Sofosbuvir and velpatasvir for HCV genotype 1, 2, 4, 5, and 6 infection. *N Engl J Med*. 2015;373(27): 2599–607.

53 Foster GR, Afdhal N, Roberts SK, Bräu N, Gane EJ, Pianko S, et al. Sofosbuvir and velpatasvir for HCV genotype 2 and 3 infection. *N Engl J Med*. 2015;373(27): 2608–2617.

54 von Felden J, Vermehren J, Ingiliz P, Mauss S, Lutz T, Simon KG, et al. High efficacy of sofosbuvir/velpatasvir and impact of baseline resistance-associated substitutions in hepatitis C genotype 3 infection. *Aliment Pharmacol Ther.* 2018;47(9): 1288–1295.

55 Curry MP, O'Leary JG, Bzowej N, Muir AJ, Korenblat KM, Fenkel JM, et al. Sofosbuvir and velpatasvir for HCV in patients with decompensated cirrhosis. *N Engl J Med.* 2015;373(27): 2618–2628.

56 Wyles D, Bräu N, Kottilil S, Daar ES, Ruane P, Workowski K, et al. Sofosbuvir and velpatasvir for the treatment of hepatitis C virus in patients coinfected with human immunodeficiency virus type 1: An open-label, phase 3 study. *Clin Infect Dis.* 2017;65(1): 6–12.

57 Younossi ZM, Stepanova M, Feld J, Zeuzem S, Jacobson I, Agarwal K, et al. Sofosbuvir/velpatasvir improves patient-reported outcomes in HCV patients: Results from ASTRAL-1 placebo-controlled trial. *J Hepatol.* 2016;65(1): 33–39.

58 von Felden J, Vermehren J, Ingiliz P, Mauss S, Lutz T, Simon KG, et al. High efficacy of sofosbuvir/velpatasvir and impact of baseline resistance-associated substitutions in hepatitis C genotype 3 infection. *Aliment Pharmacol Ther.* 2018;47, 1288–1295.

59 Mangia A, Milligan S, Khalili M, Fagiuoli S, Shafran SD, Carrat F, et al. Global real-world evidence of sofosbuvir/velpatasvir as simple, effective HCV treatment: Analysis of 5552 patients from 12 cohorts. *Liver Int.* 2020;40(8): 1841–1852. https://doi.org/10.1111/liv.14537.

60 Chan A, Patel K, Naggie, S. Genotype 3 infection–the last stand of hepatitis c virus drugs. *Drugs.* 2017;77(2): 131–144.

61 Tapper EB, Afdhal NH. Is 3 the new 1: perspectives on virology, natural history and treatment for hepatitis C genotype 3. *J Viral Hepat.* 2013;20(10): 669–677.

62 Solomon SS, Wagner-Cardoso S, Smeaton L, Sowah LA, Wimbish C, Robbins G, et al. A Minimal Monitoring Approach for the Treatment of Hepatitis C Virus (HCV) Infection: A Phase 4, Open-Label, Single-Arm Trial (ACTG 5360 [MINMON] Trial). *Lancet Gastroenterol Hepatol.* 2022;7(4): 307–317.

63 Younossi ZM, Park H, Saab S, Ahmed A, Dieterich D, Gordon SC, et al. Cost-effectiveness of all-oral ledipasvir/sofosbuvir regimens in patients with chronic hepatitis C virus genotype 1 infection. *Aliment Pharmacol Ther.* 2015;41(6): 544–563.

64 Solomon SS, Wagner-Cardoso S, Smeaton L, Sowah LA, Wimbishoi G, et al. A minimal monitoring approach for the treatment of hepatitis C virus (HCV) infection: A phase 4, open-label, single-arm trial (ACTG 5360 [MINMON] Trial). *Lancet Gastroenterol Hepatol.* 2022;7(4): 307–317.

65 World Health Organization. Elimination of hepatitis by 2030. Retrieved February 19, 2024, from https://www.who.int/health-topics/hepatitis/elimination-of-hepatitis-by-2030#tab=tab_1

Part IV

Respiratory Viruses

9

Presatovir: A Once-daily Oral Respiratory Syncytial Virus Fusion Inhibitor for the Treatment of Respiratory Syncytial Virus Infection in Adults and Infants

Richard L. Mackman, Jason Chien, Dustin S. Siegel

Gilead Sciences Inc., Foster City, CA 94044, USA

9.1 Introduction to Respiratory Syncytial Virus

Presatovir **1** (Figure 9.1) is a potent inhibitor of the respiratory syncytial virus (RSV) fusion protein [1]. The candidate compound entered clinical development in 2013 and was the first small molecule RSV inhibitor to complete both a human adult RSV challenge study and efficacy evaluation across multiple geographically diverse adult populations infected with RSV [2–6]. Presatovir is not yet approved for the treatment of RSV and is no longer in active development. However, it arguably paved a path of learning for other RSV small molecule therapeutics to follow in the subsequent years, including other promising RSV fusion inhibitors.

RSV, a member of the pneumovirus family, was first identified around 1955. RSV primarily infects the ciliated cells of the respiratory epithelium in addition to Type I and Type II pneumocytes in the alveoli [7]. Of note, RSV can also fuse adjacent epithelial cells on the apical surface, forming multi-nucleated cells known as syncytia. The viral infection ultimately leads to immune dysregulation, inflammation, and symptoms of bronchiolitis and pneumonia [8, 9]. Severe infection is associated with high morbidity and some mortality risk in young infants, immunocompromised patients, the elderly, and patients with underlying cardiopulmonary conditions such as chronic obstructive pulmonary disease [10–12]. For example, in infants less than 1 year of age severe RSV infection is associated with significantly more deaths than influenza [12]. In immunosuppressed populations, e.g. transplant patients, RSV infection also causes significant morbidity and mortality [13, 14].

Trends in Antiviral Drug Development, First Edition. Edited by Michael J. Sofia and Zhengqiang Wang.

The only FDA-approved RSV treatment option is inhaled ribavirin, which was approved for infants in 1985, more than 2 decades prior to the initiation of the presatovir program [15–17]. Ribavirin demonstrated a modest clinical benefit, but carried the disadvantage of significant safety concerns due to its mutagenic potential, along with its inconvenient aerosolized route of administration. Therefore, the goal of the presatovir program was to develop the first safe and efficacious oral treatment for RSV infection in infants and adults.

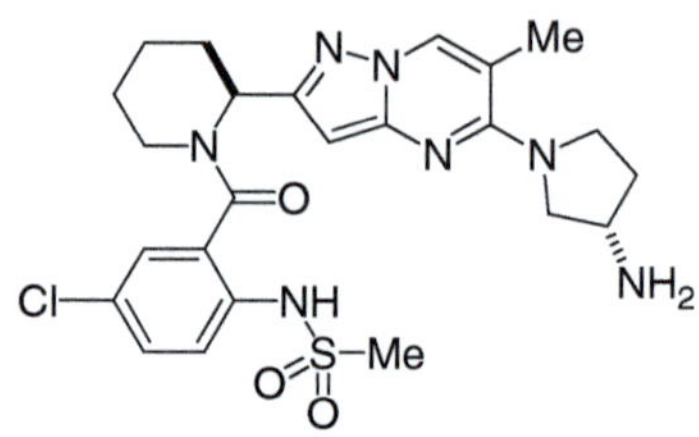

Presatovir (1)

Figure 9.1 RSV fusion inhibitor. Average RSV EC_{50} = 0.43 nM across 75 RSV A and B isolates. *Source*: R Mackman.

RSV is an enveloped negative-sense RNA virus that has 10 genes encoding 11 proteins, three proteins on the surface of the virion, and eight internal proteins (Figure 9.2) [7, 18, 19]. There are two main RSV subtypes, A and B, and within each subtype a variety of strains. On the virion surface, the glycoprotein (G) is responsible for attachment to the target cells in the respiratory airway. Following attachment, the fusion protein (F) facilitates the virus entry into the target cell to deliver the viral RNA. In addition, the fusion protein can also promote the fusion of neighboring cells leading to multi-nucleated cells, or syncytia, a hallmark of the RSV infection. Prior to this program, several small molecule fusion inhibitors from different groups had been reported in the literature but to our knowledge, none had progressed into clinical studies of natural RSV infection [20–24]. Palivizumab, a monoclonal antibody targeting the F protein, was the first antibody approved for prophylactic use in infants at high risk of severe RSV infection [25]. More recently, an improved antibody, nirsevimab, that targets the prefusion conformation of the F protein, has achieved FDA approval for prophylaxis of infants up to 2 years and neonates [26].

The internalized RSV RNA is replicated by the replication complex that includes the N, P, and L (RNA-dependent RNA polymerase) proteins [7]. Several of these

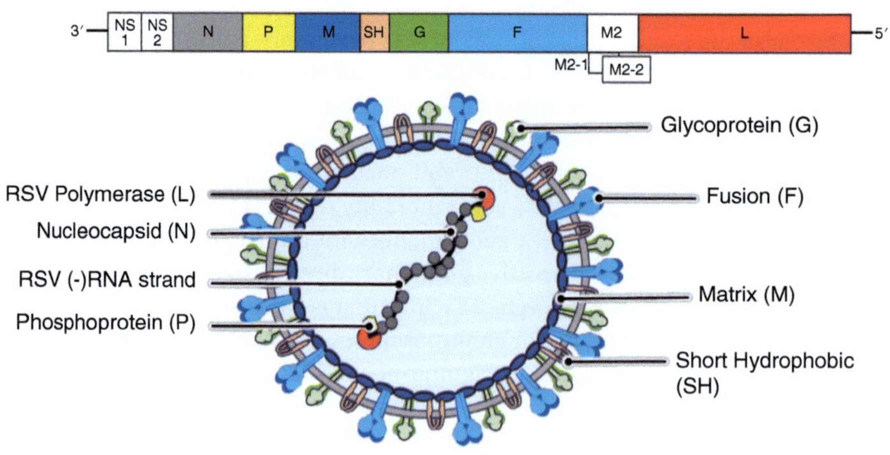

Figure 9.2 RSV virion architecture and gene encoding 10 mRNA and 11 proteins. *Source*: R. Mackman.

proteins have also been the target for small molecule interventions, but similar to the fusion target, none had progressed into broad-based RSV efficacy studies in patients before the presatovir program was initiated. The N-protein inhibitor RSV604 was evaluated in a small study of RSV infection in post-stem cell transplant patients but was subsequently discontinued due to variable exposures [27]. However, in a subset of subjects with the highest exposures a trend towards greater viral reduction in the nasopharyngeal swab analysis was observed. Recently a compound from the same structural class, EDP-938, was evaluated in otherwise healthy RSV-infected adult patients [28–30]. No clinical benefit was observed in this population, and additional trials including a pediatric trial are underway [30].

Some of the challenges encountered by the early inhibitors were used to inform the presatovir program strategy irrespective of the different mechanisms of action. Multiple inhibitors were plagued with poor physical properties that limited oral properties, exhibited a poor safety profile, or lacked broad RSV subtype coverage. Our strategy therefore began with the aim of discovering a novel chemical lead (hit) through a mechanism-agnostic, phenotypic high-throughput screen. An optimal lead would have favorable physical properties and good solubility at the outset since potency optimization generally results in increased lipophilicity, and reduced solubility. High potency when adjusted for plasma protein binding was also targeted in order to lower the systemic exposure required for efficacy, thereby minimizing the risk of off-target toxicity. The candidate profile consisted of a potent, once-daily, oral compound that had broad activity across >95% of RSV clinical isolates. The target plasma trough concentrations of the drug were refined as the program evolved and was ultimately set at ~4–5-fold above the 95% inhibitory concentration, calculated from the average RSV antiviral EC_{50} activity across a panel of clinical isolates, with adjustment for human plasma protein binding. Given the target populations of infants, elderly, and immunocompromised patients, an excellent safety profile with minimal drug–drug interactions was an integral component of the candidate profile.

9.2 Discovery of Presatovir

9.2.1 Lead Identification

Many of the optimized RSV inhibitors that were reported in the literature originated from high-throughput screening campaigns that employed cell-based and biochemical methods. Since none of these earlier inhibitors had successfully generated efficacy in large RSV natural infection studies, the respective mechanisms of action were not considered by the team to be clinically validated [20]. Therefore, the program developed a phenotypic, high-throughput RSV A2-cytopathic screen in HEp-2 cells that was then used to identify hits without bias toward any specific mechanism of action [1,31]. Compounds that protected the HEp-2 cells from the RSV A2-mediated cytopathic effect at a single concentration in duplicate were then progressed to confirmatory RSV A2 dose-response, and secondary assays. This strategy enabled the team to identify the best RSV hits across the Gilead libraries irrespective of mechanism.

Two Gilead proprietary libraries were screened, the first a ~400 K compound collection that comprised Gilead's general small molecule library, and the second, a smaller collection of proprietary nucleoside analogs that were expected to target the RNA-dependent RNA polymerase (RdRp) activity of the RSV L protein. In the latter screen, the hepatitis *C*-nucleoside lead GS-441524 was discovered as a more effective RSV inhibitor than its original HCV RdRp target. It was subsequently optimized for RSV using a prodrug strategy and this resulted in the invention of the phosphoramidate prodrug remdesivir, which was not developed further for RSV due to its intravenous route of delivery [32]. Remdesivir was later shown to possess broad activity toward multiple RNA viruses and ultimately gained FDA approval as the first treatment for SARS-CoV-2 infection [33]. From the larger 400 K compound library the top ~1% of hits were subjected to confirmatory dose-response screening together with a suite of secondary assays to identify the most promising selective hits. These secondary assays included a medium throughput RSV replicon assay to identify RSV replication inhibitors and an RSV B subtype assay to evaluate cross-subtype activity. To remove nonselective compounds with cytotoxic or cytostatic properties, the rapidly proliferating MT-4 leukemia cell line was chosen since this had been reliably used to exclude nonselective compounds in other internal antiviral programs. The most promising ~100 hits from the secondary screening were then profiled in a low throughput time-of-addition (TOA) assay and toward a known RSV fusion inhibitor mutant, D486N, to further define a potential mechanism of action for the hit [34]. In the TOA assay, compounds with substantially weaker activity when added 4 h after infection compared to 0 h, at the time of infection, were classified as entry inhibitors as opposed to replication or viral egress inhibitors. The screening process proved to be very effective in identifying multiple hits with good selectivity and classifying them into different antiviral mechanisms.

The vast majority of the most potent hits lost activity towards the known D486N RSV mutant virus, lacked RSV replicon activity, and lost potency at 4 h in the TOA assay suggesting they were likely fusion inhibitors. Notably, these hits displayed a diverse array of chemical structures affording multiple potential starting points for inhibitor optimization. In general, the fusion hits also had a low log D <2.0 compared to the best postentry inhibitors, an observation that was clearly consistent with the extracellular fusion mechanism in which cell entry was not required for activity.

From the many promising fusion hits that were identified in the secondary screening assays, racemate **2** was one of the most potent (Figure 9.3a). Racemate **2** demonstrated sub-micromolar activity, a low molecular weight, and good kinetic solubility due to its low log D. More importantly, when separated into its individual stereoisomers using chiral chromatography, the *S*-isomer **3** was >70-fold more potent than the *R*-isomer **4**. This strongly suggested that the hit was likely interacting with a defined protein pocket on the fusion protein rather than inhibiting RSV through nonspecific binding. Small molecule X-ray crystallography of **3** (Figure 9.3b), combined with an enantioselective synthesis, confirmed the *S*-isomer stereochemistry. Other key features of the hit included potent activity toward an RSV B strain supporting broad RSV subtype coverage and good stability in human microsomes (data not shown). Taken together, the properties of **3** as a screening hit were very compelling

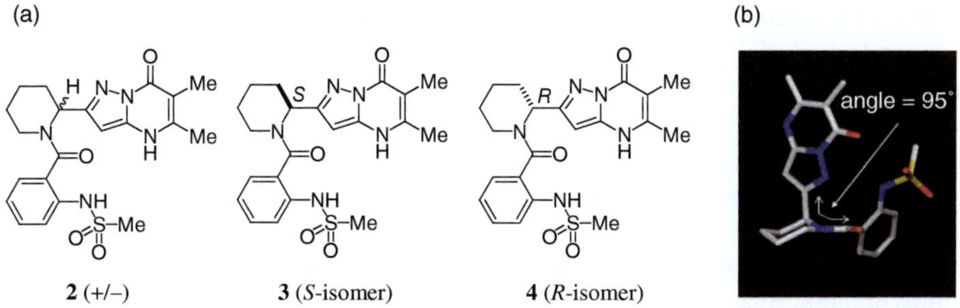

2 (+/–) RSV A2 EC_{50} = 202 nM Log D = 1.3

3 (*S*-isomer) RSV A2 EC_{50} = 65 nM

4 (*R*-isomer) RSV A2 EC_{50} = 4700 nM

3

Figure 9.3 (a) Structure and RSV activity of racemic hit **2** and its respective stereoisomers **3** and **4**. (b) X-Ray structure of **3** indicating a 95° dihedral angle between the heterocycle and the amide bond. *Source*: Reproduced from Mackman et al. [1] with permission.

given its high potency and attractive three-dimensional structure that provided many sites for further optimization of the properties. Thus, the RSV fusion hit **3** was selected as the lead transitioning the program into a lead optimization phase.

9.2.2 Lead Optimization

The fusion mechanism was prone to resistance development so a viable candidate compound required very high potency, together with excellent pharmacokinetic properties, in order to effectively inhibit the fusion protein at all times during an oral dosing regimen. The antiviral potency was measured in RSV A2-infected HEp-2 cells suspended in 10% fetal-bovine serum (FBS) and not human plasma, so an adjustment in the antiviral EC_{50} to accommodate the increased effects of human plasma protein binding on free-drug concentrations was required. The inclusion of human plasma protein binding effects has been widely adopted in the development of other prominent antiviral therapeutics, e.g. HIV, based on clinical data, but was not known for RSV therapeutics due to the lack of clinical data. Nevertheless, the program reasoned that a conservative approach was prudent and therefore the target exposure in human plasma for the program was based on the human plasma protein adjusted EC_{50} ($paEC_{50}$). To determine the $paEC_{50}$ compounds were subjected to dialysis between human plasma and 10% FBS separated by a permeable membrane and the relative concentrations on each side represented the fold-shift due to human plasma (paEC50 = fold-shift × EC_{50}). Alternatively, the ratio of dialysis between human plasma and buffer compared to dialysis between 10% FBS and buffer could be used to determine the fold-shift (fold-shift = human plasma-to-buffer ratio/10% FBS to buffer ratio). For lead **3**, a modest 13-fold human plasma protein shift was observed, since the compound was only ~95% bound to human proteins, leading to a calculated paEC50 of ~850 nM. It was clear that the $paEC_{50}$ needed to be significantly reduced during the early optimization.

A crystal structure complex of the hit **3** bound to the RSV fusion protein was not available during the early optimization stages, so only the small molecule X-ray structure of **3** (Figure 9.3b) could be used to guide early SAR. From this structure, it was evident that the tautomeric amide bond of the acylated piperidine ring forced the pyrazolopyrimidinone heterocycle to adopt a ~95° dihedral angle relative to the amide bond due to A1,3 conformational strain (Figure 9.3b). The equivalent 5-membered pyrrolidine analog of **3** was synthesized and found to be less active due to its smaller dihedral angle, calculated at ~70°, that oriented the pyrazolo-heterocycle into a different trajectory. In contrast, the larger and more flexible 7-membered azepine ring analog, which could project the pyrazolo-heterocycle into a similar geometry as the piperidine, had comparable potency. Removal of the amide bond through preparation of a benzyl instead of benzoyl derivative or removal of the piperidine ring conformational restriction by using an alanine linkage, both resulted in reduced potency. Taken together, it was clear from the SAR that the 95° spatial orientation enforced by the amide bond in the context of the 6-membered piperidine was important for potency. Although the azepine analog had equivalent potency, the ready access to cheap, stereo-chemically pure synthetic piperidine precursors or differentially substituted 2-carboxy pyridines that could be readily reduced to substituted piperidines resulted in progressing forward with the piperidine core. Further piperidine analogs included a methyl-walk around the piperidine core in an attempt to enforce a more rigid three-dimensional conformation, but this also did not result in potency improvements.

Exploration of phenyl ring substituents on lead **3** was readily performed through standard coupling reactions with different aromatic acids (Figure 9.4). In general, the SAR was very narrow, but a significant advance in potency was soon realized. Removal of the sulfonamide, analog **5**, reduced potency, while appending structurally small, nonpolar groups at the para-position (4-position; see Figure 9.4) to the

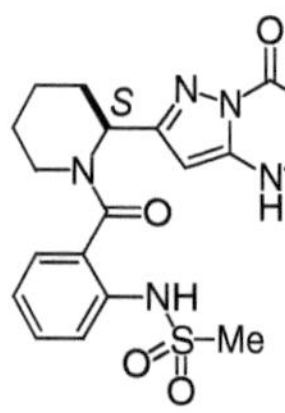

3
RSV A2 EC_{50} = 65 nM
Hu. plasma %free = 6.5
Hu. plasma fold-shift = 13
$paEC_{50}$ = 845 nM

Log D = 1.3
Caco-2 AB × 10^{-6} cm/s = 0.33
Efflux Ratio = 7
Rat Oral F% = 1.5

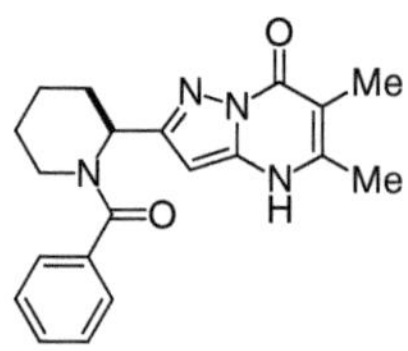

5
RSV A2 EC_{50} = 586 nM

Log D = 1.4
Caco-2 AB × 10^{-6} cm/s = 3.5
Efflux Ratio = 7

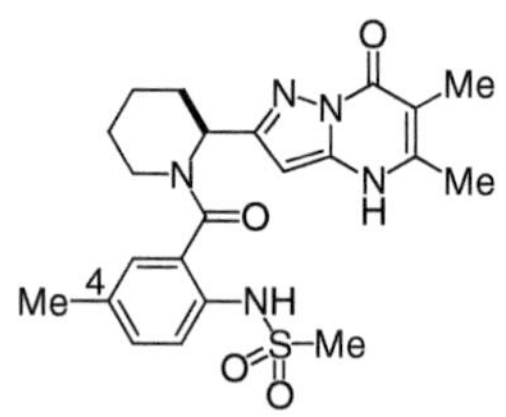

6
RSV A2 EC_{50} = 5.0 nM
Hu. plasma %free = 16
Hu. plasma fold-shift = 5
$paEC_{50}$ = 25 nM

Log D = 1.7
Caco-2 AB × 10^{-6} cm/s < 0.1
Efflux Ratio =>49

Figure 9.4 SAR of aryl substitutions on **3**. AB, apical to basolateral permeability; efflux ratio calculated as BA/AB permeability in Caco-2 cells. Hu, human.

sulfonamide, e.g. methyl **6**, or the corresponding chlorine, significantly improved potency, e.g. by > 10-fold in the case of **6**. Larger substitutions, such as phenyl, or the addition of heteroatoms into the ring, or switching to nonaromatic ring systems, all reduced potency. It was hypothesized that the phenyl ring was likely directed into a defined lipophilic region of the fusion protein, while the critical sulfonamide was engaged in hydrogen bond interactions. Another appealing property of **6** was its lower plasma-protein shift compared to **3**, leading to an overall $paEC_{50} = 25$ nM, more than 30-fold improved over **3**. The advancement in the $paEC_{50}$ provided the impetus to begin addressing the pharmacokinetic limitations of the series in parallel with driving potency.

The oral candidate needed to possess good permeability properties, not only for optimal intestinal absorption during oral delivery, but also to permit effective penetration from the plasma compartment into the lung tissue in order to reach the RSV target on the apical airway surface. Lead **3** had poor oral bioavailability of only 1.5% in rat, which was thought to be due to its very poor permeability, as demonstrated in the Caco-2 assay (Figure 9.4). The more potent lead **6** also demonstrated poor permeability and evidence of high efflux. The strategy to improve permeability focused on the two polar hydrogen bond donor groups, namely the sulfonamide and the tautomeric pyrimidinone ring within the bicyclic heterocycle. Focusing first on the sulfonamide, improved permeability was noted for **5**, suggesting that modifications to the sulfonamide group may potentially alleviate some of the poor permeability if potency could be retained. However, despite multiple attempts, including *N*-methylation to remove the hydrogen bond donor, obtaining high potency and improved permeability proved quite elusive.

Attention was turned to the other hydrogen bond donor embedded in the tautomeric pyrimidinone ring. The pyrimidinone was responsible for driving high polarity, since removal of the sulfonamide in analog **5** still resulted in a compound with a log D < 2.0. Replacement of the pyrazolopyrimidinone heterocycle with a pyrazolopyrimidine heterocycle, analog **7**, was found to have a dramatic effect on the physicochemical properties (Figure 9.5). Excellent EC_{50} potency, within twofold of **6**, was found, while the lipophilicity significantly increased to log D > 3 and Caco-2 forward permeability (apical to basolateral) was substantially improved by more than 100-fold. The increased passive permeability was also accompanied by significant improvements in oral PK. In rat, compound **7** demonstrated almost complete absorption and an oral bioavailability of 86%.

While the rat oral PK of **7** was excellent, some properties were clearly disadvantaged by the high lipophilicity. These included greater human plasma protein binding, leading to a $paEC_{50}$ sevenfold higher than lead **6**, and the observation of greater metabolic instability in microsomes. The protein binding shift was not immediately concerning, since multiple positions on the pyrazolopyrimidine ring were yet to be explored for improved overall potency. Indeed, it was quickly discovered that position 5 on the pyrimidine was tolerant of ring systems, such as azetidine, that could improve antiviral potency. The increased metabolism, however, was problematic, since it led to a safety concern regarding the potential to form reactive metabolites. Oxidation of the para-methyl group could be followed by the elimination of water to form a

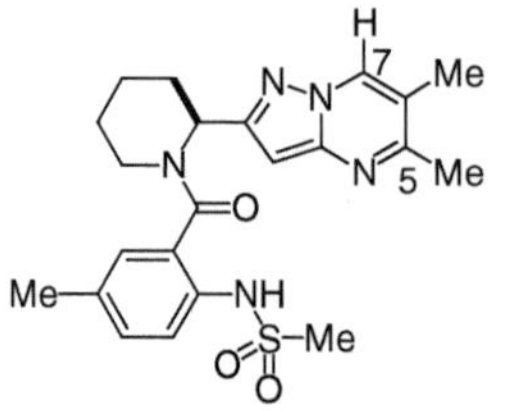

7

RSV A2 EC_{50} = 9.4 nM
Hu. plasma fold-shift = 18
paEC_{50} = 169 nM

Log D = 3.1
Caco-2 AB × 10^{-6} cm/s = 42
Efflux Ratio <=1
Rat F% = 86

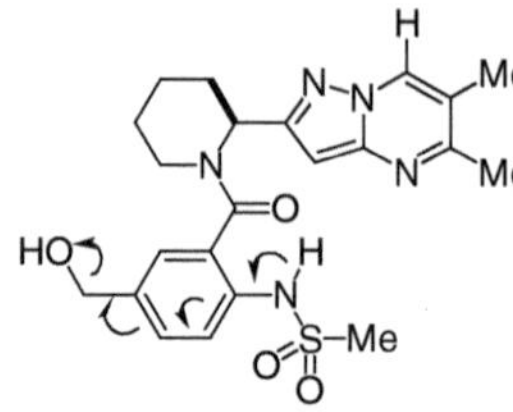

8

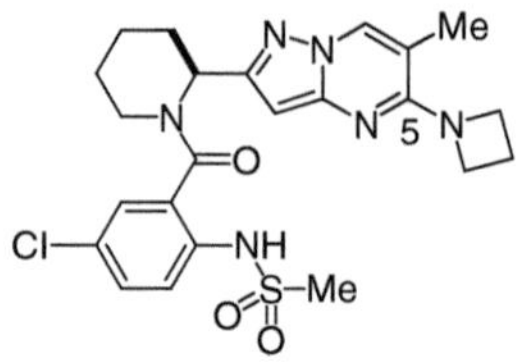

9

RSV A2 EC_{50} = 1.1 nM
Hu. plasma fold-shift = 21
paEC_{50} = 23 nM

Log D = 3.8
Caco-2 AB × 10^{-6} cm/s = 15
Efflux Ratio <=1
MS CL r/d/hu (L/h/kg) = 3.4/1.2/0.66

Figure 9.5 SAR on pyrazolopyrimidine analogs. AB, apical to basolateral permeability; efflux ratio calculated as BA/AB permeability in Caco-2 cells. MS, microsomes; r, rat; d, dog; hu, human.

reactive arene-sulfonimine, e.g. **8**. For this reason, a strategic decision was made to replace the para-methyl with the isosteric para-chloro in all future compounds, even though this resulted in a modest potency loss. Combining the azetidine and the para-chloro changes on **7** resulted in a new lead, **9**, that regained impressive potency with a paEC_{50} = 23 nM, while simultaneously maintaining excellent permeability.

The final phase of optimization to arrive at presatovir was directed at improving the clearance of **9.** The log D for **9** remained high, and the predicted clearance from human microsomes was >50% of liver blood flow. The strategy was to lower the log D by introducing polar groups, especially in the regions of the molecule that were prone to metabolism. Lowering the log D was anticipated to reduce permeability, but given the excellent properties of **9**, it was reasoned that sacrificing some permeability to obtain increased metabolic stability was a worthwhile trade-off. The azetidine was one point of weakness identified using metabolite studies, so analogs containing hydroxyl, or the primary amine **10** were designed, in addition to expanding the ring size to permit further modifications to be introduced (Figure 9.6). The amino azetidine **10** and *S*-amino pyrrolidine **1** (presatovir) both displayed significantly improved *in vitro* metabolism in human microsomes and maintained good oral bioavailability in rat. Moreover, presatovir demonstrated picomolar potency for the first time in the program, leading to an paEC_{50} = 8.1 nM. Compared to the *R*-isomer amine **11** and the corresponding *S*-hydroxy analog **12**, presatovir had a potency that was improved by 8-fold and 20-fold respectively, suggesting the amine was likely participating in a key interaction in the target binding site. Although further profiling was needed to affirm presatovir as a candidate, the key structural changes of introducing the para-chloro, switching the heterocycle to a pyrazolopyrimidine, followed by introduction of the chiral amino-pyrrolidine, resulted in >100-fold improved paEC_{50} compared to the screening hit **2**.

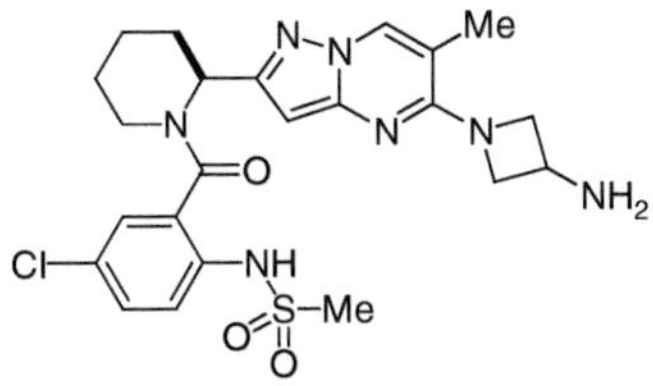

10
RSV A2 EC_{50} = 1.5 nM
Hu. plasma %free = 5.6
Hu. plasma fold-shift = 13
paEC_{50} = 20 nM

Log D = 2.1
Caco-2 AB × 10^{-6} cm/s = 3.1
Efflux Ratio = 4
Hu. MS CL (L/h/kg) = <0.09
Rat F% = 67

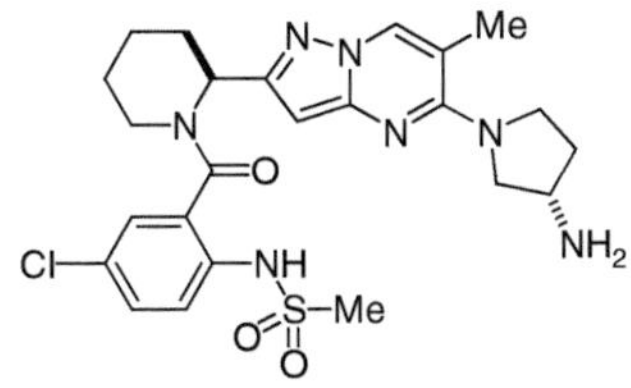

1
RSV A2 EC_{50} = 0.37 nM
Hu. plasma %free = 4.4
Hu. plasma fold-shift = 22
paEC_{50} = 8.1 nM

Log D = 2.0
Caco-2 AB × 10^{-6} cm/s = 6.4
Efflux Ratio = 3
Hu. MS CL (L/h/kg) = <0.16
Rat F% = 46

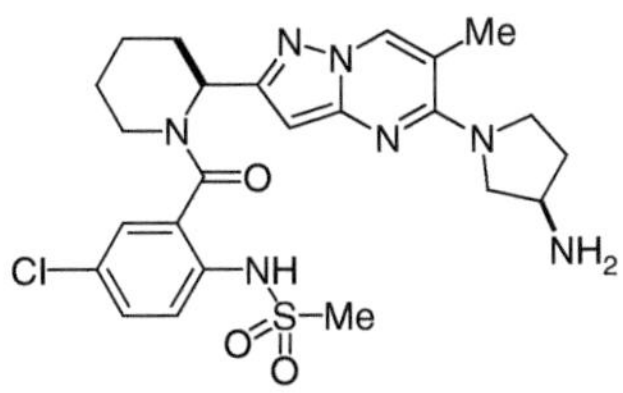

11
RSV A2 EC_{50} = 2.6 nM
Hu. plasma %free = 4.3
Hu. plasma fold-shift = 16
paEC_{50} = 42 nM

Log D = 2.0
Caco-2 AB × 10^{-6} cm/s = 1.5
Efflux Ratio = 6
Hu. MS CL (L/h/kg) = 0.13

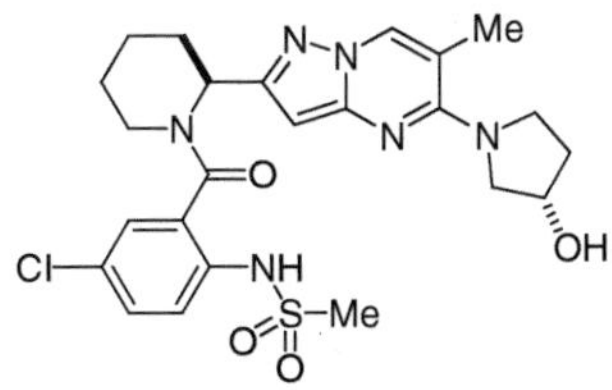

12
RSV A2 EC_{50} = 7.5 nM
Hu. plasma %free = 2.9
Hu. plasma fold-shift = 10
paEC_{50} = 75 nM

Log D = 2.9
Caco-2 AB × 10^{-6} cm/s = 43
Efflux Ratio = 1
Hu. MS CL (L/h/kg) = 0.46

Figure 9.6 SAR on the 5-substitution of the pyrazolopyrimidine analogs. AB, apical to basolateral permeability; efflux ratio calculated as BA/AB permeability in Caco-2 cells. MS, microsomes; Hu, human.

9.3 Pharmacological Profiling

9.3.1 Mechanism of Action Studies

The RSV entry process is initiated by attachment of the RSV virion to target cells through interaction of the G protein with receptors on the target cell, followed by cell entry mediated by the fusion (F) protein (Figure 9.7) [36]. The HTS hit **2** was initially classified as a fusion inhibitor based on the data generated in the TOA assay coupled with its loss of activity toward the D486N fusion mutant generated by the fusion inhibitor TMC-353121 [37]. Presatovir was therefore evaluated in the TOA

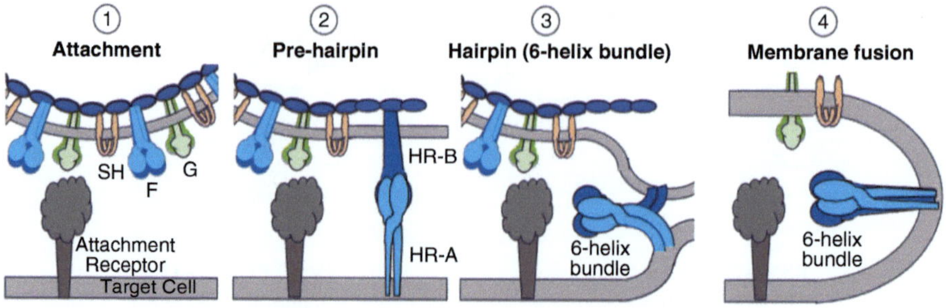

Figure 9.7 Model of the fusion mechanism involving the three surface proteins, glycoprotein (G), fusion protein (F), and short hydrophobic protein (SH) Adapted from [22, 35]. Step 1 involves attachment of the virus particle via the G protein to a receptor on the target cell. The F protein is in a pre-fusion state and exists on the viral surface as a trimer. Step 2 illustrates the conformational change in F in which the three heptad-repeat A (HR-A) units form a 3-helix bundle, the tail of which inserts into the target cell membrane. The fusion protein remains anchored to the virion through three heptad-repeat B (HR-B) helices. Step 3 illustrates the conformational transition of the three HR-A helices associating with the three HR-B helices to form a 6-helix bundle or hairpin, and through this conformational change the membranes are brought into proximity. Step 4 illustrates the hairpin conformation of the post-fusion protein following membrane fusion. *Source*: R. Mackman.

assay and also found to be >2000-fold less active when added 4 h after RSV infection compared to 0 h, consistent with the original hit **2** and an early entry mechanism of RSV inhibition [38]. The attachment process is temperature independent, while fusion is dynamic and requires temperature dependent conformational changes of the F protein from its pre-fusion to post-fusion state (see Figure 9.7). In HEp-2 cells at 4 °C, presatovir was not effective at blocking RSV attachment compared to heparin, a known attachment inhibitor, but was effective in blocking the fusion process at 37°C, further supporting its mechanism as a fusion inhibitor [38, 39].

Resistance selection in cell-culture is a viable method for generating supporting data for a specific mechanism of action of a direct-acting antiviral. *In vitro* resistance selection was conducted in HEp-2 cells in the presence of increasing concentrations of presatovir [38]. Sequencing of the resistant strains identified mutations in the F protein such as L138F in subtype A and F488L in subtype B along with some double mutants. Reverse-engineering of the mutants into the RSV wild-type strains and evaluation of antiviral activity confirmed reduced susceptibility to presatovir. Encouragingly, an analysis of >300 published RSV F gene sequences across both RSV A and B subtypes concluded that none of the presatovir-associated mutations were present suggesting that pre-existing resistance to presatovir was unlikely. Nevertheless, since resistance to presatovir was readily generated *in vitro*, it was highly desirable for the program to minimize the potential for resistance in the clinic through targeting a high trough plasma concentration that could effectively inhibit the virus at all times during the dosing regimen.

Mutations in the F protein as a result of exposure to presatovir suggested that there is most likely a direct interaction between presatovir and the F protein. To provide more conclusive evidence of this interaction, significant efforts were invested toward solving an X-ray structure of the presatovir and F protein complex. Unfortunately, this proved quite challenging, so in parallel, indirect evidence of binding to the fusion protein was sought using alternative methods. RSV-mediated membrane fusion leads to the formation of multi-nuclear fused cells called syncytia, a process that can be modeled in 293T cells transfected with a truncated F protein construct. Using microscopy, presatovir demonstrated a dose-dependent reduction in syncytia formation when compared to vehicle controls and resulted in a calculated $EC_{50} = 2.2$ nM, in close agreement with the antiviral activity [38]. Fusion protein-mediated entry also involves a conformational change in the F protein from a pre-fusion to a post-fusion state as described in Figure 9.7. This transformation was modeled using a soluble, transmembrane truncated RSV F protein that in low ionic strength buffer promotes the pre- to post-fusion conformational change resulting in its aggregation and the formation of rosette-like structures (Figure 9.8) [40]. Presatovir prevented the formation of the rosettes, observed by electron microscopy, in a dose-dependent manner at concentrations up to six-fold in excess of the protein concentration. This result suggests that presatovir binds to the fusion protein and blocks the pre- to post-fusion conformational change. Taken together, the *in vitro* pharmacological data provided strong evidence that the mechanism of action for presatovir is most likely binding to the RSV F protein and preventing the conformational change from a pre- to post-fusion state.

9.3.2 RSV Animal Models

The generation of supportive data for presatovir in RSV animal models was considered valuable proof-of-concept data prior to the nomination of a candidate for clinical development. Animal models can also be used to generate a plasma drug exposure-efficacy relationship to support or refine the human plasma exposures projected for efficacy. However, despite many animal models reported in the literature across many

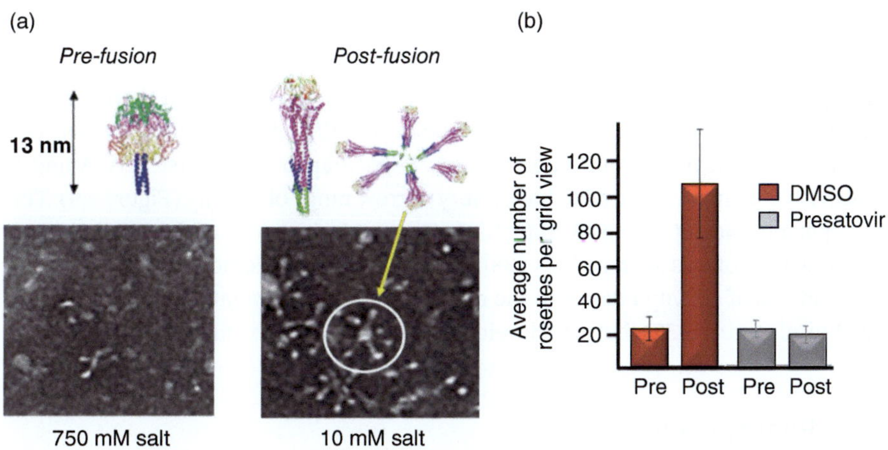

Figure 9.8 Presatovir inhibits the pre- to post-fusion transformation of a trans-membrane truncated F protein construct. (a) The post-fusion conformation is induced in low salt conditions and leads to the formation of rosette-like structures observable using electron microscopy. (b) The average number of rosettes per grid view declined in the presence of presatovir. *Source*: R. Mackman.

different species, including mice, cotton rat, lamb, bovine, and African green monkey, the translational value of these models had not been rigorously demonstrated [41, 42]. Therefore, the animal models chosen were primarily used to establish proof-of-concept data for presatovir and included the cotton rat and bovine models. The cotton rat infection model was successfully applied to earlier fusion inhibitors so an initial prophylaxis study was designed for presatovir in this model using the RSV A2 virus [43, 44]. Presatovir was dosed at three different escalating doses by intraperitoneal injection, 1 h prior to infection by nasal inoculation in the rats. After 4 days, the RSV virus in lung lavage samples was quantified by quantitative reverse-transcriptase polymerase chain reaction and a clear dose response was observed with the highest dose resulting in a ~2 log reduction in the lung viral titer compared to vehicle control [1]. This confirmed that presatovir was effective *in vivo* at preventing the RSV virus from progressing from the nasal cavity down into the lower lung.

The cotton rat model is only semi-permissive to human RSV infection and does not result in significant pathogenesis, whereas calves are susceptible to bovine RSV (bRSV) and exhibit symptoms of cough, fever, and nasal mucus formation, very similar to the clinical signs observed in RSV-infected infants. Moreover, lung lesions can be observed histologically in calves following severe bRSV infections [45, 46]. A model compound, structurally related to presatovir, was used in this proof-of-concept treatment model. Two randomized, blinded, placebo-controlled studies were designed with IV dosing of the model compound beginning 24 and 72 h post-infection. A clear clinical benefit with respect to viral load reduction, symptoms scores, respiration rates, and lung lesion pathology was observed supporting the progression of presatovir into the clinic [46]. In totality, the cotton rat and bovine model data provided significant confidence in the ability of presatovir to effectively inhibit the progression of RSV infection from the nasal cavity into the lower lung and prevent severe disease pathology.

9.3.3 Spectrum of Activity

An important component in the selection of the candidate included a thorough profiling of the spectrum of activity across the different RSV A and B virus strains. The potency of presatovir was evaluated towards 75 geographically diverse RSV A and B clinical isolates and was found to have a very narrow range of activity (Figure 9.9). The mean potency was $EC_{50} = 0.43$ nM and the vast majority, ~95% of the isolates, were within twofold of this average EC_{50} [38]. Only three isolates demonstrated more than a twofold loss of activity relative to the mean, and the other isolate was the Memphis 37b clinical strain (M37) that was developed for use in human challenge studies.

9.4 Pharmacokinetics

9.4.1 Preclinical Intravenous and Oral Pharmacokinetics

During the *in vitro* optimization leading to presatovir, rat was used as the primary species to gauge oral potential of an analog. The more promising compounds were

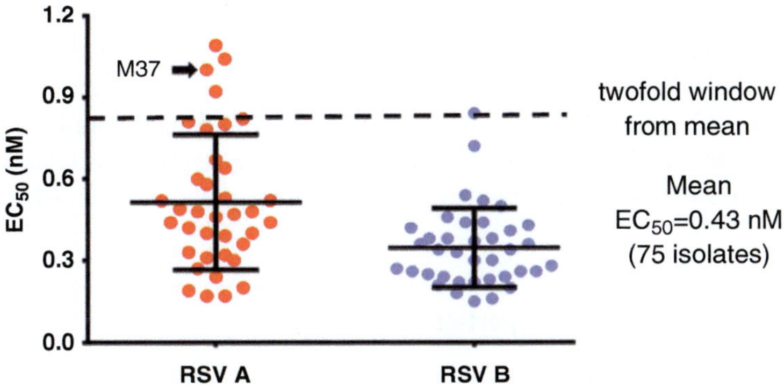

Figure 9.9 Antiviral activity of presatovir toward 75 geographically diverse RSV isolates. *Source*: Adapted from [1, 38].

also tested in higher species of dog and cynomolgus monkey to generate a more robust cross-species pharmacokinetic (PK) profile to enable a human prediction. In the case of presatovir, the predicted clearance in dog and monkey using microsomes was close to the observed clearance in these species following IV administration, suggesting microsomes, were accurately predicting *in vivo* clearance (Table 9.1) [1]. In contrast, the *in vivo* clearance in rat was much higher than predicted using microsomes suggesting a species-specific extrahepatic mechanism of clearance. Upon further investigation, it was found that plasma *N*-acetylation of the basic primary amine (pKa 8.3) was occurring in rat matrices but was minimal, or not observed, in the higher species including humans. Therefore, the primary route of clearance for presatovir in humans was predicted using preclinical data to be hepatic metabolism, leading to a clearance estimate of <0.11 L/h/kg based on the *in vitro* human microsome data.

The basic amine group on the pyrrolidine of presatovir also increased the volume of distribution across the preclinical species beyond that of blood volume. This was not surprising since basic amines embedded in lipophilic, permeable compounds often result in high volumes due to their accumulation in acidic compartments, e.g. lysosomes. The combination of higher volume of distribution and low clearance

Table 9.1 Preclinical intravenous and oral pharmacokinetics of presatovir [1].

	Sprague Dawley rat	Beagle dog	Cynomolgus monkey	Human adult
Microsome CL (L/h/kg)	0.41	0.26	0.22	<0.16
Observed CL (L/h/kg)	1.93	0.23	0.36	
Terminal $t_{1/2}$ (h)	3.4	8.9	10	35
V_{ss} (L/kg)	9.75	2.91	4.95	
Oral F%	46 ± 7	130 ± 36	60 ± 8	

Source: Adapted from [1].

both favored a long mean residence time for presatovir in human, supporting the potential for once-daily administration. Oral bioavailability was >46% across all species and >60% in higher species, suggesting that a 50% oral bioavailability in human was a reasonable prediction. Overall, presatovir was well-behaved orally across preclinical species and demonstrated a good *in vitro in vivo* clearance correlation in the higher species that supported the potential for a once-daily oral administration in human.

9.4.2 Lung Distribution Properties

Effective penetration into lung tissue was considered an important property for any candidate compound. Presatovir concentrations in lung tissue and lung epithelial lining fluid (ELF) were measured over 12 h following IV administration in rat and compared to the plasma concentrations (Figure 9.10). High concentrations relative to plasma were observed in both lung tissue and ELF, and the half-life of the compound in each tissue compartment was comparable to that in plasma [1]. Similar high ratios of lung tissue or ELF to plasma concentrations were also observed in dog and calves but at more limited timepoints (data not shown). The favorable lung penetration properties were demonstrated to be the direct result of the basic amine group by performing a comparable experiment with the hydroxyl analog **12**. Under

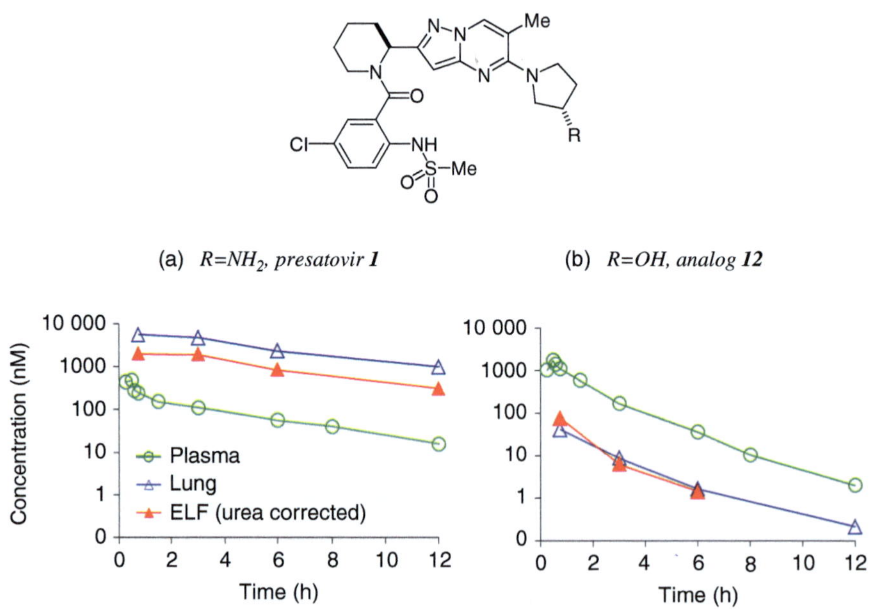

Figure 9.10 Lung tissue and epithelial lining fluid (ELF) drug concentrations compared to plasma concentrations following IV administration of compounds at 1 mg/kg in Sprague-Dawley rat. (a) Presatovir concentrations in lung tissue and ELF compared to plasma. (b) Lung tissue, ELF, and plasma concentrations of a presatovir analog, **12**, which replaces the basic amine group with a hydroxyl. Lung tissue and ELF concentrations are lower than plasma concentrations for the hydroxyl analog. *Source*: Adapted from [1, 31].

the same experimental conditions **12** showed low lung tissue and ELF concentrations well below plasma levels (Figure 9.10b). Thus, the introduction of the basic amine in presatovir not only drove high antiviral potency but also favorable lung tissue distribution properties.

9.4.3 Human Dose Prediction

Projecting effective plasma exposures for clinical efficacy was challenging given the absence of clinical efficacy data from earlier RSV inhibitors to guide the program, and the unknown translational value of the preclinical models. In addition, a further complication was the plasma compartment is the only tissue that can be readily sampled in the clinic, not the lung tissue where RSV replicates. Therefore, the program leaned on experience from a prominent respiratory small molecule antiviral, Tamiflu for influenza, in order to estimate target human exposures in plasma. Tamiflu is an approved oral influenza treatment that has an extracellular target similar to the fusion protein target of presatovir and also contains a basic amine group. Plasma trough concentrations of tamiflu at its approved clinical dose are approximately 21-fold greater than its average paEC_{50} toward different strains of influenza [1]. Therefore, a similar 21-fold increase over the RSV isolate mean EC_{50} = 0.43 nM, adjusting for human plasma protein binding (21-fold shift), was used as the preliminary trough plasma concentration target (~190 nM). Furthermore, based on the antiviral EC_{50} dose-response curves this concentration was estimated to be approximately five-fold above the 95% effective inhibition concentration, a concentration deemed suitably high to effectively inhibit the RSV virus and reduce the risk of resistance development. A low human daily dose of <50 mg was projected to achieve this concentration based on the preclinical pharmacokinetic studies.

Lipophilic compounds containing ionizable amine groups can frequently be associated with significant hERG inhibition and increased cardiovascular risk [47]. In the case of presatovir, there was no hERG signal *in vitro* up to 30 μM and standard preclinical cardiovascular studies in animals to support the IND did not highlight any cardiovascular risks. Furthermore, 28-day safety studies conducted in preclinical adult animal species to support the IND indicated that presatovir had a good safety profile at the targeted exposures in the clinic, supporting its selection for development.

9.5 Clinical Development

9.5.1 Regulatory Considerations

When the presatovir clinical development strategy was being formulated, there was no published guidance regarding the development of RSV antiviral therapies from regulatory agencies. Based on Gilead's experience in developing the influenza antiviral therapy Tamiflu, together with an appreciation for developing drugs for pediatric use such as tenofovir, and numerous interactions with regulatory agencies and key opinion leaders, a development strategy was initiated. Ultimately, the strategy was consistent with future guidance issued by the US Food & Drug Administration (FDA) and the European Medicines Agency (EMA) in October 2017 and July 2018 respectively [48, 49]. Given RSV's epidemiology, the at-risk population for severe RSV disease was identified as term infants younger than 6 months of age, preterm infants (≤35 weeks of gestation), older or elderly adults, patients with underlying disease such as chronic lung disease (e.g. bronchopulmonary dysplasia, cystic fibrosis, chronic obstructive pulmonary disease) or cardiac disease (e.g. hemodynamically significant congenital heart disease, congestive heart failure), and immunocompromised patients (e.g. hematopoietic stem cell transplant [HSCT] recipients and lung transplant patients). RSV infections primarily occur in young and vulnerable pediatric populations; therefore, the traditional approach of first accumulating safety, pharmacokinetic, and clinical efficacy data in adults was challenged by the agencies. However, while both agencies indicated that *in vivo* antiviral activity in well-characterized animal models might support the initiation of pediatric programs, the preference for demonstrating clinical proof-of-concept in humans led our program to pursue a development path that included Phase 1 and Phase 2 adult studies to evaluate tolerability, pharmacokinetics, and potential clinical benefit. As a general guideline, the FDA recommended that at least 100 adults be exposed to presatovir in clinical trials before initiating a pediatric program.

RSV infection manifests differently depending upon the target patient population. Therefore, the definition of clinical benefit varies. For instance, in clinical trials enrolling outpatients with mild disease, the endpoints are distinct from trials enrolling hospitalized patients with severe disease. Another challenge was the variable manifestations of RSV infection-related symptoms across different age groups. Physical and physiologic differences among children can also have a significant impact on not only what and how symptoms manifest, but also hospitalization rates and the need for supportive care (e.g. supplemental oxygen and hydration). The integrity of the immune response to RSV can also significantly impact the symptom profile of an infected individual, such that lack of symptoms in a severely immunocompromised long-term RSV shedder can represent a harbinger of worse outcomes. Finally, and perhaps the most challenging aspect of RSV clinical trials is that acceptable clinical endpoints were not well-defined, and the clinical tools for evaluating signs and symptoms, such as patient-reported outcomes, were not validated. The FDA now provides clear guidance citing that indications should be supported by two adequate, well-controlled trials, but the option of conducting a "single persuasive trial" may be viable. Furthermore, two trials that differ in design parameters and patient populations would be substantially more useful than two identically designed trials.

9.5.2 Phase 1 Program

A total of four Phase 1 studies were performed with presatovir in healthy adults. The first-in-human (FIH), a food effect study, a mass balance study (results unpublished), and a drug–drug interaction study [50, 51]. These studies demonstrated that presatovir was well tolerated and plasma exposures increased in an approximately dose-proportional manner. Moreover, the half-life of the compound was ~35 h, supporting continued development as a once-daily or less frequent dosing regimen. Target trough plasma exposures 24 h following dosing were reached at a dose of ~20 mg, consistent with the human prediction. Administration with a high-calorie high-fat meal did not alter the rate or extent of presatovir absorption. In the drug–drug interaction study, the data supported coadministration of presatovir with inhibitors of P-gp, BCRP, OATP1B1/1B3 or CYP3A, but not with moderate or strong CYP3A4 inducers [51].

9.5.3 Phase 2 Program

A randomized, double-blind, placebo-controlled, RSV challenge study was conducted to establish proof-of-concept for presatovir treatment in healthy adults experimentally infected with the M37 strain of RSV (Study NCT017564829) [2]. Participants were inoculated with RSV through the nose, then monitored daily with a PCR-based RSV assay. At the time of a positive test for RSV or 5 days after inoculation, whichever occurred first, participants were randomly assigned to receive presatovir or placebo in one of seven sequential cohorts evaluating various dose regimens (Figure 9.11).

Participants who received presatovir treatment had lower viral load (adjusted mean, 250.7 vs. 757.7 log10 plaque-forming-unit equivalents [PFUe] × hours per milliliter; P <0.001) (Figure 9.11), lower total mucus weight (mean, 6.9 vs. 15.1 g; P = 0.03), and a more significant change from baseline in symptom score AUC (adjusted mean, −20.2 vs. 204.9 × hours; P = 0.005). thereby establishing clinical proof-of-concept for presatovir treatment.

During preparations for the pediatric program, four Phase 2b studies were initiated to explore the clinical effects and tolerability of presatovir in high-risk patient populations that included hospitalized adults with cardiopulmonary disease [5], lung transplant patients [6], and HSCT patients [3, 4]. Among HSCT patients, two studies were conducted to evaluate treatment of upper respiratory tract infection (URTI, NCT02254408) and treatment of lower respiratory tract infection (LRTI, NCT02254421) [3, 4]. In the URTI study, 189 HSCT recipients who were infected with RSV in the outpatient setting and treated with presatovir (n = 95) or placebo (n = 90) had a numerically lower time-weighted average decline in the RSV viral load (treatment difference −0.33 $\log_{10}$ copies/mL, 95% CI −0.64 to −0.02; P = 0.040) or lower rate of progression to a lower respiratory tract complication (presatovir 11.2% vs. placebo 19.5%, odds ratio 0.50; 95% CI: 0.022–1.18; P = 0.11) (Figure 9.12). Unfortunately, because these were coprimary endpoints, the viral load effect was not considered statistically significant. In the LRTI study, among 60 HSCT patients who were hospitalized and infected with RSV in the lower respiratory tract and treated with presatovir (n = 30) or placebo (n = 29), there was no meaningful reduction in the viral load (treatment difference −0.02 $\log_{10}$ copies/mL, 95% CI: −0.62 to 0.057; P = 0.94), median supplemental oxygen-free days (26 vs. 28 days; P = 0.84), incident respiratory failure (10.3% vs. 10.7%, P = 0.98), or all-cause mortality (0 vs. 7.1%; P = 0.19). Despite these results, an important observation was made in a prespecified subgroup analysis. Among patients with baseline lymphopenia defined as

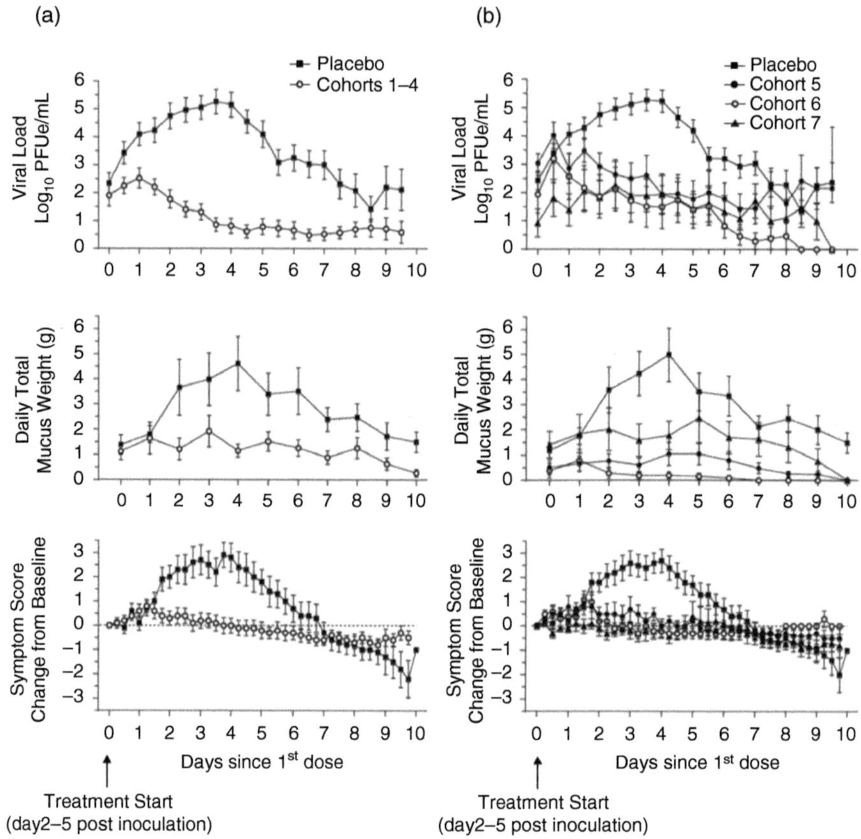

Figure 9.11 RSV challenge study participants were nasally inoculated with the M37b challenge strain. Subjects became infected, as determined using a quantitative reverse-transcriptase polymerase chain reaction assay of nasal wash samples collected twice daily, on different days so treatment began on different study days postinoculation. The virus plots represent participant plaque forming unit (PFU) data on nasal wash samples collected twice daily starting from the first measurement following the first dose. Mucus weights were plotted at the time corresponding to the beginning of each day's mucus weight collection period (i.e. weight plotted for the first day of treatment is from facial tissues collected between the morning of the first day of study treatment and the morning of the second day). The symptom diary was completed four times daily and total symptom scores began with the last assessment made before the first dose of the study drug was administered (at baseline). Values are ±SE. (a) Cohorts 1–4 received a first dose of 50 mg, then 25 mg daily for the next 4 days. (b) Cohorts 5–7: Cohort 5 received a first dose of 50 mg, then 25 mg daily for the next 2 days; Cohort 6 received a single dose of 100 mg; Cohort 7 received a first dose of 10 mg, then 5 mg daily for the next 4 days. *Source*: Reproduced from Ref. [2] with permission as co-authors of Ref. [2].

<200 cells/μL, treatment with presatovir was associated with a near absolute 50% reduction in the development of lower respiratory tract complications (treatment difference −0.51%, 95% CI: −77.8 to −13.1; P = 0.008). This observation suggested that in a setting where the immune system is unable to clear the RSV infection, antiviral treatment with presatovir may be effective in preventing progression from the upper to the lower respiratory tract. Furthermore, these results provide evidence

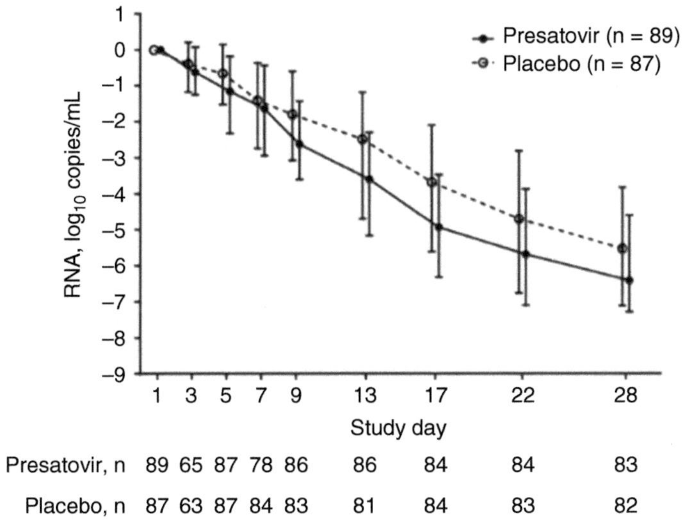

Figure 9.12 Median change from baseline nasal RSV RNA in HSCT patients. *Source*: Reproduced from Ref. [4] with permission as co-authors of Ref. [4].

of how immunologic integrity, or lack thereof, might serve as a biomarker to enrich clinical trial enrollment for clinical events.

The remaining two Phase 2b studies in lung transplant patients (NCT02534350) and hospitalized adults (NCT02135614) did not demonstrate a clinical benefit [5, 6]. Neither study demonstrated a significant effect on nasal RSV load, respiratory infection symptoms, lung function (lung transplant study), or the duration of hospital stay (hospitalized adult study). Given that the challenge study and HSCT URTI study indicate that a direct-acting antiviral may be effective when treatment is implemented early in infection or in immunocompromised settings, the totality of these data suggests that delayed treatment, when the disease has already progressed to the lower respiratory tract, or after the patient has been hospitalized, may not be effective for an RSV fusion inhibitor.

9.5.4 Treatment Emergent Mutations

For all of the Phase 2 programs, RSV RNA from nasal samples was monitored for the development of presatovir resistance. In the experimentally infected study, 28 treatment emergent F protein substitutions were identified [52]. The frequency of phenotypically resistant substitutions was higher with regimens containing lower presatovir doses and shorter treatment duration. In the natural infection studies, the frequency of mutations was highest among HSCT patients, where participants were exposed to presatovir for approximately 21 days (HSCT recipients, 6/29 (20.7%) with upper or lower respiratory tract infection, 1/35 (2.9%) lung transplant recipients, and 1/80 (1.3%) hospitalized patients treated with presatovir). Participants with phenotypic presatovir resistance tended to have higher nasal viral load and smaller decreases in viral load during treatment relative to those without, but no impact on clinical manifestations of RSV disease was observed in any of these studies. Importantly, no presatovir-associated mutations altered RSV susceptibility to palivizumab or ribavirin.

9.6 Conclusion

Presatovir was discovered following optimization of a high-throughput screening hit into a potent, orally bioavailable, and broadly active RSV fusion inhibitor toward 75 RSV A and B clinical isolates. Key to the optimization was the modification of a tautomeric pyrazolopyrimidinone heterocycle to a non-tautomeric pyrazolopyrimidine that dramatically improved permeability enabling high oral bioavailability. In addition, the inclusion of an amino-pyrrolidine on the pyrazolopyrimidine heterocycle lowered the log D and hepatic metabolism, provided exquisite picomolar potency, and led to favorable lung tissue distribution properties *in vivo*. Presatovir's mechanism of action was found to be inhibition of the pre- to post-fusion conformational change of the F protein. Prior to clinical development, presatovir and a model compound demonstrated antiviral and clinical effects in the cotton rat and bovine RSV models. Phase 1 clinical studies indicated presatovir was safe, well-tolerated, and demonstrated a long terminal half-life of 35 h, supporting multiple dosing strategies across the Phase 2 efficacy studies. In a human challenge study, presatovir was found to be effective at reducing nasal viral load and symptoms when dosed once daily following RSV detection in the nasal cavity. However, across multiple Phase 2 studies, presatovir treatment failed to demonstrate a significant clinical effect other than in HCST patients with confirmed upper respiratory tract infection, suggesting that fusion inhibition may only be effective when initiated early in infection. Despite the lack of progress since these studies for presatovir, multiple other RSV inhibitors have entered the clinic. It is anticipated that the results and learnings from presatovir as the first inhibitor to progress deep into the clinic for RSV will help provide valuable insights for these future programs and may lead to the first new RSV treatment approval in almost 40 years.

References

1 Mackman RL, Sangi M, Sperandio D, Parrish JP, Eisenberg E, Perron M, et al. Discovery of an oral respiratory syncytial virus (RSV) fusion inhibitor (GS-5806) and clinical proof of concept in a human RSV challenge study. *J Med Chem*. 2015;58(4):1630–43.

2 DeVincenzo JP, Whitley RJ, Mackman RL, Scaglioni-Weinlich C, Harrison L, Farrell E, et al. Oral GS-5806 activity in a respiratory syncytial virus challenge study. *N Engl J Med*. 2014;371(8):711–22.

3 Marty FM, Chemaly RF, Mullane KM, Lee DG, Hirsch HH, Small CB, et al. A Phase 2b, randomized, double-blind, placebo-controlled multicenter study evaluating antiviral effects, pharmacokinetics, safety, and tolerability of presatovir in hematopoietic cell transplant recipients with respiratory syncytial virus infection of the lower respiratory tract. *Clin Infect Dis*. 2020;71(11):2787–95.

4 Chemaly RF, Dadwal SS, Bergeron A, Ljungman P, Kim YJ, Cheng GS, et al. A Phase 2, randomized, double-blind, placebo-dontrolled trial of presatovir for the treatment of respiratory syncytial virus upper respiratory tract infection in hematopoietic-cell transplant recipients. *Clin Infect Dis*. 2020;71(11):2777–86.

5 Hanfelt-Goade D, Maimon N, Nimer A, Riviere F, Catherinot E, Ison M, et al. A Phase 2b, randomized, double-blind, placebo-controlled trial of presatovir (GS-5806), a novel oral RSV fusion inhibitor, for the treatment of respiratory syncytial virus (RSV) in hospitalized adults. *Am J Respir Crit Care Med*. 2018;197:A4457.

6 Gottlieb J, Torres F, Haddad T, Dhillon G, Dilling DF, Knoop C, et al. A randomized controlled trial of presatovir for respiratory syncytial virus after lung transplant. *J Heart Lung Transplant*. 2023;42(7):908–16.

7 Collins PL, Fearns R, Graham BS. Respiratory syncytial virus: virology, reverse genetics, and pathogenesis of disease. *Curr Top Microbiol Immunol*. 2013;372:3–38.

8 Johnson JE, Gonzales RA, Olson SJ, Wright PF, Graham BS. The histopathology of fatal untreated human respiratory syncytial virus infection. *Mod Pathol*. 2007;20(1):108–19.

9 Bergeron HC, Tripp RA. Immunopathology of RSV: an updated review. *Viruses*. 2021;13(12):2478.

10 Hall CB, Weinberg GA, Iwane MK, Blumkin AK, Edwards KM, Staat MA, et al. The burden of respiratory syncytial virus infection in young children. *N Engl J Med*. 2009;360(6):588–98.

11 Falsey AR, Hennessey PA, Formica MA, Cox C, Walsh EE. Respiratory syncytial virus infection in elderly and high-risk adults. *N Engl J Med*. 2005;352(17):1749–59.

12 Thompson WW, Shay DK, Weintraub E, Brammer L, Cox N, Anderson LJ, et al. Mortality associated with influenza and respiratory syncytial virus in the United States. *Jama*. 2003;289(2):179–86.

13 Englund JA, Sullivan CJ, Jordan MC, Dehner LP, Vercellotti GM, Balfour HH Jr. Respiratory syncytial virus infection in immunocompromised adults. *Ann Intern Med*. 1988;109(3):203–8.

14 Boeckh M, Englund J, Li Y, Miller C, Cross A, Fernandez H, et al. Randomized controlled multicenter trial of aerosolized ribavirin for respiratory syncytial virus upper respiratory tract infection in hematopoietic cell transplant recipients. *Clin Infect Dis*. 2007;44:245–9.

15 Taber LH, Knight V, Gilbert BE, McClung HW, Wilson SZ, Norton HJ, et al. Ribavirin aerosol treatment of bronchiolitis associated with respiratory syncytial virus infection in infants. *Pediatrics*. 1983;72(5):613–8.

16 Hall CB, McBride JT, Walsh EE, Bell DM, Gala CL, Hildreth S, et al. Aerosolized ribavirin treatment of infants with respiratory syncytial viral infection. A randomized double-blind study. *N Engl J Med*. 1983;308(24):1443–7.

17 Smith DW, Frankel LR, Mathers LH, Tang AT, Ariagno RL, Prober CG. A controlled trial of aerosolized ribavirin in infants receiving mechanical ventilation for severe respiratory syncytial virus infection. *N Engl J Med*. 1991;325(1):24–9.

18 Collins PL, Melero JA. Progress in understanding and controlling respiratory syncytial virus: still crazy after all these years. *Virus Res*. 2011;162(1–2):80–99.

19 Shang Z, Tan S, Ma D. Respiratory syncytial virus: from pathogenesis to potential therapeutic strategies. *Int J Biol Sci*. 2021;17(14):4073–91.

20 Sperandio D, Mackman R. Respiratory syncytial virus (RSV) fusion inhibitors. In: Desai MC, Meanwell NA, editors. *Successful Strategies for the Discovery of Antiviral Drugs*. 1st ed. Oxford: Royal Society of Chemistry; 2013. p. 29–62.

21 Bonfanti JF, Roymans D. Prospects for the development of fusion inhibitors to treat human respiratory syncytial virus infection. *Curr Opin Drug Discov Devel*. 2009;12(4):479–87.

22 Sun Z, Pan Y, Jiang S, Lu L. Respiratory syncytial virus entry inhibitors targeting the F protein. *Viruses*. 2013;5(1):211–25.

23 Bergeron HC, Tripp RA. Emerging small and large molecule therapeutics for respiratory syncytial virus. *Expert Opin Investig Drugs*. 2020;29(3):285–94.

24 Olszewska W, Openshaw P. Emerging drugs for respiratory syncytial virus infection. *Expert Opin Emerg Drugs*. 2009;14(2):207–17.

25 The IMpact-RSV study group. Palivizumab, a humanized respiratory syncytial virus monoclonal antibody, reduces hospitalization from respiratory syncytial virus infection in high-risk infants. *Pediatrics*. 1998;102(3):531–7.

26 Hammitt LL, Dagan R, Yuan Y, Baca Cots M, Bosheva M, Madhi SA, et al. Nirsevimab for prevention of RSV in healthy late-preterm and term infants. *N Engl J Med*. 2022;386(9):837–46.

27 Chapman J, Cockerill GS. Discovery and development of RSV604. In: Kazmierski WM, editor. *Antiviral Drugs: from Basic Discovery Through Clinical Trials*. 1st ed. New Jersey: Wiley; 2011. p. 367.

28 Ahmad A, Eze K, Noulin N, Horvathova V, Murray B, Baillet M, et al. EDP-938, a respiratory syncytial virus inhibitor, in a human virus challenge. *N Engl J Med*. 2022;386(7):655–66.

29 Rhodin MHJ, McAllister NV, Castillo J, Noton SL, Fearns R, Kim IJ, et al. EDP-938, a novel nucleoprotein inhibitor of respiratory syncytial virus, demonstrates potent antiviral activities in vitro and in a non-human primate model. *PLoS Pathog*. 2021;17(3):e1009428.

30 Enanta Pharmaceuticals (Internet). Enanta pharmaceuticals reports topline data from the RSVP study of EDP-938 in otherwise healthy adults with community-acquired respiratory syncytial virus (RSV) [Internet]. Watertown, Massachussetts: 2022 May 18 [cited 2024 Jun 27]. Available from: https://ir.enanta.com/news-releases/news-release-details/enanta-pharmaceuticals-reports-topline-data-rsvp-study-edp-938

31 Mackman R, Sangi M, Sperandio D, Parrish J, Eisenberg E, Perron M, et al. Oral respiratory syncytial virus (RSV) fusion inhibitor GS-5806: potency and oral pharmacokinetics optimization. Poster presented at: 28th International Conference on Antiviral Research, Poster 69; 2015 May 11–15; Rome, Italy.

32 Mackman RL, Hui HC, Perron M, Murakami E, Palmiotti C, Lee G, et al. Prodrugs of a 1′-CN-4-aza-7,9-dideazaadenosine C-nucleoside leading to the discovery of remdesivir (GS-5734) as a potent inhibitor of respiratory syncytial virus with efficacy in the african green monkey model of RSV. *J Med Chem*. 2021;64(8):5001–17.

33 Beigel JH, Tomashek KM, Dodd LE, Mehta AK, Zingman BS, Kalil AC, et al. Remdesivir for the treatment of Covid-19 - final report. *N Engl J Med.* 2020;383(19):1813–26.

34 Douglas JL, Panis ML, Ho E, Lin K-Y, Krawcyzk SH, Grant, DM, et al. Small molecules VP-14637 and JNJ-2408068 inhibit respiratory syncytial virus fusion by similar mechanisms. *Antimicrob Agents Chemother.* 2005;49:2460–66.

35 Cadena-Cruz C, Villarreal Camacho JL, De Ávila-Arias M, Hurtado-Gomez L, Rodriguez A, San-Juan-Vergara H. Respiratory syncytial virus entry mechanism in host cells: a general overview. *Mol Microbiol.* 2023;120(3):341–50.

36 Feng Z, Xu L, Xie Z. Receptors for respiratory syncytial virus infection and host factors regulating the life cycle of respiratory syncytial virus. *Front Cell Infect Microbiol.* 2022;12:858629.

37 Roymans D, De Bondt HL, Arnoult E, Geluykens P, Gevers T, Van Ginderen M, et al. Binding of a potent small-molecule inhibitor of six-helix bundle formation requires interactions with both heptad-repeats of the RSV fusion protein. *Proc Natl Acad Sci.* 2010;107(1):308–13.

38 Perron M, Stray K, Kinkade A, Theodore D, Lee G, Eisenberg E, et al. GS-5806 inhibits a broad range of respiratory syncytial virus clinical isolates by blocking the virus-cell fusion process. *Antimicrob Agents Chemother.* 2016;60(3):1264–73.

39 Hallak LK, Spillmann D, Collins PL, Peeples ME. Glycosaminoglycan sulfation requirements for respiratory syncytial virus infection. *J Virol.* 2000;74(22):10508–13.

40 Samuel D, Xing W, Niedziela-Majka A, Wong JS, Hung M, Brendza KM, et al. GS-5806 inhibits pre- to postfusion conformational changes of the respiratory syncytial virus fusion protein. *Antimicrob Agents Chemother.* 2015;59(11):7109–12.

41 Taylor G. Animal models of respiratory syncytial virus infection. *Vaccine.* 2017;35(3):469–80.

42 Bem RA, Domachowske JB, Rosenberg HF. Animal models of human respiratory syncytial virus disease. *Am J Physiol Lung Cell Mol Physiol.* 2011;301(2):L148–56.

43 Rouan M-C, Gevers T, Roymans D, de Zwart L, Nauwelaers D, De Meulder M, et al. Pharmacokinetics-pharmacodynamics of a respiratory syncytial virus fusion inhibitor in the cotton rat model. *Antimicrob Agents Chemother.* 2010;54(11):4534–39.

44 Cianci C, Yu K-L, Combrink K, Sin N, Pearce B, Wang A, et al. Orally active fusion inhibitor of respiratory syncytial virus. *Antimicrob Agents Chemother.* 2004;48:413–22.

45 Gershwin LJ, Schelegle ES, Gunther RA, Anderson ML, Woolums AR, Larochelle DR, et al. A bovine model of vaccine enhanced respiratory syncytial virus pathophysiology. *Vaccine.* 1998;16(11–12):1225–36.

46 Jordan R, Shao M, Mackman, RL, Perron M, Cihlar T, Lewis SA, et al. Antiviral efficacy of a respiratory syncytial virus (RSV) fusion inhibitor in a bovine model of RSV infection. *Antimicrob Agents Chemother.* 2015;59(8):4889–900.

47 Garrido A, Lepailleur A, Mignani SM, Dallemagne P, Rochais C. hERG toxicity assessment: useful guidelines for drug design. *Eur J Med Chem.* 2020;195:112290.

48 Respiratory Syncytial Virus Infection: Developing antiviral drugs for prophylaxis and treatment: Guidance for Industry [Internet]. Washington. Food and Drug Administration. 2017 October [updated 2018 Aug 24; cited 2024 Jun 27]. Available from: https://www.fda.gov/regulatory-information/search-fda-guidance-documents/respiratory-syncytial-virus-infection-developing-antiviral-drugs-prophylaxis-and-treatment-guidance

49 Guideline on the clinical evaluation of medicinal products indicated for the prophylaxis or treatment of respiratory syncytial virus (RSV) disease [Internet]. Amsterdam. European Medicines Agency. 2018 Oct 18 [cited 2024 Jun 27]. Available from: https://www.ema.europa.eu/en/treatment-prophylaxis-respiratory-syncytial-virus-rsv-disease-scientific-guideline

50 German P, Xin Y, Chien JW, Weng W, Mackman R, Lewis SA, et al. Phase 1 first-in-human, single- and multiple-ascending dose, and food effect studies to assess the safety, tolerability, and pharmacokinetics of presatovir for the treatment of respiratory syncytial virus infection. *J Clin Pharmacol.* 2018;58(8):1025–34.

51 Xin Y, Weng W, Murray BP, Eisenberg EJ, Chien JW, Ling J, et al. The drug-drug interaction profile of presatovir. *J Clin Pharmacol.* 2018;58(6):771–80.

52 Porter DP, Guo Y, Perry J, Gossage DL, Watkins TR, Chien JW, et al. Assessment of drug resistance during Phase 2b clinical trials of presatovir in adults naturally infected with respiratory syncytial virus. *Antimicrob Agents Chemother.* 2020;64(9):e02312–19.

10

Remdesivir for the Treatment of COVID-19 and Other Viral Infections with Pandemic Potential

Tomas Cihlar, Meghan S. Vermillion, Joe Llewellyn, Charlotte Hedskog, Richard L. Mackman

Gilead Sciences, Foster City, CA, USA

10.1 Introduction

COVID-19 became the most significant global viral pandemic of the past 100 years, resulting in almost 800 million infections and more than 7 million deaths globally over the course of 4 years [1, 2]. Today, it stands out as an urgent reminder of how critically important the timely availability of effective countermeasures is in our globally connected world. These countermeasures include diagnostics, vaccines, and therapeutics against viruses with pandemic potential, together with readiness for timely clinical testing, scaled manufacturing, and effective deployment supporting global access. The proactive management of emerging viral threats using the countermeasures is critically important for minimizing their severe impact on global health, economies, and the stable functioning of societies.

SARS-CoV-2 (severe acute respiratory syndrome coronavirus 2), the causative agent of the COVID-19 pandemic, emerged in the human population in late 2019 as a new member of the Coronaviridae family [3]. This family includes two other related pathogenic coronaviruses, SARS-CoV and MERS-CoV (Middle East respiratory syndrome coronavirus), which emerged in 2003 and 2012, respectively [4, 5]. In addition to coronaviruses, several other RNA virus families have been associated with significant outbreaks of global concern in recent years. These include filovirus outbreaks in West Africa, the Democratic Republic of the Congo, and, most recently, Uganda; outbreaks of flaviviruses such as dengue, Zika, and yellow fever in multiple low- and middle-income countries; and several outbreaks of influenza, including the 2009 H1N1 swine influenza pandemic [6–10]. Despite the significant burden of these viral infections and their high risk of triggering future outbreaks with pandemic potential, there are limited effective treatment options available against most of these pathogenic viruses.

Trends in Antiviral Drug Development, First Edition. Edited by Michael J. Sofia and Zhengqiang Wang.

RDV (Veklury® [Figure 10.1]) was approved by the US FDA as the first antiviral for the treatment of COVID-19 [11]. It represents a rather unique example of the preemptive approach initiated several years before the emergence of SARS-CoV-2 with a specific goal to advance this molecule for the treatment of emerging and neglected viral infections [12, 13]. FDA granted COVID-19 emergency use authorization for RDV in early May 2020, more than 6 months before the first COVID-19 vaccine, Comirnaty, reached the same regulatory status [14] and more than 18 months before the emergency use authorization was issued for Paxlovid and Lagevrio, two additional antivirals for the treatment of COVID-19 [15, 16]. The rapid advancement of RDV ahead of other coronavirus countermeasures was a result of more than a decade of extensive work focused on the design, preclinical testing, and clinical development of this direct-acting antiviral agent [17].

RDV is a mono-phosphoroamidate prodrug of an adenosine analog and belongs to a class of nucleoside and nucleotide small molecule antivirals [18, 19]. Multiple members of this drug class have been developed and are being widely utilized as successful antivirals for the treatment and/or prevention of HIV, viral hepatitis, and herpesvirus infections [20–22]. These agents resemble and structurally mimic the natural nucleotides, which are universal substrates and building blocks of nucleic acids including viral DNA and RNA. As such, nucleoside and nucleotide analogs often exhibit activity against several viral families and thus are uniquely poised for development as broad-spectrum pandemic countermeasures with efficacy against multiple pathogenic viruses.

Potent *in vitro* activity and *in vivo* efficacy of RDV across a broad range of animal models have been established against multiple families of pathogenic RNA viruses including filoviruses, pneumoviruses, paramyxoviruses, flaviviruses, and coronaviruses [17, 19, 23, 24]. Following the demonstration of clinical safety and efficacy across different stages of COVID-19 disease in multiple patient populations, including pediatrics [25–27], RDV became a standard of care for the treatment of COVID-19 in hospitalized patients. Although safe and efficacious, RDV has limited oral availability and requires daily administration via intravenous infusion, which restricts its broader use, particularly in outpatient settings. A novel antiviral molecule, obeldesivir, has been specifically designed to address this limitation [28, 29]. ODV (GS-5245) is an oral prodrug of GS-441524, the parent nucleoside of RDV. ODV forms the same active metabolite as RDV inside cells and therefore acts through an identical mechanism as RDV. In 2022, ODV entered Phase 3 clinical testing for the oral treatment of both standard- and high-risk COVID-19 patients (NCT05715528 and NCT05603143, respectively).

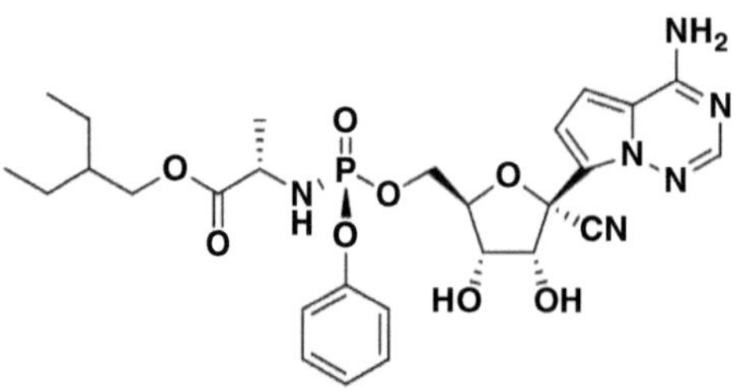

Figure 10.1 Structure of remdesivir. *Source*: R. Mackman.

Because of their potent broad-spectrum antiviral activity, clinical safety established across broad patient populations, and complementary routes of administration, both RDV and ODV are likely to play significant roles in future antiviral pandemic preparedness and outbreak response strategies.

10.2 Discovery of Remdesivir

10.2.1 Discovery of 1′-Modified C-Nucleosides

The 2′-β-Me *C*-nucleoside adenosine analog **1** is an inhibitor of the hepatitis C virus (HCV) RNA-dependent RNA polymerase (RdRp) and was identified by Gilead and, independently, Biocryst Pharmaceuticals (Table 10.1) [30, 31]. Nucleoside analogs are intracellularly metabolized by kinases into their respective triphosphate metabolites that, in turn, are inhibitors of the viral RdRp enzymes. Despite the promising antiviral activity of **1** in cells and the submicromolar potency of the active **1-TP**

Table 10.1 RSV and HCV SAR of 1′-modified 4-aza-7,9-dideazaadenosine *C*-nucleosides.

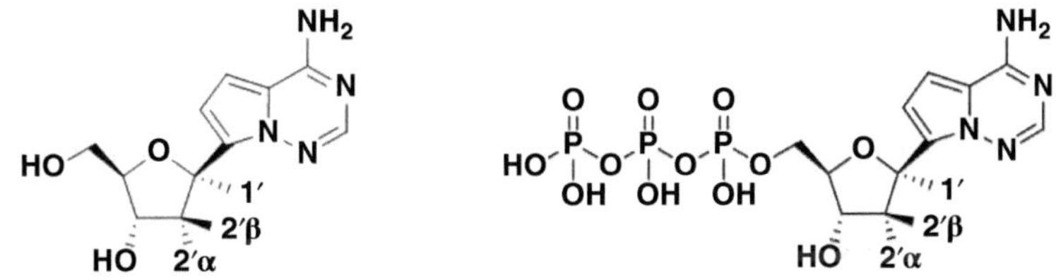

Cpd.	1′	2′-α	2′-β	HCV Huh-7 EC_{50} (μM)	RSV HEp-2 EC_{50} (μM)	MT-4 CC_{50} (μM)	HCV RdRp IC_{50} (μM)	RSV RdRp IC_{50} (μM)
	2′-β-Me compounds							
1	H	OH	Me	2.0			0.31	
2	CN	OH	Me	>44	>100	>57	0.37	38
	2′-β-H compounds							
3[a]	CN	OH	H	4.1	0.53	>57	5.6	1.1
4	Me	OH	H	38	5.5	4.5	20	1.5
5	C≡CH	OH	H	>88	>200	120	232	3.0
6	CN	F	H	>44	>100	32		0.8
7	H	OH	H	0.0035		<0.01		

CC_{50}, 50% inhibition of MT-4 T lymphoid cell line viability; EC_{50}, 50% inhibition of viral replication in the specified cells; IC_{50}, 50% inhibition of RNA synthesis by the triphosphates in an enzyme assay containing the RdRp.

a. GS-441524.

Source: R. Mackman.

metabolite toward HCV RdRp, it demonstrated significant toxicity in rats that prevented its advancement. This was thought to be due to incorporation of the active triphosphate metabolite by the mitochondrial RNA polymerase (POLRMT) into host mitochondrial RNA [32]. In the search for improved selectivity for *C*-nucleosides, 1′-modifications were explored, leading to the identification of the 1′-CN, 2′-β-Me analog **2**. The POLRMT selectivity was significantly improved, and the active metabolite **2-TP** was found to be a potent HCV RdRp inhibitor. However, **2** was not antivirally active in cells due to its poor metabolism to the triphosphate metabolite, leading to its eventual development as the phosphoramidate prodrug GS-6620 [32].

In parallel to the HCV program's efforts on *C*-nucleosides, a project targeting the respiratory syncytial virus (RSV) was initiated and included a screen of the Gilead nucleoside library in a phenotypic RSV A2 human epithelial (HEp-2) cell-based assay. Nucleoside **3** (also known as GS-441524), which bears only the 1′-CN modification on the ribose, was found to be ~10-fold more potent in the RSV cell-based antiviral assay compared with HCV, while the sterically larger 1′-modifications (e.g. **4**, **5**) had reduced antiviral activity. The 2′-F and 2′-β-Me modifications, **6** and **2**, respectively, were also inactive (Table 10.1), as were the natural *N*-nucleoside adenine or cytosine base analogs of **2** (data not shown) [19]. To assess selectivity, the compounds were evaluated in the rapidly proliferating MT-4 cell-based assay, which established that **3** was not cytotoxic, while the 1′-methyl analog **4** had no selectivity between the MT-4 assay and the HEp-2 RSV assay. The 1′-CN modification was clearly essential for selectivity since the 1′-unmodified analog **7** was highly toxic. Altogether, the antiviral structure-activity relationship (SAR) for the 1′-modifications on the 4-aza-7,9-dideaza adenosine C-nucleoside was quite steep, and the 1′-CN analog **3** was identified as the most promising inhibitor of RSV in cell culture [19].

The inherent potency of the 1′-modified triphosphate metabolites was also evaluated toward RSV RdRp using a biochemical assay. Consistent with the ~10-fold improved antiviral activity of **3** toward RSV compared with HCV, the active metabolite **3-TP** was found to be ~5-fold more potent as an RSV RdRp inhibitor than an HCV RdRp inhibitor. The triphosphate SAR for the 2′-β-Me analogs (e.g. **2**) clearly diverged between RSV and HCV RdRp enzymes, since the triphosphate **2-TP** that was 15-fold more potent than **3-TP** toward HCV RdRp was found to be 35-fold less potent toward RSV RdRp. The lack of tolerance by the RSV RdRp for the 2′-β-Me group was rationalized, using structural modeling studies, to be the result of steric effects in the RSV RdRp active site that presumably hinders the binding event prior to RNA incorporation [18]. Taken together, these studies established that the 1′-CN group together with the *C*-nucleobase were critical structural features for RSV potency and selectivity. The activity of **3** for RSV and HCV quickly prompted additional profiling toward several other respiratory RNA viruses, including hPIV-3 (human parainfluenza virus) and SARS-CoV, to establish it as a broad-spectrum antiviral [33].

10.2.2 Phosphoramidate Prodrug Optimization for RSV

While the antiviral activity of **3** for RSV was encouraging, we hypothesized that cell-based potency might be further improved through delivery of the monophosphate metabolite **3-MP** using phosphate prodrugs. Therefore, several different monophosphate prodrugs were evaluated, including bis-thioesters (SATE) [34], phosphordiamides (bisamidates) [35], and phosphoramidates. The phosphoramidate class proved to be the most effective for masking the charge of the phosphate, enabling entry into cells and rapid breakdown to the monophosphate **3-MP** [19, 36]. The metabolic pathway from the prodrug to the monophosphate is described in Section 10.3.1 [19, 36, 37]. The amidate prodrug optimization on **3** was conducted within the RSV program using a cervical HEp-2 cell line, but importantly, was reconfirmed in lung cells, including primary normal bronchial epithelial cells (NHBE) and human airway epithelial cultures (HAE).

The first prodrug evaluated was the isopropyl *Sp* (*S* stereochemistry at phosphorus) monoalanine prodrug (Table 10.2, entry 3). In HEp-2 cells, a disappointing 1.5-fold improved potency was discovered (Table 10.2, entry 3). Since the measured log D was still quite low, it was hypothesized that cell permeability and efficient recognition and cleavage by esterases may be hindering activation of the prodrug.

Table 10.2 Antiviral activity of monophenol mono-(*L*)-alanine ester phosphoramidate prodrugs.

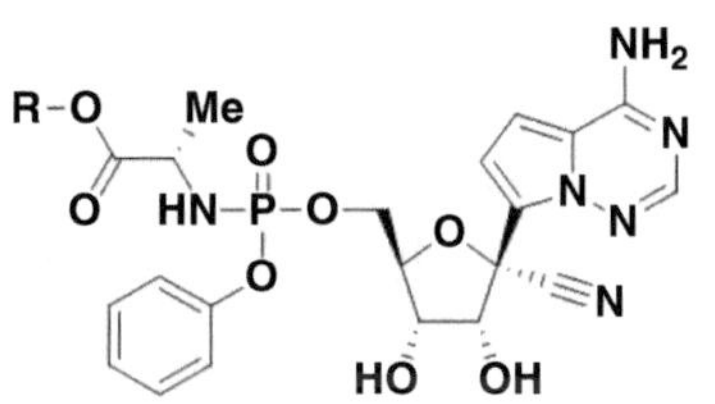

R[a]	P isomer	Log D	RSV HEp-2 EC_{50}/CC_{50} (μM)	RSV NHBE EC_{50} (μM)	EBOV HeLa EC_{50} (μM)	EBOV MΦs EC_{50} (μM)	MT-4 CC_{50} (μM)
Cpd. **3**	N/A	<0.3	0.53/>100	1.85		>10	>50
Ethyl	mix	0.6	0.30/>73	0.85	2.42		13
Isopropyl	Sp	1.1	0.36/>100	2.5	1.85	0.30	21
t-butyl	mix	1.7	4.5/>100	>5.0			>100
Cyclo-pentyl	mix	1.8	0.17/>80	0.35		0.12	8.8
CH_2-c-butyl	mix	2.1	0.024/23	0.13			
Cyclo-octyl	mix	2.3	0.028/48				
2-Ethylbutyl	mix	2.5	0.027/9.2	0.069	0.17	0.10	2.0
2-Ethylbutyl[b]	Sp	2.4	0.015/6.1	0.049	0.10	0.086	1.7
2-Ethylbutyl	Rp	2.5	0.028/26	0.065	0.08	0.11	3.0

a. Diastereomeric mixture at phosphorus unless stated otherwise. c, cyclo; N/A, not applicable; NHBE, normal human bronchial epithelial cells; MΦs, macrophages; *Sp*, *S* stereochemistry at phosphorus; *Rp*, *R* stereochemistry at phosphorus.
b. GS-5734, remdesivir
Source: R. Mackman.

Therefore, more lipophilic prodrugs were targeted through modification of the ester, including the 2-ethylbutyl ester (Table 10.2, entry 8), which, as its diastereomeric mixture at phosphorus, demonstrated a more impressive ~20-fold and 26-fold improved potency in HEp-2 cells and NHBE cells, respectively. In general, as lipophilicity increased, the potency also increased across a range of lower alkyl ester analogs (Table 10.2). When the measured log D was above 2, many of the prodrugs demonstrated RSV activity below 50 nM in the HEp-2 cell assay (Table 10.2, entries 6–8). Separation of the 2-ethylbutyl ester phosphorus diastereomers followed by X-ray crystallography allowed identification of the *Sp* isomer (designated as GS-5734 or RDV) as the most potent prodrug, with a significant 35-fold and 38-fold improved potency over parent nucleoside **3** in HEp-2 and NHBE cells, respectively (Table 10.2, entry 9) [19]. In human airway epithelial cells, a culture model that best represents human lung cells *in vitro*, the RSV potency of RDV was ~40-fold improved over parent nucleoside **3** [19].

Soon after the discovery of RDV for RSV, the West Africa Ebola outbreak occurred. It was quickly discovered through screening in partnership with the CDC and USAMRIID that RDV was highly potent toward Ebola in macrophages, a major site of early viral replication during human infection (Table 10.2) [12]. Prompted by this discovery, additional phosphoramidate prodrugs to those prepared during the RSV program were synthesized to expand the SAR. However, no significantly differentiated compounds beyond RDV were identified in this second campaign [18]. Similar to RSV, the potency of the *S*p isomer RDV for Ebola virus (EBOV), was superior to the *R*p isomer across multiple cell types including macrophages; therefore, RDV was advanced into *in vivo* studies for Ebola.

10.2.3 Synthesis

The development of a scalable and robust diastereoselective synthesis of RDV was paramount for rapid advancement into efficacy studies and the clinic to respond to the Ebola epidemic. This was achieved with the synthesis described in Figure 10.2 that began with the benzyl-protected ribolactone [18, 33]. The first step of introducing the C-nucleobase to generate the 1′-hydroxyl intermediate proved quite challenging to scale due to side reactions. Yields were improved by the use of the iodinated 4-aza-7,9-dideaza adenine base instead of the bromine-modified base, and the *i*-PrMgCl•LiCl Grignard reagent. Later, the addition of $LaCl_3$•2LiCl was discovered by Gilead process teams to suppress some of the main side reactions, further improving yields of the 1′-hydroxyl intermediate [38]. Treatment of the 1′-hydroxyl intermediate with TMSCN under Lewis acid conditions introduced the 1′-CN group with a favorable >95:5 β:α anomeric ratio. This was thought to be due to the coordination of the TMSCN reagent with the 2′ or 3′ oxygens on the lower face of the ribose ring [39]. Benzyl deprotection and crystallization provided nucleoside **3** with an overall yield of 29% over three steps. Attempts to directly couple the *p*-nitrophenolate 2-ethylbutyl-L-alaninate prodrug reagent with **3** were met with low yields, so a three-step procedure involving protection/deprotection of the 2′,3′ hydroxyl groups with an acetonide was developed. Also, the scale-up of RDV was further enabled by the facile crystallization of the *p*-nitrophenolate 2-ethylbutyl-L-alaninate

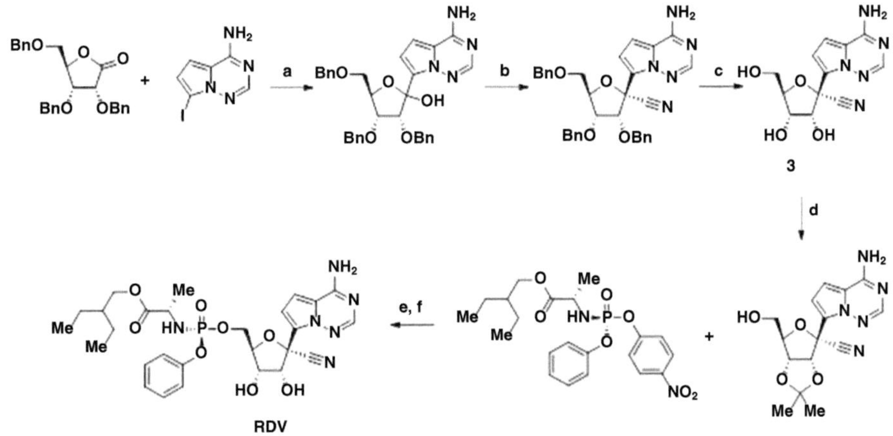

Figure 10.2 Synthesis of RDV. (a) TMSCl, PhMgCl, *i*-PrMgCl•LiCl, THF, −20°C, 40%; (b) TMSCN, TfOH, TMSOTf, CH_2Cl_2, −78°C, 85%; (c) BCl_3, CH_2Cl_2, −20°C, 86%; (d) 2,2-dimethoxypropane, H_2SO_4, acetone, rt, 90%; (e) $MgCl_2$, $(i\text{-Pr})_2NEt$, MeCN, 50°C, 70%; (f) 37% HCl, THF, rt, 69%. *Source*: R. Mackman.

coupling partner from its mixture of diastereoisomers, which enabled RDV to be prepared diastereoselectively. The optimized synthesis was used to prepare >200 g of RDV that supported efficacy and toxicity assessments for regulatory submission. The manufacturing process of RDV was later substantially shortened to 6 months by the time of FDA approval to meet increasing demand during the pandemic [40].

10.3 Pharmacokinetics

10.3.1 *In Vitro* Metabolism

To exert its antiviral activity, RDV needs to undergo intracellular metabolic conversion to its pharmacologically active metabolite, **3-TP** (GS-441524 triphosphate, GS-443902) [12, 19] (Figure 10.3). Following entry into cells, RDV is converted to the monophosphate metabolite **3-MP** via a combination of enzymatic and hydrolytic steps. It is proposed that the 2-ethylbutyl ester is hydrolyzed by carboxyesterases and/or cathepsin A to release the acid. Intramolecular cyclization of the acid onto phosphorous results in the ejection of the phenol, which, following spontaneous hydrolysis of the cyclic intermediate, is then further metabolized by histidine triad nucleotide-binding protein 1 [37] to cleave the P–N bond and release the monophosphate metabolite. In this metabolic pathway, the phosphorylation of **3** to its monophosphate metabolite **3-MP**, a rate-limiting step for the parent nucleoside activation, is effectively bypassed using the monophosphoramidate prodrug. The monophosphate metabolite is then further converted via two additional phosphorylation steps to the active triphosphate **3-TP**. Overall, the 2-ethylbutyl-*L*-alanine mono-phosphoramidate prodrug moiety present in RDV increased both the cellular permeability and the efficient conversion to the triphosphate, which then acts as a potent inhibitor of multiple viral RdRps [19].

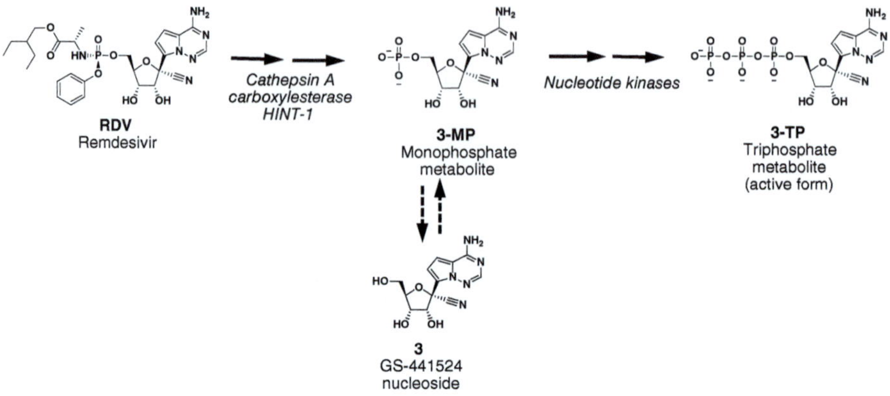

Figure 10.3 Intracellular metabolism of remdesivir. *Source*: R. Mackman.

The *in vitro* metabolism of RDV to its triphosphate metabolite **3-TP** was studied across several different cell types. This confirmed a correlation of antiviral activity with the **3-TP** concentrations inside cells and also provided an assessment of the half-life of the active **3-TP** metabolite. Traditional *in vitro* pharmacokinetic analyses such as permeability and hepatic stability assessments across preclinical species were not useful for RDV due to the rapid metabolism of the prodrug. The efficient conversion of RDV to its active triphosphate **3-TP** was shown across different cell types including primary lung cells and macrophages using LC-MS/MS techniques (Figure 10.4 and Table 10.3). Continuous incubation of RDV and parent nucleoside **3** in NHBE cells demonstrated rapid **3-TP** formation from RDV and steady-state concentrations that were clearly superior to those generated by **3** [19]. The average **3-TP** concentrations were >20-fold higher, consistent with the 38-fold improved antiviral activity toward RSV in NHBE cells. The average **3-TP** concentrations over 48 h were calculated to be 0.14 μM at the RSV EC_{50}, and 0.11 μM at the SARS-CoV-2 EC_{50}. In effect, these **3-TP** concentrations reflect the 50% inhibition of virus replication in cells and therefore are an alternative method to quantify the IC_{50} of **3-TP** toward a

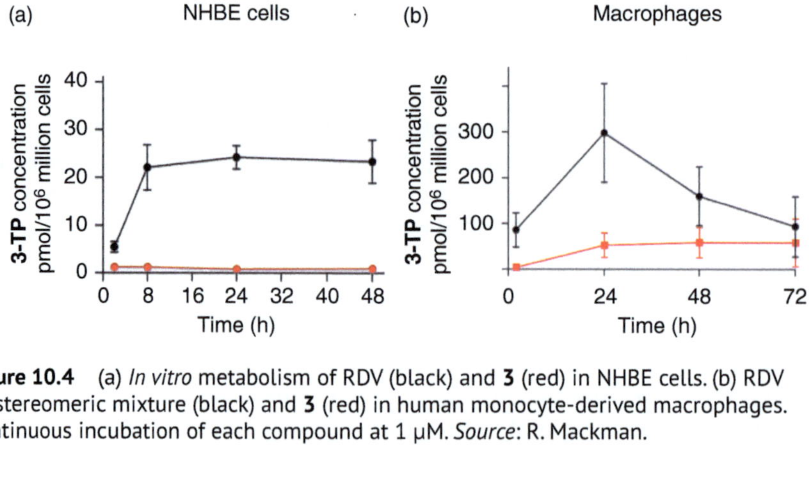

Figure 10.4 (a) *In vitro* metabolism of RDV (black) and **3** (red) in NHBE cells. (b) RDV diastereomeric mixture (black) and **3** (red) in human monocyte-derived macrophages. Continuous incubation of each compound at 1 μM. *Source*: R. Mackman.

Table 10.3 Estimated **3-TP** concentrations at 50% inhibition for different RNA viruses.

Cpd.[a]	Virus	Cell type	EC_{50} (μM)	Average [3-TP][b] (pmol/10^6 cells)	Cell volume (pL)	Calc. [3-TP] at EC_{50} (μM)	3-TP $T_{1/2}$ (h)[c]
RDV	RSV	NHBE, d1[f]	0.049	21	7	0.14	14
3	RSV	NHBE, d1[f]	1.85	0.9	7	0.22	-
RDV	SARS-CoV-2	NHBE, d2[f]	0.037	-	7	0.11[d]	-
RDV-mix	EBOV	Macrophage	0.100	156[e]	2	7.8	16

a. Continuous incubation with compounds at a concentration of 1 μM.
b. Average [**3-TP**] calculated over 48 h from a 1 μM incubation unless noted otherwise.
c. Half-life determined from a 2 h pulse experiment of RDV or its diastereomeric mixture at 1 μM.
d. Calculated using [**3-TP**] from NHBE, d1.
e. Average [**3-TP**] determined over 72 h.
f. d1, donor 1; d2, donor 2.
Source: R. Mackman.

viral polymerase. Since the antiviral activity of RDV toward RSV and SARS-CoV-2 in NHBE cells was equivalent, it follows that **3-TP** is presumably equally potent toward these two viral polymerases. Finally, the **3-TP** half-life in NHBE cells was estimated to be ~14 h using a pulse experiment of RDV and monitoring the decay of **3-TP** after removal of RDV from the cell culture after 2 h. In macrophages, similar experiments showed rapid metabolism of RDV, a ~16 h half-life of **3-TP**, and a calculated average **3-TP** concentration of 7.8 μM at 50% inhibition of EBOV replication (Figure 10.4b and Table 10.3) [12]. The long intracellular half-life of the active metabolite **3-TP** in macrophages and NHBE cells supported once-daily administration *in vivo* for EBOV, RSV, and SARS-CoV-2.

10.3.2 *In Vivo* Pharmacokinetics

In cynomolgus macaques, RDV was found to have very poor oral bioavailability (<1%) as the intact prodrug [19]. This was not unexpected since the amino acid esters of the prodrug are rapidly metabolized by hydrolases in the gut enterocytes and liver hepatocytes, leading to significant pre-systemic cleavage during the first pass. The decision was made to develop RDV as an intravenous (IV) formulation for Ebola due to the urgency of the West Africa epidemic. The IV pharmacokinetics were evaluated across preclinical species including a measurement of the **3-TP** formation in specific tissues collected at various timepoints (Table 10.4). Following a 10 mg/kg IV infusion of RDV in rhesus macaque, the plasma half-life of the intact prodrug was short (23 min), but despite this, rapid distribution of RDV into tissues and rapid intracellular metabolism was observed, leading to **3-TP** concentrations of ~45 μM in peripheral blood mononuclear cells (PBMCs) after only 2 h. This concentration was well above the estimated EBOV IC_{50} of 7.8 μM and suggested that IV administration would quickly inhibit the virus [12]. In addition, the *in vivo* half-life

Table 10.4 Plasma IV pharmacokinetics of RDV and lung **3-TP** levels in nonrodent species.

Cpd	Species	Dose IV (mg/kg)	IV CL (L/h/kg)	IV $T_{1/2}$ (min)	IV V_{ss} (L/kg)	IV 1 $AUC_{0\text{-inf.}}$ (μM.h)	IV 2 $AUC_{0\text{-last}}$ (μM.h)	IV lung 3-TP (nmol/g)[a]	Oral F%
	Ferret	10	2.6	11	0.34	7.0	18.2	1.28	
	Rhesus	10		23					
RDV	Cyno	10	1.92	24	0.45	8.8	8.7	1.39	<1[c]
	AGM	10	2.13	48	0.94	7.9	8.4	1.03	
	Human	2.5[b]	0.86	60	1.22	5.0	7.5		
3	Cyno	20	0.56	161	1.03		123[f]	0.41	5.4[d]
	AGM	20	0.49	2.63	0.89		147[f]	0.23	8.6[d]
ODV	AGM						129[f]	0.31	36[e]

a. **3-TP** measured per gram lung tissue at 24 h post dose.
b. Assumes a 60 kg adult.
c. Dose = 20 mg/kg.
d. Dose = 5 mg/kg.
e. Dose = 60 mg/kg.
f. $AUC_{0\text{-}24h}$.
V_{ss}, volume of distribution at steady state; inf, infinity; AGM, African green monkey; Cyno, cynomolgus macaque; ODV, obeldesivir.
Source: R. Mackman.

in PBMCs was ~24 h, consistent with the *in vitro* data, and fully supportive of a once-daily administration and a loading dose strategy.

In the human Phase 1 studies, the clearance of RDV from plasma and the steady-state volume of distribution were comparable to the values in rhesus macaques and other nonhuman primates (NHPs). In addition, the AUC exposure of intact prodrug from a 200 mg human IV dose was in a similar range to the NHP exposure from an IV 10 mg/kg dose. Taken together, these data supported the decision to use NHPs as a translational species for assessing the human pharmacokinetics of RDV prior to clinical development, and that a 200 mg human dose approximates a 10 mg/kg dose in NHPs [41].

10.3.3 *In Vivo* Lung Metabolism

Similar *in vivo* studies following IV administration were used to evaluate lung tissue concentrations of **3-TP** for the treatment of RSV and potentially other respiratory pathogens. From the same 10 mg/kg IV dose of RDV in cynomolgus macaques and African green monkeys, **3-TP** concentrations in the lung were measured at 24 h post-dose (Table 10.4). A short plasma half-life of intact prodrug was observed across these species, consistent with the half-life in rhesus macaques. Lung tissue samples extracted at 24 h had **3-TP** concentrations of ~1.03–1.39 nmol/g lung tissue (estimated 3–4 μM concentration), a value that was well above the calculated intracellular **3-TP** concentrations for 50% inhibition of RSV and SARS-CoV-2 replication (Table 10.4). Taken together, these data supported the potential for IV administration of RDV to effectively treat RSV and SARS-CoV-2 as well as other respiratory pathogens if **3-TP** was determined to be an effective inhibitor.

The observation that more durable exposures of **3** were generated in plasma following IV dosing of RDV raised the possibility that the **3-TP** observed in the lung at 24 h was driven by **3** and not the short exposure of RDV. To address this possibility, **3** was dosed IV at 20 mg/kg in cynomolgus macaques, a 4-fold higher molar dose than 10 mg/kg RDV. The resulting **3-TP** concentrations measured at 24 h were 3.8-fold lower. Thus, compared with amidate RDV, parent **3** at an equivalent molar dose is ~15-fold less efficient at delivery of **3-TP** into lungs in cynomolgus macaques, a result that was also replicated in African green monkeys and ferrets [19, 36, 42]. The short-lived amidate prodrug exposure clearly leads to the majority of the lung tissue **3-TP** metabolite.

10.3.4 Oral Delivery and the Discovery of Obeldesivir

The goal of achieving oral delivery of **3-TP** was not attainable with the amidate prodrug RDV due to the high first-pass metabolism effect. However, for many RNA viruses that can be inhibited by RDV, including the SARS-CoV-2 virus, the availability of an oral option to deliver **3-TP** into lung tissues was considered the next step for the development program. The parent nucleoside **3**, although less efficiently metabolized to **3-TP** *in vivo*, was not surprisingly found to be much more stable than RDV in hepatocytes *in vitro*. It was hypothesized that delivery of high systemic exposures of **3** similar to the IV 20 mg/kg experiment described above could drive sufficient **3-TP** concentrations for efficacy. Unfortunately, the oral properties of **3** in preclinical species were similar to the amidate prodrug RDV, with oral bioavailability in NHPs below 10% and only marginally improved to 32% in rats due to poor permeability (Table 10.4) [28]. It was therefore projected that multigram daily oral doses of **3** would be required for effective **3-TP** formation in tissues and comparable efficacy to RDV. This was clearly impractical, so a prodrug campaign was launched to improve the oral delivery of **3**, leading to the development of the 5′-isobutyryl prodrug ODV, which has significantly improved oral pharmacokinetic properties compared with **3** (Table 10.4). Oral delivery of ODV provides some advantages over IV RDV because it may facilitate early treatment of patients and potentially be used for prophylaxis or post-exposure prophylaxis.

10.4 Antiviral Pharmacology

10.4.1 Mechanism of Action

RdRp enzymes are essential for replication of a vast majority of RNA viruses, as they catalyze both the replication of viral RNA genome and the transcription of viral genes [43–45]. RdRps utilize the intracellular pools of natural ribonucleotide triphosphates (NTPs) to synthesize viral RNA. **3-TP** structurally mimics ATP and, as such, can compete for binding in the active site of RdRps. Binding of **3-TP** to the SARS-CoV-2 RdRp has been recently elucidated using cryo-electron microscopy, with the collected data underscoring the critical role of the 1′-CN group of **3-TP** and its interaction with a hydrophilic pocket of the RdRp active site [46]. Following its binding to the RdRp active site, **3-TP** is incorporated into the newly synthesized RNA in place of ATP. Multiple viral RdRps recognize **3-TP** as an alternative substrate for RNA incorporation, including those of RSV, EBOV, SARS-CoV, MERS-CoV, and SARS-CoV-2 [12, 47–49]. Interestingly, diverse molecular mechanisms can contribute to the inhibition of viral RNA synthesis following the incorporation of **3-TP** into nascent RNA. In RSV and EBOV, the newly synthesized RNA strand is extended by five additional nucleotides following the incorporation of **3-TP** before the viral RNA synthesis is terminated [12, 47]. In contrast, the RdRps of SARS-CoV-2 and other coronaviruses incorporate only three additional nucleotides before the RNA termination takes place following the **3-TP** incorporation [48, 49]. In the case of SARS-CoV-2 RdRp, at a certain frequency depending on the concentration of natural NTPs, viral RNA synthesis can continue following the incorporation of **3-TP**, leading to synthesis of full-length viral RNA molecules that contain the internally incorporated RDV metabolite. However, this "modified" RNA is not capable of serving as a properly functioning template for the next round of viral RNA replication, leading to a specific template-dependent inhibition [50]. In general, the most common mechanism of action for many antiviral nucleoside and nucleotide analogs involves conversion to their corresponding nucleoside triphosphate metabolites and mimicking their cognate NTPs for incorporation into RNA.

10.4.2 Spectrum of Antiviral Activity

The activity of RDV has been extensively profiled against many other viruses since its original discovery for RSV and development for EBOV (Table 10.5). In addition to potent activity against SARS-CoV-2 and all its variants of concern [51, 52], RDV is active against other human endemic CoVs (e.g. HCoV 229E and HCoV OC43), zoonotic CoVs (e.g. SARS and MERS), and "pre-pandemic" CoVs isolated from bats that are poised for emergence based on their ability to infect human cells [23, 53–55]. RDV is also a potent inhibitor of CoVs of veterinary importance, including porcine delta CoV [54] and feline infectious peritonitis viruses [56]. Using cell-based assays, RDV has been shown to inhibit many other emerging and neglected virus families (Table 10.5), including other filoviruses (e.g. Marburg, and Sudan virus, including both historic and contemporary isolates), flaviviruses (e.g. dengue serotypes 1–4, West Nile, yellow fever, Zika, Japanese encephalitis, and tick-borne encephalitis viruses), picornaviruses (e.g. enterovirus D68 and A71, coxsakievirus B3, and human rhinoviruses), paramyxoviruses (e.g. Nipah, Hendra, measles, and parainfluenza viruses), and pneumoviruses (e.g. respiratory syncytial virus as described above) [12, 18, 19, 23, 24, 57–59]. RDV demonstrated weak *in vitro* activity against Lassa fever and Andes viruses and was found to be inactive against other bunyaviruses (e.g. Crimean Congo hemorrhagic fever and Rift Valley fever viruses) [24]. The compound is also inactive against orthomyxoviruses (influenza A and B viruses), togaviruses (e.g. Chikungunya virus and Venezuelan equine encephalitis virus), rhabdoviruses

Table 10.5 *In vitro* and *in vivo* antiviral activity spectrum of RDV.

Virus family	Virus	RDV EC_{50} (μM)[a]	RDV efficacy *in vivo*	ODV efficacy *in vivo*	References
Coronaviridae	SARS-CoV-2	0.01–0.1	√	√	[28, 51, 52, 60–62, 65–67]
	MERS-CoV	0.07	√	√	[53, 64, 66]
	SARS-CoV	0.07	√	√	[53, 55, 66]
	HCoV OC43	0.07–0.1			[23, 54]
	HCoV 229E	0.02–0.09			[23, 54]
	FIPV	0.03–0.2			[56]
	PDCoV	0.02			[54]
Filoviridae	Ebola	0.003–0.1	√		[12, 18, 23, 24, 65, 67]
	Sudan	0.06–0.1	√	√	[23, 68, 69]
	Bundibugyo	0.2			[23]
	Marburg	0.01–0.04	√		[23, 24, 67, 70, 71]
	Ravn	0.07			[23]
Paramyxoviridae	Nipah	0.03–0.12	√		[24, 67, 72, 73]
	Hendra	0.06–0.16			[24, 67]
	Measles	0.03–0.04	+/−		[24, 67, 74]
	Mumps	0.2–0.8			[24, 67]
	Parainfluenza	0.02–0.03			[24, 67]
Pneumoviridae	Respiratory syncytial	0.02–0.08	√		[12, 18, 19, 24, 67]
Flaviviridae	Hepatitis C	0.07–0.09			[18, 23]
	Dengue	0.1–0.3			[23]
	Yellow fever	0.2–1	√		[23, 57, 67, 75]
	Zika	0.3–1			[23]
	West Nile	0.05			[23]
	Japanese encephalitis	0.3			[23]
	Tick-borne encephalitis	0.2–2			[24, 57]
Picornaviridae	EV-D68	0.05			[23]
	EV-A71	0.1–1			[23, 59]
	Coxsackievirus B3	0.1			[59]
	HRV A2	0.5			[23]
	HRV 14	0.4			[23]
	HRV 16	0.8			[23]

(Continued)

Table 10.5 (Continued)

Virus family	Virus	RDV EC_{50} (μM)[a]	RDV efficacy *in vivo*	ODV efficacy *in vivo*	References
Areneviridae	Junín	0.5			[12]
	Lassa	1–5			[12, 24, 67]
Hantaviridae	Andes	7			[24]
Togaviridae	Chikungunya	>20			[12]
	VEEV	>20			[12]
Bunyaviridae	Crimean-Congo	>50			[24, 67]
	Rift Valley fever	>50			[24]
Rhabdoviridae	Vesicular stomatitis	>50			[24]
Orthomyxoviridae	Influenza A	>50			[23]
	Influenza B	>50			[23]

a. Range of published EC_{50} values from relevant cell-based assays.
√ Reduced viral load and prevention or mitigation of clinical disease were observed with a clinically relevant RDV/ODV dose in a relevant animal challenge model.
+/− Reduced viral load without significant improvement of clinical disease was observed with a clinically relevant RDV/ODV dose in a relevant animal challenge model.
FIPV, feline infectious peritonitis virus; PDCoV, porcine delta coronavirus; EV, enterovirus; HRV, human rhinovirus; VEEV, Venezuelan equine encephalitis virus.
Source: M. Vermillion.

(e.g. vesicular stomatitis virus), and hepadnaviruses B, D, and E, based on published data from cell-based assays [12, 23, 24].

10.4.3 Preclinical Antiviral Efficacy in Animal Models

Based on the *in vitro* antiviral activity profile, the *in vivo* efficacy of RDV was systematically assessed in multiple relevant animal infection models. Preclinical *in vivo* efficacy was first demonstrated in the African green monkey model of RSV, where RDV was evaluated as a daily IV infusion at 10 mg/kg based on achieving the desired levels of intracellular **3-TP** in lung tissue. Prophylactic RDV treatment resulted in a significant reduction in peak viral load in bronchoalveolar lavage fluid samples over the course of infection [19]. RDV has also demonstrated efficacy as both a prophylaxis and therapeutic against SARS-CoV-2 in mice that lack carboxyesterase 1c (*Ces1c*−/−) to increase RDV stability in rodent serum. In this model, twice-daily treatment with a human equivalent dose (25 mg/kg) of RDV decreased SARS-CoV-2 titers in lung and attenuated clinical disease and functional lung damage [60]. In rhesus macaques and African green monkeys, RDV delivered at a 10 mg/kg loading dose and 5 mg/kg maintenance dose approximates the approved clinical dose for COVID-19 based on plasma prodrug exposures. Therapeutic RDV in these SARS-CoV-2 challenge models consistently reduced viral loads across the lower respiratory tract and decreased the incidence and severity of lung pathology based on microscopic examination [61–63]. To maximize delivery of RDV to the respiratory epithelium and minimize systemic drug exposures, the pharmacokinetics and efficacy of inhaled RDV was evaluated *in vivo*. In African green monkeys, inhaled RDV doses that achieved comparable lung triphosphate levels to intravenous RDV demonstrated similar efficacy against SARS-CoV-2 challenge with reduced systemic exposures [62].

Consistent with comparable *in vitro* potency of RDV across betacoronavirus species, parenterally administered RDV was similarly efficacious against SARS-CoV in *Ces1c−/−* mouse challenge models [55]. RDV was also efficacious against MERS-CoV challenge in rhesus macaques, where both prophylactic and therapeutic administration at 5 mg/kg/day reduced lung viral load and attenuated clinical disease [64].

To date, there are no approved direct-acting antiviral therapies for filoviruses, and the approved monoclonal antibodies for Ebola virus do not cross-react with Sudan or Marburg virus. The potential of RDV as a pan-filovirus therapeutic has been extensively characterized in NHP challenge models. RDV (10/5 mg/kg) treatment, initiated at 3–5 days post-challenge and continued for 12 days, provided a significant survival benefit against lethal Ebola, Sudan, and Marburg virus infection, with similar efficacy observed across filoviruses. Survival correlated with significant reductions in plasma viremia and decreased serum biomarkers of hepatocellular damage and coagulation abnormalities associated with advanced filovirus disease [12, 65, 68, 70, 71]. Moreover, the preclinical efficacy of RDV was enhanced by combination with experimental monoclonal antibodies against either Ebola or Marburg virus, where the combined treatment extended the therapeutic window and provided a significant clinical benefit over either treatment alone [68, 70]. Although persistent Ebola virus infection in sanctuary sites (e.g. testes, eyes, and brain) is not reproduced in NHPs, radiolabeled-RDV tissue distribution studies in rhesus macaques suggest that the RDV parent nucleoside **3** is efficiently delivered to these sites [12]. These tissue distribution data supported the subsequent clinical testing of RDV in the PREVIAL IV study in Ebola survivors (Section 10.5.2).

Nipah virus and measles virus, representing paramyxoviruses with no approved antiviral treatments, are susceptible to RDV *in vitro* [24]. In the African green monkey model of lethal Nipah virus challenge, RDV administered at 10 mg/kg over 12 days resulted in near-complete suppression of detectable viremia and 100% survival when treatment was initiated 1 day post-challenge [72]. Survival was reduced to 67% when treatment was delayed to 3 days post-challenge, however, and viral RNA was detectable in serum, lung, and central nervous system tissue of RDV-treated animals. These data suggest that early RDV treatment may be critical for the treatment of Nipah virus disease, although the translation of the efficacious treatment window from NHPs to humans is not well characterized [73]. In a rhesus macaque model of measles virus infection, the efficacy of RDV (10/5 mg/kg for 12 days) was evaluated as either post-exposure prophylaxis (PEP, initiated 3 days post-challenge) or treatment (initiated 11 days post-challenge). RDV as PEP resulted in a significant reduction in viral RNA detected in PBMCs during the treatment period, but the virus rebounded after treatment was stopped, and there was no appreciable improvement in clinical disease. RDV as treatment (i.e. initiated 11 days post-challenge) did not have a significant antiviral effect or clinical efficacy in this model [74]. In a single clinical case report of a patient with subacute sclerosing panencephalitis – caused by persistent measles virus infection of the central nervous system – treatment with RDV was associated with transient clinical improvement but incomplete resolution of disease following three treatment courses (2-week duration each) [76].

Flaviviruses, including dengue and yellow fever, are rising public health threats with no approved antiviral treatments. In a hamster model of yellow fever virus infection, RDV was evaluated as both a prophylaxis and treatment, where a subcutaneous dose of 30 mg/kg for 7 days effectively prevented or treated disease when administered as late as 4 days post-challenge. RDV-treated animals demonstrated increased survival and decreased viremia and liver enzymes in serum. Moreover, the treatment effect was superior to the ribavirin positive control [75]. The preclinical efficacy of RDV against dengue virus and other flaviviruses has not yet been reported. Similarly, RDV *in vivo* efficacy has not been reported for enteroviruses, but a clinical case has been reported associating RDV treatment with the clearance of chronic vaccine-derived poliovirus infection in an immunosuppressed patient [77]. Further, *in vitro* studies have identified that RDV + pleconaril + rupintrivir is a synergistic triple-drug combination with potent, broad-spectrum enterovirus activity and a high barrier to resistance that may be considered for future preclinical and clinical evaluation [78].

Despite the potential for RDV as a treatment for many emerging and neglected viral diseases, access to IV treatments remains a challenge in the resource-poor regions that are disproportionately affected by these viral outbreaks. The antiviral activity of orally delivered ODV has been demonstrated across both coronaviruses and filoviruses. ODV maintains potent *in vitro* activity across SARS-CoV-2 variants of concern [28, 79], zoonotic (SARS, MERS) and pre-pandemic bat CoVs [66], as well as against Ebola, Marburg, and Sudan virus isolates [68]. The *in vivo* efficacy of ODV against SARS, MERS, and a SARS-like bat CoV was similar to SARS-CoV-2 in relevant mouse models, where oral dosing twice daily at 30 mg/kg resulted in significantly reduced lung viral load and clinical disease metrics [66]. In a cynomolgus macaque model of lethal Sudan virus challenge, ODV was evaluated as a post-exposure prophylaxis (i.e. initiated 1 day post-challenge) at a clinically relevant dose (i.e. 100 mg/kg once daily). In this study, animals treated with ODV for 10 days experienced 100% survival, no evidence of clinical disease, and near-complete suppression of detectable viremia [68]. Based on these collective data, ODV, like RDV, has the potential to address a critical unmet medical need in coronavirus and filovirus outbreak preparedness and as potential treatments for other emerging and neglected viral diseases.

10.5 Clinical Studies

10.5.1 Phase 1

The Phase 1 safety and pharmacokinetic assessment of RDV in healthy adults demonstrated a dose-dependent pharmacokinetic profile and favorable safety with reversible low-grade liver enzyme elevations as the main adverse effect following repeat 7- to 14-day intravenous dosing [41]. The systemic exposure of RDV was short with $t_{1/2}$ ~60 min, consistent with observations in NHPs. Distribution and metabolism in cells resulted in high intracellular **3-TP** levels in PBMCs with a prolonged half-life in excess of 40 h, supporting once-daily administration [80].

10.5.2 Ebola Clinical Development

Toward the end of the West African Ebola outbreak in late 2015, RDV was administered under compassionate use to two patients, one with recurrent Ebola virus disease (EVD) and the other a neonate born with EVD [81, 82]. Both patients survived, and the neonatal case was the first documented example of a neonate surviving acute EVD [83]. Another clinical investigation of RDV against EBOV was the PREVAIL IV study in men. Up to 30% of men who survive an EBOV infection have persistent EBOV RNA in the semen that can last for years [84]. In a randomized, double-blind, Phase 2 trial enrolling from 2016 to 2018, investigators evaluated the efficacy of RDV compared with placebo in reducing EBOV RNA in the semen. Participants who were randomized to RDV 100 mg/day for 5 days had a mean assay negativity rate of 96% versus 81% for placebo ($P = 0.041$) during a long-term 5-month follow-up period [85]. Although future studies are needed, the encouraging results of PREVAIL IV suggest that RDV may be an effective intervention to help end future outbreaks more quickly by preventing sexual transmission of EBOV as well as improve longer-term clinical outcomes. During the Ebola outbreak in the Democratic Republic of Congo in 2018–2019, an open-label, Phase 2/3 randomized, active-control trial (PALM Study) was conducted to compare three different monoclonal antibody treatments and RDV as the only small-molecule antiviral [13]. The active control group of ZMapp antibody treatment demonstrated a 50% mortality rate, which was comparable to the group treated with IV RDV once daily with a 200 mg loading dose on day 1 followed by a 100 mg maintenance dose for an additional 9 days (200/100 mg), and lower than the overall outbreak mortality rate of 69%. However, the other two tested antibodies (mAb114 and REGN-EB3) demonstrated improved efficacy with mortality rates of 35% and 34%, respectively, leading to their regulatory approval as the first therapeutics for the treatment of Ebola.

10.5.3 COVID-19 Clinical Development

The first US case of COVID-19 was diagnosed on 20 January 2020, and the patient fully recovered following treatment with 200/100 mg RDV and supportive care [86]. This first case opened the international compassionate use program by Gilead to provide access to RDV [87], which was followed by the first completed randomized, double-blind, placebo-controlled clinical study. The multicenter Adaptive COVID-19 Treatment Trial (ACTT-1) was initiated on 21 February 2020 in the United States and several other countries. ACTT-1 enrolled 1062 adult hospitalized patients with moderate to severe COVID-19. In this study, treatment with 200/100 mg IV RDV for up to 10 days reduced the time to recovery to 10 days from 15 days observed in the placebo group ($P < 0.001$). The Kaplan–Meier estimates of mortality by day 15 and 29 for RDV versus placebo were 6.7% versus 11.9% (hazard ratio 95% CI = 0.36–0.83) and 11.4% versus 15.2% (hazard ratio 95% CI = 0.52–1.03) [25]. Serious adverse events observed in the RDV treatment group were less frequent compared with the placebo group and were largely consistent with severe COVID-19 disease symptoms. In parallel to the ACTT-1 study, two randomized open-label, multi-center, randomized SIMPLE studies demonstrated the clinical utility of a shorter 5-day treatment course of RDV using the same loading and maintenance dosing regimen of 200/100 mg. This was a critical finding during the early months of the pandemic when the drug supply was limited [88, 89]. Results of the ACTT-1 and SIMPLE studies led to US FDA approval of 200/100 mg RDV for the treatment of hospitalized COVID-19 patients in October 2020.

It is generally thought that the maximal benefit of direct-acting antivirals is achieved during the early stages of infection, i.e. prior to a patient needing hospitalization. A recent double-blind, placebo-controlled randomized study in 562 nonhospitalized, high-risk participants with confirmed COVID-19 (PINETREE) demonstrated that a short 3-day treatment with 200/100 mg IV RDV once daily significantly reduced the COVID-19 hospitalization rate by 87% compared with placebo (hazard ratio [HR] = 0.13, 95% CI = 0.03–0.59, P = 0.008) [26]. The 3-day course was safe and well tolerated, leading to FDA approval for use in high-risk nonhospitalized patients with mild-to-moderate COVID-19 [90].

In 2024, 4 years after the first identified case of COVID-19 in the United States, the COVID-19 endemic continues to cause significant morbidity through transmission of continually emerging variants of SARS-CoV-2. The global population's broad exposure to SARS-CoV-2 and available vaccines has resulted in reduced overall mortality; however, several recent reports demonstrate that RDV administration to hospitalized patients still provides a clinical benefit versus those who do not receive RDV. Garibaldi et al. reported that among 36 656 matched individuals, the median time to clinical improvement was faster for those who received RDV versus those who did not (7 vs. 9 days; adjusted HR = 1.19, 95% CI = 1.16–1.22) [91]. Furthermore, RDV recipients on low-flow oxygen (LFO) supplementation were less likely to die (adjusted HR = 0.85, 95% CI = 0.77–0.92; 28-day mortality of 8.4% vs. 12.5%, respectively). The authors concluded that there is a clinical benefit to treating hospitalized patients who are not on supplemental oxygen or receiving LFO, while there may not be a benefit in patients receiving more intensive oxygen supplementation. Another retrospective review assessed the impact of RDV on 67 582 patients receiving LFO, 34 857 receiving high flow or noninvasive ventilation (HF or NIV), and 4164 patients receiving invasive mechanical ventilation (IMV) or extracorporeal membrane oxygenation (ECMO) between December 2020 and April 2022 [92]. Patients were propensity score matched 1 : 1 to those who did not receive RDV. In this study, unadjusted mortality rates were significantly lower for RDV-treated patients at day 14 across all groups (LFO: 6.4% vs. 8.8%; HF/NIV: 16.8% vs. 19.4%; IMV/ECMO: 27.8% vs. 35.3%). The benefit was maintained out to day 28 and after adjustment for several variables, including different time periods of the pandemic. Finally, a meta-analysis of 9 randomized controlled trials with total individual data for 10 480 patients demonstrated that 12.5% in the RDV-treated group versus 14.1% in the non-RDV-treated group died by day 28 (adjusted odds ratio = 0.88, 95% CI = 0.78–1.00; P = 0.045) [93].

These analyses complement the findings from ACTT-1 and support the clinical use of RDV as a foundational treatment for hospitalized patients with COVID-19. RDV was the first and is currently the only approved antiviral to treat both hospitalized and nonhospitalized patients with COVID-19 [94].

10.5.4 Obeldesivir Clinical Development

A Phase 1 study of ODV examined the safety and pharmacokinetics of a single dose and repeated doses of the drug administered orally to healthy adults [95]. It was demonstrated that plasma exposure of the parent nucleoside **3** increased in a dose-dependent manner and could reach target therapeutic exposures, based on the described antiviral efficacy demonstrated in animal models [28, 66]. Overall, ODV was generally safe and well tolerated [95]. Based on the Phase 1 study results, a 350 mg dose twice daily was selected for Phase 3 clinical efficacy studies in COVID-19 patients. In late 2022, two global Phase 3 studies in COVID-19 patients were initiated. The OAKTREE study (NCT05715528) assessed the safety and efficacy of oral ODV in standard-risk COVID-19 patients, while the BIRCH study (NCT05603143) examined ODV safety and efficacy in COVID-19 patients with risk factors for progression to severe COVID-19.

In the OAKTREE Phase 3 trial, 5-day treatment with oral ODV was well tolerated across all patient groups; however, the study did not meet the prespecified primary endpoint of improvement in time to symptom alleviation [96], in part due to fast resolution of COVID-19 disease in the placebo group. At the time of this writing, the BIRCH Phase 3 study in patients with risk factors was under the final analysis, with data to be reported in due course.

10.6 Antiviral Resistance

The high mutation rate of RNA viruses enables quick adaptation to environmental changes such as selection pressure from endogenous as well as therapeutic neutralizing antibodies and, to some extent, antiviral treatment. Determination of pathways and mechanisms leading to viral resistance is important to enable effective surveillance for resistance-associated substitutions and clinical management. To understand RDV resistance pathways, *in vitro* resistance selection experiments were conducted on EBOV and SARS-CoV-2. A single amino acid residue in the EBOV RdRp, F548S, was identified that conferred a low-level resistance of 4.1-fold to RDV after 35 passages in cell culture [97]. Homology modeling suggests that the polymerase F548 residue is located in the fingers subdomain of the EBOV RdRp active site (motif F), a region where RDV resistance substitutions to the coronavirus murine hepatitis virus (MHV) were previously identified [53]. For SARS-CoV-2, two *in vitro* resistance selection experiments have been conducted. In one experiment using the parent nucleoside analog of RDV (**3**), several amino acid substitutions emerged in the SARS-CoV-2 RdRp encoded by the nsp12 gene with minimal to low reduced susceptibility to RDV (2.7- to 10.4-fold): V166A, N198S, S759A, V792I, and C799F/R [98]. In another experiment using RDV, V166L emerged after 9 passages, conferring a low to minimal reduced susceptibility to RDV (2.3-fold) [99]. In a structural model, these substitutions were clustered in three general locations: (1) in motif C in the palm domain at the RdRp catalytic site close to the incoming NTP (S759A); (2) on or adjacent to motif D in the palm domain with possibility of impacting the dynamics of NTP incorporation (V166A/L, V792I, and C799F/R); and (3) on the protein surface at a position having no predicted or known role in RNA synthesis (N198S) (Figure 10.5). The emergence of substitutions with reduced susceptibility to RDV in similar structural motifs around the active site of RdRp across EBOV, MHV, and SARS-CoV-2 indicates a consistent interaction of RDV with the active sites of viral RdRp and its mechanism of action across genetically divergent RNA viruses. Two mechanisms of resistance have been identified: (1) reduced preference for the active triphosphate **3-TP** of RDV as a substrate, as observed for SARS-CoV-2 RdRp encoding S759A, and (2) diminished uridine-triphosphate concentration needed to overcome template-dependent inhibition associated with RDV as observed for SARS-CoV-2 RdRp encoding V792I [98].

As RDV has been evaluated in clinical trials for treatment of COVID-19, emergence of drug resistance to RDV has been analyzed. In the ACTT-1 clinical trial conducted early in the COVID-19 pandemic, demonstrating clinical benefit of RDV by shortening time to recovery in adult participants hospitalized with COVID-19 (median, 10 vs. 15 days) [25], among 61 participants with baseline and postbaseline sequencing data available, a similar rate of emerging nsp12 substitutions was observed in the RDV and placebo arms. Two participants treated with RDV had emergence of substitutions previously identified in *in vitro* resistance selection experiments (nsp12 V792I in one and C799F in the other). These substitutions were associated with 2.2- and 2.5-fold reduced susceptibility, respectively, based on assessments of clinical isolates compared with the wild-type reference strain [100]. In the SIMPLE clinical trials, RDV treatment was evaluated in participants hospitalized with moderate and severe COVID-19. Emergence of RDV resistance was not observed in the 19 participants with baseline and postbaseline sequencing data [101]. In a separate clinical trial demonstrating safety and tolerability of RDV in hospitalized pediatric patients,

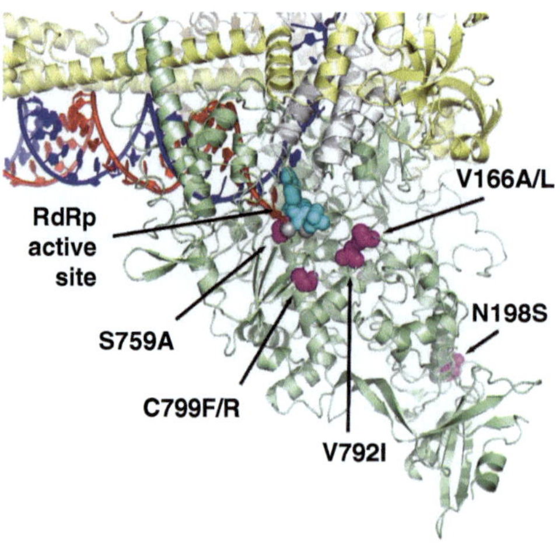

Figure 10.5 Structural model of SARS-CoV-2 RdRp. The structure is a composite model of SARS-CoV-2 RdRp derived from cryo-EM structures 7UO4 [46] and 7RDX [106]. Nsp12 is shown in green, Nsp7 in white, and Nsp8 (2 subunits) in yellow. The template RNA strand is shown in blue and nascent RNA strand in red. The pre-incorporated active triphosphate **3-TP** of RDV is shown in cyan. Amino acid substitutions observed in *in vitro* resistance selection experiment were mapped on the model and shown in magenta. S759A is in the active site, whereas V166A/L, V792A, and C799F/R are adjacent to the active site. *Source*: Adapted from [98].

no substitutions associated with reduced susceptibility to RDV were observed in the 23 participants with baseline and postbaseline sequencing data available [27]. Resistance analyses from these studies support a high barrier to RDV resistance development in COVID-19 patients.

The evolution and spread of new SARS-CoV-2 variants with increased transmissibility and/or pathogenicity such as Delta and Omicron has continued, prolonging the COVID-19 pandemic. The new variants of concern contain significant genetic changes in the spike protein, which is the target for neutralizing antibodies as well as vaccines [102]. In contrast, the RdRp gene (nsp12) of the newly emerged SARS-CoV-2 variants remained highly conserved throughout the 3 years of pandemic [103]. Only a few amino acid changes in RdRp have been observed in circulating SARS-CoV-2 variants such as P323L in Alpha, G671S in Delta, and Y273H in Omicron subvariants. RDV has been shown to retain antiviral activity against all clinical isolates tested representing the Alpha, Beta, Gamma, Delta, Epsilon, Zeta, Iota, Kappa, Lambda, and Omicron (B.1.1.529/BA.1, BA.2, BA.2.12.1, BA.2.75, BA.4, BA.4.6, BA.5, BF.5, BQ.1.1, XBB) variants, as well as against the nsp12 substitutions Y273H, P323L, and G671S introduced in the replicon system by site-directed mutagenesis [28, 51, 79, 104, 105]. These data suggest that RDV is likely to maintain antiviral activity against other variants of SARS-CoV-2 that might emerge in the future.

10.7 Summary

The discovery of RDV with potent activity against RSV and its development as the first antiviral approved for COVID-19 spanned more than a decade of intensive work focused on the mapping and testing of the compound's potential as a novel broad-spectrum antiviral active against multiple RNA virus families. The journey started with a breakthrough discovery of **3**, the parent nucleoside of both RDV and ODV, as a novel direct-acting antiviral with a unique substitution in 1′-position of the ribose ring and a replacement of the N-glycosidic bond with a carbon–carbon linker. In studies focused on profiling the antiviral activity, **3** was identified as a potent and selective inhibitor of RSV. The design of a monophosphate prodrug of **3** to further improve the RSV activity yielded the RDV molecule, an optimized phosphoramidate prodrug with improved cell permeability and efficient conversion to the pharmacologically active triphosphate metabolite. At the time of writing this chapter in early 2024, RDV was the only direct-acting antiviral approved for the treatment of hospitalized patients with COVID-19. Based on additional clinical data with a once-daily 3-day administration of RDV that reduced hospitalization and medically attended visits in high-risk nonhospitalized COVID-19 patients, the FDA approval in January 2022 has been expanded to include outpatient administration. More recently, RDV received regulatory approval for the treatment of patients with renal insufficiency as well as full approval for use in pediatric patients, including neonates [107–109].

Close collaboration across multiple functional expert teams at Gilead played a critical role in bringing RDV to patients. In addition to the significant R&D efforts, the development of a stable lyophilized RDV formulation and the rapid scale-up of the manufacturing process were essential for regulatory approval and subsequent large-scale availability of the drug product during the COVID-19 pandemic. Full global access to the drug has been enabled by an extensive voluntary licensing program introduced by Gilead early during the course of the pandemic. In addition, the role of many important partnerships and collaborations with academic and government organizations that have supported the evaluation and testing of RDV deserves credit and strong recognition. These efforts provided important preclinical and clinical data that enabled a rapid launch of the clinical testing program shortly after the identification of SARS-CoV-2 as the virus associated with COVID-19.

In addition to the efficacy against coronaviruses, the extensive characterization of RDV across multiple animal models of pathogenic viral infections, including filoviruses, paramyxoviruses, and flaviviruses, confirmed its broad antiviral activity spectrum *in vivo*. While RDV showed inferior efficacy to two monoclonal antibodies as a treatment for EBOV infection in the PALM clinical trial, synergistic efficacy was demonstrated in NHP filovirus challenge models when RDV was combined with a corresponding anti-filovirus monoclonal antibody. These results support clinical testing of the combination of direct antiviral agents and virus-neutralizing monoclonal antibodies as proposed in the WHO-sponsored Solidarity Partners trial protocol [110]. In addition, the clinical testing of RDV against paramyxoviruses is of interest based on its demonstrated *in vitro* and *in vivo* activities. Among paramyxoviruses, Nipah virus in particular has been associated with high mortality rates in multiple outbreaks across South-East Asia and the Pacific Islands. The antiviral activity of RDV against measles virus was demonstrated *in vitro* and further characterized in a post-exposure prophylaxis study in juvenile rhesus macaques, with data suggesting the potential utility of RDV as the world is facing a concerning surge of measles outbreaks due to declining rates of vaccination and exceptionally high transmissibility of the measles virus.

While RDV has the potential for targeting a wide spectrum of emerging viruses, its major limitation is the required IV route of administration. Therefore, an oral prodrug, ODV, that delivers sufficient systemic exposures of the parent nucleoside and forms the identical active triphosphate in cells and tissues as RDV was developed. ODV has demonstrated *in vivo* preclinical efficacy against multiple RNA viruses, including SARS-CoV-2. In a Phase 1 clinical study, twice-daily oral ODV delivered systemic levels of the parent nucleoside **3**, consistent with those shown to be efficacious in the preclinical *in vivo* models. Hence, ODV represents an important next step in the evolution toward broad-spectrum oral antiviral therapeutic leveraging the unique mechanism and pharmacology of RDV.

The entire process leading to the rapid regulatory approval of RDV provides several important lessons that may be more broadly applicable for building readiness and timely responses to future viral outbreaks and pandemic threats: (1) the antiviral profiles of RDV and ODV support further systemic exploration of nucleosides and nucleotides as a unique drug class with broad-spectrum antiviral potential that should be included in broader pandemic preparedness R&D strategies; (2) the fast initiation of efficacy studies with new antivirals during viral outbreaks needs to be enabled by robust preclinical and early-stage clinical development activities including pre-positioning of protocols for randomized clinical trials together with the development of adequate manufacturing processes; and (3) collaborations and partnerships across all critical sectors including the pharmaceutical and biotech industry, academia, government, nongovernment organizations, and philanthropic organizations should be established early during the R&D process and sustained throughout the entire development path.

Based on consistent results from multiple clinical trials as well as many large real-world observational cohorts, RDV was one of the key medical countermeasures in the global response to end the COVID-19 pandemic together with efficacious vaccines and a few other antivirals and host-directed therapeutics. During the course of 4 years after the Emergency Use Authorization by the FDA in May 2020, more than 14 million people across all segments of societies and populations were treated with RDV worldwide, underscoring the contribution of this drug to alleviating the devastating impact of one of the largest infectious disease pandemics in our history.

Acknowledgments

We thank Jason K. Perry for providing an updated figure for the structural model of SARS-CoV-2 RdRp. Medical writing and editorial support were provided by Becky Norquist, funded by Gilead Sciences, Inc.

References

1 Our World in Data [Internet]; [cited 2024 March 12]. Available from: https://ourworldindata.org/explorers/coronavirus-data-explorer?facet=none&Metric=Confirmed+cases&Interval=Cumulative&Relative+to+Population=false&Color+by+test+positivity=false&country=~OWID_WRL.
2 Our World in Data [Internet]; [cited 2024 March 12]. Available from: https://ourworldindata.org/explorers/coronavirus-data-explorer?facet=none&Metric=Confirmed+deaths&Interval=Cumulative&Relative+to+Population=false&Color+by+test+positivity=false&country=~OWID_WRL.
3 Zhu N, Zhang D, Wang W, Li X, Yang B, Song J, et al. A novel coronavirus from patients with pneumonia in China, 2019. *N Engl J Med.* 2020;382(8): 727–33.
4 Zumla A, Hui DS, Perlman S. Middle East respiratory syndrome. *Lancet.* 2015;386(9997): 995–1007.
5 Lee N, Hui D, Wu A, Chan P, Cameron P, Joynt GM, et al. A major outbreak of severe acute respiratory syndrome in Hong Kong. *N Engl J Med.* 2003;348(20): 1986–94.
6 World Health Organization. Zika virus [Internet]; [updated 2022 December 08; cited 2024 March 12]. Available from: https://www.who.int/news-room/fact-sheets/detail/zika-virus.
7 World Health Organization. Chikungunya [Internet]; [updated 2022 December 08; cited 2024 March 12]. Available from: https://www.who.int/en/news-room/fact-sheets/detail/chikungunya.

8 World Health Organization (WHO). Ebola virus disease [Internet]; [updated 2023 April 20; cited 2024 March 12]. Available from: https://www.who.int/news-room/fact-sheets/detail/ebola-virus-disease.

9 World Health Organization. Dengue and severe dengue [Internet]; [updated 2023 March 17; cited 2024 March 12]. Available from: https://www.who.int/news-room/fact-sheets/detail/dengue-and-severe-dengue.

10 Widdowson MA, Bresee JS, Jernigan DB. The global threat of animal influenza viruses of zoonotic concern: then and now. *J Infect Dis.* 2017;216(suppl_4):S493–S498.

11 US Food and Drug Administration (FDA). Coronavirus (COVID-19) update: FDA issues emergency use authorization for potential COVID-19 treatment [Internet]. Press Release, 2020 May 01 [cited 2024 March 12]. Available from: https://www.fda.gov/news-events/press-announcements/coronavirus-covid-19-update-fda-issues-emergency-use-authorization-potential-covid-19-treatment.

12 Warren TK, Jordan R, Lo MK, Ray AS, Mackman RL, Soloveva V, et al. Therapeutic efficacy of the small molecule GS-5734 against Ebola virus in rhesus monkeys. *Nature.* 2016;531(7594): 381–5.

13 Mulangu S, Dodd LE, Davey RT Jr, Tshiani Mbaya O, Proschan M, Mukadi D, et al. A randomized, controlled trial of Ebola virus disease therapeutics. *N Engl J Med.* 2019;381(24): 2293–303.

14 US FDA. FDA takes key action in fight against COVID-19 by issuing emergency use authorization for first COVID-19 vaccine [Internet]. Press Release, 2020 December 11 [cited 2024 March 12]. Available from: https://www.fda.gov/news-events/press-announcements/fda-takes-key-action-fight-against-covid-19-issuing-emergency-use-authorization-first-covid-19.

15 US FDA. Coronavirus (COVID-19) update: FDA authorizes first oral antiviral for treatment of COVID-19 [Internet]. Press Release, 2021 December 22 [cited 2024 March 12]. Available from: https://www.fda.gov/news-events/press-announcements/coronavirus-covid-19-update-fda-authorizes-first-oral-antiviral-treatment-covid-19.

16 US FDA. Coronavirus (COVID-19) update: FDA authorizes additional oral antiviral for treatment of COVID-19 in certain adults [Internet]. Press Release, 2021 December 23 [cited 2024 March 12]. Available from: https://www.fda.gov/news-events/press-announcements/coronavirus-covid-19-update-fda-authorizes-additional-oral-antiviral-treatment-covid-19-certain.

17 Cihlar T, Mackman RL. Journey of remdesivir from the inhibition of hepatitis C virus to the treatment of COVID-19. *Antivir Ther.* 2022;27(2): 1359653522 1082773

18 Siegel D, Hui HC, Doerffler E, Clarke MO, Chun K, Zhang L, et al. Discovery and synthesis of a phosphoramidate prodrug of a pyrrolo[2,1-f][triazin-4-amino] adenine C-nucleoside (GS-5734) for the treatment of Ebola and emerging viruses. *J Med Chem.* 2017;60(5): 1648–61.

19 Mackman RL, Hui HC, Perron M, Murakami E, Palmiotti C, Lee G, et al. Prodrugs of a 1′-CN-4-Aza-7,9-dideazaadenosine C-nucleoside leading to the discovery of remdesivir (GS-5734) as a potent inhibitor of respiratory syncytial virus with efficacy in the African green monkey model of RSV. *J Med Chem.* 2021;64(8): 5001–17.

20 Seley-Radtke KL, Yates MK. The evolution of nucleoside analogue antivirals: a review for chemists and non-chemists. Part 1: Early structural modifications to the nucleoside scaffold. *Antiviral Res.* 2018;154: 66–86.

21 Yates MK, Seley-Radtke KL. The evolution of antiviral nucleoside analogues: a review for chemists and non-chemists. Part II: Complex modifications to the nucleoside scaffold. *Antiviral Res.* 2019;162: 5–21.

22 Geraghty RJ, Aliota MT, Bonnac LF. Broad-spectrum antiviral strategies and nucleoside analogues. *Viruses.* 2021;13(4): 667.

23 Radoshitzky SR, Iversen P, Lu X, Zou J, Kaptein SJF, Stuthman KS, et al. Expanded profiling of remdesivir as a broad-spectrum antiviral and low potential for interaction with other medications in vitro. *Sci Rep.* 2023;13(1): 3131.

24 Lo MK, Jordan R, Arvey A, Sudhamsu J, Shrivastava-Ranjan P, Hotard AL, et al. GS-5734 and its parent nucleoside analog inhibit Filo-, Pneumo-, and Paramyxoviruses. *Sci Rep.* 2017;7: 43395.

25 Beigel JH, Tomashek KM, Dodd LE, Mehta AK, Zingman BS, Kalil AC, et al. Remdesivir for the treatment of Covid-19 – final report. *N Engl J Med.* 2020;383(19): 1813–26.

26 Gottlieb RL, Vaca CE, Paredes R, Mera J, Webb BJ, Perez G, et al. Early remdesivir to prevent progression to severe Covid-19 in outpatients. *N Engl J Med.* 2022;386(4): 305–15.

27 Ahmed A, Munoz FM, Muller WJ, Agwu A, Kimberlin DW, Galli L, et al. Remdesivir for COVID-19 in hospitalized children: a Phase 2/3 study. *Pediatrics.* 2024:e2023063775.

28 Mackman RL, Kalla RV, Babusis D, Pitts J, Barrett KT, Chun K, et al. Discovery of GS-5245 (obeldesivir), an oral prodrug of nucleoside GS-441524 that exhibits antiviral efficacy in SARS-CoV-2-infected African green monkeys. *J Med Chem.* 2023;66(17): 11701–17.

29 Schäfer A, Martinez DR, Won JJ, Meganck RM, Moreira FR, Brown AJ, et al. Therapeutic treatment with an oral prodrug of the remdesivir parental nucleoside is protective against SARS-CoV-2 pathogenesis in mice. *Sci Transl Med.* 2022;14(643): eabm3410.

30 Beard RL, Yuan H. 6-Substituted indole-3-carboxylic acid amide compounds having sphingosine-1-phosphate (S1P) receptor antagonist biological activity [Internet]. Patent no. WO/2008/089015 [cited 2024 April 03]. Available from: https://patentscope.wipo.int/search/en/detail.jsf?docId=WO2008089015.

31 Cho A, Zhang L, Xu J, Babusis D, Butler T, Lee R, et al. Synthesis and characterization of 2′-C-Me branched C-nucleosides as HCV polymerase inhibitors. *Bioorg Med Chem Lett.* 2012;22(12): 4127–32.

32 Cho A, Zhang L, Xu J, Lee R, Butler T, Metobo S, et al. Discovery of the first C-nucleoside HCV polymerase inhibitor (GS-6620) with demonstrated antiviral response in HCV infected patients. *J Med Chem.* 2014;57(5): 1812–25.

33 Cho A, Saunders OL, Butler T, Zhang L, Xu J, Vela JE, et al. Synthesis and antiviral activity of a series of 1′-substituted 4-aza-7,9-dideazaadenosine C-nucleosides. *Bioorg Med Chem Lett.* 2012;22(8): 2705–7.

34 Egron D, Imbach JL, Gosselin G, Aubertin AM, Périgaud C. S-acyl-2-thioethyl phosphoramidate diester derivatives as mononucleotide prodrugs. *J Med Chem.* 2003;46(21): 4564–71.

35 McGuigan C, Madela K, Aljarah M, Bourdin C, Arrica M, Barrett E, et al. Phosphorodiamidates as a promising new phosphate prodrug motif for antiviral drug discovery: application to anti-HCV agents. *J Med Chem*. 2011 Dec 22;54(24): 8632–45.

36 Mackman RL. Phosphoramidate prodrugs continue to deliver, the journey of remdesivir (GS-5734) from RSV to SARS-CoV-2. *ACS Med Chem Lett*. 2022;13(3): 338–47.

37 Li R, Liclican A, Xu Y, Pitts J, Niu C, Zhang J, et al. Key metabolic enzymes involved in remdesivir activation in human lung cells. *Antimicrob Agents Chemother*. 2021;65(9): e0060221.

38 Neville S. Process development of GS-5734 – an antiviral nucleotide analog for the treatment of Ebola. 254th ACS National Meeting, Washington DC, August 23rd, 2017.

39 Metobo SE, Xu J, Saunders OL, Butler T, Aktoudianakis V, Cho A, et al. Practical synthesis of 1′-substituted tubercidin C-nucleoside analogs. *Tetrahedron Lett*. 2012;53: 484–6.

40 Gilead Sciences update on Veklury® (remdesivir) manufacturing network [Internet]. Press release, 2020 August 06 [cited 2024 March 12]. Available from: https://www.gilead.com/news-and-press/company-statements/gilead-sciences-update-on-veklury-remdesivir-manufacturing-network.

41 Humeniuk R, Mathias A, Cao H, Osinusi A, Shen G, Chng E, et al. Safety, tolerability, and pharmacokinetics of remdesivir, an antiviral for treatment of COVID-19, in healthy subjects. *Clin Transl Sci*. 2020;13(5): 896–906.

42 Cox RM, Wolf JD, Lieber CM, Sourimant J, Lin MJ, Babusis D, et al. Oral prodrug of remdesivir parent GS-441524 is efficacious against SARS-CoV-2 in ferrets. *Nat Commun*. 2021;12(1): 6415.

43 Choi KH. Viral polymerases. *Adv Exp Med Biol*. 2012;726: 267–304.

44 Ramaswamy K, Rashid M, Ramasamy S, Jayavelu T, Venkataraman S. Revisiting viral RNA-dependent RNA polymerases: insights from recent structural studies. *Viruses*. 2022;14(10): 2200.

45 Ouizougun-Oubari M, Fearns R. Structures and mechanisms of nonsegmented, negative-strand RNA virus polymerases. *Annu Rev Virol*. 2023;10(1): 199–215.

46 Malone BF, Perry JK, Olinares PDB, Lee HW, Chen J, Appleby TC, et al. Structural basis for substrate selection by the SARS-CoV-2 replicase. *Nature*. 2023;614(7949): 781–7.

47 Tchesnokov EP, Feng JY, Porter DP, Götte M. Mechanism of inhibition of Ebola virus RNA-dependent RNA polymerase by remdesivir. *Viruses*. 2019;11(4): 326.

48 Gordon CJ, Tchesnokov EP, Feng JY, Porter DP, Götte M. The antiviral compound remdesivir potently inhibits RNA-dependent RNA polymerase from Middle East respiratory syndrome coronavirus. *J Biol Chem*. 2020;295(15): 4773–9.

49 Gordon CJ, Tchesnokov EP, Woolner E, Perry JK, Feng JY, Porter DP, et al. Remdesivir is a direct-acting antiviral that inhibits RNA-dependent RNA polymerase from severe acute respiratory syndrome coronavirus 2 with high potency. *J Biol Chem*. 2020;295(20): 6785–97.

50 Tchesnokov EP, Gordon CJ, Woolner E, Kocinkova D, Perry JK, Feng JY, et al. Template-dependent inhibition of coronavirus RNA-dependent RNA polymerase

by remdesivir reveals a second mechanism of action. *J Biol Chem*. 2020;295(47): 16156–65.

51 Pitts J, Li J, Perry JK, Du Pont V, Riola N, Rodriguez L, et al. Remdesivir and GS-441524 retain antiviral activity against Delta, Omicron, and other emergent SARS-CoV-2 variants. *Antimicrob Agents Chemother*. 2022;66(6): e0022222.

52 Xie X, Muruato AE, Zhang X, Lokugamage KG, Fontes-Garfias CR, Zou J, et al. A nanoluciferase SARS-CoV-2 for rapid neutralization testing and screening of anti-infective drugs for COVID-19. *Nat Commun*. 2020;11(1): 5214.

53 Agostini ML, Andres EL, Sims AC, Graham RL, Sheahan TP, Lu X, et al. Coronavirus susceptibility to the antiviral remdesivir (GS-5734) is mediated by the viral polymerase and the proofreading exoribonuclease. *mBio*. 2018;9(2): e00221–18.

54 Brown AJ, Won JJ, Graham RL, Dinnon KH 3rd, Sims AC, Feng JY, et al. Broad spectrum antiviral remdesivir inhibits human endemic and zoonotic deltacoronaviruses with a highly divergent RNA dependent RNA polymerase. *Antiviral Res*. 2019;169: 104541.

55 Sheahan TP, Sims AC, Graham RL, Menachery VD, Gralinski LE, Case JB, et al. Broad-spectrum antiviral GS-5734 inhibits both epidemic and zoonotic coronaviruses. *Sci Transl Med*. 2017;9(396): eaal3653.

56 Cook S, Wittenburg L, Yan VC, Theil JH, Castillo D, Reagan KL, et al. An optimized bioassay for screening combined anticoronaviral compounds for efficacy against feline infectious peritonitis virus with pharmacokinetic analyses of GS-441524, remdesivir, and molnupiravir in cats. *Viruses*. 2022 Nov 1;14(11): 2429.

57 Binderup A, Galli A, Fossat N, Fernandez-Antunez C, Mikkelsen LS, Rivera-Rangel LR, et al. Differential activity of nucleotide analogs against tick-borne encephalitis and yellow fever viruses in human cell lines. *Virology*. 2023;585: 179–85.

58 Konkolova E, Dejmek M, Hrebabecky H, Sala M, Böserle J, Nencka R, et al. Remdesivir triphosphate can efficiently inhibit the RNA-dependent RNA polymerase from various flaviviruses. *Antiviral Res*. 2020;182: 104899.

59 Ye W, Yao M, Dong Y, Ye C, Wang D, Liu H, et al. Remdesivir (GS-5734) impedes enterovirus replication through viral RNA synthesis inhibition. *Front Microbiol*. 2020 Jun 12;11: 1105.

60 Martinez DR, Schaefer A, Leist SR, Li D, Gully K, Yount B, et al. Prevention and therapy of SARS-CoV-2 and the B.1.351 variant in mice. *Cell Rep*. 2021;36(4): 109450.

61 Pitts J, Babusis D, Vermillion MS, Subramanian R, Barrett K, Lye D, et al. Intravenous delivery of GS-441524 is efficacious in the African green monkey model of SARS-CoV-2 infection. *Antiviral Res*. 2022;203: 105329.

62 Vermillion MS, Murakami E, Ma B, Pitts J, Tomkinson A, Rautiola D, et al. Inhaled remdesivir reduces viral burden in a nonhuman primate model of SARS-CoV-2 infection. *Sci Transl Med*. 2022;14(633): eabl8282.

63 Williamson BN, Feldmann F, Schwarz B, Meade-White K, Porter DP, Schulz J, et al. Clinical benefit of remdesivir in rhesus macaques infected with SARS-CoV-2. *Nature*. 2020;585(7824): 273–6.

64 de Wit E, Feldmann F, Cronin J, Jordan R, Okumura A, Thomas T, et al. Prophylactic and therapeutic remdesivir (GS-5734) treatment in the rhesus macaque model of MERS-CoV infection. *Proc Natl Acad Sci U S A*. 2020;117(12): 6771–6.

65 Warren TK, Kane CD, Wells J, Stuthman KS, Van Tongeren SA, Garza NL, et al. Remdesivir is efficacious in rhesus monkeys exposed to aerosolized Ebola virus. *Sci Rep*. 2021 Sep 30;11(1): 19458.

66 Martinez DR, Moreira FR, Catanzaro NJ, Diefenbacher MV, Zweigart MR, Gully KL, et al. The oral nucleoside prodrug GS-5245 is efficacious against SARS-CoV-2 and other endemic, epidemic, and enzootic coronaviruses. *Sci Transl Med*. 2024 May 22;16(748): eadj4504.

67 Lo MK, Shrivastava-Ranjan P, Chatterjee P, Flint M, Beadle JR, Valiaeva, N, et al. Broad-spectrum in vitro antiviral activity of ODBG-P-RVn: an orally-available, lipid-modified monophosphate prodrug of remdesivir parent nucleoside (GS-441524). *Microbiol Spectr*. 2021 Nov-Dec;9(3): e01537–21.

68 Cross RW, Bornholdt ZA, Prasad AN, Woolsey C, Borisevich V, Agans KN, et al. Combination therapy with remdesivir and monoclonal antibodies protects nonhuman primates against advanced Sudan virus disease. *JCI Insight*. 2022;7(10): e159090.

69 Cross RW, Woolsey C, Chu VC, Babusis D, Bannister R, Vermillion MS, et al. Oral administration of obeldesivir protects nonhuman primates against Sudan ebolavirus. *Science*. 2024;383(6688): eadk6176.

70 Cross RW, Bornholdt ZA, Prasad AN, Borisevich V, Agans KN, Deer DJ, et al. Combination therapy protects macaques against advanced Marburg virus disease. *Nat Commun*. 2021;12(1): 1891.

71 Porter DP, Weidner JM, Gomba L, Bannister R, Blair C, Jordan R, et al. Remdesivir (GS-5734) is efficacious in cynomolgus macaques infected with Marburg virus. *J Infect Dis*. 2020;222(11): 1894–901.

72 Lo MK, Feldmann F, Gary JM, Jordan R, Bannister R, Cronin J, et al. Remdesivir (GS-5734) protects African green monkeys from Nipah virus challenge. *Sci Transl Med*. 2019;11(494): eaau9242.

73 de Wit E, Williamson BN, Feldmann F, Goldin K, Lo MK, Okumura A, et al. Late remdesivir treatment initiation partially protects African green monkeys from lethal Nipah virus infection. *Antiviral Res*. 2023;216: 105658.

74 Peart Akindele NA, Katamoni LD, Brockhurst J, Ghimire S, Suwanmanee S, Pieterse L, et al. Effect of remdesivir post-exposure prophylaxis and treatment on pathogenesis of measles in rhesus macaques. *Sci Rep*. 2023;13(1): 6463.

75 Julander JG, Bunyan E, Jordan R, Porter DP. Remdesivir efficacy against yellow fever in a hamster model. *Antiviral Res*. 2022;203: 105331.

76 Schmitz KS, Handrejk K, Liepina L, Bauer L, Haas GD, van Puijfelik F, et al. Functional properties of measles virus proteins derived from a subacute sclerosing panencephalitis patient who received repeated remdesivir treatments. *J Virol*. 2024;98(3): e0187423.

77 Bermingham WH, Canning B, Wilton T, Kidd M, Klapsa D, Majumdar M, et al. Case report: clearance of longstanding, immune-deficiency-associated, vaccine-derived polio virus infection following remdesivir therapy for chronic SARS-CoV-2 infection. *Front Immunol*. 2023;14: 1135834.

78 Ianevski A, Frøysa IT, Lysvand H, Calitz C, Smura T, Schjelderup Nilsen HJ, et al. The combination of pleconaril, rupintrivir, and remdesivir efficiently inhibits enterovirus infections in vitro, delaying the development of drug-resistant virus variants. *Antiviral Res.* 2024;224: 105842.

79 Rodriguez L, Li J, Han D, Martin R, Moshiri J, Peinovich N, et al. 545. Remdesivir and obeldesivir retain potent activity against SARS-CoV-2 omicron variants. *Open Forum Infect Dis.* 2023;10(Suppl 2):ofad500.614.

80 Humeniuk R, Mathias A, Kirby BJ, Lutz JD, Cao H, Osinusi A, et al. Pharmacokinetic, pharmacodynamic, and drug-interaction profile of remdesivir, a SARS-CoV-2 replication inhibitor. *Clin Pharmacokinet.* 2021;60(5): 569–83.

81 Jacobs M, Rodger A, Bell DJ, Bhagani S, Cropley I, Filipe A, et al. Late Ebola virus relapse causing meningoencephalitis: a case report. *Lancet.* 2016;388(10043): 498–503.

82 Dörnemann J, Burzio C, Ronsse A, Sprecher A, De Clerck H, Van Herp M, et al. First newborn baby to receive experimental therapies survives Ebola virus disease. *J Infect Dis.* 2017;215(2): 171–4.

83 Lyman M, Mpofu JJ, Soud F, Oduyebo T, Ellington S, Schlough GW, et al. Maternal and perinatal outcomes in pregnant women with suspected Ebola virus disease in Sierra Leone, 2014. *Int J Gynaecol Obstet.* 2018;142(1): 71–7.

84 PREVAIL III Study Group; Sneller MC, Reilly C, Badio M, Bishop RJ, Eghrari AO, et al. A longitudinal study of Ebola sequelae in Liberia. *N Engl J Med.* 2019;380(10): 924–34.

85 Higgs ES, Gayedyu-Dennis D, Fischer Ii WA, Nason M, Reilly C, Beavogui AH, et al. PREVAIL IV: a randomized, double-blind, 2-phase, Phase 2 trial of remdesivir vs placebo for reduction of Ebola virus RNA in the semen of male survivors. *Clin Infect Dis.* 2021;73(10): 1849–56.

86 Holshue ML, DeBolt C, Lindquist S, Lofy KH, Wiesman J, Bruce H, et al. First case of 2019 novel coronavirus in the United States. *N Engl J Med.* 2020;382(10): 929–36.

87 Grein J, Ohmagari N, Shin D, Diaz G, Asperges E, Castagna A, et al. Compassionate use of remdesivir for patients with severe Covid-19. *N Engl J Med.* 2020;382(24): 2327–36.

88 Goldman JD, Lye DCB, Hui DS, Marks KM, Bruno R, Montejano R, et al. Remdesivir for 5 or 10 days in patients with severe Covid-19. *N Engl J Med.* 2020;383(19): 1827–37.

89 Spinner CD, Gottlieb RL, Criner GJ, Arribas López JR, Cattelan AM, Soriano Viladomiu A, et al. Effect of remdesivir vs standard care on clinical status at 11 days in patients with moderate COVID-19: a randomized clinical trial. *Jama.* 2020;324(11): 1048–57.

90 US FDA. FDA takes actions to expand use of treatment for outpatients with mild-to-moderate COVID-19 [Internet]. Press Release, 2022 January 21 [cited 2024 March 12]. Available from: FDA Takes Actions to Expand Use of Treatment for Outpatients with Mild-to-Moderate COVID-19 (prnewswire.com)https://www.fda.gov/news-events/press-announcements/fda-takes-actions-expand-use-treatment-outpatients-mild-moderate-covid-19.

91 Garibaldi BT, Wang K, Robinson ML, Betz J, Caleb Alexander G, Andersen KM, et al. Real-world effectiveness of remdesivir in adults hospitalized with coronavirus disease 2019 (COVID-19): a retrospective, multicenter comparative effectiveness study. *Clin Infect Dis.* 2022;75(1): e516–24.

92 Mozaffari E, Chandak A, Gottlieb RL, Chima-Melton C, Read SH, Jiang H, et al. Remdesivir reduced mortality in immunocompromised patients hospitalized for COVID-19 across variant waves: findings from routine clinical practice. *Clin Infect Dis.* 202b3;77(12): 1626–34.

93 Amstutz A, Speich B, Mentré F, Rueegg CS, Belhadi D, Assoumou L, et al. Effects of remdesivir in patients hospitalised with COVID-19: a systematic review and individual patient data metaanalysis of randomised controlled trials. *Lancet Respir Med.* 2023 May;11(5): 453–64.

94 Veklury (remdesivir) US Prescribing Information [Internet]. Revised 02/2004 [cited 2024 April 03]. Available from: https://www.gilead.com/-/media/files/pdfs/medicines/covid-19/veklury/veklury_pi.pdf.

95 Anoshchenko O, Abdelghany M, Lichtman A, Duan R, Chen H, Shaik NA, et al. Pharmacokinetics, mass balance, safety, and tolerability of obeldesivir in healthy participants. *Clin Pharmacol Ther.* 2024;116: 1231–9. [Online ahead of print.]

96 Gilead Sciences. Gilead Sciences Announces Fourth Quarter and Full Year 2023 Financial Results [Internet]. Press release, 2024 February 06 [cited 2024 August 03]. Available from: https://www.gilead.com/news-and-press/press-room/press-releases/2024/2/gilead-sciences-announces-fourth-quarter-and-full-year-2023-financial-results.

97 Lo MK, Albariño CG, Perry JK, Chang S, Tchesnokov EP, Guerrero L, et al. Remdesivir targets a structurally analogous region of the Ebola virus and SARS-CoV-2 polymerases. *Proc Natl Acad Sci U S A.* 2020;117(43): 26946–54.

98 Stevens LJ, Pruijssers AJ, Lee HW, Gordon CJ, Tchesnokov EP, Gribble J, et al. Mutations in the SARS-CoV-2 RNA dependent RNA polymerase confer resistance to remdesivir by distinct mechanisms. *Sci Transl Med.* 2022;14(656): eabo0718.

99 Checkmahomed L, Carbonneau J, Du Pont V, Riola NC, Perry JK, Li J, et al. In vitro selection of remdesivir-resistant SARS-CoV-2 demonstrates high barrier to resistance. *Antimicrob Agents Chemother.* 2022;66(7): e0019822.

100 Hedskog C, Rodriguez L, Roychoudhury P, Huang ML, Jerome KR, Hao L, et al. Viral resistance analyses from the remdesivir phase 3 adaptive COVID-19 treatment trial-1 (ACTT-1). *J Infect Dis.* 2023;228(9): 1263–73.

101 Hedskog C, Spinner CD, Protzer U, Hoffmann D, Ko C, Gottlieb RL, et al. No remdesivir resistance observed in the Phase 3 severe and moderate COVID-19 SIMPLE trials. *Viruses.* 2024;16(4): 546.

102 Tao K, Tzou PL, Nouhin J, Gupta RK, de Oliveira T, Kosakovsky Pond SL, et al. The biological and clinical significance of emerging SARS-CoV-2 variants. *Nat Rev Genet.* 2021;22(12): 757–73.

103 Martin R, Li J, Parvangada A, Perry J, Cihlar T, Mo H, et al. Genetic conservation of SARS-CoV-2 RNA replication complex in globally circulating isolates and recently emerged variants from humans and minks suggests minimal pre-existing resistance to remdesivir. *Antiviral Res.* 2021;188: 105033.

104 Vangeel L, Chiu W, De Jonghe S, Maes P, Slechten B, Raymenants J, et al. Remdesivir, molnupiravir and nirmatrelvir remain active against SARS-CoV-2 Omicron and other variants of concern. *Antiviral Res.* 2022:198: 105252.

105 Rosales R, McGovern BL, Rodriguez ML, Rai DK, Cardin RD, Anderson AS, et al. Nirmatrelvir, molnupiravir, and remdesivir maintain potent in vitro activity against the SARS-CoV-2 Omicron variant. *bioRxiv [Online]* 2022.

106 Chen J, Wang Q, Malone B, Llewellyn E, Pechersky Y, Maruthi K, et al. Ensemble cryo-EM reveals conformational states of the nsp13 helicase in the SARS-CoV-2 helicase replication-transcription complex. *Nat Struct Mol Biol.* 2022;29(3): 250–60.

107 Gilead Sciences. Veklury® (remdesivir) is first and only approved treatment for pediatric patients under 12 years of age with COVID-19 [Internet]. Press release, 2022 April 25 [cited 2024 April 24]. Available from: https://www.gilead.com/news-and-press/press-room/press-releases/2022/4/veklury-remdesivir-is-first-and-only-approved-treatment-for-pediatric-patients-under-12-years-of-age-with-covid19.

108 Gilead Sciences. FDA approves Veklury® (remdesivir) for COVID-19 treatment in patients with severe renal impairment, including those on dialysis [Internet]. Press release, 2023 July 14 [cited 2024 April 24]. Available from: https://www.gilead.com/news-and-press/press-room/press-releases/2023/7/fda-approves-veklury-remdesivir-for-covid19-treatment-in-patients-with-severe-renal-impairment-including-those-on-dialysis.

109 Gilead Sciences. FDA approves Veklury® (remdesivir) to treat COVID-19 in people with mild to severe hepatic impairment with no dose adjustment [Internet]. Press release, 2023 August 24 [cited 2024 April 24]. Available from: https://www.gilead.com/news-and-press/press-room/press-releases/2023/8/fda-approves-veklury-remdesivir-to-treat-covid19-in-people-with-mild-to-severe-hepatic-impairment-with-no-dose-adjustment.

110 Solidarity Partners. Core Trial Protocol [Internet]. 2024 March 08 [cited 2024 April 04]. Available from: Solidarity Partners – Platform adaptive randomized trial for new and repurposed Filovirus treatments – Core Trial Protocol (who.int).

11

Discovery and Clinical Development of Ensitrelvir: An Oral SARS-CoV-2 3CL Protease Inhibitor for the Treatment of COVID-19

Yuki Tachibana[1], Ryosuke Shimizu[2], Haruaki Nobori[1], Yuko Tsuge[2], Masaharu Shinkai[3]

[1] *Shionogi Pharmaceutical Research Center, Shionogi & Co., Ltd., Osaka, Japan*
[2] *Drug Development and Regulatory Science Division, Shionogi & Co., Ltd., Osaka, Japan*
[3] *Department of Respiratory Medicine, Tokyo Shinagawa Hospital, Tokyo, Japan*

11.1 Introduction

The COVID-19 pandemic caused by the SARS-CoV-2 virus continues to spread globally, with over 770 million people infected and 7 million deaths as of November 2023 [1]. SARS-CoV-2 is particularly dangerous for the elderly and individuals with underlying health conditions, often leading to severe acute respiratory distress syndrome. During the early phase of the COVID-19 outbreak (before the emergence of the Delta strain), there were some clinical trials using public funds to evaluate repurposing existing medicines that are widely available and have reasonably well-understood safety profiles. In Japan, clinical trials of ivermectin and ciclesonide were conducted, but these trials did not demonstrate efficacy in COVID-19 patients compared with placebo [2, 3]. Favipiravir, which has been approved for influenza treatment, was a promising drug for COVID-19 treatment. In the first Phase 3 clinical trial conducted in the early phase of the COVID-19 pandemic, treatment with favipiravir showed a significant improvement in the symptoms of hospitalized patients, who had mostly not been vaccinated or had not previously experienced SARS-CoV-2 infection [4].

A possible trend for viral clearance in hospitalized patients after early administration of favipiravir has also been observed [5]. Several clinical trials of favipiravir have been subsequently conducted, but no evidence that favipiravir can suppress severe COVID-19 development has been established [6, 7]. This lack of evidence may be because of the rapid decline in the pathogenicity of the SARS-CoV-2 variants

Trends in Antiviral Drug Development, First Edition. Edited by Michael J. Sofia and Zhengqiang Wang.

after the Omicron pandemic and the decline in the frequency of severe illness because of the host immunity as a result of vaccination and widespread SARS-CoV-2 infection.

Currently, three oral antiviral drugs (Table 11.1) and intravenous remdesivir have been granted an emergency or a special approval for COVID-19 treatment in Japan. Molnupiravir and nirmatrelvir were granted special approvals based on the risk reduction of hospitalization or death in patients at high risk of developing severe COVID-19 in clinical trials during the Delta pandemic period. Ensitrelvir was granted an emergency regulatory approval in Japan in November 2022 and a

Table 11.1 Oral COVID-19 agents available in Japan.

	Ensitrelvir fumaric acid	**Molnupiravir**	**Nirmatrelvir/ritonavir**
Approval	Emergency regulatory approval 22 November 2022 Standard approval 5 March 2024	Special approval 24 December 2021	Special approval 10 February 2022
Dosage and administration	375 mg once daily on day 1, 125 mg once daily on days 2–5	800 mg twice daily for 5 days	300/100 mg twice daily for 5 days
Indications and usage	Mild-to-moderate symptoms, such as high fever, strong cough, and severe sore throat	Mild-to-moderate symptoms in patients at high risk for progression to severe COVID-19	Mild-to-moderate symptoms in patients at high risk for progression to severe COVID-19
Days from COVID-19 onset to start of administration	Within 3 days	Within 5 days	Within 5 days
Pediatric use	12 years or older	18 years or older	12 years or older with body weight ≥ 40 kg
Dosage in patients with renal impairment	No dosage adjustment; contraindicated in patients using colchicine	No dosage adjustment	eGFR ≥ 30 to < 60 mL/min: reduce the dosage to 150/100 mg; eGFR < 30 mL/min: not recommended for use; contraindicated in patients using colchicine
Pregnancy	Contraindicated	Contraindicated	Should be administered only if the therapeutic benefits are considered to outweigh the risks

Table 11.1 (Continued)

	Ensitrelvir fumaric acid	Molnupiravir	Nirmatrelvir/ritonavir
Lactation	Not recommended	Consider continuing or discontinuing based on the therapeutic benefits and benefits of breastfeeding	Consider continuing or discontinuing based on the therapeutic benefits and benefits of breastfeeding
Contraindicated drugs	35 drugs are contraindicated	None	43 drugs are contraindicated
Risk reduction of hospitalization or death	No data	Molnupiravir: 6.8% Placebo: 9.7% (30% reduction)	Nirmatrelvir: 0.745% Placebo: 6.801% (89% reduction)
Symptom improvement	Time to resolution of five COVID-19 symptoms[a] Ensitrelvir: 167.9 h Placebo: 192.2 h (−24.3 h difference) (Omicron major)[d]	Time to first resolution of symptoms[b] Molnupiravir: 18 days Placebo: 20 days Time to first alleviation of symptoms[b] Molnupiravir: 13 days Placebo: 15 days (before Omicron)[e]	Study discontinued, no reviewed results available[c]

a. The following five symptoms: (1) malaise or fatigue; (2) feverishness or fever; (3) runny or stuffy nose; (4) sore throat; and (5) cough reported by patients were evaluated (remaining for at least 24 h).

b. Secondary and *post hoc* analyses in nonhospitalized, unvaccinated adults with mild-to-moderate, laboratory-confirmed COVID-19. Time to first resolution of symptoms and time to first alleviation of symptoms were defined as the median time to resolution/alleviation of the five characteristic symptoms of COVID-19.

c. Clinicaltrials.gov: NCT05011513.

d. Ensitrelvir common technical document: https://www.pmda.go.jp/drugs/2022/P20220719001/index.html (accessed 5 Dec 2023).

e. Yanfen Guan et al. Impact of Molnupiravir Treatment on Patient-Reported COVID-19 Symptoms in the Phase 3 MOVe-OUT Trial: A Randomized, Placebo-Controlled Trial, *Clin Infect Dis*. 2023;77(11), 1521–30.

standard approval on 5 March 2024 based on the COVID-19 symptom improvement in patients without high-risk factors. In the clinical trial, approximately 90% of the patients had received COVID-19 vaccination during the Omicron pandemic period.

SARS-CoV-2 belongs to the genus *Betacoronavirus*, an enveloped virus with a single-stranded RNA genome [8]. The virus enters host cells by binding to angiotensin-converting enzyme 2 (ACE2) through the spike glycoprotein. Once inside the cell, the SARS-CoV-2 viral RNA is released and translated into two large polyproteins, pp1a and pp1ab, which are then processed into individual nonstructural proteins. One of these nonstructural proteins, known as Nsp5 or 3C-like protease ($3CL^{pro}$), is a cysteine protease responsible for cleaving 11 distinct sites of the SARS-CoV-2 polyproteins to transform the polyproteins into mature functional proteins. 3CL^{pro} plays a crucial role in viral replication, and inhibition of 3CL^{pro} prevents the formation of replication-essential enzymes, including RNA-dependent RNA polymerase, thus inhibiting viral replication [9]. No close human analogs of the SARS-CoV-2 3CL^{pro} are known, suggesting that selective 3CL^{pro} inhibitors should avoid safety issues caused by nonselectivity, and this makes 3CL^{pro} an attractive target for developing oral therapeutics for COVID-19 (Figure 11.1).

Viral proteases have been successfully targeted for the treatment of other viral infections, such as HIV (human immunodeficiency virus) and HCV (hepatitis C virus) [10]. At the time we initiated our research that led to the development of ensitrelvir, peptide-like 3CL^{pro} inhibitors with reactive "warheads" had shown promising antiviral activity *in vitro* and had been shown to reduce viral loads in human ACE2 transgenic mouse models infected with SARS-CoV-2 [11]. In addition, GC376, a broad-spectrum antiviral inhibitor developed by Kansas State University [12], and PF-00835231, which was developed during the SARS outbreak in 2003 by Pfizer, were already known to be efficient inhibitors of the 3CL^{pro} of SARS-CoV-2 [13]. Starting from PF-00835231, Pfizer realized the first-in-class clinical drug, nirmatrelvir, which is administered with ritonavir as a PK booster [14]. The chemical structures of these molecules are very similar, mimicking the peptide substrate containing covalent warheads that can bind to the active site cysteine residue (Figure 11.2). There are challenges associated with improving the selectivity and PK profiles of peptide-like inhibitors because of their potential reactivity, general low membrane permeability, and low metabolic stability [15–17]. To avoid these intrinsic issues, nonpeptidic, noncovalent, small-molecule inhibitors have garnered attention as potential alternatives, but at the time we started our research, the potency and PK profiles needed further optimization [18–22]. We decided to initiate drug discovery research aimed

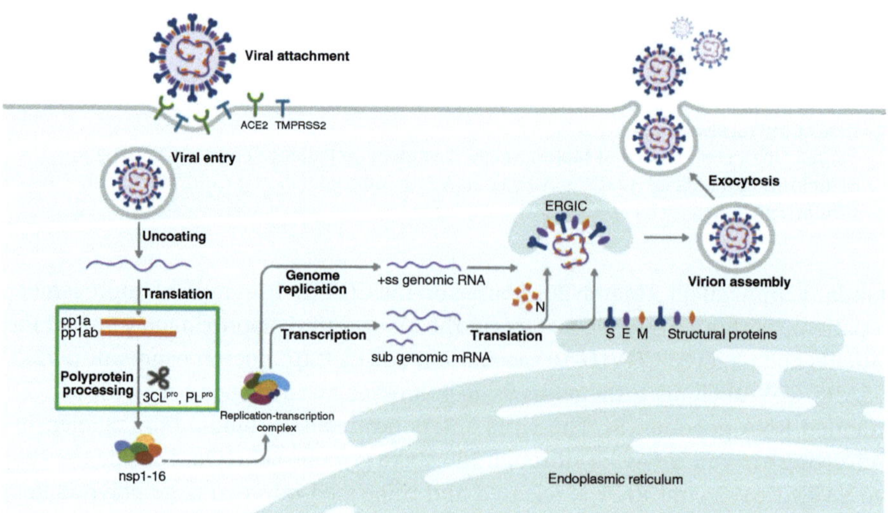

Figure 11.1 Lifecycle of SARS-CoV-2. ERGIC: endoplasmic-reticulum–Golgi intermediate compartment.

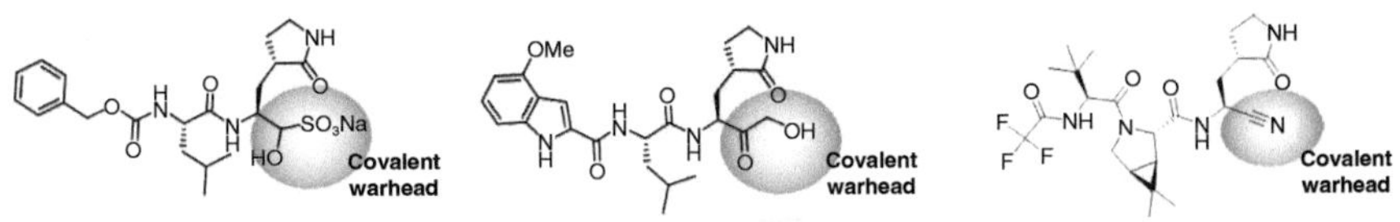

GC376	PF-00835231	Nilmatrelvir
$3CL^{pro}$ IC_{50} : 0.030 μM	$3CL^{pro}$ Ki : 0.00027 μM	$3CL^{pro}$ Ki : 0.0031 μM
Cell EC_{50} : 3.37 μM[a]	Cell EC_{50} : 0.231 μM[b]	Cell EC_{50} : 0.075 μM[b]
	Rat F_{oral} : 1.4±0.8%	Rat F_{oral} : 34±19%

Figure 11.2 Structures of SARS-CoV-2 $3CL^{pro}$ inhibitors. Conditions: [a]Vero 76 cells; [b]Vero E6 cells in the presence of 2 μM efflux inhibitor CP-100356.

at developing a nonpeptidic, noncovalent, orally administrable, small-molecule inhibitor of SARS-CoV-2 $3CL^{pro}$ that has potential as a clinical candidate for the treatment of COVID-19.

Here, we describe the discovery and preclinical and clinical characterization of ensitrelvir, the first nonpeptidic, noncovalent SARS-CoV-2 $3CL^{pro}$ inhibitor for treating COVID-19 [23]. Ensitrelvir displayed antiviral activity *in vitro* toward a range of SARS-CoV-2 variants and coronavirus families, a favorable drug metabolism and pharmacokinetics (DMPK) profile for an oral agent, and dose-dependent antiviral efficacy *in vivo*. Ensitrelvir showed antiviral efficacy in a human clinical study and received emergency regulatory approval in Japan for the treatment of SARS-CoV-2 infection in November 2022.

11.2 Discovery of Ensitrelvir

To quickly find a potential clinical candidate for a noncovalent inhibitor of SARS-CoV-2 $3CL^{pro}$, we decided to utilize structure-based virtual screening, followed by high-throughput screening using mass spectrometry, and not fluorescence-based assays, which typically have been used for the screening of viral protease inhibitors. By performing biological screening only for the compounds narrowed down by virtual screening, the period required for the screening can be considerably shortened. In addition, using mass spectrometry as a readout made it possible to eliminate false positives caused by the autofluorescence and/or quenching effects of the compounds, minimizing the numbers of false positives and false negatives, with the aim of obtaining a small number of high-quality hit compounds.

$3CL^{pro}$ is a cysteine protease with a Cys145-His41 catalytic dyad in the active site, which recognizes P1 Gln and P2 Leu/Met/Phe/Val as substrates [11]. Substrate-like substructures are also found in potent peptide-like inhibitors, such as GC-376 [12] and N3 [24], which have a glutamine-mimicking lactam moiety in the S1 pocket and a leucine-mimicking hydrophobic moiety in the S2 pocket. Noncovalent, small-molecule inhibitors, such as ML188 [18], ML300 [19, 20], and the 3-aminopyridine-like compound (TRY-UNI-714a-706-6) from the PostEra COVID moonshot project, exhibit similar binding interactions (Figure 11.3) [21, 22]. These inhibitors form a hydrogen bond with the side-chain NH of His163 in the S1 pocket and have lipophilic moieties that fit into the S2 pocket. Furthermore, the Glu166 main-chain NH

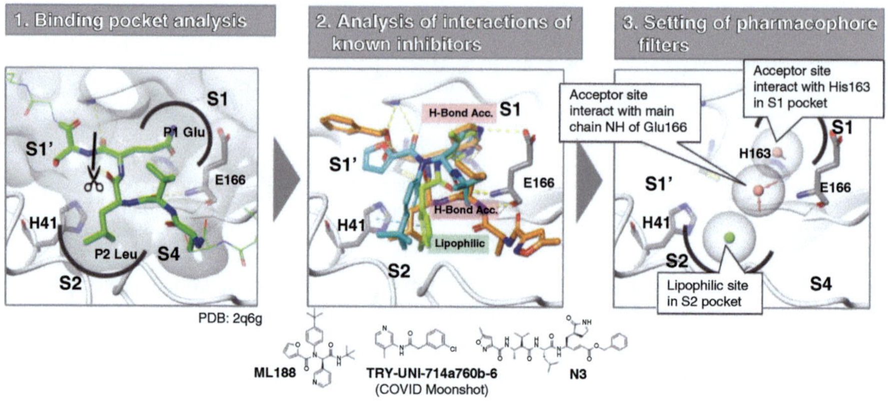

Figure 11.3 Binding pocket analysis and the process for the identification of the pharmacophore features. ML188 is shown in cyan, TRY-UNI-714a760b-6 is shown in lime green, and N3 is shown in orange, respectively.

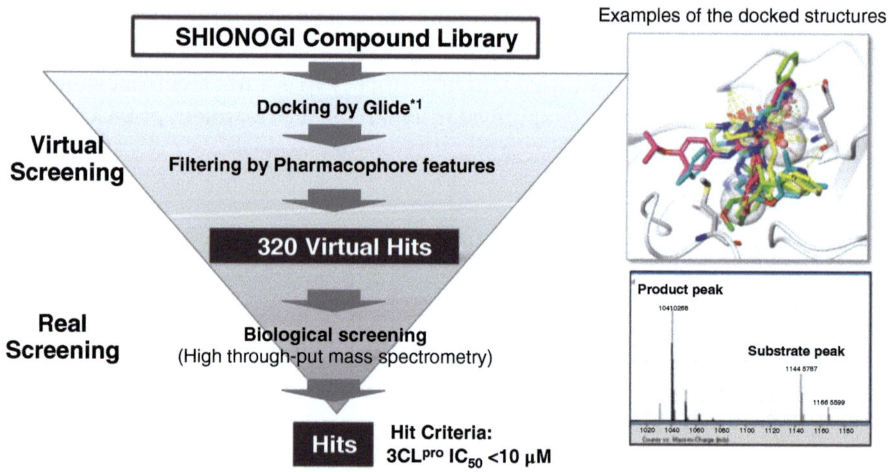

Figure 11.4 Schematic flow of the screening campaign. *1: Virtual screening was conducted using 6W63 (PDB code) as a template.

forms a hydrogen bond with the P2 main-chain carbonyl of the natural substrate, and this interaction is also found in the known inhibitors. From these interactions, we identified three pharmacophore features, namely 1) the acceptor site with the side-chain NH donor of His163 in the S1 pocket, 2) the lipophilic site in the S2 pocket, and 3) the acceptor site with the Glu166 main-chain NH, as crucial interactions for the binding of noncovalent ligands (Figure 11.3).

The screening flow to obtain hit compounds is shown in Figure 11.4. First, docking-based virtual screening was conducted against structures in the SHIONOGI compound library to facilitate the process, before virtual screening. As a template structure for the protein, the crystal structure of $3CL^{pro}$ bound to an ML188-like noncovalent small molecule (Protein Data Bank [PDB] code: 6W63 [25]) was selected because this structure has a high resolution (2.10 Å) and distinct electron density for both the compound and the active site residues. Each docking pose was then subjected to the pharmacophore filters described earlier. Using this process, we identified the 300 compounds with the highest docking scores. Then, those top-scoring compounds were subjected to screening using enzymatic assays and high-throughput mass spectrometry to avoid the false positives commonly observed in fluorescence-based assays. This thorough evaluation enabled us to identify several hit compounds that exhibited IC_{50} values < 10 μM. Then, those hit compounds were further profiled for their ADME (absorption, distribution, metabolism, excretion) characteristics, including *in vivo* PKs. In our experience, optimization of the *in vivo* PK profile and the biological activity at the same time is a very challenging and time-consuming task, and we believe that it is better to start with a hit compound that shows a good PK profile even if the compound has relatively low activity. Triazine 1 (Figure 11.5), which had a moderate IC_{50} value (8.6 μM against $3CL^{pro}$) with a potent ADME profile for oral dosing, was selected as the starting point for a structure–activity relationship (SAR) study after the extensive profiling of hit compounds (Figure 11.5a). In the ADME profile, triazine 1 showed *in vitro* metabolic stability, measured after 30 min of incubation, in human and rat microsomes of 97% and 71%, respectively, suggesting good metabolic stability, especially in humans. An *in vivo* PK study in rats indicated that triazine 1 had a favorable profile for an oral agent, a low clearance of 7.3 mL/min/mg, with an oral bioavailability (*F*) of 111%; thus, if activity optimization could be achieved while keeping this ADME profile, a candidate for a once-daily oral agent could potentially be produced. In addition, because this compound was from the SHIONOGI compound library, we had knowledge of the profile and chemistry of this chemotype, which should be a great advantage in identifying a clinical candidate within a short period of time. Furthermore, we were fortunate to succeed in the structural analysis of the X-ray

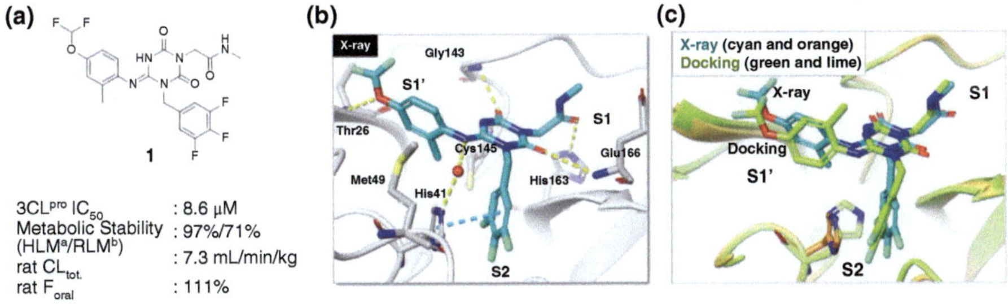

Figure 11.5 Structure of the hit compound 1 (triazine 1) and X-ray co-structure with $3CL^{pro}$. (a) Structure and profile of the hit compound 1. [a] Percentage remaining in human liver microsomes after 30-min incubation; [b] percentage remaining in rat liver microsomes after 30-min incubation. (b) Close-up view of 1 (cyan) in the binding pocket. The water molecules are shown as red spheres; hydrogen bonds are indicated as yellow dashed lines, and π–π stacking is indicated as a cyan dashed line. (c) Comparison of a predicted docking pose and the X-ray crystal structure of 1. Near the S2 pocket, the side chain of His41 was rotated to form a face-to-face π interaction with the 3,4,5-trifluorobenzene moiety of 1. The docked structure is shown in lime green and the X-ray structure is in cyan (1) and orange (protein residues).

complex of triazine 1 and 3CL^{pro} with a resolution of 2.0 Å soon after the initiation of the SAR studies (Figure 11.5b). As expected, the binding mode of triazine 1 in the X-ray structure was similar to that obtained in the docking study (Figure 11.5c), and this result was of great value because we could focus the direction of the SAR studies. However, the precise details of the interactions, which can greatly influence the design of new molecules, were slightly different. The S1 and S2 pockets were filled with the methyl amide and 3,4,5-trifluorobenzene moieties, respectively, as expected. The 4-difluoromethoxy-2-methylbenzene subunit was positioned in the S1′ pocket. The 2-carbonyl oxygen of the central triazine moiety formed a hydrogen bond with the main-chain NH of Glu166, and the 4-carbonyl oxygen was bound in the oxyanion hole of the protease and formed two hydrogen bonds with the main-chain NH groups of Gly143 and Cys145. The methyl amide moiety was positioned in the S1 pocket, and the carbonyl oxygen interacted with the side-chain NH of His163. The protease exhibited an interesting conformational change in the S2 pocket: the side chain of the catalytic His41 was rotated and formed a face-to-face π interaction with the 3,4,5-trifluorobenzene moiety of 1, whereas the predicted docking pose had an edge-to-face π interaction. Along with this side-chain flip of His41, the 4-difluoromethoxy-2-methylbenzene fragment was in a slightly different site than that in the predicted docking pose, and the oxygen of the difluoromethoxy moiety formed a hydrogen bond with the main-chain NH of Thr26. The imine linker formed a water-mediated hydrogen bond with the His41 side chain, indicating that this linker contributed to the affinity.

To enable fast optimization, the SAR strategy was organized by fully implementing the information from the X-ray structure. Figure 11.6a presents the result of the grid inhomogeneous solvation theory (GIST) analysis to evaluate the hydration properties of the binding pocket [26]. GIST is a computational method that calculates the hydration structure and thermodynamics of the water molecules around a biomolecule using data obtained from molecular dynamics simulations. GIST divides the

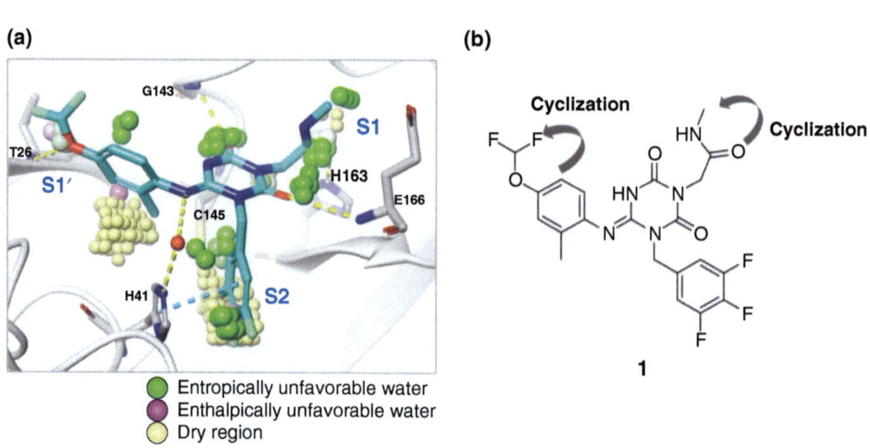

Figure 11.6 Hydration site analysis of the X-ray co-structure of 1. (a) Unfavorable water molecule occupancy calculated by a molecular dynamics simulation based on grid inhomogeneous solvation theory. (b) Strategy for the optimization.

three-dimensional space around a biomolecule into a grid and calculates both the water molecule density and the water molecule energy at each grid point. For example, water molecules that form strong hydrogen bonds with the hydrophilic residues of proteins are enthalpically favored, and displacement of such water molecules can result in disadvantageous protein-ligand binding. By contrast, a water molecule surrounded by hydrophobic residues of a protein that cannot form proper hydrogen bonds is enthalpically disadvantaged, and displacing such unfavorable water molecules results in an enthalpic contribution in the binding free energy [27, 28]. In Figure 11.6a, green spheres represent water molecules that were predicted to be entropically unfavorable molecules that can be replaced by hetero atoms that can form interactions, pink spheres represent water molecules that were predicted to be enthalpically unfavorable molecules that can be replaced by hydrophobic moieties, and yellow spheres represent a predicted dry region that water molecules cannot access. In the S1 and catalytic regions, the two core carbonyl groups replace the entropically unfavorable water molecules and form interactions with the protein, and the carboxamide side chain also replaces entropically unfavorable water. In the S2 site, the fluorobenzyl moiety replaces the enthalpically unfavored water molecules and also occupies the dry region. In the S1′ site, the ether oxygen replaces the entropically unfavorable water and forms a hydrogen bond with the main-chain NH of Thr26, and the methyl group occupies the dry region. This analysis suggested that triazine 1 already forms important hydrophilic and hydrophobic interactions with the protein. Therefore, the strategy for the structure optimization was very straightforward: investigate compounds that maintain the current interactions with the protein, combined with cyclization of the 4-difluoromethoxy-2-methylbenzene fragment in the S1′ pocket and the methyl carbamoyl unit in the S1 pocket to improve the activity by minimizing the energy loss because of the loss in entropy (Figure 11.6b).

Starting from hit compound 1 (triazine 1), we performed multiparameter optimization while preserving the hydrogen bonds indicated in the structure of the X-ray complex (Figure 11.7). Initially, we focused on optimizing the P1′ ligand to improve

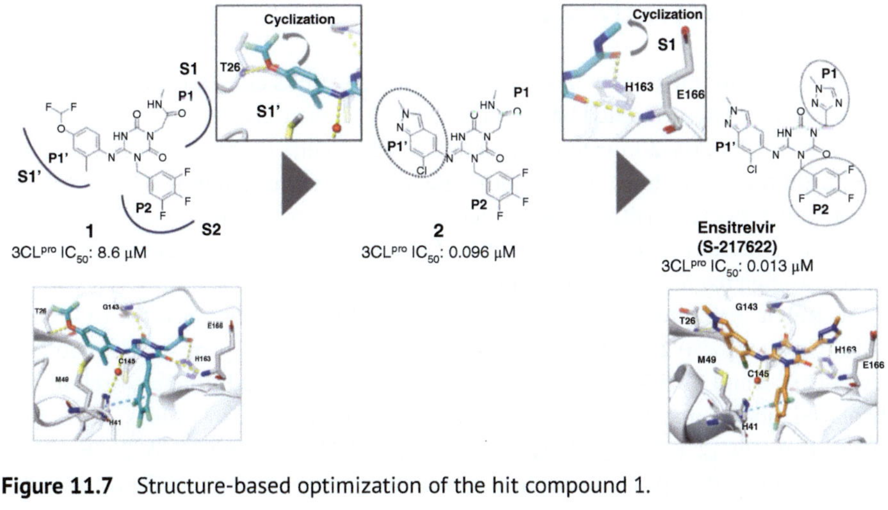

Figure 11.7 Structure-based optimization of the hit compound 1.

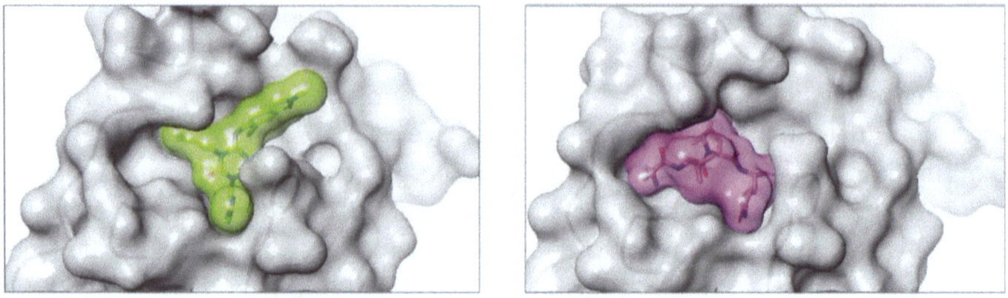

Figure 11.8 The co-crystal structures of ensitrelvir (green) and nirmatrelvir (purple) with 3CL protease.

compatibility with the S1′ pocket while maintaining the hydrogen bond with Thr26. This strategy led to the development of compound 2, which incorporated a 6-chloro-2-methyl-2H-indazole as the P1′ ligand. Compound 2 exhibited a remarkable 90-fold increase in enzymatic inhibitory activity while retaining a favorable DMPK profile.

Next, we replaced the P1 methyl amide moiety with various heterocyclic compounds, ultimately resulting in the synthesis of compound 3, now known as ensitrelvir, which exhibited an activity enhancement of over 600-fold, compared with compound 1, within four months of SAR development starting from compound 1. Ensitrelvir demonstrated potent biochemical activity with an IC_{50} value of 0.013 μM, as well as antiviral activity with an EC_{50} value of 0.37 μM.

The co-crystal structures of ensitrelvir (green) and nirmatrelvir (purple) with $3CL^{pro}$ are shown in Figure 11.8. The most obvious difference between the structures is that nirmatrelvir forms a covalent bond with Cys145 of the protein, while ensitrelvir does not. In addition, nirmatrelvir occupies site S4, while ensitrelvir does not, and ensitrelvir, but not nirmatrelvir, uses the S1′ site to boost activity. Thus, the key interactions of ensitrelvir and nirmatrelvir are different, indicating that these two molecules may be complementary, which would be useful if a viral strain develops resistance to either of these compounds.

11.3 Antiviral Activity of Ensitrelvir

The *in vitro* antiviral activity of ensitrelvir against various clinically isolated SARS-CoV-2 variants was evaluated based on the ability of ensitrelvir to inhibit cytopathic effects in SARS-CoV-2-infected Vero E6/TMPRSS2 cells (Table 11.2). Because no significant mutations have been reported near the catalytic center of $3CL^{pro}$ in these variants of concern, ensitrelvir was effective against most of the major strains in a similar manner to that expected, with half-maximal effective concentration (EC_{50}) values ranging from 0.31 to 0.50 μM [29].

Ensitrelvir exhibited comparable antiviral activity against SARS-CoV-2 and SARS-CoV (EC_{50}: 0.21 μM) because the sequence homology of $3CL^{pro}$ between SARS-CoV-2 and SARS-CoV is well conserved (>95%). In addition, ensitrelvir also displayed potent antiviral activity against other coronavirus families, including MERS-CoV (EC_{50}: 1.4 μM), HcoV-OC43 (EC_{90}: 0.074 μM), and HcoV-229E (EC_{50}: 5.5 μM), suggesting possible applications of this compound or its derivatives for

Table 11.2 Profile of ensitrelvir.

Anti-SARS-CoV-2 activity[a]	$3CL^{pro}$ IC_{50} (μM)	0.013
	Ancestral Strain_EC_{50} (μM)[b] (WK-521)	0.37
	Alpha Strains_EC_{50} (μM)[b] (QHN001/QHN002/QK002)	0.31/0.46/0.33
	Beta Strain_EC_{50} (μM)[b] (TY8-612)	0.4
	Gamma Strains_EC_{50} (μM)[b] (TY7-501/TY7-503)	0.50/0.43
	Delta Strain_EC_{50} (μM)[b] (TY11-927)	0.41
	Omicron Strain_EC_{50} (μM)[b] (BA.1/BA.1.1/BA.2)	0.29/0.36/0.52
Coronavirus family antiviral activity	SARS-CoV_EC_{50} (μM)[b]	0.21
	MERS-CoV_EC_{50} (μM)[b]	1.4
	HCoV-229E_EC_{50} (μM)[c]	5.5
	HCoV-OC43_EC_{90} (μM)[c]	0.074
Selectivity	Caspase 2_IC_{50} (μM)	>100
	Chymotrypsin_IC_{50} (μM)	>100
	Cathepsin B/D/G/L_IC_{50} (μM)	>100
	Thrombin_IC_{50} (μM)	>100
DMPK	Metabolic stability (%) HLM[d]/RLM[e]	96/88
	CL (mL/min/kg) rat/monkey/dog	1.7/0.29/0.17
	$T_{1/2}$ (h) rat/monkey/dog	2.4/10.0/29.5
	F_{oral} (%) rat/monkey/dog	97/106/65

IC_{50}: half maximal (50%) inhibitory concentration.
EC_{50}: half-maximal (50%) effective concentration.
a. Enzymatic and antiviral activity against various viral strains known to date.
b. Vero E6/TMPRSS2 cells.
c. MRC-5 cells.
d. Percentage remaining in human liver microsomes after 30 min incubation.
e. Percentage remaining in rat liver microsomes after 30 min incubation.

future pandemics caused by species from other coronavirus families. Importantly, ensitrelvir exhibited no inhibitory activity against various host-cell proteases, including caspase-2, chymotrypsin, cathepsin B/D/G/L, and thrombin, even at concentrations up to 100 μM. These results suggest that ensitrelvir has high selectivity toward coronavirus proteases.

In addition, ensitrelvir demonstrated desirable DMPK properties for oral dosing, including high metabolic stability (96% and 88% in human and rat liver microsomes, respectively), excellent oral absorption (97%), and low clearance (1.70 mL/min/mg) in rats. Remarkably, ensitrelvir exhibited more favorable DMPK profiles in monkeys and dogs, compared with rats, with low clearance; long elimination half-lives ($t_{1/2}$) of approximately 10 and 30 h in monkeys and dogs, respectively; and high oral bioavailability in all tested animals, indicating the potential for once-daily treatment of COVID-19 without the need for a PK booster.

11.3.1 *In Vivo* Antiviral Efficacy of Ensitrelvir

Next, an assessment of the antiviral efficacy of ensitrelvir was conducted using mice infected with the SARS-CoV-2 Gamma strain [23]. This strain possesses specific mutations in the receptor-binding domain of the spike protein that enhances the interaction of the virus with mouse ACE2 receptors. Five-week-old BALB/c mice were administered ensitrelvir orally in the form of a 0.5% methylcellulose suspension immediately after infection and again after 12 h. After 24 h, the mice were euthanized, and the viral titers in lung homogenates were measured. Administration of ensitrelvir resulted in a dose-dependent reduction of viral titers in the lungs (Figure 11.9). There was a significant decrease in the mean viral titer in the ensitrelvir treatment groups compared with the control group. Specifically, at doses of 2, 8, 16, and 32 mg/kg, the viral titers were significantly lower compared with the control group. At doses of 16 and 32 mg/kg, the viral titers approached the lower limit of quantification.

11.3.1.1 Delayed-treatment Model

In addition, we evaluated the inhibitory effect of ensitrelvir on SARS-CoV-2 replication in mice using a delayed-treatment model. Five-week-old mice were infected with the Gamma strain and then orally administered various doses of ensitrelvir or vehicle, twice daily, starting 24 h post-infection. While all the mice survived, the subsequent analysis revealed that ensitrelvir had a dose-dependent antiviral effect on the viral titers of the SARS-CoV-2 gamma strain in the lungs of infected mice (Figure 11.10). Treatment with both 32 and 64 mg/kg ensitrelvir resulted in a significant reduction in the viral titers, 2 and 3 days post-infection, compared with the vehicle treatment. In addition, treatment with 8 and 16 mg/kg ensitrelvir significantly reduced the lung viral titers at 3 days post-infection.

To further increase the severity of the SARS-CoV-2 infection in our test system, aged (approximately 40-week-old) mice were infected with the mouse-adapted SARS-CoV-2 MA-P10 strain. Then, the infected mice were orally administered various

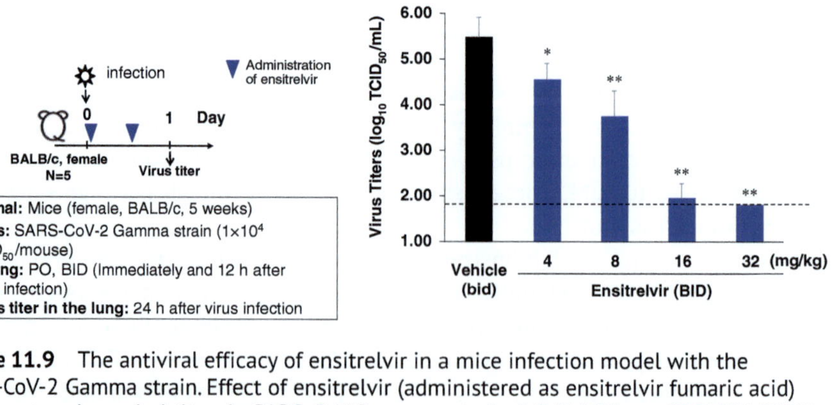

Figure 11.9 The antiviral efficacy of ensitrelvir in a mice infection model with the SARS-CoV-2 Gamma strain. Effect of ensitrelvir (administered as ensitrelvir fumaric acid) treatment on lung viral titers in SARS-CoV-2 gamma strain (hCoV-19/Japan/TY7-501/2021)-infected mice. The broken line represents the lower limit of quantification (1.80 $\log_{10}$ TCID$_{50}$/mL). The following P-values were calculated using Dunnett's test: *P <0.001, **P <0.0001 vs. vehicle; TCID$_{50}$, 50% tissue culture infectious dose; BID, twice daily.

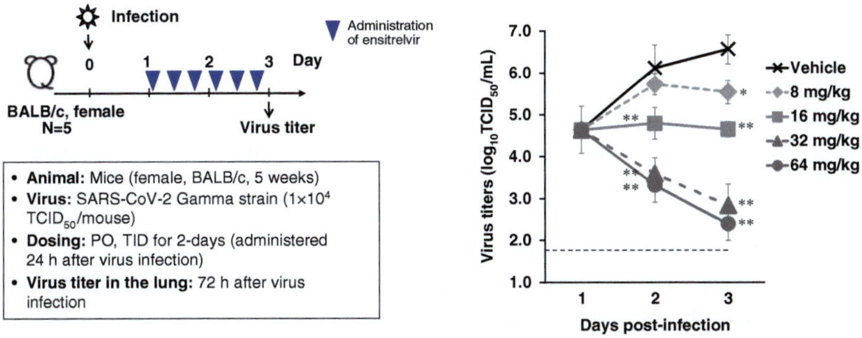

Figure 11.10 Antiviral effect of ensitrelvir on the viral titers of SARS-CoV-2 in the lungs of infected mice. Viral titers in the lungs of mice were determined using Vero E6/TMPRSS2 cells. Each point represents the mean ± SD of five mice. The dashed line represents the lower limit of quantification (LLOQ; 1.80 $\log_{10}$ TCID50/mL). P values were calculated by Dunnett's method vs. vehicle. *P <0.01, **P <0.0001. TID, three times daily.

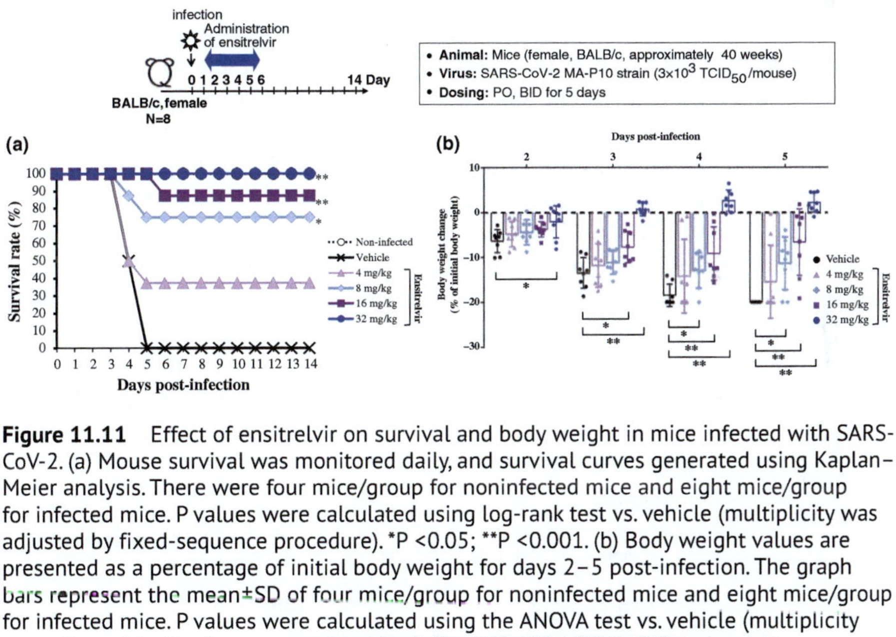

Figure 11.11 Effect of ensitrelvir on survival and body weight in mice infected with SARS-CoV-2. (a) Mouse survival was monitored daily, and survival curves generated using Kaplan–Meier analysis. There were four mice/group for noninfected mice and eight mice/group for infected mice. P values were calculated using log-rank test vs. vehicle (multiplicity was adjusted by fixed-sequence procedure). *P <0.05; **P <0.001. (b) Body weight values are presented as a percentage of initial body weight for days 2–5 post-infection. The graph bars represent the mean±SD of four mice/group for noninfected mice and eight mice/group for infected mice. P values were calculated using the ANOVA test vs. vehicle (multiplicity was adjusted by fixed-sequence procedure). *P <0.01; **P <0.001. BID, twice daily.

doses of ensitrelvir or vehicle twice daily for 5 days starting 24 h post-infection. The mice were then monitored for survival and body weight loss. The results showed that ensitrelvir improved the survival rates of mice in a dose-dependent manner (Figure 11.11). The survival time was significantly increased for infected mice treated with 8, 16, and 32 mg/kg ensitrelvir. Ensitrelvir treatment also suppressed the body weight loss observed in mice treated with vehicle in a dose-dependent manner. Treatment of SARS-CoV-2 MA-P10-infected 40-week-old mice, 24 h post-infection with ensitrelvir, also resulted in a dose-dependent suppression of the lung weight increase and the production of inflammatory cytokines and chemokines, including TNF-α, IL-1β, IL-6, and MCP-1, 4 days post-infection. Consistent with the results for cytokine/chemokine production and lung weight, lung histopathology revealed degeneration and necrosis of alveolar, bronchial, and bronchiolar epithelium cells, and vascular endothelial cells, as well as the presence of inflammation and thrombus, in vehicle-treated mice infected with SARS-CoV-2 MA-P-10 [30].

11.3.1.2 Pharmacokinetics and Pharmacodynamics

The relationship between the PK parameters and the pharmacodynamics (PD) effects of ensitrelvir was evaluated and correlated with the ensitrelvir *in vivo* efficacy. The PD effects of ensitrelvir were evaluated in a delayed-treatment mouse model system using mice infected with the SARS-CoV-2 Gamma strain. Mice were orally administered ensitrelvir at various doses and various timings, 24 h after infection, and the viral titers were measured at 72 h post-infection, which was 48 h after the first administration of ensitrelvir (Figure 11.12).

To investigate the relationship between the antiviral activity of ensitrelvir and the PK parameters, PK/PD analyses were conducted using a nonlinear regression model. The values of the PK parameters for ensitrelvir in the plasma of SARS-CoV-2-infected mice following oral treatment are presented in Table 11.3. The results are shown for single dosing (one shot) and repeated dosing (once daily for 2 days, twice daily for 2 days, and three times daily for 2 days). The maximum plasma concentration (C_{max}) for PA-EC_{50}, AUC_{0-48h} for PA-EC_{50}, C_{48h} for PA-EC_{50}, and $Time_{High}$ (the times for ensitrelvir levels to exceed 1×, 3×, 5×, and 10× the PA-EC_{50} levels [1×, 3×, 5×, and 10× PA-EC_{50}]) values were estimated. The results from the four-parameter logistic equation (model 205) that was used to calculate the relationship between the PD data of viral titer reduction (log_{10} 50% tissue culture infectious dose [$TCID_{50}$/mL]) and PK parameters, and between the PD data and $Time_{High}$, are shown in Figure 11.13. The coefficients of determination (r^2) values of $Time_{High}$ (10× PA-EC_{50}), AUC_{0-48h}/PA-EC_{50}, C_{48h}/PA-EC_{50}, $Time_{High}$ (5× PA-EC_{50}), $Time_{High}$

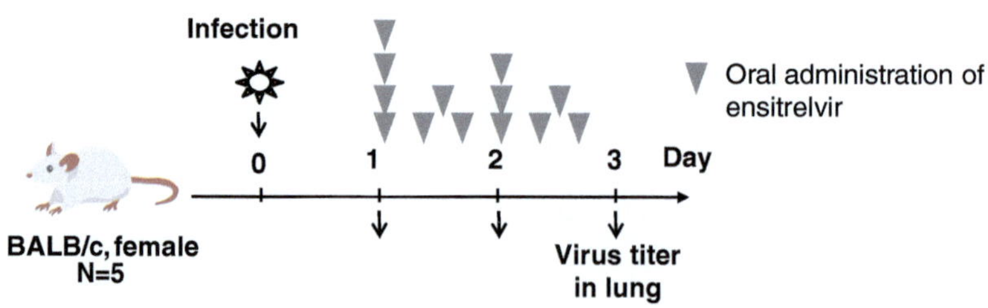

- **Animal:** Mice (female, BALB/c, 5 weeks old)
- **Virus:** SARS-CoV-2 Gamma strain (1 × 10^4 $TCID_{50}$/mouse)
- **Dosing:** 1-shot, QD, BID, or TID for 2 days (administered 24 h after virus infection)
- **Virus titer in the lung:** evaluated 3 days after virus infection

Figure 11.12 Delayed-treatment model system using mice infected with the SARS-CoV-2 Gamma strain. QD, Once daily; BID, two times daily; TID, three times daily.

Table 11.3 Pharmacokinetic parameters of ensitrelvir in the plasma of SARS-CoV-2-infected mice.

Dosing schedule	Dose (mg/kg)	PK parameters						
		C_{max}/PA-EC_{50}	AUC_{0-48h}/PA-EC_{50}	C_{48h}/PA-EC_{50}	$Time_{High}$ 1×PA-EC_{50} (h)	$Time_{High}$ 3×PA-EC_{50} (h)	$Time_{High}$ 5×PA-EC_{50} (h)	$Time_{High}$ 10×PA-EC_{50} (h)
One shot	32	34.6	244	0.0000283	15.8	13.3	12.1	9.09
	64	76.1	641	0.000531	21.7	17.1	15.7	13.8
Once daily for 2 days	16	18.6	227	0.0363	26.7	19.9	16	10.1
	32	34.7	488	0.1	31.6	26.5	24.2	18.2
	64	76.6	1280	0.617	43.4	34.2	31.4	27.5
Twice daily for 2 days	8	8.66	213	0.88	45.7	28.7	20.1	0
	16	19.8	450	1.8	48	40.1	32.3	20.6
	32	38.1	962	5.45	48	47.9	47.9	36.9
	64	90	2500	20.1	48	48	48	47.9
Three times daily for 2 days	8	10.3	314	2.75	47.9	45.2	32.7	2.13
	16	22.8	665	5.55	48	47.9	47.9	33.5
	32	45.9	1420	14.4	48	47.9	47.9	47.8
	64	111	3660	42.2	48	48	48	47.9

AUC_{0-48h}, area under the curve plasma concentrations 0 through 48 h post-first administration; C_{48h}, plasma concentration at 48 h post-first administration; C_{max}, maximum concentration; PA-EC_{50}, protein-adjusted 50% efficacy concentration; PK, pharmacokinetic; $Time_{High}$, total time above the target plasma concentration; one shot, single dosing; once daily, every 24 h; twice daily, every 12 h; three times daily, every 8 h.

(3× PA-EC_{50}), C_{max}/PA-EC_{50}, and $Time_{High}$ (1× PA-EC_{50}) against viral titer reduction were 0.775, 0.735, 0.734, 0.647, 0.499, 0.311, and 0.310, respectively. $Time_{High}$ (10× PA-EC_{50}), AUC_{0-48h}/PA-EC_{50}, and C_{48h}/PA-EC_{50} demonstrated r^2 values > 0.7. Both the $Time_{High}$ (10× PA-EC_{50}) and C_{48h}/PA-EC_{50} parameters indicated a good correlation with the PD data, suggesting that maintaining a certain level of plasma concentration was important for ensitrelvir to exert an antiviral effect in mice infected with the SARS-CoV-2 Gamma strain after administration of ensitrelvir 24 h after virus infection [31].

Using this SARS-CoV-2-infected mouse model and a mouse PA-EC_{50} value against the SARS-CoV-2 Gamma stain of 2090 ng/mL (3.93 μmol/L), the C_{48h} values required to reduce the viral titers by 1-log_{10}, 2-log_{10}, and 3-log_{10} $TCID_{50}$/mL, compared with the mean viral titer of the vehicle treatment group, 48 h after the first administration of ensitrelvir, were calculated to be 1610, 7900, and 30 700 ng/mL, respectively (Table 11.4). Similarly, using a human PA-EC_{50} value of 1610 ng/mL (3.02 μmol/L), the C_{48h} values required to reduce the viral titers compared with the vehicle treatment group by 1-log_{10}, 2-log_{10}, and 3-log_{10} $TCID_{50}$/mL in humans were calculated to be 1240, 6090, and 23 700 ng/mL, respectively [31].

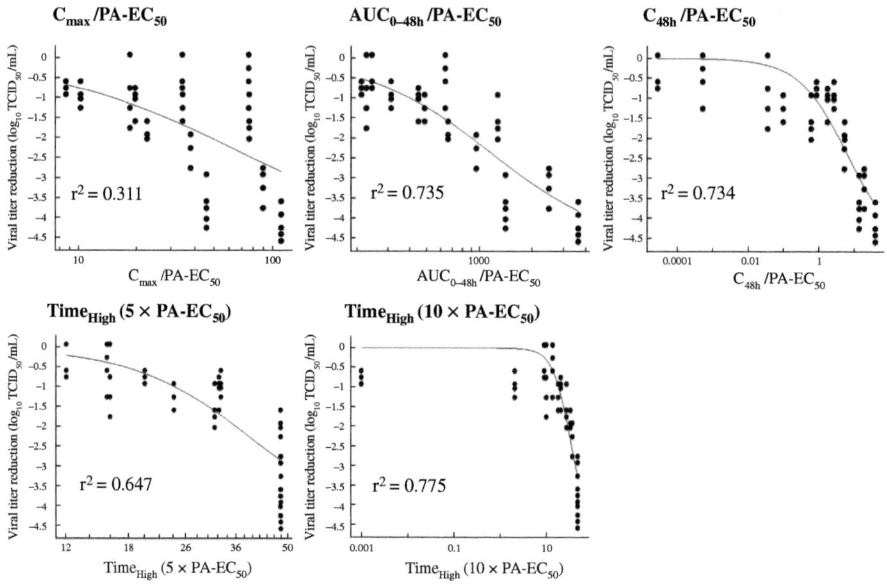

Figure 11.13 Pharmacokinetic and pharmacodynamic analysis of ensitrelvir in SARS-CoV-2-infected mice. The PK parameters maximum concentration (C_{max}), AUC0−48h post-first administration, concentration at 48 h post-first administration (C_{48h}), and total time above the target plasma concentration ($Time_{High}$) were calculated using the noncompartmental method. The parameters relative to protein-adjusted 50% efficacy concentration (PA-EC_{50}) are plotted vs. the virus titer reduction in the graphs above. Fitting curves of PD data and PK parameters using a nonlinear regression model are depicted. The coefficients of determination (r^2) are indicated. The PA-EC_{50} value was based on a previous report [30]. AUC_{0-48h}, area under the curve plasma concentrations 0 through 48 h post-first administration

Table 11.4 Pharmacokinetic analysis of ensitrelvir in SARS-CoV-2-infected mice, showing the C_{48h} values required to reduce the viral titers compared with the vehicle treatment group by 1-log_{10}, 2-log10, and 3-log_{10} $TCID_{50}$/mL.

Viral titer reduction ($TCID_{50}$/mL)	C_{48h}/ PA-EC_{50}	Mouse PA-EC_{50} (ng/mL)	Mouse C_{48h} (ng/mL)	Human PA-EC_{50} (ng/mL)	Human C_{48h} (ng/mL)
1-log10	0.769	2090	1610	1610	1240
2-log10	3.78	2090	7900	1610	6090
3-log10	14.7	2090	30 700	1610	23 700

C_{48h}, plasma concentration at 48 h post-first administration; PA-EC_{50}, protein-adjusted 50% efficacy concentration.

11.3.2 Pharmacokinetics

11.3.2.1 Pharmacokinetic Profiles

Following single-dose administration of ensitrelvir (as a suspension) at 20–2000 mg in fasted, healthy Japanese adult participants, the median time to maximum plasma

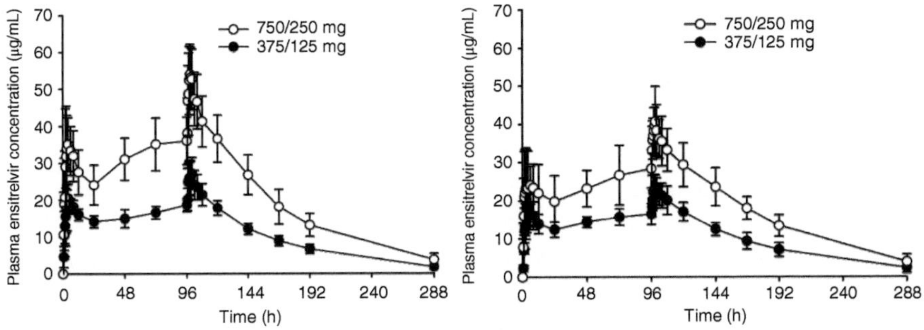

Figure 11.14 Mean (standard deviation) plasma concentration profile of ensitrelvir following administration of ensitrelvir at 375/125 or 750/250 mg. Left: healthy Japanese adult female participants; right: healthy White adult participants.

concentration (T_{max}) values of ensitrelvir ranged from 1.50 to 4.00 h and the geometric mean terminal elimination half-life ($t_{1/2,z}$) values of ensitrelvir were 42.2–48.1 h. The plasma exposure (C_{max}) and area under the concentration-time curve extrapolated to infinity ($AUC_{0\text{-}inf}$) values of ensitrelvir increased in an almost dose-proportional manner across the dose ranges after single-dose administration, although the upper limit of the 95% confidence intervals (CIs) for both parameters was < 1. In addition, urinary recovery of ensitrelvir was observed, and the geometric mean of the fraction of the dose excreted in the urine from time 0–144 h ($Feu_{0\text{-}144}$) of ensitrelvir ranged from 12.9% to 21.8% across the dose groups (20–1000 mg) [32]. Considering the long $t_{1/2,z}$ value of ensitrelvir and the need to exceed the target plasma concentration (6090 ng/mL) from the first dose, a loading dose of 375 mg on day 1, followed by a maintenance dose of 125 mg on days 2–5 (375/125 mg), and a loading dose of 750 mg on day 1 followed by a maintenance dose of 250 mg on days 2–5 were selected as the dose regimens for a multiple-dose study and a Phase 2/3 study. The PK profiles in healthy Japanese female and White adults following administration of ensitrelvir (tablet formulation) at 375/125 mg and 750/250 mg are shown in Figure 11.14. The plasma concentration at 24 h (C_{24}) after the initial dose in the 375/125 mg group exceeded the target plasma concentration (6.09 μg/mL) in all populations [33].

11.3.2.2 Effect of Food

The effect of food on the PKs of ensitrelvir was evaluated following single-dose administration of ensitrelvir 375 mg (3 ensitrelvir 125 mg tablets) in fasted and fed (high-fat/high-calorie) states. The geometric least squares (GLS) mean ratios (90% Cis) of the C_{max} and $AUC_{0\text{-}inf}$ values for ensitrelvir (fed/fasted) were 0.9320 (0.8134–1.0679) and 1.2447 (1.1396–1.3596), respectively. The 90% CIs for the C_{max} values of ensitrelvir were completely contained within the range of 0.8000–1.2500, suggesting that the C_{max} values of ensitrelvir were comparable between the fasted and fed states. The $AUC_{0\text{-}inf}$ values of ensitrelvir were increased by approximately 25% in the fed state compared with those in the fasted state, and the T_{max} value was delayed from 2.50 h in the fasted state to 6.00 h in the fed state. Food intake delayed the T_{max} of ensitrelvir in the fed state compared with the fasted state and led to a relatively small increase in the AUC value but did not affect the corresponding C_{max} values, indicating that there was no clinically meaningful difference in the exposure to ensitrelvir between the fasted and fed states [33].

11.3.2.3 Drug–Drug Interactions

In an *in vitro* study, ensitrelvir showed a time-dependent NADPH-mediated inhibition of cytochrome P450 3A (CYP3A) rather than a reversible inhibitory effect on CYP3A4. Therefore, the effect of ensitrelvir on the PKs of midazolam, a typical CYP3A substrate, was evaluated. Ensitrelvir (375/125 mg) was administered orally in the fasted state, and midazolam (2 mg/mL syrup) was administered in the fasted state alone on day 2 and concomitantly with ensitrelvir on day 5. When midazolam was co-administered with ensitrelvir (125 mg) on day 5, the GLS mean ratios for the C_{max}, $AUC_{0\text{–last}}$, and $AUC_{0\text{-inf}}$ values were 2.80-, 6.90-, and 6.77-fold higher, respectively, compared with the corresponding values for midazolam alone on day 2, suggesting that ensitrelvir was a strong CYP3A inhibitor in this dose regimen. In addition, the effects of multiple doses (once daily for 5 days) of ensitrelvir (tablet 750/250 mg) administered orally in the fasted state on the PKs of dexamethasone and prednisolone were investigated. Dexamethasone (1 mg tablet) or prednisolone (10 mg tablet) was administered on days 2, 5 (coadministered with ensitrelvir), 9 (5th day after the last ensitrelvir dose), and 14 (10th day after the last ensitrelvir dose) in the fasted state. The AUC ratio of dexamethasone on day 5 was 3.47-fold higher than the corresponding value for dexamethasone alone on day 2 and the effect diminished over time after the last dose of ensitrelvir. No clinically meaningful effect was observed for prednisolone [34].

Ensitrelvir was found to be a substrate for P-glycoprotein (P-gp) and breast cancer resistance protein (BCRP), and it inhibited P-gp, BCRP, organic anion transporting polypeptide (OATP) 1B1, OATP1B3, organic cation transporter (OCT) 1, and organic anion transporter (OAT) 3 in an *in vitro* study. The effect of P-gp and BCRP inhibitors would be limited in increasing the absorption of ensitrelvir because the oral absorption of ensitrelvir should be high in humans, considering the high bioavailability of ensitrelvir in laboratory animals [23]. Therefore, a clinical DDI study to evaluate the effect of ensitrelvir on the PKs of P-gp, BCRP, OATP1B1, OATP1B3, and OCT1 substrates was conducted using digoxin (a P-gp substrate), rosuvastatin (a BCRP, OATP1B1, and OATP1B3 substrate), and metformin (an OCT1 substrate) using a cocktail approach. A cocktail of the above drugs was administered first, and after a washout period, the cocktail and ensitrelvir 500 mg were co-administered. The exposures of digoxin and rosuvastatin were slightly increased when co-administered with ensitrelvir and the exposures of metformin did not change. Therefore, ensitrelvir needs to be administered with caution when coadministered with P-gp, BCRP, OATP1B1, or OATP1B3 substrates with a narrow therapeutic range [35]. As an additional evaluation, the plasma levels of coproporphyrin-I, an endogenous biomarker for OATP1B, did not differ with or without, administration of ensitrelvir in a drug interaction study with a cocktail of substrates, indicating that ensitrelvir did not exhibit any inhibitory effect on OATP1B [36].

11.3.3 Clinical Trial of Ensitrelvir

The multicenter, double-blind, Phase 2a part of a Phase 2/3 study was conducted during the Delta pandemic period in Japan to assess the efficacy and safety of ensitrelvir in patients with mild-to-moderate COVID-19 or asymptomatic SARS-CoV-2 infection (jRCT2031210350). Sixty-nine patients were randomized (1 : 1 : 1) to orally receive ensitrelvir fumaric acid for 5 days (375 mg on day 1, followed by 125 mg daily, or 750 mg on day 1 followed by 250 mg daily) or placebo and were followed up until day 28. A total of 16, 14, and 17 patients in the ensitrelvir 125 mg, ensitrelvir 250 mg, and placebo groups, respectively, were included in the intention-to-treat (ITT) population (mean age: 38.0–40.4 years).

The SARS-CoV-2 viral titer, measured as $\log_{10}$ $TCID_{50}$/mL, decreased with time up to day 4 and remained stable at levels at the lower limit of detection (0.8 $\log_{10}$ $TCID_{50}$/mL) until day 21 in all groups. On day 4, the change from baseline in the viral titer in patients with a positive viral titer and viral RNA at baseline was greater with ensitrelvir 125 mg (mean [standard deviation], −2.42 [1.42]; P = 0.0712) and 250 mg (−2.81 [1.21]; P = 0.0083) compared with placebo (−1.54 [0.74]), and ensitrelvir treatment reduced the viral RNA by −1.4 to −1.5 $\log_{10}$ copies/mL versus placebo. The viral titer and viral RNA were similar across groups on day 6 until the end of the follow-up period (Figure 11.15).

The median time to infectious viral clearance (first negative SARS-CoV-2 viral titer) was significantly shorter in the ensitrelvir 125 mg (61.3 h; difference from the placebo, −49.8 h; P = 0.0159) and 250 mg (62.7 h; difference from the placebo, −48.4 h; P = 0.0205) groups versus the placebo group (111.1 h). Infectious viral clearance was achieved in all patients in the ensitrelvir 125 and 250 mg groups in less than 120 h.

The mean total score of 12 COVID-19 symptoms composed of stuffy or runny nose, sore throat, shortness of breath (difficulty breathing), cough, low energy or tiredness, muscle or body aches, headache, chills or shivering, feeling hot or feverish, nausea (feeling like you wanted to throw up), vomiting (throwing up), and diarrhea (loose

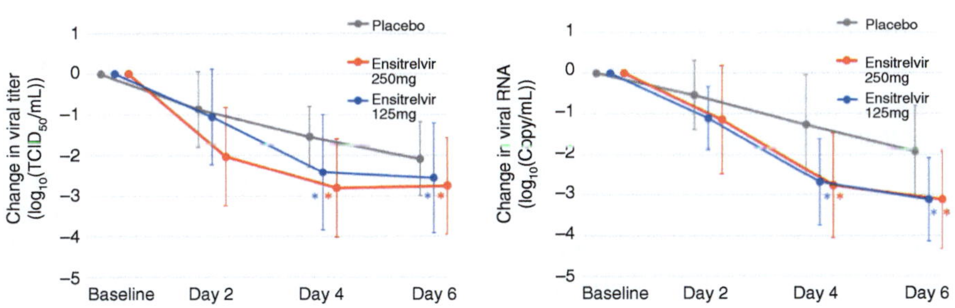

Number of patients	Day 2	Day 4	Day 6
Placebo	14	14	13
Ensitrelvir 125mg	15	15	15
Ensitrelvir 250mg	14	14	14

Number of patients	Day 2	Day 4	Day 6
Placebo	17	17	16
Ensitrelvir 125mg	15	15	15
Ensitrelvir 250mg	14	14	14

Figure 11.15 (a) SARS-CoV-2 viral titer as a change from baseline (mITT population) and (b) SARS-CoV-2 viral RNA level as a change from baseline (ITT population). Data are presented as mean ± standard deviation. *P < 0.05 vs. placebo; ITT, intention-to-treat; mITT, modified intention-to-treat; $TCID_{50}$, 50% tissue culture infectious dose.

or watery stools) in patients with mild-to-moderate COVID-19 showed a decreasing trend with time after treatment in all the groups. Generally, the mean change from baseline in the total score of the 12 COVID-19 symptoms was numerically greater in the ensitrelvir 125 and 250 mg groups compared with the placebo group.

Treatment-emergent adverse events (TEAEs) were reported in 52.4% (11/21), 69.6% (16/23), and 37.5% (9/24) of patients in the ensitrelvir 125 mg, 250 mg, and placebo groups, respectively. No serious TEAEs, deaths, or TEAEs leading to treatment discontinuation were reported, and all were mild to moderate in severity. The most common TEAE was a decrease in high-density lipoprotein (HDL: 14.3% [3/21], 52.2% [12/23], and 8.3% [2/24] of patients in the ensitrelvir 125 mg, 250 mg, and placebo groups, respectively). Transient decreases and increases in HDL and blood triglyceride levels, respectively, were observed in the ensitrelvir groups following treatment but were resolved without additional treatment. Additionally, increases in the total bilirubin and iron levels were observed in patients receiving ensitrelvir. The total bilirubin increases in two patients and an increase in the iron level in one patient in the ensitrelvir 250 mg group were observed as TEAEs. These changes were asymptomatic and transient and resolved without additional treatment. No obvious changes compared with placebo were observed in haptoglobin, reticulocytes, and low-density lipoprotein in the ensitrelvir groups, and no laboratory or clinical signs of hemolysis were observed.

Oral administration of ensitrelvir for 5 days demonstrated rapid clearance of SARS-CoV-2 and was well tolerated in patients with mild-to-moderate COVID-19 or asymptomatic SARS-CoV-2 infection. In the Phase 2a part of the Phase 2/3 trial, the proof of concept of ensitrelvir was achieved successfully [37].

11.4 Outlook

Following Phase 2a, the Phase 2b and Phase 3 (SCORPIO-SR) parts of the trial were conducted during the Omicron pandemic period. In Phase 2b part, we confirmed the rapid and favorable antiviral efficacy of ensitrelvir, compared with placebo, against Omicron [38], mainly the BA.1 variant, and from these results we selected 125 mg as the target efficacy dose in SCORPIO-SR [39], then we demonstrated the efficacy of ensitrelvir for COVID-19 symptom improvement, and ensitrelvir was granted emergency approval in November 2022.

To investigate the ensitrelvir efficacy in different patient populations, additional clinical trials are being conducted, including SCORPIO-HR (NCT05305547) for patients regardless of high-risk factors, STRIVE (NCT05605093) for inpatients, and a pediatric study (jRCT2031230140).

11.5 Conclusions

Here, we have described the discovery and development of ensitrelvir, the first non-peptidic, noncovalent, oral 3CL^{pro} inhibitor clinical candidate for treating COVID-19. When this discovery program was started, most of the known inhibitors of 3CL^{pro}

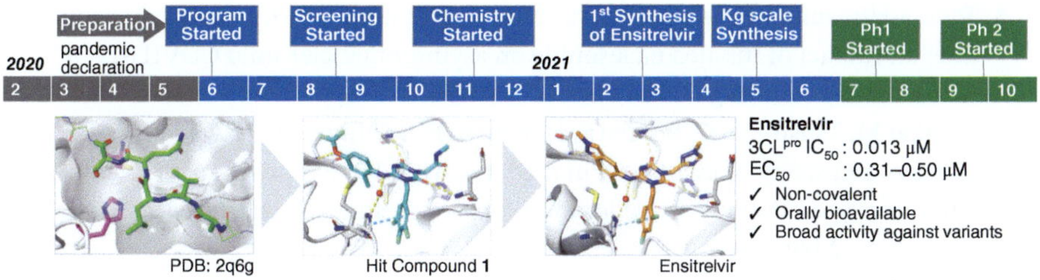

Figure 11.16 The timeline of the discovery of ensitrelvir.

were peptide substrate mimetics with covalent warheads that bound covalently to Cys145 in the active site of $3CL^{pro}$. We used a screening program focused on our in-house library of compounds to realize the fast discovery of a lead molecule. Although the activity of the initial hit compound was not ideal, the DMPK profile showed the potential for once-daily oral administration. Structure-based drug design facilitated the rapid optimization of the activity while maintaining or improving the favorable DMPK profile. The optimization finally resulted in ensitrelvir, which had promising antiviral activity against known variants of concern, a long elimination half-life *in vivo*, especially in monkeys and dogs, excellent oral bioavailability, and potent efficacy in *in vivo* mouse models of SARS-CoV-2 infections.

In the Phase 1 clinical trial that started in Japan in July 2021, ensitrelvir exhibited an excellent PK profile, including a half-life of elimination from the blood of over 40 h [33]. Clear antiviral effects were observed in Phase 2a/2b clinical trials [37, 38], and after obtaining the topline results of a Phase 3 trial, ensitrelvir was approved in Japan for the treatment of SARS-CoV-2 infection under the emergency regulatory approval system on 22 November 2022, with the brand name Xocova®. Figure 11.16 shows the discovery of ensitrelvir in chronological order. Four months after obtaining a screening hit, the first batch of ensitrelvir was synthesized in March 2021, and then GLP (Good Laboratory Practice) testing was completed in less than four months. On 22 July 2021, approximately eight months after starting the SAR optimization study, the first-in-human Phase 1 trial was initiated.

The speed toward the approval of ensitrelvir was truly beyond our expectations, something we had never imagined when we first started this drug discovery program. Looking back to the beginning of this project, we were caught up in our own common sense and assumptions, but by thinking about "how to make it possible," we realized that we could achieve things that we previously thought were impossible. We hope that this experience will bring about a paradigm shift in the conventional, fixed concepts of small molecule drug discovery in the pharmaceutical industry.

References

1 WHO Coronavirus (COVID-19) Dashboard: https://covid19.who.int (accessed 5 Dec 2023).

2 Ivermectin Press Release: https://www.kowa.co.jp/news/2022/press20220926.html (accessed 5 Dec 2023).

3 Terada-Hirashima J, Suzuki M, Tsujimoto Y, Hamamoto Y, Uemura Y, Tsushima K, et al. Impact of inhaled ciclesonide on asymptomatic or mild COVID-19: a randomized trial. *Drug Discov Ther*. 2022;16(5): 225–32.

4 Shinkai M, Tsushima K, Tanaka S, Hagiwara E, Tarumoto N, Kawada I, et al. Efficacy and safety of favipiravir in moderate COVID-19 pneumonia patients without oxygen therapy: a randomized, phase III clinical trial. *Infect Dis Ther*. 2021;10(4): 2489–509.

5 Doi Y, Hibino M, Hase R, Yamamoto M, Kasamatsu Y, Hirose M, et al. A prospective, randomized, open-label trial of early versus late favipiravir therapy in hospitalized patients with COVID-19. *Antimicrob Agents Chemother*. 2020;64(12): e01897–20.

6 Avigan® Press Release: https://www.fujifilm.com/jp/en/news/hq/8698 (accessed 5 Dec 2023).

7 Iwata S, Kobayashi O, Kurashima K, Doi Y, Kunishima H, Shinkai M, et al. Findings from a discontinued clinical trial of favipiravir in high-risk patients with early-onset COVID-19. *J Infect Chemother*. 2024;30(3): 219–27. (Online ahead of print).

8 Hu B, Guo H, Zhou P, Shi ZL. Characteristics of SARS-CoV-2 and COVID-19. *Nat Rev Microbiol*. 2021;19(3): 141–54.

9 Ullrich S, Nitsche C. The SARS-CoV-2 main protease as drug target. *Bioorg Med Chem Lett*. 2020;30(17): 127377.

10 Agbowuro AA, Huston WM, Gamble AB, Tyndall JDA. Proteases and protease inhibitors in infectious diseases. *Med Res Rev*. 2018;38(4): 1295–331.

11 Qiao J, Li YS, Zeng R, Liu FL, Luo RH, Huang C, et al. SARS-CoV-2 M^{pro} inhibitors with antiviral activity in a transgenic mouse model. *Science*. 2021;371(6536): 1374–8.

12 Dampalla CS, Zheng J, Perera KD, Wong LR, Meyerholz DK, Nguyen HN, et al. Postinfection treatment with a protease inhibitor increases survival of mice with a fatal SARS-CoV-2 infection. *Proc Natl Acad Sci U S A*. 2021;118(29): e2101555118.

13 Hoffman RL, Kania RS, Brothers MA, Davies JF, Ferre RA, Gajiwala KS, et al. Discovery of ketone-based covalent inhibitors of coronavirus 3CL proteases for the potential therapeutic treatment of COVID-19. *J Med Chem*. 2020;63(21): 12725–47.

14 Owen DR, Allerton CMN, Anderson AS, Aschenbrenner L, Avery M, Berritt S, et al. An oral SARS-CoV-2 M^{pro} inhibitor clinical candidate for the treatment of COVID-19. *Science*. 2021;374(6575): 15861593.

15 Pelly S, Liotta D. Potent SARS-CoV-2 direct-acting antivirals provide an important complement to COVID-19 vaccines. *ACS Cent Sci*. 2021;7(3): 396–99.

16 Zhang L, Lin D, Sun X, Curth U, Drosten C, Sauerhering L, et al. Crystal structure of SARS-CoV-2 main protease provides a basis for design of improved α-ketoamide inhibitors. *Science*. 2020;368(6489): 409–12.

17 Mengist HM, Mekonnen D, Mohammed A, Shi R, Jin T. Potency, safety, and pharmacokinetic profiles of potential inhibitors targeting SARS-CoV-2 main protease. *Front Pharmacol*. 2021;11: 630500.

18 Jacobs J, Grum-Tokars V, Zhou, Y. Turlington M, Saldanha SA, Chase P, et al. Discovery, synthesis, and structure-based optimization of a series of

N-(tert-butyl)-2-(N-arylamido)-2-(yridine-3-yl) acetamides (ML188) as potent noncovalent small molecule inhibitors of the severe acute respiratory syndrome coronavirus (SARS-CoV) 3CL protease. *J Med Chem.* 2013;56(2): 534–46.

19 Turlington M, Chun A, Tomar S, Eggler A, Grum-Tokars V, Jacobs J, et al. Discovery of N-(benzo[1,2,3]triazol-1-yl)-N-(benzyl)acetamido)phenyl) carboxamides as severe acute respiratory syndrome coronavirus (SARS-CoV) 3cLpro inhibitors: identification of ML300 and noncovalent nanomolar inhibitors with an induced-fit binding. *Bioorg Med Chem Lett.* 2013;23(22): 6172–77.

20 Han SH, Goins CM, Arya T, Shin WJ, Maw J, Hooper A, et al. Structure-based optimization of ML300-derived, noncovalent inhibitors targeting the severe acute respiratory syndrome coronavirus 3CL protease (SARS-CoV-2 3cLpro). *J Med Chem.* 2022;65(4): 2880–904.

21 Douangamath A, Fearon D, Gehrtz P, Krojer T, Lukacik P, Owen CD, et al. Crystallographic and electrophilic fragment screening of the SARS-CoV-2 main protease. *Nat Commun.* 2020;11(1): 5047.

22 Boby ML, Fearon D, Ferla M, Filep M, Koekemoer L, Robinson MC, The COVID Moonshot Consortium, et al. Open science discovery of oral non-covalent SARS-CoV-2 main protease inhibitor therapeutics. *bioRxiv.* 2021.

23 Unoh Y, Uehara S, Nakahara K, Nobori H, Yamatsu Y, Yamamoto S, et al. Discovery of S-217622, a noncovalent oral SARS-CoV-2 3CL protease inhibitor clinical candidate for treating COVID-19. *J Med Chem.* 2022;65(9): 6499–512.

24 Yang H, Xie W, Xue X. Yang K, Ma J, Liang W, et al. Design of wide-spectrum inhibitors targeting coronavirus main proteases. *pLoS Biol.* 2005;3(10): e324.

25 Structure of COVID-19 main protease bound to potent broad-spectrum non-covalent inhibitor X77; NCBI Structure Summary MMDB: https://www.ncbi.nlm.nih.gov/Structure/mmdb/mmdbsrv.cgi?dps=0&uid=6W63 (accessed 22 Dec 2023).

26 Nguyen CN, Young TK, Gilson MK. Grid inhomogeneous solvation theory: hydration structure and thermodynamics of the miniature receptor cucurbit[7]uril. *J Chem Phys.* 2012;137(4): 044101.

27 Jung SW, Kim M, Ramsey S, Kurtzman T, Cho AE. Water pharmacophore: designing ligands using molecular dynamics simulations with water. *Sci Rep.* 2018;8(1): 10400.

28 Ichihara O, Shimada Y, Yoshidome D. The importance of hydration thermodynamics in fragment-to-lead optimization. *ChemMedChem.* 2014;9(12): 2708–17.

29 Kawashima S, Matsui Y, Adachi T, Morikawa Y, Inoue K, Takebayashi S, et al. Ensitrelvir is effective against SARS-CoV-2 3CL protease mutants circulating globally. *Biochem Biophys Res Commun.* 2023;645: 132–6.

30 Nobori H, Fukao K, Kuroda T, Anan N, Tashima R, Nakashima M, et al. Efficacy of ensitrelvir against SARS-CoV-2 in a delayed-treatment mouse model. *J Antimicrob Chemother.* 2022;77(11): 2984–91.

31 Fukao K, Nobori H, Kuroda T, Baba K, Matsumoto K, Tanaka Y, et al. Pharmacokinetic and pharmacodynamic analysis of the 3CL protease inhibitor ensitrelvir in a SARS-CoV-2 infection mouse model. *Viruses.* 2023;15(10): 2052.

32 Shimizu R, Sonoyama T, Fukuhara T, Kuwata A, Matsuo Y, Kubota R. Safety, tolerability, and pharmacokinetics of the novel antiviral agent ensitrelvir rumaric acid, a SARS-CoV-2 3CL protease inhibitor, in healthy adults. *Antimicrob Agents Chemother*. 2022;66(10): e00632–22.

33 Shimizu R, Sonoyama T, Fukuhara T, Kuwata A, Matsuo Y, Kubota R. A. Phase 1 study of ensitrelvir fumaric acid tablets evaluating the safety, pharmacokinetics and food effect in healthy adult populations. *Clin Drug Investig*. 2023;43(10): 785–97.

34 Shimizu R, Sonoyama T, Fukuhara T, Kuwata A, Matsuzaki T, Matsuo Y, et al. Evaluation of the drug-drug interaction potential of ensitrelvir fumaric acid with cytochrome P450 3A substrates in healthy Japanese adults. *Clin Drug Investig*. 2023;43(5): 335–46.

35 Shimizu R, Matsuzaki T, Oka R, Sonoyama T, Fukuhara T, Kuwata A, et al. Evaluation of drug-drug interactions of ensitrelvir, a SARS-CoV-2 3CL protease inhibitor, with transporter substrates based on in vitro and clinical studies. *J Clin Pharmacol*. 2023;63(8): 918–27.

36 Watari R, Sawada H, Hashimoto H, Kasai Y, Oka R, Shimizu R, et al. Utility of coproporphyrin-I determination in first-in-human study for early evaluation of OATP1B inhibitory potential based on investigation of ensitrelvir, an oral SARS-CoV-2 3C-like protease inhibitor. *J Pharm Sci*. 2023;S0022–3549(23)00375–1 (in press).

37 Mukae H, Yotsuyanagi H, Ohmagari N, Doi Y, Imamura T, Sonoyama T, et al. A randomized phase 2/3 study of ensitrelvir, a novel oral SARS-CoV-2 3C-like protease inhibitor, in Japanese patients with mild-to-moderate COVID-19 or asymptomatic SARS-CoV-2 infection: results of the phase 2a part. *Antimicrob Agents Chemother*. 2022;66(10): e0069722.

38 Mukae H, Yotsuyanagi H, Ohmagari N, Doi Y, Sakaguchi H, Sonoyama T, et al. Efficacy and safety of ensitrelvir in patients with mild-to-moderate coronavirus disease 2019: the phase 2b part of a randomized, placebo-controlled, phase 2/3 study. *Clin Infect Dis*. 2023;76(8): 1403–11.

39 Yotsuyanagi H, Ohmagari N, Doi Y, Imamura T, Sonoyama T, Ichihashi G, et al. A phase 2/3 study of S-217622 in participants with SARS-CoV-2 infection (Phase 3 part). *Medicine (Baltimore)*. 2023;102(8): e33024.

12

Discovery and Development of VV116: A Novel Oral Nucleoside Anti-SARS-CoV-2 Drug

Huilong Wang[1,2], *Yuanchao Xie*[3], *Tianwen Hu*[4], *Jingshan Shen*[1,2]

[1] *State Key Laboratory of Drug Research, Shanghai Institute of Materia Medica, Chinese Academy of Sciences, Shanghai, China*
[2] *University of Chinese Academy of Sciences, Beijing, China*
[3] *Lingang Laboratory, Shanghai, China*
[4] *Vigonvita Life Sciences Co., Ltd., Suzhou, China*

12.1 Introduction

Direct antiviral agents that act on viral proteins rather than host factors are the backbone for treating virus infectious diseases and dominate the field of antiviral drug discovery [1]. Generally, antiviral drug targets mainly come from viral proteins that are fundamental to the life cycle of a virus, such as those associated with the entry, the transcription/replication, the assembly, the maturation, or the releasing event [2]. Structure-based drug design could greatly facilitate the drug discovery process if the three-dimensional structure of a druggable viral protein is solved [3, 4]. For a newly emerged pathogenic virus that humans have limited knowledge about, drug target identification is usually a prerequisite for the drug discovery campaign. Polymerase, an essential enzyme in nearly all DNA and RNA viruses, is responsible for genome transcription and replication and is regarded as one of the most important targets for antiviral drug development, even if the "target" in the newly emerged virus is not yet identified at the time. This is not only due to its crucial biological functions, highly genetic conservation in evolution, but also the unique antiviral mechanism of nucleos(t)ide antiviral drugs [5, 6].

Nucleos(t)ide analogs, which have a long history in the field of medicinal chemistry, represent a very important class of antiviral agents. At present, more than thirty nucleos(t)ide antiviral drugs have been launched around the world and play a significant role in the treatment of viral diseases, including acquired immune deficiency syndrome, herpesvirus infection, hepatitis B, hepatitis C, influenza,

Trends in Antiviral Drug Development, First Edition. Edited by Michael J. Sofia and Zhengqiang Wang.

and SARS-CoV-2 infection [7, 8]. Mechanistically, nucleos(t)ide analogs need to be converted into their active triphosphate forms to inhibit viral replication. The nucleoside triphosphates bind to the active center of viral polymerases by serving as the enzyme's natural substrates to be incorporated into the nascent viral DNA/RNA chains, which leads to chain termination or lethal mutagenesis [9]. In fact, the active triphosphates do not inhibit the polymerase's function but are mistakenly utilized by this enzyme, leading to defective viral nucleic acid. For the discovery of antiviral nucleos(t)ide analogs, it could be achieved through a design of mimicking the natural nucleosides with no need of knowing the polymerase structure or even conducting biological activity assay at the molecular level. Moreover, these kinds of antivirals that target highly conserved polymerases have the potential to be broad-spectrum antiviral agents, in contrast to those targeting other viral proteins or the non-nucleos(t)ide polymerase inhibitors [10, 11].

The coronavirus disease 2019 (COVID-19) was caused by a new human coronavirus named SARS-CoV-2. Early in 2020, this virus quickly spread around the world and caused a global pandemic, posing a great threat to human health. To find effective therapeutics as soon as possible, great effort toward the discovery of anti-SARS-CoV-2 agents by screening of various compound libraries was made [12]. Among those potential antiviral molecules, nucleos(t)ide antiviral agents became the focus of attention [12, 13]. Remdesivir (RDV), an intravenously administered nucleotide analog, received an emergency use authorization (EUA) for treating hospitalized patients with COVID-19 in May 2020 [14]. VV116, an oral anti-SARS-CoV-2 nucleoside drug derived by modifying RDV, was approved for clinical use for COVID-19 in Uzbekistan in December 2021 and conditional approval in China for use in patients with mild-to-moderate COVID-19 in January 2023. In January 2025, VV116 got regular approval in China. VV116 was originally designed by Shen and coworkers, employing an ester prodrug approach for oral administration and deuteration modification to improve its pharmacokinetic (PK) properties, different from Gilead's RDV, which is based on phosphoramidate prodrug technology [15]. In preclinical studies, VV116 exhibited remarkable oral bioavailability, potent antiviral activity against SARS-CoV-2, and a favorable safety profile [16]. These promising results have been proven in subsequent clinical studies, affirming VV116's safety and efficacy [17–20]. Moreover, further studies on VV116 are in progress to expand its antiviral applications [21, 22].

12.2 Discovery

In the early stages of the COVID-19 pandemic, Shen and coworkers conducted a screening of a variety of nucleos(t)ide analogs against SARS-CoV-2 in Vero E6 cells, identifying two active compounds, RDV and its parent nucleoside GS-441524 (Figure 12.1) [16]. The two antivirals were also discovered by some other groups at approximately the same time when SARS-CoV-2 began to spread around the world [23–25]. RDV is a phosphoramidate prodrug of GS-441524 that bears a ribose sugar and a pyrrolotriazine base, connected via a β-C-glycosidic linkage [26]. Because RDV had been investigated in clinic for the treatment of Ebola virus infection and proved to have an acceptable safety profile [27], this drug candidate attracted considerable

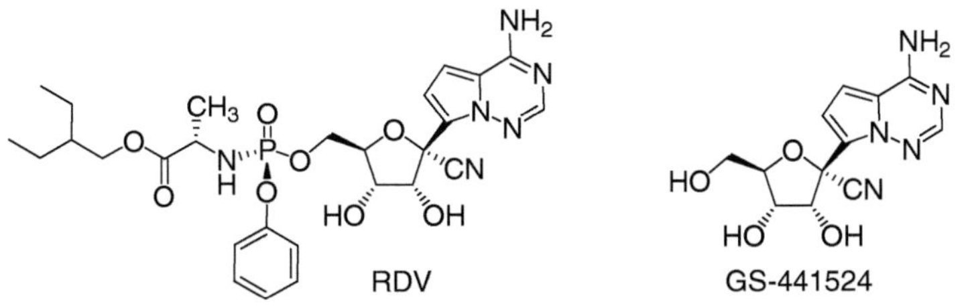

Figure 12.1 Structures of RDV and GS-441524.

attention and was fast advanced to clinical trials for the treatment of COVID-19 in many countries [28–30].

The phosphoramidate prodrug approach adopted in RDV is a practical technique for the intracellular delivery of nucleoside monophosphates or monophosphonates to efficiently generate the active nucleoside triphosphates because of the property of bypassing the rate-limiting first phosphorylation step (Figure 12.2) [31]. This prodrug approach is mainly applied in the field of antiviral drugs against hepatotropic viruses, contributing to the success of two well-known drugs (sofosbuvir and tenofovir alafenamide) and quite a few clinical drug candidates [32–34]. Because the enzymes required for activating the phosphoramidate prodrug mainly exist in the liver, nucleos(t)ide analogs decorated with a phosphoramidate moiety are typically endowed with strong liver-targeting properties, which align with its application scope [35]. RDV is administered by intravenous infusion, which could avoid the first-pass metabolism, but the liver-targeting issues remain. At the safety dose of RDV, the lung, which is the primary target organ of SARS-CoV-2, may not attain the desired level of drug exposure to sufficiently suppress viral replication [36, 37]. Though RDV had been granted EUA and subsequently officially approved for the treatment of hospitalized patients by the FDA, it was far insufficient to meet the demand for COVID-19 treatment. Oral antiviral drugs that could be taken at home enable a timely treatment for people in the early stage of infection, and are regarded as the most efficient therapeutic solution to combat COVID-19 to decrease the burden on the healthcare system caused with the infection.

GS-441524 shares the same antiviral mechanism with RDV, but their activation pathways as well as the *in vivo* PK behaviors are fundamentally different. As an

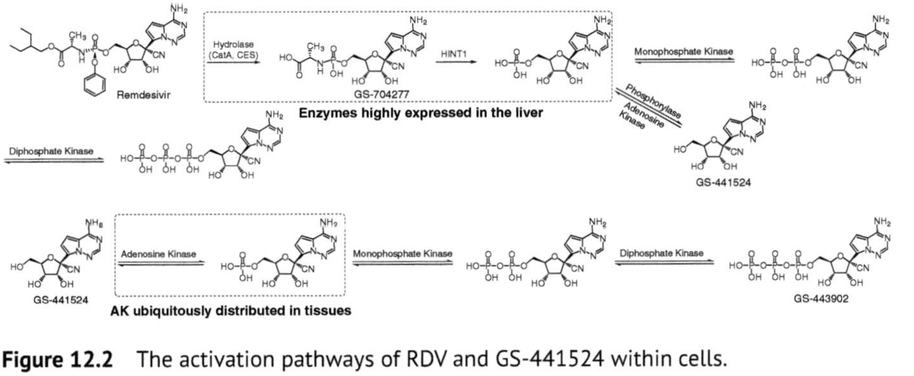

Figure 12.2 The activation pathways of RDV and GS-441524 within cells.

analog of adenosine, GS-441524 undergoes mono-phosphorylation, likely by adenosine kinase (AK), which is a nucleotide salvage pathway enzyme ubiquitously distributed in tissues (Figure 12.2) [35]. Mono-phosphorylation is perceived as the rate-limiting step for GS-441524 activation *in vivo* [38]. Once the monophosphate is formed, it will be easily trapped in cells, and the subsequent di- and triphosphorylation generate the active triphosphate [39]. For RDV, though it has high efficiency in being converted to the active triphosphate form, in principle, this is tissue-dependent, and the uneven drug distribution because of the liver-targeting properties may cause low anti-SARS-CoV-2 efficacy and even side effects [28, 36]. After the outbreak of SARS-CoV-2, GS-441524 also attracted a lot of attention, since it had been successfully used for the treatment of feline infectious peritonitis, a fatal cat disease caused by a coronavirus infection with favorable efficacy and safety. [40, 41]. As the pandemic worsened in 2020, some scientists advocated for the clinical advancement of GS-441524 as an oral therapeutic for outpatients [35]. Subsequently, the National Institutes of Health initiated the preclinical study of this nucleoside and continuously updated the research results on its website [42]. In a PK assessment conducted in mice, GS-441524 exhibited adequate distribution in the lungs and efficient conversion into its active triphosphate form [38]. However, the varied oral bioavailability observed across preclinical species (39% in mice, 33% in rats, 85% in dogs, and 8.3% in monkeys) raises concerns about the suitability of GS-441524 as an oral antiviral candidate for treating SARS-CoV-2 infection [42]. This variability implies that achieving effective plasma concentrations in humans would necessitate a high dosage, presenting challenges for its clinical application.

With the knowledge of the strengths and weaknesses about the phosphoramidate prodrugs and the parent nucleoside, Shen's group quickly designed drug discovery strategies based on the structure of GS-441524 to find the potential oral anti-SARS-CoV-2 drug candidates [43]. A number of GS-441524 derivatives with modifications on the ribose or the pyrrolotriazine base moiety were designed and synthesized (Figure 12.3). The *in vitro* anti-SARS-CoV-2 activity assay showed that all the compounds (1a–1k) with a group replacing the C1′-cyano group or with modifications at the C2′ position lost the antiviral activity or exhibited increased cytotoxicity. Among compounds 2a–2e that bore a C7 substituent on the base moiety and 2f with a C6-hydroxyl group, only the fluoro-substituted compound 2a showed a mild anti-SARS-CoV-2 activity in Vero E6 cells ($EC_{50} = 3.3\ \mu M$), but was much less potent than GS-441524 ($EC_{50} = 0.48\ \mu M$) [43]. Coronaviruses, including SARS-CoV-2, have a proofreading function mediated by nonstructural protein 14 (nsp14), i.e. a function of removing the mis-incorporated nucleotide during the transcription and replication processes [44]. This function is a great barrier to the discovery of nucleos(t)ide antiviral agents and may be an underlying cause of the loss of antiviral activity for most GS-441524 derivatives, as well as a limited number of anti-SARS-CoV-2 nucleoside analogs identified by now. According to studies published in 2021 and 2022, the cyano group at the C1′ position of RDV or GS-441524 played an important role in the delayed chain termination mechanism and did not confer resistance to the exonuclease activity of NSP14 [45, 46]. To enhance the oral bioavailability of GS-441524, Shen's group designed and synthesized some isobutyric acid-GS-441524 conjugates, such as 3a that demonstrated dramatically improved bioavailability in mice (F = 86.6%), as well as 3b (F = 98.7%, hydrobromide) [43].

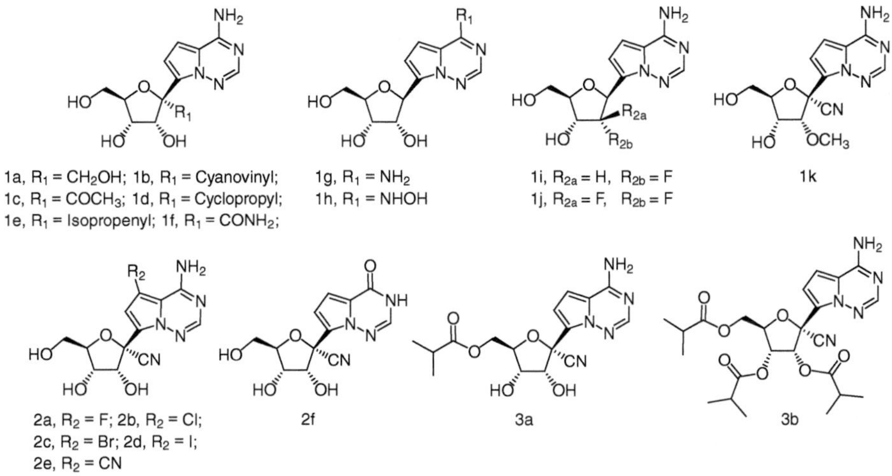

Figure 12.3 The GS-441524 derivatives and two prodrugs.

In animal models, the two compounds exhibited excellent anti-SARS-CoV-2 efficacy reported by other groups [47–50].

Deuteration, in some cases, is an effective approach for improving drug-like properties of a lead compound [51]. For GS-441524, several hydrogens on the base or the ribose moiety are amenable to deuteration. Among the deuterated GS-441524 derivatives, X1 bearing a deuterium atom at the C7 position of the base moiety was the easiest to synthesize (Figure 12.4). Some other derivatives contained two to five deuterium atoms positioned at C3′, C4′, C5′, and C7 carbons were also reported, and their synthesis involved complicated steps [52]. Compound X1 displayed potent anti-SARS-CoV-2 activity in Vero E6 cells with an EC_{50} value of 0.39 ± 0.08 μM, equivalent to that of GS-441524. In view of the low oral bioavailability of X1, the application of the ester prodrug approach by conjugating X1 with isobutyric acid afforded five ester compounds (X2–X6) that had improved lipophilicity (Figure 12.4). PK studies showed that the oral bioavailability of the two mono-ester prodrugs X2 and X3 was moderate (F = 32.6% and 46.5%, respectively) in rats, significantly higher than that of X1, but was found to be low in monkeys (F = 14.2% and 18.4%, respectively) [16]. As for the di- and tri-ester prodrugs (X4–X6), X6 bearing three isobutyryl ester groups was identified as the optimum prodrug in terms of the PK properties. Moreover, this compound was prepared most easily with no need to protect the hydroxyl group of the sugar moiety [16, 43].

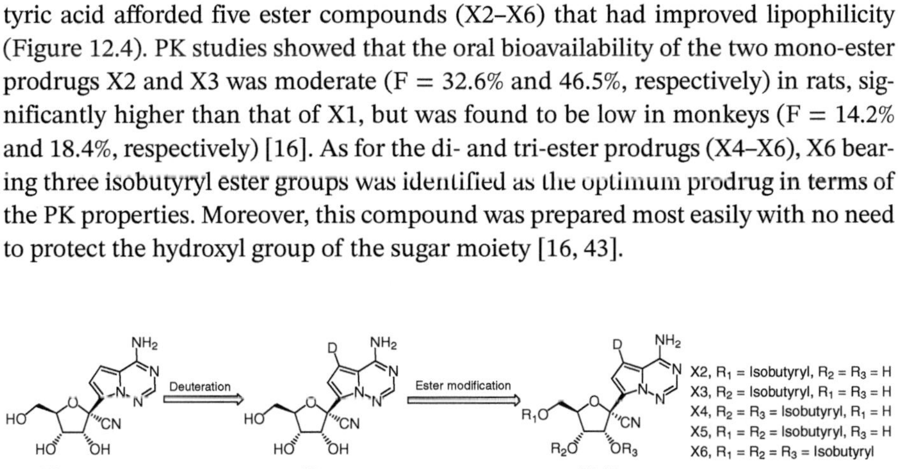

Figure 12.4 The deuterated GS-441524 derivative (X1) and the isobutyrate ester prodrugs.

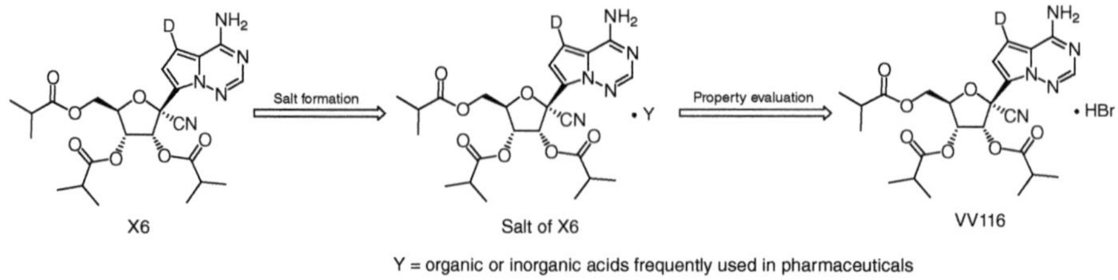

Figure 12.5 Identification of the hydrobromide salt of X6 as the drug candidate VV116.

As a tri-isobutyrate ester prodrug, X6 had a poor aqueous solubility (0.029 mg/mL), which may affect its absorption in the gastrointestinal tract and its release from a dosage form. Given that, an extensive salt screening was conducted for X6 across organic and inorganic acids frequently used in pharmaceuticals, identifying three potential salts (the hydrobromide, the hydrochloride, and the maleate salts) [53]. Among them, the hydrobromide salt of X6 was characterized by significantly improved solubility (0.41 mg/mL), no hygroscopicity even at high humidity conditions, and favorable chemical stability in solid crystalline form and in dosage form. Therefore, this salt was designated as the drug candidate, VV116 (Figure 12.5) [16].

12.3 Synthesis

VV116 is characterized by a deuterium atom at the base moiety and three isobutyryl ester groups at the sugar moiety. The biggest challenge for its synthesis, especially in the large-scale production, is how to incorporate the C7-deuterium atom simply and cost-efficiently, meanwhile with high deuteration rate.

The synthetic route of VV116 is summarized in Scheme 12.1, consisting of seven steps [16]. (2R,3R,4R,5R)-2-(4-aminopyrrolo[2,1-f][1,2,4]triazin-7-yl)-3,4-bis

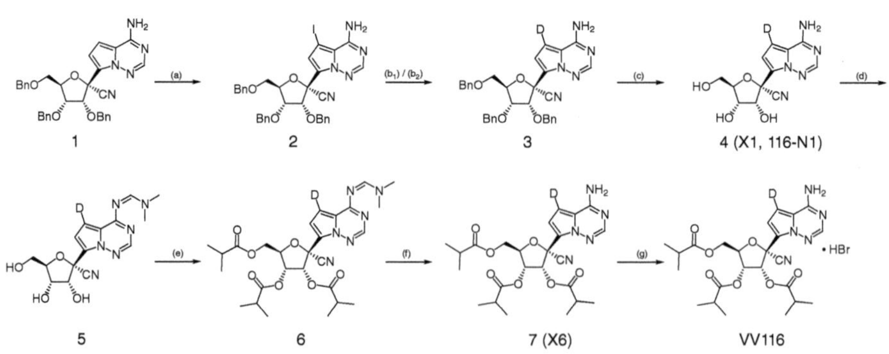

Scheme 12.1 Synthesis of VV116. Reagents and conditions: (a) NIS, TFA, DMF, 50 °C, 88%; (b_1) $PdCl_2(dppf)CH_2Cl_2$, TMEDA, $NaBD_4$, THF/D_2O, 55%; (b_2) Pd/C, D_2, 92%; (c) BCl_3, CH_2Cl_2, −30 °C, 53%; (d) DMF-DMA, DMF, CH_3OH, 60 °C, 80%; (e) isobutyryl chloride, Et_3N, DMAP, CH_2Cl_2, room temperature; (f) AcOH, EtOH, 50 °C, 75% over two steps; (g) 40% HBr, CH_3CN, 78%.

(benzyloxy)-5-((benzyloxy)methyl)tetrahydrofuran-2-carbonitrile (**1**), a key intermediate in the synthesis of RDV, was taken as the starting material. Iodination of **1** with *N*-iodosuccinimide (NIS) in *N*,*N*-dimethylformamide (DMF) afforded **2** in a high yield of 88%. The C7-iodine atom of **2** was converted to the deuterium atom using sodium borodeuteride ($NaBD_4$) through a metal-catalyzed coupling reaction in a tetrahydrofuran-deuterium oxide (THF-D_2O) mixture. This approach was hardly applicable to the large-scale synthesis of **3**, because of the limited supply of $NaBD_4$ as well as the poor atom economy. The benzyl-protecting group of **3** was easily removed by boron trichloride (BCl_3) to give the nucleoside (**4**, X1, 116-N1). For installing the three isobutyryl groups at the ribose moiety, the C6-amino group of **4** could be protected at first in order to increase the reaction selectivity. The reaction of **4** with *N*,*N*-dimethylformamide dimethyl acetal (DMF-DMA) in DMF followed by treatment with methanol (CH_3OH) afforded **5** that bore an imino group at the C6 position. Intermediate **5** was then acylated with isobutyryl chloride in the presence of triethylamine (Et_3N) and 4-dimethylaminopyridine (DMAP) to give the tri-ester **6**. Compound **7** (X6) was obtained by the acetic acid-promoted alcoholysis of **6** in ethanol (EtOH) under the heating condition and then treated with 40% hydrobromic acid (HBr) in acetonitrile (CH_3CN) to give the final product VV116.

The key point for the scalable synthesis of VV116 using the above route was the iodine-to-deuterium conversion of the second step (from **2** to **3**). This conversion was achievable via the palladium (Pd/C)-catalyzed reduction using deuterium gas (D_2). As described in a preparation patent, the reaction of **2** (600 g) with D_2, Pd/C, and Et_3N in THF under the pressure of 1.0 MPa gave **3** (451 g) in 92% yield with a deuteration rate of at least 99% [54]. With this significant improvement, the large-scale production of VV116 was achieved.

In addition, a patent disclosed a more straightforward method for the synthesis of intermediate **7** without protecting the C6-amino group of **4**, as shown in Scheme 12.2 [55]. Direct reaction of **4** with isobutyric anhydride and DMAP in *N*-methylpyrrolidone (NMP) at 0 °C gave **7** with a yield of 90%. It seemed that the reaction conditions had a great influence on the regioselectivity in the acylation reaction of **4**. From a process perspective, the optimized method for synthesizing the tri-isobutyryl ester intermediate outweighed the original protocol because two steps would be saved.

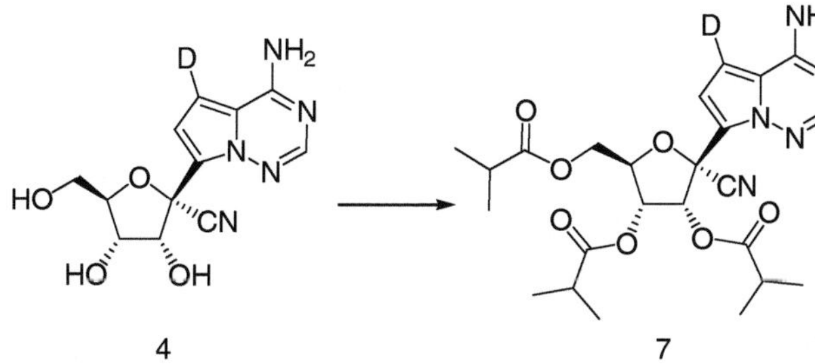

Scheme 12.2 Direct synthesis of 7 from 4. Reagents and conditions: isobutyric anhydride, DMAP, NMP, 0 °C, 90%.

12.4 Preclinical Study

12.4.1 Virology

In Vero E6 cells, VV116 and its parent nucleoside 116-N1 displayed significant inhibitory activities against various SARS-CoV-2 strains, as shown in Table 12.1 [16, 22]. The EC_{50} values of VV116 ranged from 0.15 to 0.45 μM, similar to that of 116-N1, which suggested that this tri-isobutyryl ester prodrug was efficiently metabolized to 116-N1 in the assay. RDV is also a potent anti-SARS-CoV-2 agent and has been used as the positive drug by many studies [56, 57]. As a phosphoramidate prodrug, RDV required a complicated path to be converted into the triphosphate active form, and its antiviral activity was found to be cell-dependent [58].

The *in vivo* antiviral efficacy of VV116 against SARS-CoV-2 was assessed in human angiotensin-converting enzyme 2 (hACE2)-transduced mice, with EIDD-2801 as the positive control [16]. Balb/c mice infected with SARS-CoV-2 were treated orally with VV116 at three different doses (25, 50, 100 mg/kg) twice daily for 2 or 5 days. The results showed that VV116 reduced viral RNA levels and titers in the lungs of mice in a dose-dependent manner. Treatment with VV116 50 mg/kg for 5 days or with VV116 100 mg/kg for 2 days reduced the virus titers below the detection limit. On day 2 post-infection (p.i.), VV116 100 mg significantly ameliorated the pathological changes of the lung tissues. Nirmatrelvir, a potent 3CL protease inhibitor, is the active ingredient of Paxlovid (nirmatrelvir/ritonavir), which was granted an EUA for the treatment of mild-to-moderate COVID-19 in the end of 2021 [59]. A drug combination screening found that VV116 had synergistic anti-SARS-CoV-2 activity with nirmatrelvir in cells. Furthermore, the antiviral efficacy of VV116, nirmatrelvir/ritonavir, and the combination of VV116 with nirmatrelvir/ritonavir against the SARS-CoV-2 Delta variant was evaluated in the K18-hACE2 mouse model [22]. After a 4-day treatment, 50 mg/kg of VV116 was more potent than 100 mg/kg of nirmatrelvir plus 50 mg/kg of ritonavir in the reduction of lung viral titers. Notably, the combination of the two drugs produced a significantly enhanced antiviral efficacy compared with the single treatment regimen.

VV116 could also inhibit the replication of other human coronaviruses (HCoVs) such as HCoV-OC43 and HCoV-229E in Huh-7 cells with EC_{50} values of 2.0 ± 0.085 μM and 1.4 ± 0.047 μM, respectively (Table 12.1). The antiviral efficacy against HCoV-OC43 has been validated in suckling Balb/c mice [22]. Additionally, VV116 was found to exhibit inhibitory effects on animal coronaviruses, including mouse hepatitis virus (MHV), feline infectious peritonitis virus (FIPV), feline enteric coronavirus (FECV), and canine coronavirus (CCoV), with EC_{50} values ranging from 0.665 ± 0.022 to 1.498 ± 0.017 μM in NCTC clone 929 and CRFK cells. An *in vivo* study using Balb/c mice confirmed the anti-MHV activity of VV116 [60]. Of note, VV116 was identified as a potential antiviral drug for treating respiratory syncytial virus (RSV) infection [21]. RSV is an RNA virus of the Paramyxoviridae family and a leading cause of acute respiratory tract infection, mainly in infants and young children [61]. VV116 and 116-N1 showed potent anti-RSV activities in A549, HEp-2, and NHBE cells. The *in vivo* anti-RSV efficacy study of VV116 was conducted in 6–8 weeks old Balb/c mice. A twice-daily treatment for 4 days at an oral dose of 50 mg/

Table 12.1 Antiviral potency of VV116 and RDV against SARS-CoV-2 strains and other HCoVs [16, 22].

Compound	EC_{50} (μM)						
	SARS-CoV-2 strains (Vero E6 cells)					Other HCoVs (Huh-7 cells)	
	Original strain[a]	Beta	Delta	Omicron BA.1	Omicron BA.5	OC43	229E
VV116	0.35 ± 0.09[b]	0.24 ± 0.05	0.45 ± 0.035	0.17 ± 0.016	0.15 ± 0.012	2.0 ± 0.085	1.4 ± 0.047
116-N1	0.39 ± 0.08	-	0.83 ± 0.071	0.30 ± 0.027	0.46 ± 0.004	1.1 ± 0.061	0.57 ± 0.070
RDV	1.71 ± 0.60	-	2.20 ± 0.11	0.54 ± 0.078	0.13 ± 0.011	-	0.041
GS-441524	0.59 ± 0.10	-	0.61 ± 0.079	0.28 ± 0.011	0.10 ± 0.011	1.4 ± 0.055	0.56 ± 0.082

a. 2019-nCoV-WIV04 strain.
b. The EC_{50} value of X6 (the base form of VV116).
Adapted from [16, 22].

kg VV116 reduced the lung virus titers below the detection limit and alleviated lung injury caused by the virus. Currently, VV116 has entered Phase II clinical trials in China for the treatment of RSV infection (NCT06206720) [62].

12.4.2 Mechanism of Antiviral Activity of VV116

Generally, nucleos(t)ide antiviral drugs have two acting mechanisms: chain termination and lethal mutagenesis. Among the marketed antiviral drugs, the vast majority of them belong to chain terminators, and only a few are viral mutagenetic agents, like favipiravir and molnupiravir [63–65]. By now, the detailed molecular mechanism of VV116 has not been reported, but there are many biochemistry and structural biology studies conducted for elucidating the molecular antiviral mechanism of RDV. Structurally, RDV monophosphate (RMP) resembles adenosine monophosphate (AMP), capable of forming base pairs with uridine monophosphate (UMP). Once incorporated by viral RNA-dependent RNA-polymerases (RdRp) into the RNA, RMP allowed the addition of three more nucleotides before encountering a translocation barrier that was mainly attributed to the ribose 1′-CN. Therefore, RDV was considered as the delayed chain terminator for SARS-CoV-2 viral RNA replication [45].

An *in vitro* fluorescence, polymerase-catalyzed, polymerization reaction-based assay was used for evaluating the inhibitory activity of VV116 nucleoside triphosphate (116-NTP) against the SARS-CoV-2 RdRp-mediated RNA synthesis [16]. In this assay, 116-NTP had an IC_{50} value of 0.67 ± 0.24 μM.

In mice, VV116 was observed to undergo enzymatic biotransformation after a single oral administration, yielding the corresponding nucleoside monophosphate (116-NMP) and 116-NTP, as illustrated in Figure 12.6 [66]. The deuterium atom at the C7 position remains stable in the parent nucleoside and phosphate metabolites and would not interchange with the labile hydrogen atoms of substances in the physiological environment. From a chemical perspective, the active form of VV116 is different from that of remdesivir and 3a (GS-5245, developed by Gilead Sciences), which was incorrectly described in a previous comment [67].

12.4.3 Preclinical Pharmacokinetics

VV116 showed favorable PK properties in preclinical species [16]. Because of first-pass metabolism, VV116 was rapidly metabolized to its nucleoside (116-N1) after oral absorption and was hardly detected in the plasma. The parent nucleoside 116-N1 had a time to maximum plasma concentration (T_{max}) of ~1.0 h in SD rats and

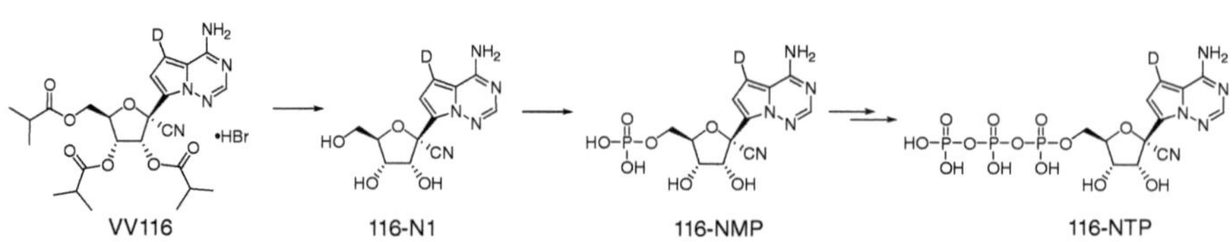

Figure 12.6 The enzymatic biotransformation of VV116.

beagle dogs at different doses, suggesting that VV116 was absorbed quickly. There was a good linear correlation between the exposure (C_{max}: maximum observed plasma concentration; AUC: area under the plasma concentration–time curve) and the dose. For the three oral doses (10, 30, and 90 mg/kg) in SD rats, VV116 had an oral bioavailability of 79.2–86.4%, and in beagle dogs, the oral bioavailability was even higher (F = 87.4–101.7%). The half-life ($t_{1/2}$) of 116-N1 was a little prolonged as the dose of VV116 increased. At the dose of 10 mg/kg, the $t_{1/2}$ was 1.43 ± 0.27 h and 4.53 ± 1.53 h in SD rats and beagle dogs, respectively. In the multiple-dose study, there was no accumulation of 116-N1 upon repeated doses of 30 mg/kg VV116 in SD rats or 20 mg/kg VV116 in beagle dogs once daily for 7 days. The tissue distribution study showed that 116-N1 was widely distributed in most organs of rats, which was significantly different from the biodistribution profile of RDV [15, 36].

The PK properties of VV116 were also investigated in mice [21]. VV116 had a high oral bioavailability of 110.2% in ICR mice, calculated on the plasma exposure of 116-N1. Balb/c mice were used for evaluating the *in vivo* antiviral efficacy of VV116 against SARS-CoV-2 and RSV infection, as well as for the PK studies of VV116 at the same doses (25, 50, and 100 mg/kg). Oral administration of VV116 exhibited a dose-dependent increase in the plasma exposure of 116-N1. 116-N1 had a $t_{1/2}$ ranging from 2.3 to 4.25 h in Balb/c mice, which supported a twice-daily dosing regimen for the *in vitro* antiviral efficacy studies. VV116 exhibited a favorable anti-SARS-CoV-2 or anti-RSV effect at the dose of 25 mg/kg twice daily for 4 or 5 days, significantly decreasing virus titers in the lungs of mice. At this dose, the AUC from time zero to the time of the last measurable concentration (AUC_{0-t}) of 116-N1 was 11461 ± 1013 h*ng/mL, which could be considered as the effective drug exposure. VV116 had a synergistic anti-SARS-CoV-2 effect with nirmatrelvir/ritonavir, which was illustrated above [22]. It was found that the combined use of the two drugs caused a slight decrease (13.5% and 19.8%, respectively) in the AUC of 116-N1 and nirmatrelvir compared with the single drug.

Notably, VV116 had better PK characteristics compared with its non-deuterated counterpart (VV116-H) [21, 68]. The chemical structures of VV116 and VV116-H were shown in Figure 12.7. At a single oral dose of 30 mg/kg, both the C_{max} and the AUC_{0-t} were obviously higher in VV116 group than that in VV116-H group (C_{max}: 2710 vs. 2060 ng/mL; AUC_{0-t}: 9259 vs. 7664 h*ng/mL), while the PK parameters were similar between the two groups when administered intravenously (5 mg/kg).

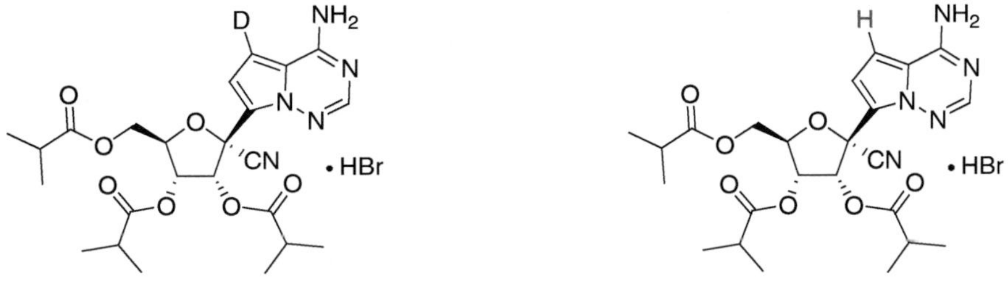

VV116 VV116-H

Figure 12.7 Structures of VV116 and its non-deuterated counterpart (VV116-H).

The better oral PK performance of VV116 resulting from the deuteration may be the foundation of its pharmacological innovation and clinical value, and the rationale behind this observation deserves further investigation.

12.4.4 Toxicology

The preclinical toxicology assessment of VV116 included but was not limited to cytotoxicity study, single-dose toxicity study, repeat-dose toxicity study, and genetic toxicology study [16, 21, 60]. VV116 and 116-N1 demonstrated minimal cytotoxicity in Vero E6 cells, with the mean CC_{50} values of 280.19 μM and over 500 μM, respectively. Additionally, VV116 showed CC_{50} values approaching or exceeding 100 μM in other cells. The maximal tolerated single doses of VV116 were at least 2.0 and 1.0 g/kg in rats and beagle dogs, respectively, and for the 14-day repeat-dose toxicity studies, the no-observed-adverse-effect levels (NOAELs) were 200 and 30 mg/kg, respectively. VV116 was negative in the Ames test, utilizing histidine-dependent *Salmonella typhimurium* strains (TA97a, TA98, TA100, TA1535) and the tryptophan-dependent *Escherichia coli* strain (WP2). In the chromosome aberration test conducted on Chinese hamster lung (CHL) cells, VV116 had no effect on inducing chromosomal damage. The micronucleus assay in rats revealed that VV116 exhibited no ability to induce any increase in micronucleated polychromatic erythrocytes.

12.5 Clinical Study

12.5.1 Pharmacokinetics

Three Phase I clinical studies of VV116, single ascending-dose study (Study 1), multiple ascending-dose study (Study 2), and food-effect study (Study 3), were conducted in healthy subjects [17]. Studies 1 and 2 were randomized, double-blind, placebo-controlled to evaluate the safety, tolerability, and PK of single and multiple ascending oral doses of VV116. Study 3 was a randomized, open-label, three-cycle, cross-over study to investigate the food effect of VV116. The main PK parameters are shown in Table 12.2.

VV116 was hydrolyzed rapidly to its metabolite 116-N1 after oral administration without being detected in human plasma, so 116-N1 was used for calculating the PK parameters. In Study 1, initially, eight ascending-dose groups (25, 50, 100, 200, 400, 600, 800, and 1000 mg) were designed. Based on the safety results of the 25 mg dose group and the clinical trial in Uzbekistan, the dose group settings were adjusted to 25, 200, 400, 800, and 1200 mg. In the dose range of 25–800 mg, the AUC and C_{max} of 116-N1 increased in an approximately dose proportional manner, but did not show significant change from 800 to 1200 mg. VV116 might suffer from absorption saturation at doses greater than 800 mg. 116-N1 reached peak plasma concentration quickly after oral administration of VV116 tablets with median T_{max} of 1.00–2.50 h. 116-N1 was predominantly excreted through the kidneys, with mean $t_{1/2}$ of 4.80–6.95 h, suggesting a twice-daily dosing regimen of VV116 for the clinical treatment of COVID-19.

Table 12.2 The main PK parameters of single and multiple ascending oral doses of VV116 in healthy subjects [17].

	Dose (mg)		T_{max} (h)	C_{max} (ng/mL)	$t_{1/2}$ (h)	AUC_{0-t} (ng·h/mL)
Single ascending-dose study	25		1.00 (0.50, 1.50)	165 (74.0)	4.80 (0.492)	744 (244)
	200		1.00 (0.75, 1.50)	1096 (412)	5.48 (0.430)	6631 (1603)
	400		1.50 (1.00, 2.00)	1898 (701)	6.15 (1.08)	12759 (2747)
	800		2.50 (1.50, 6.00)	2796 (225)	6.75 (1.36)	25886 (5904)
	1200		2.00 (1.50, 3.00)	3086 (778)	6.95 (0.659)	28057 (5145)
Multiple ascending-dose study	5.5 days × 200	Day 1	1.50 (0.750, 1.50)	858 (186)	4.72 (1.02)	4610 (864)
		Day 6	1.00 (0.750, 1.50)	1131 (231)	7.56 (1.59)	9384 (1880)
	5.5 days × 400	Day 1	1.50 (0.750, 1.50)	1968 (670)	4.88 (0.718)	10351 (2843)
		Day 6	1.00 (0.330, 3.00)	2304 (851)	8.12 (0.903)	20774 (8321)
	5.5 days × 600	Day 1	1.50 (1.00, 6.00)	2418 (708)	5.41 (0.808)	12871 (3309)
		Day 6	1.50 (1.00, 3.00)	2842 (617)	7.85 (0.692)	25077 (7100)

Note: Data are expressed as mean (SD), except for T_{max}, which is shown as median (min, max).

In Study 2, three dose groups (200, 400, and 600 mg) were designed; VV116 tablets or placebo was administered twice a day for 5.5 days (days 1–6). The results showed that VV116 had a slight accumulation upon repeated dosing with increased C_{max} and AUC at day 6 compared to day 1 for all the groups. Meanwhile, the mean $t_{1/2}$ of 116-N1 was longer at day 6 than day 1. In the 200 mg group, the plasma trough concentrations of 116-N1 at day 5 and day 6 were within 242–345 ng/mL (0.83–1.19 μM), which exceeded the EC_{90} against SARS-CoV-2. The dosage regimen of 200 mg twice a day and above was projected to maintain the effective antiviral concentration for the clinical studies in patients with COVID-19.

In Study 3, the PK parameters of 116-N1 after a single oral dose of 400 mg VV116 were investigated under different diet conditions that included fasting, standard meal, and high-fat meal. Compared with the fasting condition, the fed conditions prolonged the T_{max} of 116-N1 with no influence on the C_{max}. High-fat meal caused a slight increase (26.32%) in AUC_{0-t} compared with the fasting condition. Therefore, for treating COVID-19, VV116 could be administered under fasting or fed conditions.

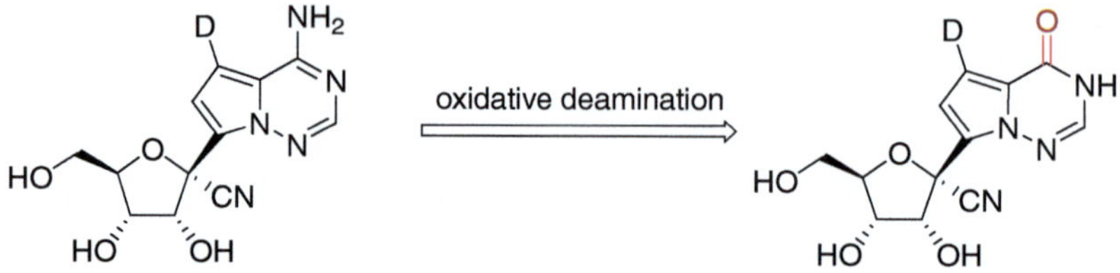

Figure 12.8 The nucleoside metabolite 116-N1 and the oxidative deamination metabolite M2.

The metabolites of VV116 in plasma, urine, and feces of healthy Chinese males have been investigated [69]. Besides 116-N1, the predominant metabolite of VV116, 6 Phase I metabolites and 11 Phase II metabolites derived from the subsequent metabolism of 116-N1 were also detected. The oxidative deamination metabolite M2 formed likely by adenosine deaminase was mainly found in feces and identified using the synthesized standard compound M2 (Figure 12.8).

12.5.2 Safety and Tolerability

The result of Phase I clinical studies showed that VV116 demonstrated a satisfactory safety profile in healthy subjects [17]. In Study 1, the incidence of adverse events (AEs) in the VV116 group was lower than the placebo group (39.3% vs. 50%), with all the AEs being Grade 1, except one case of Grade 2 neutropenia. In Study 2, the incidences of AEs were comparable (51.9% vs. 55.6%) between VV116 and the placebo groups. All AEs observed in VV116-administered subjects were Grade 1 and could be recovered without any treatment.

During the Omicron wave in March of 2022, an open, prospective cohort study was conducted to evaluate VV116's safety and effectiveness in Chinese participants [18]. Among the enrolled 136 hospitalized nonsevere patients, 60 patients receiving VV116 (300 mg orally every 12 h for 5 days) had a shorter viral shedding time than the control group. Moreover, VV116 exhibited favorable safety with a total of 9 AEs and no serious AEs in this study.

In healthy or COVID-19 volunteers, RDV treatment would be frequently accompanied by transaminase elevation, which is not common for VV116 [70]. As discussed in the Discovery part, RDV has a liver-targeting prodrug moiety and its activating pathway is greatly different from that of the parent nucleoside GS-441524 or VV116, which presumably underlies their different PK characteristics, dosing regimens, and the safety profiles.

12.5.3 Phase III Studies

From April to May of 2022, during which time the SARS-CoV-2 Omicron variant spread in Shanghai, China, a Phase III noninferiority, observer-blinded, randomized trial (NCT05341609) was conducted to compare the efficacy and safety of VV116 with Paxlovid in patients with mild-to-moderate COVID-19 [19]. The primary endpoint was the time to sustain clinical recovery (alleviation of all COVID-19-related symptoms to a total symptom score of 0 or 1 for 2 consecutive days) through day 28. This primary endpoint was chosen mainly because of the low incidences of omicron variant-related hospitalization or mortality [71]. A total of 822 participants with a median age of 53 years and at least one risk factor for progression to severe COVID-19 were randomized to the VV116 or the Paxlovid group. Among the participants, 92.1% had mild COVID-19, and 75% got vaccinated against SARS-CoV-2. Since 51 participants did not receive any medication, in the end, the VV116 group included 384 participants receiving an oral dose of 600 mg every 12 h on day 1 and 300 mg every 12 h on days 2–5, and the Paxlovid group included 387 participants receiving 300 mg of nirmatrelvir plus 100 mg of ritonavir twice daily for 5 days. This trial showed that VV116 was noninferior to Paxlovid in shortening the time (4 vs. 5 days) to sustained clinical recovery in the full analysis population. No participants died or had progression to severe or critical COVID-19 in the two groups, and VV116 was similar to Paxlovid in the secondary endpoints of time to sustained symptom resolution and to the first negative SARS-CoV-2 test. Regarding safety, VV116 with a low risk of drug–drug interactions was better than Paxlovid and was associated with fewer AEs (67.4% vs. 77.3%) and fewer Grade 3 or 4 AEs (2.6% vs. 5.7%).

A following Phase III clinical trial that was a multicenter, double-blind, randomized, placebo-controlled clinical study (NCT05582629) was conducted in China to investigate the efficacy and safety of VV116 in patients with mild-to-moderate COVID-19 [20]. In this study, the primary endpoint was the time to clinical symptom resolution for 2 consecutive days, and there was no requirement to the high-risk factors for progression to severe COVID-19. A total of 1369 patients were enrolled and assigned to VV116 and the placebo groups. The full analysis set included 1347 patients receiving either VV116 (n = 674) or the placebo (n = 673). The participants had a median age of 35 years and a high vaccination rate. In VV116 and the placebo groups, 44.0% and 42.2% had a high-risk factor for severe COVID-19, respectively. This trial showed that compared with placebo, a 5-day treatment with VV116 (600 mg every 12 h on day 1 and 300 mg every 12 h on days 2–5) significantly shortened the time to sustained clinical symptom in patients (10.9 vs. 12.9 days), regardless of the presence of high-risk factors for progression to severe COVID-19. These trends in favor of VV116 were observed across the subgroups. The incidence of AEs was similar between the two groups, suggesting that VV116 exhibited a favorable safety profile.

Based on the results of preclinical and clinical studies, VV116 received conditional approval for the treatment in adults with mild-to-moderate SARS-CoV-2 infection under the name of deuremidevir hydrobromide tablets (brand name: Mindvy®) by the National Medical Products Administration (NMPA) in 28 January 2023, and got regular approval in 10 January 2025.

12.6 Future Direction

Currently, two primary classes of small molecule antiviral drugs are used in the treatment of COVID-19: nucleos(t)ide drugs targeting SARS-CoV-2 RdRp and 3CL protease inhibitors. In addition to Veklury® (remdesivir), which is prescribed for hospitalized patients with moderate-to-severe COVID-19, several other anti-SARS-CoV-2 drugs have also received approval or condition approval for clinical use, including Paxlovid® (nirmatrelvir/ritonavir), Lagevrio® (molnupiravir, RdRp inhibitor), Xocova® (ensitrelvir fumarate, 3CL protease inhibitor, Japan), Xiannuoxin® (simnotrelvir/ritonavir, 3CL protease inhibitor, China, discovered also by Shen's group and partners), Leruiling® (leritrelvir, 3CL protease inhibitor, China), Taizhongding® (atilotrelvir/ritonavir, 3CL protease inhibitor, China), etc., but their application is limited to patients with mild-to-moderate infection [72–77]. In particular, VV116 (Mindvy) was approved for treating mild, moderate, and severe COVID-19 in adults in December 2021 in Uzbekistan.

Since the beginning of 2022, Omicron variants with lower pathogenicity than the previous variants (Delta or Alpha strains) have been the main epidemic strains and typically cause asymptomatic or mild infection. Clinically, it became difficult to assess the efficacy of antivirals in patients with severe COVID-19 due to the few cases of severe infection. However, SARS-CoV-2 is constantly mutating, and new variants with enhanced pathogenicity may emerge in the future [78]. Therefore, developing oral antiviral drugs for the treatment of severe infectious diseases is still valuable.

VV116 exhibited broad-spectrum antiviral activities [10, 21, 22, 60]. Besides SARS-CoV-2, VV116 is active against other HCoVs and animal coronaviruses. Currently, a clinical study is underway in China to evaluate the safety and effectiveness of VV116 for the treatment of RSV infection [62]. RSV, classified in the Paramyxoviridae family, shares this classification with other significant human pathogens, such as parainfluenza virus, measles virus, human metapneumovirus, and Nipah virus. In light of this, it is essential to further explore VV116's potential for preventing and treating these viral infections.

Combination therapy should also be an effective strategy for combating COVID-19. By co-administering antivirals targeting distinct viral targets, the potential of enhanced or synergistic antiviral effects, as well as reduced side effects, is an attractive proposition [79]. The synergistic potency of the VV116 and nirmatrelvir combination against HCoV-OC43 and SARS-CoV-2 have been verified *in vitro* and *in vivo* studies, highlighting its significant prospects for clinical application [22]. Further investigation into the therapeutic potential of VV116 in combination with other drugs is meaningful, as it may offer a promising approach to treating newly emerging viral infections and addressing drug-resistant viral strains.

12.7 Conclusions

VV116 was developed and proven as a safe and effective oral deuterated nucleoside antiviral drug for the treatment of COVID-19 both in Uzbekistan and China. Being an ester nucleoside prodrug, VV116 addressed the weakness of the intravenous administration associated with remdesivir. The incorporation of deuterium in VV116 represented a pharmacological innovation with improved PK properties. In several clinical trials involving healthy volunteers and patients, VV116 consistently demonstrated notable efficacy and safety, with no genotoxicity and a low risk of drug–drug interactions. These characteristics of VV116 render it a promising therapeutic option for

COVID-19. Given the broad-spectrum antiviral activities, VV116 could play a dramatic role in the fight against infectious diseases caused by other RNA viruses in the future.

References

1 Li G, De Clercq E. Overview of antiviral drug discovery and development: viral versus host targets. In: Muñoz-Fontela C, Delgado R, editors. *Antiviral Discovery for Highly Pathogenic Emerging Viruses*, The Royal Society of Chemistry. 2021.

2 Liu C, Hu L, Dong G, Zhang Y, Ferreira da Silva-Júnior E, Liu X, et al. Emerging drug design strategies in anti-influenza drug discovery. *Acta Pharm Sin B.* 2023;13(12):4715-32.

3 von Itzstein M, Wu W-Y, Kok GB, Pegg MS, Dyason JC, Jin B, et al. Rational design of potent sialidase-based inhibitors of influenza virus replication. *Nature.* 1993;363(6428):418–23.

4 Hoffman RL, Kania RS, Brothers MA, Davies JF, Ferre RA, Gajiwala KS, et al. Discovery of ketone-based covalent inhibitors of coronavirus 3CL proteases for the potential therapeutic treatment of COVID-19. *J. Med Chem.* 2020;63(21):12725–47.

5 Tsai C-H, Lee P-Y, Stollar V, Li M-L. Antiviral therapy targeting viral polymerase. *Curr Pharm Des.* 2006;12(11):1339–55.

6 Venkataraman S, Prasad BVLS, Selvarajan R. RNA dependent RNA polymerases: insights from structure, function and evolution. *Viruses.* 2018;10(2):76.

7 Clercq ED, Li G. Approved antiviral drugs over the past 50 years. *Rev Clin Microbiol.* 2016;29(3):695–747.

8 Roy V, Agrofoglio LA. Nucleosides and emerging viruses: a new story. *Drug Discov Today.* 2022;27(7):1945–1953.

9 Pruijssers AJ, Denison MR. Nucleoside analogues for the treatment of coronavirus infections. *Curr Opin Virol.* 2019;35:57–62.

10 Lo MK, Jordan R, Arvey A, Sudhamsu J, Shrivastava-Ranjan P, Hotard AL, et al. GS-5734 and its parent nucleoside analog inhibit filo-, pneumo-, and paramyxoviruses. *Sci Rep.* 2017;7(1):43395.

11 Caroline AL, Powell DS, Bethel LM, Oury TD, Reed DS, Hartman AL. Broad spectrum antiviral activity of favipiravir (T-705): protection from highly lethal inhalational rift valley fever. *PLOS. Negl Trop.* 2014;8(4):e2790.

12 Bakowski MA, Beutler N, Wolff KC, Kirkpatrick MG, Chen E, Nguyen T-TH, et al. Drug repurposing screens identify chemical entities for the development of COVID-19 interventions. *Nat Commun.* 2021;12(1):3309.

13 Borbone N, Piccialli G, Roviello GN, Oliviero G. Nucleoside analogs and nucleoside precursors as drugs in the fight against SARS-CoV-2 and other coronaviruses. Mole. 2021;26(4):986.

14 Lamb YN. Remdesivir: first approval. *Drugs.* 2020;80(13):1355–63.

15 Wang Z, Yang L, Song X-q. Oral GS-441524 derivatives: next-generation inhibitors of SARS-CoV-2 RNA-dependent RNA polymerase. *Front immunol.* 2022;13:1015355.

16 Xie Y, Yin W, Zhang Y, Shang W, Wang Z, Luan X, et al. Design and development of an oral remdesivir derivative VV116 against SARS-CoV-2. *Cell Res.* 2021;31(11):1212–4.

17 Qian H-j, Wang Y, Zhang M-q, Xie Y-c, Wu Q-q, Liang L-y, et al. Safety, tolerability, and pharmacokinetics of VV116, an oral nucleoside analog against SARS-CoV-2, in Chinese healthy subjects. *Acta Pharmacol Sin*. 2022;43(12):3130–8.

18 Shen Y, Ai J, Lin N, Zhang H, Li Y, Wang H, et al. An open, prospective cohort study of VV116 in Chinese participants infected with SARS-CoV-2 omicron variants. *EMI*. 2022;11(1):1518–23.

19 Cao Z, Gao W, Bao H, Feng H, Mei S, Chen P, et al. VV116 versus nirmatrelvir–ritonavir for oral treatment of covid-19. *N Engl J Med*. 2022;388(5):406–17.

20 Fan X, Dai X, Ling Y, Wu L, Tang L, Peng C, et al. Oral VV116 versus placebo in patients with mild-to-moderate COVID-19 in China: a multicentre, double-blind, phase 3, randomised controlled study. *Lancet Infect Dis*. 2024;24(2):129–39.

21 Zhang R, Zhang Y, Zheng W, Shang W, Wu Y, Li N, et al. Oral remdesivir derivative VV116 is a potent inhibitor of respiratory syncytial virus with efficacy in mouse model. *Signal Transduct*. 2022;7(1):123.

22 Zhang Y, Sun Y, Xie Y, Shang W, Wang Z, Jiang H, et al. A viral RNA-dependent RNA polymerase inhibitor VV116 broadly inhibits human coronaviruses and has synergistic potency with 3CLpro inhibitor nirmatrelvir. *Signal Transduct*. 2023;8(1):360.

23 Pruijssers AJ, George AS, Schäfer A, Leist SR, Gralinksi LE, Dinnon KH, et al. Remdesivir inhibits SARS-CoV-2 in human lung cells and chimeric SARS-CoV expressing the SARS-CoV-2 RNA polymerase in mice. *Cell Rep*. 2020;32(3):107940.

24 Williamson BN, Feldmann F, Schwarz B, Meade-White K, Porter DP, Schulz J, et al. Clinical benefit of remdesivir in rhesus macaques infected with SARS-CoV-2. *Nature* 2020;585(7824):273–276.

25 Li Y, Cao L, Li G, Cong F, Li Y, Sun J, et al. Remdesivir metabolite GS-441524 effectively inhibits SARS-CoV-2 infection in mouse models. *J Med Chem*. 2022;65(4):2785–93.

26 Siegel D, Hui HC, Doerffler E, Clarke MO, Chun K, Zhang L, et al. Discovery and synthesis of a phosphoramidate prodrug of a pyrrolo[2,1-f][triazin-4-amino] adenine C-nucleoside (GS-5734) for the treatment of ebola and emerging viruses. *J Med Chem*. 2017;60(5):1648–61.

27 Mulangu S, Dodd LE, Davey RT, Tshiani Mbaya O, Proschan M, Mukadi D, et al. A randomized, controlled trial of ebola virus disease therapeutics. *N Engl J Med*. 2019;381(24):2293–303.

28 Beigel JH, Tomashek KM, Dodd LE, Mehta AK, Zingman BS, Kalil AC, et al. Remdesivir for the treatment of Covid-19 – final report. *N Engl J Med*. 2020;383(19):1813–26.

29 Wang Y, Zhang D, Du G, Du R, Zhao J, Jin Y, et al. Remdesivir in adults with severe COVID-19: a randomised, double-blind, placebo-controlled, multicentre trial. *Lancet*. 2020;395(10236):1569–78.

30 Goldman JD, Lye DCB, Hui DS, Marks KM, Bruno R, Montejano R, et al. Remdesivir for 5 or 10 days in patients with severe Covid-19. *N Engl J Med*. 2020;383(19):1827–37.

31 Mackman RL, Hui HC, Perron M, Murakami E, Palmiotti C, Lee G, et al. Prodrugs of a 1′-CN-4-Aza-7,9-dideazaadenosine C-nucleoside leading to the discovery of remdesivir (GS-5734) as a potent inhibitor of respiratory syncytial virus with efficacy in the African Green Monkey model of RSV. *J Med Chem.* 2021;64(8):5001–17.

32 Sofia MJ, Bao D, Chang W, Du J, Nagarathnam D, Rachakonda S, et al. Discovery of a β-d-2′-deoxy-2′-α-fluoro-2′-β-C-methyluridine nucleotide prodrug (PSI-7977) for the treatment of hepatitis C virus. *J Med Chem.* 2010;53(19):7202–18.

33 Chapman H, Kernan M, Prisbe E, Rohloff J, Sparacino M, Terhorst T, et al. Practical synthesis, separation, and stereochemical assignment of the PMPA pro-drug GS-7340. *Nucleos Nucleot Nucl.* 2001;20(4–7):621–8.

34 Serpi M, Pertusati F. An overview of ProTide technology and its implications to drug discovery. *Expert Opin Drug Discov.* 2021;16(10):1149–61.

35 Yan VC, Muller FL. Advantages of the parent nucleoside GS-441524 over remdesivir for Covid-19 treatment. *ACS Med Chem Lett.* 2020;11(7):1361–6.

36 Hu W-j, Chang L, Yang Y, Wang X, Xie Y-c, Shen J-s, et al. Pharmacokinetics and tissue distribution of remdesivir and its metabolites nucleotide monophosphate, nucleotide triphosphate, and nucleoside in mice. *Acta Pharmacol Sin.* 2021;42(7):1195–200.

37 Yan VC, Muller FL. Why remdesivir failed: preclinical assumptions overestimate the clinical efficacy of remdesivir for COVID-19 and ebola. *Antimicrob Agents Chemother.* 2021;65(10):e01117-21.

38 Xie J, Wang Z. Can remdesivir and its parent nucleoside GS-441524 be potential oral drugs? An in vitro and in vivo DMPK assessment. *Acta Pharm Sin B.* 2021;11(6):1607–16.

39 Van Rompay AR, Johansson M, Karlsson A. Phosphorylation of nucleosides and nucleoside analogs by mammalian nucleoside monophosphate kinases. *Pharm Ther.* 2000;87(2):189–98.

40 Murphy BG, Perron M, Murakami E, Bauer K, Park Y, Eckstrand C, et al. The nucleoside analog GS-441524 strongly inhibits feline infectious peritonitis (FIP) virus in tissue culture and experimental cat infection studies. *Vet Microbiol.* 2018;219:226–33.

41 Pedersen NC, Perron M, Bannasch M, Montgomery E, Murakami E, Liepnieks M, et al. Efficacy and safety of the nucleoside analog GS-441524 for treatment of cats with naturally occurring feline infectious peritonitis. *J Feline Med Surg.* 2019;21(4):271–81.

42 Wang AQ, Hagen NR, Padilha EC, Yang M, Shah P, Chen CZ, et al. Preclinical pharmacokinetics and in vitro properties of GS-441524, a potential oral drug candidate for COVID-19 treatment. *Front pharmacol.* 2022;13:918083.

43 Wei D, Hu T, Zhang Y, Zheng W, Xue H, Shen J, et al. Potency and pharmacokinetics of GS-441524 derivatives against SARS-CoV-2. *Bioorg Med Chem.* 2021;46:116364.

44 Liu C, Shi W, Becker ST, Schatz DG, Liu B, Yang Y. Structural basis of mismatch recognition by a SARS-CoV-2 proofreading enzyme. *Science.* 2021;373(6559):1142–6.

45 Kokic G, Hillen HS, Tegunov D, Dienemann C, Seitz F, Schmitzova J, et al. Mechanism of SARS-CoV-2 polymerase stalling by remdesivir. *Nat Commun.* 2021;12(1):279.

46 Dangerfield TL, Johnson KA. Substrate specificity and kinetics of RNA hydrolysis by SARS-CoV-2 NSP10/14 exonuclease. *ACS Bio & Med Chem Au*. 2022;2(6):600–6.

47 Cao L, Li Y, Yang S, Li G, Zhou Q, Sun J, et al. The adenosine analog prodrug ATV006 is orally bioavailable and has preclinical efficacy against parental SARS-CoV-2 and variants. *Sci Transl. Med*. 2022;14(661):eabm7621.

48 Mackman RL, Kalla RV, Babusis D, Pitts J, Barrett KT, Chun K, et al. Discovery of GS-5245 (Obeldesivir), an oral prodrug of nucleoside GS-441524 that exhibits antiviral efficacy in SARS-CoV-2-infected African Green Monkeys. *J Med Chem*. 2023;66(17):11701–17.

49 Cox RM, Wolf JD, Lieber CM, Sourimant J, Lin MJ, Babusis D, et al. Oral prodrug of remdesivir parent GS-441524 is efficacious against SARS-CoV-2 in ferrets. *Nat Commun*. 2021;12(1):6415.

50 Schäfer A, Martinez DR, Won JJ, Meganck RM, Moreira FR, Brown AJ, et al. Therapeutic treatment with an oral prodrug of the remdesivir parental nucleoside is protective against SARS-CoV-2 pathogenesis in mice. *Sci Transl Med*. 2022;14(643):eabm3410.

51 Timmins GS. Deuterated drugs; updates and obviousness analysis. *Expert Opin Ther Pat*. 2017;27(12):1353–61.

52 Zheng W, Hu T, Zhang Y, Wei D, Xie Y, Shen J. Synthesis and anti-SARS-CoV-2 activity of deuterated GS-441524 analogs. *Tetrahedron Lett*. 2022;104:154012.

53 Shen J, Xie Y, Zhang L, Xiao G, Wang Z, Jiang H, et al. Salt of nucleoside analog and crystal form thereof, pharmaceutical composition and use. 2022, WO2022089302A1.

54 Tian G, Li J, Xie Y, Xu W. Preparation method of an antiviral nucleoside analogue. 2023, CN115716833A.

55 Hu T, Xie Y, Zhu F, Shen J. Preparation method for nucleoside analogue VV116. 2023, WO2023143630A1.

56 Villamarín M, Márquez-Algaba E, Esperalba J, Perelló M, Los Arcos I, Campany D, et al. Preliminary clinical experience of molnupiravir to prevent progression of COVID-19 in kidney transplant recipients. *Transplant*. 2022;106(11):2200–4.

57 Ticinesi A, Tuttolomondo D, Nouvenne A, Parise A, Cerundolo N, Prati B, et al. Co-administration of remdesivir and azithromycin may protect against intensive care unit admission in COVID-19 pneumonia requiring hospitalization: a real-life observational study. *Antibiotics*. 2022;11(7):941.

58 Li J, Sun S, Liu Y, Wang W, Zhu H-J. Understanding the cell-dependent activation of ProTides and its implications in optimizing the nucleoside antiviral prodrug design for treating pulmonary infections. *J Pharmacol Exp Ther*. 2023;385(S3):185.

59 Lamb YN. Nirmatrelvir plus ritonavir: first approval. *Drugs*. 2022;82(5):585–91.

60 Liu W, Zhang M, Hu C, Song H, Mei Y, Liu Y, et al. Remdesivir derivative VV116 Is a potential broad-spectrum inhibitor of both human and animal coronaviruses. *Viruses*. 2023;15(12):2295.

61 Borchers AT, Chang C, Gershwin ME, Gershwin LJ. Respiratory syncytial virus—a comprehensive review. *Clin Rev Allergy Immunol*. 2013;45(3):331–79.

62 Sciences VL. A Study of Deuremidevir Hydrobromide for Suspension in Chinese Infants Hospitalized With RSV. Available from: https://classic.clinicaltrials.gov/ct2/show/record/NCT06206720.

63 De Clercq E, Neyts J. Antiviral agents acting as DNA or RNA chain terminators. In: Kräusslich H-G, Bartenschlager R, editors. *Antiviral Strategies*. Berlin, Heidelberg: Springer Berlin Heidelberg; 2009. 53–84.

64 Furuta Y, Gowen BB, Takahashi K, Shiraki K, Smee DF, Barnard DL. Favipiravir (T-705), a novel viral RNA polymerase inhibitor. *Antiviral Res*. 2013;100(2):446–54.

65 Kabinger F, Stiller C, Schmitzová J, Dienemann C, Kokic G, Hillen HS, et al. Mechanism of molnupiravir-induced SARS-CoV-2 mutagenesis. *Nat Struct Mol Biol*. 2021;28(9):740–6.

66 Yin W, Luan X, Li Z, Xie Y, Zhou Z, Liu J, et al. Structural basis for repurpose and design of nucleoside drugs for treating COVID-19. *bioRxiv*. 2020;363812.

67 Gottlieb RL, Paredes R. Oral and intravenous 1′-cyano-substituted adenosine-like antivirals for early COVID-19. *Lancet Infect Dis*. 2024;24(2):108–10.

68 Jiang M, Gao Y, Hu Z. Pharmacological innovation and clinical value of VV116. *Lancet Infect Dis*. 2024;24(4):e212.

69 Zhou A, Wang Z, Diao X, Zhong D. Characterization of in-vivo human metabolites of the oral nucleoside anti-COVID-19 drug VV116 using UHPLC-Orbitrap-MS. *J Pharm Biomed Anal*. 2023;228:115340.

70 Humeniuk R, Mathias A, Cao H, Osinusi A, Shen G, Chng E, et al. Safety, tolerability, and pharmacokinetics of remdesivir, an antiviral for treatment of COVID-19, in healthy subjects. *Clin Transl Sci*. 2020;13(5):896–906.

71 Xue L, Jing S, Zhang K, Milne R, Wang H. Infectivity versus fatality of SARS-CoV-2 mutations and influenza. *Int J Infect Dis*. 2022;121:195–202.

72 Syed YY. Ensitrelvir fumaric acid: first approval. *Drugs*. 2024;84(6):721–8.

73 Cao B, Wang Y, Lu H, Huang C, Yang Y, Shang L, et al. Oral simnotrelvir for adult patients with mild-to-moderate Covid-19. *N Engl J Med*. 2024;390(3):230–41.

74 Zhu K-W. Deuremidevir and simnotrelvir–ritonavir for the treatment of COVID-19. *ACS Pharmacol Transl Sci*. 2023;6(9):1306–9.

75 Zhan Y, Lin Z, Liang J, Sun R, Li Y, Lin B, et al. Leritrelvir for the treatment of mild or moderate COVID-19 without co-administered ritonavir: a multicentre randomised, double-blind, placebo-controlled phase 3 trial. *EClinicalMedicine*. 2024;67:102359.

76 Zhang H, Zhou J, Chen H, Mao J, Tang Y, Yan W, et al. Phase I study, and dosing regimen selection for a pivotal COVID 19 trial of GST-HG171. *Antimicrob Agents Chemother*. 2024;68(1):e01115–23.

77 Jiang X, Su H, Shang W, Zhou F, Zhang Y, Zhao W, et al. Structure-based development and preclinical evaluation of the SARS-CoV-2 3C-like protease inhibitor simnotrelvir. *Nat Commun*. 2023;14(1):6463.

78 Carabelli AM, Peacock TP, Thorne LG, Harvey WT, Hughes J, de Silva TI, et al. SARS-CoV-2 variant biology: immune escape, transmission and fitness. *Nat Rev Microbiol*. 2023;21(3):162–77.

79 Shyr ZA, Cheng Y-S, Lo DC, Zheng W. Drug combination therapy for emerging viral diseases. *Drug Discov Today*. 2021;26(10):2367–76.

13

Molnupiravir

Richard K. Plemper

Center for Translational Antiviral Research, Georgia State University Institute for Biomedical Sciences, Atlanta, Georgia, USA

13.1 Introduction

When the Medicines and Healthcare products Regulatory Agency (MHRA) of the United Kingdom granted emergency approval to molnupiravir in November 2021, it became the first orally available antiviral for the treatment of severe acute respiratory syndrome coronavirus 2 (SARS-CoV-2) infection during the coronavirus disease 2019 (COVID-19) pandemic [1]. Subsequently, the drug was approved for emergency use in, among others, the United States and the European Union, giving healthcare providers an option to treat outpatients prior to advance to severe disease and hospitalization. By 2024, a second oral antiviral, Paxlovid, the ritonavir-boosted M^{pro} inhibitor nirmatrelvir, was FDA-approved for the treatment of at-risk adults and pediatric patients 12 years of age and older [2], whereas emergency authorization of molnupiravir was rescinded in the European Union. The drug remains authorized in the United States for adult outpatients who are at high risk of progressing to severe COVID-19. In the 2.5 years since the original emergency use authorization, the debate surrounding the clinical use of molnupiravir has largely focused on three areas: the potential for drug incorporation into the host genome, the risk of drug-enhanced viral evolution, and the therapeutic benefit for the patient. This chapter will outline the molecular background for this discussion and summarize the experimental and clinical data that have emerged since the first emergency authorization of the drug.

Trends in Antiviral Drug Development, First Edition. Edited by Michael J. Sofia and Zhengqiang Wang.

13.2 Mechanism of Antiviral Activity

Molnupiravir is an ester prodrug of the nucleoside analog N^4-hydroxycytidine (NHC) that was developed to address the low bioavailability of orally delivered NHC in nonhuman primates [3, 4]. NHC was first identified over 50 years ago as a bacterial mutagen [5, 6], with little use in eukaryotic systems. In the 2000s, the inhibitory activity of NHC against members of two RNA virus families, the flaviviruses and the coronaviruses, was reported in cell culture [7, 8], which was subsequently discovered to apply equally to a broad range of major human pathogens representing diverse RNA virus families, including members of the orthomyxoviruses, pneumoviruses, filoviruses, and alphaviruses, both in cell culture and in animal models of viral infection [9–11]. The antiviral activity of NHC is considered to be based on its ability to undergo spontaneous tautomeric conversions, resulting in either Watson–Crick conforming NHC:G base pairing or alternative noncanonical NHC:A pairing (Figure 13.1) [12, 13].

Groundbreaking early investigation of the mechanism of picornavirus inhibition by the FDA-approved mutagenic antiviral drug ribavirin revealed that RNA viruses are extremely sensitive to a slightly increased frequency of random mutations

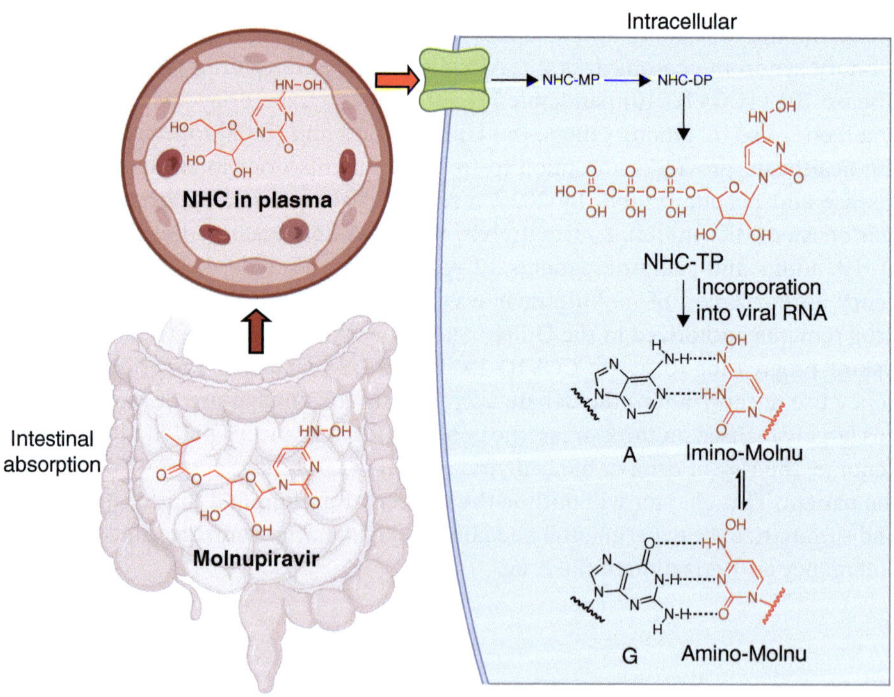

Figure 13.1 Uptake and mechanism of action of molnupiravir. Upon intestinal absorption, molnupiravir is metabolized to the free ribobase NHC, which is the only form detectable in plasma. Subsequent to cellular uptake, NHC is anabolized by host kinases to the bioactive triphosphate form NHC-TP, which, due to tautomeric conversion, can base-pair when incorporated into viral RNA as cytidine or uridine.

inserted into their genomes [14–16]. Statistically, polioviruses experience 1.5 random mutations per genome and replication cycle in the absence of any drug, which is close to the threshold of minimal genetic integrity required to sustain the long-term survival of the population. A ribavirin-mediated moderate increase to, on average, approximately seven random mutations per genome and replication cycle reduced relative viral fitness to approximately 5% of that of untreated, which rapidly caused the collapse of the virus population due to error catastrophe [16]. Random mutations accumulate rapidly in viral RNA genomes upon virus replication in the presence of a mutagenic ribonucleotide analog, since acute-disease RNA viruses are highly replication-active, RNA virus genome replication proceeds through synthesis of antigenomic RNA that subsequently serves as a template for progeny genomic RNA, and ribonucleotide analogs can be incorporated into both antigenomes and genomes. In addition to ribavirin, equivalent random mutagenesis is also considered to represent a major mechanism of the antiviral activity of favipiravir, which is conditionally approved in Japan for the treatment of pandemic influenza virus infections [17, 18].

In the case of molnupiravir, whole genome sequencing of virus populations grown in the presence of sub-sterilizing drug concentrations demonstrated that NHC incorporation into viral RNA genomes causes characteristic C-to-U and G-to-A transition mutations, which were consistent with the alternative base pairings after tautomeric conversion of NHC [3]. In addition, exposure of animals or cultured cells to very high NHC concentrations resulted in a direct destructive effect on viral and host message due to tautomeric conversion of the analog after incorporation into mRNAs. Since mRNAs are short-lived and a ribonucleoside analog cannot be directly incorporated into the host DNA genome, however, NHC had no detectable direct cytotoxic activity in cultured cells and human organoids and did not cause adverse effects in animal models at sterilizing antiviral concentrations [3, 19]. Molnupiravir was well tolerated with minimal acute side effects in subsequent clinical trials [20, 21], underscoring the predictive power of the *ex vivo* and *in vivo* models.

13.3 Potential for NHC-TP Incorporation into the Host Genome

Independent of the safety data available, the potential for long-term host mutagenic effects resulting from the metabolic conversion of NHC to its deoxy form and the incorporation of dNHC into host DNA remains a topic of intense discussion. After hydrolysis of the molnupiravir ester prodrug during intestinal absorption and release of the free NHC ribobase into the host circulation, the compound must be intracellularly anabolized by host cell kinases to its triphosphate form, NHC-TP, to become biologically active, which in a stepwise process requires conversion to the 5′ mono, -di, and finally triphosphate. The first phosphorylation step is mediated by host uridine-cytidine kinases 1 and 2 (Uck1/Uck2), followed by anabolism to the NHC diphosphate form (NHC-DP) by cytidine monophosphate kinase 1 (Cmpk1). No direct transribosylation of NHC to its deoxyribose form has been observed [22], but 5′-ribonucleoside-diphosphates stand at a junction point of the ribonucleotide and 2′-deoxyribonucleotide biosynthetic pathways, since they can be either converted to ribonucleotide triphosphates through host kinases or reduced to 2′-deoxyribonucleotide diphosphates through the host ribonucleotide reductase enzyme (RNR) (Figure 13.2). Whereas NHC-DP is clearly and efficiently anabolized to NHC-TP, the efficiency of recognition and deoxygenation by RNR is unknown.

Some studies have reported incorporation of NHC into the genome of cultured cells [23, 24], whereas others did not detect any NHC in host DNA [25, 26]. Since lack of detection is no proof of absence, however, some conversion of NHC-DP to its deoxy form by RNR should be assumed and is very likely in its efficiency directly influenced by the intracellular NHC-DP concentration. Indeed, blocking the pyrimidine salvage pathway through knockout of Uck2 reportedly lowered the mutagenic activity of NHC, whereas Uck2 overexpression had the opposite effect [27]. Importantly, Uck2 levels are strongly restricted in normal tissues [28], but upregulated in cancer cells and immortalized cell lines, such as those examined in studies reporting host genome mutations [23, 24], to meet the increased metabolic demands of rapidly proliferating cells. Differential activation of the pyrimidine salvage pathway in different cell systems should consequently be considered as a major contributing factor to the

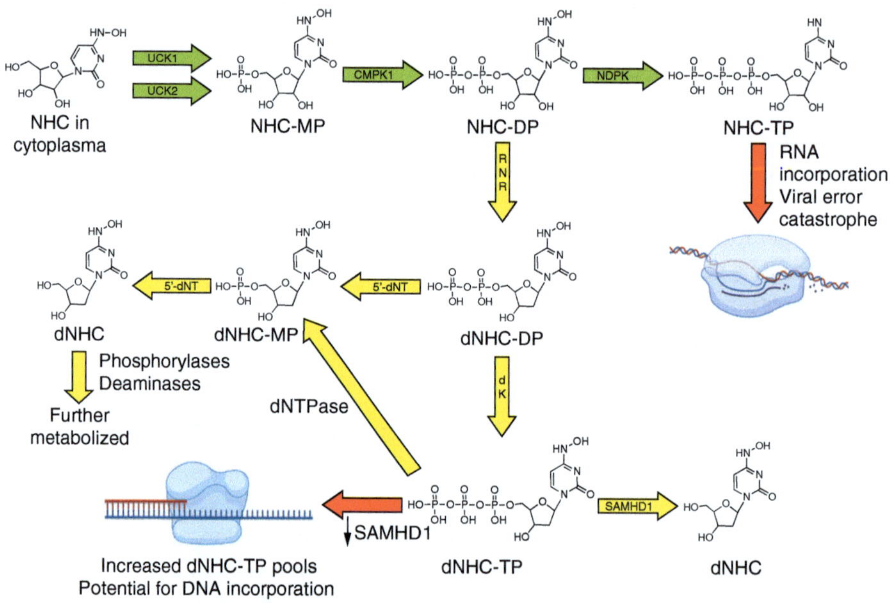

Figure 13.2 Conversion of ribonucleotide and deoxyribonucleotide synthesis pathways. NHC is anabolized stepwise to NHC-MP, NHC-DP, and NHC-TP by host kinases. NHC-DP can alternatively be converted by host RNR to the corresponding deoxyribonucleotide, dNHC-DP, which may subsequently be metabolized or anabolized to dNHC-TP. Incorporation into host DNA may occur if a sufficiently high dNHC-TP pool is formed, or dNHC-TP may be rapidly metabolized through SAMHD1 activity. Intracellular levels of dNHC-DP and dNHC-TP have not been reported, but appear to be higher in cancer cells and immortalized cell lines with high Uck2 activity than in normal tissues. UCK1/2, uridine-cytidine kinase 1/2; CMPK1, cytidine/uridine monophosphate kinase 1; NDPK, nucleoside-diphosphate kinase; RNR, ribonucleotide reductase; dK, deoxynucleoside kinase; 5-dNT, 5′-deoxynucleotidase; dNTPase, 5′-triphosphate nucleotidohydralase; SAMHD1, sterile alpha motif and HD domain-containing protein 1.

mixed literature reports and underscores that data obtained from *ex vivo* systems must be vetted carefully for their physiological relevance. Another current unknown is the fate of 2-deoxy-NHC-DP once produced by RNR, as it could be anabolized to the corresponding triphosphate or converted to the monophosphate and metabolized. If anabolism occurs, the intracellular stability of 2′-deoxy-NHC-TP, which is currently likewise unknown, will determine the likelihood of incorporation into the host genome. Long-term exposure of animals to molnupiravir in models specifically designed to assess the mutagenic potential of test articles, such as the Big Blue rodent assays [29], did not reveal an increased mutation rate, leading the FDA to conclude that the genotoxicity of molnupiravir is low [30]. However, these safety assay results have not been subjected to peer review and are not publicly available, preventing the independent evaluation of the datasets.

Acute adverse effects of molnupiravir are relatively rare and overall benign in nature, attesting to the good tolerability of the drug in clinical use [4, 31]. Spontaneous mutations, for instance, through UV exposure, cosmic radiation, or medical procedures, are furthermore common and typically trigger DNA repair pathways [32]. Although there is no safe threshold for acceptable mutation frequencies, an extremely low incorporation into somatic host cells may have no measurable late consequences. Long-term effects of molnupiravir are nevertheless not known currently and will not be for the foreseeable future. A responsible risk assessment of therapeutic benefit versus potential late effects, which will naturally be unique for each patient, must therefore guide the clinical use of the drug.

13.4 Risk of Molnupiravir-enhanced Viral Evolution

No SARS-CoV-2 escape from molnupiravir has yet been observed in the field or experimentally. These observations are consistent with earlier findings that, despite expensive viral adaptation efforts in the laboratory, none of the diverse viral targets tested against NHC has mounted viable resistance to the drug [3, 33, 34]. This unusually high genetic barrier blocking viral escape is most likely due to the close chemical similarity of NHC and cytidine, which makes it seemingly impossible for the viral polymerase complexes to distinguish between both nucleotides and reject NHC-TP while maintaining the ability to efficiently recognize CTD. Whereas molnupiravir, therefore, does not enable the evolution of viral populations with reduced susceptibility to the drug, in contrast to most other approved antivirals, its mechanism of action may inherently enhance the genetic variability of the viral population present in a treated patient [35]. This effect could be most pronounced in immunocompromised hosts, which experience longer periods of viral replication in the presence of drug due to slower viral clearance [36].

Extensive passaging of influenza viruses in the presence of sub-sterilizing NHC concentrations, followed by whole viral genome sequencing, confirmed indeed the appearance of randomly distributed transition mutations characteristic of molnupiravir, but none of these mutations became allele-dominant and changed the overall genetic integrity of the virus population over the course of the study [3]. Consistent with the extremely high sensitivity of RNA viruses to even slightly increased mutation rates discussed above, this observation suggests that molnupiravir-induced transition mutations rapidly destroy the viability of the affected individual viral genomes, preventing them from reaching dominance in the population.

In contrast to studies in cell culture, however, circulating virus populations are subject to multiple additional layers of selective pressure arising from, among other factors, a requirement to efficiently transmit between hosts and the need to navigate naturally acquired or vaccine-induced antiviral immunity, as was seen in the advanced stages of the SARS-CoV-2 pandemic. Relevant, therefore, is not whether NHC introduces low-frequency transition mutations into a viral genome that are detectable in laboratory and clinical samples – it definitely does in laboratory samples – but the question of whether these mutations expand the genetic landscape of the virus in an unusual way and/or accelerate the evolution of variants. SARS-CoV-2 genomes collected in a global database in 2022, when molnupiravir was first used on a large scale, revealed a higher frequency of G-to-A and C-to-T transition mutations in genomes sampled from geographical regions with higher molnupiravir prescription rates compared to those with lower use of the drug [35]. Although a correlation between molnupiravir use and transition mutations in the viral genome remained circumstantial, higher in-host genomic variability of the virus population is consistent with the mechanism of action of NHC and was confirmed for small sample sizes in human patients [37]. Echoing laboratory results obtained for other viral targets tested against NHC, individual transition mutations remained low in allele frequency in this study, varied in nature between different sampling time points, and did not result in the selection of a drug-resistant variant.

After several years of molnupiravir use in large geographical regions, there is no evidence that the drug has contributed to SARS-CoV-2 evolution in any impactful way. Rather, the inherent genetic variability of SARS-CoV-2 has fueled the high frequency with which viral variants and subvariants have emerged during the pandemic in response to the increasing prevalence of preexisting antiviral immunity in the human population. Apparently, SARS-CoV-2 does not require pharmacological aid to explore its genetic landscape to the fullest.

13.5 Therapeutic Benefit of Molnupiravir

The decision to administer a drug must be guided by an assessment of the anticipated therapeutic benefit for the patient versus the possible side effects. Numerous clinical reports have been released for SARS-CoV-2 therapeutics, including molnupiravir and multiple clinical trials, for instance [38–46], have been carried out since early 2021. The interpretation of the clinical results of molnupiravir is complicated by two factors in particular. First is the change in patient populations during the pandemic from initially immunologically naïve to harboring widespread hybrid immunity due to prior SARS-CoV-2 disease history and vaccination. Accordingly, the initially major risk, especially among older adults and the immunocompromised, to progress to severe COVID-19 associated with hospitalization and high case fatality rates was reduced substantially at later stages of the pandemic, despite continued high viral prevalence in all geographical areas and large case numbers [47, 48]. This shift in disease presentation compromised the power of assessing hospitalization and mortality rates as primary measures of therapeutic success, since rates in later clinical trials were increasingly low, also in reference patient groups [39, 49].

Secondly, quantifying infectious particle load directly from human throat swab samples is challenging [50], which has established viral RNA copies as a surrogate readout for viral burden and, by extension, a primary indicator of therapeutic success [46, 51]. Although widely applied, viral RNA copy numbers were shown in animal studies to underrepresent the therapeutic benefit of molnupiravir compared to a Paxlovid-like combination of nirmatrelvir and ritonavir, due to the distinct mechanisms of action of the drugs [52]. In two different animal models of SARS-CoV-2 infection, infectious particle load and viral RNA copy numbers were reduced proportionally after treatment with the main protease (M^{pro}) inhibitor Paxlovid, whereas molnupiravir had an overproportional inhibitory effect on infectious virus load compared to the reduction in viral RNA copies. This discrepancy was directly attributed to the viral error catastrophe mechanism of activity of molnupiravir, which in the first instance eliminates bioactivity of progeny viral genomes but does not directly suppress the synthesis of viral RNA. Accordingly, the reduction in physical viral RNA copies remained moderate for several days after treatment initiation, but the titer of infectious progeny particles was reduced by several orders of magnitude, indicating that the bioactivity of viral RNAs synthesized while animals were treated with molnupiravir was severely impaired [52]. If an equivalently disproportional reduction of infectious virus load versus viral RNA copy numbers applies to human patients, clinical trials based on reduction of viral RNA copies could underestimate the full antiviral effect of molnupiravir.

In the first months of the COVID-19 pandemic, a newly emerged respiratory virus, capable of efficient human-to-human transmission, spread through an immunologically naïve population. This constellation was transient, however, and the first virus wave was rapidly followed by a scenario of naturally acquired and vaccine-induced hybrid immunity in large parts of the population. Since oral antiviral therapeutics first emerged nearly 2 years after the onset of the pandemic, it remains unknown whether a nascent viral pandemic could have been silenced at its earliest stages through broad administration of antivirals to outpatients. In contrast, numerous clinical trials were carried out at later stages of the pandemic, when hybrid immunity was widely established.

The largest COVID-19 trial in the community was the UK-based PANORAMIC study, which was a national multicenter, open-label, multigroup trial of the oral COVID-19 treatments molnupiravir and Paxlovid (nirmatrelvir and ritonavir) in primary care [53]. Now closed with over 29 000 participants total, results have so far been reported only for the comparison of molnupiravir plus usual care versus usual care alone in adults at higher risk of developing severe disease [39]. Primary outcomes were hospitalization and death. Because over 90% of participants had received three COVID vaccine doses before the trial started, and far less than 10% were in the highest risk group of adults aged 75 years and older, hospitalization and death rates were very low in the reference group. As one might expect, molnupiravir did not significantly improve the primary outcome further in this relatively young patient group. However, treatment lowered virus load in patient samples and shortened the time to recovery by 4 days, which is remarkable when compared to the only 25 h shorter time to alleviation provided by the approved influenza drug oseltamivir in patients with influenza virus infection [54]. Australian data assessing COVID-19 patients aged 70 years and older revealed substantial therapeutic benefit for this highest risk group, provided treatment was started within 1 day of diagnosis [49, 55]. Early treatment with molnupiravir lowered the risk of hospitalization and death by 37% and 63%, respectively, whereas treatment initiated 4 days after diagnosis or later provided no benefit. Paxlovid provided the same risk reduction when started within 1 day of diagnosis but cannot be safely administered to many patients in this age group due to incompatibility with underlying conditions such as kidney or liver disease and multiple drug interactions of its ritonavir component with many medications commonly used by this patient group [55]. Faster recovery and lower viral burden in treated patients may furthermore provide epidemiological benefit when high-risk people are concentrated in confined settings such as assisted living facilities. Indeed, studies with animal transmission models have demonstrated that molnupiravir rapidly suppressed the spread of SARS-CoV-2 from infected and treated sources to their untreated contacts [52, 56].

13.6 Conclusions

Every pharmacological intervention requires an evaluation of anticipated therapeutic benefit versus possible undesirable side effects of treatment. The outcome of this assessment varies based on the specific needs of individual patient subgroups and can typically not be generalized. Even in populations with widespread vaccination coverage, molnupiravir reportedly provides substantial therapeutic benefit to patients aged 70 years and older, who are at the greatest risk of severe disease, hospitalization, and death, when treatment is started early after diagnosis. The drug is generally well tolerated, with minimal immediate adverse effects, and is compatible with unrelated medications commonly used by this patient group. The risk of genotoxicity was deemed low by the FDA. Although the potential long-term effects of molnupiravir use will not be fully known for years, the benefit to the highest risk patient populations is substantial. Beyond the SARS-CoV-2 indication, molnupiravir has a unique status as an orally available antiviral with an exceptionally high genetic barrier to resistance and demonstrated broad efficacy in animal models against highly pathogenic viruses of pandemic potential, such as pandemic influenza viruses [3, 9, 19, 57], and highly pathogenic alphaviruses, such as Venezuelan equine encephalitis virus and Eastern equine encephalitis virus [33]. Using it against these high-consequence viral pathogens to contain and potentially silence a nascent pandemic shifts the risk-benefit balance fundamentally in favor of treatment. Molnupiravir constitutes an important, and currently possibly the only, true first-line of defense drug that can be readily deployed to outpatients to reduce case-fatality rates, slow viral spread, prevent the overburdening of healthcare systems, and protect healthcare workers when another pandemic RNA virus emerges.

Acknowledgments

I thank J.D. Wolf for assistance with figure preparation. This work was supported, in part, by Public Health Service grants AI171403 and AI141222.

Conflict of Interest

The author reports contract testing from Enanta Pharmaceuticals, Atea Pharmaceuticals, and Icosagen Biosciences, and research support from Gilead Sciences, outside of the described work.

References

1 GOV.UK First oral antiviral for COVID-19, Lagevrio (molnupiravir), approved by MHRA. 2021. https://www.gov.uk/government/news/first-oral-antiviral-for-covid-19-lagevrio-molnupiravir-approved-by-mhra.

2 FDA: Fact sheet for healthcare providers: emergency use authorization for paxlovidtm. 2021. (https://www.google.com/url?sa=t&rct=j&q=&esrc=s&source=web&cd=&ved=2ahUKEwih_6mA1N_2AhWHMd8KHf2nAboQFnoECA8QAw&url=https%3A%2F%2Fwww.fda.gov%2Fmedia%2F155050%2Fdownload&usg=AOvVaw0YT1Xt-pO03Cft8l81eGMr).

3 Toots M, Yoon JJ, Cox RM, Hart M, Sticher ZM, Makhsous N, et al. Characterization of orally efficacious influenza drug with high resistance barrier in ferrets and human airway epithelia. *Sci. Transl. Med.* 2019;11(515): eaax5866.

4 Painter WP, Holman W, Bush JA, Almazedi F, Malik H, Eraut N, et al. Human safety, tolerability, and pharmacokinetics of molnupiravir, a novel broad-spectrum oral antiviral agent with activity against SARS-CoV-2. *Antimicrob Agents Chemother.* 2021;65(5): 10–128

5 Salganik RI, Vasjunina EA, Poslovina AS, Andreeva IS. Mutagenic action of N4-hydroxycytidine on Escherichia coli B cyt. *Mutat Res.* 1973;20: 1–5.

6 Popowska E, Janion C. N4-hydroxycytidine-a new mutagen of a base analogue type. *Biochem Biophys Res Commun.* 1974;56: 459–66.

7 Stuyver LJ, Whitaker T, McBrayer TR, Hernandez-Santiago BI, Lostia S, Tharnish PM, et al. Ribonucleoside analogue that blocks replication of bovine viral diarrhea and hepatitis C viruses in culture. *Antimicrob Agents Chemother.* 2003;47: 244–54.

8 Barnard DL, Day CW, Bailey K, Heiner M, Montgomery R, Lauridsen L, et al. Evaluation of immunomodulators, interferons and known in vitro SARS-coV inhibitors for inhibition of SARS-coV replication in BALB/c mice. *Antivir Chem Chemother.* 2006;17: 275–84.

9 Yoon JJ, Toots M, Lee S, Lee ME, Ludeke B, Luczo JM, et al. Orally efficacious broad-spectrum ribonucleoside analog inhibitor of influenza and respiratory syncytial viruses. *Antimicrob Agents Chemother.* 2018;62(8): 10–128.

10 Reynard O, Nguyen XN, Alazard-Dany N, Barateau V, Cimarelli A, Volchkov VE. Identification of a new ribonucleoside inhibitor of ebola virus replication. *Viruses* 2015;7: 6233–40.

11 Ehteshami M, Tao S, Zandi K, Hsiao HM, Jiang Y, Hammond E, et al. Characterization of beta-d-N(4)-Hydroxycytidine as a Novel Inhibitor of Chikungunya Virus. *Antimicrob Agents Chemother.* 2017;61(4): 10–128.

12 Janion C, Glickman BW. N4-hydroxycytidine: a mutagen specific for AT to GC transitions. *Mutat Res.* 1980;72: 43–47.

13 Bessi I, Stiller C, Schroeder T, Schad B, Grune M, Dietzsch J, et al. The Tautomeric State of N(4)-Hydroxycytidine within base-paired RNA. *ACS Cent Sci.* 2024;10: 1084–93.

14 Crotty S, Andino R. Implications of high RNA virus mutation rates: lethal mutagenesis and the antiviral drug ribavirin. *Microbes Infect.* 2002;4: 1301–07.

15 Crotty S, Cameron CE, Andino R. RNA virus error catastrophe: direct molecular test by using ribavirin. *Proc Natl Acad Sci U S A.* 2001;98: 6895–6900.

16 Graci JD, Cameron CE. Quasispecies, error catastrophe, and the antiviral activity of ribavirin. *Virology* 2002;298: 175–80.

17 Baranovich T, Wong SS, Armstrong J, Marjuki H, Webby RJ, Webster RG, et al. T-705 (favipiravir) induces lethal mutagenesis in influenza A H1N1 viruses in vitro. *J Virol.* 2013;87: 3741–51.

18 Davidson S. Treating influenza infection, from now and into the future. *Front Immunol.* 2018;9: 1946.

19 Toots M, Plemper RK. Next-generation direct-acting influenza therapeutics. *Transl Res.* 2020;220: 33–42.

20 Painter WP, Holman W, Bush JA, Almazedi F, Malik H, Eraut N, et al. Human safety, tolerability, and pharmacokinetics of molnupiravir, a novel broad-spectrum oral antiviral agent with activity against SARS-CoV-2. *Antimicrob Agents Chemother.* 2021;65(5): 10–128

21 Liang Y, Ma L, Wang Y, Zheng J, Su L, Lyu J. Adverse events associated with molnupiravir: a real-world disproportionality analysis in food and drug administration adverse event reporting system. *Front Pharmacol.* 2023;14: 1253799.

22 Sticher ZM, Lu G, Mitchell DG, Marlow J, Moellering L, Bluemling GR, et al. Analysis of the potential for N(4)-hydroxycytidine to inhibit mitochondrial replication and function. *Antimicrob Agents Chemother* 2020;64(2): 10–128.

23 Zhou S, Hill CS, Sarkar S, Tse LV, Woodburn BMD, Schinazi RF, et al. beta-d-N4-hydroxycytidine inhibits SARS-CoV-2 through lethal mutagenesis but is also mutagenic to mammalian cells. *J Infect Dis.* 2021;224: 415–19.

24 Miranda JA, McKinzie PB, Dobrovolsky VN, Revollo JR. Evaluation of the mutagenic effects of Molnupiravir and N4-hydroxycytidine in bacterial and mammalian cells by HiFi sequencing. *Environ Mol Mutagen.* 2022;63: 320–28.

25 Urbanowicz K, Opielka M, Stegmann KM, Dickmanns A, Dobbelstein M, Peters GJ, et al. Evaluation of N4-hydroxycytidine incorporation into nucleic acids of SARS-CoV-2-infected host cells by direct measurement with liquid chromatography-mass spectrometry. *Nucleosides Nucleotides Nucleic Acids.* 2024;43(8): 813–21.

26 Githaka JM. Molnupiravir does not induce mutagenesis in host lung cells during SARS-CoV-2 treatment. *Bioinform Biol Insights*. 2022;16: 11779322221085077.

27 Xu Z, Flensburg C, Bilardi RA, Majewski IJ. Uridine-cytidine kinase 2 potentiates the mutagenic influence of the antiviral beta-d-N4-hydroxycytidine. *Nucleic Acids Res*. 2023;51: 12031–42.

28 Van Rompay AR, Norda A, Linden K, Johansson M, Karlsson A. Phosphorylation of uridine and cytidine nucleoside analogs by two human uridine-cytidine kinases. *Mol Pharmacol*. 2001;59: 1181–86.

29 Young R. posting date. Big Blue® Transgenic Rodent (TGR) Mutation Assay;2014. (file:///Users/richard/Downloads/CR6_7-2_YOUNG-1.PDF). [Online]

30 Molnupiravir - US Food and Drug Administration;2021. (https://www.fda.gov/media/154472/download).

31 Kimata M, Watanabe A, Yanagida Y, Kinoshita D, Maekawa S. Safety and effectiveness of molnupiravir (LAGEVRIO((R))) capsules in Japanese patients with COVID-19: interim report of post-marketing surveillance in Japan. *Infect Dis Ther*. 2023;12: 1119–36.

32 Chatterjee N, Walker GC. Mechanisms of DNA damage, repair, and mutagenesis. *Environ Mol Mutagen*. 2017;58: 235–63.

33 Urakova N, Kuznetsova V, Crossman DK, Sokratian A, Guthrie DB, Kolykhalov AA, et al. beta-d-N (4)-hydroxycytidine is a potent anti-alphavirus compound that induces a high level of mutations in the viral genome. *J Virol*. 2018; 92.

34 Strizki JM, Gaspar JM, Howe JA, Hutchins B, Mohri H, Nair MS, et al. Molnupiravir maintains antiviral activity against SARS-CoV-2 variants and exhibits a high barrier to the development of resistance. *Antimicrob Agents Chemother*. 2024;68: e00953–23.

35 Sanderson T, Hisner R, Donovan-Banfield I, Hartman H, Lochen A, Peacock TP, et al. A molnupiravir-associated mutational signature in global SARS-CoV-2 genomes. *Nature*. 2023;623: 594–600.

36 Lieber CM, Kang HJ, Sobolik EB, Sticher ZM, Ngo VL, Gewirtz AT, et al. Efficacy of late-onset antiviral treatment in immunocompromised hosts with persistent SARS-CoV-2 infection. *J Virol*. 2024;98(9): e00905–24.

37 Alteri C, Fox V, Scutari R, Burastero GJ, Volpi S, Faltoni M, et al. A proof-of-concept study on the genomic evolution of Sars-Cov-2 in molnupiravir-treated, paxlovid-treated and drug-naive patients. *Commun Biol*. 2022;5: 1376.

38 Nakamura K, Fujimoto K, Hasegawa C, Aoki I, Yoshitsugu H, Ugai H, et al. A phase I, randomized, placebo-controlled study of molnupiravir in healthy Japanese to support special approval in Japan to treat COVID-19. *Clin Transl Sci*. 2022;15: 2697–708.

39 Butler CC, Hobbs FDR, Gbinigie OA, Rahman NM, Hayward G, Richards DB, et al. Molnupiravir plus usual care versus usual care alone as early treatment for adults with COVID-19 at increased risk of adverse outcomes (PANORAMIC): an open-label, platform-adaptive randomised controlled trial. *Lancet*. 2023;401: 281–93.

40 Guan Y, Puenpatom A, Johnson MG, Zhang Y, Zhao Y, Surber J, et al. Impact of molnupiravir treatment on patient-reported COVID-19 symptoms in the phase 3 MOVe-OUT Trial: a randomized, placebo-controlled trial. *Clin Infect Dis*. 2023;77: 1521–30.

41 Jayk Bernal A, Gomes da Silva MM, Musungaie DB, Kovalchuk E, Gonzalez A, Delos Reyes V, et al. Molnupiravir for oral treatment of Covid-19 in nonhospitalized patients. *N Engl J Med*. 2022;386: 509–20.

42 Alpizar SA, Accini J, Anderson DC, Eysa B, Medina-Pinon I, Ohmagari N, et al. Molnupiravir for intra-household prevention of COVID-19: the MOVe-AHEAD randomized, placebo-controlled trial. *J Infect*. 2023;87: 392–402.

43 Weil C, Bergroth T, Eisenberg A, Whiteside YO, Caraco Y, Tene L, et al. Real-world utilization of molnupiravir during the COVID-19 omicron surge in Israel. *Epidemiologia (Basel)*. 2023;4: 309–21.

44 Schilling WHK, Jittamala P, Watson JA, Boyd S, Luvira V, Siripoon T, et al. Antiviral efficacy of molnupiravir versus ritonavir-boosted nirmatrelvir in patients with early symptomatic COVID-19 (PLATCOV): an open-label, phase 2, randomised, controlled, adaptive trial. *Lancet Infect Dis*. 2024;24: 36–45.

45 Salmanton-Garcia J, Marchesi F, Koehler P, Weinbergerova B, Colovic N, Falces-Romero I, et al. Molnupiravir compared to nirmatrelvir/ritonavir for COVID-19 in high-risk patients with haematological malignancy in Europe. A matched-paired analysis from the EPICOVIDEHA registry. *Int J Antimicrob Agents*. 2023;62: 106952.

46 Fischer 2nd WA, Eron Jr JJ, Holman W, Cohen MS, Fang L, Szewczyk LJ, et al. A phase 2a clinical trial of molnupiravir in patients with COVID-19 shows accelerated SARS-CoV-2 RNA clearance and elimination of infectious virus. *Sci Transl Med*. 2022;14:eabl7430.

47 Hornsby H, Nicols AR, Longet S, Liu C, Tomic A, Angyal A, et al. Omicron infection following vaccination enhances a broad spectrum of immune responses dependent on infection history. *Nat Commun*. 2023;14: 5065.

48 Bobrovitz N, Ware H, Ma X, Li Z, Hosseini R, Cao C, et al. Protective effectiveness of previous SARS-CoV-2 infection and hybrid immunity against the omicron variant and severe disease: a systematic review and meta-regression. *Lancet Infect Dis*. 2023;23: 556–67.

49 Kidd MR, Kelly PM. PANORAMIC: important insights into molnupiravir use in COVID-19. *Lancet*. 2023;401: 250–51.

50 Wolfel R, Corman VM, Guggemos W, Seilmaier M, Zange S, Muller MA, et al. Virological assessment of hospitalized patients with COVID-2019. *Nature*. 2020;581: 465–69.

51 Puhach O, Meyer B, Eckerle I. SARS-CoV-2 viral load and shedding kinetics. *Nat Rev Microbiol*. 2023;21: 147–61.

52 Cox RM, Lieber CM, Wolf JD, Karimi A, Lieberman NAP, Sticher ZM, et al. Comparing molnupiravir and nirmatrelvir/ritonavir efficacy and the effects on SARS-CoV-2 transmission in animal models. *Nat Commun*. 2023;14: 4731.

53 Gbinigie O, Ogburn E, Allen J, Dorward J, Dobson M, Madden TA, et al. Platform adaptive trial of novel antivirals for early treatment of COVID-19 In the community (PANORAMIC): protocol for a randomised, controlled, open-label, adaptive platform trial of community novel antiviral treatment of COVID-19 in people at increased risk of more severe disease. *BMJ Open*. 2023;13: e069176.

54 Dobson J, Whitley RJ, Pocock S, Monto AS. Oseltamivir treatment for influenza in adults: a meta-analysis of randomised controlled trials. *Lancet*. 2015;385: 1729–37.

55 West D, Parta I, Canevari K, Haslett N, Wollersheim D, Martinie M, et al. posting date. Paxlovid is Australia's first-line COVID antiviral but Lagevrio also prevents severe disease in over-70s;2022. (https://theconversation.com/paxlovid-is-australias-first-line-covid-antiviral-but-lagevrio-also-prevents-severe-disease-in-over-70s-195349). [Online]

56 Cox RM, Wolf JD, Plemper RK. Therapeutically administered ribonucleoside analogue MK-4482/EIDD-2801 blocks SARS-CoV-2 transmission in ferrets. *Nat Microbiol.* 2021;6: 11–18.

57 Toots M, Yoon JJ, Hart M, Natchus MG, Painter GR, Plemper RK. Quantitative efficacy paradigms of the influenza clinical drug candidate EIDD-2801 in the ferret model. *Transl Res.* 2020;218: 16–28.

Part V

Tropical Disease Viruses

14

Development of a Novel Class of Highly Potent Dengue Virus Inhibitors: From Hit to the Clinical Candidate Mosnodenvir

Suzanne J. F. Kaptein[1], Bart Kesteleyn[2], Olivia Goethals[3], Oliver Ackaert[4], Dorothée Bardiot[5], Dominik Kiemel[6], Guillermo Herrera-Taracena[7], Ralf Bartenschlager[6,8], Marnix Van Loock[3], Tim H. M. Jonckers[2], Arnaud Marchand[5], Patrick Chaltin[5,9], Johan Neyts[1]

[1] *KU Leuven Department of Microbiology, Immunology and Transplantation, Rega Institute for Medical Research, Virology, Antiviral Drug & Vaccine Research Group, Leuven, Belgium*
[2] *Janssen Research & Development, Janssen Pharmaceutica NV, A Johnson & Johnson Company, Beerse, Belgium*
[3] *Communicable Diseases, Janssen Pharmaceutica NV, A Johnson & Johnson Company, Beerse, Belgium*
[4] *Janssen Clinical Pharmacology and Pharmacometrics, Janssen Pharmaceutica NV, A Johnson & Johnson Company, Beerse, Belgium*
[5] *Cistim Leuven vzw, Bioincubator 2, Leuven, Belgium*
[6] *Heidelberg University, Medical Faculty Heidelberg, Department of Infectious Diseases, Molecular Virology, Center for Integrative Infectious Disease Research, Heidelberg, Germany*
[7] *Communicable Diseases, Janssen Research and Development, LLC, A Johnson & Johnson Company, Horsham, PA, USA*
[8] *German Centre for Infection Research, Heidelberg Partner Site, Heidelberg, Germany*
[9] *Centre for Drug Design and Discovery (CD3), KU Leuven, Bioincubator 2, Leuven, Belgium*

14.1 Introduction

Dengue viruses (DENV) pose a serious global health threat, which infect nearly 400 million people annually, with approximately 96 million progressing to dengue disease that can be characterized by high fever, headache, retro-orbital pain, myalgia, and arthralgia, among others [1, 2]. An estimated 0.5% of the people with dengue disease (i.e. ~500 000 cases) are affected by severe dengue, such as dengue hemorrhagic fever (DHF) and dengue shock syndrome (DSS). Severe dengue is characterized by plasma leakage, hemorrhage, and thrombocytopenia, which is a potentially life-threatening condition as evidenced by the ~12 500 deaths attributed to dengue each year [2]. The actual burden may be underestimated due to misclassifications, often mistaken for febrile illnesses caused by other pathogens, such as malaria or influenza [3, 4]. The prevalence of dengue has surged, increasing approximately

Trends in Antiviral Drug Development, First Edition. Edited by Michael J. Sofia and Zhengqiang Wang.

30-fold in the last 50 years and in particular in the last years, as evidenced by the current dengue outbreak in Brazil [5], making it endemic in ~130 countries in (sub-) tropical regions. This upsurge is attributed to the expansion of the mosquito vector territories (with *Aedes aegypti* and *Aedes albopictus* as the main dengue vectors), urbanization, globalization, and climate change [6–8]. Currently, an estimated 3.9 billion people live in areas at risk of infection, with projections indicating an increase to 6.1 billion people by 2080 [8]. Dengue is therefore considered as one of the top 10 global health threats by the World Health Organization [9].

DENV comprises four serotypes (DENV-1 to DENV-4), which are further classified into genotypes (or subtypes). The four serotypes are increasingly co-circulating in dengue-endemic regions, leading to hyperendemicity. This favors sequential infections by different DENV serotypes, increasing the risk of developing severe dengue *via* a process called antibody-dependent enhancement of infection (ADE) [10, 11]. ADE occurs when preexisting antibodies from a primary DENV infection bind to infecting DENV particles during a subsequent infection with a different serotype. These cross-reactive, heterotypic antibodies cannot or only partially neutralize the virus. Instead, the infectious antibody–virus complexes attach to Fcγ-receptor-bearing cells, ultimately resulting in an increased number of infected cells, more virus produced per cell, and a faster dissemination throughout the body [12]. This may in turn trigger the development of more severe dengue disease [13, 14].

Clearly, dengue remains a pressing global health challenge, requiring comprehensive countermeasures encompassing vector control, vaccines, and antiviral drugs. Vector control strategies, such as insecticides, repellents, bed nets, and mosquito traps, have proven insufficient to halt the global spread of dengue. The use of more innovative approaches, such as the release of *Aedes* mosquitoes transinfected with certain *Wolbachia* strains that were found to interfere with and thereby impair DENV replication in these mosquitoes, is being explored to limit virus transmission [15]. Although two tetravalent, live-attenuated dengue vaccines became available in recent years, they both have challenges [reviewed in 16]. CYD-TDV (Dengvaxia, Sanofi Pasteur) has an overall efficacy of ~60% against all dengue serotypes, albeit not equally balanced over all four serotypes, with a markedly lower efficacy against DENV-2 and DENV-1. In addition, the overall efficacy in naïve recipients was considerably less than in seropositive recipients, with more cases of hospitalized dengue in the first group. The vaccine has therefore been licensed only for use in individuals aged 9–16 years who live in endemic countries and have a previously confirmed DENV infection. The recently approved TAK-003 (Qdenga, Takeda) is available in a number of countries (including Indonesia, Argentina, Brazil, the United Kingdom, and countries within the European Union), and its use is recommended in children aged 6–16 years. TAK-003 was also found to mount an imbalanced immune response against the four serotypes, with the lowest efficacy against DENV-3 (the efficacy against DENV-4 remains inconclusive due to a low number of DENV-4 cases in the trial). Moreover, vaccine efficacy waned between 1–1.5 and 3 years after vaccination, especially against DENV-3 in seronegative participants at baseline [17]. Despite these limitations, the dengue vaccines can contribute to controlling dengue outbreaks, provided they are applied properly.

Additional interventions, such as treatment and prophylaxis with potent direct-acting antiviral (DAA) drugs, are thus urgently needed. DAAs can be employed therapeutically to lower viremia levels in DENV-infected individuals and, thus, to halt the development of severe disease. Alternatively, antiviral drugs can also be taken prophylactically. Such drugs could, for example, be taken by household members of DENV-positive index cases, who are at higher risk of infection, to reduce the risk of infection. Moreover, antiviral drugs can also be used preemptively by people traveling to or working in dengue-endemic regions. This approach would be comparable to malaria chemoprophylaxis. This may not only largely reduce the risk of becoming infected, but it may also prevent the autochthonous spread of DENV to receptive regions. Hitherto, no specific antiviral drugs are available for the prevention or treatment of DENV infections. While several DENV inhibitors have been studied, none have progressed to (advanced) clinical trials because of various challenges, including limited efficacy, solubility issues, and pharmacokinetic (PK) concerns, leading to discontinuations of various drug discovery programs [reviewed in 18].

In an effort to address the lack of dengue antivirals, the Laboratory of Virology and Chemotherapy at the KU Leuven Rega Institute and the KU Leuven Centre for Drug Design and Discovery (CD3) joined their forces. The Wellcome Trust endorsed this initiative through its Seeding Drug Discovery Strategic Awards (grants 089328/Z/09 and 106327/Z/14). The collaboration led to the identification of novel and potent antiviral molecules against DENV [19]. In 2013, Janssen Pharmaceuticals Inc. joined this initiative [20], which ultimately led to the development of a first-in-class drug candidate for the prevention and treatment of DENV infections [21]. This chapter describes the trajectory of the development of the DENV inhibitor **Mosnodenvir (JNJ-1802)**, currently in Phase 2 clinical trials, under this collaborative initiative. It summarizes the work (medicinal chemistry and biological characterization) published in recent years [19, 21–26; also see Table 14.1].

14.2 Discovery and Optimization of New Dengue Drugs for Prophylaxis

14.2.1 Identification of the Hit CIM020928

The endeavor started in 2009, when the KU Leuven launched a drug discovery initiative entitled the "*Discovery of new drugs for the prophylaxis and treatment of dengue virus infections in humans*," leveraging the expertise of the KU Leuven Laboratory of Virology and Chemotherapy and that of the translational drug discovery unit of CD3. To this end, a medium-throughput phenotypic antiviral screening campaign was conducted, employing the CD3 small molecule diversity library. The phenotypic-based antiviral assay was relatively straightforward: African green monkey kidney (Vero) cells were infected with the DENV-2 RL strain (GenBank MW741553), either in the presence or absence (i.e. virus controls) of a serial dilution of the antiviral compounds. After 1 week of incubation, the (potential) inhibitory effect of the compounds on virus-induced cytopathic effect (CPE) was assessed by microscopic evaluation of the cells, which were fixed and then stained using a

Table 14.1 Compounds discussed in this chapter with references to the respective publications.

Compound number/name	Publication
1	[19]
12	[19, 24]
12a	[19]
100	[19]
100a	[19]
101	[19]
7a	[22]
9a	[22]
23a	[22]
24a	[22]
28a	[22]
13a	[23]
Mosnodenvir (JNJ-1802; 24a)	[21, 23, 26]
JNJ-A07	[24, 25]

methylene blue solution [19]. Compounds capable of inhibiting virus-induced CPE by at least 70% at nontoxic concentrations were considered potential hits, which were retested at least three times in the CPE-based assay. Compounds of which the antiviral activity was confirmed in the repeat assays (also named "confirmation assays") were next tested in virus yield reduction assays, in which the effect on viral RNA yield was assessed by quantitative RT-PCR (RT-qPCR) [19]. Compounds that lowered viral RNA levels by at least 2 $\log_{10}$ at nontoxic concentrations as compared to the virus controls (i.e. viral RNA load in the supernatant of untreated, infected cells) were considered genuine hits. Following that detailed characterization, the most interesting hits were taken into a hit explosion stage whereby commercial analogs were selected and tested to delineate potential early structure-activity relationship (SAR). During that phase, compounds were only evaluated in the virus-yield reduction assays (RT-qPCR readout); initially only against DENV-2, but later also against the other three DENV serotypes (for the most promising candidates, as based on the antiviral activity against DENV-2). More extensive biological characterization, including time-of-drug addition, mode of action, and *in vitro* resistance selection experiments, was only performed for the best (preclinical and clinical) candidates at a later stage of lead optimization.

After this hit explosion phase, our attention was dedicated to 2-((3,4-dimethoxyphenyl)amino)-1-(1*H*-indol-3-yl)-2-phenylethan-1-one, compound **1** (Figure 14.1) in Bardiot et al. [19]. Compound **1**, also denoted CIM020928, exhibited potent *in vitro* antiviral activity against DENV-2 (EC_{50} = 0.078 μM; Figure 14.2a). The compound also had some adverse effect on the host cell, albeit at significantly higher

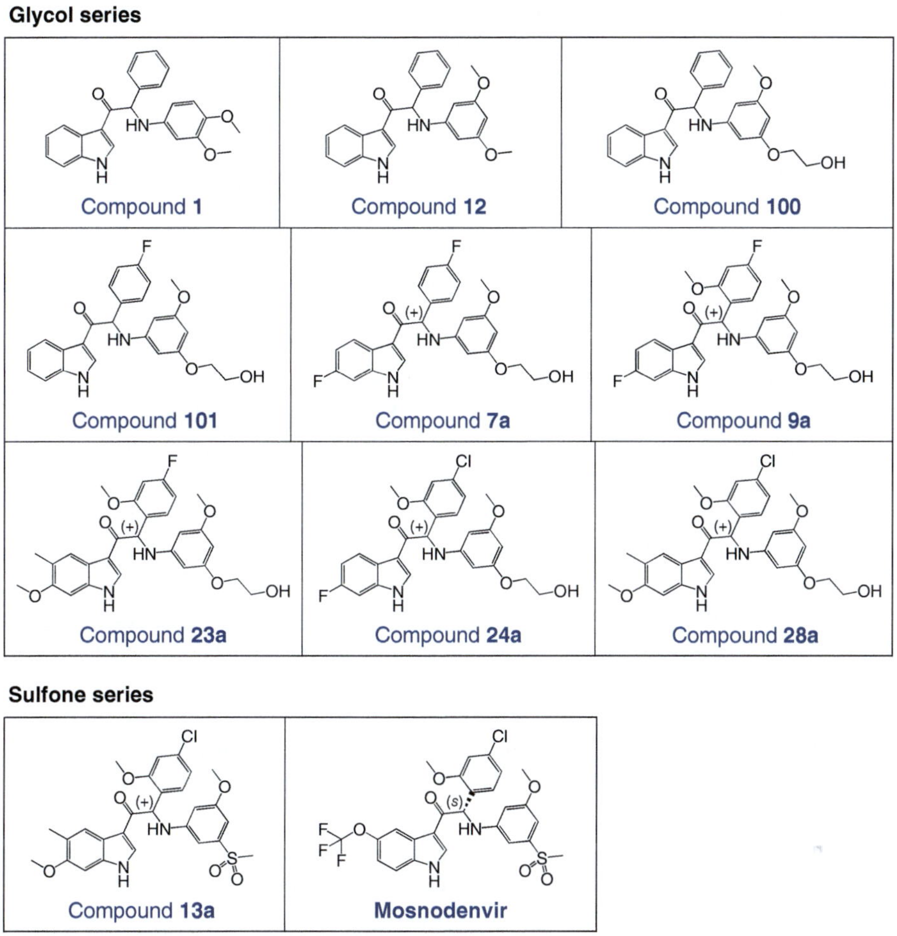

Figure 14.1 Structure of the compounds from the glycol (top panels) and sulfone (lower panels) series discussed in this chapter.

concentrations (CC_{50} = 29 μM), demonstrating that the antiviral activity is highly selective (SI = 372). In addition, the compound significantly (i.e. 3.8 $\log_{10}$-fold) inhibited viral replication in virus yield reduction assays. By contrast, the hit only displayed modest (DENV-1 and DENV-4) to weak (DENV-3) antiviral activity against the other three DENV serotypes (EC_{50} values in the μM range; Figure 14.2a). Altogether, the promising antiviral profile against DENV-2 opened avenues for further optimization of the antiviral potency, selectivity, and physicochemical properties of the hit.

14.2.2 Early SAR Exploration

The SAR exploration began by conceptually dividing compound **1** into three structural subunits: the aniline, the phenyl head part, and the indole core. Modifications to the aniline, phenyl head part, and indole core were then systematically investigated [19]. Monosubstituted aniline derivatives in which the methoxy in the *meta*-position was replaced by a trifluoromethoxy, cyano, fluorine, or methyl sulfone displayed decreased antiviral activity, demonstrating the significance of the methoxy group in the *meta*-position for optimal antiviral activity. Modifying the 3,4-dimethoxyaniline (compound **1**) to a 3,5-dimethoxyaniline moiety, as seen in compound **12** (Figure 14.1), resulted in a 10-fold increase in antiviral potency against DENV-2 (Figure 14.2a). Combining the *meta*-methoxy with other substituents in the 5-position on the aniline (such as an ethoxy or a fluorine) was well tolerated, highlighting the crucial role of at least one methoxy in the *meta*-position for obtaining double-digit nanomolar potency. By contrast, various replacements of the aniline moiety, including heteroaromatic amines and a cyclopentylamine, were associated with a decrease in antiviral activity.

Modifications to the top phenyl moiety showed that placing a methyl, fluorine, or cyano in *para*-position increased (~threefold) antiviral activity. At this stage, *ortho*- and *meta*-substitutions were also explored, but only with a methyl substituent. *Ortho*- and *meta*-substitution of the top phenyl moiety by a methyl substituent resulted, however, in respectively a 4- and 12-fold drop in antiviral activity compared to that of the *para*-methyl derivative, demonstrating that the *para*-substitution of the methyl was preferred. Replacing the phenyl ring with six-membered, five-membered, or bicyclic heteroaromatic rings revealed that the best improvement in antiviral potency was obtained with the bicyclic benzoxazole-2-yl (sevenfold) and pyrazolopyridine-2-yl (threefold) derivatives, both displaying low-nanomolar potencies. The 3- and 4-pyridyl (six-membered heteroaromatic ring) derivatives were slightly less active (~twofold) than the phenyl derivative. Other modifications to the phenyl head part that were investigated were less tolerated.

Systematic alterations of the indole core demonstrated that the introduction of halogens, methoxy, or methyl groups at different positions (i.e. positions 4, 5, 6, and 7) was allowed in most cases, slightly improving the antiviral activity (~twofold). At positions 4 and 6, the addition of a chlorine atom was preferred over a fluorine, enhancing antiviral activity threefold to fivefold. Combining the best-performing aniline (i.e. 3,5-dimethoxyaniline) with modifications to the phenyl head or with a substituted indole yielded compounds with (≥fivefold) improved potency [19]. After this first round of SAR optimizations, attention turned to the chiral center since it may impact antiviral activity. To this end, racemate **12**, which was ~10-fold more potent than the hit (compound 1) (Figure 14.2a), was selected for chiral separation. When assessing the activity against DENV-2, remarkable potency differences were observed between the two enantiomers. The (+)-rotating enantiomer **12a** was found to be the active enantiomer, while no significant differences in cytotoxicity were noted. This finding clearly underscores the importance of stereochemistry for that compound series. Interestingly, the potent (i.e. single-digit nanomolar) antiviral activity of compound **12a** against DENV-2 was mirrored by a markedly enhanced low nanomolar potency against DENV-1 and DENV-3 as compared to compound **1** (Figure 14.2a). The antiviral activity against DENV-4, however, did not improve substantially as it remained in the micromolar range. This trend was confirmed for a small selection of compounds exhibiting an improved anti-DENV-2 profile. Overall, this marked a significant advancement in pan-serotype coverage.

14.2.3 Early Evaluation of *In Vitro* ADME-Tox and *In Vivo* Pharmacokinetic Properties

To assess the suitability of compounds for *in vivo* testing, the *in vitro* Absorption, Distribution, Metabolism, Excretion, and Toxicity (ADME-Tox) profile of a small selection of derivatives was first determined. The selection was based on the (improved) pan-serotype antiviral activity and chemical diversity. Compound **12** and its enantiomers emerged as potentially interesting, displaying a comparable ADME-Tox profile: high metabolic stability in human liver microsomes (HLMs) and low micromolar inhibition of certain cytochrome P450 enzymes, however, with low to moderate solubility and permeability. Nevertheless, the (+)-enantiomer **12a** was favored as it exhibited slightly better metabolic stability in rat liver microsomes (RLMs), albeit still moderate. The other compounds tested were not further considered because of (among others) poor metabolic stability (especially in RLMs) and/or pronounced hERG inhibition at low concentrations, potentially signaling a cardiotoxicity risk.

Compound **12a** was selected for PK assessment in rats to benchmark this novel chemotype and to learn about areas for improvement in the next round of optimization. Surprisingly, results of the *in vivo* PK experiments did not match well with those obtained in the *in vitro* ADME experiments, particularly with respect to the clearance after intravenous (IV) administration. Despite a moderate turnover in rat liver microsomes, compound **12a** was rapidly cleared after IV administration. Additional experiments using rat hepatocytes were therefore conducted, demonstrating that the metabolic stability of the compound was low in these cells, potentially explaining the high clearance seen in rats. Oral dosing of compound **12a** led to a very low maximum plasma concentration and systemic exposure, resulting in low oral bioavailability, which was hypothesized to be due to poor absorption of the compound, thus matching with the *in vitro* ADME result (i.e. poor solubility and moderate permeability). This prompted a strategic shift toward further optimization of the compound series, particularly to improve the oral bioavailability.

14.2.4 Improving Oral Bioavailability

The chemical diversity explored so far, with compound **12a** as the chemotype representative, suffered from a low aqueous solubility, which could be one of the reasons for the low absorption and potentially the low oral bioavailability in rats. The next optimization therefore aimed at increasing the aqueous solubility of the series. Because adjustments to the indole core were generally well accepted, as described earlier, this moiety was selected for introducing polar alkyl chains that contained hydroxy, amino, or sulfonyl groups. Of all tried substituents on the indole, none of them led to a significant increase in solubility and were unfortunately accompanied by a slight decrease in antiviral activity and metabolic stability. Changes to the indole were thus not further pursued. Instead, focus moved to the aniline moiety, which also appeared to accept certain modifications. Introducing an elongated polar chain to the aniline also containing a *meta*-methoxy (essential for potent antiviral activity) resulted in derivatives that generally exhibited higher aqueous solubility.

These derivatives were also less cytotoxic to the host cell while retaining potent antiviral activity against DENV-2 and maintaining high metabolic stability in HLMs. The metabolic stability in RLMs was, however, variable (low to moderate) for the newly synthesized derivatives. The best performing substitution on the aniline was a glycol, which not only gave the best increase in aqueous solubility (sevenfold) but also resulted in an eightfold drop in the antiviral activity (compared to that of compound **12**), as seen for compound **100** (Figures 14.1 and 14.2a). This could, in part, be countered by adding a fluorine in *para*-position to the phenyl head part, as seen for compound **101** (Figures 14.1 and 14.2a).

Because enantiomers were earlier shown to differ in antiviral potency and ADME properties, racemate **100** was separated into its enantiomers by chiral chromatography. The (+)-enantiomer **100a** was indeed more potent than the racemate (Figure 14.2a). More importantly, it displayed a better metabolic stability (both in HLMs and RLMs) and a significant improvement in aqueous solubility (eightfold) than compound **12a**. The improvement in aqueous solubility and metabolic stability translated in a lower clearance following IV dosing, and higher maximum plasma concentrations, systemic exposure and bioavailability after oral dosing, as assessed for compound **100a**. These results demonstrate that for achieving adequate exposure levels *in vivo*, an acceptable aqueous solubility and high metabolic stability are paramount for this compound series. Notwithstanding the improvements in ADME and PK properties, compound **100a** failed to show efficacy in a DENV-2 infection mouse model, necessitating further optimization.

14.2.5 Identification of Preclinical Candidates

Continuing building on compound **100a** and compound **101**, further optimization efforts aimed to enhance pan-serotype coverage (mainly improving activity against DENV-4; also see Figure 14.2a), ADME and PK properties (systemic exposure and oral bioavailability). In addition to human metabolic stability, mouse, rat, and dog metabolic stabilities were also monitored for the most interesting compounds in order to warrant good exposures for pharmacology and toxicity assessments *in vivo*. Considering its better overall profile, compound **101** was selected as the new starting point for the subsequent round of SAR exploration [22]. To address the metabolic hotspot on the indole core, fluorines were introduced at various positions. When introduced at position 6 of the indole core moiety (compound **7a**; Figure 14.1), a better metabolic stability, both in RLMs and MLMs, was observed than when added on position 5; good metabolic stability in HLMs was maintained in both cases. Moreover, the addition of fluorines on the indole was accompanied by a slight boost in antiviral activity (as shown for compound **7a**; Figure 14.2a), as was described earlier.

To address the possibility of racemization of the chiral center, different substitutions at the *ortho*-position of the phenylglycine motif were evaluated. The *ortho*-position was selected because substituents at this position could block racemization of the chiral center by increasing steric hindrance of the intermediate enol form. An *ortho*-methoxy substituent was highly effective in stabilizing the chiral center of the compound and also outperformed the other studied substituents in terms of increasing the antiviral activity. Although the resulting compound **9a** (Figure 14.1) exhibited chiral stability, improved antiviral potency (Figure 14.2a), and stability in HLMs (over compound **100a**), the compound still presented suboptimal ADME and PK properties (such as moderate metabolic stability in RLMs and MLMs, lower exposure, and oral bioavailability), requiring additional optimization in light of future efficacy testing in the DENV-2 infection mouse model.

Starting from compound **9a**, various substitutions on different positions (i.e. 5, 6, or 7) of the indole were evaluated, including halogens (chlorines and fluorines) as well as methyl and methoxy groups. The presence of a substituent on position 6 of the indole was shown to be important for maintaining metabolic stability in HLMs, with three derivatives consistently exhibiting good metabolic stability across species [22]. Of these, compound **23a** (Figure 14.1) was selected for additional modifications, together with compound **9a**, which had a slightly better solubility than compound **23a**. In an attempt to obtain low nanomolar potencies against all four DENV serotypes, further medicinal chemistry optimization was performed on the phenylglycine part. The *ortho*-methoxy was maintained as it was essential for chiral stability. Additionally, *para*-substitutions to the phenylglycine moiety were implemented as these were generally well tolerated, as described earlier. Exchanging the *para*-fluorine for a *para*-chlorine in compounds **9a** and **23a** yielded the preclinical candidates **24a** and **28a**, respectively (Figure 14.1), which exhibited increased pan-serotype antiviral activities (Figure 14.2a). The picomolar anti-DENV-2 activity together with the favorable ADME and PK properties (both in rat and mice) warranted efficacy testing against DENV-2 in the AG129 mouse model for both compounds. Compounds **24a** and **28a** were extremely effective in inhibiting DENV-2 replication in mice treated twice-daily (BID) by oral gavage, as determined on day 3 post-infection (pi). At the highest dose (60 mg/kg/day, BID), viral RNA levels in the plasma were undetectable in 62.5% of the infected mice treated with compound **28a** [22].

14.2.6 Identification of the Clinical Candidate

In parallel to the glycol series, a structurally related sulfone series, in which the glycol side chain was replaced by a methylsulfone, was developed [23]. Interestingly, derivatives containing a *meta*-sulfone substituent replacing the glycol moiety on the aniline were generally more potent across the DENV serotypes. This was also observed for the sulfone counterpart of the preclinical candidate **28a** of the glycol series. The sulfone derivative **13a** (Figure 14.1) was approximately twofold more potent against all four DENV serotypes than its glycol equivalent (Figure 14.2a). Although the metabolic stability in HLMs and human hepatocytes was comparable for both derivatives, the *in vivo* clearance and oral bioavailability in rats were higher for compound **13a**, while the half-life and systemic exposure were higher for compound **28a**. The antivirally more potent sulfone derivative **13a** was selected for further optimization, particularly aiming at increasing the aqueous solubility. Various solubilizing substitutions to the phenylglycine head part proved to be unsuccessful as these were not accompanied by an increase in solubility or even had a dramatic impact on antiviral activity, metabolic stability, or oral bioavailability.

Focus then returned to the indole moiety. Low-nanomolar potencies against DENV-4 and DENV-3, respectively, were obtained when introducing fluorinated substituents on position 5 of the indole (thus replacing the methyl at this position in compound **13a**). In terms of antiviral potency, derivatives bearing a trifluoromethoxy group slightly outperformed the derivative that contains a trifluoromethyl group on this position. The presence of a methoxy or methyl group on positions 6 and 7, respectively, did not substantially further increase antiviral activity. The sulfone derivative **Mosnodenvir (JNJ-1802**, also named compound **24a** in Kesteleyn et al. [23]; Figure 14.1), having a trifluoro-methoxy group on position 5 with no other substituents on positions 6 and 7 and exhibiting nano- to picomolar pan-serotype antiviral activity (Figure 14.2a), was selected for further PK assessment in various animal species (mouse, rat, dog, and monkey). Intravenous administration of **Mosnodenvir** resulted in a low clearance, a moderate to high distribution volume, and long half-times (6–12 h) in the studied species. Oral dosing led to good oral bioavailability and high exposures in mouse, rat, and monkey. The good PK properties of **Mosnodenvir** together with the absence of any safety concerns, as demonstrated in various *in vitro* assays (hERG, Ames II, CYP450), paved the way for *in vivo* efficacy testing in mice and nonhuman primates (NHPs), which will be described more in detail in the next section. Altogether, these results underscore the importance of optimizing a compound series through a thorough medicinal chemistry approach to achieve a desirable PK profile in animals.

14.3 Biological Characterization

Mosnodenvir is extremely potent against the four DENV serotypes. Except for one strain belonging to DENV-4 genotype 3 (EC_{50} = 45 nM), **Mosnodenvir** exhibits double-digit picomolar to single-digit nanomolar potencies (EC_{50} < 0.04 – 1.8 nM) against the other tested strains that represent the different genotypes of DENV1–4 [21]. The antiviral activity against the other orthoflaviviruses, such as Zika virus, West Nile virus, and Japanese encephalitis virus, is about 4000–20 000 times lower than that against DENV1-4, demonstrating high specificity for DENV. Similar high potencies were obtained using various other cell types, proving that the antiviral activity is not cell type-dependent.

Experiments aimed at eliciting *in vitro* drug resistance resulted in virus strains having multiple mutations in NS4B that were not present in the wild-type virus cultured in parallel without the compound [21]. Mutations that were present in nearly all (i.e. ≥99% of the) quasi-species at the endpoint (week 42) were S85L, V91A, and T108I. Mutations V91A and T108I were also found in almost all quasi-species at endpoint (week 50) following a second, independent exercise. In addition, these quasi-species also carried the L94F mutation. The V91A, L94F, and T108I mutations were also present in quasi-species that were obtained following *in vitro* resistance selection using the structurally related **JNJ-A07** compound or the previously described early analog compound **12** [24]. Selection experiments with **JNJ-A07** also identified the T216N mutation in NS4B in one of the two (independently) obtained resistant viruses. The presence of multiple mutations in NS4B (which had appeared sequentially, with the first mutation that remained in ≥99% of the viral population emerging at ~20 weeks of virus passaging [21]) points to a high genetic barrier to *in vitro* resistance. Antiviral assessment using single NS4B mutants (i.e. subgenomic DENV, each carrying a different mutation in NS4B) demonstrated that those containing the V91A or L94F mutation became markedly less sensitive to **Mosnodenvir**, resulting in a more than 180- and 500-fold increase in the EC_{50} value, respectively [21]. Of all the drug-associated mutations identified in NS4B, only V91A and T108I are also found in DENV clinical strains present in the Virus Pathogen Database (ViPR), albeit in a low percentage (0.3–0.5%), and none of the clinical strains contains a combination of V91A and T108I.

Intriguingly, the replication competence of the resistant virus strains was heavily impaired when cultured on C6/36 mosquito cells, as was shown for the **JNJ-A07**-resistant virus strains selected during the *in vitro* resistance experiments (i.e. DENV-2 harboring the L94F, T108I, and T216N mutations; and DENV-2 harboring the V91A, L94F, and T108I mutations) [24]. By contrast, no differences in replication potential between the drug-resistant and wild-type virus were noted in Vero cells. The impaired replication fitness in mosquito cells may indicate that the propagation of viruses resistant to **JNJ-A07** (and potentially **Mosnodenvir**) may also be compromised in *Aedes* mosquitoes, which would halt any further transmission of these resistant virus strains by these mosquito vectors.

Mechanistic experiments that employed a photoaffinity labeling compound that is structurally related to **JNJ-A07** were able to show direct binding of the compound to mature NS4B as well as its cleavage intermediate NS4A-2K-NS4B, clearly pointing to an interaction between the compound and the mature/precursor form of NS4B [25]. Additional pulse-chase and pull-down experiments assessed the impact of **JNJ-A07** on the cleavage kinetics of the NS4A-2K-NS4B precursor and on the interaction between the NS2B/NS3 protease/helicase complex and mature NS4B as well as NS4A-2K-NS4B, which is instrumental for DENV replication [27, 28]. Although no evidence was found for interference with the cleavage kinetics of NS4A-2K-NS4B, **JNJ-A07** profoundly prevented the complex formation between NS2B/NS3 and the NS4A-2K-NS4B precursor (rather than that between NS2B/NS3 and mature NS4B). In contrast, already formed complexes were not substantially affected. **JNJ-A07** was additionally demonstrated to negatively impact the NS4B-dependent formation of vesicle packets (VPs), organelles that are suspected to accommodate DENV replication [29]. In cells transfected with wild-type DENV polyprotein (containing NS1–NS5), the formation of the VPs was almost completely perturbed by **JNJ-A07**, whereas VP formation was virtually unaffected when the polyprotein contained mutated NS4B carrying one of the drug-resistant mutations, at least when using **JNJ-A07** at a concentration of ~50× the EC_{50} value (i.e. 35 nM) as determined in antiviral assays. Intriguingly, by deploying NS4B variants that harbor mutations proven to abolish viral replication by hindering the interaction between NS2B/NS3 and NS4A-2K-NS4B (i.e. Q134A and M142A) [27], a functional relationship was found between this interaction and the formation of vesicle packets. Taken together, the mechanistic experiments revealed that the formation of VPs relies on the binding of NS4A-2K-NS4B to NS2B/NS3, which is blocked by **JNJ-A07**.

14.4 Efficacy of Mosnodenvir in Preclinical DENV Infection Models

Analogous to the strong *in vitro* antiviral potency, **Mosnodenvir** was also highly effective in a DENV-2 infection model using AG129 mice, which are interferon-α/β and -γ receptor deficient [21]. In this model, the mice were orally treated with vehicle or different doses of **Mosnodenvir** (60, 20, 6, 2, 0.6, or 0.2 mg/kg/day, twice-daily dosing), starting 1 h before infection with a high inoculum. Peak viral RNA levels were quantified on day 3 pi. Compared to the vehicle-treated mice, significantly lower viral RNA levels in the plasma were found in mice treated with the compound, except for mice that had received the lowest dose. Most mice treated with either of the two highest doses remained negative for viral RNA (only shown for the 60 mg/kg dose in Figure 14.2b). In mice treated with 6, 2, or 0.6 mg/kg/day, viremia was on average 2.5 $\log_{10}$ lower than in the control mice (Figure 14.2b; data for 2 mg/kg are not depicted). **Mosnodenvir** remained effective when administered once daily, at least at 30 and 3 mg/kg/day (4.2 $\log_{10}$ and 3.4 $\log_{10}$ lower viral load on day 3 pi, respectively). Moreover, **Mosnodenvir** also manifested strong antiviral activity against the other three DENV serotypes (DENV-1, DENV-3, and DENV-4) in the AG129 mice (Figure 14.2b) [21]. Mice infected with any of the three DENV serotypes and treated with 60 or 6 mg/kg/day **Mosnodenvir** (twice-daily dosing) had significantly lower viral RNA levels on day 3 pi compared to the vehicle-treated mice. **Mosnodenvir** (60 or 6 mg/kg/day, twice-daily dosing) also protected all the mice infected with any of the three DENV serotypes against virus-induced disease until the end of the experiment (day 6 pi). By contrast, none (for DENV-3 and -4) or less than half (40% for DENV-1) of the mice treated with vehicle survived the lethal challenge. Comparably high protection levels (80–100% at end point on day 25 pi) were found for DENV-2-infected mice treated with **Mosnodenvir** (10–0.3 mg/kg/day, twice-daily dosing), where none of the animals of the vehicle-treated group survived to day 21. More importantly, the compound also significantly inhibited viral replication in delayed-treatment studies. The viremic period in mice lasted 3 to 2 days shorter when **Mosnodenvir** treatment (60 mg/kg/day, twice-daily dosing) was first commenced either on day 4 or 5 pi, respectively, compared to that in the vehicle-treated mice.

Mosnodenvir's favorable PK profile and the absence of adverse effects in multiple animal species, including cynomolgus monkeys, warranted additional efficacy studies in NHPs. To cover efficacy testing against more than one serotype, NHPs (i.e. rhesus macaques) were infected with either DENV-1 or DENV-2 [21]. Animals received once-daily, oral treatments with vehicle or **Mosnodenvir** at a dose of 6 mg/kg (DENV-1 study), or with 3, 0.18, or 0.01 mg/kg (DENV-2 study). Treatment started 3 days or 1 day before infection with DENV-1 or DENV-2, respectively, and continued until day 10 pi. In DENV-1-infected NHPs, the compound fully suppressed viral replication, as evidenced by undetectable

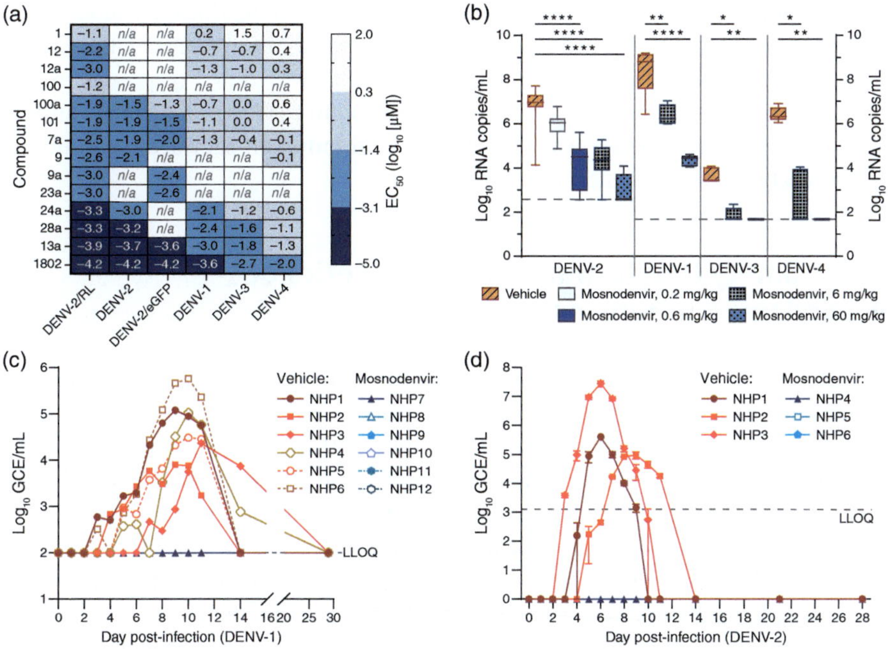

Figure 14.2 *In vitro* antiviral activity and *in vivo* efficacy in preclinical models. (a) Heatmap showing the antiviral activity of the tested compounds against four serotypes of DENV, i.e. DENV-1 strain TC974#666, DENV-3 strain H87, and DENV-4 strain H241. Antiviral activity against DENV-2 was tested against different strains: DENV-2 Rega labstrain ("DENV-2 RL"), DENV-2 strain 16681 ("DENV-2"), and DENV-2 strain 16681 expressing the enhanced green fluorescent protein ("DENV-2/eGFP"). The antiviral activity is depicted as the $\log_{10}$-transformed 50% effective concentration (EC_{50}), as determined on Vero cells. **Mosnodenvir** is depicted as "1802". (b) *In vivo* effect of **Mosnodenvir** on viremia on day 3 pi in AG129 mice infected with DENV-2 RL, DENV-1 West Pac, DENV-3 C0360/94, or DENV-4 703/4. Mice were treated prophylactically (i.e. 1 h before infection) with different doses of **Mosnodenvir**. Data depicted are from mice treated twice daily by oral gavage with vehicle or **Mosnodenvir** at doses of 60, 6, 0.6, or 0.2 mg/kg/day (for DENV-2), or 60 or 6 mg/kg/day (for DENV-1, DENV-3, and DENV-4). *$P < 0.05$; **$P < 0.01$; ****$P < 0.0001$. (c, d) *In vivo* effect of **Mosnodenvir** on viremia on day 3 pi in nonhuman primates (NHPs) infected with DENV-1/45AZ5 (c) or DENV-2 16681 (d). Treatment started 1 day (for DENV-2) or 3 days (for DENV-1) before infection and continued until day 10 pi. Data depicted are DENV genome copy equivalents (GCE) per mL plasma from NHPs treated once daily orally with vehicle or **Mosnodenvir** at a dose of 3 mg/kg/day (for DENV-2) or 6 mg/kg/day (for DENV-1). The depicted graphs in (b), (c), and (d) are (slightly) adapted from those published in [21] / Springer Nature / CC BY 4.0, which are licensed under a Creative Commons Attribution 4.0 International License (http://creativecommons.org/licenses/by/4.0/).

viral RNA levels in plasma throughout the duration of the study (endpoint at day 29 pi) (Figure 14.2c). All **Mosnodenvir**-treated animals also remained negative for anti-DENV IgG and IgM antibodies. By contrast, all vehicle-treated NHPs infected with DENV-1 experienced viremia and seroconverted. Similar virological and immunological outcomes were obtained for NHPs infected with DENV-2 in the highest dosing group (3 mg/kg **Mosnodenvir**): no detectable levels of viral RNA (Figure 14.2d), DENV NS1, IgM, and IgG in serum. One animal, though, became positive for IgM and NS1, but only at the end of the study (day

28 pi). Animals treated with the lowest dose did not differ significantly in viremia, NS1 expression, and IgM/IgG seroconversion compared to those treated with vehicle. The intermediate dose resulted in delayed and/or more blunted responses in the animals during the study period. Altogether, the potent *in vitro* pan-serotype antiviral activity of **Mosnodenvir** was reflected by a profound efficacy against multiple DENV serotypes, both in mice and NHPs.

14.5 Safety Profile and Pharmacokinetics of Mosnodenvir in Humans

The safety, tolerability, and PK of **Mosnodenvir** were assessed for the first time in humans during a double-blind, randomized, placebo-controlled clinical study [26]. In the single-dose escalation phase of the study, 39 eligible healthy participants were randomized to receive oral **Mosnodenvir** single doses (50, 150, 240, 400, and 1200 mg; n = 29) or placebo (n = 10) under fasting conditions. In the multiple-dose escalation phase, 47 eligible healthy participants were randomized to receive **Mosnodenvir** once daily (50, 150, 400, or 560 mg for 10 consecutive days or 400 mg for 31 days; n = 38) or placebo (n = 9) under fasting conditions.

Based on the results of this first-in-human study, it was concluded that **Mosnodenvir** was generally safe and well tolerated. No significant changes were observed in laboratory tests, electrocardiograms, or vital signs. No deaths, serious adverse events, or adverse events causing the discontinuation of the study drug were reported. Most adverse events were mild to moderate in severity (grade 1 or 2), with only one grade 3 adverse event of depression reported, which was not considered to be related to the study drug. In the multiple-dose part of the study, two grade 2 rash events were considered very likely related to **Mosnodenvir** by the investigator, but these resolved 4–7 weeks after treatment discontinuation. Rash is known to be one of the symptoms of dengue disease, and a rash assessment tool has been developed to further evaluate rash during clinical development in human challenge trials as well as in later phases of development.

Following single- and multiple-dose administration (10 days daily), a (close to) dose-proportional increase in exposure was observed between 50 and 150 mg **Mosnodenvir**, and a less-than-dose-proportional increase was noted following higher doses up to 1200 mg (single dose) and 560 mg (multiple dose) [26]. **Mosnodenvir** showed a biphasic decline with a long terminal half-life (6.3–9.2 days), which reduces the risk of a significant decrease in exposure in the case of a missed dose. In addition, the long terminal half-life suggests that exposure can persist long after the cessation of dosing. **Mosnodenvir** was excreted minimally in urine (<0.001% of the dose), which indicates that the predominant elimination is via metabolic clearance in the liver. Nonetheless, **Mosnodenvir** was the predominant circulating entity in human plasma (>97%).

In summary, this Phase 1 first-in-human clinical study showed that the antiviral was safe and well tolerated, with a promising PK profile, which encourages further investigation of its efficacy and safety for the prophylaxis and treatment of dengue.

14.6 Conclusions

This chapter describes the hit-to-lead optimization that resulted in the identification of extremely potent, pan-serotype DENV compounds that target the NS3–NS4B interaction. The hit-to-lead optimization process not only focused on increasing the antiviral activity against all four DENV serotypes but also on improving the drug-likeness of the compound series, resulting in compounds with a favorable PK profile, improved oral bioavailability, and high efficacy in preclinical models. It provides insight into the delicate balance between enhancing antiviral activity against all four DENV serotypes on the one hand, and adjusting the ADME and PK properties on the other. The journey from early SAR explorations to the development of the clinical candidate **Mosnodenvir** highlights the iterative nature of drug development, with challenges in potency, broad-spectrum efficacy, and oral bioavailability, which were systematically addressed.

Acknowledgments

The original work was supported by a Seeding Drug Discovery Award from the Wellcome Trust (grant numbers 089328/Z/09 and 106327/Z/14) and received funding from the Flanders Agency Innovation & Entrepreneurship (VLAIO O&O grants IWT 150863 and HBC.2019.2906). Part of the original research work was conducted using the 'Caps-It' research infrastructure (project ZW13-02) that was financially supported by the Hercules Foundation (FWO) and Rega Foundation, KU Leuven. Studies conducted at the University of Texas Medical Branch were undertaken with US Federal funds from the Division of Microbiology and Infectious Diseases, National Institute of Allergy and Infectious Diseases, National Institutes of Health, Department of Health and Human Services, under contract no. HHSN272201700040I.

References

1 Bhatt S, Gething PW, Brady OJ, Messina JP, Farlow AW, Moyes CL, et al. The global distribution and burden of dengue. *Nature*. 2013;496(7446): 504–7.
2 World Health Organization. Dengue and severe dengue. Factsheet 23 April 2024. Geneva, Switzerland: WHO;2024. https://www.who.int/news-room/fact-sheets/detail/dengue-and-severe-dengue (accessed 30 June 2024).
3 Castro MC, Wilson ME, Bloom DE. Disease and economic burdens of dengue. *Lancet Infect Dis*. 2017;17(3): e70–78.
4 Lim SP. Dengue drug discovery: progress, challenges and outlook. *Antiviral Res*. 2019;163: 156–78.
5 Vancini RL, Andrade MS, de Lira CAB. Dengue's economic impact in Brazil. *Science*. 2024;385(6706): 266.
6 Kilpatrick AM, Randolph SE. Drivers, dynamics, and control of emerging vector-borne zoonotic diseases. *Lancet*. 2012;380(9857): 1946–55.

7 Murray NE, Quam MB, Wilder-Smith A. Epidemiology of dengue: past, present and future prospects. *Clin Epidemiol.* 2013;5: 299–309.

8 Messina JP, Brady OJ, Golding, N. Kraemer MUG, Wint GRW, Ray SE, et al. The current and future global distribution and population at risk of dengue. *Nat Microbiol.* 2019;4(9): 1508–15.

9 World Health Organization. Ten threats to global health in 2019. Geneva, Switzerland: WHO;2019. https://www.who.int/news-room/spotlight/ten-threats-to-global-health-in-2019 (accessed 30 June 2024).

10 Halstead SB. Pathogenesis of dengue: dawn of a New Era. *F1000Res.* 2015;4: F1000 Faculty Rev-1353.

11 Katzelnick LC, Gresh L, Halloran ME, Mercado JC, Kuan G, Gordon A, et al. Antibody-dependent enhancement of severe dengue disease in humans. *Science.* 2017;358(6365): 929–32.

12 Narayan R, Tripathi S. Intrinsic ADE: the dark side of antibody dependent enhancement during dengue infection. *Front Cell Infect Microbiol.* 2020;10: 580096.

13 Vaughn DW, Green S, Kalayanarooj S, Innis BL, Nimmannitya S, Suntayakorn S, et al. Dengue viremia titer, antibody response pattern, and virus serotype correlate with disease severity. *J Infect Dis.* 2000;181(1): 2–9.

14 Whitehorn J, Yacoub S, Anders KL, Macareo LR, Cassetti MC, Nguyen Van VC, et al. Dengue therapeutics, chemoprophylaxis, and allied tools: state of the art and future directions. *PLoS Negl Trop Dis.* 2014;8(8): e3025.

15 Utarini A, Indriani C, Ahmad RA, Tantowijoyo W, Arguni E, Ansari MR, et al. Efficacy of wolbachia-infected mosquito deployments for the control of dengue. *N Engl J Med.* 2021;384(23): 2177–86.

16 Paz-Bailey G, Adams LE, Deen J, Anderson KB, Katzelnick LC. Dengue. *Lancet.* 2024;403(10427): 667–82.

17 Rivera L, Biswal S, Sáez-Llorens X, Reynales H, López-Medina E, Borja-Tabora C, et al. Three-year efficacy and safety of takeda's dengue vaccine candidate (TAK-003). *Clin Infect Dis.* 2022;75(1): 107–17.

18 Bifani AM, Chan KWK, Borrenberghs D, Tan MJA, Phoo WW, Watanabe S, et al. Therapeutics for flaviviral infections. *Antiviral Res.* 2023;210: 105517.

19 Bardiot D, Koukni M, Smets W, Carlens G, McNaughton M, Kaptein S, et al. Discovery of indole derivatives as novel and potent dengue virus inhibitors. *J Med Chem.* 2018;61(18): 8390–401.

20 Wellcome Trust Press release. Wellcome Trust and KU Leuven announce collaboration with Janssen for development of dengue antiviral drugs;2013. https://wellcome.org/press-release/wellcome-trust-and-ku-leuven-announce-collaboration-janssen-development-dengue (accessed 30 June 2024).

21 Goethals O, Kaptein SJF, Kesteleyn B, Bonfanti JF, Van Wesenbeeck L, Bardiot D, et al. Blocking NS3-NS4B interaction inhibits dengue virus in non-human primates. *Nature.* 2023;615(7953): 678–86.

22 Kesteleyn B, Bardiot D, Bonfanti JF, De Boeck B, Goethals O, Kaptein SJF, et al. Discovery of acyl-indole derivatives as pan-serotype dengue virus NS4B inhibitors. *J Med Chem.* 2023;66(13): 8808–21.

23 Kesteleyn B, Bonfanti JF, Bardiot D, De Boeck B, Goethals O, Kaptein SJF, et al. Discovery of JNJ-1802, a first-in-class pan-serotype dengue virus NS4B inhibitor. *J Med Chem.* 2024;67(5): 4063–82.

24 Kaptein SJF, Goethals O, Kiemel D, Marchand A, Kesteleyn B, Bonfanti JF, et al. A pan-serotype dengue virus inhibitor targeting the NS3-NS4B interaction. *Nature.* 2021;598(7881): 504–9.

25 Kiemel D, Kroell AH, Denolly S, Haselmann U, Bonfanti JF, Andres JI, et al. Pan-serotype dengue virus inhibitor JNJ-A07 targets NS4A-2K-NS4B interaction with NS2B/NS3 and blocks replication organelle formation. *Nat Commun.* 2024;15(1): 6080.

26 Ackaert O, Vanhoutte F, Verpoorten N, Buelens A, Lachau-Durand S, Lammens L, et al. Safety, tolerability, and pharmacokinetics of JNJ-1802, a pan-serotype dengue direct antiviral small molecule, in a phase 1, double-blind, randomized, dose-escalation study in healthy volunteers. *Clin Infect Dis.* 2023;77(6): 857–65.

27 Chatel-Chaix L, Fischl W, Scaturro P, Cortese M, Kallis S, Bartenschlager M, et al. A combined genetic-proteomic approach identifies residues within dengue virus NS4B critical for interaction with NS3 and viral replication. *J Virol.* 2015;89(14): 7170–86.

28 Zou J, Xie X, Wang QY, Dong H, Lee MY, Kang C, et al. Characterization of dengue virus NS4A and NS4B protein interaction. *J Virol.* 2015;89(7): 3455–70.

29 Welsch S, Miller S, Romero-Brey I, Merz A, Bleck CK, Walther P, et al. Composition and three-dimensional architecture of the dengue virus replication and assembly sites. *Cell Host Microbe.* 2009;5(4): 365–75.

Part VI

Orthopoxvirus

15

Discovery and Development of Brincidofovir

E. Randall Lanier[1], Scott Foster[2], Graciela Andrei[3]

[1] *Independent Virology Consultant*
[2] *Chimerix, Inc. 2505 Meridian Parkway, Suite 100 Durham, NC 27713 USA*
[3] *Molecular, Structural, and Translational Virology, Department of Microbiology, Immunology and Transplantation, Rega Institute for Medical Research, KU Leuven, Leuven, Belgium*

15.1 Discovery

Brincidofovir (BCV) is a lipid conjugate of cidofovir (CDV) that was designed by Karl Hostetler and colleagues to orally deliver sufficient intracellular concentrations of the active antiviral cidofovir-diphosphate (CDV-PP) for antiviral activity, while avoiding CDV's dose-limiting nephrotoxicity and requirement for intravenous (IV) dosing [1]. BCV has broad antiviral activity *in vitro* against double-stranded DNA (dsDNA) viruses, was approved by the US FDA for the treatment of smallpox in 2021, and is currently in clinical development for the treatment of adenovirus (AdV) and cytomegalovirus (CMV) [2]. From a strict *in vitro* perspective, BCV acts primarily as a more potent version of CDV. Once BCV is inside cells, CDV is liberated by phospholipase cleavage of the lipid ester and activated by two successive phosphorylation events, first to cidofovir monophosphate (CDV-P) and then to CDV-PP [3]. However, the clinical potential of BCV is much greater than that of CDV for reasons that go beyond potency differences. One simple example is the indication of oral BCV for the treatment of smallpox (caused by variola virus infection). In the event of a natural reemergence, accidental or deliberate release of variola virus, the recommended dosage of BCV for symptomatic adults is two (100 mg) oral tablets on day 1 and two more on day 8. This simple oral dosing regimen is clearly preferable to multiple in-patient administrations of IV CDV, which typically also includes prehydration and probenecid, especially in the setting of a widespread outbreak involving large numbers of patients. Less obvious examples include the increased potential for IV BCV to preempt or treat brain pathologies due to substantially higher drug concentrations in the central nervous system (CNS) following IV administration of BCV as compared to oral BCV or IV CDV. The main purpose of this review is to discuss the existing data and current development activities for BCV in the context

Trends in Antiviral Drug Development, First Edition. Edited by Michael J. Sofia and Zhengqiang Wang.

of CDV's potential, much of which is unrealized due to inherent issues with absorption, distribution, and toxicity.

CDV is an acyclic nucleoside phosphonate (ANP) that has been studied extensively since the late 1980s for a wide variety of possible antiviral and anticancer indications [4]. It was approved by the US FDA in 1996 for the treatment of CMV retinitis in patients with acquired immunodeficiency syndrome (AIDS) but is most often used off-label to treat systemic viral infections in immunocompromised individuals [5–7]. Due to the stable phosphonate bond, CDV circulates as a dianion in blood and typically penetrates cells poorly resulting in low concentrations of CDV in most anatomic compartments; however, it is rapidly taken up by dianion transporters (human organic anion transporter 1 [hOAT1]) which are selectively expressed on renal proximal tubule epithelial cells (RPTECs) of the kidney, frequently causing kidney toxicity [8]. The inability of IV CDV to achieve clinically meaningful concentrations in most target cells without inducing nephrotoxicity is a primary reason it has failed to realize much of the promise suggested by its diverse *in vitro* activities.

The Hostetler group reasoned that covalent attachment of a monoacyl lipid with the appropriate structure to CDV could yield a molecule that would be readily absorbed like natural monoacyl-phospholipids (e.g. lysophosphatidylcholine) and might require cell penetration prior to enzymatic cleavage of the lipid to release CDV, thereby greatly reducing circulating CDV in blood [1, 9]. After exploring a variety of alkyl chain lengths and linkers, BCV (hexadecyloxypropyl-CDV or phosphonic acid, [(S)-2-(4-amino-2-oxo-1(2H)-pyrimidinyl)-1-(hydroxymethyl) ethoxy]methyl] mono[3-(hexadecyloxy)propyl] ester) appeared to have the desired properties and was chosen for further development [1, 10]. Early studies confirmed that the dramatic increases (typically >100-fold) observed for antiviral activity *in vitro* with BCV versus CDV were attributable to increased intracellular concentrations of CDV-PP. Furthermore, orally delivered BCV circulated intact in mice, resulting in broader/higher distribution to all tissues examined except for the kidney, where much lower concentrations were seen versus CDV [11, 12]. These studies set the stage for over two decades of ongoing effort to develop BCV for multiple indications.

15.1.1 Key Points

- BCV is typically >100-fold more potent than CDV *in vitro*, may be dosed orally or IV, and eliminates the nephrotoxicity associated with CDV.
- Short-term oral dosing of BCV was approved by the US FDA for the treatment of smallpox on 4 June 2021.
- BCV has demonstrated antiviral activity against multiple dsDNA viruses *in vitro*, in animal models and clinical studies.
- *In vitro* and animal model studies suggest BCV has utility as a selective antiproliferative agent for cancer.
- Clearance of AdV viremia with BCV appears independent of immune reconstitution in allogeneic hematopoietic cell transplant (HCT) recipients and all species/types of AdV appear sensitive to BCV.

- BCV has a favorable resistance profile for orthopoxviruses, adenoviruses, and CMV.
- BCV is not appropriate for self-limiting or mild infections due to toxicities, which must be considered for each indication and dosing regimen.
- IV administration of BCV substantially increases exposure in the CNS and may improve the safety profile compared to IV CDV or longer term oral BCV.

15.2 Summary of Current and Potential Indications for BCV

BCV has the potential to address a surprisingly diverse set of unmet medical needs as a dsDNA virus antiviral and as an anticancer agent. Mechanistically, this is primarily attributable to the competition of CDV-PP with dCTP (deoxycytidine triphosphate) in different biochemical settings and the consequences of those events in different cellular contexts. This same property is likely responsible for its toxicity profile, which must be considered for each indication due to a relatively narrow therapeutic window. For example, oral BCV is approved for the treatment of smallpox in part because of the severity of smallpox disease and in part because the entire dosing regimen is one week [2]. As described in Section 15.5, longer term oral dosing has been associated with toxicities, especially gastrointestinal, that prompted a switch to IV dosing for current clinical programs where longer dosing is warranted. In most cases, potential indications were first described for CDV, but the majority cannot be practically or optimally addressed with CDV due to issues with distribution, cell penetration, and/or toxicity.

As an antiviral, BCV may hold the most promise to address the unmet needs of specific groups among immunocompromised patients who are at high risk for serious disease from reactivations or new infections with common dsDNA viruses. Allogeneic hematopoietic cell transplant (allo-HCT) patients are especially vulnerable shortly after transplantation (prior to immune recovery) when a surprising percentage (33–62%) have 2 or more concomitant dsDNA viral reactivations/infections [13–15]. Approximately half of the allo-HCT patients will have detectable BKV, CMV, EBV, or HHV6 in plasma during the first 6 months after transplant with AdV viremia detected in 16–23% of pediatric allo-HCT recipients, and 78% of these patients also affected by one or more of BKV, CMV, EBV, or HHV6 [16, 17]. Unsurprisingly, increased dsDNA virus infections are associated with increased mortality in this setting [13–15, 18]. The activity of BCV against all these viruses may make it especially attractive in patients who have serious disease attributable to any one of them. Improved penetration of the CNS with BCV, particularly IV BCV, also makes it appealing for indications where CNS infection is of concern [19]. Most clinical trials have evaluated BCV as an adenovirus (AdV) or CMV antiviral in immunocompromised populations, while the approved indication for treatment of smallpox was based on efficacy studies in animal models.

BCV has also generated significant interest as an anticancer agent and for several important neurodegenerative diseases. Table 15.1 lists selected targets/indications currently under study, ranked by stage of development. Key references include historical data, and proof-of-concept studies with CDV in addition to BCV studies.

Table 15.1 Current and selected potential indications for BCV.

Treatment indication	Target	Stage of development dosing route	Key references
Smallpox	Variola virus	Approved by US FDA 2021 Oral	[2]
AdV	AdV	Phase 2 Intravenous	[20–22]
BKVAN and BKV HC	BKV	Phase 2 Intravenous	[5, 23]
CMV	CMV	Phase 2 Intravenous	[24, 25]
Glioblastoma	Cancer cells	Preclinical	[26]
Head and neck cancer	Cancer cells	Preclinical	[27, 28]
Mutiple sclerosis	EBV	Preclinical	[29, 30]
Alzheimer's disease	HSV-1, VZV	Preclinical	[31, 32]
Lymphomas	Tumor cells, EBV?	Clinical	[33–35]

BKVAN, BKV-associated nephropathy; HC, hemorrhagic cystitis.

15.3 Antiviral Activity

15.3.1 Antiviral Activity Against Double-stranded DNA Viruses

While BCV has a similar spectrum of antiviral activity to CDV against dsDNA viruses, it is typically >100-fold more potent. For example, the median concentration that effectively reduced *in vitro* replication of wild-type human CMV (AD169) by 50% (EC_{50}) was 0.001 μM for BCV, 0.4 μM for CDV and 3.8 μM for GCV in a direct comparison [36]. The enhanced potency of BCV versus CDV is attributable to increased uptake into cells mediated by the lipid moiety and resultant increased intracellular concentrations of the active antiviral, CDV-PP [11]. Chemaly et al. recently performed a systematic review of antivirals with activity against dsDNA viruses known to cause significant human pathology [37]. Table 15.2 displays shows individual values for key approved antivirals reflecting the overall spectrum of activity; green indicates values that should be clinically relevant, yellow indicates values that may be achievable, and red indicates no activity or values that exceed clinically achievable concentrations [36]. Several comprehensive reviews of published values are available and in agreement with the overall picture shown below [37, 38].

There are some notable omissions from the EC_{50} values presented in Table 15.2. Papillomavirus was not included as the data available for BCV are based on cytotoxicity of HPV-transformed cells and not true measures of viral replication inhibition. Despite this gap in our knowledge, it seems clear that BCV and CDV have the potential to address papillomavirus-associated neoplasia and cancers as described in Section 15.7.1.1. For adenovirus, only one of >100 types is represented (AdV species C type 5). However, the substantial available data indicates all species/types are

Table 15.2 Overview of *in vitro* activity for selected antivirals against dsDNA viruses (EC_{50}s in μM).

Family	Virus	BCV	CDV	GCV	FOS[a]	ACV	MBV	LTMV
Herpes	CMV	0.001	0.4	3.8	50–800	5.5	0.31	0.005
	EBV	0.03	66	0.9	<500	6.2	0.63	>10
	HHV-6A	0.003	2.7	5.9	16	10	-	>10
	HHV-8	0.02	2.6	2.7	177	>100	Inactive	-
	HSV-1	0.01	3.0	0.7	92–95	3.8	Inactive	>10
	HSV-2	0.02	5.3	2.5	91–96	4.4	Inactive	>10
	VZV	0.0004	0.5	1.3	>50	3.6	Inactive	>10
Adenovirus	AdVC5	0.02	1.3	4.5–33	Inactive	>100	-	>10
Polyoma	BKV	0.13	115	-	-	-	-	-
	JCV	0.045	>0.1	-	-	-	-	-
Pox	VARV	0.1	11	-	-	-	-	-
	RPXV	0.5	39	-	-	-	-	
	ECTV	0.5	12	-	-	-	-	-

CMV, cytomegalovirus; EBV, Epstein–Barr virus; HHV-6A, human herpesvirus 6A; HHV-8, human herpesvirus 8; HSV-1, herpes simplex virus 1; HSV-2, herpes simplex virus 2; VZV, varicella zoster virus; AdVC5, adenovirus species C, type 5; BKV, BK virus; JCV, JC virus; VARV, variola virus; RPXV, rabbitpox virus; ECTV, ectromelia virus; GCV, ganciclovir; FOS, foscarnet; ACV, acyclovir; MBV, maribavir; LTMV, letermovir.

a. Foscarnet data not highlighted as it has atypically high EC50s.

sensitive to BCV with EC_{50}'s <0.03 μM; the single exception in the literature was reported for AdVA31 (EC_{50} = 0.28 μM) in 2005, but subsequent investigation showed an EC_{50} of 0.02 μM for this species/type and all patients harboring AdVA31 (N = 13) responded virologically to BCV as described in Section 15.5.2.2 [39, 40]. The most comprehensive single manuscript describing BCV *in vitro* activity against AdV determined EC_{50}'s for 17 types, selected for clinical prevalence [41]. The mean EC_{50}'s for BCV ranged from 0.001 to 0.007 μM, while those for CDV ranged from 0.3 to 1.6 μM.

Overall, BCV has demonstrated broad and potent activity against dsDNA viruses associated with human disease, prompting clinical evaluation for multiple indications as described in more detail below.

15.3.1.1 Antiviral MOA

15.3.1.1.1 Viruses with Viral DNA Polymerases

Biochemical, mechanistic, and resistance studies have been used to delineate the mechanism(s) of action for BCV and CDV against pathogenic viruses. The most studied and well-understood indications for BCV have viral DNA polymerase as the target (AdV, CMV, Pox) where CDV-PP acts as a competitive, alternative substrate inhibitor of dCTP for the virus-encoded DNA-dependent DNA polymerase.

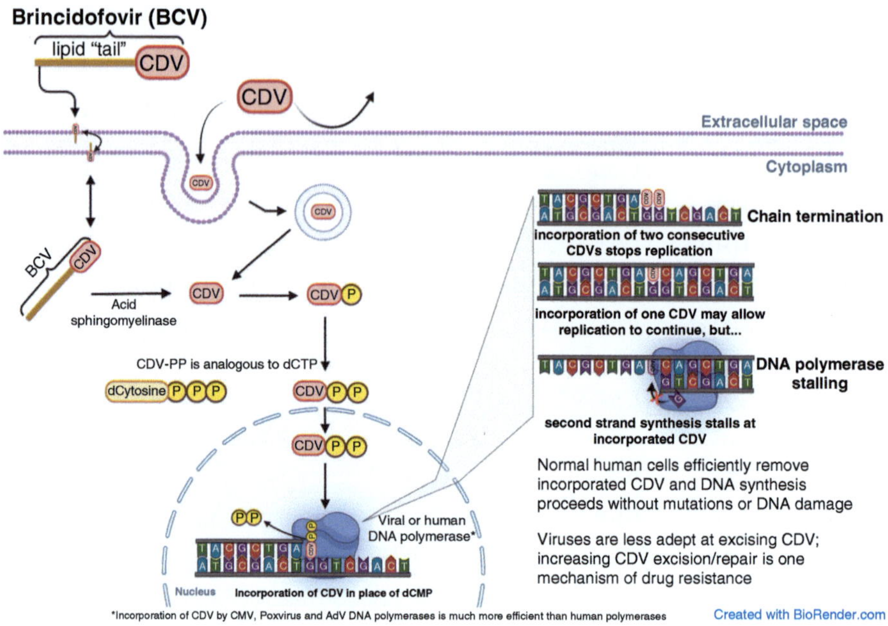

Figure 15.1 Mechanism of action for BCV against dsDNA viruses encoding viral DNA polymerases.

Incorporation of CDV into the growing viral DNA chain effectively inhibits viral DNA synthesis, which prevents the production of new infectious virions.

Biochemical studies with CMV and orthopoxvirus (vaccinia virus, VACV) DNA polymerases found that CDV-PP causes nonobligate chain termination [42, 43]. CDV was incorporated opposite dGMP as expected, and the incorporated CDV at the 3′ terminus was not efficiently excised/corrected by CMV DNA polymerase. The incorporation of one CDV molecule decreased the rate of addition for the next natural nucleotide, while the incorporation of two consecutive CDV molecules essentially terminated CMV DNA synthesis.

Incorporation of CDV in the full-length template strand effectively inhibited second-strand synthesis; both CMV and VACV DNA polymerases stalled at the site of integrated CDV [42, 43]. Therefore, CDV-PP inhibits these viral DNA polymerases at two steps: when incorporated into the nascent strand (nonobligate chain termination) and when incorporated into the template strand (polymerase stalling) as illustrated in Figure 15.1. Similar biochemical studies describe similar effects for the adenovirus DNA polymerase [44]. Resistance-associated mutations (RAMs) for BCV and CDV in these viruses (AdV, CMV, VACV) map to the respective viral DNA polymerases, providing further evidence that BCV acts on/through these enzymes [44–47].

15.3.1.1.2 Viruses Without Viral DNA Polymerases

There has been considerable interest in understanding why CDV/BCV are active against polyomaviruses (PyVs) and human papillomaviruses (HPVs), which do not encode for their own DNA polymerases, but use host cell DNA polymerases for

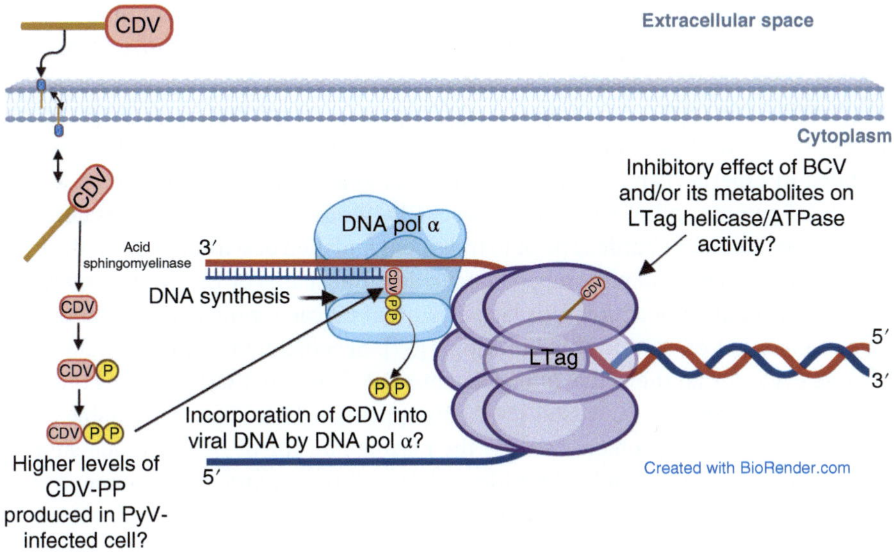

Figure 15.2 Hypothetical CDV/BCV mechanism(s) of action against PyVs.

viral DNA replication [48, 49]. Cell culture studies demonstrated that CDV does not affect early steps in BKV infection (e.g. entry, initial protein expression). However, intracellular viral DNA replication and late viral mRNA/protein expression could be reduced by ≥90% in infected RPTECs at concentrations of CDV, which inhibited cellular DNA replication by 7% in uninfected cells [50]. There is also evidence for selective cellular toxicity with infected cells being more sensitive than uninfected cells. This selectivity has been attributed to multiple factors, which are not mutually exclusive. For example, virus infection can enhance both the intracellular concentration of CDV-PP and cell proliferation [3, 51, 52]. Interactions between cellular proteins and viral proteins (e.g. PyV LTag or HPV E6/E7 with p53 and pRb) may also sensitize cells to CDV-PP [52]. Notably, mutations in the SV40 LTag have been associated with resistance to CDV [38, 53]. A schematic showing several of these possible mechanisms is shown in Figure 15.2.

15.3.2 Antiviral Activity Against RNA Viruses

15.3.2.1 Ebola

The Ebola virus (EBOV) outbreak of 2013–2015 in West Africa prompted evaluation against EBOV of virtually every antiviral in clinical development. As part of this effort, the US Centers for Disease Control and Prevention (CDC) and the NIH conducted *in vitro* activity assays for BCV using live EBOV. Initial studies in Huh7 cells showed EC_{50}s ranging from 0.31 to 3.29 μM with selectivity indices from 14 to 133. Given the urgency of the ongoing outbreak, these data were rapidly expanded to include multiple cell lines and EBOV variants [54].

Subsequently, BCV was made available under EIND (Emergency Investigational New Drug) applications for 4 patients in the United States, two of whom were nurses who provided care for one of the other two patients. The index patient in this small cluster died after receiving 2 doses of BCV administered late in the course of disease, and the other 3 patients made full recoveries. Attribution of benefit to any agent was not possible due to the simultaneous administration of multiple potentially active interventions in all cases [55, 56]. A phase 2 trial of BCV for EBOV disease was initiated in Liberia (Pan African Clinical Trials Registry 201411000939962), which enrolled 4 patients, all of whom had severe EBOV disease (hemorrhagic symptoms and/or very high viral burden) prior to treatment and died or withdrew from the trial within 1 week. The trial was stopped due to declining cases of EBOV and emerging data on mechanism of action plus lack of animal efficacy, which cast doubt on the potential benefit of BCV in this indication [57]. Specifically for the mechanism, it became evident that the active moiety against EBOV was not CDV-PP as is the case for dsDNA viruses, but BCV itself [54]. This appears to be an insurmountable clinical/pharmacokinetic problem because the plasma half-life of BCV is much shorter (hours) than the intracellular half-life of CDV-PP (days) and dose regimens estimated to provide adequate BCV exposure would greatly exceed toxicology coverage.

15.3.2.2 SARS-CoV-2

Several published reports have suggested BCV as a candidate antiviral for SARS-CoV-2 infection. One tested nucleoside triphosphate analogs in biochemical primer extension reactions using the SARS-CoV-2 RNA-dependent RNA polymerase. CDV-PP could be incorporated and showed delayed termination, i.e. RNA synthesis stopped after the addition of one base following the addition of CDV ("n+1 termination"). The authors noted this mechanism was also delineated for CDV-PP with poxvirus DNA polymerase and is similar to that of remdesivir with SARS-CoV-2 RNA polymerase [43, 58, 59]. A second manuscript reported in silico molecular docking studies of FDA-approved drugs on the SARS-CoV-2 ACE2 receptor and protease, concluding that BCV had the most potent binding interactions, but did not report any cell culture or other studies to confirm [60]. A third manuscript suggested BCV may be useful in treating reactivated AdV and potentially other dsDNA virus infections (e.g. EBV and HHV-6) in long COVID [61]. In the absence of published cell culture and/or animal model data, no firm conclusions can be drawn regarding the activity of BCV as an antiviral against SARS-CoV-2, and the safety profile would need to be considered carefully to ensure a favorable benefit/risk profile for the population in question. For example, newly SARS-CoV-2 infected patients are likely to recover without intervention, and hospitalized COVID-19 patients are often past the stage of active viral replication where antiviral intervention may provide benefit.

15.4 Key Clinical Pharmacology and Safety Considerations

One critical parameter for successful clinical development of any drug is delivering enough of the active moiety to the correct tissues/cells/subcellular compartment(s) for efficacy without delivering too much to the site(s) of toxicity. This concept has proven especially dominant in the history of CDV and BCV. BCV was developed in part to circumvent the dose-limiting renal toxicity of IV CDV which is attributable to hyperconcentration of CDV in kidney proximal tubule cells via organic anion transporter 1 (OAT1) [62, 63]. The other major early goal for BCV was to allow oral delivery as it was originally envisioned as a countermeasure for the threat of bioterror with smallpox [1]. The ability to treat large numbers of patients rapidly and easily with a convenient oral dosing regimen is a key feature of this indication and is facilitated by the very short treatment duration required (two doses of BCV, one week apart) [2, 64, 65]. However, oral dosing of BCV over substantially longer time periods for prevention of CMV in allo-HCT patients proved too toxic when initiated shortly after transplant, a time when these patients often have compromised GI tracts due to mucositis associated with conditioning regimens [25]. IV dosing of BCV prevents the excessive accumulation of BCV and its metabolites in the GI tract and has an improved therapeutic index for longer term treatment. It is currently in clinical development for the treatment of AdV and CMV in immunocompromised patients (ClinicalTrials.gov identifier NCT04706923). Several key clinical pharmacology characteristics of CDV and BCV which clarify these observations and shed light on animal models are delineated in this section. More comprehensive safety and pharmacology data are available in other publications [2, 65].

Multiple animal and human studies have evaluated the absorption, distribution, metabolism, and excretion (ADME) of BCV delivered by different routes (e.g. oral or IV) as compared to CDV. As expected, these studies have largely shown parallels between anatomical sites of highest drug concentrations and sites of concern for toxicity. CDV is highly concentrated in kidneys by OAT1 and mostly eliminated unchanged in urine; the dose-limiting toxicity of CDV is renal [66]. BCV circulates intact in blood and is not a substrate for OAT1, characteristics which effectively circumvent the dose-limiting toxicity of IV CDV irrespective of the dosing route for BCV [12, 19]. Orally delivered BCV has a more even distribution throughout the body than CDV, but is concentrated in the GI tract and, to a lesser extent, the liver [12, 19, 67]. The dose-limiting toxicities for oral BCV are gastrointestinal and, to a lesser extent, hepatic. It should be noted that the primary clinical safety data for oral BCV are derived from HCT patients who are "predisposed to GI and hepatic toxicities" [[2, 24, 25], TEMBEXA® (BCV) package insert]. These toxicity concerns were found acceptable for the treatment of smallpox in part due to the short-term dosing used for this indication; however, IV BCV has been proposed as a potentially safer alternative for longer treatment courses.

Quantitative whole-body autoradiography of IV versus oral BCV in rats showed that the amount of drug/metabolites in the small intestine following IV administration was approximately 10% of that observed with oral administration [19]. These findings are reflected in the lower incidence and severity of GI findings in toxicology studies and clinical studies to date, suggesting longer term use of IV BCV may be well-tolerated [21, 68]. Notably, 10 mg IV BCV delivered plasma drug levels equivalent to 100 mg of oral BCV and liver enzyme increases, when observed, were reversible and not associated with hyperbilirubinemia [21, 68]. Overall, IV BCV may retain the safety advantages over CDV (e.g. no nephrotoxicity or myelotoxicity), greatly reduce GI toxicity associated with longer term oral BCV, and substantially improve penetration into brain tissues versus oral BCV (approximately 6-fold), enabling development for indications where oral BCV may not provide sufficient exposure for efficacy (e.g. viral infections of CNS, glioblastoma) [19, 21, 69].

15.4.1 Metabolism and Animal Models

Species specific metabolism of BCV is a significant factor influencing animal model interpretation. While the general routes of metabolism appear shared amongst species (where studied), the rate of formation of inactive short chain lipid metabolites and the intracellular concentrations of active drug can vary widely [70, 71]. The two major intracellular routes for BCV are (i) hydrolysis of the phosphoester bond to release CDV from the lipid (e.g. by acid sphingomyelinase), followed by phosphorylation to produce CDV-PP (the active species); and (ii) omega oxidation at the terminal carbon of the lipid by cytochrome P450 (CYP4F2) to produce a terminal carboxylate, followed by multiple cycles of beta-oxidation. Pathway 1 appears unrestricted by cell type, although the rate of BCV uptake, hydrolysis, and/or CDV-PP formation depends on multiple factors (e.g. rate of cell division) [72]. The major products of pathway 2, which is believed to occur mostly in the liver, are virologically inactive short chain esters of the lipid attached to CDV ([73], TEMBEXA® (BCV) package insert).

Early studies with monkeypox (mpox) in cynomolgus macaques revealed efficacy for IV CDV and essentially no efficacy for BCV delivered by several routes. This failure of BCV to prevent mortality from lethal orthopoxvirus infections in monkeys was not generally reflected in other models (e.g. mice and rabbits, see Section 15.5.1), although lower efficacy than initially expected was also seen in prairie dogs [74]. As illustrated in Figure 15.3, plasma exposure to BCV (Cmax shown) correlated with survival in three animal models of lethal mpox. Notably, all species had substantially lower Cmax than humans on a dose-normalized basis. Mice had the highest exposure in this comparison (20% of human) and all survived, while monkeys had the lowest (2% of human) and all died. Prairie dogs had intermediate exposure and

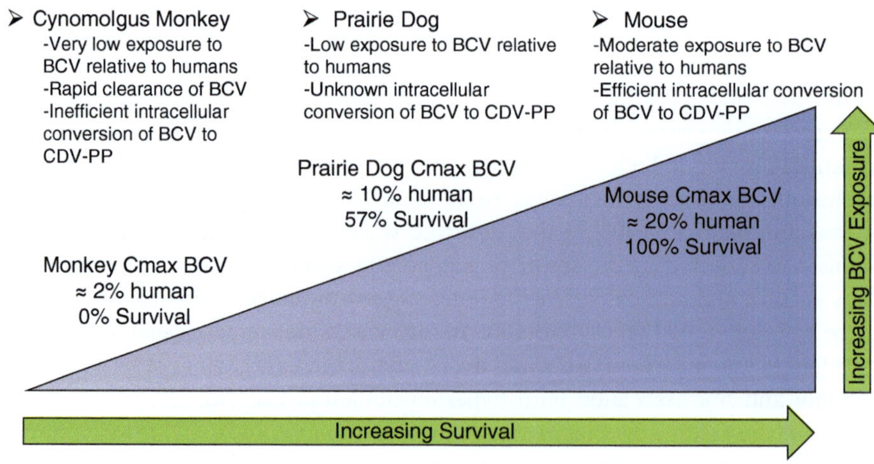

Figure 15.3 Species-specific exposure to BCV predicts efficacy in lethal models of mpox.

intermediate survival [74]. Subsequent experiments revealed that the metabolism of BCV to inactive esters was dramatically faster in monkey liver S9 fractions than in human liver S9, with only 16.7% remaining as BCV after 90 minutes in monkey fractions versus 83.3% in human fractions [75]. This was reflected in the amount of CDV-PP recovered from PBMCs where monkeys treated with BCV had very low CDV-PP compared to monkeys treated with CDV or rabbits treated with BCV [70].

15.5 Clinical Development

BCV has been administered orally to thousands of individuals in clinical studies, expanded access protocols, and through emergency/compassionate use pathways. The first patient treated, with US Food and Drug Administration (FDA) authorization under the EIND program, was a US marine with complications from vaccinia virus administered as a vaccine for smallpox [76]. Since then, oral BCV has been approved by the US FDA for the treatment of smallpox and late-stage clinical studies have been conducted for CMV and AdV infection. BCV has been primarily administered to prevent, preempt or treat disease associated with dsDNA viral infections in at-risk subjects who have undergone HCT or are otherwise immunocompromised with serious comorbidities. Two open-label expanded access protocols for the treatment of patients with one or more life-threatening or serious dsDNA viral infections have been completed (CMX001-350 and CMX001-351). Completed clinical efficacy studies of oral BCV are summarized in Table 15.3.

During the development of oral BCV, it became evident that there is a medical need for BCV in an IV formulation for patients with compromised oral intake and/or compromised GI function, to allow longer term dosing with reduced GI toxicity, and potentially to enable treatment of disease in the CNS. An ongoing clinical trial (ATHENA ClinicalTrials.gov identifier NCT04706923) is evaluating IV BCV for safety and antiviral efficacy against AdV and CMV (extension) as described in Section 15.5.2.

15.5.1 Poxviruses

Orthopoxviruses are a diverse group of enveloped dsDNA viruses with relatively large genomes (170–250 kb). One unusual feature of this genus among the dsDNA viruses is that viral replication occurs in the cytoplasm rather than the nucleus. The orthopoxvirus species variola virus (VARV) caused smallpox, which was responsible for an estimated 300 million deaths in the twentieth century alone. Although smallpox was declared eradicated in 1980, a combination of factors has kept countermeasures a high-priority for the US government and made development of effective antivirals a major public health goal [2, 77–80]. First and foremost is that while smallpox disease is currently eradicated, variola virus is not. Legal stocks exist in the United States and in Russia, unrecognized (viable) samples have been found in laboratories, clandestine stocks are believed to exist and a live version of the closely related orthopoxvirus "horsepox" was created de novo using synthetic DNA [81–84]. The historical death toll and incidence of mortality in smallpox outbreaks,

Table 15.3 Completed clinical efficacy studies for oral BCV.

Study number/name	Description
CMX001-201 ClinTrials.gov NCT00942305	Phase 2 dose escalation study to assess safety and efficacy of oral BCV once or twice weekly for prevention of CMV infection in CMV seropositive adult HCT recipients (R+) N = 230 (randomized 3:1 BCV:PBO) Cohort 1: 40 mg QW (N = 25) Cohort 2: 100 mg QW (N = 27) Cohort 3: 200 mg QW (N = 39) Cohort 4: 200 mg BIW (N = 30) Cohort 4A: 100 mg BIW (N = 50) Placebo (pooled; N = 59) Subjects received treatment for 9–11 weeks (through post-transplant week 13)
CMX001-301 SUPPRESS ClinTrials.gov NCT01769170	Phase 3 study to assess the safety and efficacy of BCV for the prevention of CMV infection in adult CMV seropositive allo-HCT recipients (R+) N = 303 oral BCV 100 mg BIW; N = 149 placebo Subjects initiated the study drug at a median of 15 days after HCT. BCV exposure was 54 days (range, 1–99)
CMX001-202 HALT ClinTrials.gov NCT01241344	Phase 2 study to assess the safety and efficacy of preemptive treatment with oral BCV, compared to placebo, for the prevention of AdV disease in HCT recipients with asymptomatic AdV viremia N = 48 (randomized 2:1 BCV:PBO)
CMX001-304 AdVise ClinTrials.gov NCT02087306	Phase 3 open-label, non-randomized, multicenter study of oral BCV administered twice weekly for the treatment of AdV infection detected during asymptomatic AdV viremia or symptomatic AdV infection Oral BCV 100 mg or 2 mg/kg BIW Subjects received treatment for 12 weeks. **Cohort A:** Adult or pediatric allogeneic HCT recipients at risk of progression to AdV disease **Cohort B:** Adult or pediatric allogeneic HCT recipients with disseminated AdV disease **Cohort C:** immunocompromised and non-immunocompromised patients with disseminated AdV disease, immunocompromised patients at risk for progression to disseminated AdV disease, allogenic HCT recipients >120 kg Total enrolled: N = 201 Cohort A: N = 66 Cohort B: N = 92 Cohort C: N = 43 Adult: N = 71 Pediatric (<18 years): N = 130

Table 15.3 (Continued)

Study number/name	Description
CMX001-350, ClinTrials.gov NCT01143181	Open-Label expanded access study of subjects with serious dsDNA viral infections without other treatment options • N = 215; average 8 weeks of dosing • Focus on treatment of refractory CMV and AdV in transplantation CMV: N = 107 (51%) AdV: N = 61 (29%) Other: N = 42 (20% [BKV, HSV, JCV, EBV, HHV-6, HPV, molluscum contagiosum virus]) HCT: N = 152 (72%) SOT: N = 33 (16%) Other disease: N = 25 (12%)
CMX001-351 ClinTrials.gov NCT02596997	Expanded access protocol to provide patients with serious AdV infection or disease access to treatment with BCV No data in public domain
EINDs/NPP	Emergency INDs and Named Patient Program for Compassionate Use • Treatment of CMV, AdV, BKV, EBV, HSV, JCV, HHV6, VZV, HPV, Orthopox

typically >30% even in populations where it was endemic, make it especially worrisome as a biothreat in populations where vaccination is uncommon, which includes most of the world [78, 80].

The US FDA approved BCV for the treatment of smallpox on 4 June 2021, under the Medical Countermeasure Initiative using the FDA Animal Rule. The Animal Rule provides a pathway to approval for drugs to treat serious or life-threatening conditions in cases where human trials are not ethical and field trials are not feasible. To use the Animal Rule as a pathway to approval, several requirements must be met. Briefly, the pathophysiology of the condition and the mechanism of action for the drug must be reasonably well-understood, the desired effect must be demonstrated in more than one species (unless a single species sufficiently predicts the response in humans), the study endpoint must be clearly related to the desired benefit in humans (e.g. survival), and data on the pharmacokinetics and pharmacodynamics of the drug must be provided to allow selection of an effective dose in humans. In addition to animal efficacy studies, the use of the Animal Rule requires that safety is established in human trials. For BCV, these data are summarized in two publications [2, 65].

15.5.1.1 Introduction to BCV Animal Models

No single animal model of smallpox is accepted to sufficiently predict treatment response in humans, so at least two animal models have been required for approval of antivirals for this indication under the Animal Rule. The ideal animal model would use variola virus (VARV); however, VARV infection and smallpox disease are highly host-specific to humans and "animal models using variola virus, including non-human primate (NHP) models, are not consistently reproducible, require unnaturally high viral challenge doses, and do not mimic human smallpox disease" [2].

Other available models deemed acceptable for the development of smallpox treatments included the monkeypox model (mpox in cynomolgus monkeys), the rabbitpox model (RPX virus [RPXV] in rabbits), and the mousepox model (ectromelia virus [ECTV] in mice) [FDA, 2011 [85]]. Since BCV is rapidly metabolized to inactive forms in monkeys [75], the rabbitpox and mousepox models were selected to demonstrate BCV efficacy (TEMBEXA® (BCV) package insert). Prior to conducting pivotal efficacy studies, detailed natural history studies were conducted to establish the course of disease in the rabbitpox and mousepox models [86–88].

15.5.1.2 Rabbitpox Model

The rabbitpox model, as implemented for BCV development, was performed in New Zealand White (NZW) rabbits. The model employed intradermal inoculation of RPXV, which was found to result in a consistent infection and course of disease. Key features of the model that align with smallpox in humans include a low inoculum, an incubation/viral replication phase, fever, lesions, and a high mortality rate [87, 88]. In the pivotal study, dosing with BCV at 20 mg/kg was initiated on day 3, 4, 5, or 6 with doses of 5 mg/kg on subsequent treatment days. A statistically significant survival benefit was demonstrated for each treatment initiation day (Table 15.4).

By day 3 post-infection, 97% (139/144) of animals had quantifiable viral DNA levels in blood, and by day 4, fever was observed in 98% of the rabbits. Therefore, BCV treatment at day 4 post-infection was considered to represent treatment of symptomatic disease.

Table 15.4 Survival results of BCV pivotal rabbitpox model study.

BCV treatment initiation day (day post-infection)	Survival % (n/N)	P-value[a]	Survival rate difference (95% CI)[b]
3	100% (29/29)	<0.0001	71% (51%, 87%)
4	90% (26/29)	<0.0001	61% (36%, 79%)
5	69% (20/29)	0.0014	40% (12%, 63%)
6	69% (20/29)	0.0014	40% (12%, 63%)
Placebo	29% (8/28)	-	-

a. P-value is from one-sided Boschloo's test compared with placebo.
b. Survival percentage at end of study (PID 42) with BCV-treated rabbits minus survival percentage in placebo-treated rabbits.

15.5.1.3 Mousepox Model

The mousepox model, as implemented for BCV development, employed intranasal infection of Balb/c mice. Key features of the model that align with smallpox in humans include a respiratory route of infection, a low inoculum, an incubation/viral replication phase, and a high mortality rate [86]. In the pivotal study, dosing with BCV at 10 or 20 mg/kg was initiated on day 4, 5, or 6 followed by 2 doses of 5 mg/kg on subsequent dosing days. Although a clear benefit with BCV treatment was observed in all groups (Table 15.5), the pharmacokinetic exposure in the 10/5/5 groups aligned better with the exposure in humans at the proposed dose.

Unlike rabbits, mice in the mousepox model do not reliably present with clinical signs to permit evaluation of disease progression. Therefore, to assess disease progression in the mousepox model, a satellite group of mice was serially harvested at specific days post-infection for viral load analysis. All mice in the viral load satellite group harvested at day 4 post-infection showed detectable viral DNA in blood, liver, and spleen, thus demonstrating disease progression at the time of BCV initiation. Therefore, BCV treatment at day 4 post-infection was considered to represent the treatment of symptomatic disease.

15.5.1.4 Scaling to Humans

Under the FDA Animal Rule, dose selection is based on delivering exposures in humans that exceed the efficacious exposures in animal models. The human BCV dose for smallpox was based on modeling among rabbit efficacy/exposures, human exposures, and mouse efficacy/exposures (www.accessdata.fda.gov, 214460Orig1s000). BCV in plasma and intracellular CDV-PP in peripheral blood mononuclear cells from infected and healthy animals were considered in the scaling analysis. A complication for the dose scaling analysis is that animal models of orthopoxvirus infection result in a more rapid disease course and shorter time to mortality than smallpox in humans. Therefore, comparisons between rabbit and human profiles were adjusted for time course differences between the animal model and humans.

Table 15.5 Survival results of BCV pivotal mousepox model study (10/5/5 dose regimen).

BCV treatment initiation (day post-infection)	Survival % (n/N)	P-value[a]	Survival rate difference (95% CI)[b]
4	78% (25/32)	<0.0001	66% (44%, 82%)
5	66% (21/32)	<0.0001	53% (29%, 72%)
6	34% (11/32)	0.023[c]	22% (1%, 43%)
Placebo	13% (4/32)	-	-

a. P-value is from one-sided Boschloo test compared with placebo.
b. Survival percentage with BCV-treated mice minus survival percentage in placebo-treated mice; CI, confidence interval.
c. Not significant at the one-sided alpha of 0.0125.

Overall, the analyses demonstrated that the BCV dose regimen of 200 mg once weekly for 2 doses delivers BCV and CDV-PP Cmax and AUC in healthy adults that are greater than the efficacious exposures seen in the rabbitpox model, supporting the clinical dose recommendations. BCV and CDV-PP Cmax and AUC in healthy adults exceeded or were similar to exposures in the mousepox model, providing additional support for the clinical dose.

15.5.1.5 Vaccine Interaction Studies

Experiments to assess the effect of BCV co-administration with smallpox vaccination were carried out in mice vaccinated with different smallpox vaccines and subsequently challenged with ectromelia virus at approximately 50 days post-vaccination [89]. With replication-competent vaccines (Dryvax and ACAM2000), BCV reduced the severity of vaccination-associated lesion development, as expected. BCV also reduced the degree of antibody response as measured by ELISA, but despite this, protective immunity as measured by survival following challenge with ectromelia virus was maintained. Experiments with non-replicating smallpox vaccine ACAM3000 (modified vaccinia Ankara or MVA) revealed no significant difference in survival following ECTV challenge and no apparent difference in antibody response measured by ELISA [89]. Overall, these results are consistent with the ability of BCV to inhibit viral DNA replication and suggest that BCV co-administration may cause some dampening of immune responses to replication-competent vaccines.

15.5.1.6 Resistance

Resistance is an important, potentially critical, aspect of antiviral treatment for smallpox. Generating resistant VARV is not permitted, so most studies use VACV. CDV resistance to orthopoxviruses *in vitro* is typically slow to emerge and is accompanied by viral attenuation [90, 91]. *In vivo*, there is no clear genetic evidence of BCV resistance selection in animal model studies to date, and no resistance to BCV has been detected in human studies of other orthopoxvirus infections [76]. Additional *in vitro* and animal studies of orthopoxvirus resistance to the parent of BCV (i.e. CDV) have made clear three important traits: (i) resistance to BCV/CDV in orthopoxviruses requires mutations in the highly conserved viral DNA polymerase; (ii) high-level resistance is not conferred by single mutations; and (iii) in all cases examined, BCV/CDV-resistant orthopoxvirus have diminished virulence *in vivo* and still respond to treatment [90, 91]. These data suggest BCV may retain activity against VARV which has been engineered or selected for resistance to the drug and that high-level resistance to BCV may be incompatible with a highly virulent orthopoxvirus due to the apparent need for changes in the conserved nucleotide-binding region of orthopoxvirus DNA polymerases.

15.5.1.7 Other Poxviruses (VACV, Mpox)

15.5.1.7.1 Vaccinia virus

The most comprehensive analysis of an immunocompromised patient who received treatment for severe progressive vaccinia was reported by Lederman et al. [76]. In this case, the patient was initially treated with escalating doses of tecovirimat and vaccinia immune globulin IV. BCV was added after the appearance of new satellite lesions prompted concerns about ongoing viral replication. Shortly after the initiation of BCV, viable virus became undetectable. While the patient ultimately survived, the complexity of treatments in this case makes attribution of success to one intervention impossible, although the temporal relationship between the addition of BCV and clinical improvement in lesions as well as improvement in virologic parameters suggests that BCV was an important component of the treatment. Upon completion of treatment, no resistance to BCV was detected; however, high-level resistance to tecovirimat was identified. Lederman et al. concluded that "in such severely ill patients, combination therapy may best be initiated at the outset, which might reduce the viral load and subsequent development of antiviral resistance mutations."

15.5.1.7.2 Monkeypox virus

BCV has been tested in a mouse model of mpox disease and showed 100% protection from mortality [92]. BCV was also tested in a prairie dog model of mpox where it showed modest efficacy. Notably, prairie dogs had intermediate BCV exposure compared to other animal models as shown in Figure 15.3, and approximately 10% of human exposure at equivalent doses [2, 74, 93]. These results highlight the importance of evaluating drug exposure in animal models when assessing efficacy. BCV has also been used in three clinical mpox cases reviewed by Adler [94]. The patients who received BCV showed consistent virologic declines in multiple compartments after BCV administration and all three fully recovered clinically, despite substantially delayed treatment, higher lesion counts (3–15-fold), and higher baseline viremia (10–30-fold) compared with the single tecovirimat-treated patient included in the review. Simultaneous use of both BCV and tecovirimat should be considered in serious cases as these agents have different mechanisms of action, different pathways to resistance, and animal model data supporting combination use. The importance of having multiple drugs to treat poxvirus infections is underscored not only by the case of tecovirimat-resistant vaccinia virus described above [94], but also by multiple cases of mpox that developed confirmed tecovirimat resistance [95]. In one well-described case, the patient was treated with tecovirimat, did not clear the infection, and progressed to a serious infection. Two mutations in the virus were identified that are known to confer high-level tecovirimat resistance and, upon adding CDV and subsequently BCV, the patient recovered [96].

15.5.2 Adenovirus

Human adenoviruses (AdVs) are nonenveloped, double-stranded DNA viruses, divided into seven major species based on clinical, virologic, and DNA sequence properties. Species are designated A through G, with >100 distinct types, e.g. AdV species C type 5 (AdVC5). AdV infection is normally mild and self-limiting, but it can cause severe disease and death in immunocompromised patients, especially among allo-HCT recipients prior to immune reconstitution. While there is wide variability in the reported incidence (3–42%) and severity of AdV infection for allo-HCT recipients, depending on multiple risk factors (e.g. age, graft type, T-cell depletion), mortality as high as 60–80% has been reported [18, 97–101]. PCR quantitation of AdV in blood is increasingly used in clinical practice to guide treatment decisions for these patients and AdV ≥ 1000 copies/mL in plasma is emerging as a clinically meaningful threshold [16, 17, 102–105]. For example, one study (N = 238) reported that survival in pediatric allo-HCT recipients was virtually identical among patients who never had AdV detected and those with AdV viremia <1000 c/mL. However, patients with peak viremia ≥1000 c/mL died more frequently and "blood HAdV loads >10 000 copies/mL (6.7% of all patients) were significant and independent risk factors for poor survival" [106].

A European study of 241 pediatric allo-HCT recipients receiving standard of care therapy (primarily CDV) who had ≥1000 copies/mL of AdV in blood concluded that "measures of AdV viral peak, duration, and overall burden of infection consistently correlate with all-cause mortality" [105]. A related US study evaluated >3000 allo-HCT recipients and reported higher 6-month mortality among patients who had ≥1000 copies/mL of AdV in plasma versus those without AdV in plasma: 23% versus 16% in adults and 17% versus 10% in pediatric patients, respectively [17]. The majority of adult (87%) and pediatric (78%) allo-HCT recipients with AdV in plasma during the first 6 months after transplant also were positive for infection with ≥1 additional dsDNA virus [17].

There are no approved therapies for AdV. The current standard of care for allo-HCT patients with AdV viremia ≥1000 copies/mL in plasma remains the reduction of immune suppression (if possible) and off-label antivirals, typically IV CDV [17, 98, 107]. Clinical guidelines and prevalent clinical practice include routine monitoring of lymphopenic allo-HCT patients (especially pediatric) for AdV viremia and intervention (typically including IV CDV) in the context of AdV viremia ≥1000 copies/mL in blood [16, 22, 107, 108].

15.5.2.1 Clinical Experience with Oral BCV for AdV

Numerous publications have reported reductions in AdV viremia for populations of immunocompromised patients following treatment with BCV [20–22, 103, 109–115]. One retrospective evaluation of longer term oral BCV (≥12 weeks) in liver transplantation (isolated liver [n = 3] or multi-organ recipients [n = 10]) analyzed subjects from CMX001-304, CMX001-350, and EINDs. BCV was stopped for a treatment-emergent adverse event (diarrhea) in 1/13 patients and the authors noted stable or declining aminotransferase levels with no liver-related AEs during BCV therapy. The median decrease in AdV viremia on therapy was −1.3 $\log_{10}$ copies/mL and the median time to maximum reduction was 15 days. The authors concluded that "treatment with BCV can reduce viral burden and is associated with favorable survival" [110]. Another early report evaluated oral BCV in 13 immunocompromised patients (11 allo-HCT) who had received prior CDV but remained viremic [111]. The median change in AdV viremia from the baseline of these patients on CDV was 0.08 $\log_{10}$ while the median change on BCV was −2.71 $\log_{10}$ (P = 0.001). Overall, 9/13 patients had a virologic response to BCV (2 $\log_{10}$ decrease in AdV viremia or undetectable AdV) by week 8; the median survival in patients with this response was 196 versus 54 days in non-responders (P = 0.04) [111].

An additional retrospective cohort study in pediatric allo-HCT recipients treated patients (N = 27) with confirmed AdV ≥1000 copies/mL using IV CDV. Oral BCV was substituted for CDV in patients with renal impairment at baseline or on CDV and for lack of virologic response to CDV after 2 weeks. There was 1 dosing interruption on BCV (severe diarrhea) and no nephrotoxicity versus 39% of CDV-treated patients having nephrotoxicity (mild-moderate). AdV was cleared from plasma in the majority (80%) of BCV-treated patients with a median time to undetectable of 4 weeks versus 35% clearance with CDV at a median time of 9 weeks. Notably, the median lymphocyte count in BCV-treated patients at the time of viral clearance (n = 13/16) was 320 cells/μL (range, 0–3000) versus 910 cells/μL (range, 380–2000) for patients who cleared AdV on CDV (n = 8/23). The authors concluded that "brincidofovir is well-tolerated and highly efficacious in controlling adenoviremia during the lymphopenic phase of HCT" [22].

The first randomized controlled trial (RCT) evaluating BCV for the treatment of AdV was HALT (CMX001-202). Allo-HCT recipients with asymptomatic AdV viremia (≥100 DNA copies/mL) were randomized to once-weekly (QW) oral BCV (200 mg or 4 mg/kg if <50 kg, n = 16), twice-weekly (BIW) BCV (100 mg or 2 mg/kg if <50 kg, n = 14), or placebo (n = 18) for up to 3 months, followed by 1 month of follow-up [20].

BCV BIW-treated patients had numerically higher rates of reaching undetectable AdV viremia, accompanied by lower rates of treatment failure and all-cause mortality, but the differences were not statistically significant (Table 15.6). A subsequent analysis of patients with baseline AdV ≥1000 copies/mL showed a statistically different virologic response as measured by undetectable AdV after 1 week of BCV therapy (86% for BCV BIW vs. 25% for Placebo [P = 0.04]). Treatment-emergent adverse events associated with the GI tract were higher in BCV-treated patients (BIW > QW > Placebo), but there was no evidence of increased neutropenia or kidney toxicity.

The largest clinical study of BCV for the treatment of AdV (AdVise;CMX001-304) enrolled 201 patients with, or at risk for, disseminated AdV disease in three cohorts shown in Table 15.3. Disseminated AdV disease was defined as: (i) AdV detection in ≥ 2 anatomical compartments other than blood and symptomatic in at least one or with unexplained fever (>100.4°F); (ii) probable/definitive AdV disease in one AdV-positive compartment and AdV in plasma ≥ 1000 copies/mL. Patients received

Table 15.6 Summary of key data from HALT (CMX001-202) AdV study [20].

Parameter	BCV BIW (n = 14)	BCV QW (n = 16)	Placebo (n = 18)
Baseline AdV, median	3.1 $\log_{10}$ copies/mL	4.0 $\log_{10}$ copies/mL	2.7 $\log_{10}$ copies/mL
Undetectable AdV after 1 week	67%	29%	33%
Treatment failure[a]	21%	38%	33%
All-cause mortality	14%	31%	39%
Acute GVHD of GI	36%	19%	11%
Diarrhea	57%	38%	28%

a. Treatment failure was defined as progression to probable or definitive AdV disease or a confirmed rise in AdV of ≥1 $\log_{10}$ copies/mL while on therapy.

Table 15.7 Virologic response to BCV is associated with higher survival for allo-HCT recipients with disseminated AdV disease [103].

Pediatric (N = 100)			Adults (N = 58)		
Proportion of responders[a]	**Mortality in responders**	**Mortality in nonresponders**	**Proportion of responders[a]**	**Mortality in responders**	**Mortality in nonresponders**
Cohort A (asymptomatic or localized infection; N = 65)					
17/25 (68%)	5/17 (29%)	2/8 (25%)	8/13 (62%)	4/8 (50%)	4/5 (80%)
Cohort B (disseminated disease; N = 93)					
28/41 (68%)	7/28 (25%)	7/13 (54%)	10/24 (42%)	5/10 (50%)	13/14 (93%)
	P = 0.031[b]			P = 0.0004[b]	

A Cox model incorporating age group was used to compare mortality at 36 weeks among responders and non-responders.

a. Responders are defined as subjects who achieved undetectable viremia at week 6; only patients with baseline AdV viremia still on study at week 6 included.

b. Responders versus nonresponders in Cohort B, Cox model for time to death.

open-label oral dosing with 100 mg BCV BIW (2 mg/kg BIW if < 50 kg) for 12 weeks and could reinitiate BCV treatment for up to 12 additional weeks if they experienced a recurrence of AdV infection or disease during the follow-up period.

The final 36-week results have been presented for the allo-HCT patients enrolled in Cohorts A and B. Pediatric patients overall received BCV sooner after HCT than adults (median 59 days from HCT in pediatric recipients vs. 99 days in adults), and 53% of children had documented prior CDV use versus 29% of adults. The most common adverse events leading to treatment discontinuation were GI, and their rate was lower in pediatric patients (7%) versus adults (14%). All-cause mortality at week 36 in pediatric patients was 42% versus 69% in adults [103]. Survival at 36 weeks was significantly higher in adult and pediatric patients with disseminated disease (Cohort B) who had a complete virologic response at week 6 as compared to non-responders, as shown in Table 15.7. There was no difference in survival based on virologic response for asymptomatic patients or those with localized AdV infection (Cohort A). While there are important caveats to studies of this nature (e.g. lack of placebo or standard of care control, time-biased response cutoff) it suggests clearance of AdV viremia with BCV may promote survival in these populations.

15.5.2.2 Adenovirus Type and Response to BCV

Different AdV types may preferentially infect different tissue types, and specific types have been associated with increased virulence [116]. While rarely investigated or used clinically, the genetic and biological diversity of AdV suggests the potential for differential activity of antivirals depending on AdV type. BCV has similar activity *in vitro* against a wide array of AdV types [39–41]. Virologic response to BCV has also been evaluated clinically based on AdV species and type. AdV was typed from plasma samples at baseline from 172 patients enrolled in studies CMX001-202

Table 15.8 Virologic response at last time on BCV therapy by AdV species (N = 148 patients).

AdV species (N = 148)	A (n = 17)	B (n = 34)	C (n = 97)	D(n = 2)	F (n = 8)
Median response to BCV ($\log_{10}$)	−1.8	−1.9	−1.6	−2.0	−2.3

(HALT; n = 42), CMX001-350 (Expanded Access; n = 34), or CMX001-304 (AdVise; n = 96) by sequencing the hypervariable region of the AdV hexon gene. Notably, 6 patients had 2 different AdV species detected in plasma, and 2 patients had 3 different AdV species detected. Overall frequencies of AdV species in order of prevalence were C, B, A, F, and D, at 59%, 23%, 13%, 3%, and 2%, respectively. To evaluate virologic response, patients with detectable baseline viremia and at least 2 weeks of treatment with BCV (N = 148) were assessed at the last timepoint on BCV therapy. All AdV species responded to BCV with a >97% median viral load reduction across AdV species (Table 15.8), and all AdV types appeared sensitive to BCV. Notably, there was no obvious reduction in virologic response for patients with species A in whom 13/15 evaluated were AdVA31, which had an elevated EC_{50} relative to other types in one report, but not in a follow-up study [39, 40]. These data suggest that BCV has antiviral activity against all species and types of AdV that have been detected in clinical trials of immunocompromised patients to date [40].

15.5.2.3 AdV Resistance to BCV

BCV and CDV-resistant isolates of AdVC5 have been selected *in vitro* [44, 46, 117]. BCV selected for V303I in the AdV DNA polymerase gene after several months and subsequently for T87I at approximately 6 months. Reintroduction of T87I and V303I mutations into a wild-type AdVC5 resulted in a 2.1-fold increase in the EC_{50} of BCV and a 1.9-fold increase in the EC_{50} of CDV, relative to control [44]. The only mutation detected in both CDV and BCV passaging studies was V303I. The emergence of AdV resistance on BCV therapy was studied in the HALT (CMX001202) and AdVise (CMX001-304) studies, which together enrolled 249 patients. Two subjects had a mutation detected in the AdV polymerase that was associated with *in vitro* resistance to BCV or CDV. The V303I mutation and 6 uncharacterized amino acid changes in the AdV DNA polymerase gene were detected in one subject from the HALT study at the time of virologic failure; V303V was confirmed at baseline. Notably, the viral load profile was consistent with resistance selection, i.e. a rapid decrease in viremia of >10-fold after initiation of BCV, followed by approximately 6 weeks of viremia around 1000 copies/mL before an increase in viremia to baseline levels [20, 40, 118]. One patient in AdVise had T87I detected at baseline in conjunction with 3 other amino acid changes in the AdV DNA polymerase gene. This patient had very high baseline viremia (820 000 copies/mL), which decreased 1.5 $\log_{10}$ by week 2 of oral BCV, but then increased to 1.7 $\log_{10}$ above baseline by week 8. Switching oral BCV to IV CDV did not decrease AdV viremia [40, 118]. Together, the available *in vitro* and clinical data suggest a relatively high barrier to resistance for BCV in AdV and that the current RAM profile is likely incomplete.

15.5.2.4 ATHENA (IV BCV for AdV)

The potential for IV BCV to preempt or treat severe disseminated AdV in immunocompromised patients is being studied in the open-label dose ascending Phase 2a ATHENA study (BCV-PA01; ClinTrials.gov NCT04706923). The completed arms of ATHENA enrolled immunocompromised pediatric and adult patients (primarily pediatric allo-HCT recipients) with disseminated AdV disease, plasma AdV > 10 000 copies/mL, or rising plasma AdV with both values >1000 copies/mL. BCV was administered by IV BIW for up to 14 weeks or until AdV in plasma was undetectable. Patients received BCV as follows: Cohort 1 (0.2 mg/kg or 10 mg/dose if ≥ 50 kg); Cohort 2 (0.3 mg/kg or 15 mg/dose if ≥ 50 kg); Cohort 3 (0.4 mg/kg or 20 mg/dose if ≥ 50 kg). Additional cohorts are ongoing to evaluate once-weekly IV BCV regimens.

Topline safety and efficacy data from Cohorts 1–3 are shown below in Table 15.9. There was a clear dose-dependent increase in virologic efficacy, with AdV clearance in blood observed in 25%, 33%, and 100% of patients from Cohorts 1, 2, and 3, respectively. Cohort 3 (0.4 mg/kg IV BCV BIW) patients also had faster clearance of AdV, with 90% having undetectable AdV in blood within the first 4 weeks of BCV treatment. It is important that clearance of AdV was maintained through the 30-day post-treatment follow-up period, and the mean duration of BCV therapy was just over 5 weeks in Cohort 3, suggesting that twice-weekly IV BCV at 0.4 mg/kg may effectively and sustainably clear AdV in these high-risk patients with a relatively short duration of therapy. Treatment-related AEs were infrequent, and all resolved during the follow-up period [21].

In summary, there is a high unmet need for an effective AdV antiviral to prevent excess mortality in allo-HCT recipients with AdV plasma viremia ≥1000 copies/mL, and existing data support AdV clearance from plasma as a marker of survival. BCV appears effective against AdV in severely lymphopenic patients who are not effectively treated with standard-of-care therapy (including IV CDV). Additionally,

Table 15.9 Summary of key data from ATHENA (BCV-PA01) AdV study [21].

Parameter	Cohort 1 (0.2 mg/kg BIW) n = 8	Cohort 2 (0.3 mg/kg BIW) n = 9	Cohort 3 (0.4 mg/kg BIW) n = 10
Patients with viral clearance (<25 copies/mL)	2 (25%)	3 (33%)	10 (100%)
Viral clearance with ≤4-week IV BCV	1 (13%)	1 (11%)	9 (90%)
Mean duration of IV BCV, weeks (range)	8.6 (3.4–13.7)	10.7 (5.4–13.4)	5.1 (2.6–10.9)
Treatment-related AE			
• Diarrhea, Grade 2	1	-	-
• Elevated AST, Grade 3	-	-	1
• Transaminitis, Grade 3	-	1	-

IV BCV appears to effectively clear AdV from blood in allo-HCT recipients at well-tolerated doses that are not associated with the renal toxicity observed with CDV or the GI toxicity associated with longer term oral BCV.

15.5.3 Cytomegalovirus

CMV is an enveloped, double-stranded DNA virus that infects most people worldwide by adulthood and establishes a latent infection. It remains a common cause of morbidity and mortality in immunocompromised populations despite recent advances in prevention and treatment, especially among HCT and solid organ transplant (SOT) recipients [119–122]. CMV replication can directly cause invasive end-organ disease and indirectly contribute to graft-versus-host disease (GVHD), graft failure, and/or other opportunistic infections [123–125].

Recent approvals for drugs with different molecular targets have led to changes in guidelines and the strategic management of CMV in transplant recipients, including advances for prevention and preemption of disease in the highest-risk groups and treatment of patients with refractory or resistant CMV [120, 126–130]. With these new agents available, the largest current unmet need for CMV antivirals is arguably in resistant/refractory (R/R) CMV disease [37].

15.5.3.1 Clinical Experience with Oral BCV for CMV

BCV has been studied extensively as a CMV antiviral due to a combination of characteristics. Specifically, the apparent absence of nephrotoxicity and myelotoxicity, which limit use of FOS, CDV, and GCV, proven antiviral efficacy, and a relatively high barrier to resistance converge to maintain interest in this area despite a high-profile failure of oral BCV in the phase 3 SUPPRESS trial for CMV prevention, discussed in more detail below. Oral BCV (100 mg twice a week (BIW) for 9–11 weeks) demonstrated efficacy in a placebo-controlled Phase 2 clinical trial (CMX001-201, ClinTrials.gov NCT00942305) for CMV prevention in primarily antiviral naïve HCT patients after engraftment. The number of patients with CMV events in the placebo group (37%) significantly exceeded those with CMV events in the 100 mg BIW BCV group (10%; $P = 0.002$) [24]. Unacceptable levels of severe diarrhea were seen at 200 mg BIW. Analysis of plasma-derived CMV from patients with viral rebound or incomplete suppression failed to detect any known UL97 or UL54 RAMs in subjects taking BCV, suggesting that BCV is unlikely to select for CMV resistance in this setting [131].

Oral BCV prophylaxis was subsequently evaluated in a Phase 3 study (SUPPRESS, CMX001-301, ClinicalTrials.gov NCT01769170) of adult CMV-seropositive allo-HCT recipients without detectable plasma CMV at baseline [25]. Patients who were no more than 28 days post-HCT were randomized 2:1 to receive 100 mg BCV twice a week or placebo for 14 weeks and were then followed through week 24 post-HCT. The primary endpoint was CMV disease or CMV in plasma ≥151 copies/mL for patients at high risk for CMV disease (e.g. cord blood transplant, T-cell depleted) and ≥1000 copies/mL for those at lower risk. While more patients receiving placebo had detectable CMV than those receiving BCV on therapy through week 14 (42% vs. 26%; $P<0.001$), the percent with clinically significant CMV infection was essentially identical at week 24 (51% vs. 52%), when the primary endpoint of CMV event prevention was assessed. This result has been attributed to a substantially higher (8-fold) use of immunosuppressants in the BCV arm for the treatment of presumptive GVHD, which is difficult to distinguish from GI toxicity of oral BCV. The higher use of immunosuppressants in the BCV arm was proposed as a major contributing factor in the higher all-cause mortality (15.5%) in the BCV arm versus the placebo arm (10.1%). As in the Phase 2 trial, no mutations associated with BCV resistance were detected. Earlier initiation of BCV in SUPPRESS (median day 15 post-HCT) versus the Phase 2 study (median day 24 post-HCT) "resulted in frequent initiation of BCV while patients were still experiencing conditioning regimen-associated mucositis" and "likely contributed to synergistic enterotoxicity" [25]. The authors end by suggesting that "development of an IV formulation of BCV that does not result in GI accumulation but retains the drug's spectrum, antiviral activity, and lack of nephrotoxicity or myelotoxicity could provide a more effective and safer prevention and treatment of CMV and the multiple dsDNA viruses that frequently affect allogeneic HCT recipients" [25, 67, 68].

These data, combined with approval of letermovir (LTMV, PREVYMIS®), a well-tolerated inhibitor of the CMV DNA terminase complex (UL51,UL56, UL89), for prophylaxis of CMV, shifted the risk/benefit calculation for BCV in CMV to later lines of therapy (PREVYMIS®, package insert, [127]). The main weakness of LTMV is a low barrier to resistance, primarily associated with single mutations in UL56, but this issue is least problematic in prophylaxis, where viral load is typically absent or very low [132–135]. Additionally, the absence of cross-resistance for UL56 mutations to other available CMV antivirals (see Table 15.10) preserves current options for preemption and treatment. LTMV is not antagonistic with CDV, BCV, FOS, or GCV, although a drug-drug interaction may preclude combination with BCV as discussed in more detail below ([136], PREVYMIS®, package insert).

The choice of which antiviral to use for CMV preemption is more complicated than prevention, in part because the risk of resistance is higher due to ongoing CMV replication. For example, in one relatively large D+R− kidney SOT study of GCV/VGCV (N = 112), significantly higher virologic failure was observed for patients treated preemptively versus prophylactically (31% vs. 3%); these results were mirrored by higher detection of GCV resistance associated mutations in UL97 or UL54 (16% with preemptive vs. 3% with prophylactic). Unsurprisingly, high CMV viral burden was significantly associated with resistance emergence [137]. Specific toxicities are also a major hurdle for many transplant patients. For example, hematologic toxicity/myelosuppression associated with GCV/VGCV reduces benefit in HCT recipients while the renal toxicity of FOS and CDV limits their utility in certain populations, e.g. kidney SOT [121, 138, 139].

Patients with refractory and/or resistant CMV pose the greatest challenge for successful antiviral therapy due to higher viral burdens, more advanced disease, and resistance. Although Maribavir (MBV, LIVTENCITY®) was recently approved for post-transplant patients with refractory CMV disease, only 55.7% of MBV-treated patients had confirmed undetectable CMV at week 8 of therapy, and only 18.7% maintained undetectable CMV and symptom control through week 16 (MBV stopped at week 8).

Resistance was common with virologic failure [126]. CDV, FOS, GCV, and other modalities (e.g. adoptive T-cells) are options, but none have proven broadly useful due to different issues (e.g. toxicity, resistance, and unproven efficacy).

15.5.3.2 CMV Resistance to BCV

CMV resistance to antivirals remains a significant issue, as exemplified by a recent study that examined the prevalence of CMV antiviral RAMs in 2750 patient samples submitted to Eurofins Viracor between 1 January 2021, and 31 January 2023 [121, 129, 140]. RAMs associated with phenotypic resistance to GCV, CDV, FOS, MBV, and/or LTMV were determined for UL54, UL56, and/or UL97. Almost one-third (30.04%) had RAMs for one or more antivirals, with 27.64% of samples having a UL97 RAM associated with GCV and 9.96% having a UL97 RAM associated with MBV. UL54 RAMs were less common, with only 6.11%, 5.98%, and 1.76% of samples having GCV, CDV, and/or FOS RAMs, respectively. For LTMV, RAMs in UL56 were present in 7.17% of samples. RAMs for two CMV antivirals were identified in 7.82% of samples and 3 or more in 1.27% of samples. The authors note that the high prevalence of RAMS in these samples reflects their origin from clinicians who suspected their patients might harbor antiviral-resistant CMV strains; however, this would be the case for some preemption patients and most R/R patients [140].

Resistance to CDV and BCV maps predominantly to the exonuclease domain located between amino acids 300 and 545 of the CMV DNA polymerase (UL54) gene. *In vitro* passage of CMV with BCV by two groups identified mutations selected in UL54. James et al. found BCV selected a unique mutation (D542E) in UL54 that was associated with slower CMV replication and resistance to BCV and CDV, but no cross-resistance to GCV or FOS [141]. Chou et al. found a wider array of UL54 RAMs, including known RAMs N408K and V812L and the novel RAMs D413Y, E303D, or G; these mutations conferred ganciclovir and CDV resistance with 6- to 17-fold resistance to BCV [142]. The published *in vitro* selection experiments with CDV [143, 144] or BCV [141, 142] suggest that resistance to BCV is limited to changes in the UL54 gene, may be slow to emerge, and is associated with reduced viral replication kinetics. FOS resistance is most often associated with the UL54 palm domain at amino acids 588 to 821, while GCV resistance can map to the CMV phosphotransferase (UL97) or UL54. Notably, the vast majority (up to 90%) of RAMs associated with GCV/VGCV occur in UL97 without detectable UL54 mutations [144–148]. While a thorough review of CMV resistance and cross-resistance is beyond the scope of this review, two key points for BCV are that clinically relevant cross-resistance to BCV is uncommon and is not conferred by mutations selected with LTMV or MBV, as shown graphically in Table 15.10 [129, 149].

Viral genes associated with drug resistance mutations found in the clinic are highlighted in red. Drug resistance mutations in viral genes described *in vitro* but with uncertain clinical relevance are depicted in yellow, while genes not implicated in drug resistance for a given drug are shown in green.

Consistent with these data, an evaluation of CMV resistance and virologic response to BCV in patients participating in the expanded access study CMX001-350 (Table 15.3) revealed that mutations in UL97 did not substantially affect BCV activity, while mutations in UL54 (CMV polymerase gene) associated with CDV and

Table 15.10 Clinically relevant cross-resistance among CMV antivirals.

CMV gene	BCV	CDV	FOS	GCV/vGCV	LTMV	MBV
UL27	green	green	green	green	green	yellow
UL51	green	green	green	green	yellow	green
UL54	red	red	red	red	green	green
UL56	green	green	green	green	red	green
UL89	green	green	green	green	yellow	green
UL97	green	green	green	red	green	red

Source: Adapted from [129, 149].

GCV resistance reduced BCV activity (Table 15.11) as predicted [150]. CMX001-350 enrolled 215 patients with immediately life-threatening or serious diseases/conditions caused by one or more dsDNA viruses, including CMV (n = 107). Subjects received oral BCV once weekly (QW) or twice weekly (BIW) for up to 3 months until the target virus was cleared or clinical disease resolved. The majority (77%) were HCT recipients. The most frequent reasons given for having no alternative therapy available was viral resistance to available CMV antivirals (36%), lack of efficacy of available CMV antivirals (26%), or toxicity of available CMV antivirals (24%). Reported prior use of CMV antivirals included FOS (47%), GCV (31%), VGCV (21%), and CDV (14%).

The population evaluable for assessing BCV efficacy (N = 70) met the following criteria: (i) CMV quantified and ≥200 copies/mL at baseline; (ii) received BCV for at least 4 weeks; and (iii) CMV quantified on-treatment for at least 4 weeks. Plasma samples were evaluated for CMV genotype (UL54 and UL97) by Viracor-IBT Labs (5600 Cytomegalovirus Antiviral Resistance) and/or Nextgen Sequencing from subjects with ≥ 4 weeks of CMV viral load testing while receiving BCV who had confirmed CMV viremia ≥ 500 copies/mL. CMV genotypes obtained for subjects with at least 4 weeks of BCV and detectable CMV at baseline (N = 63) revealed BCV/CDV-associated RAMs in UL54 at baseline or on therapy for 14 subjects while 21 subjects had UL97 RAMs. As expected, UL97 RAMs had no discernible effect on the response to BCV. However, BCV/CDV RAMs in UL54 were associated with a diminished response to BCV. For example, at the last time on therapy, patients without BCV/CDV RAMs had a mean decrease in CMV viremia of −0.81 $\log_{10}$ (53% <200 copies/mL) while those with CDV RAMs had a mean CMV decrease of −0.43 $\log_{10}$ and none reached <200 copies/mL.

Notably, subjects with CDV RAMs detected had approximately twice the plasma viral load at baseline of subjects without a CDV RAM, and the virologic response lessened over time (mean −0.77 $\log_{10}$ at week 5 and −0.43 $\log_{10}$ at Last Value). By comparison, subjects without any BCV/CDV RAM detected had an improved mean antiviral response at the later time (mean −0.64 $\log_{10}$ at week 5 and −0.81 $\log_{10}$ at Last Value). Detection of UL54 BCV/CDV RAMs significantly reduced the proportion of subjects who achieved a viral load of ≤200 copies/mL at week 5 (P = 0.028) and at the last value (P = 0.0002).

Table 15.11 CMX001-350: Summary of CMV viremia (log_{10} copies/mL) by detection of UL54 CDV RAM at baseline or on treatment [150].

	CDV-R UL54 mutation[a]	No CDV-R UL54 mutation	Not tested
N	14	49	7
Mean (SD)	4.04 (0.78)	3.74 (0.69)	2.66 (0.42)
Change from baseline to week 5			
N	14	49	7
Mean (SD)	−0.77 (1.03)	−0.64 (0.97)	−0.51 (0.64)
P-value[b]	-	0.59	-
Proportion of subjects ≤200 copies/mL at week 5	0/14 (0.0%)	15/49 (30.6%)	6/7 (85.7%)
P-value[c]	-	0.028	-
Proportion of subjects 0.5 log drop or undetectable at week 5	8/14 (57.1%)	27/49 (55.1%)	5/7 (71.4%)
P-value[c]	-	1.0	-
Change from baseline to last value			
N	14	49	7
Mean (SD)	−0.43 (1.00)	−0.81 (1.25)	−0.67 (0.42)
P-value[b]	-	0.21	-
Proportion of subjects ≤200 copies/mL at last value	0/14 (0.0%)	26/49 (53.1%)	7/7 (100.0%)
P-value[c]	-	0.0002	-
Proportion of subjects 0.5 log drop or undetectable at last value	6/14 (42.9%)	32/49 (65.3%)	7/7 (100%)
P-value[c]	-	0.21	-

a. Two UL54 CDV RAMs were detected at baseline by standard sequencing and two additional by Nextgen at <2% representation; 10 were treatment-emergent.
b. Wilcoxon rank sum test (two-sided *t* approximation) vs. UL54 mutation group.
c. Fisher's exact test (two-sided) vs. UL54 mutation group.

The use of combination antiviral therapy for CMV has been reduced by several factors, including overlapping toxicities, antagonistic drug-drug interactions, shared resistance profiles, and lack of controlled efficacy/safety data. *In vitro*, BCV combinations with ACV, FOS, GCV, MBV, or LTMV were at least additive. However, the combination of BCV with LTMV may be contraindicated since LTMV inhibits organic anion-transporting polypeptides (OATP1B1/3) and BCV is a substrate of OATP1B1/3. [136, 151–153]. Therefore, co-administration of LTMV and BCV has the potential to significantly increase plasma concentrations of BCV and increase toxicity (TEMBEXA® (BCV) and PREVYMIS® (letermovir), Package inserts).

Overall, there remains a substantial unmet need for improved CMV antivirals, especially for patients with recurrent and/or resistant CMV. BCV-delivered IV is promising in this area for several reasons: (i) oral BCV has clinically proven antiviral activity against CMV; (ii) BCV has promise in combination with most approved

CMV antivirals; (iii) BCV has a relatively high barrier to resistance and a predictable/well-understood cross-resistance profile; and (iv) IV delivery of BCV significantly reduces the GI toxicity associated with oral BCV. IV BCV is being evaluated in an extension of the ATHENA trial for antiviral efficacy against CMV (ClinicalTrials.gov identifier NCT04706923).

15.5.4 Polyomaviruses

Polyomaviruses (PyVs) are non-enveloped viruses characterized by a circular double-stranded DNA genome of approximately 5.0 kb, bearing a noncoding control region (NCCR) with the origin of replication, transcription, and regulatory elements [49]. Four human PyVs species BKV, JCV, Trichodysplasia Spinulosa Polyomavirus (TSPyV), and Merkel cell polyomavirus (MCPyV) are well-known for their roles in human diseases, mainly in immunocompromised patients [154]. Post-transplant BKV infections within adult and childhood kidney transplant recipients can lead to graft dysfunction due to BKV-associated nephropathy (BKVAN), making BKV one of the leading causes of kidney graft dysfunction [155]. BKV is commonly linked to hemorrhagic cystitis (HC) and other complications in HCT recipients [156]. JCV is mostly associated with progressive multifocal leukoencephalopathy (PML), a rare but fatal brain disease characterized by progressive demyelination [157]. PML occurs mainly in immunocompromised individuals and has increased with potent immunomodulatory therapies for diseases such as multiple sclerosis (MS) [158]. MCPyV is associated with approximately 80% of Merkel cell carcinoma (MCC), an aggressive neuroendocrine carcinoma of the skin with high mortality [159, 160]. Antiviral development for PyVs has been hampered by the absence of optimal animal models since human PyVs are species-specific; however, a mouse model using murine polyomavirus (MuPyV) showed virologic efficacy of BCV treatment, including a 10-fold decrease in kidney viral load [161]. While there are no specific agents approved to treat human PyV-associated diseases, reduction of immunosuppressive agents or other means to improve host immunity (e.g. optimization of HIV antiviral therapy, cellular immunotherapy) have been used, sometimes in conjunction with off-label agents such as CDV, BCV, leflunomide and/or quinolones [166–175]. Current public data with BCV is limited to BKV and JCV which will be the focus here.

15.5.4.1 BKV Clinical Data with CDV/BCV

There are several published clinical reports of CDV or BCV in the treatment of BKV (PVAN in kidney transplants or HC in HCT recipients) and for JCV in patients with PML; as expected, the main CDV-related toxicity was renal failure [162]. Several reports describe the antiviral efficacy of BCV as measured by reductions in BKV-DNAemia and clinical improvement of PVAN in transplant patients without renal toxicity [163, 164]. An early dose escalating PK study of BCV enrolled subjects with BKV but the top dose evaluated (40 mg) was too low for efficacy (Clintrial.gov identifier NCT00793598).

Higher oral BCV doses (100–200 mg) were used in CMX001-201 (ClinicalTrials.gov Identifier: NCT00942305, Table 15.3) a Phase 2 study in HCT recipients focused on the prevention of CMV disease [24]. At baseline, 44% of subjects (n = 101/230)

Table 15.12 Exploratory analysis of BCV effect on BKV viruria-associated hematuria and renal impairment.

	BKV + urine BCV (N = 94)	BKV + urine placebo (N = 32)	BKV − urine BCV (N = 77)	BKV − urine placebo (N = 27)
Treatment emergent hematuria[a]	6.4%	25.0%	7.8%	3.7%
Creatinine increase to >120 μM/L	14.9%	34.4%	13.0%	11.1%
Renal Impairment[b]	7.5%	18.8%	7.8%	7.4%

a. Confirmed microscopic hematuria.
b. Creatinine >120 μM/L and ≥25% increase from baseline at last measure.
Source: Adapted from Ref. [23].

had BKV viremia while 55% (n = 126/230) had BKV in urine at some point during treatment [24]. As shown in Table 15.12, placebo patients with BKV detected in urine at baseline had higher incidences of hematuria, creatinine increases, and renal impairment compared to placebo recipients without BKV in urine at baseline while BCV-treated patients had similar rates irrespective of BKV detection. The incidence of treatment-emergent hematuria was significantly (P = 0.008) higher for placebo-treated patients who had BKV in urine (viruria) at baseline (25%) when compared to patients with BKV viruria receiving BCV (6.4%). Similarly, creatinine increases above 120 μM/L were more frequent (P = 0.02) in subjects with BKV viruria who received a placebo (34.4%) as compared to those treated with BCV (14.9%). While signs of renal impairment were also more than twice as frequent in placebo-treated patients with BKV in urine than for those receiving BCV, the difference was not significant (P = 0.09) [23].

Consistent with these findings, laboratory signs of end-organ disease were significantly (P = 0.003) lower for BCV-treated patients (20%, n = 19/94) versus placebo (50%, n = 16/32) when BKV was detected in urine, but these differences were not observed in patients without BKV viruria. Further analysis showed a significantly improved glomerular filtration rate (GFR) for subjects receiving 100 mg twice a week compared to placebo by week 8 of dosing [165].

Overall, the BKV subanalysis from the clinical trial CMX001-201 suggested BCV may be useful in preventing or treating BKV-associated renal and bladder complications post-HCT [23].

15.5.4.2 JCV Clinical Data with CDV/BCV

The use of CDV for the treatment of PML has shown controversial results with some reports showing significant clinical improvement [166–169], while others report discouraging outcomes [170–172]. Oral BCV has been evaluated in several PML cases. One case reported JCV viral load reduction and stabilized neurological function with BCV, following unsuccessful treatment with risperidone, IV CDV, and mefloquine [173]. Another report of two lung transplant patients with PML showed initial improvement of neurologic function using the Kurtzke disability scale with BCV treatment but both patients eventually declined [174]. A failure with BCV therapy has also been reported in 2 non-HIV PML patients who had a significant improvement after initiation of BCV, but the disease progressed with a fatal outcome [175]. Additional case studies of patients with PML related to natalizumab (for multiple sclerosis) have shown some promise with evidence for decreased JCV viral load in the CSF during CDV or BCV treatment with mirtazapine and neurologic recovery in some patients but also noted GI toxicity [176]. Early initiation of BCV with concurrent mirtazapine may have potential to arrest the progression of natalizumab-related PML and help in neurologic recovery. The improved CNS penetration of BCV administered IV may improve efficacy for this indication relative to IV CDV or oral BCV. More data are needed to support BCV as a valid option for the treatment of PML.

15.6 Neurodegenerative Diseases

As described above for PML, the use of CDV and BCV for the treatment of viral and neoplastic disease in the CNS has been hampered by relatively poor penetration into the brain with IV CDV and oral BCV. Development of the IV formulation of BCV combined with mounting evidence for herpesvirus involvement in serious neurodegenerative diseases has increased enthusiasm for testing BCV as a novel therapeutic in several indications, including multiple sclerosis (related to EBV) and Alzheimer's disease (related to HSV and VZV).

15.6.1 EBV and Multiple Sclerosis

EBV is one of the most common herpesviruses with a prevalence of around 90% in adults where it persists in memory B cells. Typical disease is limited to infectious mononucleosis in children and adolescents. However, it is clearly causative for some lymphoproliferative disorders and recent studies support a causal relationship between Multiple Sclerosis (MS) and EBV [29, 177]. The proposed mechanism(s) by which EBV might cause MS are varied, but the most important questions for this section center on the potential utility of BCV as a therapeutic intervention in MS. As background, there are publications providing evidence of active EBV replication in the CNS of MS patients and restriction of lytic EBV infection detection to MS lesions in adult brain tissue [178, 179]. Recent reports have also shown reduced relapsing-remitting MS (RRMS) in HIV patients on antiviral therapy for HIV (highly active antiretroviral therapy [HAART]), suggesting the potential benefit of repurposing HIV antivirals for MS, although the link to specific antivirals and potential mechanisms is not clear [180]. Additionally, a 20-year study that followed 10 million active-duty members of the US military found that MS risk increased over 30-fold following EBV infection, but not after infection with other viruses (including CMV) and concluded that "these findings cannot be explained by any known risk factor for MS and suggest EBV as the leading cause of MS" [29].

Taken together, the available data suggest that antivirals active against EBV are good candidates for treatment of MS and support the rationale behind an ongoing clinical study with tenofovir disoproxil fumarate (TDF), administered with emtricitabine as Truvada, in MS patients (ClinicalTrials.gov identifier NCT05957913). While there is evidence of EBV activity in the clinic for several antivirals (e.g. ACV, GCV, and CDV [Table 15.2]) for indications other than MS (e.g. PTLD, lymphomas), potency, safety, and penetration of relevant tissues are limiting factors that have diminished enthusiasm for these to date in MS and acyclovir/valacyclovir failed to demonstrate benefit in MS [181]. While tenofovir has a safety advantage over BCV, BCV may have an efficacy advantage due to higher potency and better CNS penetration, especially when administered IV. Although TDF and tenofovir alafenamide (TAF) are relatively potent antivirals against EBV lytic infection *in vitro*, TDF is rapidly hydrolyzed to TFV in blood and TFV ((R)-9-(2-phosphonomethoxypropyl)-adenine [(R)-PMPA]) is not a potent inhibitor of EBV (EC50s >10 μg/mL) as monitored by inhibition of EBV DNA replication and virus capsid antigen (VCA) expression [182, 183]. These data combined with the reported low penetration of TAF in cerebrospinal fluid suggest current prodrugs for TFV may be suboptimal for inhibition of EBV in the CNS [184], (ClinicalTrials.gov Identifier: NCT04880577). The rationale for evaluating BCV as a potential treatment for worsening MS is based on the high morbidity of the disease, evidence for a causal association between active EBV infection and MS, BCV potency against EBV (EC_{50} 0.03 μM, Table 15.2), and enhanced penetration of the CNS with IV BCV [19].

15.6.2 HSV and Alzheimer's Disease

HSV-1 is another common herpesvirus that establishes latency, primarily in the peripheral nervous system. The most common symptoms of HSV-1 are self-resolving vesicular lesions in and around the mouth, although it is associated with multiple conditions and may penetrate the CNS where it can cause encephalitis, especially in neonates and immunocompromised individuals. A possible link between HSV-1 and Alzheimer's disease (AD) was postulated in part due to similarities in anatomical sites and symptoms observed for HSV-1 encephalitis and AD. The majority of HSV-1 in brain tissue from patients with AD was associated with amyloid beta (Aβ) plaques and epidemiology studies correlated HSV-1 infection and reactivation with AD. For example, a large retrospective study found a 2.5-fold increased risk of AD associated with HSV infection and a large (10-fold) decrease in this risk with antiviral therapy [32]. Like EBV in MS, the exact (potentially causal) mechanism remains somewhat elusive but appears to involve reactivation of HSV-1 (perhaps stimulated by VZV) and the host response to this reactivation (e.g. tau phosphorylation, upregulation of extracellular Aβ and entrapment of virions in plaques); APOE-ε4 carriers appear especially susceptible [185–188]. Valacyclovir studies are ongoing (ClinicalTrials.gov NCT03282916 and NCT02997982) in symptomatic AD patients, but questions around the timing of therapy with respect to disease state, penetration of the antiviral to the relevant anatomical site, and potency remain [187, 189].

The interest in BCV for preemption and/or treatment of AD is based on its potency against HSV-1 (EC_{50} 0.01μM) and VZV (EC_{50} 0.0004 μM) as shown in Table 15.2, improved virologic efficacy in the brain versus ACV against HSV in a mouse model [67], synergistic antiviral activity with ACV *in vitro* and in a lethal mouse model of HSV-1 [151], enhanced penetration of the CNS with IV BCV [19], and preliminary clinical data supporting virologic efficacy against HSV-1 and VZV. There are several published case reports of successful treatment of ACV-resistant HSV with BCV [190–192]. Additionally, BCV was successful in managing disseminated recurrent/refractory VZV in an HCT patient where ACV and CDV were ineffective [193].

15.6.3 HHV-6

Human herpesvirus 6 (HHV-6) infection occurs early in childhood and typically remains latent in immunocompetent individuals; however, HHV6-B commonly reactivates following HCT and may (infrequently) cause severe disease including fatal encephalitis [194, 195]. The SUPPRESS study described in Section 15.5.3 was primarily designed to evaluate BCV prophylaxis of CMV but it also provided an opportunity to study BCV prevention of HHV-6B reactivation. Briefly, 153 subjects (92 BCV: 61 placebo) from SUPPRESS who received at least 6 doses of BCV had plasma samples evaluated for HHV-6B. "The cumulative incidence of HHV-6B plasma detection through day 42 post-HCT was significantly lower among patients receiving BCV (14.2%) compared with placebo (32.4%; log-rank, 0.019)" [196]. Although HHV-6-associated encephalitis is uncommon, it is strongly associated with plasma HHV-6 viremia above 10,000 copies/mL [197]. While the low incidence of HHV-6B encephalitis complicates the conduct of clinical studies specific to HHV-6, the common reactivation of these viruses in patients with severe AdV infections (approximately 50%) may permit additional evaluation in case studies of HHV-6 during ongoing clinical trials of IV BCV in patients with severe AdV [17, 198].

15.7 Antiproliferative Activity

CDV and BCV are best known as antivirals, but they have also demonstrated significant antineoplastic activities in multiple systems, including against tumors of viral and non-viral origin. For example, there is evidence for the activity of CDV against PyV-associated tumors (e.g. hemangioma and hemangiosarcoma) both *in vitro* and in animal models [199, 200]. A virus-independent example is the efficacy of CDV in a mouse xenograft model of vascular tumors originating from basic fibroblast growth factor (FGF2)-overexpressing endothelial cells (FGF2-T-MAE) [200]. Effects associated with CDV treatment included inhibition of FGF2 expression, S-phase arrest, up-regulation of p53, induction of apoptosis, and inhibition of metastasis [201, 202]. CDV also demonstrated significant activity in a mouse model of aggressive melanoma unrelated to oncogenic viruses and there is a clinical case report of regression of a cutaneous melanoma metastasis after intralesional CDV [203, 204]. Unravelling the mechanisms underlying antitumor activities of CDV, and to a lesser extent BCV per se, has been the aim of numerous research studies. Where evaluated, BCV and CDV share intracellular mechanisms of action, suggesting proof-of-concept studies with CDV will translate to BCV and provide substantial support for further evaluation of BCV as a therapeutic for several cancers. Here we will focus on glioblastoma and tumors commonly associated with HPV and EBV.

15.7.1 Papillomavirus-associated Tumors

15.7.1.1 *In Vitro* Activity Against Papillomavirus-associated Tumors and MOA

Since the first report describing CDV activity against human papillomavirus (HPV) positive cells in 1998, several studies have shown its potential for the treatment of HPV-induced benign and malignant diseases. CDV hindered the proliferation of HPV-positive cells in a dose- and time-dependent manner compared to normal primary human keratinocytes (PHKs) [52]. A follow-up study by Johnson and Gangemi also showed that CDV irreversibly and selectively suppressed the proliferation of HPV-16 transformed cells in a dose-dependent manner compared to PHKs [51]. Compared to uninfected control PHKs, CDV-treated HPV-16 transformed cells displayed a decrease in thymidine incorporation and S-phase arrest, pointing to an inhibitory effect of host DNA polymerase activity. These results are consistent with incorporated CDV presenting a block to first and second-strand synthesis as observed with CMV and orthopoxvirus polymerases and, potentially, diminished ability of transformed cells to excise/repair incorporated CDV as discussed in more detail below. One example of this is the 4–9-fold higher incorporation of CDV into cellular DNA by papillomavirus-transformed cell lines (SiHa, HeLa, HaCat) as compared to primary human keratinocytes [205].

HPV-transformed cervical carcinoma cells do not need to produce infectious virus because an integrated HPV genome expressing E6 and E7 viral oncoproteins is sufficient for neoplasia. Hence, CDV inhibitory effects can be explained by an antiproliferative effect on HPV-transformed cells and do not require an antiviral effect [52]. The antiproliferative effect of CDV on HPV-transformed cells may be explained, at least in part, by a selective inhibitory effect on rapidly dividing cells, and the HPV genome in transformed cells might enhance the sensitivity to CDV considering that CDV showed antiproliferative effects against HPV(-) tumor cells such as human melanomas, lung carcinomas, colon carcinomas, breast carcinomas albeit with reduced potency compared to HPV(+) cells [52]. In HPV-transformed cells, CDV treatment increased levels of the tumor suppressor proteins p53 and pRb (which are degraded by the HPV oncoproteins E6 and E7, respectively), increased levels of p21/WAF1, led to accumulation of cells in S-phase and induced apoptosis. [206, 207]. Notably, genes involved in cell cycle regulation and double-strand breaks (DSB) repair mechanisms (i.e. ATM [ataxia telangiectasia mutated kinase] signaling and DSB repair by homologous recombination) were activated in CDV-exposed PHKs but not in transformed cells, suggesting DDR and/or cell cycle regulation deficiencies may be responsible for selective accumulation of DSBs following CDV exposure in HPV-transformed cells versus PHKs.

Similar studies demonstrated that CDV incorporation into cellular DNA, phospho-ATM levels, and γ-H2AX (a quantitative marker of DSBs) were all selectively higher in HPV(+) and HPV(-) cancer cell lines compared to normal cells [27]. In aggregate these studies suggest CDV is selectively toxic to transformed cells in part by inducing DSBs following failure of DNA repair subsequent to incorporation. Normal keratinocytes appear to detect CDV incorporation as DNA damage, activate cell cycle regulation, and remove the incorporated CDV via excision repair or homologous recombination; this results in higher genomic stability and survival following CDV exposure.

While these studies were all conducted using CDV, it is assumed that BCV will act via the same mechanisms but at lower systemic drug exposures. One study compared BCV and other alkoxyalkyl esters of CDV to CDV for their antiproliferative effects against both HPV-negative [HPV(-)] and HPV-positive [HPV(+)] cervical cancer cells [208]. The HPV(+) CaSki (HPV-16), HeLa (HPV-18), and Me-180 (HPV-18, 39, and 68) cells, the cervical cancer cell line HPV(-) C33A, and the nonmalignant primary human foreskin fibroblast cells were investigated. BCV was 18 to 1,230 times more active than CDV depending on the cell line.

15.7.1.2 Animal Studies

CDV significantly reduced cervical carcinoma tumor burden in an athymic nude mouse xenograft model, accompanied by enhanced tumor protein levels of p53 and p-pRb and increased apoptosis [209, 210]. Additionally, a double human cervical xenograft model in nude mice demonstrated that reduction of the primary tumor with local CDV treatment also had a temporary far-reaching effect resulting in delayed growth of a distal tumor. This effect was associated with reduced tumor-driven inflammation and enhanced anti-tumor immune response, suggesting CDV killing of primary tumor cells might influence the immune system to fight cancer, possibly by exposing novel tumor antigens [211]. Another study combined CDV with irradiation both *in vitro* and in xenografted nude mice, resulting in a marked radiosensitization of HPV-positive cells in the presence of CDV, which was not observed in HPV-uninfected cells [207].

15.7.1.3 Clinical Studies

Clinical studies evaluating the efficacy of CDV in the treatment of HPV-related diseases have been reviewed in recent publications [212–216]. Briefly, several trials demonstrated the potential of CDV to treat patients with cervical intraepithelial neoplasia (CIN), an aberrant cellular proliferation clearly associated with HPV. In one, topical CDV (1% gel) was applied three times, every other day, on the cervix of 15 women with biopsy-proven CIN III [217]. Surgical removal of the cervix was carried out within 1 month after the start of treatment and HPV PCR was performed. Complete response by histological evaluation was found in 7 of the 15 patients, which was confirmed by PCR in 4 of the 7 patients. Five patients had a partial response and the drug proved non-toxic to the normal epithelium, i.e. CDV was selectively toxic to the dysplastic epithelium.

A subsequent randomized controlled trial evaluated topical CDV treatment for CIN II and III (CIN II+) [218]. Patients (N = 53) with a biopsy-proven CIN II+ were randomly assigned either 3 applications of 2% CDV or placebo 6 weeks before surgical excision of the cone-shaped portion of the cervix bearing the transformation zone. Forty-eight out of the 53 patients were treated (23 CDV and 25 placebo) and followed according to the protocol. In the CDV group, 14 of the 23 cones were free of any CIN (60.8%) versus only 5 of 25 cones (20%) in the placebo group. CDV had no systemic toxicity and cervicovaginal side effects were limited and comparable to the placebo group. Although this study showed that CDV cannot replace conization, CDV could be considered as a candidate for topical chemotherapy of CIN II+ lesions.

Lastly, a Phase I trial evaluated CDV in combination with chemoradiation in cervical cancer patients to determine safety and the recommended Phase II dose [28]. A total of 15 patients were treated with CDV (1, 2.5, 5, or 6.5 mg/kg) administered intravenously weekly for two weeks, and then every 2 weeks from the start of chemoradiotherapy until initiation of utero-vaginal brachytherapy. Women had tumor assessment at inclusion and during the follow-up. In 8 of 9 evaluable patients, complete responses (89%) were seen. At the first evaluation (end of treatment), one patient experienced progressive disease. The 4-year overall survival was 93%, and 2-year progression-free survival was 76%. Immunohistochemical analyses of HPV-related markers in sequential tumor biopsy samples demonstrated a relative decrease in p16 expression, and a relative increase in p53 and pRb levels at week 4 compared with baseline. This study showed that CDV at a dose of 5 mg/kg in combination with chemoradiotherapy was tolerable and afforded tumor regression, though the recommended Phase II dose could not be determined because of premature trial discontinuation.

Oral BCV was evaluated in a pilot study of three patients suffering from HPV-related laryngeal diseases [219]. Two patients had laryngeal recurrent respiratory papillomatosis due to infection with the low-risk types HPV-6 and HPV-11 and the third suffered from recurring aryepiglottic fold carcinoma in situ associated with the high-risk type HPV-16. Papilloma burden was reduced by BCV in 1 patient and a marked attenuation of the rate and magnitude of recurrence was noted in the other patients. After surgery, Patient 1 remained disease-free at 10 years and Patient 2 was symptom-free at 8 years. Patient 3, suffering from recurring carcinoma in situ, needed less frequent resections and the specimens showed lower levels of dysplasia (present only in islands within normal mucosa at 8 years), with no evidence of disease at 21 months from last intervention.

An additional area of high unmet need under this broader category is head and neck cancer squamous cell carcinoma (HNSCC) which is associated with approximately 330,000 deaths annually. The current standard of care for HNC patients is ionizing radiation plus cisplatin. Although this treatment is relatively effective, it is associated with high morbidity and is contraindicated for patients with renal insufficiency. Treatment of recurrent/metastatic HNSCC is considerably less effective and more diverse [220]. BCV has been proposed as a more tolerable alternative chemotherapy to cisplatin and as an adjunct to standard of care for recurrent/refractory HNSCC. Considering its likely mechanism(s) of action, BCV could be effective against HNC when combined with agents that cause DNA damage, inhibit DNA repair, and inhibit immune checkpoints.

15.7.2 EBV-associated Tumors

Over 25 years ago Neyts et al. demonstrated CDV inhibition of EBV-associated tumor nasopharyngeal carcinoma growth in an experimental mouse model by a mechanism unrelated to its antiviral activity [221]. Subsequent studies showed that CDV treatment resulted in downregulation of the EBV oncoprotein latent membrane protein 1 (LMP1) with a concomitant decrease of the antiapoptotic Bcl-2 protein and an increase of the proapoptotic Bax protein in a Burkitt lymphoma cell line

(Raji) and a nasopharyngeal carcinoma cell line (C15) [34]. In the Burkitt lymphoma BL2 cell line infected with the B95.8 strain of EBV, CDV significantly reduced LMP1 and EBV nuclear protein 2 (EBNA2) transcripts and protein expression, inhibited cell proliferation, stimulated apoptosis, and decreased Bcl-2 expression. Moreover, CDV in combination with ionizing radiation proved synergistic both *in vitro* and *in vivo*. In combination with the ribonucleotide reductase inhibitors hydroxyurea and didox (3,4-dihydroxybenzohydroxamic acid), CDV-induced apoptosis in EBV-transformed epithelial cells and in EBV-positive nasopharyngeal carcinoma xenografts [222]. Enhanced apoptosis through the proteolytic cleavage of the death effectors caspase-9 and -3 was specifically induced by the combined treatment in EBV-positive cells compared to their negative counterparts. In two human EBV-positive cancer xenograft models (Raji and C15 cells) in nude mice, the combined treatment led to complete tumor remission without increasing toxicity. Clinically, successful CDV treatment of locally recurrent EBV-associated nasopharyngeal carcinoma was reported in two patients [223].

More recently Chan et al. tested BCV against a panel of natural killer/T-cell lymphoma (NKTCL) lines which are associated with EBV infection. BCV was >1000-fold more active than CDV against these lines, demonstrating significant *in vitro* and *in vivo* (mouse xenograft) activity independent of EBV [35]. Two of the most sensitive lines (KAI-3 and NK-S1) were evaluated thoroughly, revealing S-phase arrest, downregulation of MYC target pathways, upregulation of PDL-1, activation of p53, increased DSBs, activation of STING, and increased calreticulin expression. Together, these data support the development of novel anticancer strategies utilizing BCV to augment current treatments, including immunotherapies, in EBV-related cancers.

15.7.3 Glioblastoma

Glioblastoma is the most common malignant brain tumor in adults. It is a type of astrocytoma arising from the glial cells or their progenitors in the CNS and has a dismal prognosis (median survival <2 years) [224]. The standard approach for glioblastoma is surgery followed by radiation and chemotherapy; however, recurrence is almost universal, highlighting a clear need for novel therapeutic options. Biological factors, including distinct tumor genetics, immunological microenvironments and the blood–brain barrier, are important challenges in the development of new therapies.

The potential of BCV as an alternative therapeutic option for glioblastomas is based in large part on a study reported by Hadaczek [26]. CDV cytotoxicity against the glioblastoma cells U87MG and primary SF7796 was evaluated both *in vitro* and *in vivo* in an intracranial xenograft model. CDV was preferentially incorporated in U87MG tumor cell DNA versus non-transformed controls (MRC5 fibroblasts) despite similar doubling times when MRC5s were in log growth, leading to increased double-stranded DNA breaks and preferential apoptosis in tumor cells. Unsurprisingly, CDV incorporation was inversely proportional to in vitro CC_{50}, with more incorporation corresponding to a lower CC_{50}. CDV also augmented radiation-induced DNA damage, increasing a marker for DSB in DNA (P-H2AX) by 21-fold in U87MG cells as compared to radiation alone. Additionally, CDV promoted tumor cell death in mice intracerebrally implanted with human glioblastoma cells and significantly extended survival. Notably, CDV-induced glioblastoma cell death was independent of CMV infection. Considering this mechanism of cell killing by CDV, combining BCV with ionizing radiotherapy (standard care for glioblastoma) is a promising strategy. SymBio is evaluating the potential of BCV as a treatment for refractory brain tumors and further elucidate its mode of action. As previously described in Section 15.4, IV administration of BCV substantially increases brain exposure relative to oral BCV or IV CDV while retaining the respective GI and renal safety advantages [19, 21, 69].

15.7.4 Antiproliferative MOA

The selective cytotoxicity of CDV to transformed cells versus normal cells has been attributed in part to higher incorporation of CDV into cellular DNA. This could occur via increased intracellular formation of CDV-PP and/or enhanced retention in transformed cell genomes due to the reduced/altered DNA damage repair and/or cell cycle control observed in many transformed cells. The end result is increased double-stranded DNA breaks and cell death [27, 38, 205]. Additional effects, some of which may be unrelated to incorporation of CDV into host cell DNA, have been described in different systems as discussed above. Based on published studies in viral-induced and non-viral induced tumors, the selective antiproliferative activity of CDV/BCV against transformed cells may be different depending on cell-specific properties (e.g. genetics, epigenetics, metabolism, mode of transformation) and may include one or more of the following mechanisms:

1. Increased production of the active drug (CDV-PP) in transformed cells versus normal cells.
2. Increased incorporation/retention of CDV into cellular DNA and increased DNA damage (e.g. double-stranded DNA breaks) in cancer cells compared to normal cells (Figure 15.4).
3. Increased levels of tumor suppressor proteins (e.g. pRb and p53) in treated cancer cells versus untreated cancer cells.
4. Selective or exclusive activation in immortalized cells of several pathways associated with host control of neoplasia following exposure to CDV or BCV, including Rho GTPase and STING, among others.

15.8 Summary and Future Directions

BCV is a lipid conjugate (hexadecyloxypropyl-cidofovir) that dramatically alters the ADME profile of CDV while enabling oral delivery and effectively eliminating the nephrotoxicity of CDV. Oral BCV is approved for the treatment of smallpox. The utility of BCV as a countermeasure for potential bioterrorism with smallpox relies on several characteristics, including the ability to treat large numbers of patients rapidly and easily with a convenient 1 week oral dosing regimen, the long intracellular half-life of the active antiviral (CDV-PP), synergy with tecovirimat, and BCV's

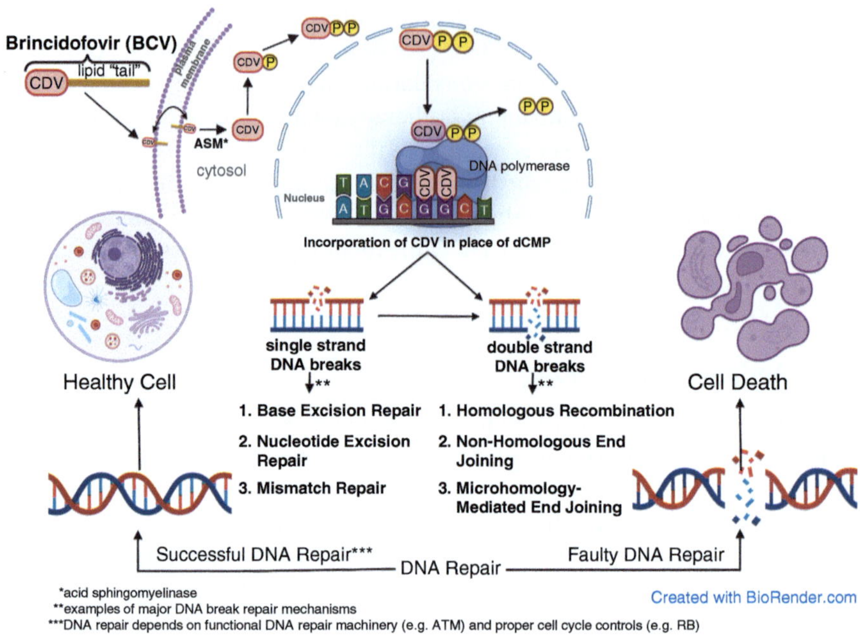

Figure 15.4 Proposed mechanism for selective induction of apoptosis by BCV in cancer cells with altered DDR.

high barrier to resistance for orthopoxviruses [64]. The lipid conjugate-mediated increase in cell uptake for BCV enhances its potency relative to CDV, typically by >100-fold, while retaining CDV's broad-spectrum antiviral activity against dsDNA viruses and selective antiproliferative activity in transformed cells.

The most advanced clinical programs for BCV (beyond the approved smallpox indication) are evaluating IV BCV as a treatment for AdV and CMV infection [21]. IV administration of BCV appears to substantially reduce the gastrointestinal toxicity associated with longer term use of oral BCV [21, 25, 68]. The increased concentration of BCV in the CNS with IV administration may enable development for indications that are presumed to require effective concentrations in brain tissue (e.g. glioblastoma, HSV/VZV in Alzheimer's disease, EBV in Multiple Sclerosis). While the evidence base for these indications is currently preclinical, studies are progressing to enable clinical proof of concept. One important aspect of data pertaining to CDV/BCV use in cancer indications is that viral infection/transformation is not a requirement for efficacy in vitro or in animal models. Another is that the precise mechanism(s) of action may be different for different tumor types and even amongst cancers of the same tumor type.

Covalently attaching the hexadecyloxpropyl lipid moiety to the phosphonate of CDV to form BCV improved the distribution, activity, and safety of CDV, enabling indications that cannot be addressed practically with CDV. Dose regimen and route of administration remain critical elements to optimize the risk-benefit profile of BCV for each indication. Overall, BCV has the potential to address significant unmet needs as both an antiviral and anti-cancer therapeutic for critically ill patients.

Acknowledgments

The authors wish to thank Odin Naderer, Heidi Colton, Roy Ware, Marion Morrison, and Tim Tippin for valuable insights and assistance in the preparation of this review.

Data presented in this review is restricted to what was available in the public domain in mid-2024.

References

1 Hostetler KY. Synthesis and early development of hexadecyloxypropylcidofovir: an oral antipoxvirus nucleoside phosphonate. *Viruses*. 2010;2(10):2213–25.
2 Chan-Tack K, Harrington P, Bensman T, Choi SY, Donaldson E, O'Rear J, et al. Benefit-risk assessment for brincidofovir for the treatment of smallpox: U.S. food and drug administration's evaluation. *Antiviral Res*. 2021;195:105182.
3 Cihlar T, Chen MS. Identification of enzymes catalyzing two-step phosphorylation of cidofovir and the effect of cytomegalovirus infection on their activities in host cells. *Mol Pharmacol*. 1996;50(6):1502–10.
4 De Clercq E, Holy A. Acyclic nucleoside phosphonates: a key class of antiviral drugs. *Nat Rev Drug Discov*. 2005;4(11):928–40.
5 Cesaro S, Pillon M, Tridello G, Aljurf M, Martino R, Schroyens W, et al. Relationship between clinical and BK virological response in patients with late hemorrhagic cystitis treated with cidofovir: a retrospective study from the European group for blood and marrow transplantation. *Bone Marrow Transplant*. 2013;48(6):809–13.
6 Riggsbee DL, Alali M, Kussin ML. Cidofovir for viral infections in immunocompromised children: guidance on dosing, safety, efficacy, and a review of the literature. *Ann Pharmacother*. 2024;58(3):286–304.
7 Snoeck R, De Clercq E. Role of cidofovir in the treatment of DNA virus infections, other than CMV infections, in immunocompromised patients. *Curr Opin Investig Drugs*. 2002;3(11):1561–6.
8 Connelly MC, Robbins BL, Fridland A. Mechanism of uptake of the phosphonate analog (S)-1-(3-hydroxy-2-phosphonylmethoxypropyl)cytosine (HPMPC) in Vero cells. *Biochem Pharmacol*. 1993;46(6):1053–7.
9 Hostetler KY. Alkoxyalkyl prodrugs of acyclic nucleoside phosphonates enhance oral antiviral activity and reduce toxicity: current state of the art. *Antiviral Res*. 2009;82(2):A84–98.
10 Wan WB, Beadle JR, Hartline C, Kern ER, Ciesla SL, Valiaeva N, et al. Comparison of the antiviral activities of alkoxyalkyl and alkyl esters of cidofovir against human and murine cytomegalovirus replication in vitro. *Antimicrob Agents Chemother*. 2005;49(2):656–62.
11 Aldern KA, Ciesla SL, Winegarden KL, Hostetler KY. Increased antiviral activity of 1-O-hexadecyloxypropyl-[2-(14)C]cidofovir in MRC-5 human lung fibroblasts is explained by unique cellular uptake and metabolism. *Mol Pharmacol*. 2003;63(3):678–81.

12 Ciesla SL, Trahan J, Wan WB, Beadle JR, Aldern KA, Painter GR, et al. Esterification of cidofovir with alkoxyalkanols increases oral bioavailability and diminishes drug accumulation in kidney. *Antiviral Res.* 2003;59(3):163–71.

13 Hill JA, Mayer BT, Xie H, Leisenring WM, Huang ML, Stevens-Ayers T, et al. The cumulative burden of double-stranded DNA virus detection after allogeneic HCT is associated with increased mortality. *Blood.* 2017;129(16):2316–25.

14 Hill JA, Moon SH, Chandak A, Zhang Z, Boeckh M, Maziarz RT. Clinical and economic burden of multiple double-stranded DNA viral infections after allogeneic hematopoietic cell transplantation. *Transplant Cell Ther.* 2022;28(9):619 e1–8.

15 Huang YT, Kim SJ, Lee YJ, Burack D, Nichols P, Maloy M, et al. Co-infections by double-stranded DNA viruses after ex vivo t cell-depleted, CD34(+) selected hematopoietic cell transplantation. *Biol Blood Marrow Transplant.* 2017;23(10):1759–66.

16 Sedlacek P, Petterson T, Robin M, Sivaprakasam P, Vainorius E, Brundage T, et al. Incidence of adenovirus infection in hematopoietic stem cell transplantation recipients: findings from the advance study. *Biol Blood Marrow Transplant.* 2019;25(4):810–518.

17 Papanicolaou GA, Dvorak CC, Dadwal S, Maron G, Prasad VK, Giller R, et al. Practice patterns and incidence of adenovirus infection in allogeneic hematopoietic cell transplant recipients: multicenter survey of transplant centers in the United States. *Transpl Infect Dis.* 2020;22(4):e13283.

18 Cesaro S, Porta F. Adenovirus infection in pediatric hematopoietic cell transplantation: a challenge still open for survival. *J Clin Med.* 2022;11(16):4827.

19 Tippin T, Srnka A, Savina P, Van Sickle K, Naderer O. Tissue distribution of radioactivity after intravenous and oral administration of brincidofovir to rats. Poster presented at the American Association of Pharmaceutical Scientists Annual Meeting and Exposition November 2016 Denver, CO, USA.

20 Grimley MS, Chemaly RF, Englund JA, Kurtzberg J, Chittick G, Brundage TM, et al. Brincidofovir for asymptomatic adenovirus viremia in pediatric and adult allogeneic hematopoietic cell transplant recipients: a randomized placebo-controlled phase ii trial. *Biol Blood Marrow Transplant.* 2017;23(3):512–21.

21 Grimley M, Maron G, Gomez CA, Prasad VK, Dara J, Papanicolaou GA, et al. Preliminary results of a phase 2a clinical trial to evaluate safety, tolerability and antiviral activity of intravenous brincidofovir (BCV IV) in immunocompromised patients with adenovirus infection. *Blood.* 2023;142.

22 Hiwarkar P, Amrolia P, Sivaprakasam P, Lum SH, Doss H, O'Rafferty C, et al. Brincidofovir is highly efficacious in controlling adenoviremia in pediatric recipients of hematopoietic cell transplant. *Blood.* 2017;129(14):2033–7.

23 Mommeja-Marin H, Marty F, Boeckh M, Winston D, Rowley SD, Godkin S, et al. CMX001, a novel broad spectrum antiviral, may mitigate signs of bk virus (BKV) associated bladder and kidney end-organ damage: 2576. *Transplantation.* 2012;94(10S):576.

24 Marty FM, Winston DJ, Rowley SD, Vance E, Papanicolaou GA, Mullane KM, et al. CMX001 to prevent cytomegalovirus disease in hematopoietic-cell transplantation. *N Engl J Med.* 2013;369(13):1227–36.

25 Marty FM, Winston DJ, Chemaly RF, Mullane KM, Shore TB, Papanicolaou GA, et al. A Randomized, double-blind, placebo-controlled phase 3 trial of oral brincidofovir for cytomegalovirus prophylaxis in allogeneic hematopoietic cell transplantation. *Biol Blood Marrow Transplant*. 2019;25(2):369–81.

26 Hadaczek P, Ozawa T, Soroceanu L, Yoshida Y, Matlaf L, Singer E, et al. Cidofovir: a novel antitumor agent for glioblastoma. *Clin Cancer Res*. 2013;19(23):6473–83.

27 Mertens B, Nogueira T, Stranska R, Naesens L, Andrei G, Snoeck R. Cidofovir is active against human papillomavirus positive and negative head and neck and cervical tumor cells by causing DNA damage as one of its working mechanisms. *Oncotarget*. 2016;7(30):47302–18.

28 Deutsch E, Haie-Meder C, Bayar MA, Mondini M, Laporte M, Mazeron R, et al. Phase I trial evaluating the antiviral agent Cidofovir in combination with chemoradiation in cervical cancer patients. *Oncotarget*. 2016;7(18):25549–57.

29 Bjornevik K, Cortese M, Healy BC, Kuhle J, Mina MJ, Leng Y, et al. Longitudinal analysis reveals high prevalence of Epstein-Barr virus associated with multiple sclerosis. *Science*. 2022;375(6578):296–301.

30 Lanz TV, Brewer RC, Ho PP, Moon JS, Jude KM, Fernandez D, et al. Clonally expanded B cells in multiple sclerosis bind EBV EBNA1 and GlialCAM. *Nature*. 2022;603(7900):321–7.

31 Letenneur L, Peres K, Fleury H, Garrigue I, Barberger-Gateau P, Helmer C, et al. Seropositivity to herpes simplex virus antibodies and risk of Alzheimer's disease: a population-based cohort study. *PLoS One*. 2008;3(11):e3637.

32 Tzeng NS, Chung CH, Lin FH, Chiang CP, Yeh CB, Huang SY, et al. Anti-herpetic medications and reduced risk of dementia in patients with herpes simplex virus infections-a nationwide, population-based cohort study in Taiwan. *Neurotherapeutics*. 2018;15(2):417–29.

33 Shannon-Lowe C, Rickinson AB, Bell AI. Epstein-barr virus-associated lymphomas. *Philos Trans R Soc Lond B Biol Sci*. 2017;372(1732):20160271.

34 Abdulkarim B, Sabri S, Zelenika D, Deutsch E, Frascogna V, Klijanienko J, et al. Antiviral agent cidofovir decreases Epstein-Barr virus (EBV) oncoproteins and enhances the radiosensitivity in EBV-related malignancies. *Oncogene*. 2003;22(15):2260–71.

35 Chan JY, Lee ECY, Chai KXY, Lee JY, Kannan B, Lim BY, et al. Therapeutic repurposing of brincidofovir in natural killer/t-cell lymphoma reveals potent induction of replication stress, sting pathway activation and immunogenic cell death. *Blood*. 2022;140(Supplement 1):2256–7.

36 Williams-Aziz SL, Hartline CB, Harden EA, Daily SL, Prichard MN, Kushner NL, et al. Comparative activities of lipid esters of cidofovir and cyclic cidofovir against replication of herpesviruses in vitro. *Antimicrob Agents Chemother*. 2005;49(9):3724–33.

37 Chemaly RF, Hill JA, Voigt S, Peggs KS. In vitro comparison of currently available and investigational antiviral agents against pathogenic human double-stranded DNA viruses: a systematic literature review. *Antiviral Res*. 2019;163:50–8.

38 Andrei G, Topalis D, De Schutter T, Snoeck R. Insights into the mechanism of action of cidofovir and other acyclic nucleoside phosphonates against polyoma- and papillomaviruses and non-viral induced neoplasia. *Antiviral Res.* 2015;114:21–46.

39 Hartline CB, Gustin KM, Wan WB, Ciesla SL, Beadle JR, Hostetler KY, et al. Ether lipid-ester prodrugs of acyclic nucleoside phosphonates: activity against adenovirus replication in vitro. *J Infect Dis.* 2005;191(3):396–9.

40 Bae A, Sethna P, Brundage TM, Momméja-Marin H, Nichols G, Lanier R. Role of adenovirus species and type on virologic response to brincidofovir. *Biology of Blood and Marrow Transplantation.* 2016;22(3):S55.

41 Hanaoka N, Hazama M, Fukushima K, Fujimoto T. Sensitivity of human mastadenovirus, the causal agent of pharyngoconjunctival fever, epidemic keratoconjunctivitis, and hemorrhagic cystitis in immunocompromised individuals, to brincidofovir. *Microbiol Spectr.* 2022;10(1):e0156921.

42 Xiong X, Smith JL, Chen MS. Effect of incorporation of cidofovir into DNA by human cytomegalovirus DNA polymerase on DNA elongation. *Antimicrob Agents Chemother.* 1997;41(3):594–9.

43 Magee WC, Hostetler KY, Evans DH. Mechanism of inhibition of vaccinia virus DNA polymerase by cidofovir diphosphate. *Antimicrob Agents Chemother.* 2005;49(8):3153–62.

44 Chamberlain JM, Sortino K, Sethna P, Bae A, Lanier R, Bambara RA, et al. Cidofovir diphosphate inhibits adenovirus 5 dna polymerase via both nonobligate chain termination and direct inhibition, and polymerase mutations confer cidofovir resistance on intact virus. *Antimicrob Agents Chemother.* 2019;63(1):10–1128.

45 Chou S, Lurain NS, Thompson KD, Miner RC, Drew WL. Viral DNA polymerase mutations associated with drug resistance in human cytomegalovirus. *J Infect Dis.* 2003;188(1):32–9.

46 Kinchington PR, Araullo-Cruz T, Vergnes JP, Yates K, Gordon YJ. Sequence changes in the human adenovirus type 5 DNA polymerase associated with resistance to the broad spectrum antiviral cidofovir. *Antiviral Res.* 2002;56(1):73–84.

47 Magee WC, Aldern KA, Hostetler KY, Evans DH. Cidofovir and (S)-9-[3-hydroxy-(2-phosphonomethoxy)propyl]adenine are highly effective inhibitors of vaccinia virus DNA polymerase when incorporated into the template strand. *Antimicrob Agents Chemother.* 2008;52(2):586–97.

48 Fanning E, Zhao K. SV40 DNA replication: from the A gene to a nanomachine. *Virology.* 2009;384(2):352–9.

49 DeCaprio JA, Garcea RL. A cornucopia of human polyomaviruses. *Nat Rev Microbiol.* 2013;11(4):264–76.

50 Bernhoff E, Gutteberg TJ, Sandvik K, Hirsch HH, Rinaldo CH. Cidofovir inhibits polyomavirus BK replication in human renal tubular cells downstream of viral early gene expression. *Am J Transplant.* 2008;8(7):1413–22.

51 Johnson JA, Gangemi JD. Selective inhibition of human papillomavirus-induced cell proliferation by (S)-1-[3-hydroxy-2-(phosphonylmethoxy)propyl]cytosine. *Antimicrob Agents Chemother.* 1999;43(5):1198–205.

52 Andrei G, Snoeck R, Piette J, Delvenne P, De Clercq E. Antiproliferative effects of acyclic nucleoside phosphonates on human papillomavirus (HPV)-harboring cell lines compared with HPV-negative cell lines. *Oncol Res.* 1998;10(10):523–31.

53 Topalis D, Lebeau U, Fiten P., Opdenakker G., Andrei G., and Snoeck R. Is the large T antigen a target for the inhibition of SV40 replication?. *Antiviral Res.* 2010;1(86):A27–8.

54 McMullan LK, Flint M, Dyall J, Albarino C, Olinger GG, Foster S, et al. The lipid moiety of brincidofovir is required for in vitro antiviral activity against Ebola virus. *Antiviral Res.* 2016;125:71–8.

55 Liddell AM, Davey Jr RT, Mehta AK, Varkey JB, Kraft CS, Tseggay GK, et al. Characteristics and clinical management of a cluster of 3 patients with ebola virus disease, including the first domestically acquired cases in the united states. *Ann Intern Med.* 2015;163(2):81–90.

56 Florescu DF, Kalil AC, Hewlett AL, Schuh AJ, Stroher U, Uyeki TM, et al. Administration of brincidofovir and convalescent plasma in a patient with ebola virus disease. *Clin Infect Dis.* 2015;61(6):969–73.

57 Dunning J, Kennedy SB, Antierens A, Whitehead J, Ciglenecki I, Carson G, et al. Experimental treatment of ebola virus disease with brincidofovir. *PLoS One.* 2016;11(9):e0162199.

58 Gordon CJ, Tchesnokov EP, Woolner E, Perry JK, Feng JY, Porter DP, et al. Remdesivir is a direct-acting antiviral that inhibits RNA-dependent RNA polymerase from severe acute respiratory syndrome coronavirus 2 with high potency. *J Biol Chem.* 2020;295(20):6785–97.

59 Jockusch S, Tao C, Li X, Anderson TK, Chien M, Kumar S, et al. A library of nucleotide analogues terminate RNA synthesis catalyzed by polymerases of coronaviruses that cause SARS and COVID-19. *Antiviral Res.* 2020;180:104857.

60 Hussien MA, Abdelaziz AEM. Molecular docking suggests repurposing of brincidofovir as a potential drug targeting SARS-CoV-2 ACE2 receptor and main protease. *Netw Model Anal Health Inform Bioinform.* 2020;9(1):56.

61 Hannestad U, Apostolou E, Sjogren P, Bragee B, Polo O, Bertilson BC, et al. Post-COVID sequelae effect in chronic fatigue syndrome: sars-CoV-2 triggers latent adenovirus in the oral mucosa. *Front Med (Lausanne).* 2023;10:1208181.

62 Ho ES, Lin DC, Mendel DB, Cihlar T. Cytotoxicity of antiviral nucleotides adefovir and cidofovir is induced by the expression of human renal organic anion transporter 1. *J Am Soc Nephrol.* 2000;11(3):383–93.

63 Cihlar T, Lin DC, Pritchard JB, Fuller MD, Mendel DB, Sweet DH. The antiviral nucleotide analogs cidofovir and adefovir are novel substrates for human and rat renal organic anion transporter 1. *Mol Pharmacol.* 1999;56(3):570–80.

64 Foster SA, Parker S, Lanier R. The role of brincidofovir in preparation for a potential smallpox outbreak. *Viruses.* 2017;9(11):320.

65 Chittick G, Morrison M, Brundage T, Nichols WG. Short-term clinical safety profile of brincidofovir: a favorable benefit-risk proposition in the treatment of smallpox. *Antiviral Res.* 2017;143:269–77.

66 Caruso Brown AE, Cohen MN, Tong S, Braverman RS, Rooney JF, Giller R, et al. Pharmacokinetics and safety of intravenous cidofovir for life-threatening viral infections in pediatric hematopoietic stem cell transplant recipients. *Antimicrob Agents Chemother.* 2015;59(7):3718–25.

67 Quenelle DC, Lampert B, Collins DJ, Rice TL, Painter GR, Kern ER. Efficacy of CMX001 against herpes simplex virus infections in mice and correlations with drug distribution studies. *J Infect Dis*. 2010;202(10):1492–99.

68 Wire MB, Morrison M, Anderson M, Arumugham T, Dunn J, Naderer O. Pharmacokinetics (PK) and safety of intravenous (IV) brincidofovir (BCV) in healthy adult subjects. *Open Forum Infect Dis*. 2017;4(1):S311.

69 Naderer O, Schuck V, Morrison M, Anderson M, Arumugham T, Dunn J. 1421. IV Brincidofovir (BCV): pharmacokinetics (PK) and safety of multiple ascending doses (MAD) in healthy subjects. *Open Forum Infectious Diseases*. 2018;5(suppl_1):S438–9.

70 Lanier R, Trost L, Tippin T, Lampert B, Robertson A, Foster S, et al. Development of CMX001 for the Treatment of Poxvirus Infections. *Viruses*. 2010;2(12):2740–62.

71 Ruiz JC, Beadle JR, Aldern KA, Keith KA, Hartline CB, Kern ER, et al. Synthesis and antiviral evaluation of alkoxyalkyl-phosphate conjugates of cidofovir and adefovir. *Antiviral Res*. 2007;75(1):87–90.

72 Sethna P, Selleseth D, Bae A, Keilholz L, Lampert B, Lanier R. Brincidofovir (BCV, CMX001) delivers high intracellular concentrations of cidofovir diphosphate. 27th International Conference on Antiviral Research. 2014;152.

73 van Engen CE, Ofman R, Dijkstra IM, van Goethem TJ, Verheij E, Varin J, et al. CYP4F2 affects phenotypic outcome in adrenoleukodystrophy by modulating the clearance of very long-chain fatty acids. *Biochim Biophys Acta*. 2016;1862(10):1861–70.

74 Hutson CL, Kondas AV, Mauldin MR, Doty JB, Grossi IM, Morgan CN, et al. Pharmacokinetics and efficacy of a potential smallpox therapeutic, brincidofovir, in a lethal monkeypox virus animal model. *Msphere*. 2021;6(1):10–1128.

75 Tippin T, Almond M, Painter G, Lampert B. In vitro hepatic metabolism of CMX001, a Lipid Conjugate of Cidofovir, *Drug Metabolism Reviews*. 16th North American Regional International Society for the Study of Xenobiotics Meeting. 2009.

76 Lederman ER, Davidson W, Groff HL, Smith SK, Warkentien T, Li Y, et al. Progressive vaccinia: case description and laboratory-guided therapy with vaccinia immune globulin, ST-246, and CMX001. *J Infect Dis*. 2012;206(9):1372–85.

77 Fenner F. Smallpox: emergence, global spread, and eradication. *Hist Philos Life Sci*. 1993;15(3):397–420.

78 Fenner F, Henderson DA, Arita I, Jezek Z, Ladnyi ID, World Health O. Smallpox and its eradication / F. Fenner [et al.]. Geneva: World Health Organization; 1988.

79 Henderson DA, Inglesby TV, Bartlett JG, Ascher MS, Eitzen E, Jahrling PB, et al. Smallpox as a biological weapon: medical and public health management. Working Group on Civilian Biodefense. *JAMA*. 1999;281(22):2127–37.

80 Merchlinsky M, Albright A, Olson V, Schiltz H, Merkeley T, Hughes C, et al. The development and approval of tecoviromat (TPOXX((R))), the first antiviral against smallpox. *Antiviral Res*. 2019;168:168–74.

81 Arita I. Discovery of forgotten variola specimens at the National Institutes of Health in the USA. *Expert Rev Anti Infect Ther*. 2014;12(12):1419–21.

82 Gellman H. 4 nations thought to possess smallpox: iraq, N. Korea named, two officials say. Washington Post, 5 November 2001. 2001:A1.

83 Kupferschmidt K. Labmade smallpox is possible, study shows. *Science*. 2017;357(6347):115–6.

84 Noyce RS, Lederman S, Evans DH. Construction of an infectious horsepox virus vaccine from chemically synthesized DNA fragments. *PLoS One*. 2018;13(1):e0188453.

85 FDA. U.S. Food and Drug Administration. Summary Minutes of the Antiviral Drugs Advisory Committee Meeting. 2011:14–15.

86 Garver J, Weber L, Vela EM, Anderson M, Warren R, Merchlinsky M, et al. Ectromelia virus disease characterization in the Balb/C mouse: a surrogate model for assessment of smallpox medical countermeasures. *Viruses*. 2016;8(7):203.

87 Adams MM, Rice AD, Moyer RW. Rabbitpox virus and vaccinia virus infection of rabbits as a model for human smallpox. *J Virol*. 2007;81(20):11084–95.

88 Rice AD, Adams MM, Wallace G, Burrage AM, Lindsey SF, Smith AJ, et al. Efficacy of CMX001 as a post exposure antiviral in New Zealand White rabbits infected with rabbitpox virus, a model for orthopoxvirus infections of humans. *Viruses*. 2011;3(1):47–62.

89 Parker S, Crump R, Foster S, Hartzler H, Hembrador E, Lanier ER, et al. Co-administration of the broad-spectrum antiviral, brincidofovir (CMX001), with smallpox vaccine does not compromise vaccine protection in mice challenged with ectromelia virus. *Antiviral Res*. 2014;111:42–52.

90 Andrei G, Gammon DB, Fiten P, De Clercq E, Opdenakker G, Snoeck R, et al. Cidofovir resistance in vaccinia virus is linked to diminished virulence in mice. *J Virol*. 2006;80(19):9391–401.

91 Duraffour S, Andrei G, Topalis D, Krecmerova M, Crance JM, Garin D, et al. Mutations conferring resistance to viral DNA polymerase inhibitors in camelpox virus give different drug-susceptibility profiles in vaccinia virus. *J Virol*. 2012;86(13):7310–25.

92 Stabenow J, Buller RM, Schriewer J, West C, Sagartz JE, Parker S. A mouse model of lethal infection for evaluating prophylactics and therapeutics against Monkeypox virus. *J Virol*. 2010;84(8):3909–20.

93 Hutson CL, Carroll DS, Gallardo-Romero N, Weiss S, Clemmons C, Hughes CM, et al. Monkeypox disease transmission in an experimental setting: prairie dog animal model. *PLoS One*. 2011;6(12):e28295.

94 Adler H, Gould S, Hine P, Snell LB, Wong W, Houlihan CF, et al. Clinical features and management of human monkeypox: a retrospective observational study in the UK. *Lancet Infect Dis*. 2022;22(8):1153–62.

95 Mitja O, Alemany A, Marks M, Lezama Mora JI, Rodriguez-Aldama JC, Torres Silva MS, et al. Mpox in people with advanced HIV infection: a global case series. *Lancet*. 2023;401(10380):939–49.

96 Contag CA, Mische L, Fong I, Karan A, Vaidya A, McCormick DW, et al. Treatment of Mpox with suspected tecovirimat resistance in immunocompromised patient, United States, 2022. *Emerg Infect Dis*. 2023;29(12):2520–3.

97 Howard DS, Phillips IG, Reece DE, Munn RK, Henslee-Downey J, Pittard M, et al. Adenovirus infections in hematopoietic stem cell transplant recipients. *Clin Infect Dis*. 1999;29(6):1494–501.

98 Lion T. Adenovirus infections in immunocompetent and immunocompromised patients. *Clin Microbiol Rev*. 2014;27(3):441–62.

99 Lion T, Baumgartinger R, Watzinger F, Matthes-Martin S, Suda M, Preuner S, et al. Molecular monitoring of adenovirus in peripheral blood after allogeneic bone marrow transplantation permits early diagnosis of disseminated disease. *Blood*. 2003;102(3):1114–20.

100 Kojaoghlanian T, Flomenberg P, Horwitz MS. The impact of adenovirus infection on the immunocompromised host. *Rev Med Virol*. 2003;13(3):155–71.

101 Wold WS, Toth K. New drug on the horizon for treating adenovirus. *Expert Opin Pharmacother*. 2015;16(14):2095–99.

102 Ison MG, Hayden RT. Adenovirus. *Microbiol Spectr*. 2016;4(4):1–14.

103 Prasad VK, Papanicolaou GA, Marón GM, Vainorius E, Brundage TM, Chittick G, et al. 46 - Treatment of adenovirus (AdV) infection in allogeneic hematopoietic cell transplant (allo HCT) patients (pts) with brincidofovir: final 36 week results from the advise trial. *Biol Blood Marrow Transplant*. 2017;23(3, Supplement):S57–8.

104 Brundage TM, Vainorius E, Chittick G, Nichols G. Brincidofovir decreases adenovirus viral burden, which is associated with improved mortality in pediatric allogeneic hematopoietic cell transplant recipients. *Biol Blood Marrow Transplant*. 2018;24(3):S372.

105 Zecca M, Wynn R, Dalle JH, Feuchtinger T, Vainorius E, Brundage TM, et al. Association between adenovirus viral load and mortality in pediatric allo-HCT recipients: the multinational advance study. *Bone Marrow Transplant*. 2019;54(10):1632–42.

106 Mynarek M, Ganzenmueller T, Mueller-Heine A, Mielke C, Gonnermann A, Beier R, et al. Patient, virus, and treatment-related risk factors in pediatric adenovirus infection after stem cell transplantation: results of a routine monitoring program. *Biol Blood Marrow Transplant*. 2014;20(2):250–6.

107 Gonzalez-Vicent M, Verna M, Pochon C, Chandak A, Vainorius E, Brundage T, et al. Current practices in the management of adenovirus infection in allogeneic hematopoietic stem cell transplant recipients in Europe: the advance study. *Eur J Haematol*. 2019;102(3):210–7.

108 Matthes-Martin S, Feuchtinger T, Shaw PJ, Engelhard D, Hirsch HH, Cordonnier C, et al. European guidelines for diagnosis and treatment of adenovirus infection in leukemia and stem cell transplantation: summary of ECIL-4 (2011). *Transpl Infect Dis*. 2012;14(6):555–63.

109 Paolino K, Sande J, Perez E, Loechelt B, Jantausch B, Painter W, et al. Eradication of disseminated adenovirus infection in a pediatric hematopoietic stem cell transplantation recipient using the novel antiviral agent CMX001. *J Clin Virol*. 2011;50(2):167–70.

110 Florescu DF, Grimley M, Bourne E, Chittick G, Brundage T, Mommeja-Marin H, et al. Brincidofovir (BCV) for the treatment of adenovirus (AdV) infection in patients receiving liver transplantation (LT). *Open Forum Infectious Diseases*. 2015;2(suppl_1):1227.

111 Florescu DF, Pergam SA, Neely MN, Qiu F, Johnston C, Way S, et al. Safety and efficacy of CMX001 as salvage therapy for severe adenovirus infections in immunocompromised patients. *Biol Blood Marrow Transplant*. 2012;18(5):731–8.

112 Sulejmani N, Nagai S, Safwan M, Rizzari MD, Raoufi M, Abouljoud MS, et al. Brincidofovir as salvage therapy for adenovirus disease in intestinal transplant recipients. *Pharmacotherapy*. 2018;38(4):470–5.

113 Sudhindra P, Knoll B, Nog R, Singh N, Dhand A. Brincidofovir (CMX001) for the treatment of severe adenoviral pneumonia in kidney transplant recipient. *Cureus*. 2019;11(8):e5296.

114 Hemmersbach-Miller M, Bailey ES, Kappus M, Prasad VK, Gray GC, Alspaugh JA. Disseminated adenovirus infection after combined liver-kidney transplantation. *Front Cell Infect Microbiol*. 2018;8:408.

115 Gray GC, McCarthy T, Lebeck MG, Schnurr DP, Russell KL, Kajon AE, et al. Genotype prevalence and risk factors for severe clinical adenovirus infection, United States 2004–2006. *Clin Infect Dis*. 2007;45(9):1120–31.

116 Lion T, Wold WSM. *Fields Virology: DNA Viruses*. Seventh ed. The Netherlands: Walters Kluwer; 2021.

117 Gordon YJ, Araullo-Cruz TP, Johnson YF, Romanowski EG, Kinchington PR. Isolation of human adenovirus type 5 variants resistant to the antiviral cidofovir. *Invest Ophthalmol Vis Sci*. 1996;37(13):2774–8.

118 Bae A, Sethna P, Brundage T, Mommeja-Marin H, Nichols G. Adenovirus (ADV) virologic response to brincidofovir in patients with evidence of resistance. 55th Interscience Conference on Antimicrobial Agents and Chemotherapy 2015;Abstract #2732.

119 Teira P, Battiwalla M, Ramanathan M, Barrett AJ, Ahn KW, Chen M, et al. Early cytomegalovirus reactivation remains associated with increased transplant-related mortality in the current era: a CIBMTR analysis. *Blood*. 2016;127(20):2427–38.

120 Jakharia N, Howard D, Riedel DJ. CMV infection in hematopoietic stem cell transplantation: prevention and treatment strategies. *Curr Treat Options Infect Dis*. 2021;13(3):123–40.

121 Kotton CN, Kamar N. New insights on CMV management in solid organ transplant patients: prevention, treatment, and management of resistant/refractory disease. *Infect Dis Ther*. 2023;12(2):333–42.

122 Cho SY, Lee DG, Kim HJ. Cytomegalovirus infections after hematopoietic stem cell transplantation: current status and future immunotherapy. *Int J Mol Sci*. 2019;20(11):2666.

123 Rubin RH. The indirect effects of cytomegalovirus infection on the outcome of organ transplantation. *JAMA*. 1989;261(24):3607–9.

124 Nichols WG, Corey L, Gooley T, Davis C, Boeckh M. High risk of death due to bacterial and fungal infection among cytomegalovirus (CMV)-seronegative recipients of stem cell transplants from seropositive donors: evidence for indirect effects of primary CMV infection. *J Infect Dis*. 2002;185(3):273–82.

125 Cantoni N, Hirsch HH, Khanna N, Gerull S, Buser A, Bucher C, et al. Evidence for a bidirectional relationship between cytomegalovirus replication and acute graft-versus-host disease. *Biol Blood Marrow Transplant*. 2010;16(9):1309–14.

126 Avery RK, Alain S, Alexander BD, Blumberg EA, Chemaly RF, Cordonnier C, et al. Maribavir for refractory cytomegalovirus infections with or without resistance post-transplant: results from a phase 3 randomized clinical trial. *Clin Infect Dis*. 2022;75(4):690–701.

127 Marty FM, Ljungman P, Chemaly RF, Maertens J, Dadwal SS, Duarte RF, et al. Letermovir prophylaxis for cytomegalovirus in hematopoietic-cell transplantation. *N Engl J Med*. 2017;377(25):2433–44.

128 Ljungman P, de la Camara R, Robin C, Crocchiolo R, Einsele H, Hill JA, et al. Guidelines for the management of cytomegalovirus infection in patients with haematological malignancies and after stem cell transplantation from the 2017 European Conference on Infections in Leukaemia (ECIL 7). *Lancet Infect Dis*. 2019;19(8):e260–72.

129 Yong MK, Shigle TL, Kim YJ, Carpenter PA, Chemaly RF, Papanicolaou GA. American society for transplantation and cellular therapy series: #4 - cytomegalovirus treatment and management of resistant or refractory infections after hematopoietic cell transplantation. *Transplant Cell Ther*. 2021;27(12):957–67.

130 Yadav DK, Adhikari VP, Yadav RK, Singh A, Huang X, Zhang Q, et al. Antiviral prophylaxis or preemptive therapy for cytomegalovirus after liver transplantation?: a systematic review and meta-analysis. *Front Immunol*. 2022;13:953210.

131 Lanier ER, Foster S, Brundage T, Chou S, Prichard MN, Kleiboeker S, et al. Analysis of mutations in the gene encoding cytomegalovirus dna polymerase in a phase 2 clinical trial of brincidofovir prophylaxis. *J Infect Dis*. 2016;214(1):32–5.

132 Chou S. Rapid in vitro evolution of human cytomegalovirus UL56 mutations that confer letermovir resistance. *Antimicrob Agents Chemother*. 2015;59(10):6588–93.

133 Lischka P, Michel D, Zimmermann H. Characterization of cytomegalovirus breakthrough events in a phase 2 prophylaxis trial of letermovir (AIC246, MK 8228). *J Infect Dis*. 2016;213(1):23–30.

134 Douglas CM, Barnard R, Holder D, Leavitt R, Levitan D, Maguire M, et al. Letermovir resistance analysis in a clinical trial of cytomegalovirus prophylaxis for hematopoietic stem cell transplant recipients. *J Infect Dis*. 2020;221(7):1117–26.

135 James SH. Letermovir resistance in hematopoietic stem cell transplant recipients: the risks associated with cytomegalovirus prophylaxis. *J Infect Dis*. 2020;221(7):1036–8.

136 O'Brien MS, Markovich KC, Selleseth D, DeVita AV, Sethna P, Gentry BG. In vitro evaluation of current and novel antivirals in combination against human cytomegalovirus. *Antiviral Res*. 2018;158:255–63.

137 Couzi L, Helou S, Bachelet T, Moreau K, Martin S, Morel D, et al. High incidence of anticytomegalovirus drug resistance among D+R- kidney transplant recipients receiving preemptive therapy. *Am J Transplant*. 2012;12(1):202–9.

138 Boeckh M, Ljungman P. How we treat cytomegalovirus in hematopoietic cell transplant recipients. *Blood*. 2009;113(23):5711–9.

139 Winston DJ, Ho WG, Bartoni K, Du Mond C, Ebeling DF, Buhles WC, et al. Ganciclovir prophylaxis of cytomegalovirus infection and disease in allogeneic bone marrow transplant recipients. results of a placebo-controlled, double-blind trial. *Ann Intern Med*. 1993;118(3):179–84.

140 Kleiboeker SB. Prevalence of cytomegalovirus antiviral drug resistance in transplant recipients. *Antiviral Res*. 2023;215:105623.

141 James SH, Price NB, Hartline CB, Lanier ER, Prichard MN. Selection and recombinant phenotyping of a novel CMX001 and cidofovir resistance mutation in human cytomegalovirus. *Antimicrob Agents Chemother*. 2013;57(7):3321–5.

142 Chou S, Ercolani RJ, Lanier ER. Novel cytomegalovirus UL54 DNA polymerase gene mutations selected in vitro that confer brincidofovir resistance. *Antimicrob Agents Chemother*. 2016;60(6):3845–8.

143 Cihlar T, Fuller MD, Cherrington JM. Characterization of drug resistance-associated mutations in the human cytomegalovirus DNA polymerase gene by using recombinant mutant viruses generated from overlapping DNA fragments. *J Virol*. 1998;72(7):5927–36.

144 Cihlar T, Fuller MD, Mulato AS, Cherrington JM. A point mutation in the human cytomegalovirus DNA polymerase gene selected in vitro by cidofovir confers a slow replication phenotype in cell culture. *Virology*. 1998;248(2):382–93.

145 Chou S, Wechel LC, Marousek GI. Cytomegalovirus UL97 kinase mutations that confer maribavir resistance. *J Infect Dis*. 2007;196(1):91–4.

146 Gilbert C, Boivin G. Human cytomegalovirus resistance to antiviral drugs. *Antimicrob Agents Chemother*. 2005;49(3):873–83.

147 Shi R, Azzi A, Gilbert C, Boivin G, Lin SX. Three-dimensional modeling of cytomegalovirus DNA polymerase and preliminary analysis of drug resistance. *Proteins*. 2006;64(2):301–7.

148 Lurain NS, Chou S. Antiviral drug resistance of human cytomegalovirus. *Clin Microbiol Rev*. 2010;23(4):689–712.

149 Chou S. Advances in the genotypic diagnosis of cytomegalovirus antiviral drug resistance. *Antiviral Res*. 2020;176:104711.

150 Lanier R, Foster S, Chou S, Prichard M, James S, Brundage T, et al. CMV resistance profile of CMX001. abstracts of the 26th International Conference on Antiviral Research. 2013;San Francisco, California, May 11–15(Abstract):205.

151 Prichard MN, Kern ER, Hartline CB, Lanier ER, Quenelle DC. CMX001 potentiates the efficacy of acyclovir in herpes simplex virus infections. *Antimicrob Agents Chemother*. 2011;55(10):4728–34.

152 Selleseth DW, Talarico CL, Miller T, Lutz MW, Biron KK, Harvey RJ. Interactions of 1263W94 with other antiviral agents in inhibition of human cytomegalovirus replication. *Antimicrob Agents Chemother*. 2003;47(4):1468–71.

153 Chou S, Ercolani RJ, Derakhchan K. Antiviral activity of maribavir in combination with other drugs active against human cytomegalovirus. *Antiviral Res*. 2018;157:128–33.

154 Moens U, Krumbholz A, Ehlers B, Zell R, Johne R, Calvignac-Spencer S, et al. Biology, evolution, and medical importance of polyomaviruses: an update. *Infect Genet Evol*. 2017;54:18–38.

155 Chong S, Antoni M, Macdonald A, Reeves M, Harber M, Magee CN. BK virus: current understanding of pathogenicity and clinical disease in transplantation. *Rev Med Virol*. 2019;29(4):e2044.

156 Zhou X, Zhu C, Li H. BK polyomavirus: latency, reactivation, diseases and tumorigenesis. *Front Cell Infect Microbiol*. 2023;13:1263983.

157 Atkinson AL, Atwood WJ. Fifty years of JC polyomavirus: a brief overview and remaining questions. *Viruses*. 2020;12(9):969

158 De Gascun CF, Carr MJ. Human polyomavirus reactivation: disease pathogenesis and treatment approaches. *Clin Dev Immunol*. 2013;2013:373579.

159 Pietropaolo V, Prezioso C, Moens U. Merkel cell polyomavirus and merkel cell carcinoma. *Cancers (Basel)*. 2020;12(7):1774.

160 Dalianis T, Hirsch HH. Human polyomaviruses in disease and cancer. *Virology*. 2013;437(2):63–72.

161 Naderer O, Colton H, Jin G, Lauver M, Lukacher A. Activity of brincidofovir (BCV) against murine polyoma virus (MuPyV) in a mouse infection model. American Society of Nephrology; October; San Diego2018. SA-PO641.

162 Wu Z, Graf FE, Hirsch HH. Antivirals against human polyomaviruses: leaving no stone unturned. *Rev Med Virol*. 2021;31(6):e2220.

163 Reisman L, Habib S, McClure GB, Latiolais LS, Vanchiere JA. Treatment of BK virus-associated nephropathy with CMX001 after kidney transplantation in a young child. *Pediatr Transplant*. 2014;18(7):E227–31.

164 Papanicolaou GA, Lee YJ, Young JW, Seshan SV, Boruchov AM, Chittick G, et al. Brincidofovir for polyomavirus-associated nephropathy after allogeneic hematopoietic stem cell transplantation. *Am J Kidney Dis*. 2015;65(5):780–4.

165 Mommeja-Marin H. editor Renal Safety of Broad Spectrum Antiviral CMX001 in the Prevention of CMV Infection Post- Allogeneic Hematopoietic Cell Transplantation (HCT). BMT Tandem Meetings; 2013:13–17;Salt Lake City, Utah, United States.

166 Cardenas RL, Cheng KH, Sack K. The effects of cidofovir on progressive multifocal leukoencephalopathy: an MRI case study. *Neuroradiology*. 2001;43(5):379–82.

167 Happe S, Besselmann M, Matheja P, Rickert CH, Schuierer G, Reichelt D, et al. [Cidofovir (vistide) in therapy of progressive multifocal leukoencephalopathy in AIDS. Review of the literature and report of 2 cases]. *Nervenarzt*. 1999;70(10):935–43.

168 Zimmermann T, Stingele K, Hartmann M, Haas J, von Einsiedel R, Wildemann B. Successful treatment of aids related PML with HAART and cidofovir. *Eur J Med Res*. 2001;6(5):190–62.

169 Segarra-Newnham M, Vodolo KM. Use of cidofovir in progressive multifocal leukoencephalopathy. *Ann Pharmacother*. 2001;35(6):741–4.

170 Osorio S, de la Camara R, Golbano N, Marti E, Fedele CG, Nieto S, et al. Progressive multifocal leukoencephalopathy after stem cell transplantation, unsuccessfully treated with cidofovir. *Bone Marrow Transplant*. 2002;30(12):963–6.

171 Houston S, Roberts N, Mashinter L. Failure of cidofovir therapy in progressive multifocal leukoencephalopathy unrelated to human immunodeficiency virus. *Clin Infect Dis*. 2001;32(1):150–2.

172 De Luca A, Ammassari A, Pezzotti P, Cinque P, Gasnault J, Berenguer J, et al. Cidofovir in addition to antiretroviral treatment is not effective for AIDS-associated progressive multifocal leukoencephalopathy: a multicohort analysis. *AIDS*. 2008;22(14):1759–67.

173 Patel A, Patel J, Ikwuagwu J. A case of progressive multifocal leukoencephalopathy and idiopathic CD4+ lymphocytopenia. *J Antimicrob Chemother*. 2010;65(12):2697–8.

174 Schneider J.A, Arbogast TA, Wolfe C.R. editor Brincidofovir - Finally an Option for PML?. *J. Heart Lung Transplant*. 37(4):S358.

175 Talarmin J-P, Rezig S, Tran-Minoui A, Berthou C, Eveillard J-R. Failure of antiviral therapy with brincidofovir in non-HIV progressive multifocal leukoencephalopathy. *Acta Neurologica Belgica*. 2020;120(5):1225–6.

176 Eckert S. W-G B, Kolb C, Hojnacki D. Treatment of natalizumab-induced PML with cidofovir or CMX001 (brincidofovir) and mirtazapine. *Neurology*. 2018;90(15_supplement):P5–380.

177 Soldan SS, Lieberman PM. Epstein-Barr virus and multiple sclerosis. *Nat Rev Microbiol*. 2023;21(1):51–64.

178 Serafini B, Rosicarelli B, Franciotta D, Magliozzi R, Reynolds R, Cinque P, et al. Dysregulated epstein-barr virus infection in the multiple sclerosis brain. *J Exp Med*. 2007;204(12):2899–912.

179 Moreno MA, Or-Geva N, Aftab BT, Khanna R, Croze E, Steinman L, et al. Molecular signature of epstein-barr virus infection in MS brain lesions. *Neurol Neuroimmunol Neuroinflamm*. 2018;5(4):e466.

180 Stefanou MI, Krumbholz M, Ziemann U, Kowarik MC. Human immunodeficiency virus and multiple sclerosis: a review of the literature. *Neurol Res Pract*. 2019;1:24.

181 Andrei G, Trompet E, Snoeck R. Novel Therapeutics for Epstein(-)Barr Virus. Molecules. 2019;24(5):997.

182 Meerbach A, Holy A, Wutzler P, De Clercq E, Neyts J. Inhibitory effects of novel nucleoside and nucleotide analogues on Epstein-Barr virus replication. *Antivir Chem Chemother*. 1998;9(3):275–82.

183 Drosu NC, Edelman ER, Housman DE. Tenofovir prodrugs potently inhibit Epstein-Barr virus lytic DNA replication by targeting the viral DNA polymerase. *Proc Natl Acad Sci U S A*. 2020;117(22):12368–74.

184 Ma Q, Ocque AJ, Morse GD, Sanders C, Burgi A, Little SJ, et al. Switching to tenofovir alafenamide in elvitegravir-based regimens: pharmacokinetics and antiviral activity in cerebrospinal fluid. *Clin Infect Dis*. 2020;71(4):982–8.

185 Linard M, Letenneur L, Garrigue I, Doize A, Dartigues JF, Helmer C. Interaction between APOE4 and herpes simplex virus type 1 in Alzheimer's disease. *Alzheimers Dement*. 2020;16(1):200–8.

186 Itzhaki RF. overwhelming evidence for a major role for herpes simplex virus type 1 (HSV1) in Alzheimer's disease (AD); underwhelming evidence against. *Vaccines (Basel)*. 2021;9(6):679.

187 Wainberg M, Luquez T, Koelle DM, Readhead B, Johnston C, Darvas M, et al. The viral hypothesis: how herpesviruses may contribute to Alzheimer's disease. *Mol Psychiatry*. 2021;26(10):5476–80.

188 Cairns DM, Itzhaki RF, Kaplan DL. Potential involvement of varicella zoster virus in Alzheimer's disease via reactivation of quiescent herpes simplex virus type 1. *J Alzheimers Dis*. 2022;88(3):1189–200.

189 Protto V, Marcocci ME, Miteva MT, Piacentini R, Li Puma DD, Grassi C, et al. Role of HSV-1 in Alzheimer's disease pathogenesis: a challenge for novel preventive/therapeutic strategies. *Curr Opin Pharmacol*. 2022;63:102200.

190 El-Haddad D, El Chaer F, Vanichanan J, Shah DP, Ariza-Heredia EJ, Mulanovich VE, et al. Brincidofovir (CMX-001) for refractory and resistant CMV and HSV infections in immunocompromised cancer patients: a single-center experience. *Antiviral Res*. 2016;134:58–62.

191 Voigt S, Hofmann J, Edelmann A, Sauerbrei A, Kuhl JS. Brincidofovir clearance of acyclovir-resistant herpes simplex virus-1 and adenovirus infection after stem cell transplantation. *Transpl Infect Dis*. 2016;18(5):791–4.

192 Lee YJ, Neofytos D, Kim SJ, Cheteyan L, Huang YT, Papadopoulos EB, et al. Efficacy of brincidofovir as prophylaxis against HSV and VZV in hematopoietic cell transplant recipients. *Transpl Infect Dis*. 2018;20(6):e12977.

193 Mullane KM, Nuss C, Ridgeway J, Prichard MN, Hartline CB, Theusch J, et al. Brincidofovir treatment of acyclovir-resistant disseminated varicella zoster virus infection in an immunocompromised host. *Transpl Infect Dis*. 2016;18(5):785–90.

194 Agut H, Bonnafous P, Gautheret-Dejean A. Human herpesviruses 6A, 6B, and 7. *Microbiol Spectr*. 2016;4(3):157–76.

195 Agut H, Bonnafous P, Gautheret-Dejean A. Update on infections with human herpesviruses 6A, 6B, and 7. *Med Mal Infect*. 2017;47(2):83–91.

196 Hill JA, Nichols WG, Marty FM, Papanicolaou GA, Brundage TM, Lanier R, et al. Oral brincidofovir decreases the incidence of HHV-6B viremia after allogeneic HCT. *Blood*. 2020;135(17):1447–51.

197 Ogata M, Satou T, Kadota J, Saito N, Yoshida T, Okumura H, et al. Human herpesvirus 6 (HHV-6) reactivation and HHV-6 encephalitis after allogeneic hematopoietic cell transplantation: a multicenter, prospective study. *Clin Infect Dis*. 2013;57(5):671–81.

198 Ljungman P. Is antiviral therapy against HHV-6B beneficial? *Blood*. 2020;135(17):1413–14.

199 Liekens S, Andrei G, Vandeputte M, De Clercq E, Neyts J. Potent inhibition of hemangioma formation in rats by the acyclic nucleoside phosphonate analogue cidofovir. *Cancer Res*. 1998;58(12):2562–7.

200 Liekens S, Neyts J, De Clercq E, Verbeken E, Ribatti D, Presta M. Inhibition of fibroblast growth factor-2-induced vascular tumor formation by the acyclic nucleoside phosphonate cidofovir. *Cancer Res*. 2001;61(13):5057–64.

201 Liekens S, Gijsbers S, Vanstreels E, Daelemans D, De Clercq E, Hatse S. The nucleotide analog cidofovir suppresses basic fibroblast growth factor (FGF2) expression and signaling and induces apoptosis in FGF2-overexpressing endothelial cells. *Mol Pharmacol*. 2007;71(3):695–703.

202 Liekens S, Noppen S, Gijsbers S, Sienaert R, Ronca R, Tobia C, et al. The broad-spectrum anti-DNA virus agent cidofovir inhibits lung metastasis of virus-independent, FGF2-driven tumors. *Oncotarget*. 2015;6(7):4633–48.

203 Redondo P, Idoate M, Galofre JC, Solano T. Cidofovir inhibits growth of B16 melanoma cells in vivo. *Br J Dermatol*. 2000;143(4):741–8.

204 Calista D. Regression of a cutaneous melanoma metastasis after intralesional cidofovir. *Melanoma Res*. 2003;13(2):205–6.

205 De Schutter T, Andrei G, Topalis D, Naesens L, Snoeck R. Cidofovir selectivity is based on the different response of normal and cancer cells to DNA damage. *BMC Med Genomics*. 2013;6:18.

206 Andrei G, Snoeck R, Schols D, De Clercq E. Induction of apoptosis by cidofovir in human papillomavirus (HPV)-positive cells. *Oncol Res*. 2000;12(9–10):397–408.

207 Abdulkarim B, Sabri S, Deutsch E, Chagraoui H, Maggiorella L, Thierry J, et al. Antiviral agent Cidofovir restores p53 function and enhances the radiosensitivity in HPV-associated cancers. *Oncogene*. 2002;21(15):2334–46.

208 Hostetler KY, Rought S, Aldern KA, Trahan J, Beadle JR, Corbeil J. Enhanced antiproliferative effects of alkoxyalkyl esters of cidofovir in human cervical cancer cells in vitro. *Mol Cancer Ther*. 2006;5(1):156–9.

209 Andrei G, Snoeck R, Piette J, Delvenne P, De Clercq E. Inhibiting effects of cidofovir (HPMPC) on the growth of the human cervical carcinoma (SiHa) xenografts in athymic nude mice. *Oncol Res*. 1998;10(10):533–9.

210 Yang Y, Zhao X, Chen W, Gao Z, Liu A, Guo J, et al. Effects of cidofovir on human papillomavirus-positive cervical cancer cells xenografts in nude mice. *Oncol Res*. 2010;18(11–12):519–27.

211 Mertens B, Cristina de Araujo Nogueira T, Topalis D, Stranska R, Snoeck R, Andrei G. Investigation of tumor-tumor interactions in a double human cervical carcinoma xenograft model in nude mice. *Oncotarget*. 2018;9(31):21978–2000.

212 Gazia F, Galletti B, Freni F, Bruno R, Sireci F, Galletti C, et al. Use of intralesional cidofovir in the recurrent respiratory papillomatosis: a review of the literature. *Eur Rev Med Pharmacol Sci*. 2020;24(2):956–62.

213 Desravines N, Miele K, Carlson R, Chibwesha C, Rahangdale L. Topical therapies for the treatment of cervical intraepithelial neoplasia (CIN) 2–3: a narrative review. *Gynecol Oncol Rep*. 2020;33:100608.

214 Brogden DRL, Walsh U, Pellino G, Kontovounisios C, Tekkis P, Mills SC. Evaluating the efficacy of treatment options for anal intraepithelial neoplasia: a systematic review. *Int J Colorectal Dis*. 2021;36(2):213–26.

215 Cook MK, Hagen EM, Feldman SR. Cidofovir in the management of non-genital warts: a review. *J Drugs Dermatol*. 2023;22(10):1009–16.

216 Zagzoog FH, Mogharbel AM, Alqutub A, Bukhari M, Almohizea MI. Intralesional cidofovir vs. bevacizumab for recurrent respiratory papillomatosis: a systematic review and indirect meta-analysis. *Eur Arch Otorhinolaryngol*. 2024;281(2):601–27.

217 Snoeck R, Noel JC, Muller C, De Clercq E, Bossens M. Cidofovir, a new approach for the treatment of cervix intraepithelial neoplasia grade III (CIN III). *J Med Virol*. 2000;60(2):205–9.

218 Van Pachterbeke C, Bucella D, Rozenberg S, Manigart Y, Gilles C, Larsimont D, et al. Topical treatment of CIN 2+ by cidofovir: results of a phase II, double-blind, prospective, placebo-controlled study. *Gynecol Oncol*. 2009;115(1):69–74.

219 Richardson BE. Initial experience treating HPV-related laryngeal diseases with oral brincidofovir: a pilot study. *Ann Otol Rhinol Laryngol*. 2021;130(12):1383–91.

220 Yilmaz E, Ismaila N, Bauman JE, Dabney R, Gan G, Jordan R, et al. Immunotherapy and biomarker testing in recurrent and metastatic head and neck cancers: ASCO guideline. *J Clin Oncol*. 2023;41(5):1132–46.

221 Neyts J, Sadler R, De Clercq E, Raab-Traub N, Pagano JS. The antiviral agent cidofovir [(S)-1-(3-hydroxy-2-phosphonyl-methoxypropyl)cytosine] has pronounced activity against nasopharyngeal carcinoma grown in nude mice. *Cancer Res*. 1998;58(3):384–88.

222 Wakisaka N, Yoshizaki T, Raab-Traub N, Pagano JS. Ribonucleotide reductase inhibitors enhance cidofovir-induced apoptosis in EBV-positive nasopharyngeal carcinoma xenografts. *Int J Cancer*. 2005;116(4):640–5.
223 Yoshizaki T, Wakisaka N, Kondo S, Murono S, Shimizu Y, Nakashima M, et al. Treatment of locally recurrent Epstein-Barr virus-associated nasopharyngeal carcinoma using the anti-viral agent cidofovir. *J Med Virol*. 2008;80(5):879–82.
224 Tan AC, Ashley DM, Lopez GY, Malinzak M, Friedman HS, Khasraw M. Management of glioblastoma: state of the art and future directions. *CA Cancer J Clin*. 2020;70(4):299–312.

Index

Note: Page numbers in *italics* and **bold** refers to figures and tables, respectively.

Trends in Antiviral Drug Development, First Edition. Edited by Michael J. Sofia and Zhengqiang Wang.